The Identification of the Genetic Components of Autism Spectrum Disorders 2017

Special Issue Editor
Merlin G. Butler

MDPI • Basel • Beijing • Wuhan • Barcelona • Belgrade

MDPI

Special Issue Editor
Merlin G. Butler
The University of Kansas Medical Center
USA

Editorial Office
MDPI AG
St. Alban-Anlage 66
Basel, Switzerland

This edition is a reprint of the Special Issue published online in the open access journal *International Journal of Molecular Sciences* (ISSN 1422-0067) from 2015–2017 (available at: http://www.mdpi.com/journal/ijms/special_issues/asd).

For citation purposes, cite each article independently as indicated on the article page online and as indicated below:

Author 1; Author 2. Article title. *Journal Name* **Year**, *Article number*, page range.

First Edition 2017

ISBN 978-3-03842-520-5 (Pbk)
ISBN 978-3-03842-521-2 (PDF)

Table of Contents

About the Special Issue Editor vii

Preface to "The Identification of the Genetic Components of Autism Spectrum Disorders 2017" ..ix

Chapter I: Clinical Genetics

G. Bradley Schaefer
Clinical Genetic Aspects of Autism Spectrum Disorders
Reprinted from: *Int. J. Mol. Sci.* **2016**, *17*(2), 180; doi: 10.3390/ijms17020180..3

**Cyrille Robert, Laurent Pasquier, David Cohen, Mélanie Fradin, Roberto Canitano,
Léna Damaj, Sylvie Odent and Sylvie Tordjman**
Role of Genetics in the Etiology of Autistic Spectrum Disorder: Towards a Hierarchical
Diagnostic Strategy
Reprinted from: *Int. J. Mol. Sci.* **2017**, *18*(3), 618; doi: 10.3390/ijms18030618..17

Ewa Pisula and Karolina Ziegart-Sadowska
Broader Autism Phenotype in Siblings of Children with ASD—A Review
Reprinted from: *Int. J. Mol. Sci.* **2015**, *16*(6), 13217–13258; doi: 10.3390/ijms160613217 46

Annaëlle Charrier, Bertrand Olliac, Pierre Roubertoux and Sylvie Tordjman
Clock Genes and Altered Sleep–Wake Rhythms: Their Role in the Development of
Psychiatric Disorders
Reprinted from: *Int. J. Mol. Sci.* **2017**, *18*(5), 938; doi: 10.3390/ijms18050938..75

Juliana Magdalon, Sandra M. Sánchez-Sánchez, Karina Griesi-Oliveira and Andréa L. Sertié
Dysfunctional mTORC1 Signaling: A Convergent Mechanism between Syndromic and
Nonsyndromic Forms of Autism Spectrum Disorder?
Reprinted from: *Int. J. Mol. Sci.* **2017**, *18*(3), 659; doi: 10.3390/ijms18030659..97

José Guevara-Campos, Lucía González-Guevara and Omar Cauli
Autism and Intellectual Disability Associated with Mitochondrial Disease and
Hyperlactacidemia
Reprinted from: *Int. J. Mol. Sci.* **2016**, *16*(2), 3870–3884; doi: 10.3390/ijms16023870 124

**Antonella De Jaco, Dalila Mango, Federica De Angelis, Flores Lietta Favaloro,
Diego Andolina, Robert Nisticò, Elena Fiori, Marco Colamartino and Tiziana Pascucci**
Unbalance between Excitation and Inhibition in Phenylketonuria, a Genetic Metabolic Disease
Associated with Autism
Reprinted from: *Int. J. Mol. Sci.* **2017**, *18*(5), 941; doi: 10.3390/ijms18050941..136

**Jeffrey A. Bennett, Sandra Hodgetts, Michelle L. Mackenzie, Andrea M. Haqq and
Lonnie Zwaigenbaum**
Investigating Autism-Related Symptoms in Children with Prader-Willi Syndrome: A Case Study
Reprinted from: *Int. J. Mol. Sci.* **2017**, *18*(3), 517; doi: 10.3390/ijms18030517..146

Devin M. Cox and Merlin G. Butler
The 15q11.2 BP1–BP2 Microdeletion Syndrome: A Review
Reprinted from: *Int. J. Mol. Sci.* **2015**, *16*(2), 4068–4082; doi: 10.3390/ijms16024068 160

Sureni V. Mullegama, Joseph T. Alaimo, Li Chen and Sarah H. Elsea
Phenotypic and Molecular Convergence of 2q23.1 Deletion Syndrome with Other
Neurodevelopmental Syndromes Associated with Autism Spectrum Disorder
Reprinted from: *Int. J. Mol. Sci.* **2015**, *16*(4), 7627–7643; doi: 10.3390/ijms16047627 171

Opal Ousley, A. Nichole Evans, Samuel Fernandez-Carriba, Erica L. Smearman,
Kimberly Rockers, Michael J. Morrier, David W. Evans, Karlene Coleman and Joseph Cubells
Examining the Overlap between Autism Spectrum Disorder and 22q11.2 Deletion Syndrome
Reprinted from: *Int. J. Mol. Sci.* **2017**, *18*(5), 1071; doi: 10.3390/ijms18051071.......................................184

Chapter II: Genetics

Merlin G. Butler, Syed K. Rafi and Ann M. Manzardo
High-Resolution Chromosome Ideogram Representation of Currently Recognized
Genes for Autism Spectrum Disorders
Reprinted from: *Int. J. Mol. Sci.* **2015**, *16*(3), 6464–6495; doi: 10.3390/ijms16036464 197

Austen B. McGuire, Syed K. Rafi, Ann M. Manzardo and Merlin G. Butler
Morphometric Analysis of Recognized Genes for Autism Spectrum Disorders and Obesity in
Relationship to the Distribution of Protein-Coding Genes on Human Chromosomes
Reprinted from: *Int. J. Mol. Sci.* **2016**, *17*(5), 673; doi: 10.3390/ijms17050673.......................................221

Karen S. Ho, E. Robert Wassman, Adrianne L. Baxter, Charles H. Hensel, Megan M. Martin,
Aparna Prasad, Hope Twede, Rena J. Vanzo and Merlin G. Butler
Chromosomal Microarray Analysis of Consecutive Individuals with Autism Spectrum
Disorders Using an Ultra-High Resolution Chromosomal Microarray Optimized for
Neurodevelopmental Disorders
Reprinted from: *Int. J. Mol. Sci.* **2016**, *17*(12), 2070; doi: 10.3390/ijms17122070.....................................240

Oded Oron and Evan Elliott
Delineating the Common Biological Pathways Perturbed by ASD's Genetic Etiology: Lessons
from Network-Based Studies
Reprinted from: *Int. J. Mol. Sci.* **2017**, *18*(4), 828; doi: 10.3390/ijms18040828.......................................254

Naveen S. Khanzada, Merlin G. Butler and Ann M. Manzardo
GeneAnalytics Pathway Analysis and Genetic Overlap among Autism Spectrum Disorder,
Bipolar Disorder and Schizophrenia
Reprinted from: *Int. J. Mol. Sci.* **2017**, *18*(3), 527; doi: 10.3390/ijms18030527.......................................273

Merlin G. Butler, Syed K. Rafi, Waheeda Hossain, Dietrich A. Stephan and Ann M. Manzardo
Whole Exome Sequencing in Females with Autism Implicates Novel and Candidate Genes
Reprinted from: *Int. J. Mol. Sci.* **2015**, *16*(1), 1312–1335; doi: 10.3390/ijms16011312 284

Hee Jeong Yoo, Kyung Kim, In Hyang Kim, Seong-Hwan Rho, Jong-Eun Park, Ki Young Lee,
Soon Ae Kim, Byung Yoon Choi and Namshin Kim
Whole Exome Sequencing for a Patient with Rubinstein-Taybi Syndrome Reveals *de Novo* Variants
besides an Overt *CREBBP* Mutation
Reprinted from: *Int. J. Mol. Sci.* **2015**, *16*(3), 5697–5713; doi: 10.3390/ijms16035697 303

Yasin Panahi, Fahimeh Salasar Moghaddam, Zahra Ghasemi, Mandana Hadi Jafari,
Reza Shervin Badv, Mohamad Reza Eskandari and Mehrdad Pedram
Selection of Suitable Reference Genes for Analysis of Salivary Transcriptome in Non-Syndromic
Autistic Male Children
Reprinted from: *Int. J. Mol. Sci.* **2016**, *17*(10), 1711; doi: 10.3390/ijms17101711.....................................317

Martina Landini, Ivan Merelli, M. Elisabetta Raggi, Nadia Galluccio, Francesca Ciceri,
Arianna Bonfanti, Serena Camposeo, Angelo Massagli, Laura Villa, Erika Salvi,
Daniele Cusi, Massimo Molteni, Luciano Milanesi, Anna Marabotti
and Alessandra Mezzelani
Association Analysis of Noncoding Variants in Neuroligins 3 and 4X Genes with Autism
Spectrum Disorder in an Italian Cohort
Reprinted from: *Int. J. Mol. Sci.* **2016**, *17*(10), 1765; doi: 10.3390/ijms17101765.....................................333

Naila Al Mahmuda, Shigeru Yokoyama, Jian-Jun Huang, Li Liu, Toshio Munesue, Hideo
Nakatani, Kenshi Hayashi, Kunimasa Yagi, Masakazu Yamagishi and Haruhiro Higashida
A Study of Single Nucleotide Polymorphisms of the *SLC19A1/RFC1* Gene in Subjects with
Autism Spectrum Disorder
Reprinted from: *Int. J. Mol. Sci.* **2016**, *17*(5), 772; doi: 10.3390/ijms17050772...346

Hae Jeong Park, Su Kang Kim, Won Sub Kang, Jin Kyung Park, Young Jong Kim, Min Nam,
Jong Woo Kim and Joo-Ho Chung
Association between *IRS1* Gene Polymorphism and Autism Spectrum Disorder: A Pilot
Case-Control Study in Korean Males
Reprinted from: *Int. J. Mol. Sci.* **2016**, *17*(8), 1227; doi: 10.3390/ijms17081227.....................................355

Chapter III: Other

Costas Koufaris and Carolina Sismani
Modulation of the Genome and Epigenome of Individuals Susceptible to Autism by
Environmental Risk Factors
Reprinted from: *Int. J. Mol. Sci.* **2015**, *16*(4), 8699–8718; doi: 10.3390/ijms16048699 365

Colin Heberling and Prasad Dhurjati
Novel Systems Modeling Methodology in Comparative Microbial Metabolomics: Identifying
Key Enzymes and Metabolites Implicated in Autism Spectrum Disorders
Reprinted from: *Int. J. Mol. Sci.* **2015**, *16*(4), 8949–8967; doi: 10.3390/ijms16048949 381

Jessie M. Cameron, Valeriy Levandovskiy, Wendy Roberts, Evdokia Anagnostou,
Stephen Scherer, Alvin Loh and Andreas Schulze
Variability of Creatine Metabolism Genes in Children with Autism Spectrum Disorder
Reprinted from: *Int. J. Mol. Sci.* **2017**, *18*(8), 1665; doi: 10.3390/ijms18081665.....................................396

Zheng Wang, Minhyuk Kwon, Suman Mohanty, Lauren M. Schmitt, Stormi P. White,
Evangelos A. Christou and Matthew W. Mosconi
Increased Force Variability Is Associated with Altered Modulation of the Motorneuron Pool
Activity in Autism Spectrum Disorder (ASD)
Reprinted from: *Int. J. Mol. Sci.* **2017**, *18*(4), 698; doi: 10.3390/ijms18040698...415

Teri Smith, Susan Sharp, Ann M. Manzardo and Merlin G. Butler
Pharmacogenetics Informed Decision Making in Adolescent Psychiatric Treatment: A Clinical
Case Report
Reprinted from: *Int. J. Mol. Sci.* **2015**, *16*(3), 4416–4428; doi: 10.3390/ijms16034416 435

**Jarle Johannessen, Terje Nærland, Sigrun Hope, Tonje Torske, Anne Lise Høyland, Jana
Strohmaier, Arvid Heiberg, Marcella Rietschel, Srdjan Djurovic and Ole A. Andreassen**
Parents' Attitudes toward Clinical Genetic Testing for Autism Spectrum Disorder—Data from a
Norwegian Sample
Reprinted from: *Int. J. Mol. Sci.* **2017**, *18*(5), 1078; doi: 10.3390/ijms18051078..446

About the Special Issue Editor

Merlin G. Butler MD, PhD is Professor of Psychiatry, Behavioral Sciences and Pediatrics, as well as Director of the Division of Research and Genetics and the Genetics Clinic at the University of Kansas Medical Center, Kansas City. He received his medical degree from the University of Nebraska College of Medicine in Omaha and his doctorate in Medical Genetics from Indiana University School of Medicine in Indianapolis where he trained in both Clinical Genetics and Clinical Cytogenetics and completed a fellowship accredited by the American Board of Medical Genetics (ABMG) with certification received in 1984. He has held previous academic positions at Indiana University and Vanderbilt University. Dr. Butler is a member of several state and national committees, study sections and editorial boards. His research interests include the genetics of autism, Prader–Willi syndrome, genotype–phenotype correlations and delineation of rare disorders. He has published over 500 peer-reviewed articles, book chapters and textbooks.

Preface to "The Identification of the Genetic Components of Autism Spectrum Disorders 2017"

The textbook entitled Identification of the Genetic Components of Autism Spectrum Disorders includes themes associated with ASD and is divided into three sections (clinical, genetics, other). These sections include information on clinical description, diagnosis, treatment and characterization of ASD with current reviews on supported topics including other ASD related syndromes; the genetics of autism based on laboratory research and testing; and other factors contributing to our understanding and causation of ASD including metabolism, environmental risks, epidemiology, neurophysiology and parental attitudes. This textbook includes 28 chapters written by experts in the field of autism in research, discovery, treatment and diagnosis with 11 chapters dedicated to clinical description, characterization and related reviews of ASD; 11 chapters dedicated to laboratory genetics research and reviews with other genetic and cytogenetic syndromes; and six chapters involving other risk factors contributing to our understanding of ASD and should be a useful resource for physicians, basic scientists and clinical researchers with translation to the clinic setting and other professionals, students and healthcare providers interested in this important neurodevelopmental disorder that is on the rise in our society. Those interested in the current understanding of the scientific causes and contributing factors of ASD as well as those caring for individuals affected with this disorder should find this book useful. These should include basic laboratory researchers involved with identification of genetic factors in the causation of ASD as reported studies have indicated as many as 90% of individuals with autism have a genetic cause contributing to their clinical findings. In addition, healthcare professionals including psychiatrists, psychologists, clinical geneticists and genetic counselor, neurologists, special educators and child-life experts, developmental specialists, social workers, nurses, occupational/ physical therapists, speech therapists and pathologists, public health experts and community activists would find this resource of value in their practice. This book would also serve as a resource for parents and other family members and agencies involved with providing services and care of those with this common neurodevelopmental disorder.

Merlin G. Butler
Special Issue Editor

Chapter I:
Clinical Genetics

International Journal of
Molecular Sciences

MDPI

Review

Clinical Genetic Aspects of Autism Spectrum Disorders

G. Bradley Schaefer

University of Arkansas for Medical Sciences, Arkansas Children's Hospital, 1 Children's Way, Slot 512-22, Little Rock, AR 72202, USA; schaefergb@uams.edu; Tel.: +1-501-364-2971; Fax: +1-501-364-1564

Academic Editor: Merlin G. Butler
Received: 9 December 2015; Accepted: 26 January 2016; Published: 29 January 2016

Abstract: Early presumptions opined that autism spectrum disorder (ASD) was related to the rearing of these children by emotionally-distant mothers. Advances in the 1960s and 1970s clearly demonstrated the biologic basis of autism with a high heritability. Recent advances have demonstrated that specific etiologic factors in autism spectrum disorders can be identified in 30%–40% of cases. Based on early reports newer, emerging genomic technologies are likely to increase this diagnostic yield to over 50%. To date these investigations have focused on etiologic factors that are largely mono-factorial. The currently undiagnosed causes of ASDs will likely be found to have causes that are more complex. Epigenetic, multiple interacting loci, and four dimensional causes (with timing as a variable) are likely to be associated with the currently unidentifiable cases. Today, the "Why" is more important than ever. Understanding the causes of ASDs help inform families of important issues such as recurrence risk, prognosis, natural history, and predicting associated co-morbid medical conditions. In the current era of emerging efforts in "personalized medicine", identifying an etiology will be critical in identifying endo-phenotypic groups and individual variations that will allow for tailored treatment for persons with ASD.

Keywords: multifactorial inheritance; genetic testing; diagnostic yield; copy number variants; gene sequencing; genomics

1. Introduction

1.1. Autism as a Neuro-Genetic Disorder

Not long after autism was identified in the 1940s as a distinct developmental disorder, the question of causation was discussed. Most of the early theories as to the etiology of autism were framed in the context of the prevailing psychiatric models of the time. Thus many theorized that autism was an acquired condition associated with children raised by cold, emotionally-distant mothers. Given our current understanding of the etiology of autism spectrum disorders (which will be a large focus of this review) it is hard to imagine how these theories gained popular acceptance. One has to also appreciate the extreme guilt that this must have evoked on the parents, and this compounded by the stress of having a special needs child.

From the 1940s to the present, the understanding of the etiology of autism has mirrored the eras of scientific advancement in the field of developmental biology and genetics. Population studies in the 1960s and 1970s provided clear proof of the biologic and genetic basis of autism. Progress in genetic testing in the 1980s and 1990s began to identify specific genetic markers in some patients with autism. The progression from linkage studies to cytogenetic, molecular cytogenetic, molecular and now genomic testing have all been very congruent in identifying genomic "hot spots" associated with autism.

Current understanding recognizes autism as having a strong genetic basis with a complex inheritance pattern. Strong genetic factors are involved. As with all human medical conditions, there is environmental modulation. There is clear etiologic and genetic heterogeneity. Literally hundreds of "autism genes" have been identified. Thus, from an etiologic standpoint, it would be better stated "the autisms" rather than "autism". This understanding will be critical as the science of autism therapies moves forward. Using targeted therapies for specific identified causes of autism holds the promise of improved outcomes and reduced adverse events. For the purposes of this review I will discuss autism spectrum disorders (ASDs) unless otherwise stated. In general, from a clinical genetic standpoint, the evaluation is not different for persons on different parts of the spectrum. The focus of this review is to highlight the genetic factors in autism especially in the light of clinical applications.

1.1.1. Indisputable Evidence of the Genetic Basis of Autism Spectrum Disorders (ASDs)

As noted above, population studies have provided strong evidence of the genetic basis of ASDs [1]. In fact, when compared to other neurodevelopmental disabilities (NDD), autism is one of most—if not the most—heritable NDDs known.

From 2006 to 2008 we reviewed a large portion of the existing published information on the population genetic characteristics of autism spectrum disorders [2,3]. This information provided a very strong indication of the genetic basis of ASDs. Classic parameters of population genetics that supported this included data on twin concordance, heritability, relative risk ratio and sibling risk ratio.

In this review, I will provide a brief summary of this collected body of information. Twin concordance studies showed an estimated 70% concordance among monozygotic twins (reported ranges 36%–95%). In contrast, concordance among dizygotic twins was around 3% (range 0%–31%) or 30% if a broader phenotypic definition was used. Heritability estimates were noted to be 0.8 to 0.9. The estimated relative (sibling) risk ratio is reported to be about 150 for monozygotic twins and 8–10 for dizygotic twins and full siblings [4]. In the realm of the discipline of population genetics this is overwhelming support of a genetic basis to ASDs. These data have been replicated and validated in more recent and larger studies [4–6]. With a broader design and larger sample size, these more recent studies have also added another population parameter not previously noted. A 2–3-fold increased recurrence risk of ASD has been documented in half-siblings (both maternal and paternal) of probands with an ASD [4–7]. The increased occurrence in even less closely related individuals lends very strong support for the genetic basis of autism.

1.1.2. Why Then Is the Genetic Basis of ASDs Still Debated?

Even with such a strong body of evidence as to the genetic basis of ASDs, not everyone is convinced. Significant opposition to this notion exists. There exist multiple factions of people who do not recognize the validity of the commonly accepted medical literature. Common objections voiced include the concern over some link between ASDs and childhood immunizations. Some question how could a "genetic disorder" be increasing in frequency? This is cited in light of the continued increase in the reported incidence of ASD. Over the past 20 years the reported occurrence of ASD has almost quadrupled [8]. Current estimates put the prevalence of ASD at about 1/68 children [9]. The logical question is "How could a condition could be increasing at such a rapid rate and yet have a primary genetic origin?" As one of my partners expressed to me "It is interesting that the incidence of conditions like diabetes and asthma are similarly increasing and yet no one is questioning the genetic basis of these conditions [10].

1.2. Multifactorial Inheritance

Strictly speaking, all human medical conditions could be classified as having a "multi-factorial etiology". This implies that these conditions have both genetic and environmental factors which contribute to the overall phenotype. All conditions have genetic and environmental contributions—depending on the condition, the relative proportion of each varies considerably. *Mendelian inheritance* is the term applied to those conditions in which a single gene mutation has a major phenotypic effect with significantly less environmental influences on the phenotype and in which the inheritance follows a monogenic pattern. For more complex traits, the relative contribution of environmental and genetic factors is more balanced. Conditions are designated as having *multifactorial inheritance* if several characteristics are noted:

1. Clear genetic variability exists yet no uni-factorial mode of inheritance can be identified
2. Family studies indicate an increased risk for near relatives to be affected
3. Complicated pathophysiology or morphogenetic processes are involved
4. Biologic influences of environmental factors

Genetic susceptibility refers to a condition that is predominantly determined by environmental factors with expression and phenotypic variability affected by genetic changes that alter the susceptibility to the environmental factor [11].

Considering the above definitions, ASDs clearly fall into the category of multi-factorial inheritance. The recurrence risk pattern of multi-factorial traits typically demonstrates:

1. An increased recurrence risk in close relatives as compared to the general population frequency
2. A non-linear decrease in frequency with increasing distance of relationship—typically no increased recurrence rates are seen beyond 3rd degree relatives
3. The recurrence risk increases with the number of affected individuals
4. There is an increased risk with increased severity of the condition
5. There is an increased risk if person(s) affected are of the "rarer" gender. (In ASD a distinct gender bias of a 3- to 4-fold rate of affected males has been noted)

Recurrence risk data that had accumulated up until about 2008 matched well with this model. The overall recurrence risk for a sibling of a single proband with ASD was reported as 3%–10%. Further refinement of these risks noted that the estimated recurrence risk for a sibling of an affected person with an ASD was established as 4% if the affected individual was male and 7% if female. If there were more than one affected sibling with an ASD, the recurrence risk for future sibs increased to a remarkable 30%–50%. These numbers have been verified in multiple studies [12].

As already stated, multifactorial inheritance implies both genetic and environmental factors are at work. The established heritability rates of around 0.8 or more implies that the majority of the ASD phenotype is determined by genetic factors. Still, environmental influences need to be acknowledged. The issue of environmental contributors to autism is one of the most highly debated topics in all of the body of medical literature. A discussion of this is beyond the focus of this review.

2. Discussion

A thorough discussion of the genetic aspects of ASDs crosses the gamut of genetic principles to be considered in human diseases and disorders. In this section we will discuss each of these parameters individually. We will also review changes in the reported data in light of changes in reported information before and after about 2008. Table 1 provides a summary of the information below.

2.1. Epidemiology/Population Genetics

Epidemiologic data is extremely helpful in population genetics. Knowing the specifics about occurrence, incidence, prevalence and biases of a condition is important in defining genetic factors and the magnitude of their influence on the trait. The reported overall incidence of ASDs has steadily increased over the past 3 decades. The curve has risen sharply since the year 2000. At that point, the reported incidence of ASDs was 1/150 school-aged children. In 2014 it had risen to 1/68 (a 120% increase). Boys are estimated to be affected 4–5 times as often as girls; with the reported incidence now being 1/42 in boys and 1/189 in girls. With this notable increase in reported cases, recent studies have re-evaluated the population genetics of ASDs. Some have reinforced what has already been known. Others have challenged long-standing assumptions. We will look at each of these individually.

2.1.1. Recurrence Risk

Recent studies [6,13,14] have suggested that the recurrence for siblings in the situation of a single affected proband may be higher than the previously noted 3%–10%. These more recent studies reported risks of 10%–19%. While there has not been a formal weighted analysis yet of these changing numbers, these studies provide sufficient support of using an increased recurrence risk of 10%–20%. In the event of more than one affected sibling with an ASD recent studies have confirmed the estimated 30%–50% recurrence risk for subsequent sibs [14–16]. The newer studies have also reported relative risk ratios (RRR) for full siblings as 6.9–10.3 and significant concordance calculations with RRR estimates of 150 for monozygotic twins as compared to 8.2 for dizygotic twins [4,5].

2.1.2. Heritability

Heritability is defined as the proportion of the observed phenotype that is attributable to genetic factors. Heritability estimates over the past 20 years for ASD have been in the range of 0.7 to 0.9. One recent study [4] has given a much lower estimate of 0.4–0.6. For now these data must be cautiously interpreted in light of most other reports.

2.1.3. Sex Bias (Occurrence Gender)

Historically a reproducible and significant gender bias has been noted in the incidence of most neurodevelopmental disabilities. In general males have been found to have a 4–5-fold increased occurrence of these conditions [17]. This is true for ASDs. This difference is noted more so in those persons with ASD who have milder degrees of cognitive impairment. If individuals with mild cognitive impairment are considered, the gender bias is about 2 times increased in males. In individuals with more severe impairments the bias is 4–7 times increased [18]. The first assumption that would be made in this setting would be that X-linked genes might be associated with this observed increased occurrence in males. Interestingly it has been suggested that only about 10% of the reported male excess can be attributed to X-linked genes. A previously cited study [13] did not identify a gender bias. However, others have confirmed this finding [14,19]. For now it appears safe to assume that there is indeed a 4–5-fold increased incidence of ASDs in males.

2.1.4. Proband Gender Effect (Recurrence)

As previously mentioned, most multi-factorial conditions show a gender bias in their occurrence. Recurrence rates can then also be shown to have a gender specific recurrence rate with the rate being higher if the affected individual is of the less frequent gender. In ASDs the recurrence rate in siblings of an affected male has been stated as 4% as compared to 7% if the affected individual is a female [3]. In more recent studies, this "proband gender effect" on recurrence in siblings has been questioned. Three studies [5,14,19] found no proband gender differences. Two other recent studies [15,20] did observe this phenomenon. From a practical standpoint, if the proband gender effect is indeed real, the small difference that would be present would not be sufficiently large enough to be predicted to be of sufficient magnitude as to affect familial decision making. Thus, while this is an interesting question and worthy of additional investigations, clinical counseling could probably be best given as 10%–20% recurrence risk for either gender.

2.1.5. Reproductive Stoppage and Birth Order

Earlier observations have noted that despite a per pregnancy prediction of a 3%–10% recurrence rate of ASDs, the actual number of observed recurrences was less (2%–3%). This raised the question of reproductive "stoppage" (aka curtailment of reproduction). That is the presumption that once a couple has had a child with an ASD, they are less likely to have other children—regardless. These correlated observations suggested that stoppage might be at work, but without investigative support. Three recent reports [16,21,22] have all objectively confirmed the suspicion of stoppage in families with children with an ASD. Finally, some studies have investigated the association with birth order and ASD. Birth order is an interesting parameter to consider. In actuality birth order effects are predicted to be more of an environmental influence. Two studies have noted no birth order effect in one report [14] *versus* a decrease in ASDs in later siblings. Reported birth order effects, if real, are likely to be related in some degree to the same factors in reproductive curtailment (psycho-social environmental factors).

2.1.6. Parental Age Effects

An association with an advanced paternal age (which is felt to be associated with an increased occurrence of *de novo* mutations) has previously been reported in ASDs [23]. One recent report [14] actually reported the opposite—an association with younger fathers.

Table 1. Summarizes the current parameters of the population aspects of ASDs.

Parameter	Value	Comments
Recurrence risk	10%–20%	Value increased based on newer studies
Relative recurrence ratio		
Monozygotic twins	150	
Dizygotic twins	8	
Full siblings	7–10	
Heritability	0.7–0.9	One recent study estimate of 0.5
Occurrence gender	4–5× higher in males	Few studies have not seen this
Proband gender effect	2× increase if female	Recent studies differ on this effect
Paternal age	Increased	One recent study saw a higher occurrence in younger fathers
Reproductive curtailment (stoppage)	Appears to be real phenomenon	
Birth order	Decreased in later sibs	To be confirmed

2.2. Genetic Loci and ASDs

Given the information in the preceding section, a genetic basis for ASDs is unquestionable. As such, the next question is "Can these genetic factors be identified?" In this context a couple of terms should be defined. "*Etiology*" is a specific diagnosis that can be translated into useful clinical information for the family, including providing information about prognosis, recurrence risks, and preferred modes of available therapy [24]. This definition was further refined to include the stipulation that there is "sufficient literature evidence to make a causal relationship of the disorder . . . and if it meets the Schaefer-Bodensteiner definition" [25]. "*Diagnostic yield*" is the proportion of cases in which the etiology can be determined after a complete evaluation. Essentially it is the batting average in finding the etiology of a given condition.

Genetic testing technologies have dramatically improved over the past 3 decades. Each significant advancement in genetic testing technology has produced an increased understanding of the genetic factors that cause ASDs. A rise in the diagnostic yield in defining an etiology for ASD parallels the introduction of these new technologies. A discussion of the causes of ASDs can be framed in the context of the ontogeny of the development of each new modality.

2.2.1. Linkage

Linkage technology was developed and refined, with a further expansion into genome wide association studies. Multiple studies using this technology were reported in cohorts of patients with ASD. A summary of these results is that linkage was identified with most autosomes. A careful review of the cumulative results identified consistent reports of linkage to chromosomal regions 2q, 7q22-31 (with a parent of origin effect), 13q, 15q11-13, 16p and 17q11 (a male specific locus). Better *lod* scores were obtained if a broader phenotypic definition was used. Repeat studies with larger sample sizes did not change these results/conclusions.

Since ASDs shows a clear male predominance, a logical assumption might be that X-linked genes could be playing a major role. However, linkage studies specifically targeting the X chromosome did not identify X-linked genes as accounting for a large portion of the male predominance. Only four minor linkages to the X chromosome were noted. Stated another way, a much smaller proportion of the male preponderance is explained by X-linked genes than initially suspected [26]. This raises an interesting corollary question. If X-linked genes do not explain the male excess in cases, what does? Other possible explanations that have been proposed include (1) Mosaicism/tissue specific expression of X-linked genes; (2) Dysregulation of methylation of brain expressed genes on the X chromosome and (3) Hormonally mediated changes (*i.e.*, differential expression on the "male brain").

2.2.2. Cytogenetics

Microscopically detected chromosome abnormalities have consistently been reported in association with ASD. Refinements of chromosome analysis (prometaphase and later interphase studies) enhance detection rates. Estimated positive findings in ASDs have a reported range of 3%–12% with the best overall estimates being 3%–5% [27–29]. The most commonly seen abnormalities reported are deletions or duplications of the proximal 15q region. Other commonly reported aneuploidies include deletions of 2q37, 7q, 18q, and Xp and duplication of 22q13. An association with whole chromosome aneuploidies such as 47 XXY and 45X/46XY has also been seen.

2.2.3. Fluorescent-*in Situ*-Hybridization (FISH)

Fluorescent-*in situ*-hybridization (FISH) technology became generally available for clinical use in the 1990s. With the introduction of FISH studies, clinical geneticists had another tool beyond just karyotype analysis for identifying mutations in patients. When applied to patients with ASDs the most common single loci FISH findings were found in association with the chromosomal regions 2q, 15q, 17p, 22q11, and 22q13.

2.2.4. Chromosomal Microarray (CMA)

Around the year 2000, molecular cytogenetic techniques were introduced to clinical medicine. Chromosomal microarray (CMA) quickly became a powerful tool for identifying copy number variants (CNVs) in the human genome. When microarray studies are performed in patients with neurodevelopmental disabilities, a significant number of patients will have either small deletions or duplications identified. Currently microarray studies are still one of the highest yield diagnostic tests in these patients. In the process of the study of CNVs in ASDs, a handful of "ASD syndromes" has been identified. These syndromes are recognizable phenotypes that are associated with specific CNVs. These conditions have notable clinical features besides having ASD. Table 2 list a few of the most well-known ASD syndromes associated with small chromosomal deletions or duplications.

Table 2. "Autism Syndromes" Identified by Chromosomal Microarray.

Copy Number Variant	Incidence in Cohorts with ASDs	Eponym	Other Key Features (besides ASD)
1q21.1 del	1%	None	Congenital heart disease (30%)
2q22.3 del dup	<1%	Mowat-Wilson	Hirschprung disease, epilepsy, facial dysmorphisms
16p11.2 del/dup	1%	None	
17p11.2 dup	<1%	Potocki-Lupski	Hypotonia, slow growth, epilepsy
22q11.2 del	<1%	DiGeorge/Shprintzen	Multiple congenital anomalies
22q13.3 del	1%	Phelan-McDermid	Hypotonia, accelerated growth

Over the past 15 years, many studies have been published that have reported on the incidence of CNVs identified by microarrays in patients with ASDs. Cumulatively over 100 different clinically significant changes have been reported in patients with ASDs. Estimates of the prevalence of pathogenic CNVs in ASD range from 8% to 21%. Overall it has been concluded that CMA will identify a known pathogenic CNV in about 10% of all cases of ASD [12]. Prior to the introduction of microarray technology, the overall diagnostic yield for a complete evaluation of a person with ASD was accepted as being around 6%–12%. Thus the introduction of this one test modality greatly enhances the overall diagnostic yield. As will be discussed below, this high yield warrants making CMA a "first-tier" test in the etiologic evaluation of persons with ASD.

2.2.5. Key ASD Loci

One of the fascinating observations that can be made from the data presented in the above section is the consistency of the findings across studies regardless of the testing modality utilized. A review of this information readily identifies several "ASD hot spots". Specifically, certain loci jump out as consistently being identified by linkage, chromosomal studies, FISH and microarray studies. This aggregate information serves as a strong indicator for investigators that key genes that paly a strong role in the occurrence of ASD lie in these regions. Some of the strongest evidence points to strong genetic factors in the pathogenesis of ASD lie in chromosomal regions 2q, 7q, 15q11-13, 16p, 17q, 22q, Xp, and Xq. A more detailed listing of these areas of interest can be found in the ACMG practice guidelines reported in 2013 [12].

2.3. Clinical Genetic Evaluation of ASD

Not surprisingly, the number of requests for consultations by clinical geneticists with the referral indication of ASD has tremendously increased over the past 15 years. Of course this is in part due to the greatly increased reported incidence. In addition, there is an increased awareness among both the lay and medical communities of the recent rapid advances in genetics and genomics. As such, a large number of children with the diagnosis of ASD are being seen for an etiologic evaluation. However, many question the rationale for such evaluations. The question raised is simply "You can't cure the condition, so what good is it in figuring out what caused it?" This is an age old question that geneticists have answered almost since the inception of the discipline. It is critically important to note that an etiologic answer can be helpful to patients and their families in many different ways [30]. Examples of these would include:

(1) Genetic counseling—including providing recurrence risk information
(2) Counseling regarding the natural history of the condition
(3) Anticipation of a later associated co-morbid condition
(4) Prevention of secondary disorders
(5) Availability of prenatal diagnosis
(6) Access to public support systems
(7) Access to syndrome-specific support groups
(8) The reassurance of knowing "Why" in reliving the stress of the unknown
(9) The possibility of a specific treatment strategy—should one be available or developed in the future

2.3.1. Role of Dysmorphology/Clinical Genetics

When a patient is referred for a clinical genetic evaluation for ASD, many families want to know "Why" (discussed above) and "What will happen?" as clinical genetics is not a discipline familiar to many people [31]. A major skill set of a trained clinical geneticist is that of a "dysmorphology assessment". The geneticist has the ability to identify differences in morphology of patients and often to assemble the list of noted findings into a recognizable pattern [32]. The identification of such patterns can then aid in directing the rest of the diagnostic workup.

When a cohort of persons with ASD is evaluated, two sub-groups are easily identified [33]. An estimated 10%–20% of persons with ASD fit into the category of "complex" or syndromic' ASD. These are patients who, besides having ASD, are noted to have recognizable abnormalities of morphogenesis manifest as dysmorphic features, microcephaly, or cerebral dysgenesis [34]. As a whole, this group has a more guarded outcome with a higher degree of cognitive impairment and epilepsy and a greater occurrence of co-morbid medical problems. The second group (those with just ASD) has been referred to by several terms including "simple", "essential", "idiopathic" or "non-syndromic". As a whole this group has a higher heritability, more affected family members, and a higher male: female ratio than the complex group [33].

The initial part of the etiologic evaluation of ASD then is to determine whether or not the patient fits into the syndromic or non-syndromic category. If it is the former the next step is to see if a specific syndromic etiology can be identified. Sometimes this can be accomplished simply as part of the initial history and physical examination. Table 3 lists some of the most important syndromes associated with ASD that should be considered. A related question in this regards is when to do an etiologic evaluation for ASD in a patient who already has a known syndromic diagnosis. For some syndromes the known association with ASD is so strong that one can safely assume that the condition is indeed the cause of the ASD. In these cases no further evaluation is indicated. For other dysmorphic syndromes the reported association is not as convincing and a full evaluation should be considered. For a listing of these two groups the reader is referred to a recent review [12].

Table 3. Important Dysmorphic Syndromes to Consider in the Etiologic Evaluation of ASD.

22q11.2 Deletions (Including DiGeorge and Shprintzen Syndromes)
CHARGE syndrome
Fragile X syndrome
Opitz FG syndrome
Prader Willi/Angelman syndrome
PTEN associated disorders
Rett syndrome
Smith-Magenis syndrome
Sotos syndrome

In the discipline of dysmorphology, there is a well-known relationship between certain types of pigmentary abnormalities and neurodevelopmental disorders [35]. This association is felt to be due to the embryonic neuro-ectodermal cells that both populate the glial elements of the brain and give rise to the melanocytes in the skin. As such part of the dysmorphologic evaluation includes a careful cutaneous examination aided by the use of a Woods lamp to identify even subtle changes in skin pigmentation. In some cases abnormal pigmentation in association with ASD is indicative of a neuro-cutaneous syndrome. The strongest link here is tuberous sclerosis as a clearly identified cause of ASD [36]. Similar to the assumptions with certain dysmorphic syndromes, the co-existent diagnosis of ASD in a person with tuberous sclerosis does not warrant further investigation. Other patients may have pigmentary abnormalities noted in non-random patterns suggestive of somatic mosaicism [37]. In patients with ASD with these types of skin changes, a skin biopsy for cultured fibroblasts may be needed to identify the etiology (Figure 1).

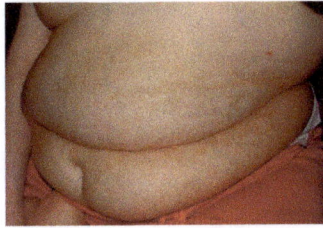

Figure 1. Pigmentary Changes in a Patient with Somatic Mosaicism. Note the linear pattern of the pigmentary changes.

Another important concept to consider is that of "expanded phenotypes". An expanded phenotype refers to the full range of phenotypes seen with mutations (variants) at a specific locus (gene). What is typically seen in the progress of clinical genetics is that mutations in a gene are initially reported in association with a known clinical disorder. Soon after this relationship is defined, the same or similar mutations at the same locus are identified with different phenotypes. Over time a broad range of phenotypes associated with changes in the same gene is defined. For example, mutations in the gene MeCP2 were initially described as the cause of Rett syndrome. As the range of the expanded phenotype was defined, mutations in MeCP2 have also been seen in patients with non-syndromic cognitive impairment and cerebral palsy [38]. Salient to these discussions, 4% of females with idiopathic ASD will have pathogenic mutations in MeCP2 [12]. This highlights a couple of very important concepts. First, it must be recognized that expanded phenotypes exist and need to be appreciated in a diagnostic workup. It is not readily intuitive that MeCP2 testing should be part of an ASD evaluation. The second point is critically important: genotype does not change phenotype. If the patient with ASD is found to have a MeCP2 mutation, the diagnosis does not change to Rett syndrome, simply the cause of the ASD has been identified.

As noted in an earlier section ASD exhibits a familial pattern of multifactorial inheritance. Thus multiple members of the same kindred may be found to have ASD. Even within families, a large degree of variable expression is typically seen. Family members with ASD may differ in their degree of social and cognitive impairment. They may also differ in co-morbid medical conditions. Along the lines of the discussions of expanded phenotypes the same causative mutation may produce different phenotypes. Thus as one investigates the kindred of a person with ASD, other neurodevelopmental and neurobehavioral phenotypes need to be considered. It is not uncommon for a family with a person with ASD to also have persons affected with other conditions. The incidence of other conditions such as depression, schizophrenia, anxiety and even Attention Deficit Hyperactivity Disorder (ADHD) are higher in relatives of persons with ASD [39]. Recent discoveries of other etiologies of ASD have identified some of these links. For example, a copy number variant—deletion 17q12—has been reported with both ASD and schizophrenia in the same family [40].

An interesting question has been raised in regard to the craniofacial appearance of patients with non-syndromic ASD. Specifically, is there an "ASD face"? In a non-published series of studies, myself and one other clinical geneticist reviewed the photographs of 33 children with ASD [41]. We then cataloged any dysmorphic features noted. These findings were tabulated for recurrent features. Those features are listed in Table 4. These photographs plus another 31 photographs of age and gender matched normal children without ASD were presented to three other clinical geneticists in a blinded fashion. They were asked to identify any of the features in the established list in the 64 photographs. Five of the nine features (shortened columella, thick scalp hair, lateral extension of the eyebrows, thickened ala nasae, and prominent nasal root) occurred more often (*p* value range 0.006–0.02) in the group of children with ASDs. While this is a very interesting observation, it is not clear what, if any, clinical significance it might have.

Table 4. Common Facial Features Noted in Non-syndromic ASD [41].

Thick scalp hair
Lateral extension of the eyebrows
Sloped forehead with prominent brow
Infraorbital hypoplasia
Prominent pre-maxilla
Short columella
Prominent (broad) nasal root
Thickened ala nasae
Prominent philtral ridges
Small ears

Bolded features were noted to occur statistically more often in children with ASD in a blinded study.

2.3.2. Guidelines for the Clinical Genetic Evaluation of ASD

With ever changing advancements in genetic testing technology comes the question of which test to use when. To answer this question the American College of Medical Genetics' Professional Practice and Guidelines Committee developed a framework to aid the geneticists in these evaluations. The initial guidelines were published in 2008 [42] and subsequently revised [12] to reflect further changes in testing technologies and emerging findings in the literature. This report proposes a "tiered" approach to the diagnostic evaluation. The earlier tiers use testing modalities that have a higher diagnostic yield and are less invasive/easier to accomplish than others. In brief, the most recent report notes an expected diagnostic yield of 30%–40%. That is to say that a specific mono-factorial etiology can be identified in patients with ASD in 30%–40% of the cases. For details of these evaluations the reader is referred to the reference provided [12]. While these numbers may seem low to many, they represent a significant increase from the 6%–12% yield reported prior to the year 2000. As such it is recommended that every person with the diagnosis of ASD be offered a clinical genetic consultation. That is not to say that every person with ASD will receive such a work up. Many details factor into the ultimate decision of whether to proceed with testing or not. It is crucial that these families receive detailed pre-evaluation informed consent. They need to understand what the tiered process entails, what the projected costs may be (especially in light of 3rd party payer policies), any potential risks of the testing and, of course, the diagnostic yield. Once the family understands what is involved that can decide whether such an evaluation is appropriate for them.

At the time of this writing, the 2013 guidelines are only a couple of years old. However, the ever advancing science of genetic testing is soon going to necessitate another revision of these recommendations. The advent of relatively inexpensive and rapid genomic testing methods has already been applied to cohorts of patients with ASD [43–46]. Currently a handful of publications have found that whole exome sequencing (WES) and whole genome sequencing (WGS) are expected to significantly increase the diagnostic yield when applied to patients with ASD. A summary of se current reports suggests that WES/WGS may identify an additional 10%–15% of causes of ASD. It is important to note that current sequencing techniques are not very good at detecting the larger copy number variants. Thus these techniques do not obviate the use of chromosomal microarray (CMA). Microarray testing itself is a high yield test identifying an etiology of ASD in about 10% of cases. Together WES and CMA alone may identify the cause of ASD in 20% of cases [45]. Finally, it is also important to note that the diagnostic yields are different for the 2 ASD sub-groups. Not surprisingly, the yield is higher (about double) in the complex/syndromic group.

2.3.3. What about Unknowns?

The recent advances in genetics and genomics have been extremely exciting and gratifying for clinical geneticists. Cases that have defied answers for literally decades are now being understood. As noted above, the increase in diagnostic yield for ASD from around 10% to almost 40% is indeed exciting. However, that still means that 60% of cases will have no answer despite a thorough evaluation. So the questions remain; "What about the unknown cases?" "What could possibly be the cause of the other cases?" We have already alluded to the fact that the application of WES/WGS testing to the diagnostic algorithm will undoubtedly increase the diagnostic yield. It is a little too early yet to definitively say by how much. However, consideration of the several early reports suggests that the application of these modalities in testing may increase the yield to 50%, maybe 60%. At that point it appears that the straight-forward, mono-factorial etiologies will mostly have been identified. The remainder of cases is likely due to more complicated mechanisms. Many such mechanisms have been suggested. Some of these would include:

(1) Mutations in non-coding DNA
(2) Epigenetic disorders including those fitting a MEGDI model—mixed genetics/epigenetics
(3) Multiple contributing loci
(4) Complex gene x gene interactions
(5) Snippets

It is highly likely that further insights into the etiology of ASD will come from investigations into these (and other) mechanisms.

2.4. Metabolic and Mitochondrial Disorders

Among the many genetic causes of ASD, metabolic and mitochondrial disorders deserve a separate discussion. These conditions have been the subject of vigorous debate among genetic and metabolic specialists over the past many years. The importance in sorting out the answers in these conditions lies in the commonly stated point that these conditions are "low yield, but high impact" [12]. Specifically, these conditions have the greatest potential for treatment among the myriad of causes of ASD. They may also present special situations of genetic susceptibility to environmental factors as well as project for a more progressive/deteriorative course.

Many metabolic (*i.e.*, inborn errors of metabolism) and mitochondrial conditions have been reported in association with ASD [47–49]. While some have been reported with the non-syndromic phenotype, many exhibit a variety of other features that stand as clues to the possible cause. Signs and symptoms that suggest such a possible etiology include poor linear growth, laboratory changes (anemia, acid-base abnormalities, lactic acidosis), cyclical vomiting, rashes, tone abnormalities, gastrointestinal problems, refractory epilepsy, changes in sensorium, neuro-regression outside of the typical speech loss of 18–24 months and microcephaly. An especially important situation to consider is developmental regression associated with fever or illness. This type of regression especially with recurrent episodes and multiple organ dysfunctions are clues to a possible mitochondrial disorder. These patients have been noted to have loss of speech after a febrile illness or immunizations with a subsequent encephalopathy. (It is particularly important to note that despite this association, immunizations do not cause ASD. There simply are no reputable studies which show any such plausible link.) In the evaluation of the etiology of a person with ASD, the presence of any of these clues should prompt an extended evaluation to include metabolic and mitochondrial causes.

3. Conclusions

A large body of solid and consistent studies identifies a clear genetic etiology of ASD. Like most complex phenotypes, ASD has a "multi-factorial etiology". This means that both genetic and environmental factors contribute to the overall phenotype. Heritability estimates suggest that genetic factors by far play the largest role. Clinical genetic evaluations currently can identify the cause of ASD in 30%–40% of cases. Early evidence suggests that the addition of genomic techniques to the standardly recommended testing protocol is likely to increase this yield to over 50%. Such evaluations should be offered to families after clear and careful informed consent. Knowledge of the cause of ASD can be very empowering for families. Simply knowing "Why" can be very reassuring and directive for many families. In addition, this provides genetic information for familial risks, prognosis, and awareness of possible associated conditions.

Most importantly, the identification of a genetic etiology of ASD is the first step in establishing "personalized medicine" for these patients. What is currently quite clear is that ASD is not a single condition. It is a collective of conditions with a common, strictly defined neuro-behavioral phenotype.

In reality it is "the ASDs" with literally hundreds of genetic causes identified. The delineation of specific subtypes (endo-phenotypes) of ASD will facilitate the investigation of differential responses to therapies. This coupled with a growing understanding of ASD as a disorder of the synapses should allow for the development of targeted therapies based upon a specific genetic diagnosis. While it is impossible to dogmatically state when this might happen, it is anticipated that this type of focused intervention is just on the horizon.

Conflicts of Interest: The author declare no conflict of interest.

References

1. Folstein, S.E.; Piven, J. Etiology of autism: Genetic influences. *Pediatrics* **1991**, *87*, 767–773. [PubMed]
2. Schaefer, G.B.; Lutz, R.E. Diagnostic yield in the clinical genetic evaluation of autism spectrum disorders. *Genet. Med.* **2006**, *8*, 549–556. [PubMed]
3. Schaefer, G.B.; Mendelsohn, N.J. Genetics evaluation for the etiologic diagnosis of autism spectrum disorders. *Genet. Med.* **2008**, *10*, 4–12. [CrossRef] [PubMed]
4. Sandin, S.; Lichtenstein, P.; Kuja-Halkola, R.; Larsson, H.; Hultman, C.M.; Reichenberg, A. The familial risk of autism. *JAMA* **2014**, *311*, 1770–1777. [CrossRef] [PubMed]
5. Gronborg, T.K.; Schendel, D.E.; Parner, E.T. Recurrence of autism spectrum disorders in full- and half-siblings and trends over time: A population-based cohort study. *JAMA Pediatr.* **2013**, *167*, 947–953. [CrossRef] [PubMed]
6. Risch, N.; Hoffmann, T.J.; Anderson, M.; Croen, L.A.; Grether, J.K.; Windham, G.C. Familial recurrence of autism spectrum disorder: Evaluating genetic and environmental contributions. *Am. J. Psychiatry* **2014**, *171*, 1206–1213. [CrossRef] [PubMed]
7. Constantino, J.N.; Todorov, A.; Hilton, C.; Law, P.; Zhang, Y.; Molloy, E.; Fitzgerald, R.; Geschwind, D. Autism recurrence in half siblings: Strong support for genetic mechanisms of transmission in ASD. *Mol. Psychiatry* **2013**, *18*, 137–138. [CrossRef] [PubMed]
8. Jensen, C.M.; Steinhausen, H.C.; Lauritsen, M.B. Time trends over 16 years in incidence-rates of autism spectrum disorders across the lifespan based on nationwide Danish register data. *J. Autism Dev. Disord.* **2014**, *44*, 1808–1818. [CrossRef] [PubMed]
9. McCarthy, M. Autism diagnoses in the US rise by 30%, CDC reports. *BMJ* **2014**. [CrossRef] [PubMed]
10. Kahler, S.G. Increasing incidence of autism and genetics. Personal Communication, 2014.
11. Schaefer, G.B.; Thompson, J.N. *Medical Genetics: An Integrated Approach*; McGraw Hill Education: New York, NY, USA, 2014.
12. Schaefer, G.B.; Mendelsohn, N.J.; Professional Practice Guideline Committee. Clinical genetics evaluation in identifying the etiology of autism spectrum disorders: 2013 Guideline revisions. *Genet. Med.* **2013**, *15*, 399–407. [CrossRef] [PubMed]
13. Constantino, J.N.; Zhang, Y.; Frazier, T.; Abbacchi, A.M.; Law, P. Sibling recurrence and the genetic epidemiology of autism. *Am. J. Psychiatry* **2010**, *167*, 1349–1356. [CrossRef] [PubMed]
14. Ozonoff, S.; Young, G.S.; Carter, A.; Messinger, D.; Yirmiya, N.; Zwaigenbaum, L.; Bryson, S.; Carver, L.J.; Constantino, J.N.; Dobkins, K.; et al. Recurrence risk for autism spectrum disorders: A baby siblings research consortium study. *Pediatrics* **2011**, *128*, e488–e495. [CrossRef] [PubMed]
15. Werling, D.M.; Geschwind, D.H. Recurrence rates provide evidence for sex-differential, familial genetic liability for autism spectrum disorders in multiplex families and twins. *Mol. Autism* **2015**, *6*, 27. [CrossRef] [PubMed]
16. Wood, C.L.; Warnell, F.; Johnson, M.; Hames, A.; Pearce, M.S.; McConachie, H.; Parr, J.R. Evidence for ASD recurrence rates and reproductive stoppage from large UK ASD research family databases. *Autism Res.* **2015**, *8*, 73–81. [CrossRef] [PubMed]
17. Jacquemont, S.; Coe, B.P.; Hersch, M.; Duyzend, M.H.; Krumm, N.; Bergmann, S.; Beckmann, J.S.; Rosenfeld, J.A.; Eichler, E.E. A higher mutational burden in females supports a "female protective model" in neurodevelopmental disorders. *Am. J. Hum. Genet.* **2014**, *94*, 415–425. [CrossRef] [PubMed]
18. Ritvo, E.R.; Jorde, L.B.; Mason-Brothers, A.; Freeman, B.J.; Pingree, C.; Jones, M.B.; McMahon, W.M.; Petersen, P.B.; Jenson, W.R.; Mo, A. The UCLA-University of Utah epidemiologic survey of autism: Recurrence risk estimates and genetic counseling. *Am. J. Psychiatry* **1989**, *146*, 1032–1036. [PubMed]

19. Messinger, D.S.; Young, G.S.; Webb, S.J.; Ozonoff, S.; Bryson, S.E.; Carter, A.; Carver, L.; Charman, T.; Chawarska, K.; Curtin, S.; *et al.* Early sex differences are not autism-specific: A Baby Siblings Research Consortium (BSRC) study. *Mol. Autism* **2015**, *6*, 32. [CrossRef] [PubMed]

20. Frazier, T.W.; Youngstrom, E.A.; Hardan, A.Y.; Georgiades, S.; Constantino, J.N.; Eng, C. Quantitative autism symptom patterns recapitulate differential mechanisms of genetic transmission in single and multiple incidence families. *Mol. Autism* **2015**, *6*, 58. [CrossRef] [PubMed]

21. Grønborg, T.K.; Hansen, S.N.; Nielsen, S.V.; Skytthe, A.; Parner, E.T. Stoppage in Autism Spectrum Disorders. *J. Autism Dev. Disord.* **2015**, *45*, 3509–3519. [CrossRef] [PubMed]

22. Hoffmann, T.J.; Windham, G.C.; Anderson, M.; Croen, L.A.; Grether, J.K.; Risch, N. Evidence of reproductive stoppage in families with autism spectrum disorder: A large, population-based cohort study. *JAMA Psychiatry* **2014**, *71*, 943–951. [CrossRef] [PubMed]

23. Durkin, M.S.; Maenner, M.J.; Newschaffer, C.J.; Lee, L.C.; Cunniff, C.M.; Daniels, J.L.; Kirby, R.S.; Leavitt, L.; Miller, L.; Zahorodny, W.; *et al.* Advanced parental age and the risk of autism spectrum disorder. *Am. J. Epidemiol.* **2008**, *168*, 1268–1276. [CrossRef] [PubMed]

24. Schaefer, G.B.; Bodensteiner, J.B. Evaluation of the child with idiopathic mental retardation. *Pediatr. Clin. N. Am.* **1992**, *39*, 929–943.

25. Moeschler, J.B.; Shevell, M.; Committee on Genetics. Comprehensive evaluation of the child with intellectual disability or global developmental delays. *Pediatrics* **2014**, *134*, e903–e918. [CrossRef] [PubMed]

26. Hallmayer, J.; Hebert, J.M.; Spiker, D.; Lotspeich, L.; McMahon, W.M.; Peterse, P.B.; Nicholas, P.; Pingree, C.; Lin, A.A.; Cavalli-Sforza, L.L.; *et al.* Autism and the X chromosome: Multipoint sib-pair analysis. *Arch. Gen. Psychiatry* **1996**, *53*, 985–989. [CrossRef] [PubMed]

27. Reddy, K.S. Cytogenetic abnormalities and fragile-X syndrome in Autism Spectrum Disorder. *BMC Med. Genet.* **2005**, *6*, 3. [CrossRef] [PubMed]

28. Shevell, M.I.; Majnemer, A.; Rosenbaum, P.; Abrahamowicz, M. Etiologic yield of autistic spectrum disorders: A prospective study. *J. Child Neurol.* **2001**, *16*, 509–512. [CrossRef] [PubMed]

29. Weidmer-Mikhail, E.; Sheldon, S.; Ghaziuddin, M. Chromosomes in autism and related pervasive developmental disorders: A cytogenetic study. *J. Intellect. Disabil. Res.* **1998**, *42*, 8–12. [CrossRef] [PubMed]

30. Johnson, C.P.; Myers, S.M. Identification and evaluation of children with autism spectrum disorders. *Pediatrics* **2007**, *120*, 1183–1215. [CrossRef] [PubMed]

31. Carter, M.T.; Scherer, S.W. Autism spectrum disorder in the genetics clinic: A review. *Clin. Genet.* **2013**, *83*, 399–407. [CrossRef] [PubMed]

32. Jones, K.L.; Jones, M.C.; del Campo, M. *Smith's Recognizable Patterns of Human Malformations*, 7th ed.; Elsevier, Saunders: Philadelphia, PA, USA, 2013.

33. Miles, J.H.; Hillman, R.E. Value of a clinical morphology examination in autism. *Am. J. Med. Genet.* **2000**, *91*, 245–253. [CrossRef]

34. Wong, V.C.; Fung, C.K.; Wong, P.T. Use of dysmorphology for subgroup classification on autism spectrum disorder in Chinese children. *J. Autism Dev. Disord.* **2014**, *44*, 9–18. [CrossRef] [PubMed]

35. Thomas, I.T.; Frias, J.L.; Cantu, E.S.; Lafer, C.Z.; Flannery, D.B.; Graham, J.G., Jr. Association of pigmentary anomalies with chromosomal and genetic mosaicism and chimerism. *Am. J. Hum. Genet.* **1989**, *45*, 193–205. [PubMed]

36. Guo, X.; Tu, W.J.; Shi, X.D. Tuberous sclerosis complex in autism. *Iran J. Pediatr.* **2012**, *22*, 408–411. [PubMed]

37. Biesecker, L.G.; Spinner, N.B. A genomic view of mosaicism and human disease. *Nat. Rev. Genet.* **2013**, *14*, 307–320. [CrossRef] [PubMed]

38. Hoffbuhr, K.; Devaney, J.M.; LaFleur, B.; Sirianni, N.; Scacheri, C.; Giron, J.; Schuett, J.; Innis, J.; Marino, M.; Philippart, M.; *et al.* MeCP2 mutations in children with and without the phenotype of Rett syndrome. *Neurology* **2001**, *56*, 1486–1495. [CrossRef] [PubMed]

39. Eriksson, M.A.; Westerlund, J.; Anderlid, B.M.; Gillber, C.; Fernell, E. First-degree relatives of young children with autism spectrum disorders: Some gender aspects. *Res. Dev. Disabil.* **2012**, *33*, 1642–1648. [CrossRef] [PubMed]

40. Moreno-De-Luca, D.; SGENE Consortium. Deletion 17q12 is a recurrent copy number variant that confers high risk of autism and schizophrenia. *Am. J. Hum. Genet.* **2010**, *87*, 618–630. [CrossRef] [PubMed]

41. Schaefer, G.B. Personal Communication of unpublished data, 2015.

42. Schaefer, G.B.; Mendelsohn, N.J. Clinical genetics evaluation in identifying the etiology of autism spectrum disorders. *Genet. Med.* **2008**, *10*, 301–305. [CrossRef] [PubMed]
43. Butler, M.G.; Dasouki, M.J.; Zhou, X.P.; Talebizadeh, Z.; Brown, M.; Takahashi, T.N.; Miles, J.H.; Wang, C.H.; Stratton, R.; Pilarski, R.; *et al.* Subset of individuals with autism spectrum disorders and extreme macrocephaly associated with germline PTEN tumour suppressor gene mutations. *J. Med. Genet.* **2005**, *42*, 318–321. [CrossRef] [PubMed]
44. Jiang, Y.H.; Yuen, R.K.; Jin, X.; Wang, M.; Chen, N.; Wu, X.; Ju, J.; Mei, J.; Shi, Y.; He, M.; *et al.* Detection of clinically relevant genetic variants in autism spectrum disorder by whole-genome sequencing. *Am. J. Hum. Genet.* **2013**, *93*, 249–263. [CrossRef] [PubMed]
45. Tammimies, K.; Marshall, C.R.; Walker, S.; Kaur, G.; Thiruvahindrapuram, B.; Lionel, A.C.; Yuen, R.K.; Uddin, M.; Roberts, W.; Weksberg, R.; *et al.* Molecular diagnostic yield of chromosomal microarray analysis and whole-exome sequencing in children with Autism Spectrum Disorder. *JAMA* **2015**, *314*, 895–903. [CrossRef] [PubMed]
46. Yuen, R.K.; Thiruvahindrapuram, B.; Merico, D.; Walker, S.; Tammimies, K.; Hoang, N.; Chrysler, C.; Nalpathamkalam, T.; Pellecchia, G.; Liu, Y.; *et al.* Whole-genome sequencing of quartet families with autism spectrum disorder. *Nat. Med.* **2015**, *21*, 185–191. [CrossRef] [PubMed]
47. Haas, R.H. Autism and mitochondrial disease. *Dev. Disabil. Res. Rev.* **2010**, *16*, 144–153. [CrossRef] [PubMed]
48. Page, T. Metabolic approaches to the treatment of autism spectrum disorders. *J. Autism Dev. Disord.* **2000**, *30*, 463–469. [CrossRef] [PubMed]
49. Zecavati, N.; Spence, S.J. Neurometabolic disorders and dysfunction in autism spectrum disorders. *Curr. Neurol. Neurosci. Rep.* **2009**, *9*, 129–136. [CrossRef] [PubMed]

International Journal of
Molecular Sciences

MDPI

Review

Role of Genetics in the Etiology of Autistic Spectrum Disorder: Towards a Hierarchical Diagnostic Strategy

Cyrille Robert [1,2], Laurent Pasquier [2], David Cohen [3], Mélanie Fradin [2], Roberto Canitano [4], Léna Damaj [2], Sylvie Odent [2] and Sylvie Tordjman [1,5,*]

[1] Pôle Hospitalo-Universitaire de Psychiatrie de l'Enfant et de l'Adolescent (PHUPEA), University of Rennes 1 and Centre Hospitalier Guillaume Régnier, 35200 Rennes, France; robertcyrille@live.fr
[2] Service de Génétique Clinique, Centre de Référence Maladies Rares Anomalies du Développement (Centre Labellisé pour les Anomalies du Développement de l'Ouest: CLAD Ouest), Hôpital Sud, Centre Hospitalier Universitaire de Rennes, 35200 Rennes, France; laurent.pasquier@chu-rennes.fr (L.P.); Melanie.FRADIN@chu-rennes.fr (M.F.); Lena.DAMAJ@chu-rennes.fr (L.D.); Sylvie.Odent@chu-rennes.fr (S.O.)
[3] Hospital-University Department of Child and Adolescent Psychiatry, Pitié-Salpétrière Hospital, Paris 6 University, 75013 Paris, France; david.cohen@psl.aphp.fr
[4] Division of Child and Adolescent Neuropsychiatry, University Hospital of Siena, 53100 Siena, Italy; r.canitano@ao-siena.toscana.it
[5] Laboratory of Psychology of Perception, University Paris Descartes, 75270 Paris, France
* Correspondence: s.tordjman@yahoo.fr; Tel.: +33-6-1538-0748; Fax: +33-2-9964-1807

Academic Editor: Merlin G. Butler
Received: 13 December 2016; Accepted: 20 February 2017; Published: 12 March 2017

Abstract: Progress in epidemiological, molecular and clinical genetics with the development of new techniques has improved knowledge on genetic syndromes associated with autism spectrum disorder (ASD). The objective of this article is to show the diversity of genetic disorders associated with ASD (based on an extensive review of single-gene disorders, copy number variants, and other chromosomal disorders), and consequently to propose a hierarchical diagnostic strategy with a stepwise evaluation, helping general practitioners/pediatricians and child psychiatrists to collaborate with geneticists and neuropediatricians, in order to search for genetic disorders associated with ASD. The first step is a clinical investigation involving: (i) a child psychiatric and psychological evaluation confirming autism diagnosis from different observational sources and assessing autism severity; (ii) a neuropediatric evaluation examining neurological symptoms and developmental milestones; and (iii) a genetic evaluation searching for dysmorphic features and malformations. The second step involves laboratory and if necessary neuroimaging and EEG studies oriented by clinical results based on clinical genetic and neuropediatric examinations. The identification of genetic disorders associated with ASD has practical implications for diagnostic strategies, early detection or prevention of co-morbidity, specific treatment and follow up, and genetic counseling.

Keywords: autism; genetic disorders; hierarchical diagnostic strategy; child psychiatric and psychological assessment; clinical genetics; neuropediatric evaluation

1. Introduction

Autism Spectrum Disorder (ASD) is considered as a neurodevelopmental disorder defined in the DSM-5 (the Diagnostic and Statistical Manual of Mental Disorders 5th ed.) [1], the most recent diagnostic classification of mental disorders, by social communication deficits associated with repetitive/stereotyped behaviors or interests with early onset (referenced in the DSM-5 as symptoms present at early developmental period and evident in early childhood). According to the historian Normand J. Carrey, the first description of autism can be traced to Itard in his 1828 report on "mutism produced by a lesion of intellectual functions" [2]. In 1943, Leo Kanner [3], American psychiatrist with Austrian origins, borrowed the term *autism* from the Swiss psychiatrist Eugen Bleuler to define a specific syndrome observed in 11 children (the term "autism", derived from the Greek autos which means "oneself", was used for the first time by Eugen Bleuler in 1911 [4] to describe social withdrawal in adult patients with schizophrenia). The Kanner syndrome was characterized by its early onset (from the first year of life) and symptomatology (social withdrawal, sameness, language impairment, stereotyped motor behaviors, and intellectual disability). It is noteworthy that Kanner did not describe autism in individuals with severe intellectual disability and known brain disorders. The same year, the Austrian psychiatrist Hans Asperger [5] distinguished "personalities with autistic tendencies" which differed from the children that Kanner had described, due to the expression of exceptional isolated talents and conserved linguistic abilities. Currently, autism is viewed as an epistatic and multifactorial disorder involving genetic factors and environmental factors (prenatal, neonatal and/or postnatal environmental factors such as gestational diabetes, neonatal hypoxia conditions, exposure to air pollution, parental immigration or sensory/social deficits; for a review, see Tordjman et al., 2014 [6]).

Several literature reviews underline the important role of genetics in the etiology of autism [7–9]. The data come from family and twin studies. Indeed, the concordance rate among monozygotic twins is high (60%–90%) compared to the concordance rate among dizygotic twins (0%–30%) [10,11]. Technological advances in epidemiological and molecular genetics have led recently to new findings in the field of the genetics of neuropsychiatric disorders. These new findings concern also the domain of the genetics of autism and have increased the state of current knowledge on the genetic disorders associated with ASD. Identified genetic causes of ASD can be classified as the cytogenetically visible chromosomal abnormalities, copy number variants (CNVs) (i.e., variations in the number of copies of one segment of DNA, including deletions and duplications), and single-gene disorders [12]. CNVs can have different sizes (small to large deletions or duplications) and therefore concern a variable number of genes according to their size. The number of known genetic disorders associated with ASD has increased with the use of array Comparative Genomic Hybridization (aCGH), also called chromosomal microarrays (CMAs) or high-resolution molecular karyotype, which is one of the most molecular cytogenetic methods used by geneticists. The accuracy of aCGH is 10 to 100 times higher than low-resolution karyotype accuracy and allows detection of small CNVs. Unfortunately, this method does not detect certain genetic anomalies such as balanced chromosome rearrangements (which are rarely involved in developmental disorders) or CNVs present in less than 10% to 20% of cells (somatic mosaicism). CNVs are usually considered to be rare variants but recent techniques have shown that they may be in fact more frequent than expected [13]. In practice, after CNV detection, the geneticist needs to determine the contribution of this variation to the patient's disorder. Most of the time, parental CNVs are also analyzed for family segregation. Among genetic causes of ASD, besides single specific genetic factors, cumulative effects of Single Nucleotide Variants (SNVs: common small variations in the DNA sequence occurring within a population) also have to be taken into consideration. Single-nucleotide polymorphism (SNP) is the term usually used for SNVs occurring in more than 1% of the general population. In the next few years, it will be probably possible to improve the identification and interpretation of SNPs in ASD using new techniques of High Throughput Sequencing, such as targeted sequencing on known genes (gene panel) or Whole Exome/Genome Sequencing. Finally, it should be highlighted that a genetic model of ASD (including cumulative genetic variance) does not exclude the role of environment (including cumulative environmental variance). It is noteworthy that heritability (h^2) is defined as $h^2 = GV/(GV + EV)$ where GV is the cumulative genetic variance and EV, the environmental variance [14].

Many susceptibility genes and cytogenetic abnormalities have been reported in ASD and concern almost every chromosome (for a review, see Miles, 2011; available online: http://projects.tcag.ca/autism/) [15]. How can we explain such genetic diversity associated with similar autism cognitive-behavioral phenotypes? Given the current state of knowledge, we can state the hypothesis that the majority of ASD-related genes are involved in brain development and functioning (such as synapse formation and functioning, brain metabolism, chromatin remodeling) [16].

Several studies have reported associations between genetic contribution (such as unbalanced chromosome abnormalities, in particular large chromosomal abnormalities), dysmorphic features and low cognitive functioning [17–19]. Indeed, large chromosomal abnormalities are more often found in children with dysmorphic features, whereas small CNVs and de novo SNPs are more often found in individuals without dysmorphic features. The concept of "syndromic autism" or "complex autism" (autism associated with genetic disorders/genetic syndromes) which qualifies individuals with at least one dysmorphic feature/malformation or severe intellectual disability, is opposed to the concept of "non-syndromic autism" or "simplex"/"pure"/idiopathic autism (isolated autism) which qualifies individuals with moderate intellectual disability to normal cognitive functioning and no other associated signs or symptoms (except the presence of seizures) [19,20]. The prevalence of epilepsy in ASD varies from 5% to 40% (compared to 0.5% to 1% in the general population) according to different variables, including syndromic vs. non-syndromic autism (higher rates of epilepsy are observed in syndromic autism compared to non-syndromic autism) [21–24] and the age of the population (the rate of seizures varies with age and increases especially during the prepubertal period). It is noteworthy that genome-wide association and genetic analyses such as aCGH have revealed hundreds of previously unknown rare mutations and CNVs linked to non-syndromic autism. Some authors [12,19] suggest that syndromic autism (SA), compared to non-syndromic autism (NSA), is associated with a poorer prognosis, a lower male-to-female sex ratio (SA: 3/1; NSA: 6/1), and a lower sibling recurrence risk (SA: 4%–6%; NSA: up to 35%) as well as a lower risk for family history of autism (SA: up to 9%; NSA: up to 20%) given that SA would be more related to accidental mutations than NSA. Similarly, the concept of syndromic intellectual disability (intellectual disability associated with dysmorphism and medical or behavioral signs and symptoms) is widely used as opposed to non-syndromic intellectual disability (intellectual disability without other abnormalities). Intellectual disability and autism spectrum disorder share other similarities given that they are both genetically heterogeneous, and a significant number of genes have been associated with both conditions. It is noteworthy that the concept of non-syndromic autism or non-syndromic intellectual disability depends on technological progress and current state of knowledge. We can foresee, as underlined by Tordjman et al. [25], that individuals with currently and apparently "non-syndromic autism" could become a few years later individuals with "syndromic autism", if new genetic disorders are identified, allowing a better understanding and more efficient identification of discrete clinical symptoms as minor physical anomalies, malformations and associated biological anomalies. For example, individuals with Smith–Magenis syndrome can show in early childhood ASD associated with subtle minor facial dysmorphic features (these discrete dysmorphic features become more evident later), mild or moderate intellectual disability, disrupted sleep patterns and repetitive self-hugging characteristic of this syndrome. The molecular discovery of Smith–Magenis syndrome is relatively recent (1982) and allowed to better know the characteristic behavioral phenotype combined with physical anomalies of Smith–Magenis syndrome (including short stature and scoliosis, not enough specific to help alone to identify the genetic syndrome), orienting towards the search for certain dysmorphic features and the possible identification of this genetic disorder. Thus, some children with Smith–Magenis syndrome could be considered in early childhood, especially before the discovery of Smith–Magenis syndrome in 1982, as individuals with apparently "non-syndromic autism" (they are more viewed as individuals with syndromic autism over their developmental trajectory). Therefore, taking into account this bias, it would be better to consider autism as a behavioral syndrome related either to known genetic disorders or to currently unknown causes.

The objective of this article is to show the diversity of the genetic disorders associated with ASD (including single-gene disorders, CNVs and other chromosomal disorders), and consequently to propose a hierarchical diagnostic strategy with a stepwise evaluation, helping general practitioners/pediatricians and child psychiatrists to collaborate with geneticists and neuropediatricians, in order to search for genetic disorders associated with ASD.

2. Known Genetic Syndromes Associated with Autism Spectrum Disorder

We chose to limit the review of literature to genetic disorders presented in Table 1. Table 1 includes one part on chromosomal disorders and another part on single-gene disorders. The present descriptions focus on autistic traits encountered in each genetic disorder and other symptoms that may be helpful for the diagnosis. Table 1 summarizes these clinical data and indicates the estimated frequency of each genetic disorder in autism and the frequency of autistic traits in each genetic disorder (when available). This table does not present an exhaustive list of genetic disorders associated with ASD. Furthermore, Table 1 includes only genetic syndromes involving a single-gene disorder or a chromosomal rearrangement, and therefore does not include polygenic causes and possible genetic disorders only related to epigenetic mechanisms. It is noteworthy that some studies suggest that the genetic background might play an important role in many ASD individuals, with in particular cumulative genetic effects related to the load of common risk variants [16].

Int. J. Mol. Sci. **2017**, *18*, 618

Table 1. Main genetic disorders associated with autistic syndrome.

Genetic Disorder [References]	Estimated Rate (%) of the Disorder in Autism	Estimated Rate (%) of Autism in the Disorder	Degree of Intellectual Disability (ID)	Possible Autistic Behaviors	Other Possible Behaviors	Other Possible Symptoms
			Chromosomal Disorders			
Maternal * 15q11-q13 duplication [26–38]	1–2	80–100	Severe	Severe autistic syndrome with severe expressive language impairment	Hyperactivity and aggression	Seizures (75%), hypotonia, genitor/urinary abnormalities
Angelman syndrome * (maternal 15q11-q13 deletion, paternal uniparental disomy, mutations of *UBE3A* that encodes an ubiquitin E3 ligase) [35,39–48]	1	48–80	Severe	No language, stereotypies, sameness	Attention Deficit with Hyperactivity Disorder (ADHD), paroxysmal laughter, tantrums	Facial dysmorphism, microcephaly, seizures (>1 year), ataxia and walking disturbance
Prader–Willi syndrome * (maternal uniparental disomy at 15q11-q13, paternal deletions) [35,49–58] Paternal 15q11-q13 duplication (only few cases reported to be associated with autism): behavioral phenotype similar to Prader–Willi syndrome [58–60]	Not Available (NA)	19–37	Mild to moderate	Motor and verbal stereotypies, rituals	Hyperphagia, obsessive-compulsive traits, temper tantrums	Obesity, growth delay and hypogonadism, facial dysmorphism, Hypotonia
Phelan–McDermid * syndrome (Inherited, de novo deletions at 22q13.3 leading to loss of *SHANK 3*) 22q13.3 duplication [20,61–70]	NA	75–84	Severe	Variable autistic syndrome with social communication impairments, including delayed or absent verbal language	Global developmental delay, atatonia in adolescence and adulthood	Dysmorphic features, hypotonia, gait disturbance, recurring upper respiratory tract infections, gastroesophageal reflux and seizures
16p11.2 duplication 16p11.2 deletion [71–77]	1	33	Severe	Severe autistic syndrome with speech impairment	Gross and fine coordination problems, SCH (Schizophrenia), anxiety, ADHD	Hypotonia (Multiple congenital anomalies are possible with more distal region)
Inverted duplication/deletion 8p21–23 [78,79]	NA	30–57	Variable	Mild to moderate autistic syndrome with absent or delayed verbal language	ADHD	Minor facial dysmorphism, hypotonia, agenesis of the corpus callosum, possible heart defect
Genetic disorder [References]	Estimated rate (%) of the disorder in autism	Estimated rate (%) of autism in the disorder	Degree of intellectual disability (ID)	Autistic behaviors	Other behaviors	Other symptoms
Down syndrome * (trisomy 21) [80–84]	2	5–10	Variable but usually severe when autism	Severe autistic syndrome	-	Facial dysmorphism, heart and intestine malformations
Smith–Magenis syndrome (17p11.2 deletion) [85–88]	<1	80–100	Variable	Self-injurious and stereotyped behaviors, sameness	Tantrums, possible social contact, sleep disturbance	Facial dysmorphism, peripheral neuropathy, hypotonia

21

Int. J. Mol. Sci. **2017**, *18*, 618

Table 1. *Cont.*

Genetic Disorder [References]	Estimated Rate (%) of the Disorder in Autism	Estimated Rate (%) of Autism in the Disorder	Degree of Intellectual Disability (ID)	Possible Autistic Behaviors	Other Possible Behaviors	Other Possible Symptoms
Potocki–Lupski syndrome (17p11.2 duplication) [89]	NA	50–100	Normal to moderate	Decreased eye contact, motor mannerisms or posturing, sensory hypersensitivity or preoccupation, repetitive behaviors or interests, lack of appropriate functional or symbolic play and lack of joint attention	Developmental delay, language impairment, and cognitive impairment	hypotonia, poor feeding and failure to thrive in infancy, oral-pharyngeal dysphagia, obstructive and central sleep apnea, structural cardiovascular abnormalities, electroencephalography (EEG), abnormalities, and hypermetropia
2q37 deletion [90–94]	<1	25–35	Mild to moderate	Severe communication impairment, stereotypies	Hypotonia, hyperactivity, Obsessive-Compulsive Disorder (OCD), aggression, sleep disturbance	Facial dysmorphism, microcephaly, growth delay/short stature, intestine and heart malformations, seizure
22q11.21 duplication and 22q11 deletion (DiGeorge/Velocardio-facial syndrome) [95–100]	NA	<10	Normal to severe ID	Autistic syndrome, Pervasive Developmental Disorder-Not Otherwise Specified PDD-NOS (ICD-10 criteria)	Learning disability, anxiety, ADHD, oppositional-defiant disorder, OCD, motor impairment	Facial dysmorphism, microcephaly, growth delay/short stature, craniofacial abnormalities/cleft palate, heart defect, hypotonia
1q21.1 Copy-Number Variation (CNV) (1q21.1 deletion/duplication) [101–104]	NA	<30	Normal to mild ID	Autistic syndrome, PDD-NOS (ICD-10 criteria)	Developmental delay, learning disability, anxiety, ADHD, aggression, SCZ and hallucination	Microcephaly (deletion) Macrocephaly (duplication)
Williams-Beuren syndrome * (7q11.23 deletion) and Reciprocal 7q11.23 duplication syndrome [105–117]	<1	<10	Mild to moderate	Autistic syndrome	Overfriendliness, over talkativeness, visual spatial deficit, hyperacusis, feeding and sleep problems	Facial dysmorphism, short stature, heart and endocrine malformations, hypercalcemia
Turner syndrome * (most common monosomy For X chromosome) [37,118,119]	NA	3	Usually normal IQ	Females monosomic for the maternal chromosome X score significantly worse on social adjustment and verbal skills	-	Short stature, skeletal abnormalities, absence of ovarian function, webbed neck, lymphedema in hands and feet, heart defects and kidney problems

Table 1. *Cont.*

Genetic Disorder [References]	Estimated Rate (%) of the Disorder in Autism	Estimated Rate (%) of Autism in the Disorder	Degree of Intellectual Disability (ID)	Possible Autistic Behaviors	Other Possible Behaviors	Other Possible Symptoms
Beckwith–Wiedemann * syndrome (abnormal expression of imprinted genes on chromosome 11p15.5 such as *IGF2* and/or *CDKN1C*) [120–123]	NA	6.8 (replication needed)	Usually normal IQ	Autistic syndrome	-	Pre- and postnatal overgrowth (hemihyperplasia, macroglossia, visceromegaly) and increased risk of embryonal tumors
Isodicentric chromosome 15 or duplication/inversion 15q11 [124,125]	NA	NA	Moderate to severe	Autistic behavior	Developmental delay and intellectual deficit, epilepsy	Early central hypotonia
Ito hypomelanosis [126,127]	NA	NA	Inconstant	Asperger syndrome(high functioning autism) or atypical autism	Psychomotor delay and cognitive deficit	hypopigmented skin lesions along the Blaschko lines, motor delay, seizures, microcephaly or macrocephaly, hypotonia, ophthalmological abnormalities
Single Gene Disorders						
CHARGE syndrome * (*CHD7*, 8q21.1) [128–133]	<1	15–50	Variable but often normal IQ	Variable autistic syndrome	Hyperactivity, obsessive-compulsive traits, tic disorders	Coloboma of the eye, Heart defects, Atresia of the nasal choane, Retardation of growth and/or development, Genital/urinary abnormalities, Ear abnormalities/deafness
Tuberous sclerosis (*TSC1*, 9q34) (*TSC2*, 16p13.3) [134–136]	1–4	25–60	Variable	Severe autistic syndrome	Learning disorder	Ectodermal anomalies, renal lesions, seizures
PTEN macrocephaly syndromes (*PTEN*, 10q23.31) [137–142]	4 in ASD with macro-cephaly	NA	Severe	Autistic syndrome and language delay	=	Progressive macrocephaly, developmental delay, macrosomy, tumors in adulthood
Rett's syndrome * (*MECP2*, Xq28) [117,143–149]	<1 in female	61–100	Severe	Stereotyped hand movements, absence of language, loss of social engagement	Stagnation stage (6–18 months) in girls, then regression stage (12–36 months), pseudostationary t stage (2–10 years), l and late motor t deterioration (>10 years)	Head growth deceleration, progressive motor neuron (gait and truncal apraxia, ataxia, decreasing mobility) and respiratory (hyperventilation, breath holding, apnea) symptoms
San Filippo syndrome # (*SGSH*, 17q25.3) [150–153]	1 replication needed	80–100	Severe	Language impairment, autistic withdrawal, stereotyped behaviors impulsivity, inappropriate affects	Progressive loss of acquisitions	Motor regression, hepatomegaly
Cerebral folate deficiency # (*FOLR1*, 11q13.4) [154–157]	NA	NA	Variable	Autistic syndrome including especially social interaction and language impairment	irritability, movement (such as tremors) and gait disturbances with ataxia, sleep problems	Psychomotor regression, epilepsy, developmental delay, deceleration of head growth, dystonia/hypotonia, visual and hearing deficit
Smith–Lemli–Opitz syndrome # (*DHCR7*, 1q12–13) [158–163]	NA	50	Variable	Self-injurious behaviors, stereotypies ("opisthokinesis") language impairment	Sensory hyper-reactivity, irritability, sleep disturbance	Facial dysmorphism, cleft palate, congenital heart disease, hypospadias, 2–3 toe syndactyly

Table 1. *Cont.*

Genetic Disorder [References]	Estimated Rate (%) of the Disorder in Autism	Estimated Rate (%) of Autism in the Disorder	Degree of Intellectual Disability (ID)	Possible Autistic Behaviors	Other Possible Behaviors	Other Possible Symptoms
Phenylketonuria # (*PAH*, 12q22-q24.1) [20,164,165]	NA	NA	Severe	Self-injurious behavior, lack of social responsiveness	Temper tantrums, hyperactivity	Eczema, hypertonia, seizures, hypo-pigmentation
Adenylosuccinate lyase deficiency # (*ASL*, 22q13.1–13.2) [166–170]	<1	80–100	Variable	Severe autistic syndrome	-	Seizures
Creatine deficiency syndrome # (*GAMT*, 19p13.3) (*CRTR*, Xq28) [171–174]	<1	80–100	Severe	Severe autistic syndrome with poor language	-	Seizures, hypotonia
SHANK 3 (22q13.3) [175–177]	<1	NA	NA	Severe autistic syndrome with no language	-	-
Neurexin family: Neurexin 1 (*NRX1*, 2p16.3) [178–182]	1	NA	Variable	Autistic syndrome	Hyperactivity, depression, learning disability, but also normal behavior	Seizures, hypotonia, facial dysmorphism?
Contactin Associated Protein-like 2 (*CNTNAP2*, 7q35) [183–185]	NA	NA	Variable	Autistic syndrome including verbal language impairment	-	Seizures
Contactin 4 (*CNTN4*, 3p26.2-3p26.3) [186–189]	<1	NA	Variable	Autistic syndrome, PDD-NOS (ICD-10 criteria)	Visual spatial impairment, regression	Facial dysmorphism, developmental delay, hypotonia, ptosis.
Cell adhesion molecule-1 (*CADM1*, 11q23-23.2) [190–192]	NA	NA	NA	Autistic syndrome with especially social communication impairment including verbal language deficit	-	-
Protocadherin 10 (*PCDH10*, 4q28) [193]	<1	NA	NA	Autistic syndrome	-	-
Neuroligin family: Neuroligin 3 (*NLG3*, Xq13) Neuroligin 4 (*NLG4*, Xq22.33) [194–196]	<1	NA	Variable	Severe autistic syndrome, PDD-NOS (ICD-10 criteria)	Regression	Tic
Fragile X * (*FMR1*, Xq27.3) [191,196–222]	0-8	0-33	Variable	Poor eye contact , social anxiety, language deficit and stereotypies	Hyperactivity with attention deficit, sensory hyper-reactivity	Facial dysmorphism, macro-orchidism
Neurofibromatosis type 1 [109,223–225]	NA	21-40	inconstant	Restrictive/repetitive behaviors and severe social-communicative impairments	Cognitive deficits and learning difficulties	Café-au-lait spots, iris Lisch nodules, axillary and inguinal freckling, multiple neurofibromas
Sotos syndrome [226,227]	NA	70 (replication needed)	Mild to severe	Self-injurious behavior, physical aggression, and destruction	-	Facial dysmorphism, overgrowth of the body in early life with macrocephaly
Aarskog syndrome [228,229]	NA	NA	-	Variable autistic syndrome	Learning and behavioral disabilities	facial, limbs and genital features, acromelic short stature
Cornelia de Lange syndrome [117,230,231]	NA	43 (replication needed)	Variable	Self-injury, excessive repetitive behaviors and expressive language deficits	Psychomotor delay, language acquisition difficulties	Facial dysmorphism, severe growth delay, abnormal hands and feet, and constant brachymetacarpia of the first metacarpus, various other malformations (heart, kidney, etc.)

Table 1. *Cont.*

Genetic Disorder [References]	Estimated Rate (%) of the Disorder in Autism	Estimated Rate (%) of Autism in the Disorder	Degree of Intellectual Disability (ID)	Possible Autistic Behaviors	Other Possible Behaviors	Other Possible Symptoms
Joubert syndrome [232]	NA	<40	severe intellectual deficit to normal intelligence	Autistic syndrome	-	Facial dysmorphism, abnormal respiratory pattern, nystagmus, hypotonia, ataxia, and delay in achieving motor milestones
Cohen syndrome [117,233–235]	NA	54 (replication needed)	variable	Atypical autism	Often sociable with a cheerful disposition, learning and behavioral problems	Microcephaly, facial dysmorphism, hypotonia, myopia, retinal dystrophy, neutropenia and truncal obesity
Lujan–Fryns syndrome [236]	NA	NA	Mild to moderate	Autistic-like disorder	Behavioral problems, emotional instability, hyperactivity and shyness, schizophrenia	Tall, marfanoid stature, facial dysmorphism, hypernasal voice and generalized hypotonia
Noonan syndrome [237–239]	NA	15 (replication needed)	Mild	ASD	Poor feeding in infancy	Short stature, facial dysmorphism and congenital heart defects
Moebius syndrome [128,240,241]	NA	0–45	Inconstant, mild	ASD	Delayed walk development	Oculofacial paralysis, strabismus
Helsmoortel–Van der Aa syndrome (ADNP-related ID/ASD) [242]	0.17	100	Mild to severe	ASD, behavioral problems, sleep disturbance	Delayed developmental milestones (walking between 19 months and 4.5 years, and speech ranging from sentences to no words)	Facial dysmorphism, hypotonia, seizures, feeding difficulties, visual problems (hypermetropia, strabismus, cortical visual impairment), and cardiac defects
Timothy syndrome (CACNA1C) [243]	NA	50–70 (replication needed)	Inconstant, mild to severe	ASD	Language, motor, and generalized cognitive impairment	Facial dysmorphism, rate-corrected QT (QTc) interval >480 ms, unilateral or bilateral cutaneous syndactyly, heart defects, seizures

Adapted from Tordjman et al. [25]. # These genetic disorders are metabolic disorders (the list is not exhaustive). * The asterisk indicates the existence of epigenetic mechanisms observed in these genetic disorders (the list is not exhaustive).

Interestingly, as underlined by Abrahams and Geschwind [7], the descriptive analysis of Table 1 indicates that each of the many known genes or genomic regions associated with ASD, taken separately, accounts usually and on average for less than 2% of autism cases, suggesting high genetic heterogeneity (the maximum average rate is 5% and concerns the Fragile X syndrome). However, taken all together, these single genetic impairments account at least for 10%–25% of ASD individuals [244] and even 35%–40% in more recent studies [245].

3. Recommended Investigations for the Identification of Genetic Disorders Associated with ASD

Facing such a diversity of genetic disorders associated with ASD (see Table 1), it seems important to clarify, as best as possible with regard to the current state of knowledge, the diagnostic steps towards identification of these genetic disorders. A hierarchical diagnostic strategy with a stepwise evaluation towards identification of genetic disorders associated with autism is proposed and developed in Figure 1.

As it appears in Table 1, the severity of the autistic syndrome is an important variable to take into consideration in the search for genetic disorders associated with ASD. Therefore, the preliminary step in the hierarchical diagnostic strategy is to confirm autism diagnosis and assess autism severity. Furthermore, based on the genetic syndromes listed in Table 1, certain variables are of particular interest such as dysmorphic features, malformations, epilepsy and intellectual disability. It indicates clearly that clinical investigation of behavioral phenotype of autism for identification of genetic disorders associated with ASD requires clinical genetic examination, neuropediatric examination and psychological assessment. In this section, we propose a hierarchal diagnostic strategy with focused investigations for the identification of genetic disorders associated with ASD, based on the history of the individual (including family history and developmental trajectory) and the clinical neurologic and genetic evaluation, but also on up-to-date knowledge and new technology. The first step of the hierarchal diagnostic strategy is a clinical investigation involving: (i) a child psychiatrist's and psychologist's evaluation to confirm autism diagnosis using different observational sources (including an extensive parental interview facilitating the study of family history) and assess autism severity (including behavioral and cognitive assessments); (ii) a neuropediatrician's evaluation to examine neurological symptoms/signs and developmental milestones; and (iii) a geneticist's evaluation to search for dysmorphic features and malformations based on a clinical physical examination. This clinical step is followed by a second step involving laboratory and if necessary neuroimaging or EEG studies oriented by clinical results based on clinical genetic and neuropediatric examinations. These two steps are described below.

3.1. First Step: Clinical Investigation

3.1.1. Autism Diagnosis Using Different Observational Sources and Autism Severity Assessment (Child Psychiatric and Psychological Evaluation)

The preliminary step requires confirming the diagnosis of autism and assessing the severity of autistic behavioral impairments with validated tools used by trained professionals in different observational situations, such as the Autism Diagnostic Interview-Revised [246] (the ADI-R scale is a parental interview) and the Autism Diagnostic Observation Schedule [247] (the ADOS scale allows a direct observation of the individual through a standardized play situation). The psychiatric appreciation of ICD-10 (International Classification of Diseases, World Health Organization, Geneve, Switzerland, 1993) and DSM-5 (American Psychiatric Association, Washington, DC, USA, 2013) diagnostic criteria of autism is also necessary and provides a clinical psychiatric judgment. This approach, combining information from multiple sources based on the clinical psychiatric judgment and the administration of the ADI-R completed by the ADOS improves the confidence in the diagnosis of ASD [248]. Interestingly, the ADI-R scale is validated to assess current behavior but also behavior during the 4–5 years old period (the ADI-R algorithm is based on the 4–5 years old period of life which is supposed to correspond to the most severe period of autistic impairments), allowing to see the evolution between these two periods of life. It happens more frequently than expected that children who fulfill the diagnostic criteria for ASD based on the ADI-R algorithm (parental interview for the 4–5 years old period) do not meet the full diagnostic criteria for ASD based on the ADOS algorithm (direct observation of the child for the current period).

The ADI-R scale is the most common assessment used to conduct an extensive semi-structured parental interview searching for medical and psychiatric family history. This clinical interview should be conducted with empathy and allow a three-generation family tree to be constructed, showing, in particular, family psychiatric history (such as, for example, schizophrenia, depression or bipolar disorder) and ASD behaviors (including delayed language development) within the family (mother's and father's sides). The construction of the pedigree can be very helpful to get specific hypotheses of genetic disorders associated with autism. More generally, higher rates of CNVs are observed in ASD individuals with a family history of psychiatric disorders or developmental disabilities, highlighting the importance to study family history [249]. It is noteworthy that family history is also investigated and analyzed by the neuropediatrician and the geneticist during their evaluation.

Furthermore, the clinician should indicate in the family tree the existence of consanguineous mating, family malformations, deceased infants and spontaneous abortions occurring in the mother but also in relatives (repeated spontaneous abortions are in favor of chromosomal rearrangement [20,250,251]). It is necessary to specify the trimester of pregnancy when spontaneous abortions occurred given that repeated miscarriages at the first trimester of pregnancy are more specifically in favor of genetic anomalies. Reproductive stoppage is rarely evoked but should be investigated because reproductive curtailment following the birth of a child with ASD has been reported [252]. In addition, the birth order of the child with autism has to be noted in the family tree given that a decrease of ASD in later siblings has been observed [253]. The maternal and paternal ages at the time of procreation should be also systematically noted. Indeed, several studies suggest that advanced parental age, in particular for the father, is a risk factor for ASD [254], but there are some discrepancies in the results [253]. Although advanced parental age may be a risk factor, it orients towards genetic testing but does not lead to definitive conclusions.

Finally, a psychological evaluation is necessary in this hierarchical diagnostic strategy to assess the level of cognitive functioning given that intellectual disability and its severity are associated, as indicated previously, with a risk for several genetic disorders in ASD individuals (see Table 1). In addition, the level of cognitive functioning should be taken into consideration and therefore assessed given that severe intellectual disability can introduce an important bias leading to a misdiagnosis of autism. Indeed, ASD is often misdiagnosed in individuals with severe intellectual disability. It should be noted that the diagnosis of autism might be not valid or relevant for genetic disorders associated with severe intellectual disability due to first, the behavioral overlap between severe intellectual disability and autistic behavioral impairments (social communication deficit as well as repeated behaviors or interests), and second, the absence of validity and reliability of autism diagnostic instruments in the context of very low IQ or mental age of less than 24 months [255]. It is noteworthy that the most currently international instruments used for autism diagnosis, such as the ADI-R and the ADOS scales, were not normed in individuals with severe intellectual disability [255]. Cognitive functioning can be assessed, whenever possible with the child, by a psychologist or a neuropsychologist if available, using the age-appropriate Wechsler intelligence scales (Wechsler Preschool and Primary Scale of Intelligence: WPPSI-IV for children between 2 and 7 years old, Wechsler Intelligence Scale for Children: WISC-5 for children between 6 and 16 years old, Wechsler Adult Intelligence Scale: WAIS-IV for individuals above 16 years old) or the Kaufman Assessment Battery for Children (K-ABC) [256]. The Kaufman K-ABC is supposed to be more adapted for nonverbal children but cognitive assessment appears in fact easier to perform using the Wechsler intelligence scales with nonverbal subtests for nonverbal ASD children. Some instruments are of particular interest for intellectual disabled children with ASD such as the Raven's Color Progressive Matrices (CPM) which is a short (20 min) nonverbal intelligence test validated for young children and individuals with intellectual disability [257].

The child psychiatric evaluation and psychological evaluation confirming autism diagnosis from different observational sources and assessing autism severity and intellectual disability are followed by a neuropediatrician's evaluation and a geneticist's evaluation (see Figure 1).

Int. J. Mol. Sci. **2017**, *18*, 618

Autism Behavioral Phenotype

• Child Psychiatric and Psychological Evaluation

Confirmed Autism Diagnosis
(based on different observational sources)
+
Assessment of Autism Severity and Intellectual Disability (ID assessment when possible)

• Neuropediatric Evaluation
• Clinical Genetic Evaluation

Non-Syndromic Autism

(Isolated Autism but possible epilepsy or mild/moderate ID)

- If family History: aCGH

- If consanguinity or selective food aversion:
Targeted Metabolic Testing

- If individuals with ID:
aCGH with DNA testing for Fragile X Syndrome

- If male individuals with no ID:
DNA testing for Fragile X Syndrome

- If no Hypothesis:
Repeated Clinical Genetic Examination
(possible appearance of dysmorphic features later)

Syndromic Autism

(Dysmorphic Features, Malformations, symptomatic or cryptogenic epilepsy, or severe ID)

- If specific Chromosomal or Point Mutation suspected:
aCGH or specific gene sequencing

- If Metabolic Disorder suspected:
Targeted Metabolic Testing

- If Mitochondrial Disorder suspected:
Targeted Mitochondrial Testing

- If no Hypothesis:
1) First: aCGH with DNA Testing for Fragile X Syndrome
2) Then, if no identified variant: Targeted NGS based on clinical investigation (ID, Epilepsy, etc.) or WES (if available)

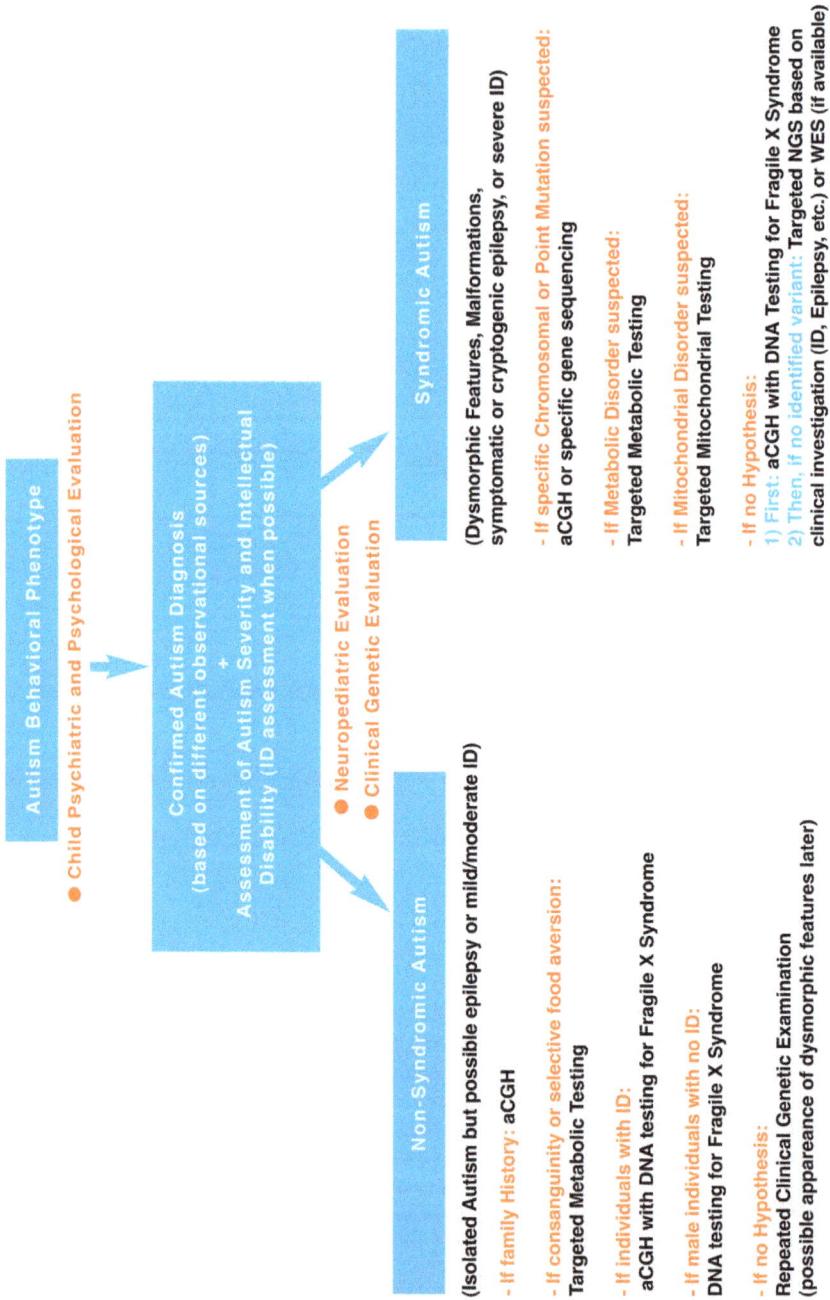

Figure 1. Hierarchical diagnostic strategy with a stepwise evaluation towards identification of genetic disorders associated with autism.

3.1.2. Developmental Milestones (Neuropediatric Evaluation)

A neuropediatric examination should be systematically conducted searching for neurological symptoms associated with ASD such as hypotonia, ataxia, abnormal movements or epilepsy. In case of clinical symptoms of epilepsy, the electroencephalogram (EEG) can be a useful tool to detect epileptic brain activity. In complement to this neuropediatric examination, it is necessary to rule out a sensory deficit, such as an auditory or visual deficit, based on ophtamological and ears-nose-throat examination (an audiogram is recommended). In addition, the neuropediatric examination investigates extensively developmental milestones for the following periods (some developmental milestones are often explored also during the child psychiatric evaluation): (i) prenatal period: course of pregnancy, fetal movements, results of ultrasound exams, treatments (in particular, in-utero exposure to thalidomide or valproate, known as teratogenic medications and risk factors for ASD), and other possible prenatal risk factors for ASD (such as gestational diabetes, gestational bleeding or multiple birth [6]); (ii) perinatal period: birth presentation (abnormal presentation in general and breech presentation in particular, have been associated with the development of ASD [258], mode of delivery, clinical status at birth (APGAR, score, weight, body length, and head circumference), premature birth (the risk for autism increased with the severity of preterm birth [259]), conditions related to hypoxia at birth reported as risk factors for ASD (such as umbilical cord complications, meconium aspiration, neonatal anaemia, ABO or Rh incompatibility, hyperbilirubinemia, or birth injury and trauma; for a review, see Gardener et al. [260]), feeding pattern and difficulties, screening tests (hypothyroidism, phenylketonuria), physical examination (apparent dysmorphism and malformations) and neurological examination (neonatal hypotonia); and (iii) postnatal period: psychomotor development, feeding pattern and possible gastrointestinal problems, sleep pattern, epilepsy, sensory dysfunction, language development, early behavioral signs of autism and other atypical behaviors. This clinical evaluation investigates developmental and learning disorders. The study of the height-weight-head circumference growth curve can show a break, decline with regression, or decrease with stagnation. For example, a decrease between 6 and 18 months of head circumference growth should orient the clinician towards the search for Rett syndrome in its first stage (stagnation stage) which occurs before the stage of fast decline of the communication skills (regression stage between 12 and 18 months of age). In addition, a developmental regression should orient the clinician towards the search for metabolic diseases. More generally, neuroimaging is recommended in individuals with ASD and microcephaly [249]. It is noteworthy that parental measures of height-weight-head circumference have also to be considered in order to determine whether the child's height-weight-head circumference measures are really abnormal compared to the parents' ones or whether a parent may be mildly affected, orienting towards a congenital or acquired abnormality.

3.1.3. Physical Examination (Genetic Evaluation)

A clinical genetics evaluation is systematically required and involves an extensive physical examination of the whole body. This physical examination should be conducted by a clinical geneticist trained in dysmorphology searching for dysmorphic features or malformations, and includes photos of the head (face and profile) with a good visibility of the ears, hands (palm and dorsal view) and feet. When clinical geneticists are not easily available, alternative options are to send, as a preliminary step, patients' pictures (face and hands pictures) to a geneticist for a specialized opinion (our group is testing the feasibility and efficiency of such a procedure and is comparing the results obtained based on the analyses of pictures by a group of geneticists to the ones obtained based on the analyses of the same pictures using the genetic search and reference mobile application Face2Gene developed by FDNA: Facial Dysmorphology Novel Analysis). In addition, the skin should be examined with a Wood lamp, in particular when tuberous sclerosis is suspected [251]. Finally, clinical genetics evaluation can greatly benefit, in this strategy of identification of genetic disorders associated with ASD, from examining repeatedly over time the child to follow his/her developmental trajectory and examining at the same time the patient, his/her parents and siblings to compare the findings.

3.2. Second Step: Complementary Investigations (Laboratory Studies)

This secondary investigation includes cytogenetic, molecular analyses, biochemical analyses. Other exams such as neuroimaging and EEG are requested based on the results of the clinical genetic and neuropediatric examinations.

The following hierarchical diagnostic strategy, in line with recent American guidelines [249], can be proposed (see Figure 1).

3.2.1. Non-Syndromic Autism (Isolated ASD)

It should be remembered that non-syndromic autism includes ASD individuals with moderate intellectual disability to normal cognitive functioning (high functioning ASD) and no other associated signs or symptoms (except the possible presence of seizures). To our knowledge, the probability to detect a mutation (CNV or point mutation) is very low in high functioning non-syndromic ASD individuals, and, in this case, we do not recommend molecular analysis. However, in the case of family history of psychiatric disorders or developmental disabilities, it could be of interest to perform aCGH and to collect samples for a research program. In addition, in the case of consanguinity or selective food aversion, it could be useful to perform targeted metabolic testing. In addition, in the case of individuals with intellectual disability or males with no intellectual disability, the search for Fragile X mutation is recommended. If there is no hypothesis, a repeated clinical genetic evaluation over time is proposed given the possible later appearance of dysmorphic features along with the developmental trajectory.

3.2.2. Syndromic Autism (ASD Associated with Dysmorphic Features and/or Malformations and/or Symptomatic or Cryptogenic Epilepsy, and/or Severe Intellectual Disability)

If a specific chromosomal or point mutation is suspected: perform aCGH or specific gene sequencing.

If a metabolic disorder is suspected based on developmental regression (regression observed on the growth curves, including macrocephaly and in particular microcephaly, but also developmental psychomotor regression or neuroregression/neurodegeneration with neurological symptoms), seizures and especially epilepsy (recurrent and refractory seizures), consanguinity and specific dysmorphy or other symptoms described below: perform targeted metabolic testing including measurements of ammonemia and lactatemia, urinary organic acids chromatography, plasmatic amino-acids chromatography, creatine metabolism, iron, vitamin D, glutathione, oxidative stress, and cerebral magnetic resonance spectroscopy combined with standard neuroimaging. A metabolic disease could also be suspected based on certain clinical symptoms, such as digestion-related symptoms (cyclic vomiting, selective eating or gastrointestinal problems), dermatologic and hair changes (rashes, pigmented skin eruptions, hypertrichosis or alopecia), lethargy with hypotonia/extrapyramidal signs (dystonia, abnormal movements), or multiple organ dysfunction (in particular, heart, liver, kidney). In addition, the search for metabolic disorders can be indicated by abnormal biological measures such as anemia or lactic acidosis, and more generally acid/base or electrolyte disturbances. Searching for metabolic disorders is important given the potential availability of treatments such as enzymotherapy for certain metabolic diseases, diet low in phenylalanine for phenylketonuria due to the enzyme phenylalanine hydroxylase deficiency, or supplementation with folinic acid for cerebral folate deficiency. Several metabolic disorders associated with ASD are listed in Table 1 using the superscript #.

If a mitochondrial disorder is suspected based on repeated regressions after three years old, neurological abnormalities (including hypotonia) and multiple organ dysfunction: perform targeted mitochondrial testing including measurement of lactatemia. Abnormal neurologic examination and/or elevated plasma lactate concentration have been indeed found in mitochonfrial diseases [260]. A mitochondrial disorder could be also suspected in case of ASD associated with loss of speech after a febrile illness or immunization with encephalopathy [261]. Finally, it is noteworthy that mitochondrial disease (rare in ASD) has to be differentiated from mitochondrial dysfunction (common in ASD) [262].

If there is no hypothesis of specific syndrome: perform aCGH and molecular diagnosis of Fragile-X syndrome. Without any identified pathogenic variant, we recommend targeted New Generation Sequencing (NGS) on known genes (for example, panels of genes related to intellectual disability or epilepsy based on the clinical examination) or Whole Exome Sequencing (WES) if available.

4. Conclusions

The finality of the hierarchal diagnostic strategy proposed in this article is to identify genetic disorders associated with ASD based on a stepwise evaluation characterized by low invasiveness and cost but high accessibility, feasibility and adaptability of current knowledge and technology to the individual and his/her family. These clinical and laboratory investigations can lead to identify genetic disorders associated with ASD in 35%–40% of individuals [225,262]. The identification of a specific genetic syndrome associated with ASD has practical clinical implications, in terms of diagnostic strategies including early detection and prevention of co-morbidity related to the known medical risks of the genetic disorder, follow up adapted to the genetic disorder with a more precise prognosis, and therapeutic strategies with access to needed services, help provided by specific family associations or specific individualized treatment in some cases. In many cases, identification of genetic disorders does not lead to therapeutic implications but there are a few cases where enzymotherapy, food depletion or supplementation in certain metabolic diseases as well as administration of melatonin associated with acebutolol in Smith–Magenis syndrome have improved considerably the quality of life of patients and their families. It should be noted that sleep problems are frequent in ASD (prevalence of insomnia: 50%–80% [263]) and decreased nocturnal and/or diurnal levels of melatonin (a sleep neurohormone) have been reported in many ASD studies; trial studies support therapeutic benefits of melatonin use in ASD (for a review, see Tordjman et al. [116,264]. Furthermore, the identification of genetic disorders associated with ASD allows unique personalized genetic counseling adapted to each family with regard to the transmission risk to the descendants and the possibility of a prenatal or preimplant diagnosis in the perspective of a future pregnancy. According to the 2013 American guidelines [249], the genetic counselor should note the type of array technology used (e.g., oligonucleotide vs. single-nucleotide polymorphism arrays) and access major databases relevant to this technology in order to provide the best possible and adapted information for the family. Many parents of children with ASD worry about the autism risk for a future pregnancy and knowing that their impaired child has a de novo mutation can be very helpful concerning the decision to have another child. It can be noted that the majority of the found mutations or rearrangements occurs de novo and are therefore not inherited from the parents or relatives [262]. In the same vein, siblings of children with ASD are very often preoccupied by the possible genetic transmission of autism to their future children. This preoccupation is legitimate considering, for instance, that the sisters of a boy with Fragile X Syndrome, even if they express very few or no symptoms, have a risk of 50% to carry the mutation or premutation and therefore a risk of 50% to have a child with Fragile X Syndrome (more severely impaired if it is a male). It is noteworthy that genetic counseling is also offered to families of ASD individuals without identified etiology, using in this case epidemiologic studies, and in particular sibling recurrence-risk updated data [249]. In any case, genetic counseling provides helpful family guidance, and parents of children with ASD are often relieved that biological factors involved in the development of ASD are considered and possibly found when some of them might feel guilty with regard to the role of family environment in ASD. Interestingly, based on our experience and the experience of geneticists or neuropediatricians working with psychoanalysts [265], knowing the underlying genetic cause of a relative's or patient's disorder, far from fixing the family or caregiver representations with regard to genetic determinism, rather creates and stimulates new dynamics within the family as well as the caregiver team centered on the ASD individual and probably related to a better known developmental trajectory with therapeutic perspectives and a parental relief. Finally, given all these implications, it is essential that all individuals (children but also adolescents and adults) with ASD can benefit from a clinical genetic examination and neuropediatric examination searching for associated signs or symptoms helping to identify known genetic disorders.

Acknowledgments: Funds from the Laboratory Psychology of Perception (LPP) were receiving for covering the costs to publish in open access.

Author Contributions: Cyrille Robert and Sylvie Tordjman wrote the manuscript; Sylvie Odent, Laurent Pasquier, David Cohen, Mélanie Fradin, Roberto Canitano and Léna Damaj revised the first draft of the paper; Sylvie Tordjman, Sylvie Odent and Laurent Pasquier modified the article according to the reviewers' comments.

Conflicts of Interest: The authors declare no conflict of interest.

References

1. American Psychiatric Association. *Diagnostic and Statistical Manual of Mental Disorders (DSM-5)*, 5th ed.; American Psychiatric Association: Washington, DC, USA, 2013.
2. Carrey, N.J. Itard's 1828 Memoire on Mutism caused by a lesion of the intellectual functions: A historical analysis. *J. Am. Acad. Child Adolesc. Psychiatry* **1995**, *34*, 1655–1661. [CrossRef] [PubMed]
3. Kanner, L. Austistic disturbances of affective contact. *Nervous Child* **1943**, *32*, 217–253.
4. Bleuler, E. *Dementia Praecox oder Gruppe der Schizophrenien*; Handbuch der Psychiatrie; Deuticke: Leipzig, Germany, 1911.
5. Asperger, H. Die autistischen psychopathen im kindesalter. *Arch. Psychiatry Nervenkrandeiten* **1944**, *117*, 73–136. [CrossRef]
6. Tordjman, S.; Somogyi, E.; Coulon, N.; Kermarrec, S.; Cohen, D.; Bronsard, G.; Bonnot, O.; Weismann-Arcache, C.; Botbol, M.; Lauth, B.; et al. Gene-environment interactions in autism spectrum disorders: Role of epigenetic mechanisms. *Front. Psychiatry* **2014**, *5*, 1–17. [CrossRef] [PubMed]
7. Abrahams, B.S.; Geschwind, D.H. Advances in autism genetics: On the threshold of a new neurobiology. *Nat. Rev. Genet.* **2008**, *9*, 341–355. [CrossRef] [PubMed]
8. Bill, B.R.; Geschwind, D.H. Genetic advances in autism: Heterogeneity and convergence on shared pathways. *Curr. Opin. Genet. Dev.* **2009**, *19*, 271–278. [CrossRef] [PubMed]
9. Muhle, R.; Trentacoste, S.V.; Rapin, I. The genetics of autism. *Pediatrics* **2004**, *113*, 472–486. [CrossRef]
10. Ronald, A.; Hoekstra, R.A. Autism spectrum disorders and autistic traits: A decade of new twin studies. *Am. J. Med. Genet.* **2011**, *156*, 255–274. [CrossRef] [PubMed]
11. Sandin, S.; Lichtenstein, P.; Kuja-Halkola, R.; Larsson, H.; Hultman, C.M.; Reichenberg, A. The familial risk of autism. *JAMA* **2014**, *311*, 1770–1777. [CrossRef] [PubMed]
12. Rosti, R.O.; Sadek, A.A.; Vaux, K.K.; Gleeson, J.G. The genetic landscape of autism spectrum disorders. *Dev. Med. Child Neurol.* **2014**, *56*, 12–18. [CrossRef] [PubMed]
13. Sebat, J.; Lakshmi, B.; Malhotra, D.; Troge, J.; Lese-Martin, C.; Walsh, T. Strong association of de novo copy number mutations with autism. *Science* **2007**, *316*, 445–449. [CrossRef] [PubMed]
14. Hegmann, J.P.; Possidente, B. Estimating genetic correlations from inbred strains. *Behav. Genet.* **1981**, *11*, 103–114. [CrossRef] [PubMed]
15. Miles, J.H. Autism spectrum disorders, a genetics review. *Genet. Med.* **2011**, *13*, 278–294. [CrossRef] [PubMed]
16. Bourgeron, T. From the genetic architecture to synaptic plasticity in autism spectrum disorder. *Nat. Rev. Neurosc.* **2015**, *16*, 551–563. [CrossRef] [PubMed]
17. Miles, J.H.; Hillman, R.E. Value of a clinical morphology examination in autism. *Am. J. Med. Genet.* **2000**, *91*, 245–253. [CrossRef]
18. Jacquemont, M.L.; Sanlaville, D.; Redon, R.; Raoul, O.; Cormier-Daire, V.; Lyonnet, S.; Amiel, J.; Le Merrer, M.; Heron, D.; de Blois, M.C.; et al. Array-based comparative genomic hybridisation identifies high frequency of cryptic chromosomal rearrangements in patients with syndromic autism spectrum disorders. *J. Med. Genet.* **2006**, *43*, 843–849. [CrossRef] [PubMed]
19. Miles, J.H.; Takahashi, T.N.; Bagby, S.; Sahota, P.K.; Vaslow, D.F.; Wang, C.H.; Hillman, R.E.; Farmer, J.E. Essential versus complex autism: Definition of fundamental prognostic subtypes. *Am. J. Med. Genet.* **2005**, *135*, 171–180. [CrossRef] [PubMed]

20. Cohen, D.; Pichard, N.; Tordjman, S.; Baumann, C.; Burglen, L.; Excoffier, E.; Lazar, G.; Mazet, P.; Pinquier, C.; Verloes, A.; et al. Specific genetic disorders and autism: Clinical contribution towards their identification. *J. Autism Dev. Disord.* **2005**, *35*, 103–116. [CrossRef] [PubMed]

21. Amiet, C.; Gourfinkel-An, I.; Bouzamondo, A.; Tordjman, S.; Baulac, M.; Lechat, P.; Cohen, D.J. Epilepsy in autism is associated with intellectual disability and gender: Evidence from a meta-analysis. *Biol. Psychiatry* **2008**, *64*, 577–582. [CrossRef] [PubMed]

22. Amiet, C.; Gourfinkel-An, I.; Lauent, C.; Bodeneau, N.; Genin, B.; Leguern, E.; Tordjman, S.; Cohen, D.J. Does epilepsy in multiplex autism define a different subgroup of clinical characteristics and genetic risk? *Mol. Autism* **2013**, *4*, 47. [CrossRef] [PubMed]

23. Amiet, C.; Gourfinkel-An, I.; Lauent, C.; Carayol, J.; Genin, B.; Leguern, E.; Tordjman, S.; Cohen, D.J. Epilepsy in simplex autism pedigrees is much lower than the rate in multiplex autism pedigrees. *Biol. Psychiatry* **2013**, *74*, 3–4. [CrossRef] [PubMed]

24. Canitano, R. Epilepsy in autism spectrum disorders. *Eur. Child Adolesc. Psychiatry* **2007**, *16*, 61–66. [CrossRef] [PubMed]

25. Tordjman, S.; Cohen, D.; Anderson, G.M.; Canitano, R.; Botbol, M.; Coulon, N.; Roubertoux, P.L. Reframing autism as a behavioral syndrome and not a specific mental disorder: Implications of genetic and phenotypic heterogeneity. *Neurosci. Biobehav. Rev.* **2017**. [CrossRef] [PubMed]

26. Cook, E.H., Jr.; Lindgren, V.; Leventhal, B.L.; Courchesne, R.; Lincoln, A.; Shulman, C.; Lord, C.; Courchesne, E. Autism or atypical autism in maternally but not paternally derived proximal 15q duplication. *Am. J. Hum. Genet.* **1997**, *60*, 928–934. [PubMed]

27. Bolton, P.F.; Dennis, N.R.; Browne, C.E.; Thomas, N.S.; Veltman, M.W.; Thompson, R.J.; Jacobs, P. The phenotypic manifestations of interstitial duplications of proximal 15q with special reference to the autistic spectrum disorders. *Am. J. Hum. Genet.* **2001**, *105*, 675–685. [CrossRef] [PubMed]

28. Cohen, D.; Martel, C.; Wilson, A.; Déchambre, N.; Amy, C.; Duverger, L.; Guile, J.M.; Pipiras, E.; Benzacken, B.; Cavé, H.; et al. Brief report: visual-spatial deficit in a 16-year-old girl with maternally derived duplication of proximal 15q. *J. Autism Dev. Disord.* **2007**, *37*, 1585–1591. [CrossRef] [PubMed]

29. Schroer, R.J.; Phelan, M.C.; Michaelis, R.C.; Crawford, E.C.; Skinner, S.A.; Cuccaro, M.; Simensen, R.J.; Bishop, J.; Skinner, C.; Fender, D.; et al. Autism and maternally derived aberrations of chromosome 15q. *Am. J. Hum. Genet.* **1998**, *76*, 327–336. [CrossRef]

30. Wandstrat, A.E.; Leana-Cox, J.; Jenkins, L.; Schwartz, S. Molecular cytogenetic evidence for a common break-point in the largest inverted duplications for chromosomes 15. *Am. J. Hum. Genet.* **1998**, *62*, 925–936. [CrossRef] [PubMed]

31. Sutcliffe, J.S.; Nurmi, E.L. Genetics of chilhood disorders: XLVII. Autism, Part 6: Duplication and inherited susceptibility of chromosome 15q11–q13 genes in autism. *J. Am. Acad. Child Adolesc. Psychiatry* **2003**, *42*, 253–256. [CrossRef] [PubMed]

32. Jacobsen, J.; King, B.H.; Leventhal, B.L.; Christian, S.L.; Ledbetter, D.H.; Cook, E.H., Jr. Molecular screening for proximal 15q abnormalities in a mentally retard population. *J. Med. Genet.* **1998**, *35*, 534–538. [CrossRef] [PubMed]

33. Steffenburg, S.; Gillberg, C.L.; Seffenburg, U.; Kyllerman, M. Autism in Angelman syndrome: A population-based study. *Pediatr. Neurol.* **1996**, *14*, 131–136. [CrossRef]

34. Peters, S.U.; Beaudet, A.L.; Madduri, N.; Bacino, C.A. Autism in Angelman syndrome: Implications for autism research. *Clin. Genet.* **2004**, *66*, 530–536. [CrossRef] [PubMed]

35. Chamberlain, S.J.; Chen, P.F.; Ng, K.Y.; Bourgois-Rocha, F.; Lemtiri-Chlieh, F.; Levine, E.S.; Lalande, M. Induced pluripotent stem cell models of the genomic imprinting disorders Angelman and Prader-Willi syndromes. *Proc. Natl. Acad. Sci. USA* **2010**, *107*, 17668–17673. [CrossRef] [PubMed]

36. Rangasamy, S.; D'Mello, S.R.; Narayanan, V. Epigenitics, autism spectrum, and neurodevelopmental disorders. *Neurotherapeutics* **2013**, *10*, 742–756. [CrossRef] [PubMed]

37. Grafodatskaya, D.; Chung, B.; Szatmari, P.; Weksberg, R. Autism spectrum disorders and epigenetics. *J. Am. Acad. Child Adolesc. Psychiatry* **2010**, *49*, 794–809. [CrossRef] [PubMed]

38. Dawson, A.J.; Cox, J.; Hovanes, K.; Spriggs, E. PWS/AS MS-MLPA confirms maternal origin of 15q11.2 microduplication. *Case Rep. Genet.* **2015**, *2015*, 474097. [CrossRef] [PubMed]

39. Huang, H.S.; Allen, J.A.; Mabb, A.M. Topoisomerase inhibitors unsilence the dormant allele of Ube3a in neurons. *Nature* **2011**, *481*, 185–189. [CrossRef] [PubMed]

40. Mabb, A.M.; Judson, M.C.; Zylka, M.J.; Philpot, B.D. Angelman syndrome: Insights into genomic imprinting and neurodevelopmental phenotypes. *Trends Neurosci.* **2011**, *34*, 293–303. [CrossRef] [PubMed]

41. Erickson, C.A.; Wink, L.K.; Baindu, B.; Ray, B.; Schaefer, T.L.; Pedapati, E.V.; Lahiri, D.K. Analysis of peripheral amyloid precursor protein in Angelman syndrome. *Am. J. Med. Genet.* **2016**, *170*, 2334–2337. [CrossRef] [PubMed]

42. Li, W.; Yao, A.; Zhi, H.; Kaur, K.; Zhu, Y.; Jia, M.; Wang, Q.; Jin, S.; Zhao, G.; et al. Angelman syndrome protein Ube3a regulates synaptic growth and endocytosis by inhibiting BMP signaling in *Drosophila*. *PLoS Genet.* **2016**, *12*, e1006062. [CrossRef] [PubMed]

43. Meng, L.; Person, R.E.; Beaudet, A.L. Ube3a-ATS is an atypical RNA polymerase II transcript that represses the paternal expression of Ube3a. *Hum. Mol. Genet.* **2012**, *21*, 3001–3012. [CrossRef] [PubMed]

44. Kishino, T.; Lalande, M.; Wasgstaff, J. UBE3A/E6-AP mutations cause Angelman syndrome. *Nat. Genet.* **1997**, *15*, 70–73. [CrossRef] [PubMed]

45. Petit, E.; Hérault, J.; Martineau, J.; Perrot, A.; Barthélémy, C.; Hameury, L.; Sauvage, D.; Lelord, G.; Müh, J.P. Association study with two markers of a human homeogene in infantile autism. *J. Med. Genet.* **1995**, *32*, 269–274. [CrossRef] [PubMed]

46. Clayton-Smith, J. Clinical research on Angelman syndrome in the United Kingdom: Observations on 82 affected individuals. *Am. J. Med. Genet.* **1993**, *46*, 12–15. [CrossRef] [PubMed]

47. Summers, J.A.; Allison, D.B.; Lynch, P.S.; Sandler, L. Behaviour problems in Angelman syndrome. *J. Intellect. Disabil. Res.* **1995**, *39*, 97–106. [CrossRef] [PubMed]

48. Walz, N.C. Parent report of stereotyped behaviors, social interaction, and developmental disturbances in individuals with Angelman syndrome. *J. Autism Dev. Disord.* **2007**, *37*, 940–947. [CrossRef] [PubMed]

49. Dykens, E.M.; Cassidy, S.B. Correlates of maladaptive behavior in children and adults with Prader-Willi syndrome. *Am. J. Med. Genet.* **1995**, *60*, 546–549. [CrossRef] [PubMed]

50. Demb, H.B.; Papola, P. PDD and Prader-Willi syndrome. *J. Am. Acad. Child Adolesc. Psychiatry* **1995**, *34*, 539–540. [CrossRef] [PubMed]

51. Dykens, E.M.; Cassidy, S.B.; King, B.H. Maladaptive behavior differences in Prader-Willi syndrome due to paternal deletion versus maternal uniparental disomy. *Am. J. Ment. Retard.* **1995**, *104*, 67–77. [CrossRef]

52. Veltman, M.W.; Thompson, R.J.; Roberts, S.E.; Thomas, N.S.; Whittington, J.; Bolton, P.F. Prader-Willi syndrome—A study comparing deletion and uniparental disomy cases with reference to autism spectrum disorders. *Eur. Child Adolesc. Psychiatry* **2004**, *13*, 42–50. [CrossRef] [PubMed]

53. Descheemaeker, M.J.; Govers, V.; Vermeulen, P.; Fryns, J.P. Pervasive developmental disorders in Prader-Willi syndrome: The Leuven experience in 59 subjects and controls. *Am. J. Med. Genet.* **2006**, *140*, 1136–1142. [CrossRef] [PubMed]

54. Hogart, A.; Wu, D.; LaSalle, J.M.; Schanen, N.C. The comorbidity of autism with the genomic disorders of chromosome 15q11.2–q13. *Neurobiol. Dis.* **2010**, *38*, 181–191. [CrossRef] [PubMed]

55. Horsthemke, B.; Wagstaff, J. Mechanisms of imprinting of the Prader-Willi/Angelman region. *Am. J. Med. Genet.* **2008**, *146*, 2041–2052. [CrossRef] [PubMed]

56. Buiting, K. Prader-Willi and Angelman syndrome. *Am. J. Med. Genet.* **2010**, *154*, 365–376. [CrossRef] [PubMed]

57. Cassidy, S.B.; Scwartz, S.; Miller, H.; Driscoll, D.J. Prader-Willi syndrome. *Genet. Med.* **2012**, *14*, 10–26. [CrossRef] [PubMed]

58. Bennett, J.A.; Germani, T.; Haqq, A.M.; Zwaigenbaum, L. Autism spectrum disorder in Prader-Willi syndrome: A systematic review. *Am. J. Med. Genet.* **2015**, *167*, 2936–2944. [CrossRef] [PubMed]

59. Veltman, M.W.; Thompson, R.J.; Craig, E.E.; Dennis, N.R.; Roberts, S.E.; Moore, V.; Brown, J.A.; Bolton, P.F. A paternally inherited duplication in the Prader-Willi/Angelman syndrome critical region: A case and family study. *J. Autism Dev. Disord.* **2015**, *35*, 117–127. [CrossRef]

60. Bolton, P.F.; Roobol, M.; Allsopp, L.; Pickles, A. Association between idiopathic infantile macrocephaly and autism spectrum disorders. *Lancet* **2001**, *358*, 726–727. [CrossRef]

61. Roberts, S.E.; Dennis, N.R.; Browne, C.E.; Willatt, L.; Woods, G.; Cross, I.; Jacobs, P.A.; Thomas, S. Characterisation of interstitial duplications and triplications of chromosome 15q11–q13. *Hum. Genet.* **2002**, *110*, 227–234. [CrossRef] [PubMed]

62. Goizet, C.; Excoffier, E.; Taine, L.; Taupiac, E.; El Moneim, A.A.; Arveiler, B.; Bouvard, M.; Lacombe, D. Case with autistic syndrome and chromosome 22q13.3 deletion detected by FISH. *Am. J. Hum. Genet.* **2000**, *96*, 839–844. [CrossRef]

63. Manning, M.A.; Cassidy, S.B.; Clericuzio, C.; Cherry, A.M.; Schwartz, S.; Hudgins, L.; Enns, G.M.; Hoyme, H.E. Terminal 22q deletion syndrome: A newly recognized cause of speech and language disability in the autism spectrum. *Pediatrics* **2004**, *114*, 451–457. [CrossRef] [PubMed]

64. Prasad, C.; Prasad, A.N.; Chodirker, B.N.; Lee, C.; Dawson, A.K.; Jocelyn, L.J.; Chudley, A.E. Genetic evaluation of pervasive developmental disorders: The terminal 22q13 deletion syndrome may represent a recognizable phenotype. *Clin.Genet.* **2000**, *57*, 103–109. [CrossRef] [PubMed]

65. Phelan, M.C.; Rogers, R.C.; Saul, R.A.; Stapleton, G.A.; Sweet, K.; McDermid, H.; Shaw, S.R.; Claytor, J.; Willis, J.; Kelly, D.P. 22q13 Deletion Syndrome. *Am. J. Hum. Genet.* **2001**, *101*, 91–99. [CrossRef]

66. Soorya, L.; Kolevzon, A.; Zweifach, J.; Lim, T.; Dobry, Y.; Schwartz, L.; Frank, Y.; Wang, A.T.; Cai, G.; Parkhomenko, E.; et al. Prospective investigation of autism and genotype-phenotype correlations in 22q13 deletion syndrome and SHANK3 deficiency. *Mol. Autism* **2013**, *4*, 18. [CrossRef] [PubMed]

67. Shcheglovitov, A.; Shcheglovitova, O.; Yazawa, M.; Portmann, T.; Shu, R.; Sebastiano, V.; Krawisz, A.; Froehlich, W.; Bernstein, J.A.; Hallmayer, J.F.; et al. SHANK3 and IGF1 restore synaptic deficits in neurons from 22q13 deletion syndrome patients. *Nature* **2013**, *503*, 267–271. [CrossRef] [PubMed]

68. Phelan, K.; McDermid, H.E. The 22q13.3 deletion syndrome (Phelan-McDermid syndrome). *Mol. Syndromol.* **2012**, *2*, 186–201. [CrossRef] [PubMed]

69. Peca, J.; Feliciano, C.; Ting, J.T.; Wang, W.; Wells, M.F.; Venkatraman, T.N.; Lascola, C.D.; Fu, Z.; Feng, G. SHANK3 mutant mice display autistic-like behaviours and striatal dysfunction. *Nature* **2011**, *472*, 437–442. [CrossRef] [PubMed]

70. Jafri, F.; Fink, J.; Higgins, R.R.; Tervo, R. 22q13.32 deletion and duplication and inversion in the same family: A rare occurrence. *ISRN Pediatr.* **2011**, *2011*, 829825. [CrossRef] [PubMed]

71. Zwanenburg, R.J.; Ruiter, S.A.J.; van Den Heuvel, E.R.; Flapper, B.C.T.; van Ravenswaaij-Arts, C.M.A. Developmental phenotype in Phelan-McDermid (22q13.3 deletion) syndrome: A systematic and prospective study in 34 children. *J. Neurodev. Disord.* **2016**, *8*, 16. [CrossRef] [PubMed]

72. Marshall, C.R.; Noor, A.; Vincent, J.B.; Lionel, A.C.; Feuk, L.; Skaug, J.; Shago, M.; Moessner, R.; Pinto, D.; Ren, Y.; et al. Structural variation of chromosomes in autism spectrum disorder. *Am. J. Hum. Genet.* **2008**, *82*, 477–488. [CrossRef] [PubMed]

73. Hanson, E.; Nasir, R.H.; Fong, A.; Lian, A.; Hundley, R.; Shen, Y.; Wu, B.L.; Holm, I.A.; Miller, D.T. 16p11.2 Study group clinicians cognitive and behavioral characterization of 16p11.2 deletion syndrome. *J. Dev. Behav. Pediatr.* **2010**, *31*, 649–657. [CrossRef] [PubMed]

74. Weiss, L.A.; Shen, Y.; Korn, J.M.; Arking, D.E.; Miller, D.T.; Fossdal, R.; Saemundsen, E.; Stefansson, H.; Ferreira, M.A.; Green, T.; et al. Consortium association between microdeletion and microduplication at 16p11.2 and autism. *New Engl. J. Med.* **2008**, *358*, 667–675. [CrossRef] [PubMed]

75. Kumar, R.A.; KaraMohamed, S.; Sudi, J.; Conrad, D.F.; Brune, C.; Badner, J.A.; Gilliam, T.C.; Nowak, N.J.; Cook, E., Jr.; Dobyns, W.B.; et al. Recurrent 16p11.2 microdeletions in autism. *Hum. Mol. Genet.* **2008**, *17*, 628–638. [CrossRef] [PubMed]

76. Fombonne, E.; Du Mazaubrun, C.; Cans, C.; Grandjean, H. Autism and associated medical disorders in a French epidemiological survey. *J. Am. Child Adolesc. Psychiatry* **1997**, *36*, 1561–1569.

77. Steinman, K.J.; Spence, S.J.; Ramocki, M.B.; Proud, M.B.; Kessler, S.K.; Marco, E.J.; Green Snyder, L.; D'Angelo, D.; Chen, Q.; Chung, W.K.; et al. 16p11.2 Deletion and duplication: Characterizing neurologic phenotypes in a large clinically ascertained cohort. *Am. J. Med. Genet.* **2016**, *9999*, 1–13. [CrossRef] [PubMed]

78. Green Snyder, L.; D'Angelo, D.; Chen, Q.; Bernier, R.; Goin-Kochel, R.P.; Wallace, A.S.; Gerdts, J.; Kanne, S.; Berry, L.; Blaskey, L.; et al. Autism spectrum disorder, developmental and psychiatric features in 16p11.2 duplication. *J. Autism Dev. Disord.* **2016**, *46*, 2734–2748. [CrossRef] [PubMed]

79. Fisch, G.S.; Grossfeld, P.; Falk, R.; Battaglia, A.; Youngblom, J.; Simensen, R. Cognitive-behavioral features of Wolf-Hirschhorn syndrome and other subtelomeric microdeletions. *Am. J. Med. Genet.* **2010**, *154*, 417–426. [CrossRef] [PubMed]

80. Fisch, G.S.; Davis, R.; Youngblom, J.; Gregg, J. Genotype-phenotype association studies of chromosome 8p inverted duplication deletion syndrome. *Behav. Genet.* **2011**, *41*, 373–380. [CrossRef] [PubMed]

81. Bregmen, J.D.; Volkmar, F.R. Autism spectrum disorders in genetic syndromes: Implications for diagnosis, intervention and understanding the wider autism spectrum disorder population. *J. Intellect. Disabil. Res.* **1988**, *53*, 852–873.

82. Mariner, R.; Jackson, A.W.; Levitas, A.; Hagerman, R.J.; Braden, M.; McBogg, P.M.; Smith, A.C.; Berry, R. Autism, mental retardation, and chromosomal abnormalities. *J. Autism Dev. Disord.* **1986**, *16*, 425–440. [CrossRef] [PubMed]

83. Ghaziuddin, M. Autism in down syndrome: Family history correlates. *J. Intellect. Disabil. Res.* **1997**, *41*, 87–91. [CrossRef] [PubMed]

84. Moss, J.; Howlin, P. Autism spectrum disorders in genetic syndromes: Implications for diagnosis, intervention and understanding the wider autism spectrum disorder population. *J. Intellect. Disabil. Res.* **2009**, *53*, 852–873. [CrossRef] [PubMed]

85. Jiang, J.; Jing, Y.; Cost, G.J.; Chiang, J.C.; Kolpa, H.J.; Cotton, A.M.; Carone, D.M.; Carone, B.R.; Shivak, D.A.; Guschin, D.Y.; et al. Translating dosage compensation to trisomy 21. *Nature* **2013**, *500*, 296–300. [CrossRef] [PubMed]

86. Smith, A.C.; McGavran, L.; Robinson, J.; Waldstein, G.; Macfarlane, J.; Zonona, J.; Reiss, J.; Lahr, M.; Allen, L.; Magenis, E. Interstitial deletion of (17) (P11.2p11.2) in nine patients. *Am. J. Hum. Genet.* **1986**, *24*, 393–414. [CrossRef] [PubMed]

87. Vostanis, P.; Harrington, R.; Prendergast, M.; Farndon, P. Case reports of autism with interstitial deletion of chromosome 17 (p11.2 p11.2) and monosomy of chromosome 5 (5pter > 5p15.3). *Psychiatr. Genet.* **1994**, *4*, 109–111. [CrossRef] [PubMed]

88. Laje, G.; Morse, R.; Richter, W.; Ball, J.; Pao, M.; Smith, A.C.M. Autism spectrum features in smith-magenis syndrome. *Am. J. Med. Genet.* **2010**, *154*, 456–462. [CrossRef] [PubMed]

89. Potocki, L.; Bi, W.; Treadwell-Deering, D.; Carvalho, C.M.; Eifert, A.; Friedman, E.M.; Glaze, D.; Krull, K.; Lee, J.A.; Lewis, R.A.; et al. Characterization of Potocki-Lupski syndrome (dup(17) (p11.2p11.2)) and delineation of a dosage-sensitive critical interval that can convey an autism phenotype. *Am. J. Med. Genet.* **2007**, *80*, 633–649. [CrossRef]

90. Ghaziuddin, M.; Burmeister, M. Deletion of chromosome 2q37 and autism: A distinct subtype? *J. Autism Dev. Disord.* **1999**, *29*, 259–263. [CrossRef] [PubMed]

91. Devillard, F.; Guinchat, V.; Moreno-De-Luca, D.; Tabet, A.C.; Gruchy, N.; Guillem, P.; Nguyen Morel, M.A.; Leporrier, N.; Leboyer, M.; Jouk, P.S.; et al. Paracentric inversion of chromosome 2 associated with cryptic duplication of 2q14 and deletion of 2q37 in a patient with autism. *Am. J. Med. Genet.* **2010**, *152*, 2346–2354. [CrossRef] [PubMed]

92. Galasso, C.; Lo-Castro, A.; Lalli, C.; Nardone, A.M.; Gullotta, F.; Curatolo, P. Deletion 2q37: An identifiable clinical syndrome with mental retardation and autism. *J. Child Neurol.* **2008**, *23*, 802–806. [CrossRef] [PubMed]

93. Falk, R.E.; Casas, K.A. Chromosome 2q37 deletion: Clinical and molecular aspects. *Am. J. Med. Genet.* **2007**, *145*, 357–371. [CrossRef] [PubMed]

94. Fisch, G.S.; Falk, R.E.; Carey, J.C.; Imitola, J.; Sederberg, M.; Caravalho, K.S.; South, S. Deletion 2q37 syndrome: Cognitive-Behavioral trajectories and autistic features related to breakpoint and deletion size. *Am. J. Med. Genet.* **2016**, *170*, 2282–2291. [CrossRef] [PubMed]

95. Lo-Castro, A.; Galasso, C.; Cerminara, C.; El-Malhany, N.; Benedetti, S.; Nardone, A.M.; Curatolo, P. Association of syndromic mental retardation and autism with 22q11.2 duplication. *Neuropediatrics* **2009**, *40*, 137–140. [CrossRef] [PubMed]

96. Lo-Castro, A.; Benvenuto, A.; Galasso, C.; Porfirio, C.; Curatolo, P. Autism spectrum disorders associated with chromosomal abnormalities. *Res. Autism Spectr. Disord.* **2010**, *4*, 319–327. [CrossRef]

97. Portnoï, M.F. Microduplication 22q11.2: A new chromosomal syndrome. *Eur. J. Med. Genet.* **2009**, *52*, 88–93. [CrossRef] [PubMed]

98. Baker, K.D.; Skuse, D.H. Adolescents and young adults with 22q11 deletion syndrome: Psychopathology in an at-risk group. *Br. J. Psychiatry* **2005**, *186*, 115–120. [CrossRef] [PubMed]

99. Wenger, T.L.; Miller, J.S.; DePolo, L.M.; de Marchena, A.B.; Clements, C.C.; Emanuel, B.S.; Zackai, E.H.; McDonald-McGinn, D.M.; Schultz, R.T. 22q11.2 duplication syndrome: Elevated rate of autism spectrum disorder and need for medical screening. *Mol. Autism* **2016**, *7*, 27. [CrossRef] [PubMed]

100. Hidding, E.; Swaab, H.; de Sonneville, L.M.J.; van Engeland, H.; Vorstman, J.A.S. The role of COMT and plasma proline in the variable penetrance of autistic spectrum symptoms in 22q11.2 deletion syndrome. *Clin. Genet.* **2016**, *90*, 420–427. [CrossRef] [PubMed]

101. Mefford, H.C.; Sharp, A.J.; Baker, C.; Itsara, A.; Jiang, Z.; Buysse, K.; Huang, S.; Maloney, V.K.; Crolla, J.A.; Baralle, D.; et al. Recurrent rearrangements of chromosome 1q21.1 and variable pediatric phenotypes. *N. Engl. J. Med.* **2008**, *359*, 1685–1699. [CrossRef] [PubMed]

102. Brunetti-Pierri, N.; Berg, J.S.; Scaglia, F.; Belmont, J.; Bacino, C.A.; Sahoo, T.; Lalani, S.R.; Graham, B.; Lee, B.; Shinawi, M.; et al. Recurrent reciprocal 1q21.1 deletions and duplications associated with microcephaly or macrocephaly and developmental and behavioral abnormalities. *Nat. Genet.* **2008**, *40*, 1466–1471. [CrossRef] [PubMed]

103. Bernier, R.; Steinman, K.J.; Reilly, B.; Wallace, A.S.; Sherr, E.H.; Pojman, N.; Mefford, H.C.; Gerdts, J.; Earl, R.; Hanson, E.; et al. Clinical phenotype of the recurrent 1q21.1 copy-number variant. *Gen. Med.* **2016**, *18*, 341–349. [CrossRef] [PubMed]

104. Torres, F.; Barbosa, M.; Maciel, P. Recurrent copy number variations as risk factors for neurodevelopmental disorders: Critical overview and analysis of clinical implications. *J. Med. Genet.* **2016**, *53*, 73–90. [CrossRef] [PubMed]

105. Berg, J.S.; Brunetti-Pierri, N.; Peters, S.U.; Kang, S.H.; Fong, C.T.; Salamone, J.; Freedenberg, D.; Hannig, V.L.; Prock, L.A.; Miller, D.T.; et al. Speech delay and autism spectrum behaviors are frequently associated with duplication of the 7q11.23 Williams-Beuren syndrome region. *Genet. Med.* **2007**, *9*, 427–441. [CrossRef] [PubMed]

106. Depienne, C.; Heron, D.; Betancur, C.; Benyahia, B.; Trouillard, O.; Bouteiller, D.; Verloes, A.; LeGuern, E.; Leboyer, M.; Brice, A. Autism, language delay and mental retardation in a patient with 7q11 duplication. *J. Med. Genet.* **2007**, *44*, 452–458. [CrossRef] [PubMed]

107. Edelmann, L.; Prosnitz, A.; Pardo, S.; Bhatt, J.; Cohen, N.; Lauriat, T.; Ouchanov, L.; González, P.J.; Manghi, E.R.; Bondy, P.; et al. An atypical deletion of the Williams-Beuren syndrome interval implicates genes associated with defective visuospatial processing and autism. *J. Med. Genet.* **2007**, *44*, 136–143. [CrossRef] [PubMed]

108. Gagliardi, C.; Bonaglia, M.C.; Selicorni, A.; Borgatti, R.; Giorda, R. Unusual cognitive and behavioural profile in a Williams syndrome patient with atypical 7q11.23 deletion. *J. Med. Genet.* **2003**, *40*, 526–530. [CrossRef] [PubMed]

109. Lincoln, A.J.; Searcy, Y.M.; Jones, W.; Lord, C. Social interaction behaviors discriminate young children with autism and Williams syndrome. *J. Am. Acad. Child Adolesc. Psychiatry* **2007**, *46*, 323–331. [CrossRef] [PubMed]

110. Sanders, S.J.; Ercan-Sencicek, A.G.; Hus, V.; Luo, R.; Murtha, M.T.; Moreno-de-Luca, D.; Chu, S.H.; Moreau, M.P.; Gupta, A.R.; Thomson, S.A.; et al. Multiple recurrent de novo CNVs, including duplications of the 7q11.23 Williams syndrome region, are strongly associated with autism. *Neuron* **2011**, *70*, 863–885. [CrossRef] [PubMed]

111. Somerville, M.J.; Mervis, C.B.; Young, E.J.; Seo, E.J.; del Campo, M.; Bamforth, S.; Peregrine, E.; Loo, W.; Lilley, M.; Pérez-Jurado, L.A.; et al. Severe expressive-language delay related to duplication of the Williams-Beuren locus. *N. Engl. J. Med.* **2005**, *353*, 1694–1701. [CrossRef] [PubMed]

112. Van der Aa, N.; Rooms, L.; Vandeweyer, G.; van den Ende, J.; Reyniers, E.; Fichera, M.; Romano, C.; Delle Chiaie, B.; Mortier, G.; Menten, B.; et al. Fourteen new cases contribute to the characterization of the 7q11.23 microduplication syndrome. *Eur. J. Hum. Genet.* **2009**, *52*, 94–100. [CrossRef] [PubMed]

113. Makeyev, A.V.; Enkhamandakh, B.; Hong, S.H.; Joshi, P.; Shin, D.G.; Bayarsaihan, D. Diversity and complexity in chromatin recognition by TFII-I transcription factors in pluripotent embryonic stem cells and embryonic tissues. *PLoS ONE* **2012**, *7*, e44443. [CrossRef] [PubMed]

114. Tordjman, S.; Anderson, G.M.; Botbol, M.; Toutain, A.; Sarda, P.; Carlier, M.; Saugier-Veber, P.; Baumann, C.; Cohen, D.; Lagneaux, C.; et al. Autistic disorder in patients with Williams-Beuren syndrome: A reconsideration of the Williams-Beuren syndrome phenotype. *PLoS ONE* **2012**, *7*, e30778. [CrossRef] [PubMed]

115. Tordjman, S.; Anderson, G.M.; Cohen, D.; Kermarrec, S.; Carlier, M.; Touitou, Y.; Saugier-Veber, P.; Lagneaux, C.; Chevreuil, C.; Verloes, A. Presence of autism, hyperserotonemia, and severe expressive language impairment in Williams-Beuren syndrome. *Mol. Autism* **2013**, *4*, 29. [CrossRef] [PubMed]

116. Tordjman, S.; Najjar, I.; Bellissant, E.; Anderson, G.M.; Barburoth, M.; Cohen, D.; Jaafari, N.; Schischmanoff, O.; Fagard, R.; Lagdas, I.; et al. Advances in the research of melatonin in autism spectrum disorders: Literature review and new perspectives. *Int. J. Mol. Sci.* **2013**, *14*, 20508–20542. [CrossRef] [PubMed]
117. Richards, C.; Jones, C.; Groves, L.; Moss, J.; Oliver, C. Prevalence of autism spectrum disorder phenomenology in genetic disorders: A systematic review and meta-analysis. *Lancet Psychiatry* **2015**, *2*, 909–916. [CrossRef]
118. Skuse, D.H.; James, R.S.; Bishop, D.V.; Coppin, B.; Dalton, P.; Aamodt-Leeper, G.; Bacarese-Hamilton, M.; Creswell, C.; McGurk, R.; Jacobs, P.A. Evidence from Turner's syndrome of an imprinted X-linked locus affecting cognitive function. *Nature* **1997**, *387*, 705–708. [CrossRef] [PubMed]
119. Sybert, V.P.; McCauley, E. Turner' syndrome. *N. Engl. J. Med.* **2004**, *351*, 1227–1238. [CrossRef] [PubMed]
120. Kent, L.; Bowdin, S.; Kirby, G.A.; Cooper, W.N.; Maher, E.R. Beckwith Wiedemann syndrome: A behavorial phenotype-genotype study. *Am. Med. Genet.* **2008**, *147*, 1295–1297. [CrossRef] [PubMed]
121. DeBaun, M.R.; Niemitz, E.L.; Feinberg, A.P. Association of in vitro fertilization with Beckwith-Wiedemann syndrome and epigenetic alterations of LIT1 and H19. *Am. J. Hum. Genet.* **2003**, *72*, 156–160. [CrossRef] [PubMed]
122. Smith, A.C.; Choufani, S.; Ferreira, J.C.; Weksberg, R. Growth regulation, imprinted genes, and chromosome 11p15.5. *Pediatr. Res.* **2007**, *61*, 43–47. [CrossRef] [PubMed]
123. Weksberg, R.; Shuman, C.; Beckwith, J.B. Beckwith-Wiedemann syndrome. *Eur. J. Hum. Genet.* **2010**, *18*, 8–14. [CrossRef] [PubMed]
124. Wolpert, C.M.; Menold, M.M.; Bass, M.P.; Qumsiyeh, M.B.; Donnelly, S.L.; Ravan, S.A.; Vance, J.M.; Gilbert, J.R.; Abramson, R.K.; Wright, H.H.; et al. Three probands with autistic disorder and isodicentric chromosome 15. *Am. J. Med. Genet.* **2000**, *96*, 365–372. [CrossRef]
125. Wolpert, C.; Pericak-Vance, M.A.; Abramson, R.K.; Wright, H.H.; Cuccaro, M.L. Autistic symptoms among children and young adults with isodicentric chromosome 15. *Am. J. Med. Genet.* **2000**, *96*, 128–129. [CrossRef]
126. Parisi, L.; di Filippo, T.; Roccella, M. Hypomelanosis of ito: Neurological and psychiatric pictures in developmental age. *Minerva Pediatr.* **2012**, *64*, 65–70. [PubMed]
127. Åkefeldt, A.; Gillberg, C. Hypomelanosis of ito in three cases with autism and autistic-like conditions. *Dev. Med. Child Neurol.* **1991**, *33*, 737–743. [CrossRef] [PubMed]
128. Johansson, M.; Gillberg, C.; Råstam, M. Autism spectrum conditions in individuals with Möbius sequence, CHARGE syndrome and oculo-auriculo-vertebral spectrum: Diagnostic aspects. *Res. Dev. Disabil.* **2010**, *31*, 9–24. [CrossRef] [PubMed]
129. Johansson, M.; Råstam, M.; Billstedt, E.; Danielsson, S.; Strömland, K.; Miller, M.; Gillberg, C. Autism spectrum disorders and underlying brain pathology in CHARGE association. *Dev. Med. Child Neurol.* **2006**, *48*, 40–50. [CrossRef] [PubMed]
130. Smith, I.M.; Nichols, S.L.; Issekutz, K.; Blake, K. Canadian paediatric surveillance program behavioral profiles and symptoms of autism in CHARGE syndrome: Preliminary Canadian epidemiological data. *Am. J. Med. Genet.* **2005**, *133*, 248–256. [CrossRef] [PubMed]
131. Hartshorne, T.S.; Grialou, T.L.; Parker, K.R. Autistic-like behavior in CHARGE syndrome. *Am. J. Med. Genet.* **2005**, *133*, 257–261. [CrossRef] [PubMed]
132. Patten, S.A.; Jacobs-McDaniels, N.L.; Zaouter, C.; Drapeau, P.; Albertson, R.C.; Moldovan, F. Role of Chd7 in zebrafish: A model for CHARGE syndrome. *PLoS ONE* **2012**, *7*, e31650. [CrossRef] [PubMed]
133. Blake, K.D.; Prasad, C. CHARGE syndrome. *Orphanet J. Rare Dis.* **2006**, *1*, 34. [CrossRef] [PubMed]
134. Bolton, P.F.; Park, R.J.; Higgins, J.N.; Griffiths, P.D.; Pickles, A. Neuro-epileptic determinants of autism spectrum disorders in tuberous sclerosis complex. *Brain* **2002**, *125*, 1247–1255. [CrossRef] [PubMed]
135. Curatolo, P.; Bombardieri, R.; Jozwiak, S. Tuberous sclerosis. *Lancet* **2008**, *372*, 657–668. [PubMed]
136. Jeste, S.S.; Varcin, K.J.; Hellemann, G.S.; Gulsrud, A.C.; Bhatt, R.; Kasari, C.; Wu, J.Y.; Sahin, M.; Nelson, C.A. Symptom profiles of autism spectrum disorder in tuberous sclerosis complex. *Neurology* **2016**. [CrossRef] [PubMed]
137. McBride, K.L.; Varga, E.A.; Pastore, M.T.; Prior, T.W.; Manickam, K.; Atkin, J.F.; Herman, G.E. Confirmation study of PTEN mutations among individuals with autism or developmental delays/mental retardation and macrocephaly. *Autism Res.* **2010**, *3*, 137–141. [CrossRef] [PubMed]
138. Lintas, C.; Persico, A.M. Autistic phenotypes and genetic testing: State-of-the-art for the clinical geneticist. *J. Med. Genet.* **2009**, *46*, 1–8. [CrossRef] [PubMed]

139. Buxbaum, J.D.; Cai, G.; Chaste, P.; Nygren, G.; Goldsmith, J.; Reichert, J.; Anckarsäter, H.; Rastam, M.; Smith, C.J.; Silverman, J.M.; et al. Mutation screening of the PTEN gene in patients with autism spectrum disorders and macrocephaly. *Am. J. Med. Genet.* **2007**, *144*, 484–491. [CrossRef] [PubMed]

140. Herman, G.E.; Butter, E.; Enrile, B.; Pastore, M.; Prior, T.W.; Sommer, A. Increasing knowledge of PTEN germline mutations: Two additional patients with autism and macrocephaly. *Am. J. Med. Genet.* **2007**, *143*, 589–593. [CrossRef] [PubMed]

141. Goffin, A.; Hoefsloot, L.H.; Bosgoed, E.; Swillen, A.; Fryns, J.P. PTEN mutation in a family with Cowden syndrome and autism. *Am. J. Med. Genet.* **2001**, *105*, 521–524. [CrossRef] [PubMed]

142. Spinelli, L.; Black, F.M.; Berg, J.N.; Eickholt, B.J.; Leslie, N.R. Functionally distinct groups of inherited PTEN mutations in autism and tumour syndromes. *J. Med. Genet.* **2015**, *52*, 128–134. [CrossRef] [PubMed]

143. Amir, R.E.; Van Den Veyver, I.B.; Wan, M.; Tran, C.Q.; Francke, U.; Zoghbi, H.Y. Rett syndrome is cause by mutations in X-linked MECP2, encoding methil-CpG-binding protein 2. *Nat. Genet.* **1999**, *23*, 185–188. [PubMed]

144. Olsonn, B.; Rett, A. A review of the Rett syndrome with a theory of autism. *Brain Dev.* **1990**, *12*, 11–15. [CrossRef]

145. Young, D.J.; Bebbington, A.; Anderson, A.; Ravine, D.; Ellaway, C.; Kulkarni, A.; de Klerk, N.; Kaufmann, W.E.; Leonard, H. The diagnosis of autism in a female: Could it be Rett syndrome? *Eur. J. Pediatr.* **2008**, *167*, 661–669. [CrossRef] [PubMed]

146. Neul, J.L. The relationship of Rett syndrome and MECP2 disorders to autism. *Dialogues Clin. Neurosci.* **2012**, *14*, 253–262. [PubMed]

147. Gadalla, K.; Bailey, M.E.; Cobb, S.R. MeCP2 and Rett syndrome: Reversibility and potential avenues for therapy. *Biochem. J.* **2011**, *439*, 1–14. [CrossRef] [PubMed]

148. Neul, J.L.; Kaufmann, W.E.; Glaze, D.G.; Christodoulou, J.; Angus, J.; Clarke, A.J.; Bahi-Buisson, N.; Leonard, H.; Bailey, M.E.S.; Schanen, C.; et al. Rett syndrome: Revised diagnostic criteria and nomenclature. *Ann. Neurol.* **2010**, *68*, 944–950. [CrossRef] [PubMed]

149. Ramocki, M.B.; Peters, S.U.; Tavyev, Y.J.; Zhang, F.; Carvalho, C.M.; Schaaf, C.P.; Richman, R.; Fang, P.; Glaze, D.G.; Lupski, J.R.; et al. Autism and other neuropsychiatric symptoms are prevalent in individuals with MeCP2 duplication syndrome. *Ann. Neurol.* **2009**, *66*, 771–782. [CrossRef] [PubMed]

150. Nidiffer, F.D.; Kelly, T.E. Developmental and degenerative patterns associated with cognitive, behavioural and motor difficulties in the Sanfilippo syndrome: An epidemiological study. *J. Ment. Defic. Res.* **1983**, *27*, 185–203. [CrossRef] [PubMed]

151. Wraith, J.E. The mucopolysaccharidoses: A clinical review and guide to management. *Arch. Dis. Child.* **1995**, *72*, 263–267. [CrossRef] [PubMed]

152. Ritvo, E.R.; Mason-Brothers, A.; Freeman, B.J.; Pingree, C.; Jenson, W.R.; McMahon, W.M.; Petersen, P.B.; Jorde, L.B.; Mo, A.; Ritvo, A. The UCLA-university of Utah epidemiologic survey of autism: The etiologic role of rare diseases. *Am. J. Psychiatry* **1990**, *147*, 1614–1621. [PubMed]

153. Shapiro, E.; King, K.; Ahmed, A.; Rudser, K.; Rumsey, R.; Yund, B.; Delaney, K.; Nestrasil, I.; Whitley, C.; Potegal, M. The neurobehavioral phenotype in mucopolysaccharidosis type IIIB: An exploratory study. *Mol. Genet. Metab. Rep.* **2016**, *6*, 41–47. [CrossRef] [PubMed]

154. Ramaekers, V.T.; Blau, N. Cerebral folate deficiency. *Dev. Med. Child Neurol.* **2004**, *46*, 843–851. [CrossRef] [PubMed]

155. Moretti, P.; Scaglia, F. Cerebral folate deficiency with developmental delay, autism, and response to folinic acid. *Neurology* **2005**, *64*, 1088–1090. [CrossRef] [PubMed]

156. Gordon, N. Cerebral folate deficiency. *Dev. Med. Child Neurol.* **2009**, *51*, 180–182. [CrossRef] [PubMed]

157. Raidah, S.A.; Mohammed, W.C. Diagnosis and management of cerebral folate deficiency. *Neurosciences* **2014**, *19*, 312–316.

158. Ryan, A.K.; Bartlett, K.; Clayton, P.; Eaton, S.; Mills, L.; Donnai, D.; Winter, R.M.; Burn, J. Smith-Lemli-Opitz syndrome: A variable clinical and biochemical phenotype. *J. Med. Genet.* **1998**, *35*, 558–565. [CrossRef] [PubMed]

159. Kelley, R.I.; Hennekam, R.C.M. The Smith-Lemli-Opitz syndrome. *J. Med. Genet.* **2000**, *37*, 321–335. [CrossRef] [PubMed]

160. Tierny, E.; Nwokoro, N.A.; Porter, F.D.; Freund, L.S.; Ghuman, J.K.; Kelley, R.I. Behavior phenotype in the RSH/ Smith-Lemli-Opitz syndrome. *Am. J. Med. Genet.* **2001**, *98*, 191–200. [CrossRef]

161. Tierny, E.; Nwokoro, N.A.; Kelley, R.I. Behavior phenotype of RSH/Smith-Lemli-Opitz syndrome. *Ment. Retard. Dev. Disabil. Res. Rev.* **2001**, *98*, 191–200.

162. Tint, G.S.; Irons, M.; Elias, E.R.; Batta, A.K.; Frieden, R.; Chen, T.S.; Salen, G. Defective cholesterol biosynthesis associated with the Smith-Lemli-Opitz syndrome. *N. Engl. J. Med.* **1994**, *330*, 107–113. [CrossRef] [PubMed]

163. Thurm, A.; Tierney, E.; Farmer, C.; Albert, P.; Joseph, L.; Swedo, S.; Bianconi, S.; Bukelis, I.; Wheeler, C.; Sarphare, G.; et al. Development, behavior, and biomarker characterization of Smith-Lemli-Opitz syndrome: An update. *J. Neurodev. Disord.* **2016**, *8*, 12. [CrossRef] [PubMed]

164. Baieli, S.; Pavone, L.; Meli, C.; Fiumara, A.; Coleman, M. Autism and phenylketonuria. *J. Autism Dev. Disord.* **2003**, *33*, 201–204. [CrossRef] [PubMed]

165. Miladi, N.; Larnaout, A.; Kaabachi, N.; Helayem, M.; Ben Hamida, M. Phenylketonuria: An underlying etiology of autistic syndrome report. *J. Child Neurol.* **1992**, *7*, 22–23. [CrossRef] [PubMed]

166. Fon, E.A.; Sarrazin, J.; Meunier, C.; Alarcia, J.; Shevell, M.I.; Philippe, A.; Leboyer, M.; Rouleau, G.A. Adenylosuccinate lyase (ADSL) and infantile autism: Absence of previously reported point mutation. *Am. J. Hum. Genet.* **1995**, *18*, 554–557. [CrossRef] [PubMed]

167. Jaeken, J.; van den Berghe, G. An infantile autistic syndrome characterized by the presence of succinylpurines in body fluids. *Lancet* **1984**, *2*, 1058–1061. [PubMed]

168. Race, V.; Marie, S.; Vincent, M.F.; van der Berghe, G. Clinical, biochemical and molecular genetic correlations in adenylosuccinate lyase deficiency. *Hum. Mol. Genet.* **2000**, *9*, 2159–2163. [CrossRef] [PubMed]

169. Stone, R.L.; Aimi, J.; Barshop, B.A.; Jaeken, J.; van den Berghe, G.; Zalkin, H.; Dixon, J.E. A mutation in adenylosuccinate lyase associated with mental retardation autistic features. *Nat. Genet.* **1992**, *1*, 59–63. [CrossRef] [PubMed]

170. Jinnah, H.A.; Sabina, R.L.; Van Den Berghe, G. Metabolic disorders of purine metabolism affecting the nervous system. *Handb. Clin. Neurol.* **2013**, *113*, 1827–1836. [PubMed]

171. Nasrallah, F.; Feki, M.; Kaabachi, N. Creatine and creatine deficiency syndromes: Biochemical and clinical aspects. *Pediatr. Neurol.* **2010**, *42*, 163–171. [CrossRef] [PubMed]

172. Póo-Argüelles, P.; Arias, A.; Vilaseca, M.A.; Ribes, A.; Artuch, R.; Sans-Fito, A.; Moreno, A.; Jakobs, C.; Salomons, G. X-linked creatine transporter deficiency in two patients with severe mental retardation and autism. *J. Inherit. Metab. Dis.* **2006**, *29*, 220–223. [CrossRef] [PubMed]

173. Schulze, A.; Bauman, M.; Tsai, A.C.; Reynolds, A.; Roberts, W.; Anagnostou, E.; Cameron, J.; Nozzolillo, A.A.; Chen, S.; Kyriakopoulou, L.; et al. Prevalence of creatine deficiency syndromes in children with nonsyndromic autism. *Pediatrics* **2016**. [CrossRef] [PubMed]

174. Campistol, J.; Díez-Juan, M.; Callejón, L.; Fernandez-De Miguel, A.; Casado, M.; Garcia Cazorla, A.; Lozano, R.; Artuch, R. Inborn error metabolic screening in individuals with nonsyndromic autism spectrum disorders. *Dev. Med. Child Neurol.* **2016**, *58*, 842–847. [CrossRef] [PubMed]

175. Durand, C.M.; Betancur, C.; Boeckers, T.M.; Bockmann, J.; Chaste, P.; Fauchereau, F.; Nygren, G.; Rastam, M.; Gillberg, I.C.; Anckarsäter, H.; et al. Mutations in the gene encoding the synaptic scaffolding protein SHANK3 are associated with autism spectrum disorders. *Nat. Genet.* **2007**, *39*, 25–27. [CrossRef] [PubMed]

176. Moessner, R.; Marshall, C.R.; Sutcliffe, J.S.; Skaug, J.; Pinto, D.; Vincent, J.; Zwaigenbaum, L.; Fernandez, B.; Roberts, W.; Szatmari, P.; et al. Contribution of SHANK3 mutations to autism spectrum disorder. *Am. J. Hum. Genet.* **2007**, *81*, 1289–1297. [CrossRef] [PubMed]

177. Guiltmatre, A.; Huquet, G.; Delorme, R.; Bourgeron, T. The emerging role of *SHANK* genes in neuropsychiatric disorders. *Dev. Neurobiol.* **2014**, *74*, 113–122. [CrossRef] [PubMed]

178. Feng, J.; Schroer, R.; Yan, J.; Song, W.; Yang, C.; Bockholt, A.; Cook, E.H., Jr.; Skinner, C.; Schwartz, C.E.; Sommer, S.S. High frequency of neurexin 1β signal peptide structural variants in patients with autism. *Neurosci. Lett.* **2006**, *27*, 10–13. [CrossRef] [PubMed]

179. Ching, M.S.; Shen, Y.; Tan, W.H.; Jeste, S.S.; Morrow, E.M.; Chen, X.; Mukaddes, N.M.; Yoo, S.Y.; Hanson, E.; Hundley, R.; et al. Deletions of NRXN1 (neurexin-1) predispose to a wide spectrum of developmental disorders. *Am. J. Med. Genet.* **2010**, *153*, 937–947. [CrossRef] [PubMed]

180. Autism Genome Project Consortium. Mapping autism risk loci using genetic linkage and chromosomal rearrangements. *Nat. Genet.* **2007**, *39*, 319–328.

181. Viñas-Jornet, M.; Esteba-Castillo, S.; Gabau, E.; Ribas-Vidal, N.; Baena, N.; San, J.; Ruiz, A.; Coll, M.D.; Novell, R.; Guitart, M. A common cognitive, psychiatric and dysmorphic phenotype in carriers of *NRXN1* deletion. *Mol. Genet. Genom. Med.* **2014**, *2*, 512–521. [CrossRef] [PubMed]

182. Dabell, M.P.; Rosenfeld, J.A.; Bader, P.; Escobar, L.F.; El-Khechen, D.; Vallee, S.E.; Dinulos, M.B.P.; Curry, C.; Fisher, J.; Tervo, R.; et al. Investigation of *NRXN1* deletions: Clinical and molecular characterization. *Am. J. Med. Genet.* **2013**, *161*, 717–731. [CrossRef] [PubMed]

183. Alarcon, M.; Abrahams, B.S.; Stone, J.L. Linkage, association, and gene-expression analyses identify *CNTNAP2* as an autism-susceptibility gene. *Am. J. Med. Genet.* **2008**, *82*, 150–159. [CrossRef] [PubMed]

184. Werling, A.M.; Bobrowski, E.; Taurines, R.; Gundelfinger, R.; Romanos, M.; Grünblatt, E.; Walitza, S. *CNTNAP2* gene in high functioning autism: No association according to family and meta-analysis approaches. *J. Neural Transm.* **2016**, *123*, 353–363. [CrossRef] [PubMed]

185. Murdoch, J.D.; Gupta, A.R.; Sanders, S.J.; Walker, M.F.; Keaney, J.; Fernandez, T.V.; Murtha, M.T.; Anyanwu, S.; Ober, G.T.; Raubeson, M.J.; et al. No evidence for association of autism with rare heterozygous point mutations in contactin-associated protein-like 2 (*CNTNAP2*), or in other contactin-associated proteins or contactins. *PLoS Genet.* **2015**, *11*, e1004852. [CrossRef] [PubMed]

186. Bakkaloglu, B.; O'Roak, B.J.; Louvi, A.; Gupta, A.R.; Abelson, J.F.; Morgan, T.M.; Chawarska, K.; Klin, A.; Ercan-Sencicek, A.G.; Stillman, A.A.; et al. Molecular cytogenetic analysis and resequencing of contactin associated protein-like 2 in autism spectrum disorders. *Am. J. Hum. Genet.* **2008**, *82*, 165–173. [CrossRef] [PubMed]

187. Fernandez, T.; Morgan, T.; Davis, N.; Klin, A.; Morris, A.; Farhi, A.; Lifton, R.P.; State, M.W. Disruption of contactin 4 (CNTN4) results in developmental delay and other features of 3p deletion syndrome. *Am. J. Hum. Genet.* **2004**, *74*, 1286–1293. [CrossRef] [PubMed]

188. Roohi, J.; Montagna, C.; Tegay, D.H.; Palmer, L.E.; DeVincent, C.; Pomeroy, J.C.; Christian, S.L.; Nowak, N.; Hatchwell, E. Disruption of contactin 4 in three subjects with autism spectrum disorder. *J. Med. Genet.* **2008**, *46*, 176–182. [CrossRef] [PubMed]

189. Karayannis, T.; Au, E.; Patel, J.C.; Kruglikov, I.; Markx, S.; Delorme, R.; Héron, D.; Salomon, D.; Glessner, J.; Restituito, S.; et al. Cntnap4 differentially contributes to GABAergic and dopaminergic synaptic transmission. *Nature* **2014**, *511*, 236–240. [CrossRef] [PubMed]

190. Zhiling, Y.; Fujita, E.; Tanabe, Y. Mutations in the gene encoding CADM1 are associated with autism spectrum disorder. *Biochem. Biophys. Res. Commun.* **2008**, *377*, 926–929. [CrossRef] [PubMed]

191. Fujita, E.; Tanabe, Y.; Imhof, B.A.; Mamsi, M.Y.; Momoi, T. A complex of synaptic adhesion molecule CADM1, a molecule related to autism spectrum disorder, with MUPP1 in the cerebellum. *J. Neurochem.* **2012**, *123*, 886–894. [CrossRef] [PubMed]

192. Kitagishi, Y.; Minami, A.; Nakanishi, A.; Ogura, Y.; Matsuda, S. Neuron membrane trafficking and protein kinases involved in autism and ADHD. *Int. J. Mol. Sci.* **2015**, *16*, 3095–3115. [CrossRef] [PubMed]

193. Morrow, E.M.; Yoo, S.Y.; Flavell, S.W.; Kim, T.K.; Lin, Y.; Hill, R.S.; Mukaddes, N.M.; Balkhy, S.; Gascon, G.; Hashmi, A.; et al. Identifying autism loci and genes by tracing recent shared ancestry. *Science* **2008**, *321*, 218–223. [CrossRef] [PubMed]

194. Jamain, S.; Quach, H.; Betancur, C.; Rastam, M.; Colineaux, C.; Gillberg, I.C.; Soderstrom, H.; Giros, B.; Leboyer, M.; Gillberg, C.; et al. Mutations of the X-linked genes encodage neuroligins NLGN3 and NLGN4 are associated with autism. *Nat. Genet.* **2003**, *34*, 27–29. [CrossRef] [PubMed]

195. Laumonnier, F.; Bonnet-Brilhault, F.; Gomot, M.; Blanc, R.; David, A.; Moizard, M.P.; Raynaud, M.; Ronce, N.; Lemonnier, E.; Calvas, P.; et al. X-linked mental retardation and autism are associated with a mutation in the *NLGN4* gene, a member of the neuroligin family. *Am. J. Hum. Genet.* **2004**, *74*, 552–557. [CrossRef] [PubMed]

196. Yan, J.; Oliveira, G.; Coutinho, A.; Yang, C.; Feng, J.; Katz, C.; Sram, J.; Bockholt, A.; Jones, I.R.; Craddock, N.; et al. Analysis of the neuroligin 3 and 4 genes in autism and other neuropsychiatric patients. *Mol. Psychiatry* **2005**, *10*, 329–332. [CrossRef] [PubMed]

197. Fisch, G.S. Is autism associated with the fragile X syndrome? *Am. J. Med. Genet.* **1992**, *43*, 47–55. [CrossRef] [PubMed]

198. Fisch, G.S. What is associated with the fragile X syndrome? *Am. J. Med. Genet.* **1993**, *48*, 112–121. [CrossRef] [PubMed]

199. Rogers, S.J.; Wehner, D.E.; Hagerman, R. The behavioral phenotype in fragile X: Symptoms of autism in very young children with fragile X syndrome, idiopathic autism, and other developmental disorders. *J. Dev. Behav. Pediatr.* **2001**, *22*, 409–417. [CrossRef] [PubMed]

200. Bailey, D.B., Jr.; Hatton, D.D.; Skinner, M.; Mesibov, G. Autistic behavior, FMR1 protein, and developmental trajectories in young males with fragile X syndrome. *J. Autism Dev. Disord.* **2001**, *31*, 165–174. [CrossRef] [PubMed]

201. Belmonte, M.K.; Bourgeron, T. Fragile X syndrome and autism at the intersection of genetic and neural netwoks. *Nat. Neurosci.* **2006**, *9*, 1221–1225. [CrossRef] [PubMed]

202. Kau, A.S.; Tierney, E.; Bukelis, I.; Stump, M.H.; Kates, W.R.; Trescher, W.H.; Kaufmann, W.E. Social behavior profile in young males with fragile X syndrome: Characteristics and specificity. *Am. J. Med. Genet.* **2004**, *126*, 9–17. [CrossRef] [PubMed]

203. Loesch, D.Z.; Bui, Q.M.; Dissanayake, C.; Clifford, S.; Gould, E.; Bulhak-Paterson, D.; Tassone, F.; Taylor, A.K.; Hessl, D.; Hagerman, R.; et al. Molecular and cognitive predictors of the continuum of autistic behaviours in fragile X. *Neurosci. Biobehav. Rev.* **2007**, *31*, 315–326. [CrossRef] [PubMed]

204. Clifford, S.; Dissanayake, C.; Bui, Q.M.; Huggins, R.; Taylor, A.K.; Loesch, D.Z. Autism spectrum phenotype in males and females with fragike X full mutation and premutation. *J. Autism Dev. Disord.* **2007**, *337*, 738–747. [CrossRef] [PubMed]

205. Harris, S.W.; Hessl, D.; Goodlin-Jones, B.; Ferranti, J.; Bacalman, S.; Barbato, I.; Tassone, F.; Hagerman, P.J.; Herman, H.; Hagerman, R.J. Autism profiles of males with fragile X syndrome. *Am. J. Ment. Retard.* **2008**, *113*, 427–438. [CrossRef] [PubMed]

206. Dölen, G.; Carpenter, R.L.; Ocain, T.D.; Bear, M.F. Mechanism-based approaches to treating fragile X. *Pharmacol. Ther.* **2010**, *127*, 78–93. [CrossRef] [PubMed]

207. McLennan, Y.; Polussa, J.; Tassone, F.; Hagerman, R. Fragile X syndrome. *Curr. Genom.* **2011**, *12*, 216–224. [CrossRef] [PubMed]

208. Bar-Nur, O.; Caspi, I.; Benvenisty, N. Molecular analysis of FMR 1 reactivation in fragile-X induced pluripotent stem cells and their neuronal derivatives. *J. Mol. Cell. Biol.* **2012**, *4*, 180–183. [CrossRef] [PubMed]

209. Hagerman, R.; Lauterborn, J.; Au, J.; Berry-Kravis, E. Fragile X syndrome and targeted treatment trials. *Results Probl. Cell Differ.* **2012**, *54*, 297–335. [PubMed]

210. Tabolacci, E.; Chiurazzi, P. Epigenetics, fragile X syndrome and transcriptional therapy. *Am. J. Med. Genet.* **2013**, *161*, 2797–2808. [CrossRef] [PubMed]

211. De Vries, B.B.; Halley, D.J.; Oostra, B.A.; Niermeijer, C. The fragile X syndrome. *Am. J. Med. Genet.* **1998**, *35*, 579–589. [CrossRef]

212. Tolmie, J. The genetics of mental retardation. *Curr. Opin. Psychiatry* **1998**, *11*, 507–513. [CrossRef]

213. Hagerman, R.J. Fragile X syndrome. *Child Adolesc. Psychiatr. Clin. N. Am.* **1996**, *5*, 895–911. [CrossRef]

214. Hagerman, R.J.; Chudley, A.E.; Knoll, J.H.; Jackson, A.W.; Kemper, M.; Ahmad, R. Autism in fragile X females. *Am. J. Med. Genet.* **1986**, *23*, 375–380. [CrossRef] [PubMed]

215. Hatton, D.D.; Sideris, J.; Skinner, M.; Mankowski, J.; Bailey, D.B., Jr.; Roberts, J.; Mirrett, P. Autistic behavior in children with fragile X syndrome: Prevalence, stability, and the impact of FMRP. *Am. J. Med. Genet.* **2006**, *140*, 1804–1813. [CrossRef] [PubMed]

216. Roberts, J.E.; Tonnsen, B.L.; McCary, L.M.; Caravella, K.E.; Shinkareva, S.V. Brief report: Autism symptoms in infants with fragile X syndrome. *J. Autism Dev. Disord.* **2016**, *46*, 3830–3837. [CrossRef] [PubMed]

217. Einfeld, S.; Molony, H.; Hall, W. Autism is not associated with the fragile X syndrome. *Am. J. Med. Genet.* **1989**, *34*, 187–193. [CrossRef] [PubMed]

218. Reiss, A.L.; Freund, L. Behavioral phenotype of fragile X syndrome; DSM-III-R autistic behavior in male children. *Am. J. Med. Genet.* **1992**, *43*, 35–46. [CrossRef] [PubMed]

219. Todd, R.D.; Hudziak, J.J. Genetics of autism. *Curr. Opin. Psychiatry* **1993**, *6*, 486–488. [CrossRef]

220. Farzin, F.; Perry, H.; Hessl, D.; Loesch, D.; Cohen, J.; Bacalman, S.; Gane, L.; Tassone, F.; Hagerman, P.; Hagerman, R. Autism spectrum disorders and attention-deficit/hyperactivity disorder in boys with the fragile X premutation. *J. Dev. Behav. Pediatr.* **2006**, *27*, 137–144. [CrossRef]

221. Hagerman, R.J.; Berry-Kravis, E.; Kaufmann, W.E.; Ono, M.Y.; Tartaglia, N.; Lachiewicz, A.; Kronk, R.; Delahunty, C.; Hessl, D.; Visootak, J.; et al. Advances in the treatment of fragile X syndrome. *Pediatrics* **2009**, *123*, 378–390. [CrossRef] [PubMed]

222. Hallmayer, J.; Hebert, J.M.; Spiker, D.; Lotspeich, L.; McMahon, W.M.; Petersen, P.B.; Nicholas, P.; Pingree, C.; Lin, A.A.; Cavalli-Sforza, L.L.; et al. Autism and the X chromosome. Multipoint sib-pair analysis. *Arch. Gen. Psychiatry* **1996**, *53*, 985–989. [CrossRef] [PubMed]

223. Constantino, J.N.; Zhang, Y.; Holzhauer, K.; Sant, S.; Long, K.; Vallorani, A.; Malik, L.; Gutmann, D.H. Distribution and Within-Family specificity of quantitative autistic traits in patients with neurofibromatosis type I. *J. Pediatr.* **2015**, *167*, 621–626. [CrossRef] [PubMed]

224. Garg, S.; Plasschaert, E.; Descheemaeker, M.J.; Huson, S.; Borghgraef, M.; Vogels, A.; Evans, D.G.; Legius, E.; Green, J. Autism spectrum disorder profile in neurofibromatosis type I. *J. Autism Dev. Disord.* **2015**, *45*, 1649–1657. [CrossRef] [PubMed]

225. Plasschaert, E.; Descheemaeker, M.J.; Van Eylen, L.; Noens, I.; Steyaert, J.; Legius, E. Prevalence of autism spectrum disorder symptoms in children with neurofibromatosis type 1. *Am. J. Med. Genet.* **2014**, *168*, 72–80. [CrossRef] [PubMed]

226. Timonen-Soivio, L.; Vanhala, R.; Malm, H.; Hinkka-Yli-Salomäki, S.; Gissler, M.; Brown, A.; Sourander, A. Brief report: Syndromes in autistic children in a Finnish birth cohort. *J. Autism Dev. Disord.* **2016**, *46*, 2780–2784. [CrossRef] [PubMed]

227. Sheth, K.; Moss, J.; Hyland, S.; Stinton, C.; Cole, T.; Oliver, C. The behavioral characteristics of Sotos syndrome. *Am. J. Med. Genet.* **2015**, *167*, 2945–2956. [CrossRef] [PubMed]

228. Assumpcao, F.; Santos, R.C.; Rosario, M.; Mercadante, M. Brief report: Autism and Aarskog syndrome. *J. Autism Dev. Disord.* **1999**, *29*, 179–181. [CrossRef] [PubMed]

229. De Wolf, V.; Crepel, A.; Schuit, F.; van Lommel, L.; Ceulemans, B.; Steyaert, J.; Seuntjens, E.; Peeters, H.; Devriendt, K. A complex Xp11.22 deletion in a patient with syndromic autism: Exploration of *FAM120C* as a positional candidate gene for autism. *Am. J. Med. Genet.* **2014**, *164*, 3035–3041. [CrossRef] [PubMed]

230. Parisi, L.; Di Filippo, T.; Roccella, M. Behavioral phenotype and autism spectrum disorders in Cornelia de Lange syndrome. *Ment. Illn.* **2015**, *7*, 5988. [CrossRef] [PubMed]

231. Srivastava, S.; Landy-Schmitt, C.; Clark, B.; Kline, A.D.; Specht, M.; Grados, M.A. Autism traits in children and adolescents with Cornelia de Lange syndrome. *Am. J. Med. Genet.* **2014**, *321*, 1400–1410. [CrossRef] [PubMed]

232. Alvarez Retuerto, A.I.; Cantor, R.M.; Gleeson, J.G.; Ustaszewska, A.; Schackwitz, W.S.; Pennacchio, L.A.; Geschwind, D.H. Association of common variants in the Joubert syndrome gene (*AHI1*) with autism. *Hum. Mol. Genet.* **2008**, *17*, 3887–3896. [CrossRef] [PubMed]

233. Ionita-Laza, I.; Capanu, M.; De Rubeis, S.; McCallum, K.; Buxbaum, J.D. Identification of rare causal variants in sequence-based studies: Methods and applications to *VPS13B*, a gene involved in Cohen syndrome and Autism. *PLoS Genet.* **2014**, *10*, e1004729. [CrossRef] [PubMed]

234. Howlin, P.; Karpf, J.; Turk, J. Behavioural characteristics and autistic features in individuals with Cohen syndrome. *Eur. Child Adolesc. Psychiatry* **2005**, *14*, 57–64. [CrossRef] [PubMed]

235. Chandler, K.E.; Moffett, M.; Clayton-Smith, J.; Baker, G.A. Neuropsychological assessment of a group of UK patients with Cohen syndrome. *Neuropediatrics* **2003**, *34*, 7–13. [CrossRef] [PubMed]

236. Lerma-Carrillo, I.; Molina, J.D.; Cuevas-Duran, T.; Julve-Correcher, C.; Espejo-Saavedra, J.M.; Andrade-Rosa, C.; Lopez-Muñoz, F. Psychopathology in the Lujan-Fryns syndrome: Report of two patients and review. *Am. J. Med. Genet.* **2006**, *140*, 2807–2811. [CrossRef] [PubMed]

237. Alfieri, P.; Piccini, G.; Caciolo, C.; Perrino, F.; Gambardella, M.L.; Mallardi, M.; Cesarini, L.; Leoni, C.; Leone, D.; Fossati, C.; et al. Behavioral profile in RASopathies. *Am. J. Med. Genet.* **2014**, *164*, 934–942. [CrossRef] [PubMed]

238. Adviento, B.; Corbin, I.L.; Widjaja, F.; Desachy, G.; Enrique, N.; Rosser, T.; Risi, S.; Marco, E.J.; Hendren, R.L.; Bearden, C.E.; et al. Autism traits in the RASopathies. *J. Med. Genet.* **2014**, *51*, 10–20. [CrossRef] [PubMed]

239. Ghaziuddin, M.; Bolyard, B.; Alessi, N. Autistic disorder in Noonan syndrome. *J. Intellect. Disabil. Res.* **1994**, *38*, 67–72. [CrossRef] [PubMed]

240. Johansson, M.; Wentz, E.; Fernell, E.; Strömland, K.; Miller, M.T.; Gillberg, C. Autistic spectrum disorders in Moëbius sequence: a comprehensive study of 25 cases. *Dev. Med. Child Neurol.* **2001**, *43*, 338–345. [CrossRef] [PubMed]

241. Briegel, W.; Schimek, M.; Kamp-Becker, I.; Hofmann, C.; Schwab, K.O. Autism spectrum disorders in children and adolescents with Moebius sequence. *Eur. Child Adolesc. Psychiatry* **2009**, *18*, 515–519. [CrossRef] [PubMed]

242. Helsmoortel, C.; Vulto-van Silfhout, A.T.; Coe, B.P.; Vandeweyer, G.; Rooms, L.; van den Ende, J.; Schuurs-Hoeijmakers, J.H.; Marcelis, C.L.; Willemsen, M.H.; Vissers, L.E.; et al. A SWI/SNF-related autism syndrome caused by de novo mutations in ADNP. *Nat. Genet.* **2014**, *46*, 380–384. [CrossRef] [PubMed]

243. Splawski, I.; Timothy, K.W.; Sharpe, L.M.; Decher, N.; Kumar, P.; Bloise, R.; Napolitano, C.; Schwartz, P.J.; Joseph, R.M.; Condouris, K.; et al. Ca(V)1.2 calcium channel dysfunction causes a multisystem disorder including arrhythmia and autism. *Cell* **2004**, *119*, 19–31. [CrossRef] [PubMed]

244. Huguet, G.; Ey, E.; Bourgeron, T. The genetic landscapes of autism spectrum disorders. *Annu. Rev. Genom. Hum. Genet.* **2013**, *14*, 191–213. [CrossRef] [PubMed]

245. Schaefer, G.B. Clinical genetic aspects of autism spectrum sisorders. *Int. J. Mol. Sci.* **2016**, *17*, 180. [CrossRef] [PubMed]

246. Lord, C.; Rutter, M.; Le Couteur, A. Autism diagnostic interview-revised: A revised version of a diagnostic interview for caregivers of individuals with possible pervasive developmental disorders. *J. Autism Dev. Disord.* **1994**, *24*, 659–685. [CrossRef] [PubMed]

247. Lord, C.; Risi, S.; Lambrecht, L.; Cook, E.; Leventhal, B.; DiValore, P.; Pickles, A.; Rutter, M. The autism diagnostic observation schedule-generic: A standard measures of social and communication deficits associated with the spectrum of autism. *J. Autism Dev. Disord.* **2000**, *30*, 205–223. [CrossRef] [PubMed]

248. Risi, S.; Lord, C.; Gotham, K.; Corsello, C.; Chrysler, C.; Szatmari, P.; Cook, E.H., Jr.; Leventhal, B.L.; Pickles, A. Combining information from multiple sources in the diagnosis of autism spectrum disorders. *J. Am. Acad. Child Adolesc. Psychiatry* **2006**, *45*, 1094–1103. [CrossRef] [PubMed]

249. Schaefer, G.B.; Mendelson, N.J. Clinical genetics evaluation in identifying the etiology of autism spectrum disorders: 2013 guideline revisions. *Genet. Med.* **2013**, *15*, 399–407. [CrossRef] [PubMed]

250. Ponsot, O.; Moutard, M.L.; Villeneuve, N. Orientation Diagnostique en Neurologie Pédiatrique. In *Neurologie Pédiatrique*, 2nd ed.; Flammarion Médecine-Sciences: Paris, France, 1998; pp. 75–87.

251. Simonoff, E.; Bolton, P.; Rutter, M. Mental retardation: Genetics findings, clinical implications and research agenda. *J. Child Psychol. Psychiatry* **1996**, *37*, 259–280. [CrossRef] [PubMed]

252. Hoffmann, T.J.; Windham, G.C.; Anderson, M.; Croen, L.A.; Grether, J.K.; Risch, N. Evidence of reproductive stoppage in families with autism spectrum disorder: A large, population-based cohort study. *JAMA Psychiatry* **2014**, *71*, 943–951. [CrossRef] [PubMed]

253. Ozonoff, S.; Young, G.S.; Carter, A.; Messinger, D.; Yirmiya, N.; Zwaigenbaum, L.; Bryson, S.; Carver, L.J.; Constantino, J.N.; Dobkins, K.; et al. Recurrence risk for autism spectrum disorders: A baby siblings research consortium study. *Pediatrics* **2011**, *128*, 488–495. [CrossRef] [PubMed]

254. Durkin, M.S.; Maenner, M.J.; Newschaffer, C.J.; Lee, L.C.; Cunniff, C.M.; Daniels, J.L.; Kirby, R.S.; Leavitt, L.; Miller, L.; Zahorodny, W.; et al. Advanced parental age and the risk of autism spectrum disorder. *Am. J. Epidemiol.* **2008**, *168*, 1268–1276. [CrossRef] [PubMed]

255. Lord, C. Diagnostic Instruments in Autism Spectrum Disorders. In *Handbook of Autism and Pervasive Developmental Disorders*, 2nd ed.; Cohen, D.J., Volkmar, F.R., Eds.; John, Wiley and Sons Inc.: New York, NY, USA, 1997; pp. 460–483.

256. Anastasi, A. *Psychological Testing*, 6th ed.; Macmillan Publishing Co Inc.: New York, NY, USA, 1988.

257. *Manual for the Raven's Color Progressive Matrices (CPM)*; Les Editions Centre de Psychologie Appliquée (ECPA): Paris, France, 2016.

258. Guinchat, V.; Thorsen, P.; Laurent, C.; Cans, C.; Bodeau, N.; Cohen, D. Pre-peri-and neonatal risk factors for autism. *Acta Obstet. Gynecol. Scand.* **2012**, *91*, 287–300. [CrossRef] [PubMed]

259. Gardener, H.; Spiegelman, D.; Buka, S.L. Perinatal and neonatal risk factors for autism: A comprehensive meta-analysis. *Pediatrics* **2011**, *128*, 344–355. [CrossRef] [PubMed]

260. Weissman, J.R.; Kelley, R.I.; Bauman, M.L.; Cohen, B.H.; Murray, K.F.; Mitchell, R.L.; Kern, R.L.; Natowicz, M.R. Mitochondrial disease in autism spectrum disorder patients: A cohort analysis. *PLoS ONE* **2008**, *3*, e3815. [CrossRef] [PubMed]

261. Rossignol, D.A.; Frye, R.E. Mitochondrial dysfunction in autism spectrum disorders: A systematic review and meta-analysis. *Mol. Psychiatry* **2012**, *17*, 290–314. [CrossRef] [PubMed]

262. Demily, C.; Assouline, M.; Boddaert, N.; Barcia, G.; Besmond, C.; Poisson, A.; Sanlaville, D.; Munnich, A. Genetic approaches in autism spectrum disorders. *Neuropsychiatr. Enfance Adolesc.* **2016**, *64*, 395–401. [CrossRef]

263. Lai, M.C.; Lombardo, M.V.; Baron-Cohen, S. Autism. *Lancet* **2014**, *383*, 896–910. [CrossRef]

264. Tordjman, S.; Anderson, G.M.; Bellissant, E.; Botbol, M.; Charbuy, H.; Camus, F.; Graignic, R.; Kermarrec, S.; Fougerou, C.; Cohen, D.; et al. Day and nighttime excretion of 6-sulphatoxymelatonin in adolescents and young adults with autistic disorder. *Psychoneuroendocrinology* **2012**, *37*, 1990–1997. [CrossRef] [PubMed]

265. Haag, G.; Tordjman, S.; Duprat, A.; Urwand, S.; Jardin, F.; Clement, M.C.; Cukierman, A.; Druon, C.; Maufras du Chatellier, A.; et al. Presentation of a diagnostic grid of the progressive stages of infantile autism as observed in treatment. *Int. J. Psychoanal.* **2005**, *86*, 335–352. [CrossRef] [PubMed]

International Journal of
Molecular Sciences

MDPI

Review

Broader Autism Phenotype in Siblings of Children with ASD—A Review

Ewa Pisula [†,*] and Karolina Ziegart-Sadowska [†]

Faculty of Psychology, University of Warsaw, Stawki 5/7, 00-183 Warsaw, Poland; karolinaziegart@gmail.com
* Author to whom correspondence should be addressed; ewa.pisula@psych.uw.edu.pl;
  Tel.: +48-22-55-49-785; Fax: +48-22-63-57-991.
† These authors contributed equally to this work.

Academic Editor: Merlin G. Butler
Received: 18 January 2015; Accepted: 19 May 2015; Published: 10 June 2015

Abstract: Although less pronounced, social, cognitive, and personality characteristics associated with autism spectrum disorders (ASD) may be present in people who do not meet ASD diagnostic criteria, especially in first-degree relatives of individuals with ASD. Research on these characteristics, referred to as broader autism phenotype (BAP), provides valuable data on potential expressions of autism-specific deficits in the context of family relations. This paper offers a review of research on BAP in siblings of individuals with ASD, focusing on reports regarding social, communication, and cognitive deficits, published from 1993 to 2014. The studies are divided into two groups based on participants' age: papers on preschool and older siblings of individuals with ASD; and publications on infants at risk for ASD. On the basis of this review, suggestions are offered for further research and its significance for our understanding of the genetic determinants of autism.

Keywords: broad autism phenotype; siblings; review; at-risk infants

---

1. Introduction

Autism spectrum disorders (ASD) are neurodevelopmental disorders characterized by deficits in social communication and the presence of repetitive patterns of behavior, activity, and interests [1,2]. An important contributing factor in ASD etiology could be genetic liability, since the occurrence rate for ASD in siblings is on average 20- or 25-fold higher than in the general population [3–5].

The genetic mechanisms involved in the development of ASD are complex and heterogeneous [4,6,7]. This complexity is reflected by the variety of clinical characteristics of ASD, involving differences in the combination and expression of symptoms and severity of disorders in affected individuals. The heterogeneous, multifaceted nature of ASD is what drives research into the many possible expressions of autism that incorporate its typical deficits. The data obtained in that research may help us understand the genetic mechanisms that contribute to the development of autism-specific functional traits and atypical developmental trajectories. Moreover, this data may be useful in identifying genetic factors specific for various autism phenotypes, both in subjects with ASD and individuals from the general population.

A clear diagnostic conceptualization of autism still has yet to be established. It has been proposed that behavioral and cognitive characteristics of ASD, which include social, communication, and cognitive processes, rigid and persistent interests, and rigid and aloof personality traits, are continuously distributed throughout the general population [8]. These characteristics are likely to be more prevalent in first-degree relatives of individuals with ASD than in other groups [6,9–15]. Frazier and colleagues [16] studied how caregivers reported autism symptoms among children in their charge diagnosed with ASD and their unaffected siblings. They obtained data supporting the view that ASD is best characterized as a category. However, they note that "these data do not discount the possibility of endophenotypes or subthreshold ASD symptoms in unaffected family members" [16]. Wheelwright and colleagues [17] point out that studying ASD in nonclinical samples may be valuable in the context of the variety of genetic factors that seem to be connected to ASD. The genetic research on autism may benefit from a more inclusive concept of genetic expression that comprises cognitive, social, and communication deficits, as well as personality traits [18,19]. Characteristics similar to autism but less severe are referred to as broader autism phenotype (BAP) [20]. The notion of BAP facilitated involvement in studies on ASD and related phenomena of a large number of individuals without a clinical diagnosis of ASD. Consequently, it became possible to apply methods of the quantitative genetics typically used in the studies of normally distributed characteristics [21]. Research on BAP may help identify more functionally homogeneous subgroups of affected individuals and pinpoint the genetic factors that contribute to the development of ASD symptoms or traits [22]. The identification of phenotype may, therefore, open up the possibility of hypothetically assigning responsibility to a candidate gene or chromosome region. It therefore seems that data obtained in studies on BAP may provide promising cues for more detailed hypotheses on the genetic background of ASD. These studies may also contribute to a better understanding of the lesser variants of autism [23]. Since the data collected to date are not uniform, their applicability to genetic studies remains limited, and indicating the phenotypes with a clear genetic connection is still a difficult question. This may be due to the fact that the BAP concept covers a range of cognitive and social abilities and personality traits. Precisely establishing which characteristics should be included in BAP is a somewhat controversial undertaking. The lack of standardized criteria for BAP complicates attempts to engage in a comparative review of research on the topic. The features generally recognized as the most typical are deficits in social functioning, pragmatic language difficulties, restricted, repetitive behaviors and interests, as well as cognitive deficits (with respect to theory of mind in particular and social cognition in general, executive function, weak central coherence) and rigid personality [23]. Therefore, this paper focuses on the phenotypic characteristics of the individuals, observed from the perspective of clinical and experimental psychology. It seems that the concise summary of studies on BAP proposed here may provide some assistance in designing more advanced future studies aimed at identifying specific genetic mechanisms. Moreover, this paper will also demonstrate the methodological diversity in this field of study, which might help us in understanding the inconsistency of results across papers being published.

Gottesman and Gould [24] pointed out that phenotypes, understood as measurable characteristics, are heritable traits that are located in the path of pathogenesis from genetic predisposition to psychopathology. These traits are found at a higher rate in unaffected relatives of diagnosed family members than in the general population. The search for phenotypes may be conducted in specific populations that carry risk genes, but remain unaffected. Therefore unaffected relatives of individuals with the diagnosis, like healthy siblings, fulfill both criteria: they are enriched in risk genes and are healthy [25].

Therefore, a number of studies on BAP have been conducted on siblings of individuals with ASD. Initial findings have already indicated a significantly higher risk of autism in siblings than in the general population. Autism was initially reported to occur in approximately 2%–3% of siblings of affected children [26,27]. Folstein and Rutter [19], studying 21 pairs of same-sex siblings of whom one of the pairs was diagnosed with autism, found a concordance for autism in 4 of 11 sets of monozygotic twins, while in 10 dizygotic twin pairs they found no coexistence of autism in both children. Bolton and colleagues [28] analyzed family histories of autism in 137 individuals whose siblings had been diagnosed with the disorder, and found that 5.8% of siblings were diagnosed with autism, atypical autism, or Asperger syndrome, while no ASD diagnoses were reported among the siblings of individuals with Down syndrome. In a study by Ghaziuddin [29], 4.3% of siblings of 114 children with Asperger syndrome or autism were also diagnosed with Asperger syndrome or autism disorder.

Rates of BAP among siblings of individuals with ASD are much higher. August, Stewart, and Tsai [18] found cognitive disabilities in 15.5% of 71 siblings of probands with autism (compared to less than 3% in the control group). Folstein and Rutter [19] put forward a hypothesis that it is not autism as such that is inherited, but rather a pervasive cognitive deficit present both in individuals with autism and their immediate family. Currently, it is estimated that characteristics typical for ASD are present in at least 10%–20% of siblings of children diagnosed with ASD [28,30]. Ozonoff and colleagues [31], whose study involved a sample of 600 children in the USA and Canada, reported that autism-characteristic symptoms developed in nearly 19% of children whose older siblings had ASD. Boys are particularly at risk, with the correlation three times higher than in girls. At least one BAP trait is found in approximately 50% of family members of individuals with ASD when parents are included [32]. Biological siblings are attracting the attention of researchers studying BAP due to the large percentage of shared genes and environmental factors affecting their development.

Over the past few decades the corpus of data on BAP in siblings of individuals with ASD has expanded significantly. However, the studies in this area are highly diverse in terms of aspects such as design, instruments used, participant groups, as well as number of participants, their ages, and functions of interest, which lead to their findings being particularly difficult to integrate. Some help in overcoming that difficulty is offered by review papers that pool information on various aspects of studies and their results [23,33–37]. These reviews, especially those on preschool or older siblings, rarely provide information on participants' age. Due to the neurodevelopmental nature of ASD and the specifics of the processes involved, the inclusion of that information would seem reasonable. In the light of increasing interest in studying infant siblings of children with ASD among researchers seeking early predictors of ASD, summarizing the data collected from siblings could provide valuable information. Moreover, the methods for collecting data from children, adolescents, and adults are different, which provides even greater encouragement to have a closer look at the studies on BAP in different groups of siblings of individuals with ASD. The present paper provides a review of research on BAP traits in siblings of individuals with ASD, wherein the body of research will be divided into two groups: preschool or older siblings, and at-risk infants. The two will be treated as distinct due to the different nature of studies on infant siblings of children with ASD compared to research on older siblings. The studies on infants and toddlers are quite often prospective, enrolling both healthy children and children later diagnosed with ASD. With repeated measures it is possible to follow the siblings' developmental trajectories, unlike in most available studies on children over three or four years of age. Studies on at-risk infants have only emerged in the last decade; earlier research focused on older children, and as such these studies will be discussed first. For the reasons mentioned above, we have not attempted to propose a systematic review as defined in the PRISMA statement [38]. The present paper rather involves the subjective perspective of the authors reflected by the choice of the reviewed papers.

2. Research on Preschool-Age or Older Siblings

Table 1 presents a list of studies on broader autism phenotype in preschool or older siblings of individuals with ASD, published in the years 1993–2014. The list includes mainly studies that enrolled healthy siblings not affected by autism, although in a handful of them some participants were diagnosed with ASD or another disorder (*i.e.*, delayed language development) during the course of the study [28,39,40]. We have rejected studies in which the results of siblings were pooled in statistical analysis with the results of parents or other relatives *i.e.*, [41,42], which was sometimes a consequence of very small sibling subgroup size (like in [43]—only four siblings). Furthermore, publications included in the list meet the following criteria: (a) Peer-reviewed articles published in English; (b) Original studies; (c) Containing a control group; and (d) The social communication, language, and cognitive characteristics of autism in siblings of individuals with autism were the objects of study.

As shown in Table 1, studies that involve healthy siblings of individuals with ASD vary in design. They differ in terms of compared groups, sibling ages, sample sizes, analyzed aspects of functioning, as well as instruments and methods.

Table 1. Social communication, language, and cognitive BAP characteristics in siblings of individuals with ASD.

Study	Journal & Year	Characteristics	Sample	Age of Siblings of Individuals with ASD	Control Group, Matching Criteria	Measures/Instruments (Samples)	Main Findings
Ozonoff, Rogers, Farnham, & Pennington [45]	*J. Autism Dev. Disord.*, 1993	Executive function, theory of mind	18 Siblings of high functioning autism (HFA) children	8–18 years	18 siblings of children with learning disability, 18 HFA individuals, 18 learning-disabled children, matched on the basis of IQ, gender, socioeconomic status (SES), and ethnic background	Wechsler Intelligence Scale for Children–Revised (WISC–R), Wechsler Adults Intelligence Scale–Revised (WAIS–R), Wisconsin Card Sorting Test (WCST), Tower of Hanoi, Second-Order Belief Attribution Task, Fox and Grapes Task, Apple–Dog Task	HFA siblings performed worse than controls in the measure of planning (Tower of Hanoi). No differences between siblings groups in: set-shifting, working memory, and inhibitory control, theory-of-mind, Verbal IQ, Performance IQ, and Full Scale IQ
Szatmari et al. [45]	*J. Am. Acad. Child Adolesc. Psychiatry,* 1993	Cognitive impairments, adaptive behavior, developmental history	Siblings (and parents) of 52 pervasive developmental disorders (PDD) probands	6–18 years	Siblings (and parents) of 33 Down syndrome and low birth weight controls, matching criteria: IQ, family size, SES	Vineland Adaptive Behavior Scales (VABS), the Revised Stanford-Binet, Wide Range Achievement Test-Revised (WRAT-R), WCST	No differences between ASD siblings compared to control siblings on the social and communication domains of the VABS. No group differences in developmental history of language delays
Bolton et al. [28]	*J. Child Psychol. Psychiatry,* 1994	Social and communication impairments	137 Siblings (and 198 parents) of ASD individuals	Younger than 8 years	Down syndrome probands relatives (64 siblings, 72 parents); matching for age, sex, social class, birth order, and maternal age	Family History Interview, Autism Diagnostic Interview	20.4% of ASD siblings (and 3.1% of control siblings) demonstrated communication atypicalities, social impairments, or restricted behaviors; 4 out of 137 siblings met ICD-10 criteria for autism (two of them were mentally retarded), a further three were classified as having atypical autism and one was diagnosed with Asperger syndrome
Fombonne et al. [46]	*J. Child Psychol. Psychiatry,* 1997	General intellectual functioning, reading and spelling skills	Siblings (and parents) of 99 autism probands	Children and adults, lack of precise data	Siblings (and parents) of 36 Down syndrome individuals, matched for age, sex, social class, birth order, and maternal age with the autism probands	Family History Schedule, WAIS-R, WISC-R, Gray Oral Reading Test (GORT), Edinburgh Reading Tests, National Adult Reading Test, The Schonell Graded Word Spelling Test-B	Slightly higher mean verbal IQ scores in siblings of ASD individuals. Only the group of siblings of ASD individuals identified as affected with the BAP had significantly lower IQ scores and poorer reading and spelling abilities than unaffected siblings
Piven, Palmer, Jacobi, Childress, and Arndt [14]	*Am. J. Psychiatry,* 1997	Social and communication deficits	12 siblings from multiple incidence autism families	4–30 years	53 siblings of Down syndrome individuals; matched by probands age; no differences in parental education level and age	Family History Interview for Developmental Disorders of Cognition and Social Functioning	Higher rates of social deficits in siblings from families with multiple-incidence autism. No differences in the rates of communication deficits or stereotyped behaviors in siblings
Folstein et al. [47]	*J. Child Psychol. Psychiatry,* 1999	Pragmatic language, verbal IQ, reading and spelling skills	87 siblings (and 166 parents) of individuals with autism	5–46 years	64 siblings (and 75 parents) of individuals with Down syndrome; very similar age of siblings and parental education	WISC-R, WAIS-R, GORT, Kaufman Battery	No differences in verbal IQ scores, reading and spelling skills
Hughes, Hughes, Plumet, & Leboyer [48]	*J. Child Psychol. Psychiatry,* 1999	Executive function: verbal fluency, planning, flexibility	31 siblings of children with autism	5 years 8 months–19 years 11 months	32 siblings of children with developmental delay, 32 children from unaffected families (with no family history of ASD; similar family backgrounds (living in a low-income area)	Cambridge Neuropsychological Test Automated Battery, Multistage set-shifting task, akin to the WCST, Corsi Block Tapping task, Tower of London, Verbal Fluency task	Superior verbal and spatial span in siblings of children with autism; higher number of autism siblings than controls performed poorly on the verbal fluency tasks, planning, and set-shifting
Briskman, Happé, and Frith [49]	*J. Child Psychol. Psychiatry,* 2001	Everyday-life preferences and activities	19 siblings of children with autism	8–18 years	13 dyslexic siblings, 11 controls; matched on the basis of SES, parental education	Parental report	No differences between ASD siblings and other groups in everyday skills and preferences (parent-rated) with the exception of two boys subsequently diagnosed on the Autism Spectrum

Table 1. Cont.

Study	Journal & Year	Characteristics	Sample	Age of Siblings of Individuals with ASD	Control Group, Matching Criteria	Measures/Instruments (Samples)	Main Findings
Happé, Briskman and Frith [50]	J. Child Psychol. Psychiatry, 2001	Information processing, "Central coherence"	19 siblings of children with autism	8-18 years	14 dyslexia siblings, 17 controls; matched on the basis of SES, parental education; no difference in chronological age between autism group probands, dyslexia group probands, or normal controls	WISC-R, WISC-III, phonological measures (i.e., the Reading tests); Experimental measures (i.e., the Embedded Figures Test, Block Design Task, Titchener Circles Illusion Task)	Intact central coherence in siblings
Pilowsky, Yirmiya, Shalev, & Gross-Tsur [51]	J. Child Psychol. Psychiatry, 2003	Language abilities	27 siblings of children with autism	6 years-15 years 1 month	23 siblings of children with mental retardation of unknown etiology, 22 siblings of children with developmental language disorders; groups matched by siblings' age, gender, birth order, family size, ethnicity, and family income and by probands' gender and mental age	Children's Evaluation of Language Fundamentals (CELF)	Higher scores in siblings of children with autism on receptive, expressive, and total language scales and on verbal IQ compared to siblings of children with developmental language disorders
Dorris, Espie, Knott, & Salt [52]	J. Child Psychol. Psychiatry, 2004	Mind-reading	27 siblings of children with Asperger syndrome (AS)	7 years 6 months-17 years	27 control children matched for age, sex, and a measure of verbal comprehension	"Eyes Test", British Picture Vocabulary Scale II	Poorer performance of AS siblings in the "Eyes Test"
Bishop, Maybery, Wong, Maley, & Hallmayer [11]	Am. J. Med. Genet. B: Neuropsychiatr. Genet., 2006	Communication deficits	43 ASD siblings	4-16 years	46 control children; matching criteria: age, sex	Parent' report on Children's Communication Checklist-2 (CCC-2)	The only difference between groups was syntax; 23.8% of ASD siblings scored 2 SD below the control mean on CCC-2, compared to 2.2% of controls. Some differences in structural language skills
Constantino et al. [6]	Am. J. Psychiatry, 2006	Subsyndromal autistic impairments	49 siblings of children with autism from multiple-incidence families, 100 siblings of children with any PDD	4-18 years	45 siblings of children with psychopathology unrelated to autism, no matching criteria discussed	Social Responsiveness Scale (SRS)	Siblings of children with autism from multiple-incidence families—the highest scores in the SRS; followed by siblings of probands with any PDD, and then siblings of the probands with psychopathology unrelated to autism
Shaked, Gamliel, & Yirmiya [53]	Autism, 2006	Theory of mind	24 siblings of children with ASD (SIBS-A)	54-57 months	24 typically developing siblings (SIBS-TD), matched by siblings' age, gender, birth order, parents' age, and education	The false belief and the strange stories tasks	No differences on both theory of mind tasks
Christ, Holt, White, & Green [54]	J. Autism Dev. Disord., 2007	Executive function: inhibitory control, processing speed	21 siblings of children with ASD	6-15 years	18 children with ASD, 25 typically developing (TD) controls, matched on age, overall IQ, and processing speed	Stroop Card Task, Stroop Computer Task, Flanker Task, Go/No-go Task	No differences between ASD siblings and controls in processing speed and inhibitory control
Chuthapisith, Ruangdaraganon, Sombuntham, & Roongpraiwan [39]	Autism, 2007	Language development	32 preschool siblings of children with autism	2-6 years	28 control children, matched by siblings' age, gender, maternal educational level and family income	The Stanford-Binet IV	Delayed language development in eight of the autism siblings. Following exclusion of siblings with ASD and developmental language disorder (DLD) diagnosis, the remaining 29 siblings' verbal IQs were not significantly different from the control group
Dalton, Nacewicz, Alexander, & Davidson [55]	Biol. Psychiatry, 2007	Face processing	12 ASD siblings	8-18 years	21 individuals with autism, 19 TD controls, matched for age and intelligence quotient (IQ)	Facial Recognition Task, eye tracking, brain functional magnetic resonance imaging	Decreased gaze fixation and brain function in response to images of human faces in ASD siblings; less time than the control group fixating the eye region in response to naturalistic photographs of both familiar and unfamiliar human faces

Table 1. Cont.

Study	Journal & Year	Characteristics	Sample	Age of Siblings of Individuals with ASD	Control Group, Matching Criteria	Measures/Instruments (Samples)	Main Findings
Gamliel, Yirmiya, & Sigman [56]	J. Autism Dev. Disord., 2007	Cognitive and language development—a prospective study	39 ASD siblings (SIBS-A)	4–54 months	39 siblings of TD children (SIBS-TD); matched at 4 months according to chronological age, sex, birth order, number of children in the family, sex of the older proband, temperament profile, and mental and motor scores	Bayley Scales of Infant Development—2nd Edition (BSID-II), Reynell Developmental Language Scales (RDLS), Kaufman Assessment Battery for Children (K-ABC), Clinical Evaluation of Language Fundamentals-Preschool (CELF-P)	A delay in cognition and/or language in 12 of the SIBS-A and only two SIBS-TD; one child subsequently diagnosed with autism. Cognitive differences disappeared by age 54 months, while some differences in receptive and expressive language abilities remained. Most SIBS-A were well-functioning
Pilowsky, Yirmiya, Gross-Tsur, & Shalev [40]	J. Autism Dev. Disord., 2007	Neurocognitive functioning (i.e., intellectual abilities, acquired knowledge and achievement, executive function, attention and distractibility, sequential and simultaneous processing), behavior problems, developmental history, language abilities	30 siblings of children with autism	6–16 years	28 siblings of children with mental retardation, 30 siblings of children with developmental language delay; matched by siblings' chronological age, gender, birth order, probands' gender, and family income	WISC-III, WRAT-III, Tower of Hanoi, CELF-III (Word Associations Test), Rapid Automatic Naming test, Attention Deficit and Hyperactivity Questionnaire, Sequences Test, Visual Perception Test, Child Behavior Checklist, Family History Questionnaire	After excluding from ASD sibling group two siblings diagnosed with PDD, there were no differences between siblings of children with autism and the other groups
Gamliel, Yirmiya, Jaffe, Manor, & Sigman [57]	J. Autism Dev. Disord., 2009	Cognitive and language development—a prospective study	37 siblings of children with ASD (SIBS-A)	4 months–7 years	47 siblings of TD children (SIBS-TD); matched at 4 months on the basis of age, sex, birth order, number of children in the family, sex of the older proband, and temperament profile	BSID-II, K-ABC, RDLS, CELF-P, WISC-III, WRAT-III, CELF-III	At 7 years, 40% of the SIBS-A (and 16% of SIBS-TD) showed cognitive, language and/or academic difficulties (this sub-group was named SIBS-A-BP). Early language scores (14–54 months) were significantly lower in SIBS-A-BP compared to the language scores of SIBS-TD. Language as a major area of difficulty for SIBS-A during the preschool years
Kawakubo et al. [58]	PLoS ONE, 2009	Verbal fluency	24 siblings of children with ASD	Children (n = 12; M age = 11.1; SD = 3); Adults (n = 12)	27 high functioning individuals with ASD, 27 unrelated healthy controls with no family history of ASD, matched for age and IQ	Letter fluency task	No differences between either the child or the adult group
Koh, Milne, & Dobkins [59]	Neuropsychologia, 2010	Motor perception	13 adolescents with siblings diagnosed with ASD	13 years–17 years 11 months	23 adolescents with ASD, 42 TD adolescents, matched for age and gender	Two experiments: "detection task", and the "motion task"	SIBS showed higher chromatic contrast sensitivity than both participants with ASD and TD participants, what authors interpreted as the possible existence of a protective factor in these individuals against developing ASD
Ben-Yizhak et al. [60]	J. Autism Dev. Disord., 2011	Pragmatic language, school related linguistic abilities	35 siblings of children with autism (SIBS-A)	9–12 years	42 siblings of TD children; matching criteria at age 4 months; chronological age, sex, birth order, number of children in the family, sex of the older proband, temperament profile, and mental and motor scores	ADOS, SCQ, WISC-III, CELF-III, WRAT-III) Diagnostic Battery for Reading Processes in Hebrew	Lower pragmatic language abilities in a subgroup of SIBS-A identified with BAP-related difficulties. No differences between groups in general linguistic measures, school achievements, and reading processes
Levy and Bar-Yuda [61]	Autism, 2011	Language performance	28 siblings of nonverbal children with autism SIBS-ANV	4–9 years	27 controls matched for age, family background, socioeconomic status, and type of school they attended	CELF, spontaneous speech samples	SIBS-ANV achieved lower scores on the Receptive Scale, Expressive Scale and the Total Language Scale of the CELF; differences in the language scores were associated with IQ. No differences between groups in the results of grammatical analysis of spontaneous speech samples

Table 1. *Cont.*

Study	Journal & Year	Characteristics	Sample	Age of Siblings of Individuals with ASD	Control Group, Matching Criteria	Measures/Instruments (Samples)	Main Findings
Sumiyoshi, Kawakubo, Suga, Sumiyoshi, & Kasai [62]	*Neurosci. Res.*, 2011	Ability to organize information, executive function	14 siblings of individuals with ASD	Adults: M age = 24.5; SD = 4.0	22 individuals with ASD, 15 age-matched control subjects	WCST, Verbal Learning Task (VLT), AQ, CARS	No differences between siblings and controls in the WCST and VLT results. Authors noticed that a linear increase of the memory organization score on the VLT was absent in siblings as well as the ASD group. More autistic traits measured by AQ and CARS in siblings than in controls
Fiorentini *et al.* [63]	*Neuropsychologia*, 2012	Face identity	8 siblings (and 20 parents) of ASD children	7 years 11 months–16 years 3 months	10 TD children, (and 20 parents of TD children); matched for age, IQ	Face identity after effect task	Face-coding mechanisms in relatives of ASD individuals similar but less efficient compared to the relatives of typical children
Warren *et al.* [64]	*J. Autism Dev. Disord.*, 2012	Neurocognitive, language, and behavior measures—a prospective study	39 younger siblings of children with ASD (SIBS-ASD)	4 years 2 months–7 years 4 months	22 younger siblings of TD children; matching criteria: gender, chronological age, and SES	Differential Ability Scales—Second Edition, NEPSY-II, CELF-P, Children's Communication Checklist-2 for Parents, ADOS, SRS, CBCL, Social Skills Rating System	Executive functioning composite, Auditory attention, Inhibition—naming—worse in SIBS-ASD compared to controls. No differences in CBCL results
Gerdts, Bernier, Dawson, & Estes [65]	*J. Autism Dev. Disord.*, 2013	Broader autism phenotype traits as measured by Broader Phenotype Autism Symptom Scale	Siblings and parents from 87 multiplex and 41 simplex ASD families	Simplex families: M age = 11.51 (SD = 3.59); Multiplex families: M age = 10.20 (SD = 4.20)	Only members of ASD families	Broader Phenotype Autism Symptom Scale	Siblings from multiplex ASD families revealed less social interest, poorer conversational skills, higher rigidity, and intense interests and were less expressive in the use of nonverbal communication than siblings from simplex ASD families
Malesa *et al.* [66]	*Autism*, 2013	Prospective study–follow-up evaluation at age 5 years	38 from 54 later-born SIBS-TD, participating in the original study by Yoder *et al.* [67] (see: Table 2)	4–7 years	23 from 31 later-born SIBS-TD, participating in the original study; matching criteria: chronological age, race, and gender	ADOS, ADI-R, CELF-P, CELF-4, Differential Ability Scales—Second Edition (DAS-II), Social Skills Rating System (SSRS)	Two SIBS-ASD received a diagnosis of ASD at follow-up; none of the SIBS-TD received a clinical diagnosis. At age five there were no differences between SIBS-ASD and SIBS-TD in performance on most social and language domains assessed with standardized measures
Oerlemans *et al.* [68]	*J. Autism Dev. Disord.*, 2013	Executive function, social cognition, local processing style	172 siblings of children with ASD	6–21 years	140 Children with ASD, 127 controls; matched on the basis of age and ethnic background	Face recognition task, the Identification of Facial Emotions task, the Prosody task, Go/No-Go task, The Response Organization Objects task, Digit Span task (from Wechsler Scale)	ASD siblings performed worse than controls in face recognition task and inhibition task (but the differences referred only to processing speed)
Pickles, St Clair, & Conti-Ramsden [69]	*J. Autism Dev. Disord.*, 2013	Communication and social deficits	134 siblings and 193 parents of probands with ASD	8–42 years	Specific Language Impairment (SLI-only), 79 siblings, 103 parents), SLI + ASD (43 siblings, 30 parents), Down syndrome (DS, 63 siblings, 70 parents), matched on the basis of age and other criteria (not listed precisely)	Family History Information (with modifications)	ASD and SLI siblings had higher levels of communication deficits in relation to DS siblings (especially in the rate of language delay, level of spelling difficulties). More social deficits in ASD relatives in comparison to DS and SLI-only relatives
Robel *et al.* [70]	*Eur. Child Adolesc. Psychiatry*, 2013	Autistic traits	24 siblings of children with autism	M age = 9.4 (SD = 1.82)	96 siblings of TD children, aged three or older, from different socioeconomic backgrounds	French Autism Quotient (two main factors: F1 corresponding to socialization and communication, F2 to imagination and rigidity)	F1 and total AQ score higher in siblings of children with autism and global scores; no differences in F2 scores

Table 1. *Cont.*

Study	Journal & Year	Characteristics	Sample	Age of Siblings of Individuals with ASD	Control Group, Matching Criteria	Measures/Instruments (Samples)	Main Findings
Gizzonio *et al.* [71]	*Exp. Brain Res.*, 2014	Cognitive profile	21 siblings of children with ASD	6–16 years	32 Affected with ASD brothers of participating siblings, 43 TD children; matching criteria not listed precisely	WISC-III, SRS	No significant differences between Verbal Intelligence Quotient and Performance Intelligent Quotient scores among groups. Not significant, predominance of performance over verbal abilities observed in siblings group. Common cognitive profile in ASD group and ASD siblings
Holt *et al.* [22]	*Psychol. Med.*, 2014	Theory of mind	40 full siblings of individuals with autism or AS	12–18 years	50 adolescents with HFA or AS, 40 TD controls; matching criteria: age, full scale IQ above 70	"Reading the Mind in the Eyes" task	No differences between siblings and controls
Oerlemans *et al.* [72]	*Eur. Child Adolesc. Psychiatry*, 2014	Recognition of facial emotion and affective prosody; verbal attention	79 siblings of children with ASD	6–13 years	90 children with ASD (43 with and 47 without ADHD), 139 controls; matched on the basis of age and ethnic background	Facial emotion and affective prosody experimental tasks	Poorer performance of unaffected siblings than controls and better than the ASD probands in recognition of facial emotion and affective prosody tasks

2.1. Comparison Groups

All papers listed in Table 1 included one or more comparison groups. One exception is the paper by Gerdts *et al.* [65], which is listed despite the fact that only ASD families were included in that study. However, the inclusion of multiplex and simplex ASD families provides valuable insight into the severity of BAP traits in families with more than one child with autism.

Analysis of the groups used in comparisons with siblings of individuals with ASD reveals several strategies adopted by researchers. The most commonly-used comparison group consists of healthy siblings of typically developing individuals, or simply typically developing individuals themselves with no family history of autism [48,54,60,63,64,70]. This strategy allows for a comparison of the development of ASD siblings with their typically developing peers from families with no autism.

In many studies, one of the comparison groups was composed of individuals diagnosed with ASD, usually high-functioning autism or Asperger syndrome [22,44,54,55,58,59,62,68,72]. This way, the results of siblings could be compared with those obtained from individuals with ASD, and it was possible to determine whether the instruments used actually identified these deficits. In some of these studies, the results achieved by siblings of individuals with ASD landed in the middle, between the scores of individuals with ASD and those achieved by typically developing controls *i.e.*, [62,71]. This could indicate qualitatively similar traits or behavioral profiles of siblings and individuals with ASD, although the results of siblings are closer to the developmental norm. Some studies evaluated affected and unaffected siblings from the same family, *i.e.*, [71]. This empirical approach has particular merit in light of the fact that siblings share on average 50% of genes with their affected brother or sister [68]. Verifying similarities between siblings in terms of autistic traits may offer new insights into genetic susceptibility to autism.

In several studies, typically older ones, the comparison group for siblings of individuals with ASD was composed of siblings of individuals with Down syndrome, *i.e.*, [14,45–47,69]. This was to control for the effects of a family member with a hereditary disability not associated with ASD on the development and functioning of siblings. Recently, however, the focus has been on the comorbidity of Autism Spectrum Disorders and Down syndrome. Lowenthal *et al.* [73] showed that the frequency of Pervasive Developmental Disorders in individuals with Down syndrome was 15.6%, including 5.58% for autism. These findings must be taken into account in the selection of siblings of individuals with Down syndrome for the group compared with the siblings of ASD individuals. The presence and severity of autistic traits and symptoms in probands with Down syndrome must also be controlled for. With the high incidence of ASD (see [74]), this should be a universally-observed principle applied in the selection of comparison groups in studies on BAP, not limited to Down syndrome participants.

In some research projects, comparison groups consisted of siblings of individuals with developmental delay [48], mental retardation of unknown etiology [40,51], or psychopathology unrelated to autism, such as ADHD, affective disorders, and anxiety disorders [6]. There is also an interesting group of studies that included siblings of children with such developmental problems as dyslexia [49,50] and specific language impairment [69]. Extensive research has recently been done on potential associations between these complex developmental disorders and autism as, similarly to autism, their incidence is greater in males, they involve a number of neurophysiological and neuroanatomical abnormalities, they are likely to have strong genetic components, and they encompass language and communication deficits [75,76].

There is no question that the type of comparison groups should be taken into consideration when interpreting results and forming generalizations regarding the presence or absence of specificity in the functioning of siblings of individuals with ASD. In the case of heritable conditions that share features with ASD (e.g., SLI), it is possible that the first-degree relatives of probands will present some characteristics that overlap with the BAP [23]. This may cause additional difficulties in comparisons and the isolation of the autism phenotype.

2.2. Participants' Age and Sample Size

In the majority of studies listed in Table 1, the participants' functioning was measured only once. The exceptions are two longitudinal studies by Gamliel and colleagues [56,57] that include children aged 4 months to 4.6 and 7 years, and studies by Warren *et al.* [64] and Malesa *et al.* [66], which were a continuation of a study by Yoder *et al.* [67]. As these studies provide information about preschool and school-age children, we decided to include them in Table 1. There is only one other study [39] in which participants were under 4 years of age. In all other works the siblings of individuals with ASD were aged 4 or older, and in a number of them the age range was quite large. In some papers it exceeded 10 years, *i.e.*, [44,49,55], and could be as high as 40 years, *i.e.*, [47,69].

The age of siblings at the time of the study is not always crucial; in designs where caregivers were asked to provide information on the child's development history, that variable is less significant *i.e.*, [46,69]. However, in those studies that measure developmental levels across various domains or mastery of specific functions, a wide age range of participants may compromise the precision of inferences. It also cannot capture any developmental delays or irregularities that may be present during a certain period. However, knowledge of the specifics of that aspect of development could help in planning support for children in ASD families. A wide sibling age range makes it difficult to determine the level of that group's functioning, especially when coupled with a small sample size, which precludes statistical analysis on subgroups more homogenous in terms of age. In some studies the ASD siblings group counted under 20 participants *i.e.*, [14,44,49,50,55,59,62,63]. There is, however, a clear trend towards increased sample size in more recent studies, *i.e.*, [68,69]. In turn, small groups tend to be more homogenous, including in respect of participants' age. One illustration is the study by Warren *et al.* [64], which enrolled 40 siblings aged 4 to 7; another comes from research on preschool and early school children aged 2 to 9 years *i.e.*, [39,61]. Ben-Yizhak *et al.* [60] assessed older children in a narrow age range from 9 to 12 years, while Koh *et al.* [59] and Holt *et al.* [22] evaluated adolescents aged 12–18. Nevertheless, the majority of studies are conducted on children and adolescents with wider age ranges: from 6–8 to 16–18 years, *i.e.*, [6,40,54,71].

As we have previously mentioned, research on siblings who are of preschool age, or on older children and adolescents, suffers from a lack of longitudinal studies. In one of a handful of reports that provide such information, Gamliel *et al.* [56] stated that cognitive deficits originally present in siblings of children with ASD disappeared during the preschool period (age: 54 months), but—compared to controls—some differences in terms of receptive and expressive language abilities remained. The results indicating differences in language development were confirmed in the next phase of the study, conducted when the children reached 7 years of age. Data of that sort is particularly relevant for understanding the specifics of developmental processes in children from ASD-affected families. While not meeting the diagnostic criteria for the disorder, these children may experience difficulties due to atypical development. Some of these difficulties are resolved, but there are some that may compromise their adjustment at a given moment in life. Thus, longitudinal studies that enable us to follow developmental processes, as well as to identify their determinants and effective methods of developmental support, are particularly valuable.

2.3. Functioning Characteristics of Interest

The studies listed in Table 1 cover a wide variety of characteristics. In some, non-affected siblings were assessed for autism symptomatology with instruments used in diagnosis (Autism Diagnostic Observation Schedule, ADOS [77] or Autism Diagnostic Interview-Revised, ADI-R [78]) or based on developmental history established from medical history or interviews other than ADI-R *i.e.*, [14,28,69]. The results of these studies suggest a greater incidence of deficits in at least one domain typical for ASD. However, their results are not fully consistent in terms of the domains affected by the deficits and the depth of the deficits in question.

Greater severity of autistic traits in the siblings of individuals with ASD compared to controls was shown in those studies that measured such traits using the Social Communication Questionnaire (SCQ [79]), Childhood Autism Rating Scale (CARS [80]), Autism Spectrum Quotient (AQ [8]) and Broader Phenotype Autism Symptom Scale (BPASS [32]) [6,62,65,79]. Another finding was that autistic traits, including less social interest, poorer conversational skills, higher rigidity, and intense interests, as well as less expressiveness in the use of nonverbal communication, are more pronounced in siblings from multiplex ASD families than in children from simplex ASD families [65]. Based on similar results obtained for social and communication domains of BAP in parents, Bernier *et al.* [81] concluded that different genetic transfer mechanisms may operate in families with one and families with several children with autism displaying many of these characteristics. It has been suggested that *de novo* mutations and non-inherited copy number variants may be particularly important risk factors in simplex ASD families, while in multiplex ASD families they are present to a much lesser degree [82]. However, this finding was not confirmed in all studies [83]. As noted by Spiker and colleagues [84], the variability of autistic phenotype expressed in multiplex families is relatively low. In their study on families with two or more children with autism, in 37 out of 44 participating families at least two children met the ADI diagnostic criteria. Out of all the children in these families, 71% met all of the criteria for autism diagnosis in ADI, 22% failed to meet any of the criteria, and 7% met one or more criteria without reaching all of the ADI cutpoints. The number of items classified as "uncertain" was small, and there was a clear-cut distinction between children affected and unaffected by autism. It should also be noted that studies involving siblings from simplex ASD families, siblings from multiplex ASD families, and controls are scarce. This reduces the possibility of identifying the scope of the differences between the three groups with respect to the BAP.

A substantial body of research into the functioning of ASD siblings concerns language and communication as part of the phenotype(s) of ASD. Findings vary, as do methods and instruments. In their grammatical analysis of spontaneous speech samples, Levy and Bar-Yuda [61] found no significant differences between ASD siblings and preschool controls. In some studies ASD siblings were found to be more likely to have language delays relative to their comparison groups [39,56,57], while in other papers such differences were not stated, *i.e.*, [45]. A frequently encountered pattern of results is one in which the group of ASD siblings as a whole does not differ from controls, but a subgroup may be distinguished that demonstrates more pronounced BAP traits, including inferior language development, *i.e.*, [39,60]. Interesting data comes from comparisons of ASD siblings with siblings of individuals with language disorders. Some researchers found less severe difficulties in language and communication in siblings of individuals with ASD than siblings of people with language disorders, *i.e.*, [51], while others demonstrated that ASD siblings, similar to specific language impairment siblings, demonstrate higher levels of communication deficits in relation to siblings of individuals with Down syndrome [69].

A number of studies analyzed siblings' intellectual abilities [44,46,47,51,71]. This is an important issue since intellectual disability (ID) and autism are highly co-morbid [85]. It is estimated that ID is present in 50%–70% of all ASD cases [85]. Precise estimation of the co-morbidity of ASD and ID is complicated due to changing criteria, diagnostic procedures, and educational policy (e.g., providing more support to certain groups of pupils) as well as the methodological problems (screening tools properties) [86]. Nevertheless, it has been shown that the number of high-functioning individuals with ASD diagnosis has increased in recent years (see [74]). It is also noteworthy that IQ may be underestimated in people with high levels of autistic traits [87]. However, the interrelationships among the level of autistic traits and intellectual abilities are still far from being fully recognized, and the results of studies on IQ in siblings of individuals with ASD are equally difficult to generalize. In some of them, no differences were found between ASD siblings and controls in terms of IQ levels (e.g., [40,44,47]), while Fombonne *et al.* [46] and Pilowsky *et al.* [51] reported even higher verbal IQ in ASD siblings. Gizzonio *et al.* [71] found no differences between ASD siblings and controls in terms of Verbal IQ and Performance IQ, but reported a slight (non-significant) predominance of performance over verbal abilities. As the authors have suggested, this could indicate the presence of a certain cognitive profile common for individuals with ASD and their siblings, but statistically non-significant differences make this a very tentative conclusion. Fombonne *et al.* [46] identified a subgroup of participants who demonstrated BAP traits (referred to as BAP+) and had significantly lower IQ scores than the group of siblings non-affected with BAP. Similar results were reported by Chuthapisith *et al.* [39].

There have been numerous studies on autism-specific cognitive deficits: in theory of mind, central coherence, and executive function. Research on theory of mind, recognition of facial emotions, and face processing indicate that they are less developed in siblings of individuals with ASD compared to controls, *i.e.*, [55,63,68,72]. By contrast, the results of a study using the "Reading the Mind in the Eyes" task were inconclusive. Dorris *et al.* [52] showed that siblings scored lower than controls, but Holt *et al.* [22] found no differences between adolescent ASD siblings and typically developing control adolescents. Similarly, no differences in theory of mind between siblings and controls were reported by Ozonoff [44]. The key in studies of this type is to take into account the age of participating siblings, which is often overlooked. It is also noteworthy that this area of study is still strongly affected by a lack of precise conceptual definitions and methodological scrutiny, which makes the discussion even more complex. It must be noted that the results of studies on deficits discussed above in individuals with ASD are mixed, and are far from conclusive. A detailed discussion of the definitional controversies relating to particular aspects of cognitive deficits typical for ASD, as well as a presentation of current opinion concerning theory of mind, emotion recognition, central coherence, and executive function in people with ASD, remain outside the scope of this paper.

A complex picture also emerges from research on executive function. Siblings of individuals with ASD were found to be no different from controls in terms of inhibitory control and processing speed [54], but performed worse in planning tasks [44,48]. Ozonoff and colleagues [44] found no differences in working memory and set-shifting, while Hughes *et al.* [48] concluded that a larger than expected proportion of siblings of individuals with autism demonstrated difficulties in set-shifting. In the study conducted by Pilowsky *et al.* [40], differences in executive function between ASD siblings and controls disappeared once two participants diagnosed with Pervasive Developmental Disorders were removed from the former group. It should be also mentioned that Happe, Briskman, and Frith [50] reported no differences in terms of weak central coherence. In recent years the field of research on siblings of individuals diagnosed with ASD has been dominated by studies focused on infants at high familial risk for ASD (see [35]). It is estimated that about 10%–20% of high-risk infants can be affected with subclinical ASD traits or other developmental problems [88]. It should be emphasized that the analyses in some of the studies on infants also included children who later received an ASD diagnosis, which is why these results should be approached with caution. Nevertheless, research on infant siblings of children with ASD can not only provide us with valuable information on early signs of autism, but also pave the way for investigation of BAP traits [35].

3. Research on High-Risk Infants

Most studies focus on infant siblings of older children diagnosed with ASD, aged from 4 to 24 or 36 months. Diagnosis of the ASD proband is usually confirmed with ADOS and ADI-R outcomes, and the age of the proband is not relevant. Infant siblings are usually participants in long-lasting longitudinal studies in which different aspects of infants' functioning are assessed, *i.e.*, [56,77,89]. The control groups mainly consist of typically developing infants without familial risk for ASD. Some research projects perform comparison analyses between high-risk infants (HR, siblings of older children diagnosed with ASD) and low-risk infants (LR, infants without familial risk for ASD) [90–94], while others strive for interpretation of infants' functioning and test performance in the context of later ASD diagnoses or BAP characteristics [30,95–98]. Many studies differ in the sizes of HR and LR infant samples involved in analysis (*i.e.*, nine HR infants in [99] to 507 HR infants in [88]), making interpretation and comparison of results problematic.

A number of research projects on infant siblings are focused on early characteristics of ASD core symptoms, as well as overall risk for developing ASD, *i.e.*, [88,94,100–102]. Macari *et al.* [101] suggest that two thirds of infants at high risk for ASD experience some kind of developmental difficulties in the second year of life. A large study conducted by Messinger *et al.* [88] indicated elevated levels of autistic traits (higher mean ADOS severity scores) in HR infants compared to LR infants. Georgiades *et al.* [100] suggested that significantly more HR infants had exhibited higher levels of autistic-like traits than LR children, and at the age of three years these children had more social communication impairments, lower cognitive abilities, and more internalizing problems than typically developing children.

As shown in Table 2, many authors report early communication and language deficits [56,89,90,94, 96,102–104], social interaction impairments [91,94,97,98,105–107], and increased levels of stereotyped behaviors [108–110] in HR infant siblings.

It should be noted that the emergence of some of the highlighted differences are probably caused by inclusion in analyses of high-risk infant siblings who later developed ASD. For instance, Rozga *et al.* [97] ascertained that high-risk infant siblings who later developed ASD exhibited lower rates of joint attention and requesting behavior than typically developing children. However, HR infant siblings without ASD outcomes did not differ in these characteristics from the control infants. Similarly, Bedford *et al.* [111] indicated that only the high-risk infants with later emerging socio-communication difficulties (ASD and atypical development) differed significantly from the control group in the gaze following task. What is more, high-risk infants without ASD outcome performed similarly to typically developing infants. Hutman *et al.* [95] also suggest social impairments only in the high-risk infants with later ASD diagnosis. Hudry *et al.* [104] observed reduced receptive vocabulary advantage in all HR infants by 14 months, but this difference was maintained through 24 months only in children with ASD outcome, while typically developing HR infants regained a more normative profile.

Some empirical data suggest the presence of deficits in quality of mother–infant interaction and differences in responses to separation events in high-risk infant siblings. For instance, Esposito *et al.* [112] found differences in cry sample patterns in HR toddlers compared to LR toddlers in expression of distress during the separation phase. However, Haltigan *et al.* [113] concluded that HR infants are not less likely to form secure affectional bonds with their caregivers than LR infants, but also mentioned that infant siblings of children with ASD are less distressed during separation and more reserved after reunion with a caregiver compared to typically developing children. Finally, a study conducted by Wan *et al.* [114] indicated that infant attentiveness to parent and positive affect were lower in the high-risk group later diagnosed with ASD. These characteristics, as well as dyadic mutuality, predicted a three-year ASD outcome.

A number of studies analyzed face processing and social visual fixation patterns as predictors of autistic traits in infants. Some HR siblings demonstrated diminished gaze to the mother's eyes relative to her mouth in the Still Face episode [93]. During the face processing task, LR infants demonstrated a preference for looking at the left side of the face (characteristic left visual field bias) that emerged by 11 months of age and was absent in HR infants at any age [115]. Hutman *et al.* [95] observed no difference in the proportion of attention to social stimuli or attention shifting during the play condition between HR and LR infants. However, children later diagnosed with ASD tended to continue looking at a toy during the distress condition despite the salience of social information. Overall, no group differences between HR and LR infants in gaze following behavior at either age was observed in the study by Bedford *et al.* [111]. Nevertheless, it should be noted that HR infants with later emerging socio-communication difficulties (ASD and atypical development) allocated less attention to a congruent object compared to typically developing HR siblings and LR controls.

Table 2. Studies on infants and toddlers at risk for ASD.

Study	Journal & Year	Characteristics	Sample	Age of Siblings of Individuals with ASD	Control Group, Matching Criteria	Measures/Instruments (Samples)	Main Findings
Goldberg et al. [105]	J. Autism Dev. Disord., 2005	Social communication behaviors	8 children diagnosed with ASD; 8 younger siblings of children with ASD	Below 3 years old	9 TD children, age and IQ controlled	ADI-R, ADOS-G, CARS, Early Social Communication Scales (ESCS)	On three of four of the ESCS subscales (Responds to Social Interaction, Initiates of Joint Attention, and Requesting Behaviors) social communicative behaviors of younger siblings differed from those of typically developing children but not from the behaviors displayed by ASD group
Zwaigenbaum et al. [102]	Int. J. Dev. Neurosci., 2005	Autistic traits, autism-specific behavior	65 siblings of children with ASD	6 to 24 months	75 low risk infants, gender, birth-order, and age-matched to high-risk infants	Novel observational scale a computerized visual orienting task, and standardized measures of temperament, cognitive and language development	Lower receptive language scores and use of fewer gestures and phrases at 24 months in non-autistic siblings. By 12 months of age, siblings who are later diagnosed with autism may be distinguished from other siblings and low-risk controls on the basis of: (1) behavioral markers, including atypicalities in eye contact or visual tracking; (2) prolonged latency to disengage visual attention; (3) a characteristic pattern of early temperament, with marked passivity and decreased activity level at six months; and (4) delayed expressive and receptive language
Landa and Garrett-Mayer [30]	J. Child Psychol. Psychiatry, 2006	Autistic traits, autism-specific behavior	60 HR infants (siblings of children with autism, SIBS-A) and 27 LR infants (no family history of autism), at 24 months of age categorized as: unaffected, ASD, or language delayed (LD)	6–24 months	27 low risk infants (no family history of autism). age, ethnic group, and SES were controlled	Language test scores, ADOS, MSEL	Lower scores on all MSEL scales in SIB-A at 14 months, compared to LD and TD children. By 14 months of age, the ASD group performed significantly worse than the unaffected group on all scales except Visual Reception. By 24 months of age, the ASD group performed significantly worse than the unaffected group in all domains, and worse than the language delayed group in Gross Motor, Fine Motor, and Receptive Language
Mitchell et al. [96]	J. Dev. Behav. Pediatr., 2006	Early language and communication development	97 SIBS-A (then part of them diagnosed with ASD)	12 to 24 months	49 control children, recruited from three regions in numbers roughly proportionate to each region's high-risk siblings	MacArthur Communicative Development Inventory (CDI), Preschool Language Scale—Third Edition, MSEL	Children with ASD showed delays in early language and communication compared with non-ASD siblings and controls. At 12 months, the ASD group was reported to understand significantly fewer phrases and to produce fewer gestures. At 18 months, they showed delays in their understanding of phrases, comprehension and production of single words, and use of gestures. Siblings not diagnosed with ASD also used fewer play-related gestures at 18 months than low-risk controls, even when children with identified language delays were excluded
Yirmiya et al. [90]	J. Child Psychol. Psychiatry, 2006	Social engagement, communication, and cognition	21 SIBS-A	4–14 months	21 TD infants, age-matched	Bayley Scales of Infant Development–2nd edition, Infant Characteristics Questionnaire (ICQ), Still-face paradigm, Name-calling responsiveness, Early Social Communication Scales (ESCS), Checklist for Autism in Toddlers (CHAT)	At 14 months, SIBS-A made fewer nonverbal requesting gestures and achieved lower language scores on the Bayley Scale. Infant SIBS-A, who showed more neutral affect to the still face and were less able to respond to their name being called by their mothers, initiated fewer nonverbal joint attention and requesting behaviors at 14 months, respectively

Table 2. *Cont.*

Study	Journal & Year	Characteristics	Sample	Age of Siblings of Individuals with ASD	Control Group, Matching Criteria	Measures/Instruments (Samples)	Main Findings
Bryson *et al.* [99]	*J. Autism Dev. Disord.*, 2007	Autistic traits, IQ	9 HR infants with older siblings with ASD, all of them diagnosed with ASD at 36 months	6–24 months, assessment every 6 months	Developmental study, no control group	ADOS, Bayley Scales of Infant Development, 2nd ed. or MSEL, CDI-Words and Gestures, Infant Temperament Scale or Toddler Behavior Assessment Questionnaire	Two groups were identified: 1st subgroup (*n* = 6) showed a decrease in IQ between 12 and 24 or 36 months; 2nd subgroup (*n* = 3) continued to obtain average IQs. Signs of autism emerged and/or were more striking earlier in the 1st group (*n* = 6). In all children early impairment in social-communicative development coexisted with atypical sensory and/or motor behaviors and temperamental profile marked by irritability/distress and dysregulated state
Cassel *et al.* [91]	*J. Autism Dev. Disord.*, 2007	Social and emotional communication	12 infant siblings of children with autism	6–18 months	19 age-matched TD control children	Face-to-face/still-face (FFSF), Early Social Communication Scale (ESCS)	Siblings smiled for a lower proportion of the FFSF than TD and lacked emotional continuity between episodes. Siblings engaged in lower rates of initiating joint attention at 15 months, lower rates of higher-level behavioral requests at 12 months, and responded to fewer joint attention bids at 18 months. Infant siblings experience subtle, inconsistent, but multi-faceted deficits in emotional expression and referential communication
Iverson *et al.* [92]	*J. Autism Dev. Disord.*, 2007	Vocal-motor development	21 infant siblings of children with autism	5–18 months	18 TD control children, maternal and parental age and levels of parental education comparable in sample and control group	Videotaping; naturalistic observation, semi-structured play, play in a Johnny Jump-Up, face-to-face interaction, and play with toys; MacArthur–Bates Communicative Development Inventory; Pervasive Developmental Disorder Screening Test-II	Infant siblings were delayed in the onset of early developmental milestones and spent significantly less time in a greater number of postures, suggestive of relative postural instability. Infant siblings demonstrated attenuated patterns of change in rhythmic arm activity around the time of reduplicated babble onset; and were highly likely to exhibit delayed language development at 18 months
Loh *et al.* [108]	*J. Autism Dev. Disord.*, 2007	Stereotyped motor behaviors	8 infant siblings of children later diagnosed with ASD, 9 infant siblings of children with autism not diagnosed with ASD	12 and 18 months	15 TD control children, same geographic area and age-matched to the high-risk infants	Measurement of Repetitive Motor Behaviors, videotaping	At 12 and 18 months the ASD group "arm waved" more frequently and at 18 months, one posture ("hands to ears") was more frequently observed in the ASD and non-diagnosed group compared to the TD
Merin *et al.* [93]	*J. Autism Dev. Disord.*, 2007	Visual fixation patterns during reciprocal social interaction	31 infant siblings of children with autism	6 months	24 Comparison infants with no autism family history, age and gender controlled	Modified Still Face paradigm, Eye tracking	Eleven infants demonstrated diminished gaze to the mother's eyes relative to her mouth during the Still Face episode; 10 of them had an older sibling with ASD
Stone *et al.* [103]	*Arch. Pediatr. Adolesc. Med.*, 2007	Communicative and cognitive development	64 siblings of children with ASD (SIBS-A)	12–23 months	42 control children with no autism family history, no matching criteria except for age range	MSEL, CARS, Screening Tool for Autism in Two-Year-Olds (STAT), MacArthur Communicative Development Inventories, Detection of Autism by Infant Sociability Interview	Younger siblings of children with ASD demonstrated poorer performance in nonverbal problem solving, directing attention, understanding words, understanding phrases, gesture use, and social-communicative interactions with parents, and had increased autism symptoms relative to control siblings

Table 2. Cont.

Study	Journal &Year	Characteristics	Sample	Age of Siblings of Individuals with ASD	Control Group, Matching Criteria	Measures/Instruments (Samples)	Main Findings
Sullivan et al. [98]	J. Autism Dev. Disord., 2007	Response to joint attention	51 infant siblings of children with autism; Outcome groups at age 3 years: 16 ASD, 8 broader autism phenotype, and 27 non-broader autism phenotype	14 and 24 months	Developmental study; no control group	Adaptation of a task described by Butterworth and Jarrett (1991) to assess response to joint attention; the Communication and Symbolic Behavior Scales Developmental Profile, MSEL, ADOS	Lower response to joint attention was observed for the ASD group at 24 months. Response to joint attention performance at 14 months predicted ASD outcome. The ASD group made minimal improvement in response to joint attention between 14 and 24 months
Toth et al. [94]	J. Autism Dev. Disord., 2007	Social, imitation, play and language abilities	42 non-autistic siblings of children with autism 20 toddlers with no family history of autism	18-27 months	20 toddlers with no family history of autism, controlled for age and ethnic group	MSEL, The Vineland Social-Emotional Early Childhood Scales, The Communication and Symbolic Behavior Scale-Developmental Profile, Imitation battery developed by Meltzoff, The Play Assessment Scale, The Early Development Interview	Siblings scored poorer in Receptive language scale (MSEL), Daily living skills, Motor and Composite (Vineland)
Yirmiya et al. [89]	J. Autism Dev. Disord., 2007	Cognitive and language profile—a prospective study	30 siblings of children with autism (SIBS-A)	24-36 months	30 siblings of typically developing children (SIBS-TD); matched on the basis of chronological age, gender, birth order, scores on mental and psychomotor indices, temperamental characteristics and number of children in the family	BSID-II, RDLS, CHAT, K-ABC, CELF-P, The Social and Communication Questionnaire (SCQ)	At 24 months: more SIBS-A demonstrated language scores one or two standard deviations below the mean compared to SIBS-TD. At 36 months: more SIBS-A displayed receptive and expressive difficulties compared to SIBS-TD. Six SIBS-A (including one diagnosed with autism) revealed language scores more than two standard deviations below the mean at both ages, a pattern not seen in the SIBS-TD
Garon et al. [116]	J. Abnorm. Child Psychol., 2009	Temperamental traits	138 HR infants with an older sibling with autistic spectrum disorder	6-36 months	73 low risk infants with no family history of ASD, no matching criteria listed except for age range	Toddler Behavior Assessment Questionnaire-Revised, MSEL, ADOS, ADI-R	HR children, who were diagnosed with ASD at 36 months, had temperament profile marked by lower positive affect, higher negative affect, and difficulty controlling attention and behavior (labelled as Effortful Emotion Regulation). This temperamental profile distinguished also HR children without ASD diagnosis at 36 months from LR children
Yoder et al. [67]	J. Autism Dev. Disord., 2009	Social impairment and ASD diagnosis	43 siblings of children with autism (SIBS-ASD)	15-34 months	24 SIBS-TD, matched on the basis of child's age and maternal education	MSEL, STAT, Responding to Joint Attention (RJA), Social Behavior Checklist (SBC), ADOS, ADI-R	Initial level of responding to joint attention and growth rate of weighted triadic communication predicted the degree of social impairment at the final measurement period of SIBS-ASD. Both predictors were associated with later ASD diagnosis, contrary to unweighted triadic communication, age of entry into the study, and initial language level, which did not predict later social impairment
Christensen et al. [117]	J. Autism Dev. Disord., 2010	Play behaviors	17 ASD siblings later diagnosed with ASD, infant siblings of children with autism with and without other delays (Other Delays and No Delays siblings; $n = 12$ and $n = 19$, respectively)	18 months	19 TD children, no matching criteria listed except for age range	Free-play task: functional, symbolic, and repeated play actions	ASD siblings showed fewer functional and more non-functional repeated play behaviors than TD children. Other Delays siblings showed more non-functional repeated play than TD controls. Group differences disappeared with the inclusion of verbal mental age

Table 2. *Cont.*

Study	Journal &Year	Characteristics	Sample	Age of Siblings of Individuals with ASD	Control Group, Matching Criteria	Measures/Instruments (Samples)	Main Findings
Holmboe *et al.* [109]	*Infant Behav. Dev.,* 2010	Executive functions, attention and inhibition, frontal cortex functioning	31 SIBS-ASD	9–10 months	33 typically developing children with no family history of autism, no matching criteria listed except for age range	Freeze-Frame task	SIBS-ASD had difficulty disengaging attention and showed less selective inhibition than controls (less difference between interesting and boring trials); however, they demonstrated selective inhibitory learning (tendency to show a larger decrease in looks to the distractors in the interesting trials than in the boring trials, whereas controls showed a similar decrease in the two trial type)
Haltigan *et al.* [113]	*J. Autism Dev. Disord.,* 2011	Attachment security	51 infant siblings of older children with ASD (SIBS-ASD)	15 months	34 typically developing children with no family history of autism (SIBS-COMP); no matching criteria listed except for age range	Strange Situation Procedure (SSP)	SIBS-ASD are not less likely to form secure affectional bonds with their caregivers than SIBS-COMP. Larger rate of B1–B2 secure subclassifications in SIBS-ASD than controls (B1–B2 infants are less distressed during separation and are more reserved after reunion with caregiver)
Ozonoff *et al.* [31]	*Pediatrics,* 2011	Recurrence risk for ASD	664 infants with an older sibling with ASD	18–36 months	No control group, developmental study	ADOS, MSEL	18.7% of the infants developed ASD. Infant gender (threefold increase in risk for male subjects) and the presence of 1 affected older sibling (twofold increase in risk) were significant predictors of ASD outcome
Paul *et al.* [118]	*J. Child Psychol. Psychiatry,* 2011	Vocal production	At 6 months: 28 high-risk (HR) infants; 37 HR infants; at 12 months: 38 HR infants; at 24 months: 24 HR infants	6–24 months	At 6 months: 20 low-risk (LR) infants; at 9 months: 29 LR infants; at 12 months: 31 LR infants; at 24 months: 21 LR infants; no matching criteria listed except for age range	ADOS, MSEL, Vocalization Sample Collection	Differences were seen between risk groups for certain vocal behaviors. Differences in vocal production in the first year of life were associated with outcomes in terms of autistic symptomatology in the second year
Rozga *et al.* [97]	*J. Autism Dev. Disord.,* 2011	Mother–infant interaction and nonverbal communication, social gaze, affect, and joint attention behaviors	17 infant siblings of older children with ASD, later diagnosed with autism; 84 infant siblings of older children with ASD without ASD diagnosis (NoASD-sib)	6–36 months	66 TD children, no matching criteria listed except for age range	Free Play Mother–Infant Interaction, Still Face Procedure, ESCS	The ASD group did not differ from the other two groups at six months in the frequency of gaze, smiles and vocalizations directed toward the caregiver, nor in their sensitivity to her withdrawal from interaction. By 12 months, infants in the ASD group exhibited lower rates of joint attention and requesting behaviors. NoASD-sibs did not differ from comparison infants on any variables of interest at 6 and 12 months
Bedford *et al.* [111]	*J. Autism Dev. Disord.,* 2012	Social gaze, communication and attentional engagement	54 HR infants	7 and 13 months	50 LR infants, no matching criteria listed except for age range	Eye-tracking	No group difference between high-risk and low-risk infants in gaze-following behavior at either age. At-risk infants with later emerging socio- communication difficulties (ASD and atypical development) allocated less attention to the congruent object compared to typically developing high-risk siblings and low-risk controls

Table 2. *Cont.*

Study	Journal &Year	Characteristics	Sample	Age of Siblings of Individuals with ASD	Control Group, Matching Criteria	Measures/Instruments (Samples)	Main Findings
Cornew *et al.* [119]	*J. Autism Dev. Disord.*, 2012	Social referencing	38 HR infants	17.7–20.6 months	44 LR infants; LR and HR groups controlled for equality of mean age and infants' maturity at birth	Social referencing procedure	Compared to both typically developing infants and high-risk infants without ASD, infants later diagnosed with ASD engaged in slower information seeking. High-risk infants, both those who were and those who were not later diagnosed with ASD, exhibited impairments in regulating their behavior based on the adults' emotional signals
Dundas *et al.* [115]	*J. Autism Dev. Disord.*, 2012	Face processing	43 HR infants	6 and 11 months	31 LR infants; gender, ethnicity, and age of infants in LR and HR groups described	Eye-tracking	Low-risk infants demonstrated a preference for looking at the left side of the face, which emerged by 11 months of age. High-risk infants did not demonstrate a left visual field bias at either age
Hutman *et al.* [95]	*J. Autism Dev. Disord.*, 2012	Social interactions, selective visual attention	81 HR infants; Outcome groups: 15 ASD; 12 Other Concerns; 59 High-Risk Typical; and 43 Low-Risk Typical	12 months	48 LR infants, no matching criteria listed except for age range	Examiner-Child Interaction, play, and distress condition	No difference in proportion of attention to social stimuli or attention shifting during the play condition between groups. Infants later diagnosed with ASD tended to continue looking at a toy during the distress condition despite the salience of social information. Emotion recognition is intact in infants who later develop autism, but the emotional value of the information appears to be less salient
Macari *et al.* [101]	*J. Autism Dev. Disord.*, 2012	Risk for ASD	53 HR infants	12, 18 and 24 months	31 LR infants; no matching criteria listed except for age range	MSEL, ADOS-Toddler	About 2/3 of infants at high risk for ASD experience some kind of developmental difficulties in the second year of life
Curtin &Vouloumanos [120]	*J. Autism Dev. Disord.*, 2013	Speech preference	31 HR infants	12–18 months	31 LR infants; no matching criteria listed except for age range	Speech/Non-Speech task, MSEL, MacArthur-Bates Communicative Development Inventories	Only low-risk infants listened significantly longer to speech than to non-speech at 12 months. In both groups, relative preference for speech correlated positively with general cognitive ability at 12 months. However, in high-risk infants only, preference for speech was associated with autistic-like behavior at 18 months, while in low-risk infants, preference for speech correlated with language abilities
Damiano *et al.* [110]	*J. Autism Dev. Disord.*, 2013	Repetitive and stereotyped movements	20 HR infants (SIBS-ASD)	15–24 months	20 typically developing siblings (SIBS-TD), differences in maternal educational between HR and LR groups were noted	STAT, Repetitive and Stereotyped Movement Scales (RSMS)	SIBS-ASD displayed higher rates of repetitive and stereotyped movements (RSM) relative to SIBS-TD; SIBS-ASD as a group demonstrated a significantly higher inventory of RSMs than controls, but this difference was no longer significant after excluding subgroup of ASD diagnosed SIBS-ASD. Different patterns of ASD diagnosed SIBS-ASD. Body RSM inventory for the high-risk groups with different diagnostic outcomes (Sibs-ASD/+ *vs.* Sibs-ASD/−)

Table 2. *Cont.*

Study	Journal &Year	Characteristics	Sample	Age of Siblings of Individuals with ASD	Control Group, Matching Criteria	Measures/Instruments (Samples)	Main Findings
Georgiades et al. [100]	JAMA Psychiatry, 2013	Autistic-like traits	170 HR infants	12 months	90 LR control subjects with no family history of ASD; no matching criteria listed except for age range	The Autism Observation Scale for Infants	Cluster 1 (n = 37), having significantly higher levels of autistic-like traits, consisted of 33 children from the siblings and only four from the control subjects. At the age of three, children from cluster 1 had more social-communication impairments, lower cognitive abilities, and more internalizing problems. Nineteen percent of HR siblings who did not meet ASD diagnostic criteria at the age of three showed autistic-like traits resembling a BAP by 12 months
Messinger et al. [88]	J. Am. Acad. Child Adolesc. Psychiatry, 2013	ASD risk, autistic traits	507 HR siblings	8–36 months	324 LR control subjects, no matching criteria listed except for age range	ADOS calibrated severity scores, and Mullen Verbal and Non-Verbal Developmental Quotients (DQ)	At three years, HR siblings without an ASD outcome exhibited higher mean ADOS severity scores and lower verbal and non-verbal DQs than LR controls. HR siblings were over-represented (21% HR vs. 7% LR) in latent classes characterized by elevated ADOS severity and/or low to low-average DQs. The remaining HR siblings without ASD outcomes (79%) belonged to classes in which they were not differentially represented with respect to LR siblings
Wan et al. [114]	J. Child Psychol. Psychiatry, 2013	Quality of interaction	At 6–10 months: 45 HR infants; at 12–15 months: 43 HR siblings	6–10 and 12–15 months	At 6–10 months: 47 LR siblings; at 12–15 months: 48 LR siblings; no matching criteria listed except for age range	Six-min videotaped episodes of parent-infant free play; Manchester Assessment of Caregiver-Infant Interaction (MACI)	At six months, infant liveliness was lower in the at-risk groups; at 12 months, infant attentiveness to parent and positive affect were lower in the at-risk group later diagnosed with ASD. Dyadic mutuality, infant positive affect and infant attentiveness to parent at 12 months predicted three-year ASD outcome, whereas infant ASD-related behavioral atypicality did not
Del Rosario et al. [121]	J. Autism Dev. Disord, 2014	Temperament trajectories	16 HR infants who were later diagnosed with ASD (SIBS-ASD), 27 HR infants who demonstrated typical patterns of development (SIBS-HR-TD)	6–36 months	No control group	Carey Temperament Scales completed by parents, MSEL	Temperament trajectories of children with ASD reflected increases over time in activity level, and decreasing adaptability and approach behaviors, relative to high-risk typically developing (HR-TD) children
Esposito et al. [112]	J. Autism Dev. Disord, 2014	Expression of distress during the separation phase	13 HR infants	15 months	14 LR infants; LR and HR groups were controlled for the age of infants and the age of older siblings	Cry samples derived from vocal recordings	HR toddlers, compared to those with LR, produced cries that were shorter and had a higher fundamental frequency (F0). Three HR toddlers later classified with an ASD at 36 months produced cries that had among the highest F0 and shortest durations
Gangi et al. [106]	J. Autism Dev. Disord, 2014	Joint attention initiation, social communication	56 HR siblings	8–12 months	26 LR siblings; LR and HR groups were controlled for age and ethnicity	Initiating joint attention (IJA) smiling patterns (i.e., anticipatory smiling, reactive smiling, and no smiling) assessed with Early Social Communication Scales	High-risk siblings produced less anticipatory smiling than low-risk siblings, suggesting early differences in communicating pre-existing positive affect. Among high-risk siblings, only IJA without smiling was associated with later ASD severity scores

Table 2. *Cont.*

Study	Journal & Year	Characteristics	Sample	Age of Siblings of Individuals with ASD	Control Group, Matching Criteria	Measures/Instruments (Samples)	Main Findings
Gliga *et al.* [122]	*Dev. Psychol.*, 2014	Spontaneous belief attribution, mental state understanding for action prediction	47 siblings of children with ASD	36 months	39 typically developing children; no matching criteria listed except for age range	Eye-tracking	In tasks demanding mental state understanding for action prediction, at-risk siblings performed at chance (contrary to control children, who performed above the chance), independently of their later clinical outcome (ASD, broader autism phenotype, or typically developing). Performance was not related to children's verbal or general IQ, nor was it explained by children "missing out" on crucial information, as shown by an analysis of visual scanning during the task
Hudry *et al.* [104]	*J. Autism Dev. Disord.*, 2014	Early language profiles, communication	54 HR infants	7–38 months	50 LR controls, no matching criteria listed except for the age range	MCDI: Words and Gestures (WG) and Words and Sentences (WS), VABS—2nd edition and MSEL	Reduced receptive vocabulary advantage was observed in HR infants by 14 months, but was maintained to 24 months only in children with ASD outcome, while typically-developing HR infants regained a more normative profile
Nichols *et al.* [107]	*J. Autism Dev. Disord.*, 2014	Social communication, social smiling	15 SIBS-ASD/AS (siblings of children with ASD, who demonstrated later ASD symptomatology), 27 SIBS-ASD/NS (siblings of children with ASD, who did not demonstrate ASD symptoms)	15 months	25 siblings of children with no family history of ASD, SIBS-TD: no matching criteria listed except for the age range	MSEL, STAT, ADOS	Both SIBS-ASD subgroups demonstrated lower levels of social smiling than SIBS-TD. Only the SIBS-ASD/AS demonstrated less eye contact and non-social smiling than SIBS-TD
Patten *et al.* [123]	*J. Autism Dev. Disord.*, 2014	Vocal patterns, vocalization frequency	37 HR infants (23 obtained the ASD diagnosis later)	9–12 and 15–18 months	14 typically developing infants with no autism family history (LR); HR and LR groups did not differ in terms of age, gender or SES	Video records	Infants later diagnosed with ASD produced low rates of canonical babbling and low volubility by comparison with the typically developing infants

Some authors have posited alternative temperament development trajectories in HR infants, characteristic for early ASD symptoms or broader autism phenotype condition. HR children who were diagnosed with ASD at 36 months had a temperament profile marked by lower positive affect, higher negative affect, and difficulty controlling attention and behavior. This temperamental profile also distinguished HR children without ASD diagnosis at 36 months from LR children [116]. Rosario *et al.* [121] compared HR infants who were or were not later diagnosed with ASD and discovered that the temperament trajectories of children with ASD reflected increases over time in activity levels and decreasing adaptability and approach behaviors relative to high-risk typically developing (TD) children.

It is also predicted that HR infants will exhibit problems with executive functions, especially attention and inhibition deficits, which are considered to be associated with frontal cortex functioning impairments [109]. HR infants had difficulty disengaging attention and showed less selective inhibition than controls (less difference between interesting and boring trials); however, they demonstrated a larger decrease in looks to the distractors in the interesting trials than in the boring trials, whereas controls showed a similar decrease in the two trial types.

Studies investigating vocalization patterns in HR infants suggest that infants later diagnosed with ASD produce low rates of canonical babbling and low volubility in comparison with typically developing infants [123]. Iverson *et al.* [92] also found impaired vocal-motor developmental behaviors in HR infants. Infant siblings demonstrated attenuated patterns of change in rhythmic arm activity around the time of reduplicated babble onset, and were highly likely to exhibit delayed language development at 18 months.

It should be stressed that the analyzed studies have many limitations. As previously mentioned, some authors include in the analysis high-risk infant siblings who later developed ASD, which can lead to overestimation of the differences between high-risk and low-risk children, *i.e.*, [95,97,111]. Furthermore, some studies do not report any clear matching criteria of the control group participants during the recruitment procedure—there is only a description provided of recruited participants in terms of different parameters (such as gender, age, ethnicity, *etc.*), *i.e.*, [110,115,119]. It should be noted that these features often vary across studies, which can raise questions about the comparability and recurrence of obtained results with other research projects. There are also studies that do not have any control group at all, because of the aim of the study, *i.e.*, comparing HR siblings who later developed ASD or did not [107,121]. These research projects provide information about differences within the HR group, but lack data from comparisons of these HR infants with typically developing children that do not have an ASD family background. There are also discrepancies between infants' age ranges in different studies. Some papers focus on describing the features of children at exactly the same age (*i.e.*, 6, 12, or 24 months [102], or other time points [30,90,96]), while others collect data covering a fixed age range (18–27 months [94], below 3 years old [105], 12–23 months [103]). These distinct approaches produce different types of data and therefore variable quality of comparisons, which should be considered during an analysis of studies.

In summary, there are many studies suggesting the existence of a broad range of impairments in infant siblings of children with ASD. Some of the differences are probably characteristic for later ASD diagnosis, but it should be noted that infant siblings are also at a high risk of developing broader autism phenotype-like traits.

4. Summary and Conclusions

The review of studies on BAP traits in siblings of individuals with ASD shows that the issue has been extensively explored. This goes hand in hand with investigation of hereditary mechanisms involved in the etiology of autism spectrum disorders. Research on BAP provides important information about the varied expressions of autism-specific traits. Integration of the results of these studies presents a challenge due to differences in methods, control groups, age of participants, and other aspects of research protocols. Since the relationships between autistic traits and other individual characteristics in a general population, including first-degree relatives of individuals with ASD, have not yet been fully elaborated, this paper focuses on the phenotypic characteristics of individuals observed mainly in psychological studies.

A number of investigators reported cognitive deficits in siblings of children with ASD, which can be a part of ASD phenotype(s). These include emotion recognition tasks [52], lower levels of efficiency in planning, attention shifting, and verbal fluency [44,48]. Some studies indicated differences in social skills development [6,14,28,69]. Difficulties for siblings of children with ASD were also found with respect to communication and language. There are reports of histories of language delay and pragmatic language deficits in this group compared with the siblings of children with Down syndrome and typically developing children, *i.e.*, [28,60,61,94,102].

Researchers have posited various BAP traits in siblings of individuals with autism as important components of the neurocognitive endophenotype for autism. For instance, Dalton *et al.* [55] suggested that social and emotional processing along with underlying neural circuitry constitute an important element of the endophenotype. Fiorentini and colleagues [63] point to face-coding mechanisms, emphasizing their role in the impairment of adaptive mechanisms, while Holt and colleagues [22] highlight the role of mentalizing deficits and atypical social cognition. The inclusion of language deficits in BAP remains controversial. Based on their findings from research on children aged 4–9 years, Levy and Bar-Yuda [61] question whether these deficits should indeed be included in BAP, while Gamliel *et al.* [56,57] consider them to be the key component in the difficulties encountered by preschool and early school individuals with ASD.

It should also be noted that some researchers found no differences between siblings of people with ASD and comparison groups with respect to BAP characteristics [40,45]. Importantly, in a number of studies subgroups demonstrating developmental deficits were distinguished from the group of healthy siblings of people with autism [39,65]. The clear distinction between ASD symptoms and BAP traits is difficult in studies involving siblings of individuals with ASD. This is especially true in studies on infant siblings, in which the risk of ASD rather than the BAP characteristics is the main concern. Prospective studies in this group of children should be continued, in order to track the developmental trajectories in these children at subsequent stages of their development.

It should be stressed that in the light of BAP research, the majority of brothers and sisters of individuals with ASD develop typically, without displaying autistic traits to a greater extent than the relevant control groups. However, traits in the siblings group are often widely dispersed, suggesting that there is much variation among these children in the course of developmental processes and their outcomes.

Some data suggest that genetic susceptibility to autism may differ among families. It is likely to be higher in families with two or more children diagnosed with ASD. Other siblings in these families demonstrate more pronounced BAP traits [65,124]. Research on these families could provide valuable insights on genetic involvement in the development of autistic traits.

Longitudinal studies are especially useful, as they enable researchers to trace developmental dynamics of ASD siblings. Despite achieving the status of the gold standard in research on infants, more longitudinal studies on children over 36 months of age are needed.

The information available at present is insufficient to formulate final conclusions regarding BAP characteristics in siblings of people with ASD. There is no doubt, however, that current research is bringing us closer to an understanding of the genetic factors involved in the etiology of this group of disorders. The studies are also of fundamental importance due to the rising numbers of ASD diagnoses and the presence of autistic characteristics in the general population.

Acknowledgments: This paper was funded by the project of the National Science Center of Poland, #UMO-2011/03/B/HS6/03326 and by the University of Warsaw.

Author Contributions: Ewa Pisula: Study design, Literature search, Manuscript preparation, Funds collection; Karolina Ziegart-Sadowska: Literature search, Manuscript preparation.

Conflicts of Interest: The authors declare no conflict of interest.

References

1. American Psychiatric Association. *Diagnostic and Statistical Manual of Mental Disorders*, 5th ed.; American Psychiatric Association: Arlington, VA, USA, 2013.
2. World Health Organization. *Manual of the International Statistical Classification of the Diseases, and Related Health Problems*, 10th ed.; World Health Organization: Geneva, Switzerland, 2002; Volume 1.
3. Lauritsen, M.B.; Pedersen, C.B.; Mortensen, P.B. Effects of familial risk factors and place of birth on the risk of autism: A nationwide register-based study. *J. Child Psychol. Psychiatry* 2005, *46*, 963–971. [CrossRef] [PubMed]
4. Persico, A.M.; Sacco, R. Endophenotypes in autism spectrum disorders. In *Comprehensive Guide to Autism*; Patel, V.B., Preedy, V.R., Martin, C.R., Eds.; Springer: New York, NY, USA, 2014; pp. 77–95.
5. Rogers, S.J. What are infant siblings teaching us about autism in infancy? *Autism Res.* 2009, *2*, 125–137. [CrossRef] [PubMed]
6. Constantino, J.N.; Lajonchere, C.; Lutz, M.; Gray, T.; Abbacchi, A.; McKenna, K.; Singh, D.; Todd, R.D. Autistic social impairment in the siblings of children with pervasive developmental disorders. *Am. J. Psychiatry* 2006, *163*, 294–296. [CrossRef] [PubMed]
7. Rutter, M.; Thapar, A. Genetics of autism spectrum disorders. In *Handbook of Autism and Pervasive Developmental Disorders*, 4th ed.; Volkmar, R., Rogers, S.J., Paul, R., Pelphrey, K.A., Eds.; Wiley & Sons: New Jersey, NJ, USA, 2014; Volume 1, pp. 411–423.
8. Baron-Cohen, S.; Wheelwright, S.; Skinner, R.; Martin, J.; Clubley, E. The autism-spectrum quotient (AQ): Evidence from Asperger syndrome/high-functioning autism, males and females, scientists and mathematicians. *J. Autism Dev. Disord.* 2001, *31*, 5–17. [CrossRef] [PubMed]
9. Bailey, A.; Palferman, S.; Heavey, L.; Couteur, A.L. Autism: The phenotype in relatives. *J. Autism Dev. Disord.* 1998, *28*, 369–392. [CrossRef] [PubMed]
10. Bishop, D.V.M.; Maybery, M.; Maley, A.; Wong, D.; Hill, W.; Hallmayer, J. Using self-report to identify the broad phenotype in parents of children with autistic spectrum disorders: A study using the Autism-Spectrum Quotient. *J. Child Psychol. Psychiatry* 2004, *45*, 1431–1436. [CrossRef] [PubMed]
11. Bishop, D.V.M.; Maybery, M.; Wong, D.; Maley, A.; Hallmayer, J. Characteristics of the broader phenotype in autism: A study of siblings using the children's communication checklist-2. *Am. J. Med. Genet. Part B* 2006, *141B*, 117–122. [CrossRef] [PubMed]
12. Losh, M.; Childress, D.; Lam, K.; Piven, J. Defining key features of the broad autism phenotype. *Am. J. Med. Genet. Part B* 2008, *147B*, 424–433. [CrossRef] [PubMed]
13. Losh, M.; Piven, J. Social-cognition and the broad autism phenotype: Identifying genetically meaningful phenotypes. *J. Child Psychol. Psychiatry* 2007, *48*, 105–112. [CrossRef] [PubMed]
14. Piven, J.; Palmer, P.; Jacobi, D.; Childress, D.; Arndt, S. Broader autism phenotype: Evidence from a family history study of multiple-incidence autism families. *Am. J. Psychiatry* 1997, *154*, 185–190. [PubMed]
15. Piven, J.; Palmer, P.; Landa, R.; Santangelo, S.; Jacobi, D.; Childress, D. Personality and language characteristics in parents from multiple-incidence autism families. *Am. J. Med. Genet.* 1997, *74*, 398–411. [CrossRef] [PubMed]
16. Frazier, T.W.; Youngstrom, E.A.; Sinclair, L.; Kubu, C.S.; Law, P.; Rezai, A.; Constantino, J.N.; Eng, C. Autism spectrum disorders as a qualitatively distinct category from typical behavior in a large, clinically ascertained sample. *Assessment* 2010, *17*, 308–320. [CrossRef] [PubMed]
17. Wheelwright, S.; Auyeung, B.; Allison, C.; Baron-Cohen, S. Defining the broader, medium and narrow autism phenotype among parents using the Autism Spectrum Quotient (AQ). *Mol. Autism* 2010, *1*. [CrossRef] [PubMed]
18. August, G.J.; Stewart, M.A.; Tsai, L. The incidence of cognitive disabilities in the siblings of autistic children. *Br. J. Psychiatry* 1981, *138*, 416–422. [CrossRef] [PubMed]
19. Folstein, S.; Rutter, M. Infantile autism: A genetic study of 21 twin pairs. *J. Child Psychol. Psychiatry* 1977, *18*, 297–321. [CrossRef] [PubMed]
20. Piven, J. The broad autism phenotype: A complementary strategy for molecular genetic studies of autism. *Am. J. Med. Genet.* 2001, *105*, 34–35. [CrossRef]
21. Schlichting, C.D.; Pigliucci, M. Gene regulation, quantitative genetics and the evolution of reaction norms. *Evol. Ecol.* 1995, *9*, 154–168. [CrossRef]

22. Holt, R.J.; Chura, L.R.; Lai, M.-C.; Suckling, J.; von dem Hagen, E.; Calder, A.J.; Bullmore, E.T.; Baron-Cohen, S.; Spencer, M.D. "Reading the Mind in the Eyes": An fMRI study of adolescents with autism and their siblings. *Psychol. Med.* **2014**, *44*, 3215–3227. [CrossRef] [PubMed]

23. Ingersoll, B.; Wainer, A. The broader autism phenotype. In *Handbook of Autism and Pervasive Developmental Disorders*, 4th ed.; Volkmar, F.R., Rogers, S.J., Paul, R., Pelphrey, K.A., Eds.; Wiley & Sons: New Jersey, NJ, USA, 2014; Volume 1, pp. 28–56.

24. Gottesman, I.I.; Gould, T.D. The endophenotype concept in psychiatry: Etymology and strategic intentions. *Am. J. Psychiatry* **2003**, *160*, 636–645. [CrossRef] [PubMed]

25. Rasetti, R.; Weinberger, D.R. Intermediate phenotypes in psychiatric disorders. *Curr. Opin. Genet. Dev.* **2011**, *21*, 340–348. [CrossRef] [PubMed]

26. Rutter, M. Concepts of autism: A review of research. *J. Child Psychol. Psychiatry* **1968**, *9*, 1–25. [CrossRef] [PubMed]

27. Smalley, S.L.; Asarnow, R.F.; Spence, M.A. Autism and genetics. A decade of research. *Arch. Gen. Psychiatry* **1988**, *45*, 953–961. [CrossRef] [PubMed]

28. Bolton, P.; Macdonald, H.; Pickles, A.; Rios, P.; Goode, S.; Crowson, M.; Bailey, A.; Rutter, M. A case-control family history study of autism. *J. Child Psychol. Psychiatry* **1994**, *35*, 877–900. [CrossRef] [PubMed]

29. Ghaziuddin, M. A family history study of Asperger syndrome. *J. Autism Dev. Disord.* **2005**, *35*, 177–182. [CrossRef] [PubMed]

30. Landa, R.; Garrett-Mayer, E. Development in infants with autism spectrum disorders: A prospective study. *J. Child Psychol. Psychiatry* **2006**, *47*, 629–638. [CrossRef] [PubMed]

31. Ozonoff, S.; Young, G.S.; Carter, A.; Messinger, D.; Yirmiya, N.; Zwaigenbaum, L.; Bryson, S.; Carver, L.J.; Constantino, J.N.; Dobkins, K.; *et al.* Recurrence risk for autism spectrum disorders: A Baby Siblings Research Consortium study. *Pediatrics* **2011**, *128*, e488–e495. [PubMed]

32. Dawson, G.; Estes, A.; Munson, J.; Schellenberg, G.; Bernier, R.; Abbott, R. Quantitative assessment of autism symptom-related traits in probands and parents: Broader Phenotype Autism Symptom Scale. *J. Autism Dev. Disord.* **2007**, *37*, 523–536. [CrossRef] [PubMed]

33. Bauminger, N.; Yirmiya, N. The functioning and well-being of SIBS-A: Behavioral-genetic and familial contributions. In *The Development of Autism: Perspectives from Theory and Research*; Burack, J.A., Charman, T., Yirmiya, N., Zelazo, P.R., Eds.; Erlbaum: Hillsdale, NJ, USA, 2001; pp. 61–80.

34. Gerdts, J.; Bernier, R. The broader autism phenotype and its implications on the etiology and treatment of autism spectrum disorders. *Autism Res. Treat.* **2011**, *2011*, e545901. [CrossRef] [PubMed]

35. Jones, E.J.H.; Gliga, T.; Bedford, R.; Charman, T.; Johnson, M.H. Developmental pathways to autism: A review of prospective studies of infants at risk. *Neurosci. Biobehav. Rev.* **2014**, *39*, 1–33. [CrossRef] [PubMed]

36. Losh, M.; Adolphs, R.; Piven, J. The broad autism phenotype. In *Autism Spectrum Disorders*; Dawson, G., Amaral, D., Geschwind, D., Eds.; Oxford University Press: Oxford, UK, 2011; pp. 457–476.

37. Sucksmith, E.; Roth, I.; Hoekstra, R.A. Autistic traits below the clinical threshold: Re-examining the broader autism phenotype in the 21st century. *Neuropsychol. Rev.* **2011**, *21*, 360–389. [CrossRef] [PubMed]

38. Moher, D.; Liberati, A.; Tetzlaff, J.; Altman, D.G. Preferred reporting items for systematic reviews and meta-analyses: The PRISMA statement. *PLoS Med.* **2009**, *6*, e1000097. [CrossRef] [PubMed]

39. Chuthapisith, J.; Ruangdaraganon, N.; Sombuntham, T.; Roongpraiwan, R. Language development among the siblings of children with autistic spectrum disorder. *Autism* **2007**, *11*, 149–160. [CrossRef] [PubMed]

40. Pilowsky, T.; Yirmiya, N.; Gross-Tsur, V.; Shalev, R.S. Neuropsychological functioning of siblings of children with autism, siblings of children with developmental language delay, and siblings of children with mental retardation of unknown genetic etiology. *J. Autism Dev. Disord.* **2007**, *37*, 537–552. [CrossRef] [PubMed]

41. Delorme, R.; Goussé, V.; Roy, I.; Trandafir, A.; Mathieu, F.; Mouren-Siméoni, M.-C.; Betancur, C.; Leboyer, M. Shared executive dysfunctions in unaffected relatives of patients with autism and obsessive-compulsive disorder. *Eur. Psychiatry* **2007**, *22*, 32–38. [CrossRef] [PubMed]

42. Pickles, A.; Starr, E.; Kazak, S.; Bolton, P.; Papanikolaou, K.; Bailey, A.; Goodman, R.; Rutter, M. Variable expression of the autism broader phenotype: Findings from extended pedigrees. *J. Child Psychol. Psychiatry* **2000**, *41*, 491–502. [CrossRef] [PubMed]

43. Hill, E.; Berthoz, S.; Frith, U. Brief report: Cognitive processing of own emotions in individuals with autistic spectrum disorder and in their relatives. *J. Autism Dev. Disord.* **2004**, *34*, 229–235. [CrossRef] [PubMed]

44. Ozonoff, S.; Rogers, S.J.; Farnham, J.M.; Pennington, B.F. Can standard measures identify subclinical markers of autism? *J. Autism Dev. Disord.* **1993**, *23*, 429–441. [CrossRef] [PubMed]

45. Szatmari, P.; Jones, M.B.; Tuff, L.; Bartolucci, G.; Fisman, S.; Mahoney, W. Lack of cognitive impairment in first-degree relatives of children with pervasive developmental disorders. *J. Am. Acad. Child Adolesc. Psychiatry* **1993**, *32*, 1264–1273. [CrossRef] [PubMed]

46. Fombonne, E.; Bolton, P.; Prior, J.; Jordan, H.; Rutter, M. A family study of autism: Cognitive patterns and levels in parents and siblings. *J. Child Psychol. Psychiatry* **1997**, *38*, 667–683. [CrossRef] [PubMed]

47. Folstein, S.E.; Santangelo, S.L.; Gilman, S.E.; Piven, J.; Landa, R.; Lainhart, J.; Hein, J.; Wzorek, M. Predictors of cognitive test patterns in autism families. *J. Child Psychol. Psychiatry* **1999**, *40*, 1117–1128. [CrossRef] [PubMed]

48. Hughes, C.; Plumet, M.H.; Leboyer, M. Towards a cognitive phenotype for autism: Increased prevalence of executive dysfunction and superior spatial span amongst siblings of children with autism. *J. Child Psychol. Psychiatry* **1999**, *40*, 705–718. [CrossRef] [PubMed]

49. Briskman, J.; Happé, F.; Frith, U. Exploring the cognitive phenotype of autism: Weak "central coherence" in parents and siblings of children with autism: II. Real-life skills and preferences. *J. Child Psychol. Psychiatry* **2001**, *42*, 309–316. [CrossRef] [PubMed]

50. Happé, F.; Briskman, J.; Frith, U. Exploring the cognitive phenotype of autism: Weak "central coherence" in parents and siblings of children with autism: I. Experimental tests. *J. Child Psychol. Psychiatry* **2001**, *42*, 299–307. [CrossRef] [PubMed]

51. Pilowsky, T.; Yirmiya, N.; Shalev, R.S.; Gross-Tsur, V. Language abilities of siblings of children with autism. *J. Child Psychol. Psychiatry* **2003**, *44*, 914–925. [CrossRef] [PubMed]

52. Dorris, L.; Espie, C.A.E.; Knott, F.; Salt, J. Mind-reading difficulties in the siblings of people with Asperger's syndrome: Evidence for a genetic influence in the abnormal development of a specific cognitive domain. *J. Child Psychol. Psychiatry* **2004**, *45*, 412–418. [CrossRef] [PubMed]

53. Shaked, M.; Gamliel, I.; Yirmiya, N. Theory of mind abilities in young siblings of children with autism. *Autism* **2006**, *10*, 173–187. [CrossRef] [PubMed]

54. Christ, S.E.; Holt, D.D.; White, D.A.; Green, L. Inhibitory control in children with autism spectrum disorder. *J. Autism Dev. Disord.* **2007**, *37*, 1155–1165. [CrossRef] [PubMed]

55. Dalton, K.M.; Nacewicz, B.M.; Alexander, A.L.; Davidson, R.J. Gaze-fixation, brain activation, and amygdala volume in unaffected siblings of individuals with autism. *Biol. Psychiatry* **2007**, *61*, 512–520. [CrossRef] [PubMed]

56. Gamliel, I.; Yirmiya, N.; Sigman, M. The development of young siblings of children with autism from 4 to 54 months. *J. Autism Dev. Disord.* **2007**, *37*, 171–183. [CrossRef] [PubMed]

57. Gamliel, I.; Yirmiya, N.; Jaffe, D.H.; Manor, O.; Sigman, M. Developmental trajectories in siblings of children with autism: Cognition and language from 4 months to 7 years. *J. Autism Dev. Disord.* **2009**, *39*, 1131–1144. [CrossRef] [PubMed]

58. Kawakubo, Y.; Kuwabara, H.; Watanabe, K.-I.; Minowa, M.; Someya, T.; Minowa, I.; Kono, T.; Nishida, H.; Sugiyama, T.; Kato, N.; *et al.* Impaired prefrontal hemodynamic maturation in autism and unaffected siblings. *PLoS ONE* **2009**, *4*, e6881. [CrossRef] [PubMed]

59. Koh, H.C.; Milne, E.; Dobkins, K. Contrast sensitivity for motion detection and direction discrimination in adolescents with autism spectrum disorders and their siblings. *Neuropsychologia* **2010**, *48*, 4046–4056. [CrossRef] [PubMed]

60. Ben-Yizhak, N.; Yirmiya, N.; Seidman, I.; Alon, R.; Lord, C.; Sigman, M. Pragmatic language and school related linguistic abilities in siblings of children with autism. *J. Autism Dev. Disord.* **2011**, *41*, 750–760. [CrossRef] [PubMed]

61. Levy, Y.; Bar-Yuda, C. Language performance in siblings of nonverbal children with autism. *Autism* **2011**, *15*, 341–354. [CrossRef] [PubMed]

62. Sumiyoshi, C.; Kawakubo, Y.; Suga, M.; Sumiyoshi, T.; Kasai, K. Impaired ability to organize information in individuals with autism spectrum disorders and their siblings. *Neurosci. Res.* **2011**, *69*, 252–257. [CrossRef] [PubMed]

63. Fiorentini, C.; Gray, L.; Rhodes, G.; Jeffery, L.; Pellicano, E. Reduced face identity aftereffects in relatives of children with autism. *Neuropsychologia* **2012**, *50*, 2926–2932. [CrossRef] [PubMed]

64. Warren, Z.E.; Foss-Feig, J.H.; Malesa, E.E.; Lee, E.B.; Taylor, J.L.; Newsom, C.R.; Crittendon, J.; Stone, W.L. Neurocognitive and behavioral outcomes of younger siblings of children with autism spectrum disorder at age five. *J. Autism Dev. Disord.* **2012**, *42*, 409–418. [CrossRef] [PubMed]

65. Gerdts, J.A.; Bernier, R.; Dawson, G.; Estes, A. The broader autism phenotype in simplex and multiplex families. *J. Autism Dev. Disord.* **2013**, *43*, 1597–1605. [CrossRef] [PubMed]

66. Malesa, E.; Foss-Feig, J.; Yoder, P.; Warren, Z.; Walden, T.; Stone, W.L. Predicting language and social outcomes at age 5 for later-born siblings of children with autism spectrum disorders. *Autism* **2013**, *17*, 558–570. [CrossRef] [PubMed]

67. Yoder, P.; Stone, W.L.; Walden, T.; Malesa, E. Predicting social impairment and ASD diagnosis in younger siblings of children with autism spectrum disorder. *J. Autism Dev. Disord.* **2009**, *39*, 1381–1391. [CrossRef] [PubMed]

68. Oerlemans, A.M.; Droste, K.; van Steijn, D.J.; de Sonneville, L.M.J.; Buitelaar, J.K.; Rommelse, N.N.J. Co-segregation of social cognition, executive function and local processing style in children with ASD, their siblings and normal controls. *J. Autism Dev. Disord.* **2013**, *43*, 2764–2778. [CrossRef] [PubMed]

69. Pickles, A.; St Clair, M.C.; Conti-Ramsden, G. Communication and social deficits in relatives of individuals with SLI and relatives of individuals with ASD. *J. Autism Dev. Disord.* **2013**, *43*, 156–167. [CrossRef] [PubMed]

70. Robel, L.; Rousselot-Pailley, B.; Fortin, C.; Levy-Rueff, M.; Golse, B.; Falissard, B. Subthreshold traits of the broad autistic spectrum are distributed across different subgroups in parents, but not siblings, of probands with autism. *Eur. Child Adolesc. Psychiatry* **2013**, *23*, 225–233. [CrossRef] [PubMed]

71. Gizzonio, V.; Avanzini, P.; Fabbri-Destro, M.; Campi, C.; Rizzolatti, G. Cognitive abilities in siblings of children with autism spectrum disorders. *Exp. Brain Res.* **2014**, *232*, 2381–2390. [CrossRef] [PubMed]

72. Oerlemans, A.M.; van der Meer, J.M.J.; van Steijn, D.J.; de Ruiter, S.W.; de Bruijn, Y.G.E.; de Sonneville, L.M.J.; Buitelaar, J.K.; Rommelse, N.N.J. Recognition of facial emotion and affective prosody in children with ASD (+ADHD) and their unaffected siblings. *Eur. Child Adolesc. Psychiatry* **2014**, *23*, 257–271. [CrossRef] [PubMed]

73. Lowenthal, R.; Paula, C.S.; Schwartzman, J.S.; Brunoni, D.; Mercadante, M.T. Prevalence of pervasive developmental disorder in Down's syndrome. *J. Autism Dev. Disord.* **2007**, *37*, 1394–1395. [CrossRef] [PubMed]

74. Biao, J. Prevalence of autism spectrum disorder among children aged 8 years—Autism and developmental disabilities monitoring network, 11 Sites, United States, 2010. *Surveill. Summ.* **2014**, *63*, 1–21.

75. Leyfer, O.T.; Tager-Flusberg, H.; Dowd, M.; Tomblin, J.B.; Folstein, S.E. Overlap between autism and specific language impairment: Comparison of Autism Diagnostic Interview and Autism Diagnostic Observation Schedule scores. *Autism Res.* **2008**, *1*, 284–296. [CrossRef] [PubMed]

76. Tsermentseli, S.; O'Brien, J.M.; Spencer, J.V. Comparison of form and motion coherence processing in autistic spectrum disorders and dyslexia. *J. Autism Dev. Disord.* **2008**, *38*, 1201–1210. [CrossRef] [PubMed]

77. Lord, C.; Rutter, M.; DiLavore, P.; Risi, S. *Autism Diagnostic Observation Schedule (ADOS) Manual*; Western Psychological Services: Los Angeles, CA, USA, 2001.

78. Rutter, M.; le Couteur, A.; Lord, C. *ADI-R: Autism Diagnostic Interview–Revised*; Western Psychological Services: Los Angeles, CA, USA, 2003.

79. Rutter, M.; Bailey, A.; Lord, C. *Social Communication Questionnaire*; Western Psychological Services: Los Angeles, CA, USA, 2003.

80. Schopler, E.; Reichler, R.J.; Renner, B.R. *The Childhood Autism Rating Scale*; Western Psychological Services: Los Angeles, CA, USA, 1988.

81. Bernier, R.; Gerdts, J.; Munson, J.; Dawson, G.; Estes, A. Evidence for broader autism phenotype characteristics in parents from multiple-incidence autism families. *Autism Res.* **2012**, *5*, 13–20. [CrossRef] [PubMed]

82. Sebat, J.; Lakshmi, B.; Malhotra, D.; Troge, J.; Lese-Martin, C.; Walsh, T.; Yamrom, B.; Yoon, S.; Krasnitz, A.; Kendall, J.; *et al.* Strong association of de novo copy number mutations with autism. *Science* **2007**, *316*, 445–449. [CrossRef] [PubMed]

83. Pinto, D.; Pagnamenta, A.T.; Klei, L.; Anney, R.; Merico, D.; Regan, R.; Conroy, J.; Magalhaes, T.R.; Correia, C.; Abrahams, B.S.; *et al.* Functional impact of global rare copy number variation in autism spectrum disorders. *Nature* **2010**, *466*, 368–372. [CrossRef] [PubMed]

84. Spiker, D.; Lotspeich, L.; Kraemer, H.C.; Hallmayer, J.; McMahon, W.; Petersen, P.B.; Nicholas, P.; Pingree, C.; Wiese-Slater, S.; Chiotti, C. Genetics of autism: Characteristics of affected and unaffected children from 37 multiplex families. *Am. J. Med. Genet.* **1994**, *54*, 27–35. [CrossRef] [PubMed]

85. Matson, J.L.; Shoemaker, M. Intellectual disability and its relationship to autism spectrum disorders. *Res. Dev. Disabil.* **2009**, *30*, 1107–1114. [CrossRef] [PubMed]

86. Newschaffer, C.J.; Falb, M.D.; Gurney, J.G. National autism prevalence trends from United States special education data. *Pediatrics* **2005**, *115*, e277–e282. [CrossRef] [PubMed]

87. Bölte, S.; Dziobek, I.; Poustka, F. Brief report: The level and nature of autistic intelligence revisited. *J. Autism Dev. Disord.* **2009**, *39*, 678–682. [CrossRef] [PubMed]

88. Messinger, D.; Young, G.S.; Ozonoff, S.; Dobkins, K.; Carter, A.; Zwaigenbaum, L.; Landa, R.J.; Charman, T.; Stone, W.L.; Constantino, J.N.; *et al.* Beyond autism: A baby siblings research consortium study of high-risk children at three years of age. *J. Am. Acad. Child Adolesc. Psychiatry* **2013**, *52*, 300–308. [CrossRef] [PubMed]

89. Yirmiya, N.; Gamliel, I.; Shaked, M.; Sigman, M. Cognitive and verbal abilities of 24- to 36-month-old siblings of children with autism. *J. Autism Dev. Disord.* **2007**, *37*, 218–229. [CrossRef] [PubMed]

90. Yirmiya, N.; Gamliel, I.; Pilowsky, T.; Feldman, R.; Baron-Cohen, S.; Sigman, M. The development of siblings of children with autism at 4 and 14 months: Social engagement, communication, and cognition. *J. Child Psychol. Psychiatry* **2006**, *47*, 511–523. [CrossRef] [PubMed]

91. Cassel, T.D.; Messinger, D.S.; Ibanez, L.V.; Haltigan, J.D.; Acosta, S.I.; Buchman, A.C. Early social and emotional communication in the infant siblings of children with autism spectrum disorders: An examination of the broad phenotype. *J. Autism Dev. Disord.* **2007**, *37*, 122–132. [CrossRef] [PubMed]

92. Iverson, J.M.; Wozniak, R.H. Variation in vocal-motor development in infant siblings of children with autism. *J. Autism Dev. Disord.* **2007**, *37*, 158–170. [CrossRef] [PubMed]

93. Merin, N.; Young, G.S.; Ozonoff, S.; Rogers, S.J. Visual fixation patterns during reciprocal social interaction distinguish a subgroup of 6-month-old infants at-risk for autism from comparison infants. *J. Autism Dev. Disord.* **2007**, *37*, 108–121. [CrossRef] [PubMed]

94. Toth, K.; Dawson, G.; Meltzoff, A.N.; Greenson, J.; Fein, D. Early social, imitation, play, and language abilities of young non-autistic siblings of children with autism. *J. Autism Dev. Disord.* **2007**, *37*, 145–157. [CrossRef] [PubMed]

95. Hutman, T.; Chela, M.K.; Gillespie-Lynch, K.; Sigman, M. Selective visual attention at twelve months: Signs of autism in early social interactions. *J. Autism Dev. Disord.* **2012**, *42*, 487–498. [CrossRef] [PubMed]

96. Mitchell, S.; Brian, J.; Zwaigenbaum, L.; Roberts, W.; Szatmari, P.; Smith, I.; Bryson, S. Early language and communication development of infants later diagnosed with autism spectrum disorder. *J. Dev. Behav. Pediatr.* **2006**, *27*, S69–S78. [CrossRef] [PubMed]

97. Rozga, A.; Hutman, T.; Young, G.S.; Rogers, S.J.; Ozonoff, S.; Dapretto, M.; Sigman, M. Behavioral profiles of affected and unaffected siblings of children with autism: Contribution of measures of mother-infant interaction and nonverbal communication. *J. Autism Dev. Disord.* **2011**, *41*, 287–301. [CrossRef] [PubMed]

98. Sullivan, M.; Finelli, J.; Marvin, A.; Garrett-Mayer, E.; Bauman, M.; Landa, R. Response to joint attention in toddlers at risk for autism spectrum disorder: A prospective study. *J. Autism Dev. Disord.* **2007**, *37*, 37–48. [CrossRef] [PubMed]

99. Bryson, S.E.; Zwaigenbaum, L.; Brian, J.; Roberts, W.; Szatmari, P.; Rombough, V.; McDermott, C. A prospective case series of high-risk infants who developed autism. *J. Autism Dev. Disord.* **2007**, *37*, 12–24. [CrossRef] [PubMed]

100. Georgiades, S.; Szatmari, P.; Zwaigenbaum, L.; Bryson, S.; Brian, J.; Roberts, W.; Smith, I.; Vaillancourt, T.; Roncadin, C.; Garon, N. A prospective study of autistic-like traits in unaffected siblings of probands with autism spectrum disorder. *JAMA Psychiatry* **2013**, *70*, 42–48. [CrossRef] [PubMed]

101. Macari, S.L.; Campbell, D.; Gengoux, G.W.; Saulnier, C.A.; Klin, A.J.; Chawarska, K. Predicting developmental status from 12 to 24 months in infants at risk for autism spectrum disorder: A preliminary report. *J. Autism Dev. Disord.* **2012**, *42*, 2636–2647. [CrossRef] [PubMed]

102. Zwaigenbaum, L.; Bryson, S.; Rogers, T.; Roberts, W.; Brian, J.; Szatmari, P. Behavioral manifestations of autism in the first year of life. *Int. J. Dev. Neurosci.* **2005**, *23*, 143–152. [CrossRef] [PubMed]

103. Stone, W.L.; McMahon, C.R.; Yoder, P.J.; Walden, T.A. EArly social-communicative and cognitive development of younger siblings of children with autism spectrum disorders. *Arch. Pediatr. Adolesc. Med.* **2007**, *161*, 384–390. [CrossRef] [PubMed]

104. Hudry, K.; Chandler, S.; Bedford, R.; Pasco, G.; Gliga, T.; Elsabbagh, M.; Johnson, M.H.; Charman, T. Early language profiles in infants at high-risk for autism spectrum disorders. *J. Autism Dev. Disord.* **2014**, *44*, 154–167. [CrossRef] [PubMed]

105. Goldberg, W.A.; Jarvis, K.L.; Osann, K.; Laulhere, T.M.; Straub, C.; Thomas, E.; Filipek, P.; Spence, M.A. Brief report: Early social communication behaviors in the younger siblings of children with autism. *J. Autism Dev. Disord.* **2005**, *35*, 657–664. [CrossRef] [PubMed]

106. Gangi, D.N.; Ibañez, L.V.; Messinger, D.S. Joint attention initiation with and without positive affect: Risk group differences and associations with ASD symptoms. *J. Autism Dev. Disord.* **2014**, *44*, 1414–1424. [CrossRef] [PubMed]

107. Nichols, C.M.; Ibañez, L.V.; Foss-Feig, J.H.; Stone, W.L. Social smiling and its components in high-risk infant siblings without later ASD symptomatology. *J. Autism Dev. Disord.* **2014**, *44*, 894–902. [CrossRef] [PubMed]

108. Loh, A.; Soman, T.; Brian, J.; Bryson, S.E.; Roberts, W.; Szatmari, P.; Smith, I.M.; Zwaigenbaum, L. Stereotyped motor behaviors associated with autism in high-risk infants: A pilot videotape analysis of a sibling sample. *J. Autism Dev. Disord.* **2007**, *37*, 25–36. [CrossRef] [PubMed]

109. Holmboe, K.; Elsabbagh, M.; Volein, A.; Tucker, L.A.; Baron-Cohen, S.; Bolton, P.; Charman, T.; Johnson, M.H. Frontal cortex functioning in the infant broader autism phenotype. *Infant Behav. Dev.* **2010**, *33*, 482–491. [CrossRef] [PubMed]

110. Damiano, C.R.; Nahmias, A.; Hogan-Brown, A.L.; Stone, W.L. What do repetitive and stereotyped movements mean for infant siblings of children with autism spectrum disorders? *J. Autism Dev. Disord.* **2013**, *43*, 1326–1335. [CrossRef] [PubMed]

111. Bedford, R.; Elsabbagh, M.; Gliga, T.; Pickles, A.; Senju, A.; Charman, T.; Johnson, M.H.; BASIS Team. Precursors to social and communication difficulties in infants at-risk for autism: Gaze following and attentional engagement. *J. Autism Dev. Disord.* **2012**, *42*, 2208–2218. [CrossRef] [PubMed]

112. Esposito, G.; del Carmen Rostagno, M.; Venuti, P.; Haltigan, J.D.; Messinger, D.S. Brief report: Atypical expression of distress during the separation phase of the strange situation procedure in infant siblings at high risk for ASD. *J. Autism Dev. Disord.* **2014**, *44*, 975–980. [CrossRef] [PubMed]

113. Haltigan, J.D.; Ekas, N.V.; Seifer, R.; Messinger, D.S. Attachment security in infants at-risk for autism spectrum disorders. *J. Autism Dev. Disord.* **2011**, *41*, 962–967. [CrossRef] [PubMed]

114. Wan, M.W.; Green, J.; Elsabbagh, M.; Johnson, M.; Charman, T.; Plummer, F.; BASIS Team. Quality of interaction between at-risk infants and caregiver at 12–15 months is associated with 3-year autism outcome. *J. Child Psychol. Psychiatry* **2013**, *54*, 763–771. [CrossRef] [PubMed]

115. Dundas, E.; Gastgeb, H.; Strauss, M.S. Left visual field biases when infants process faces: A comparison of infants at high- and low-risk for autism spectrum disorder. *J. Autism Dev. Disord.* **2012**, *42*, 2659–2668. [CrossRef] [PubMed]

116. Garon, N.; Bryson, S.E.; Zwaigenbaum, L.; Smith, I.M.; Brian, J.; Roberts, W.; Szatmari, P. Temperament and its relationship to autistic symptoms in a high-risk infant sib cohort. *J. Abnorm. Child Psychol.* **2009**, *37*, 59–78. [CrossRef] [PubMed]

117. Christensen, L.; Hutman, T.; Rozga, A.; Young, G.S.; Ozonoff, S.; Rogers, S.J.; Baker, B.; Sigman, M. Play and developmental outcomes in infant siblings of children with autism. *J. Autism Dev. Disord.* **2010**, *40*, 946–957. [CrossRef] [PubMed]

118. Paul, R.; Fuerst, Y.; Ramsay, G.; Chawarska, K.; Klin, A. Out of the mouths of babes: Vocal production in infant siblings of children with ASD. *J. Child Psychol. Psychiatry* **2011**, *52*, 588–598. [CrossRef] [PubMed]

119. Cornew, L.; Dobkins, K.R.; Akshoomoff, N.; McCleery, J.P.; Carver, L.J. Atypical social referencing in infant siblings of children with autism spectrum disorders. *J. Autism Dev. Disord.* **2012**, *42*, 2611–2621. [CrossRef] [PubMed]

120. Curtin, S.; Vouloumanos, A. Speech preference is associated with autistic-like behavior in 18-months-olds at risk for Autism Spectrum Disorder. *J. Autism Dev. Disord.* **2013**, *43*, 2114–2120. [CrossRef] [PubMed]

121. Del Rosario, M.; Gillespie-Lynch, K.; Johnson, S.; Sigman, M.; Hutman, T. Parent-reported temperament trajectories among infant siblings of children with autism. *J. Autism Dev. Disord.* **2014**, *44*, 381–393. [CrossRef] [PubMed]

122. Gliga, T.; Senju, A.; Pettinato, M.; Charman, T.; Johnson, M.H. Spontaneous belief attribution in younger siblings of children on the autism spectrum. *Dev. Psychol.* **2014**, *50*, 903–913. [CrossRef] [PubMed]

123. Patten, E.; Belardi, K.; Baranek, G.T.; Watson, L.R.; Labban, J.D.; Oller, D.K. Vocal patterns in infants with autism spectrum disorder: Canonical babbling status and vocalization frequency. *J. Autism Dev. Disord.* **2014**, *44*, 2413–2428. [CrossRef] [PubMed]
124. Schwichtenberg, A.J.; Young, G.S.; Sigman, M.; Hutman, T.; Ozonoff, S. Can family affectedness inform infant sibling outcomes of autism spectrum disorders? *J. Child Psychol. Psychiatry* **2010**, *51*, 1021–1030. [CrossRef] [PubMed]

International Journal of
Molecular Sciences

MDPI

Article

Clock Genes and Altered Sleep–Wake Rhythms: Their Role in the Development of Psychiatric Disorders

Annaëlle Charrier [1,*], Bertrand Olliac [2,3], Pierre Roubertoux [4] and Sylvie Tordjman [1,5]

1   Pôle Hospitalo-Universitaire de Psychiatrie de l'Enfant et de l'Adolescent (PHUPEA), Université de Rennes 1, Centre Hospitalier Guillaume-Régnier, 154 Rue de Châtillon, Rennes 35000, France; s.tordjman@yahoo.fr
2   Pôle Universitaire de Psychiatrie de l'Enfant et de l'Adolescent, Centre Hospitalier Esquirol, Limoges 87025, France; Bertrand.Olliac@chu-limoges.fr
3   INSERM, U1094, Tropical Neuroepidemiology, Limoges 87000, France
4   Aix Marseille Université, INSERM, GMGF UMR_S 910, Marseille 13385, France; pierre.roubertoux@univ-amu.fr
5   Laboratoire Psychologie de la Perception (LPP), Université Paris Descartes, CNRS UMR 8158, Paris 75270, France
*   Correspondence: a.charrier@ch-guillaumeregnier.fr; Tel.: +33-2-9951-0604

Academic Editor: Merlin G. Butler
Received: 19 October 2016; Accepted: 9 March 2017; Published: 29 April 2017

Abstract: In mammals, the circadian clocks network (central and peripheral oscillators) controls circadian rhythms and orchestrates the expression of a range of downstream genes, allowing the organism to anticipate and adapt to environmental changes. Beyond their role in circadian rhythms, several studies have highlighted that circadian clock genes may have a more widespread physiological effect on cognition, mood, and reward-related behaviors. Furthermore, single nucleotide polymorphisms in core circadian clock genes have been associated with psychiatric disorders (such as autism spectrum disorder, schizophrenia, anxiety disorders, major depressive disorder, bipolar disorder, and attention deficit hyperactivity disorder). However, the underlying mechanisms of these associations remain to be ascertained and the cause–effect relationships are not clearly established. The objective of this article is to clarify the role of clock genes and altered sleep–wake rhythms in the development of psychiatric disorders (sleep problems are often observed at early onset of psychiatric disorders). First, the molecular mechanisms of circadian rhythms are described. Then, the relationships between disrupted circadian rhythms, including sleep–wake rhythms, and psychiatric disorders are discussed. Further research may open interesting perspectives with promising avenues for early detection and therapeutic intervention in psychiatric disorders.

Keywords: clock genes; circadian rhythm; circadian clocks network; synchronization of oscillators; sleep-wake rhythm; psychiatric disorders; schizophrenia; autism spectrum disorder; mood disorders; attention deficit hyperactivity disorder

1. Introduction: Circadian Rhythms and Their Molecular Mechanisms

The function of the body is subject to different biological rhythms, the circadian one being keyed to a cycle of a 24-h day corresponding approximately to the 24-h light/dark cycle of the Earth's rotation. In fact, the range of the internal clock period for healthy adults is 24 h and 11 ± 16 min but it is set back to a 24-h cycle each day by exposure to morning light and to external clocks [1]. This rhythm regulates most of our biological and behavioral functions. Its dysregulation leads to sleep disorders and major physiological disturbances. In psychiatric disorders, sleep problems (especially reduced total sleep with insomnia, longer sleep latency, nocturnal and early morning awakenings) are highly prevalent symptoms. Insomnia is the most frequent sleep problem reported in psychiatric disorders [2]. In autism, a prevalence of insomnia from 50% to 80% has been reported [3–6] compared to 9–50% in age-matched typically developing children [7–11]. However, the interpretation of these results has to take into consideration the high prevalence of insomnia (30%) in typically developing individuals [12–14]. It is noteworthy that sleep disorders are not specific to autism spectrum disorder (ASD) given that they are observed in ASD individuals with intellectual disability (ID), ID individuals without ASD, and individuals with brain injury as well as ID, suggesting a possible overlap between ASD and ID. Concerning schizophrenia, a genome-wide association study (GWAS) shows genetic correlations between sleep disorders and schizophrenia and between the late evening chronotype and risk of schizophrenia [15]. No rigorous study of prevalence of sleep disorders was conducted in bipolar spectrum disorder and depression-related disorders but altered patterns of clock genes were reported in these disorders [16,17]. Conversely, high rates of sleep problems and mental disorders have been reported in shift workers, including insomnia, fatigue, anxiety and depression [18–24]. It is noteworthy that altered patterns of clock genes were also reported in shift workers [25–27].

The objective of this article is an attempt to clarify the relationships between psychiatric disorders, altered expression of clock genes and endogenous circadian rhythm disturbances (especially altered sleep–wake rhythms). Is the disruption of circadian rhythms, and therefore of sleep–wake rhythms, a prerequisite for the genesis of psychiatric disorders? What roles do the clock genes play in the disruption of endogenous circadian rhythms? After describing the molecular mechanisms of circadian rhythms, their relationships with psychiatric disorders are discussed, focusing in particular on bipolar disorder, anxiety, depression, attention deficit disorder, schizophrenia and autism spectrum disorder (a rich literature review on this topic was available for these psychiatric disorders, which was not the case for other disorders such as addictions or dementia).

1.1. Daily and Circadian Rhythms

There are different categories of physiological rhythms with different time scales, such as ultradian, tidal, circadian, lunar, and seasonal rhythms. A daily rhythm is a regular and predictable phenomenon which is defined as a series of significant physiological changes over 24 h. The biological rhythm consists of two components: the first is exogenous and modulated by environmental factors such as light–dark alternations, sleep–wake, hot–cold, and seasonal change [28–31]; the second is endogenous and linked to genetic factors (studies in identical twins showed that they had identical biological rhythms [32]). Homologous genes involved in the animal's activity–rest cycle have been described in humans. It has been observed that the individual tendency to get up and go to bed more or less early was associated with a polymorphism of the *circadian locomotor output cycles kaput* (*Clock*) gene. Socio-ecological exogenous factors modulate rhythms and are called synchronizers. Endogenous factors, genetic by nature, underpin the internal biological clock which is responsible for an internal time synchronization coordinating the circadian variations of biochemical, physiological and behavioral parameters.

Circadian rhythms are the best known biological rhythms at the molecular level. The alternation of activity and rest over a period of 24 h (or rest-activity cycle) was observed in drosophila, fish and mammals including rodents (rats, mice, hamsters) and humans. In mammals, this circadian rhythm is generated by a master central clock, mainly reset by ambient light, and is located in the suprachiasmatic nuclei (SCN) of the hypothalamus [33,34]. Circadian regulation of all biological functions is processed through direct or indirect signals (cyclic hormone production) between the suprachiasmatic nuclei and different body structures (brain regions, organs). The body has peripheral clocks located in each organ (heart, lung, liver, muscles, kidneys, retina, etc.) that optimize the function of each organ according to the environmental context allowing therefore adaptation of the organism to environmental changes. Thus, the circadian clocks network has an adaptive function [35]. Both central and peripheral clocks are detectable though both cyclic gene expression and rhythmic physiological processes. These peripheral clocks work independently but must be re-synchronized continuously through the brain's master clock acting as a real conductor. Furthermore, the neurohormone melatonin is involved in the synchronization (i.e., adjustment of the timing of existing oscillations) of peripheral oscillators; the nocturnal synthesis and release of melatonin by the pineal gland are controlled by the SCN master clock and inhibited by light exposure [36].

1.2. Clock Genes

The first gene known as being responsible for the circadian rhythm was found in drosophila; it is the *period* (*Per*) gene [37]. Subsequent research has highlighted another gene involved in the circadian rhythm: the *timeless* (*Tim*) gene [38–41].

Chemical mutagenesis performed in mammals such as mice allowed identification of the first clock genes: the *Clock* gene [42] (or *Npas2* in neuronal tissue), whose mutation was responsible for a lengthened rest–activity cycle. Other clock genes were then identified in the circadian regulation of the mouse: the *brain and muscle ARNT-like protein 1* (*Bmal1*) gene as a *Clock* partner, *Per1* and *Per2* genes, and *cryptochrome-1* (*Cry1*) and *Cry2* genes (genes encoding proteins involved in blue light reception in non-mammalian species).

The role of melatonin on the circadian oscillatory rhythms, and in particular as a neuroendocrine synchronizer of molecular oscillatory systems has been documented [36,43–46]. Studies in animal models showed that melatonin is involved in the regulation of circadian expression of several clock genes, such as *Per1*, *Per2*, *Bmal1*, *reverse erythroblastosis virus* (*REV-ERBα*), *Clock* and *Cry1*, in both central and peripheral melatonin target tissues. For example, in rat and mouse models, it has been shown that melatonin induces rhythmic expression of *Per1*, *Bmal1*, *Clock* and *Cry1* in the pituitary (*pars tuberalis*) [47,48]. It is noteworthy that *Per1* expression is undetectable in melatonin-deficient mice [49,50] and pinealectomy abolishes rhythmic expression of *Per1* in the *Pars tuberalis* (PT) and desynchronizes *Per1* and *Per2* expression in the SCN [51,52]. Moreover, melatonin synchronizes circadian oscillations in the cardiovascular system by influencing circadian rhythmic expression of both *Per1* and *Bmal1* in the rat heart [53]. In the pituitary of sheep, melatonin stimulates *Cry1* expression and suppresses other clock genes expression [54]. These effects are probably mediated by melatonin (MT) receptors, such as the MT1 receptors. Indeed, in the *pars tuberalis* of MT1 knockout mice, expression of *Per1*, *Bmal1*, *Clock* and *Cry1* was dramatically reduced but not changed in MT2 knockout mice [54]. Furthermore, in the hypothalamic SCN of rats, melatonin treatment induces phase advance of the nuclear receptor *REV-ERBα* expression [55]. In adipose tissue, melatonin synchronizes metabolic and hormonal function [56] by regulating *Per2*, *Clock* and *REV-ERBα* [57] (*REV-ERBα* is required for the daily balance of carbohydrate and lipid metabolism [58]). Finally, melatonin regulates oscillation of *Clock* genes in healthy and cancerous human breast epithelial cells [59] and induces a shift in the 24-h oscillatory expression of *Per2* and *Bmal1* in cultured fetal adrenal gland [60]. Taken together, these studies underline the major role of melatonin in the regulation of clock genes expression allowing the synchronization of central and peripheral oscillators.

In humans, homologues of *Clock*, *Bmal1*, *Per* and *Cry* were identified. The *Clock* gene, the only cloned circadian rhythm gene, is located on chromosome 4 [61]. There are other clock genes involved in circadian regulation, such as *retinoic acid receptor-related orphan receptor* (*ROR*)-*A* and *ROR-B*, *REV-ERB*, and *casein kinase-1* (*Ck1*)ε and *Ck1δ*, which control transcription of *Bmal1* gene. Additionally, the *albumin-D-site-binding protein* (*Dbp*) is also a circadian clock-controlled gene, involved in the circadian transcriptional regulation of several metabolic enzymes and some transcription factors. Some studies [25–27] report altered expression of certain clock genes in cases of shift workers (*Per1* and *Per2*), and free-running/constant conditions (*Per1*). However, most of the clock genes are affected in case of sleep-deprivation studies [62,63], highlighting the interaction between clock genes and sleep reported by Franken et al. [64,65]. Indeed, clock genes contribute to the homeostatic aspect of sleep regulation and mutations in some clock genes modify the markers of sleep homeostasis and an increase in homeostatic sleep drive alters clock genes expression in the forebrain [64]. It is noteworthy that the expression of clock genes is ubiquitous in humans (central nervous system, spleen, thymus, intestine, heart, lung, etc.), but also in animals (clock genes are found in most cell types among vertebrates).

1.3. Molecular Working of the Cellular Circadian Clock

The discoveries of the different clock genes led to the proposal of a molecular clock model based on a feedback loop running over 24 h [66]. The molecular mechanisms underlying the circadian rhythms are rather similar regardless of the species. They include enhancer elements, repressor elements and control loops involving phosphorylation-dephosphorylation, methylation, acetylation reactions and the specific protein dimerization [67,68]. In mammals, this molecular circadian system is present in the hypothalamic central clock, the suprachiasmatic nuclei, and in secondary clocks within the brain and peripheral organs. Light perceived by the retina leads to changes in transcription of certain genes of the circadian rhythm in the master clock and shows the ability of the body to adjust to a change in the cycle of the photoperiodic environment. This master clock synchronizes multiple peripheral circadian oscillators via mechanisms that remain to be better ascertained.

The molecular loop of the circadian clock seems to imply in a general way two types of mechanisms: transcriptional (transcriptional regulation of genes at the DNA level, i.e., their copy in the form of RNA messenger) and post-transcriptional (regulating steps downstream of the transcription) (see Figure 1 [69]).

Transcriptional mechanisms: these mechanisms can be summarized by autoregulatory feedback loops. The first is a primary negative feedback loop: they rely on a pair of positive elements and a pair of negative elements. In mammals, the positive elements are two proteins, CLOCK and BMAL1 transcription factors which heterodimerize each other. The two negative elements are also two proteins, PERIOD (PER) and CRYPTOCHROME (CRY). In a neuron of the suprachiasmatic nucleus, there is little PER and CRY in the morning. The CLOCK–BMAL1 complex activates maximally the transcription of *Per* and *Cry* genes. However, the PER and CRY proteins do not accumulate immediately due to their instability, which allows the cycle not to be interrupted prematurely. Following interactions with other proteins they are gradually stabilized during the day, then they heterodimerize and finally migrate into the nucleus [70], where they inhibit the transcriptional activity of the CLOCK–BMAL1 heterodimer and thus their own transcription [71,72]. The *Per* and *Cry* genes being less and less active, the PER and CRY proteins are less and less produced. Their amount reaches a maximum at the beginning of night, and then decreases. At the same time, the CLOCK–BMAL1 heterodimer gradually regains its activity during the night. Thus, the molecular mechanism of the clock consists principally in a feedback loop comprising a positive component (with the CLOCK–BMAL1 heterodimer) and a negative component (with the PER–CRY heterodimer). A complete cycle of this loop lasts about 24 h. CLOCK and BMAL1 heterodimers are also involved in a daily transcription of many clock-controlled genes (CCGs) in different peripheral tissues [73–75]. In addition to the primary feedback loop, another regulatory feedback loop is formed by the orphan nuclear receptors REV-ERBα and RORα. In the nucleus, REV-ERBα competes with RORα for binding to the ROR-responsive element (RORE) in the *Bmal1* promoter. Whereas RORα activates transcription of *Bmal1*, REV-ERBα represses it. Consequently, the cyclic expression of *Bmal1* is achieved by both positive and negative regulation of RORs and REV-ERBs, respectively. This secondary feedback loop is called the "stabilizing loop".

Post-transcriptional mechanisms: the proper functioning of the circadian loop requires that PER and CRY proteins disappear in due course, once they have served their purpose. As long as they inhibit the activity of positive elements (CLOCK and BMAL1), a new cycle cannot start. Several post-transcriptional processes thus affect the ability of PER and CRY to act on CLOCK and BMAL1. The most studied aspect is the modification of these proteins by phosphorylation and dephosphorylation. Several kinase proteins include PER and/or CRY among their targets, and each one of the two proteins can be phosphorylated on many distinct sites. They accelerate or slow down their transfer to the proteasome. Proteins to be degraded are "marked" by ubiquitin. Post-transcriptional regulations of the clock proteins of the main molecular loop make it possible to ensure intracellular traffic, functionality and degradation of clock proteins which are crucial for the functioning of the molecular loop over 24 h [76].

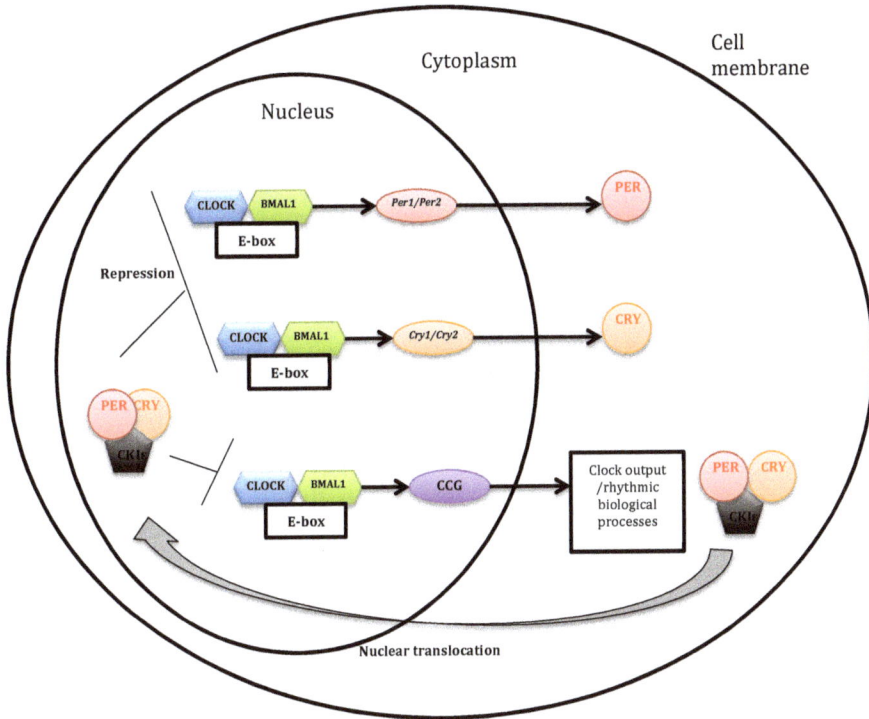

Figure 1. Model of the mammalian cell-autonomous oscillator (based on Lowrey and Takahashi, 2011) [69]. The transcriptional activators *circadian locomotor output cycles kaput* (CLOCK) and *brain and muscle ARNT-like protein 1* (BMAL1) stimulate the expression of *cryptochrome-1* (*Cry*) and *period* (*Per*) genes. The protein products of these genes are associated in the cytoplasm to form dimers that go into the core. There, they serve two functions: first, the repression of their own transcription, via the inhibition of CLOCK–BMAL1; and second, the activation of *Bmal1* gene, by a mechanism that remains to be discovered. These proteins are thus two regulating loops, one negative and the other positive. CLOCK and BMAL1 activate also the so-called clock-controlled genes (CCG) whose products transmit the rhythm information to the rest of the body via the output channels of the clock. Some proteins modulate the progression of control loops. Thus, casein kinase Iε (CKIε) can phosphorylate PER proteins, which destabilizes them and prevents their translocation into the nucleus.

2. Relationships between Circadian Rhythms and Psychiatric Disorders

Several hypotheses can be raised regarding the relationships between the development of psychiatric disorders and problems of circadian rhythms, including altered sleep–wake rhythms:

(1) First, sleep problems may lead to cognitive impairments due to the effects of sleep deprivation and fatigue on learning and attention capacities, long-term memory, language development and emotions [77–82];

(2) Recent studies in cognitive and developmental psychology have highlighted the importance of rhythmicity and synchrony of motor, emotional, and inter-personal rhythms in early development of social communication; the synchronization of rhythms allows tuning and adaptation to the external environment [83]. Impaired circadian rhythms with an absence of synchronization of

the circadian clocks network might alter the functioning of motor, emotional and interpersonal rhythms, leading to social communication impairments and vulnerability to psychiatric disorders with social communication deficit such as ASD or schizophrenia (for a review of literature on the importance of rhythmicity and synchrony of motor, emotional, and inter-personal rhythms in early development of social communication, see [83,84]). In addition, circadian rhythms involve sequences of continuities/discontinuities that might be important for typical fetal and child development in order to provide a secure environment (through stable and predictable regularities) but also variations allowing the individual to adapt to changes. Impaired circadian rhythms with no, little or irregular variability might lead to anxiety and difficulties in adapting to changes associated with restricted and repetitive interests observed in some psychiatric disorders described in this article such as ASD, schizophrenia and anxiety disorder;

(3) Clock genes control critical periods of brain development [85] and therefore, abnormal expression of clock genes might participate to neurodevelopmental disorders such as psychiatric disorders. It is noteworthy that only a few days of circadian rhythm impairments may impact the maturation and specialization of some brain structures at specific developmental periods; these abnormalities can alter the temporal organization of brain maturation and development [86];

(4) Circadian rhythm impairments may alter transcriptional and splicing regulation of Parvalbumin (PV) neurons, knowing that PV knockout mice (PV−/−) or heterozygous (PV+/−) mice showed autism behavioral phenotype [87]. More generally, circadian rhythm impairments may affect gene expression involved in synapse formation and brain maturation;

(5) Also, clock-controlled genes (CCGs) may have pleiotropic effects outside the molecular clock and have therefore more widespread impact on cognition, mood, and reward-related behaviors [88];

(6) Finally, circadian rhythm impairments (provoked or not by sleep problems) may also alter the adaptation of the individual to his/her environment and therefore his/her state of homeostasis. In this perspective, psychiatric disorders might reflect a loss of synchronization between the external environment's rhythms and the individual's internal rhythms, leading to major problems of adaptation for the individual and the appearance of psychiatric disorders. Single-nucleotide polymorphisms (SNPs) in core circadian clock genes have been associated with autism spectrum disorder [89], attention deficit hyperactivity disorder [90,91], anxiety disorder [92], major depressive disorder [93–95], bipolar disorder [95–97] and schizophrenia [98–101]. However, the causal relationship for these associations remains to be better ascertained. Circadian clock genes may affect specific aspects of psychiatric disorders through circadian control or through distinct regulation of downstream effectors.

The suprachiasmatic nuclei (SCN) coordinate the rhythms of other brain regions and peripheral organs like a pacemaker [102]. The sleep–wake cycle is the most documented example of the activity of the SCN, but endocrine, metabolic and immunological activities are also conducted by the SCN. For example, the glucocorticoids are regulated by the circadian cycle and affect the circadian rhythms of the amygdala [103,104]. Also, they regulate directly the clock genes [105,106].

We will explore the current state of research on this subject through five psychiatric disorders: bipolar disorder, anxiety and depression, attention deficit disorder, schizophrenia, and autism spectrum disorders.

2.1. Bipolar Spectrum Disorder

The mechanisms underlying the association between alterations in circadian rhythms and mood disorders are still unclear. However, current scientific studies provide food for thought. According to Jackson et al. [107], sleep disorder is the most common prodrome of mania and one of the six most common prodromes of depression in bipolar disorder (bipolar type I disorder is defined as an isolated manic episode, and bipolar type II disorder is defined as manic and depressive episodes) [108]. Between mood episodes, the sleep–wake cycle remains disrupted in patients with bipolar disorder who have some difficulty falling asleep and frequent night awakenings [109]. These patients have a significantly more eventide sleep pattern than control subjects, in other words, they go to bed later and have more difficulty getting up early compared to healthy controls [110].

McClung [111] studied the involvement of the *Clock* gene in manic episodes in mice. She compared *Clock* mutant mice to mice with a normal *Clock* gene. Mice in which the gene had been modified had mania-like behavior. Those mice were hyperactive, less anxious and less depressed. They slept less and showed greater brain activity in response to sugar water, cocaine and a mild electrical stimulation of the brain. When researchers added lithium in their water, these mice did not show their manic behavior anymore and started to act like the mice with a normal *Clock* gene. A lithium salt treatment could affect circadian rhythms by modulating the expression of clock genes [112–115]. Indeed, lithium salts might inhibit the expression of *glycogen synthase kinase 3β (GSK-3β)* gene [116], a gene involved in the brain's biological clock [117]. Kaladchibachi demonstrated that both genetic and pharmacological reduction of *GSK-3* activity in mice have a specific effect on the circadian transcriptional oscillation consisting of *mPer2* period lengthening, indicating a delay in phase [116]. Moreover, it has been shown that *GSK-3β* is rhythmically expressed in the SCN and liver of mice, and that it undergoes a daily cycle in phosphorylation in vivo. Lithium chloride treatment inhibits the *GSK-3β* expression and results in a phase delay of *Clock* gene expression in fibroblasts [118]. Bipolar spectrum disorder has been associated with variations in the *Clock* gene. A single-nucleotide polymorphism (SNP) in the 3-flanking region of the *Clock* gene (3111 T to C) is associated with a higher recurrence rate of bipolar episodes [16]. This SNP was also associated in bipolar disorder and/or antidepressant treatment with sleep problems (insomnia and decreased need for sleep) [17,119]. Some studies reported other clock genes associated with bipolar disorder (such as *Bmal1* and *Per3*) [96]. Furthermore, a SNP in *Bmal1* and a SNP in *Tim* have also been identified as having a link with bipolar disorder [99]. One South Indian study showed that the occurrence of the five repeat alleles of *Per3* may be a risk factor for bipolar disorder onset in this ethnic group [120]. The coding region of *Per3* gene contains a variable number tandem-repeat (VNTR) polymorphisms which has been associated with diurnal preference, sleep structure and sleep homeostasis in healthy individuals. In a homogeneous sample of patients with bipolar type I disorder (occurrence of one or more manic episodes or mixed episodes), they observed that *Per3* VNTR influenced age of onset: earlier onset in homozygote carriers of *Per35* variant, later onset in homozygotes for *Per34*, and intermediate onset in heterozygotes. Sjöholm et al. [121] studied four single-nucleotide polymorphisms (SNPs) of *Cry2* gene in a cohort of bipolar patients, some with fast cycles. They observed that *Cry2* is associated with rapid cycling in bipolar disorder patients. Rapid cycling in bipolar disorder is defined according to the 5th version of Diagnostic and Statistical Manual of Mental Disorders (DSM-5) as four or more mood episodes in any combination or order within any year in the course of the illness. The A allele of rs10838524 (*Cry2* SNP) was significantly overrepresented among bipolar disorder cases with rapid cycling compared to controls. There was a significant trend in the association between the A allele of rs10838524 and rapid cycling ($p = 0.0076$) and this allele increased the risk for rapid cycling both in a homozygote and a heterozygote form (that is dominant model) compared to controls. Two closely related clock genes, retinoid-related orphan receptors α (*RORA*) and β (*RORB*) are involved in a number of pathways including neurogenesis, stress response, and modulation of circadian rhythms. A study reports that four intronic *RORB* SNPs showed positive associations with the pediatric bipolar phenotype and suggests that clock genes in general and *RORB* in particular may be important candidates for further investigation in the search for the molecular basis of bipolar disorder [122]. Further studies are necessary to better assess the role of *RORB* in bipolar spectrum disorder and explore if this gene is particularly relevant for this disorder compared to other clock genes.

Several genetic studies suggest that certain polymorphisms of clock genes are more common in bipolar disorder than in the general population [123,124]. The clock genes might therefore be viewed as genetic vulnerability factors to bipolar disorder [125]. These results need to be examined further.

The litterature review on clock genes and bipolar spectrum disorder is presented in Table 1.

Table 1. Clock genes and bipolar spectrum disorder. RORα: retinoid-related orphan receptors α; RORβ: retinoid-related orphan receptors β.

Studies	Measure	Individuals with Psychiatric Disorder and/or Organisms Models (*n*)	Controls (*n*)	Results
Kaladchibachi et al. [116]	Cyclical expression of clock genes (*Per2*)	Mouse embryonic fibroblasts (MEFs)	-	Genetic depletion of glycogen synthase kinase 3 (GSK3) activity results in a significant delay in the cycling period of *Per2*.
McGrath et al. [122]	Genotyping and analysis of 312 single-nucleotide polymorphisms (SNPs) in *RORA* and 43 SNPs in *RORB*	Bipolar disorder (BD) children (*n* = 305)	Healthy parents (*n* = 306) Healthy individuals (*n* = 140)	Four intronic *RORB* SNPs showed positive associations with the pediatric bipolar phenotype.
Lavebratt et al. [94]	Assessment of *Cry2* gene expression before and after one night of sleep deprivation	BD individuals (*n* = 13)	Healthy individuals (*n* = 8)	*Cry2* mRNA levels are reduced and unresponsive to sleep deprivation in depressed patients with bipolar disorder.
Sjöholm et al. [121]	Analysis of four *Cry2* single-nucleotide polymorphisms	BD individuals in Sweden (*n* = 577); BD type I (*n* = 497); BD type II (*n* = 60); BD with rapid cycling (*n* = 155)	Healthy individuals (*n* = 1044)	Association between the circadian gene *Cry2* and rapid cycling in bipolar disorder.
Karthikeyan et al. [120]	Genotyping and analysis of *Per3* in blood samples	Bipolar type I disorder individuals in South India (*n* = 311)	Healthy individuals (*n* = 346)	The occurrence of the five repeat allele of *Per3* may be a risk factor for bipolar type I disorder onset in this ethnic group.

2.2. Anxiety and Depression-Related Disorders

2.2.1. Anxiety Disorder

A team of French researchers has evaluated a range of behaviors associated with psychiatric illnesses in mice in which two genes of the circadian clock, *Cry1* and *Cry2*, were investigated [126]. Their work highlighted the causal relationship between the disruption of encoding genes for cryptochrome proteins 1 and 2 (*Cry1* and *Cry2*) and behaviors associated with anxiety. Mice deficient for CRY 1 and 2 proteins show behavioral alterations characterized among other things by an abnormally high level of anxiety. These results indicate clearly that in addition to their critical roles in regulating the molecular clock; these proteins are directly involved in the control of emotional states.

2.2.2. Major Depressive Disorder (MDD)

Sleep problems, in particular insomnia or hyperinsomnia, are parts of depression criteria and have been hypothesized to be under genetic control. Major depressive disorder, as well as seasonal affective disorder (winter depression and summer depression) and bipolar type I disorder provide excellent models for studying the molecular mechanisms of mood disorder. An important study published in 2013 first demonstrated dysfunctioning clock genes in brains from depressed humans compared to healthy controls [127]. Involvement of the clock genes in depression is also evident from several genetic studies. It has been reported that polymorphisms of clock genes appear in depressed patients [93,128–132].

Several SNPs in the *Clock* gene (T3111C, 3117 G to T, 3125 A to G) has been reported [127,128,133] to be associated with major depression and sleep disturbances (but some studies did not found any association between T3111C and sleep problems; [134,135]). More precisely, the two rare SNPs (3117 G to T and 3125 A to G) were associated with alternating phases of good sleep and insomnia over the course of a few days [133].

A Swedish team, Lavebratt et al. [94] reported in healthy controls a marked diurnal variation in *Cry2* mRNA levels; total sleep deprivation induced a 2.0-fold increase in *Cry2* mRNA levels. In patients with depressive state of bipolar disorder, sleep deprivation induced significantly decreased *Cry2* mRNA expression compared with healthy controls. In addition, *Cry2* mRNA levels were found to be lowered in blood mononuclear cells from depressed patients with bipolar disorder after total sleep deprivation in comparison to healthy controls, and *Cry2* gene variation was associated with winter depression in both Swedish and Finnish patients [94]. Deletion of *Cry2* gene lengthened the circadian period by approximately 48 min. This study [94] suggests that a *Cry2* locus is associated with vulnerability for depression, and that mechanisms of action involve dysregulation of *Cry2* expression. Associations between the gene *Cry2* and winter depression, but also dysthymia and bipolar type I disorder as seen previously, support the view that the *Cry2* gene has a role in mood disorders.

Shi et al. [132] found genetic polymorphisms in circadian genes, especially *Clock* and *Per3*, in major depressive disorder individuals. Authors propose that the impact of the *Clock* and *Per3* SNPs on transcription and/or expression may not be on the core circadian oscillator in humans, but on global output transcriptional pathways, that are mediated sex-dependently by the circadian system.

The litterature review on clock genes and depression related disorders is presented in Table 2.

2.2.3. Familial Advanced Sleep Phase Syndrome (FASPS)

A good example of an abnormal circadian system in humans is the familial advanced sleep phase syndrome (FASPS), described in three families by Jones et al. [136]. FASPS is often associated with depression and anxiety. Affected individuals have a lead of a few hours of their sleep patterns and rhythms of temperature and melatonin, and a shorter period than healthy individuals in constant conditions. The joint efforts of several laboratories of the University of Utah in Salt Lake City have recently led to the description of *Per2* mutation in one of those families affected by FASPS. The mutation caused a substitution of serine by glycine in the *Per2* gene at position 662 [137]; this serine that can normally be phosphorylated by casein kinase Iε (CKIε). Inhibiting the action of CKIε accelerates the clock, which is indeed what is observed in FASPS patients. FASPS are also to be caused by a T44A missense mutation in human CK1δ, which causes hypophosphorylation of *Per2* [138,139]. Transgenic mice over-expressing the *hPer2* mutation exhibit shorter free-running periods, mimicking FASPS [140], and their phenotype is sensitive to CK1δ: increased dosage of CK1δ shortens their period further.

Recently, a team has identified a new mutation in the *hCry2* gene associated with FASPS. The mutation leads to replacement of an alanine residue at position 260 with a threonine. In mice, the *Cry2* mutation causes a shortened circadian period and reduced phase-shift to early–night light pulse associated with phase-advanced behavioral rhythms in the light-dark cycle [141].

2.2.4. Seasonal Affective Disorder (SAD) and Delayed Sleep Phase Syndrome (DSPS)

Environmental changes like the seasons can be associated with depressive episodes in the inability of the circadian clock to adjust appropriately. Seasonal affective disorder (SAD) is defined as recurrent depressive disorder characterized by a seasonal pattern with, generally, an appearance in the fall or winter, apart from triggering external factor, and spontaneous disappearance in the spring or summer, even in the absence of treatment. SAD affects about 5% of the general population [142]. Lewy evokes the idea of a "phase delay of the endogenous circadian oscillator in relation to sleep-wake rhythm" [143], the equivalent of a delayed sleep phase syndrome (DSPS), as the cause of SAD. Genetic variants in *Npas2*, *Per2*, and *Bmal1* have been found to combine with the development of SAD [93,144].

Iwase et al. [145] reported that the SNP T3111C in the *Clock* gene was associated with morning or evening preference for activity and its frequency was decreased in DSPS. Futures studies are required to investigate the possible contribution of T3111C to DSPS susceptibility.

Table 2. Clock genes and depression related disorders.

Studies	Measure	Individuals with Psychiatric Disorder (n)	Controls (n)	Results
Takimoto et al. [146]	Daily variation of melatonin and cortisol, and daily expression of clock genes (*Per*, *Bmal1* and *Clock*) in whole blood cells	Individuals with circadian rhythm sleep disorder (*n* = 1)	Healthy male individuals (*n* = 12)	The peak phase of *Per1*, *Per2*, and *Per3* appeared in the early morning, whereas that of *Bmal1* and *Clock* appeared in the midnight hours in healthy male individuals.
Partonen et al. [93]	Analysis of sequence variations (single-nucleotide polymorphisms) in three core clock genes: *Per2*, *Bmal*, and *Npas2*	Depressed individuals (*n* = 189)	Healthy individuals (*n* = 189)	Variations in the three circadian clock genes *Per2*, *Bmal*, and *Npas2* are associated with winter depression.
Utge et al. [147]	Analysis of 113 single-nucleotide polymorphisms of 18 genes of the circadian system	Depressed individuals (*n* = 384)	Healthy individuals (*n* = 1270)	Significant association between *Tim* variants and depression with fatigue in females, and association to depression with early morning awakening in males.
Lavebratt et al. [94]	Genotyping of single nucleotide polymorphism of the *Cry2* gene	Depressed individuals with bipolar disorder (*n* = 204)	Healthy individuals (*n* = 2017)	The *Cry2* gene was significantly associated with winter depression in both samples.
Kovanen et al. [130]	Genotyping of 48 single-nucleotide polymorphisms in *Cry1* and *Cry2* gene	Individuals with dysthymia (*n* = 136)	Healthy individuals (*n* = 3871)	Four *Cry2* genetic variants (rs10838524, rs7121611, rs7945565, rs1401419) are significantly associated with dysthymia.
Hua et al. [131]	Genotyping of single nucleotide polymorphisms (SNPs) of *Cry1* rs2287161, *Cry2* rs10838524 and *Tef* (*thyrotroph embryonic factor*) rs738499	Chinese individuals with major depressive disorder (MDD) (*n* = 105)	Chinese healthy individuals (*n* = 485)	The polymorphisms of *Cry1* rs2287161 and *Tef* rs738499 are associated to major depressive disorder.
Shi et al. [132]	Genotyping of 32 genetic variants from eight clock genes	Major depressive disorder individuals (*n* = 592)	Healthy individuals (*n* = 776)	Genetic polymorphisms in circadian genes, especially *Clock* and *Per3*, influence risk of developing depression in a sex- and stress-dependent manner.

2.3. Attention Deficit Hyperactivity Disorder (ADHD)

Many sleep disorders are associated with attention deficit hyperactivity disorder (ADHD), both in adults and children. In particular, the following sleep problems occurring at sleep–wake transition were observed: bed-time refusal, delayed sleep-onset, and early awakenings [148]. Differences in sleep problems were found as a function of ADHD subtype: children with ADHD inattentive type (ADHD-I) had the fewest sleep problems and did not differ from controls, children with ADHD combined type (ADHD-C) had more sleep problems than controls and children with ADHD-I. Daytime sleepiness was greatest in ADHD-I and was associated with sleeping more (not less) than normal [149].

Several studies reported genetic associations between polymorphism (rs1801260) at the 3′-untranslated region (3′-UTR) of the *Clock* gene and psychiatric disorders [17,90] (see Table 3). Concerning the association between the rs1801260 polymorphism and ADHD, Kissling et al. [90] found at least one T-mutation being the risk allele in Caucasians of western European origin and German background.

Xu et al. [91] conducted a study which investigated a previously reported discovery of an association between a single nucleotide polymorphism in the 3′-UTR region of the *Clock* gene (rs1801260) in two independent samples of ADHD probands from UK and Taiwan (aged 5 to 15).

They found a significant over-transmission of the T allele of rs1801260 SNP, which is associated as a risk allele for delayed sleep phase syndrome [145] in ADHD cases in the Taiwanese population. No association was observed between this polymorphism and ADHD in the UK sample. However, they did find evidence for increased transmission of the T allele of rs1801260 in the Taiwanese samples. Their findings support the hypothesis that genetic variation in the 3'-UTR region of *Clock* gene might be a risk factor for the development of ADHD, particularly in the Taiwanese sample studied. Therefore, more functional polymorphisms of this region should be investigated in other independent studies using larger samples.

Table 3. Clock genes and attention deficit hyperactivity disorder (ADHD).

Studies	Measure	Individuals with Psychiatric Disorder (*n*)	Controls (*n*)	Results
Kissling et al. [90]	Analysis of polymorphism (rs1801260) at the 3'-untranslated region of the *Clock* gene	ADHD individuals (*n* = 143)	Healthy individuals (*n* = 143)	Significant association (*p* < 0.001) between genotype and ADHD-scores of the adult ADHD assessments, and the rs1801260 polymorphism with at least one T-mutation is the risk allele.
Xu et al. [91]	Analysis of polymorphism (rs1801260) at the 3'-untranslated region of the *Clock* gene in ADHD using within-family transmission disequilibrium test	Two clinical ADHD samples: United Kingdom (UK) sample: (*n* = 180); Taiwan sample: (*n* = 212)	Both parents or mother alone or father alone UK sample: (*n* = 296); Taiwan sample: (*n* = 326)	Increased transmission of the T allele of the rs1801260 polymorphism in Tawainese samples.

2.4. Schizophrenia

Schizophrenia is often associated with sleep alterations [88,150], observed also in untreated patients [151]. Patients with schizophrenia have a desynchronization between the sleep–wake rhythms and melatonin profiles [88,152,153], the rhythms of body temperature [154], and serum levels of tryptophan and prolactin [155]. These anomalies suggest strongly disturbed circadian rhythms in these patients [156]. Studies of polymorphisms of clock genes or alterations in the regulation of these genes in schizophrenia are rare. The litterature review on clock genes and schizophrenia is presented in Table 4.

The expression of *Per1* mRNA in the temporal lobe in individuals with schizophrenia decreases significantly compared with age-matched healthy controls [98]. Another study suggested that *Per3* but not *Per2* abnormalities were associated with schizophrenia [99]. In a sample of 145 Japanese individuals with schizophrenia compared with healthy controls, it was reported that the T3111C polymorphism of the *Clock* gene presented a transmission bias. The T3111C polymorphism of the *Clock* gene might be associated with aberrant dopaminergic transmission in the suprachiasmatic nucleus, which is presumably involved in the pathophysiology of schizophrenia [157]. More recently, two studies focused on white blood cells and fibroblast from smaller samples: Sun et al. [100] reported altered expressions of *Per1*, *Per2*, *Per3* and *Npas2* in white blood cells in individuals with schizophrenia. Indeed, compared with healthy controls, schizophrenia patients presented disruptions in diurnal rhythms of the expression of *Per1*, *Per3*, and *Npas2*, accompanied by a delayed phase in the expression of *Per2* and by a decreasing in *Per3* and *Npas2* expression. Johansson et al. [101] reported a loss of rhythmic expression of *Cry1* and *Per2* in fibroblasts from individuals with schizophrenia compared to cells from healthy controls.

Table 4. Clock genes and schizophrenia. REV-ERBα: reverse erythroblastosis virus; Dbp: albumin-D-site-binding protein.

Studies	Measure	Individuals with Psychiatric Disorder (*n*)	Controls (*n*)	Results
Takao et al. [157]	Analysis of 3111C single nucleotide polymorphism of the *Clock* gene	Individuals with schizophrenia (*n* = 145)	Healthy individuals (*n* = 128)	Individuals with schizophrenia had a significantly higher frequency of the C allele compared to controls.
Sun et al. [100]	Relative expression of clock gene mRNA: *Per1*, *Per2* and *Per3* in blood samples	Individuals with schizophrenia (*n* = 13)	Healthy controls (*n* = 15)	Individuals with schizophrenia presented disruptions in diurnal rhythms of the expression of *Per1*, *Per3*, and *Npas2* compared with healthy controls, accompanied by a delayed phase in the expression of *Per2* and by a decrease in *Per3* and *Npas2* expression.
Johansson et al. [101]	Analysis of *Clock*, *Bmal1*, *Per1*, *Per2*, *Cry1*, *Cry2*, *REV-ERBα* and *Dbp* in fibroblasts from skin samples	Individuals with chronic schizophrenia under neuroleptic medication (*n* = 11)	Healthy individuals (*n* = 11)	Loss of rhythmic expression of *Cry1* and *Per2* in fibroblasts from individuals with schizophrenia compared to cells from healthy controls.

To date, the associations between clock genes and schizophrenia are not clear.

2.5. Autism Spectrum Disorder (ASD)

Surveys of parents show that the prevalence of sleep problems in ASD is 50%–80% compared to 9%–50% in age-matched typically developing children [7–14]. Recent results showing abnormal melatonin secretion in ASD children may change the initial disregard of these disorders and suggest a possible key role of the clock and circadian regulations in ASD. Several independent groups detected abnormal melatonin levels in ASD [158–163]. These studies conducted on independent autism samples and using different methodologies indicate that abnormally low melatonin level is a frequent trait in ASD. More precisely decreased nocturnal as well as diurnal levels have been reported in individuals with ASD; trial studies support therapeutic benefits of melatonin use in ASD [83,84,161]. Nevertheless, both the underlying cause of this anomaly and its relationship with ASD (cause or consequence?) remain still unexplained.

Many studies have advocated a genetic etiology for autism [164] involving in particular synaptic genes related to synaptic cell adhesion molecules NLGN3, NLGN4, and NRXN1 and a postsynaptic scaffolding protein SHANK3. This protein complex is crucial for the maintenance of functional synapses as well as the adequate balance between neuronal excitation and inhibition. Among the factors that could modulate this pathway there are genes controlling circadian rhythms. However, no direct association has been reported between clock genes and NLGN3, NLGN4, NRXN1 and SHANK3 in humans, but epistasis mechanisms cannot be ruled out. SHANK3 expression is modulated by melatonin concentration and the modulation is brain structure-dependent [165]. Concerning relationships between sleep disorders and NLGN3, NLGN4, NRXN1 and SHANK3 could be a pleiotropic effect on sleep of genes associated with brain disorganization. It is noteworthy that possible interplay of synaptic and clock genes may increase the risk of ASD [166]. The involvement of clock genes in ASD was first suggested by Wimpory et al. [167] who stated the hypothesis that anomalies in clock genes operating as timing genes in high frequency oscillator systems may underline timing deficits that could be important in the development of autism spectrum disorder, notably in autistic communication impairment. To test this hypothesis, Nicholas et al. [89] screened single-nucleotide polymorphisms in 11 clock/clock-related genes in 110 individuals with ASD and their parents.

A significant allelic association was detected for *Per1* and *Npas2*. It should be noted that it was a small population and results were not significant after correction for multiple testing. However, the association between *clock* genes and ASD was confirmed by a more recent study [168] reporting also mutations in other circadian clock genes (*Per2, Per3, Clock, Bmal1, Tim, Cry1, Cry2, Dbp* and *Ck1ε*) in ASD patients. Taken together, these findings suggest that the circadian rhythm abnormalities observed in ASD may be linked to abnormalities of the circadian clock genes. Finally, it is noteworthy that clock genes are related to the protein association network described in ASD by Roubertoux and Tordjman [169] either through direct gene × gene interactions or protein × protein interactions.

The litterature review on clock genes and autism spectrum disorder is presented in Table 5.

Table 5. Clock genes and autism spectrum disorder (ASD).

Studies	Measure	Individuals with Psychiatric Disorder (*n*)	Controls (*n*)	Results
Nicholas et al. [89]	Screening of eleven clock/clock-related genes	High-functioning ASD individuals (*n* = 110)	Healthy parents (*n* = 220)	Significant association for two single-nucleotide polymorphisms in *Per1* and in *Npas2*.
Yang et al. [168]	Direct sequencing analysis of the coding regions of 18 canonical clock genes and clock-controlled genes	ASD individuals with sleep disorders (*n* = 14); ASD individuals without sleep disorders (*n* = 14)	Healthy individuals (*n* = 23)	Mutations in circadian-relevant genes (specifically *Per1, Per2, Per3, Clock, Npas2, Bmal1, Tim, Cry1, Cry2, Dbp* and *Ck1ε*) affecting gene function are more frequent in individuals with ASD than in controls.

3. Conclusions

Circadian clocks enable organisms to anticipate temporal organization of biological functions in relation to periodic changes of the environment, and to adapt consequently their behavior. The genes *Clock, Per, Cry* and *Bmal1* are currently the major clock genes identified in humans as being involved in the rhythmicity and timing of biological rhythms at the molecular level. Their alteration involves changes to the 24-h rhythm through poor synchronization between the endogenous circadian rhythms and the sleep-wake cycle, and act especially on sleep disorders. These are often early symptoms of altered sleep–wake rhythms at the onset of psychiatric disorders, especially for mood disorders. Furthermore, impairments in the four major clock genes (*Clock, Per, Cry* and *Bmal1*) were found for bipolar disorder, depression-related disorders, autism spectrum disorder, and impairments in some of these major clock genes were also reported for schizophrenia (*Clock, Per* and *Cry*), anxiety disorder (*Cry*) and attention deficit hyperactivity disorder (*Clock*). In addition, other *clock* genes were associated with these psychiatric disorders, such as *Npas2* (winter depression, autism spectrum disorder and schizophrenia), *RORA* and *RORB* (bipolar disorder) or *Tim, Dbp* and *Ck1ε* (autism spectrum disorder). The associations of identical clock genes with these different psychiatric disorders suggest that they may share similar pathways and etiopathogenic mechanisms. It highlights the interest and need to study these mental disorders through a transnosographic and multidimensional approach focusing on depression, anxiety and stress responses.

The cascading effects resulting from altered clock genes, poorly understood so far, could participate in sleep problems and the emergence of symptoms present in certain psychiatric disorders through, as discussed in the article, impaired regulation of circadian rhythms and emotional states with neurodevelopmental effects (including impaired control of the temporal organization of brain maturation, neurogenesis, synapses formation/functioning and brain specialization at specific developmental periods). Inversely, sleep problems can alter the expression of clock genes and contribute, as seen in this article, to the development of psychiatric disorders through cognitive effects of sleep deprivation and fatigue. More generally, alteration of clock genes (directly or indirectly through sleep problems) might lead to desynchronized and abnormal circadian rhythms (including sleep/wake rhythm but also other circadian rhythms such as neuroendocrine or body temperature rhythms) impairing in turn the synchronization between external and internal rhythms and therefore the adaptation of the individual to his/her internal and external environment with the development of psychiatric disorders. Future studies are required to better ascertain the underlying mechanisms of the relationships between clock genes, sleep disturbances and psychiatric disorders. Further research may open interesting perspectives with promising avenues for early detection and therapeutic intervention in psychiatric disorders.

Author Contributions: Annaëlle Charrier, Bertrand Olliac, Pierre Roubertoux and Sylvie Tordjman wrote the paper.

Conflicts of Interest: The authors declare no conflict of interest.

References

1. Czeisler, C.A.; Duffy, J.F.; Shanahan, T.L.; Brown, E.N.; Mitchell, J.F.; Rimmer, D.W.; Ronda, J.M.; Silva, E.J.; Allan, J.S.; Emens, J.S.; et al. Stability, precision, and near-24-hour period of the human circadian pacemaker. *Science* **1999**, *284*, 2177–2181. [CrossRef] [PubMed]
2. Singh, K.; Zimmerman, A.W. Sleep in Autism Spectrum Disorder and Attention Deficit Hyperactivity Disorder. *Semin. Pediatr. Neurol.* **2015**, *22*, 113–125. [CrossRef] [PubMed]
3. Lai, M.C.; Lombardo, M.V.; Baron-Cohen, S. Autism. *Lancet* **2014**, *383*, 896–910. [CrossRef]
4. Kotagal, S.; Broomall, E. Sleep in children with autism spectrum disorder. *Pediatr. Neurol.* **2012**, *47*, 242–251. [CrossRef] [PubMed]
5. Onore, C.; Careaga, M.; Ashwood, P. The role of immune dysfunction in the pathophysiology of autism. *Brain Behav. Immun.* **2012**, *26*, 383–392. [CrossRef] [PubMed]
6. Kohyama, J. Possible neuronal mechanisms of sleep disturbances in patients with autism spectrum disorders and attention-deficit/hyperactivity disorder. *Med. Hypotheses* **2016**, *97*, 131–133. [CrossRef] [PubMed]
7. Allik, H.; Larsson, J.O.; Smedje, H. Sleep patterns of school-age children with Asperger syndrome or high-functioning autism. *J. Autism Dev. Disord.* **2006**, *36*, 585–595. [CrossRef] [PubMed]
8. Richdale, A.L.; Schreck, K.A. Sleep problems in autism spectrum disorders: Prevalence, nature, and possible biopsychosocial aetiologies. *Sleep Med. Rev.* **2009**, *13*, 403–411. [CrossRef] [PubMed]
9. Polimeni, M.A.; Richdale, A.L.; Francis, A.J. A survey of sleep problems in autism, Asperger's disorder and typically developing children. *J. Intellect. Disabil. Res.* **2005**, *49*, 260–268. [CrossRef] [PubMed]
10. Doo, S.; Wing, Y.K. Sleep problems of children with pervasive developmental disorders: Correlation with parental stress. *Dev. Med. Child. Neurol.* **2006**, *48*, 650–655. [CrossRef] [PubMed]
11. Giannotti, F.; Cortesi, F.; Cerquiglini, A. An investigation of sleep characteristics, electroencephalogram abnormalities and epilepsy in developmentally regressed and non-regressed children with autism. *J. Autism Dev. Disord.* **2008**, *38*, 1888–1897. [CrossRef] [PubMed]
12. Stoleru, S.; Nottelmann, E.D.; Belmont, B.; Ronsaville, D. Sleep problems in children of affectively ill mothers. *J. Child. Psychol. Psychiatry Allied Discip.* **1997**, *38*, 831–841. [CrossRef]
13. Brown, W.D. Insomnia: Prevalence and daytime consequences. In *Sleep: A Comprehensive Handbook*; Lee-Chiong, T., Ed.; John Wiley and Sons: Hoboken, NJ, USA, 2006; pp. 93–98.
14. Ahmed, A.E.; Al-Jahdali, H.; Fatani, A.; Al-Rouqi, K.; Al-Jahdali, F.; Al-Harbi, A.; Baharoon, S.; Ali, Y.Z.; Khan, M.; Rumayyan, A. The effects of age and gender on the prevalence of insomnia in a sample of the Saudi population. *Ethn. Health* **2016**, 1–10. [CrossRef] [PubMed]
15. Lane, J.M.; Vlasac, I.; Anderson, S.G.; Kyle, S.D.; Dixon, W.G.; Bechtold, D.A.; Gill, S.; Little, M.A.; Luik, A.; Loudon, A.; et al. Genome-wide association analysis identifies novel loci for chronotype in 100,420 individuals from the UK Biobank. *Nat. Commun.* **2016**, *7*, 10889. [CrossRef] [PubMed]
16. Benedetti, F.; Dallaspezia, S.; Fulgosi, M.C.; Lorenzi, C.; Serretti, A.; Barbini, B.; Colombo, C.; Smeraldi, E. Actimetric evidence that CLOCK 3111 T/C SNP influences sleep and activity patterns in patients affected by bipolar depression. *Am. J. Med. Genet. B Neuropsychiatr. Genet.* **2007**, *144B*, 631–635. [CrossRef] [PubMed]
17. Serretti, A.; Benedetti, F.; Mandelli, L.; Lorenzi, C.; Pirovano, A.; Colombo, C.; Smeraldi, E. Genetic dissection of psychopathological symptoms insomnia in mood disorders and CLOCK gene polymorphism. *Am. J. Med. Genet. B Neuropsychiatr. Genet.* **2003**, *121B*, 35–38. [CrossRef] [PubMed]

18. Scott, A.J.; Monk, T.H.; Brink, L.L. Shiftwork as a Risk Factor for Depression: A Pilot Study. *Int. J. Occup. Environ. Health* **1997**, *3*, S2–S9. [PubMed]
19. Bildt, C.; Michelsen, H. Gender differences in the effects from working conditions on mental health: A 4-year follow-up. *Int. Arch. Occup. Environ. Health* **2002**, *75*, 252–258. [CrossRef] [PubMed]
20. Costa, G. Shift work and occupational medicine: An overview. *Occup. Med.* **2003**, *53*, 83–88. [CrossRef]
21. Muecke, S. Effects of rotating night shifts: Literature review. *J. Adv. Nurs.* **2005**, *50*, 433–439. [CrossRef] [PubMed]
22. Bara, A.C.; Arber, S. Working shifts and mental health–findings from the British Household Panel Survey (1995–2005). *Scand. J. Work Environ. Health* **2009**, *35*, 361–367. [CrossRef] [PubMed]
23. Pallesen, S.; Bjorvatn, B.; Mageroy, N.; Saksvik, I.B.; Waage, S. Measures to counteract the negative effects of night work. *Scand. J. Work Environ. Health* **2010**, *36*, 109–120. [CrossRef] [PubMed]
24. Harma, M.; Kecklund, G. Shift work and health—How to proceed? *Scand. J. Work Environ. Health* **2010**, *36*, 81–84. [CrossRef] [PubMed]
25. James, F.O.; Cermakian, N.; Boivin, D.B. Circadian rhythms of melatonin, cortisol, and *clock* gene expression during simulated night shift work. *SLEEP* **2007**, *30*, 11–1427. [CrossRef]
26. Husse, J.; Hintze, S.C.; Eichele, G.; Lehnert, H.; Oster, H. Circadian Clock Genes *Per1* and *Per2* Regulate the Response of Metabolism-Associated Transcripts to Sleep Disruption. *PLoS ONE* **2012**, *7*, 12–52983. [CrossRef] [PubMed]
27. Taniyama, Y.; yamauchi, T.; takeuchi, S.; Kuroda, Y. *PER1* polymorphism associated with shift work disorder. *Sleep Biol. Rhythm.* **2015**, *13*, 342–347. [CrossRef]
28. Duffy, J.F.; Kronauer, R.E.; Czeisler, C.A. Phase-shifting human circadian rhythms: Influence of sleep timing, social contact and light exposure. *J. Physiol.* **1996**, *495*, 289–297. [CrossRef] [PubMed]
29. Dawson, D.; Lack, L.; Morris, M. Phase resetting of the human circadian pacemaker with use of a single pulse of bright light. *Chronobiol. Int.* **1993**, *10*, 94–102. [CrossRef] [PubMed]
30. Honman, K.; Honman, S.; Nakamura, K.; Sasaki, M.; Endo, T.; Takahashi, T. Differential effects of bright light and social cues on reentrainment of human circadian rhythms. *Am. J. Physiol.* **1995**, *268*, R528–R535.
31. Klerman, E.B.; Rimmer, D.W.; Dijk, D.J.; Kronauer, R.E.; Rizzo, J.F.I.; Czeisler, C.A. Nonphotic entrainment of the human circadian pacemaker. *Am. J. Physiol.* **1998**, *274*, R991. [PubMed]
32. Reinberg, A.; Touitou, Y.; Restoin, A.; Migraine, C.; Levi, F.; Montagner, H. The genetic background of circadian and ultradian rhythm patterns of 17-hydroxycorticosteroids: A cross-twin study. *J. Endocrinol.* **1985**, *105*, 247–253. [CrossRef] [PubMed]
33. Moore, R.Y. Circadian rhythms: Basic neurobiology and clinical applications. *Annu. Rev. Med.* **1997**, *48*, 253–266. [CrossRef] [PubMed]
34. Reppert, S.M.; Weaver, D.R. Coordination of circadian timing in mammals. *Nature* **2002**, *418*, 935–941. [CrossRef] [PubMed]
35. Johnson, C.H. Testing the Adaptive Value of Circadian Systems. *Methods Enzymol.* **2005**, *393*, 818–837. [PubMed]
36. Pevet, P.; Challet, E. Melatonin: Both master clock output and internal time-giver in the circadian clock network. *J. Physiol. Paris* **2011**, *105*, 170–182. [CrossRef] [PubMed]
37. Konopka, R.J.; Benzer, S. Clock mutants of Drosophila melanogaster. *Proc. Natl. Acad. Sci. USA* **1971**, *58*, 2112–2116. [CrossRef]
38. Edery, I.; Rutila, J.E.; Rosbash, M. Phase shifting of the circadian clock by induction of the Drosophila period protein. *Science* **1994**, *263*, 237–240. [CrossRef] [PubMed]
39. Sehgal, A.; Price, J.L.; Man, B.; Young, M.W. Loss of circadian behavioral rhythms and per RNA oscillations in the Drosophila mutant timeless. *Science* **1994**, *263*, 1603–1606. [CrossRef] [PubMed]
40. Gekakis, N.; Saez, L.; Delahaye-Brown, A.M.; Myers, M.P.; Sehgal, A.; Young, M.W.; Weitz, C.J. Isolation of timeless by PER protein interaction: Defection interaction between timeless protein and long-period mutant *per1*. *Science* **1995**, *270*, 811–815. [CrossRef] [PubMed]
41. Myers, M.; Wager-Smith, K.; Rothenfluh-Hilfiker, A.; Young, M. Light-induced degradation of TIMELESS and entrainment of the Drosophila circadian clock. *Science* **1996**, *271*, 1736–1740. [CrossRef] [PubMed]
42. Vitaterna, M.H.; King, D.P.; Chang, A.M.; Kornhauser, J.M.; Lowrey, J.D. Mutagenesis and mapping of a mouse gene, Clock, essential for circadian behavior. *Science* **1994**, *264*, 719–725. [CrossRef] [PubMed]

43. Pevet, P.; Bothorel, B.; Slotten, H.; Saboureau, M. The chronobiotic properties of melatonin. *Cell. Tissue Res.* **2002**, *309*, 183–191. [CrossRef] [PubMed]
44. Slotten, H.A.; Krekling, S.; Sicard, B.; Pévet, P. Daily infusion of melatonin entrains circadian activity rhythms in the diurnal rodent Arvicanthis ansorgei. *Behav. Brain Res.* **2002**, *133*, 11–19. [CrossRef]
45. Slotten, H.A.; Pitrosky, B.; Krekling, S.; Pévet, P. Entrainment of circadian activity rhythms in rats to melatonin administered at T cycles different from 24 hours. *Neurosignals* **2002**, *11*, 73–80. [CrossRef] [PubMed]
46. Johnston, J.D.; Messager, S.; Barrett, P.; Hazlerigg, D.G. Melatonin action in the pituitary: Neuroendocrine synchronizer and developmental modulator? *J. Neuroendocrinol.* **2003**, *15*, 405–408. [CrossRef] [PubMed]
47. Dardente, H.; Menet, J.S.; Poirel, V.J.; Streincher, D.; Gauer, F.; Vivien-Roels, B.; Klosen, P.; Pévet, P.; Masson-Mévet, M. Melatonin induces *Cry1* expression in the pars tuberalis of the rat. *Mol. Brain Res.* **2003**, *114*, 101–106. [CrossRef]
48. Von Gall, C.; Weaver, D.R.; Moek, J.; Jilg, A.; Stehle, J.H.; Korf, H.W. Melatonin plays a crucial role in the regulation of rhythmic *clock* gene expression in the mouse pars tuberalis. *Ann. N. Y. Acad. Sci.* **2005**, *1040*, 508–511. [CrossRef] [PubMed]
49. Stehle, J.H.; von Gall, C.; Korf, H.W. Organisation of the circadian system in melatonin- proficient C3H and melatonin-deficient C57BL mice: A comparative investigation. *Cell. Tissue Res.* **2002**, *309*, 173–182. [CrossRef] [PubMed]
50. Von Gall, C.; Garabette, M.L.; Kell, C.A.; Frenzel, S.; Dehghani, F.; Schumm-Draeger, P.M.; Weaver, D.R.; Korf, H.W.; Hastings, M.H.; Stehle, J.H. Rhythmic gene expression in pituitary depends on heterologous sensitization by the neurohormone melatonin. *Nat. Neurosci.* **2002**, *5*, 234–238. [CrossRef] [PubMed]
51. Messager, S.; Garabette, M.L.; Hastings, M.H.; Hazlerigg, D.G. Tissue-specific abolition of Per1 expression in the pars tuberalis by pinealectomy in the Syrian hamster. *Neuroreport* **2001**, *12*, 579–582. [CrossRef] [PubMed]
52. Agez, L.; Laurent, V.; Guerrero, H.Y.; Pévet, P.; Masson-Pévet, M.; Gauer, F. Endogenous melatonin provides an effective circadian message to both the suprachiasmatic nuclei and the pars tuberalis of the rat. *J. Pineal Res.* **2009**, *46*, 95–105. [CrossRef] [PubMed]
53. Zeman, M.; Herichova, I. Melatonin and *clock* genes expression in the cardiovascular system. *Front. Biosci.* **2013**, *5*, 743–753. [CrossRef]
54. Johnston, J.D.; Tournier, B.B.; Andersson, H.; Masson-Pevet, M.; Lincoln, G.A.; Hazlerigg, D.G. Multiple effects of melatonin on rhythmic *clock* gene expression in the mammalian pars tuberalis. *Endocrinology* **2006**, *147*, 959–965. [CrossRef] [PubMed]
55. Agez, L.; Laurent, V.; Pevet, P.; Gauer, F. Melatonin affects nuclear orphan receptors mRNA in the rat suprachiasmatic nuclei. *Neuroscience* **2007**, *144*, 522–530. [CrossRef] [PubMed]
56. Alonso-Vale, M.I.; Andreotti, S.; Mukai, P.Y.; Borges-Silva, C.D.; Peres, S.B.; Cipolla-Neto, J.; Lima, F.B. Melatonin and the circadian entrainment of metabolic and hormonal activities in primary isolated adipocytes. *J. Pineal Res.* **2008**, *45*, 422–429. [CrossRef] [PubMed]
57. Kennaway, D.J.; Owens, J.A.; Voultsios, A.; Wight, N. Adipokines and adipocyte function in Clock mutant mice that retain melatonin rhythmicity. *Obesity* **2012**, *20*, 295–305. [CrossRef] [PubMed]
58. Delezie, J.; Dumont, S.; Dardente, H.; Oudart, H.; Gréchez-Cassiau, A.; Klosen, P.; Teboul, M.; Delaunay, F.; Pévet, P.; Challet, E. The nuclear receptor REV-ERBα is required for the daily balance of carbohydrate and lipid metabolism. *FASEB J.* **2012**, *26*, 3321–3335. [CrossRef] [PubMed]
59. Xiang, S.; Mao, L.; Duplessis, T.; Yuan, L.; Dauchy, R.; Dauchy, E.; Blask, D.E.; Frasch, T.; Hill, S.M. Oscillation of clock and clock controlled genes induced by serum shock in human breast epithelial and breast cancer cells: Regulation by melatonin. *Breast Cancer Epub.* **2012**, *6*, 137–150.
60. Torres-Farfan, C.; Mendez, N.; Abarzua-Catalan, L.; Vilches, N.; Valenzuela, G.J.; Seron-Ferre, M. A circadian clock entrained by melatonin is ticking in the rat fetal adrenal. *Endocrinology* **2011**, *152*, 1891–1900. [CrossRef] [PubMed]
61. Steeves, T.D.; King, D.P.; Zhao, Y.; Sangoram, A.M.; Du, F.; Bowcock, A.M.; Moore, R.Y.; Takahashi, J.S. Molecular cloning and characterization of the human *CLOCK* gene: Expression in the suprachiasmatic nuclei. *Genomics* **1999**, *57*, 189–200. [CrossRef] [PubMed]
62. Wisor, J.P.; Pasumarthi, R.; Gerashchenko, D.; Thompson, C.; Pathak, S.; Sancar, A.; Kilduff, T.S. Sleep Deprivation Effects on Circadian *Clock* Gene Expression in the Cerebral Cortex Parallel Electroencephalographic Differences Among Mouse Strains. *J. Neurosci.* **2008**, *28*, 28–7193. [CrossRef] [PubMed]

63. Cedernaes, J.; Osler, M.E.; Voisin, S.; Broman, J.; Vogel, H.; Dickson, S.L.; Zierath, J.R.; Schiöth, H.B.; Benedict, C. Acute sleep loss induces tissue-specific epigenetic and transcriptional alterations to circadian clock genes in men. *J. Clin. Endocrinol. Metab.* **2015**, *100*, 9. [CrossRef] [PubMed]

64. Franken, P.; Thomason, R.; Heller, H.C.; O'Hara, B.F. A non-circadian role for clock-genes in sleep homeostasis: A strain comparison. *BMC Neurosci.* **2007**, *8*, 87. [CrossRef] [PubMed]

65. Franken, P.; Dijk, D.J. Circadian *clock* genes and sleep homeostasis. *Eur. J. Neurosci.* **2009**, *29*, 1820–1829. [CrossRef] [PubMed]

66. Panda, S.; Antoch, M.P.; Miller, B.H.; Su, A.I.; Schook, A.B.; Straume, M.; Schultz, P.G.; Kay, S.A.; Takahashi, J.S.; Hogenesch, J.B. Coordinated transcription of key pathways in the mouse by the circadian clock. *Cell* **2002**, *109*, 307–320. [CrossRef]

67. Hardin, P.E. Activating inhibitors and inhibiting activators: A day in the life of a fly. *Curr. Opin. Neurobiol.* **1998**, *8*, 642–647. [CrossRef]

68. Dunlap, J.C. Molecular bases for circadian clocks. *Cell* **1999**, *96*, 271–290. [CrossRef]

69. Lowrey, P.L.; Takahashi, J.S. Genetics of Circadian Rhythms in Mammalian Model Organisms. *Adv. Genet.* **2011**, *74*, 175–230. [PubMed]

70. Lee, C.; Etchegaray, J.P.; Cagampang, F.R.; Loudon, A.S.; Reppert, S.M. Posttranslational mechanisms regulate the mammalian circadian clock. *Cell* **2001**, *107*, 855–867. [CrossRef]

71. Kume, K.; Zylka, M.J.; Sriram, S.; Shearman, L.P.; Weaver, D.R. mCRY1 and mCRY2 are essential components of the negative limb of the circadian clock feedback loop. *Cell* **1999**, *98*, 193–205. [CrossRef]

72. Shearman, L.P.; Sriram, S.; Weaver, D.R.; Maywood, E.S.; Chaves, I. Interacting molecular loops in the mammalian circadian clock. *Science* **2000**, *288*, 1013–1019. [CrossRef] [PubMed]

73. Marcheva, B.; Ramsey, K.M.; Buhr, E.D.; Kobayashi, Y.; Su, H.; Ko, C.H.; Bass, J. Disruption of the clock components CLOCK and BMAL1 leads to hypoinsulinaemia and diabetes. *Nature* **2010**, *466*, 627–631. [CrossRef] [PubMed]

74. Janich, P.; Pascual, G.; Merlos-Suárez, A. The circadian molecular clock creates epidermal stem cell heterogeneity. *Nature* **2011**, *480*, 209–214. [CrossRef] [PubMed]

75. Paschos, G.K.; Ibrahim, S.; Song, W.; Kunieda, T.; Grant, G.; Reyes, T.M.; Fitzgerald, G.A. Obesity in mice with adipocyte-specific deletion of clock component Arntl. *Nat. Med.* **2012**, *18*, 12–1768. [CrossRef] [PubMed]

76. Takahashi, J.S.; Hong, H.K.; Ko, C.H.; McDearmon, E.L. The genetics of mammalian circadian order and disorder: Implications for physiology and disease. *Nat. Rev. Genet.* **2008**, *9*, 764–775. [CrossRef] [PubMed]

77. Jiang, F.; Van Dyke, R.D.; Zhang, J.; Li, F.; Gozal, D.; Shen, X. Effect of chronic sleep restriction on sleepiness and working memory in adolescents and young adults. *J. Clin. Exp. Neuropsychol.* **2011**, *33*, 892–900. [CrossRef] [PubMed]

78. Wang, G.; Grone, B.; Colas, D.; Appelbaum, L.; Mourrain, P. Synaptic plasticity in sleep: Learning, homeostasis and disease. *Trends Neurosci.* **2011**, *34*, 452–463. [CrossRef] [PubMed]

79. Berger, R.H.; Miller, A.L.; Seifer, R.; Cares, S.R.; Le Bourgeois, M.K. Acute sleep restriction effects on emotion responses in 30- to 36-month-old children. *J. Sleep Res.* **2012**, *21*, 235–246. [CrossRef] [PubMed]

80. Deliens, G.; Gilson, M.; Peigneux, P. Sleep and the processing of emotions. *Exp. Brain Res.* **2014**, *232*, 5–1403. [CrossRef] [PubMed]

81. Coogan, A.N.; Baird, A.L.; Popa-Wagner, A.; Thome, J. Circadian rhythms and attention deficit hyperactivity disorder: The what, the when and the why. *Prog. Neuropsychopharmacol. Biol. Psychiatry.* **2016**, *67*, 74–81. [CrossRef] [PubMed]

82. Seegers, V.; Touchette, E.; Dionne, G.; Petit, D.; Seguin, J.R.; Montplaisir, J.; Vitaro, F.; Falissard, B.; Boivin, M.; Tremblay, R.E. Short persistent sleep duration is associated with poor receptive vocabulary performance in middle childhood. *J. Sleep Res.* **2016**, *25*, 325–332. [CrossRef] [PubMed]

83. Tordjman, S.; Davlantis, K.S.; Georgieff, N.; Geoffray, M.M.; Speranza, M.; Anderson, G.M.; Xavier, J.; Botbol, M.; Oriol, C.; Bellissant, E.; et al. Autism as a disorder of biological and behavioral rhythms: Toward new therapeutic perspectives. *Front. Pediatr.* **2015**, *3*, 1. [CrossRef] [PubMed]

84. Tordjman, S.; Najjar, I.; Bellissant, E.; Anderson, G.M.; Barburoth, M.; Cohen, D.; Jaafari, N.; Schischmanoff, O.; Fagard, O.; Lagdas, I.; et al. Advances in the research of melatonin in autism spectrum disorders: Literature review and new perspectives. *Int. J. Mol. Sci.* **2013**, *14*, 20508–20542. [CrossRef] [PubMed]

85. Kobayashi, Y.; Ye, Z.; Hensch, T.K. Clock genes control cortical critical period timing. *Neuron* **2015**, *86*, 264–275. [CrossRef] [PubMed]

86. Geoffray, M.M.; Nicolas, A.; Speranza, M.; Georgieff, N. Are circadian rhythms new pathways to understand Autism spectrum disorder? *J. Neuropsychol* **2016**, in press.

87. Wöhr, M.; Orduz, D.; Gregory, P.; Moreno, H.; Khan, U.; Vörckel, K.J.; Wolfer, D.P.; Welzl, H.; Gall, D.; Schiffmann, S.N.; et al. Lack of parvalbumin in mice leads to behavioral deficits relevant to all human autism core symptoms and related neural. *Transl. Psychiatry* **2015**, *5*, 525. [CrossRef] [PubMed]

88. Wulff, K.; Dijk, D.J.; Middleton, B.; Foster, R.G.; Joyce, E.M. Sleep and circadian rhythm disruption in schizophrenia. *Br. J. Psychiatry J. Ment. Sci.* **2012**, *200*, 308–316. [CrossRef] [PubMed]

89. Nicholas, B.; Rudrasingham, V.; Nash, S.; Kirov, G.; Owen, M.J.; Wimpory, D.C. Association of *Per1* and *Npas2* with autistic disorder: Support for the clock genes/social timing hypothesis. *Mol. Psychiatry* **2007**, *12*, 581–592. [CrossRef] [PubMed]

90. Kissling, C.; Retz, W.; Wiemann, S.; Coogan, AN.; Clement, RM.; Hunnerkopf, R.; Conner, A.C.; Freitag, C.M.; Rösler, M.; Thome, J. A polymorphism at the 3′-untranslated region of the *CLOCK* gene is associated with adult attention-deficit hyperactivity disorder. *Am. J. Med. Genet. B Neuropsychiatr. Genet.* **2008**, *147*, 333–338. [CrossRef] [PubMed]

91. Xu, X.; Breen, G.; Chen, C.K.; Huang, Y.S.; Wu, Y.Y.; Asherson, P. Association study between a polymorphism at the 3′-untranslated region of *CLOCK* gene and attention deficit hyperactivity disorder. *Behav. Brain Funct.* **2010**, *6*, 48. [CrossRef] [PubMed]

92. Sipila, T.; Kananen, L.; Greco, D.; Donner, J.; Silander, K.; Terwilliger, J.D. An association analysis of circadian genes in anxiety disorders. *Biol. Psychiatry* **2010**, *67*, 1163–1170. [CrossRef] [PubMed]

93. Partonen, T.; Treutlein, J.; Alpman, A.; Frank, J.; Johansson, C.; Depner, M.; Aron, L.; Rietschel, M.; Wellek, S.; Soronen, P. Three circadian clock genes Per2, Arntl, and Npas2 contribute to winter depression. *Ann. Med.* **2007**, *39*, 229–238. [CrossRef] [PubMed]

94. Lavebratt, C.; Sjöholm, L.K.; Soronen, P.; Paunio, T.; Vawter, M.P.; Bunney, W.E.; Adolfsson, R.; Forsell, Y.; Wu, J.C.; Kelsoe, J.R.; et al. CRY2 Is Associated with Depression. *PLoS ONE* **2010**, *5*, e9407. [CrossRef] [PubMed]

95. Soria, V.; Martinez-Amoros, E.; Escaramis, G.; Valero, J.; Perez-Egea, R.; Garcia, C. Differential association of circadian genes with mood disorders: CRY1 and NPAS2 are associated with unipolar major depression and CLOCK and VIP with bipolar disorder. *Neuropsychopharmacology* **2010**, *35*, 1279–1289. [CrossRef] [PubMed]

96. Nievergelt, C.M.; Kripke, D.F.; Barrett, T.B.; Burg, E.; Remick, R.A.; Sadovnick, A.D. Suggestive evidence for association of the circadian genes PERIOD3 and ARNTL with bipolar disorder. *Am. J. Med. Genet. B Neuropsychiatr. Genet.* **2006**, *141B*, 234–241. [CrossRef] [PubMed]

97. Shi, J.; Wittke-Thompson, J.K.; Badner, J.A.; Hattori, E.; Potash, J.B.; Willour, V.L. Clock genes may influence bipolar disorder susceptibility and dysfunctional circadian rhythm. *Am. J. Med. Genet. B Neuropsychiatr. Genet.* **2008**, *147B*, 1047–1055. [CrossRef] [PubMed]

98. Aston, C.; Jiang, L.; Sokolov, B.P. Microarray analysis of postmortem temporal cortex from patients with schizophrenia. *J. Neurosci. Res.* **2004**, *77*, 858–866. [CrossRef] [PubMed]

99. Mansour, H.A.; Wood, J.; Logue, T.; Chowdari, K.V.; Dayal, M.; Kupfer, D.J.; Monk, T.H.; Devlin, B.; Nimgaonkar, V.L. Association study of eight circadian genes with bipolar I disorder, schizoaffective disorder and schizophrenia. *Genes Brain Behav.* **2006**, *5*, 150–157. [CrossRef] [PubMed]

100. Sun, H.Q.; Li, S.X.; Chen, F.B.; Zhang, Y.; Li, P.; Jin, M.; Sun, Y.; Wang, F.; Mi, W.F.; Shi, L.; et al. Diurnal neurobiological alterations after exposure to clozapine in first-episode schizophrenia patients. *Psychoneuroendocrinology* **2016**, *64*, 108–116. [CrossRef] [PubMed]

101. Johansson, A.S.; Owe-Larsson, B.; Hetta, J.; Lundkvist, G.B. Altered circadian clock gene expression in patients with schizophrenia. *Schizophr. Res.* **2016**, *174*, 17–23. [CrossRef] [PubMed]

102. Buhr, E.D.; Takahashi, J.S. Molecular components of the Mammalian circadian clock. *Handb. Exp. Pharmacol.* **2013**, *217*, 3–27.

103. Oster, H.; Damerow, S.; Kiessling, S.; Jakubcakova, V.; Abraham, D.; Tian, J.; Hoffmann, M.W.; Eichele, G. The circadian rhythm of glucocorticoids is regulated by a gating mechanism residing in the adrenal cortical clock. *Cell. Metab.* **2006**, *4*, 163–173. [CrossRef] [PubMed]

104. Son, G.H.; Chung, S.; Choe, H.K.; Kim, H.D.; Baik, S.M.; Lee, H.; Lee, H.W.; Choi, S.; Sun, W.; Kim, H.; et al. Adrenal peripheral clock controls the autonomous circadian rhythm of glucocorticoid by causing rhythmic steroid production. *Proc. Natl. Acad. Sci. USA* **2008**, *105*, 20970–20975. [CrossRef] [PubMed]

105. Yamamoto, T.; Nakahata, Y.; Tanaka, M.; Yoshida, M.; Soma, H.; Shinohara, K.; Yasuda, A.; Mamine, T.; Takumi, T. Acute physical stress elevates mouse PERIOD1 mRNA expression in mouse peripheral tissues via a glucocorticoid-responsive element. *J. Biol. Chem.* **2005**, *280*, 42036–42043. [CrossRef] [PubMed]

106. Dulcis, D.; Jamshidi, P.; Leutgeb, S.; Spitzer, N.C. Neurotransmitter switching in the adult brain regulates behavior. *Science* **2013**, *340*, 449–453. [CrossRef] [PubMed]

107. Jackson, A.; Cavanagh, J.; Scott, J. A systematic review of manic and depressive prodromes. *J. Affect. Disord.* **2003**, *74*, 209–217. [CrossRef]

108. American Psychiatric Association. *Diagnostic and Statistical Manual of Mental Disorders*, 5th ed.; American Psychiatric Association: Washington, DC, USA, 2015.

109. Millar, A.; Espie, C.A.; Scott, J. The sleep of remitted bipolar outpatients: A controlled naturalistic study using actigraphy. *J. Affect. Disord.* **2004**, *80*, 145–153. [CrossRef]

110. Mansour, H.A.; Wood, J.; Chowdari, K.V.; Dayal, M.; Thase, M.E.; Kupfer, D.J.; Monk, T.H.; Devlin, B.; Nimgaonkar, V.L. Circadian phase variation in bipolar I disorder. *Chronobiol. Int.* **2005**, *22*, 571–584. [CrossRef] [PubMed]

111. McClung, C.A. Role for the *Clock* gene in bipolar disorder. *Cold Spring Harb. Symp. Quant. Biol.* **2007**, *72*, 637–644. [CrossRef] [PubMed]

112. Klemfuss, H. Rhythms and the pharmacology of lithium. *Pharmacol. Ther.* **1992**, *56*, 53–78. [CrossRef]

113. Abe, M.; Herzog, E.D.; Block, G.D. Lithium lengthens the circadian period of individual suprachiasmatic nucleus neurons. *Neuroreport* **2000**, *11*, 3261–3264. [CrossRef] [PubMed]

114. Yin, L.; Wang, J.; Klein, P.S.; Lazar, M.A. Nuclear receptor REV-ERBα is a critical lithium-sensitive component of the circadian clock. *Science* **2006**, *311*, 1002–1005. [CrossRef] [PubMed]

115. Li, J.; Lu, W.Q.; Beesley, S.; Loudon, A.S.; Meng, Q.J. Lithium impacts on the amplitude and period of the molecular circadian clockwork. *PLoS ONE* **2012**, *7*, 33292. [CrossRef] [PubMed]

116. Kaladchibachi, S.A.; Doble, B.; Anthopoulos, N.; Woodgett, J.R.; Manoukian, A.S. Glycogen synthase kinase 3, circadian rythms, and bipolar disorder: A molecular link in the therapeutic action of lithium. *J. Circadian Rythm.* **2007**, *5*, 3. [CrossRef] [PubMed]

117. Gould, T.D.; Manji, H.K. Glycogen synthase kinase-3: A putative molecular target for lithium mimetic drugs. *Neuropsychopharmacology* **2005**, *30*, 1223–1237. [CrossRef] [PubMed]

118. Iitaka, C.; Miyazaki, K.; Akaike, T.; Ishida, N. A role for glycogen synthase kinase-3β in the mammalian circadian clock. *J. Biol. Chem.* **2005**, *280*, 29397–29402. [CrossRef] [PubMed]

119. Serretti, A.; Cusin, C.; Benedetti, F.; Mandelli, L.; Pirovano, A.; Zanardi, R.; Colombo, C.; Smeraldi, E. Insomnia improvement during antidepressant treatment and *CLOCK* gene polymorphism. *Am. J. Med. Genet. B Neuropsychiatr. Genet.* **2005**, *137B*, 36–39. [CrossRef] [PubMed]

120. Karthikeyan, K.; Marimuthu, G.; Ramasubramanian, C.; Arunachal, G.; Bahammam, A.; Spence, D.W.; Cardinali, D.P.; Brown, G.M.; Pandi-Perumal, S.R. Association of *Per3* length polymorphism with bipolar disorder and schizophrenia. *Neuropsychiatr. Dis. Treat.* **2014**, *10*, 2325–2330. [PubMed]

121. Sjöholm, L.; Backlund, L.; Cheteh, E.H.; Ek, I.R.; Frisén, L.; Schalling, M.; Ösby, U.; Lavebratt, C.; Nikamo, P. CRY2 is associated with rapid cycling in bipolar disorder patients. *PLoS ONE* **2010**, *5*, e12632. [CrossRef] [PubMed]

122. McGrath, C.L.; Glatt, S.L.; Sklar, P.; Le-Niculescu, H.; Kuczenski, R.; Doyle, A.E.; Biederman, J.; Mick, E.; Faraone, S.V.; Niculescu, A.B.; et al. Evidence for genetic association of RORB with bipolar disorder. *BMC Psychiatry* **2009**, *9*, 70. [CrossRef] [PubMed]

123. Kato, T. Molecular genetics of bipolar disorder and depression. *Psychiatry Clin. Neurosci.* **2007**, *61*, 3–19. [CrossRef] [PubMed]

124. Kato, T.; Kakiuchi, C.; Iwamoto, K. Comprehensive gene expression analysis in bipolar disorder. *Can. J. Psychiatry* **2007**, *52*, 763–771. [PubMed]

125. McClung, C.A. Circadian genes, rhythm and biology of mood disorder. *Pharmacol. Ther.* **2007**, *114*, 222–232. [CrossRef] [PubMed]

126. DeBundel, D.; Gangarossa, G.; Biever, A.; Bonnefont, X.; Valjent, E. Cognitive dysfunction, elevated anxiety, and reduced cocaine response in circadian clock-deficient cryptochrome knockout mice. *Front. Behav. Neurosci.* **2013**, *7*, 152.

127. Li, J.Z.; Bunney, B.G.; Meng, F.; Hagenauer, M.H.; Walsh, D.M.; Vawter, M.P.; Evans, S.J.; Choudary, P.V.; Cartagena, P.; Barchas, J.D.; et al. Circadian patterns of gene expression in the human brain and disruption in major depressive disorder. *Proc. Natl. Acad. Sci. USA* **2013**, *110*, 9950–9955. [CrossRef] [PubMed]

128. Kripke, D.F.; Nievergelt, C.M.; Joo, E.; Shekhtman, T.; Kelsoe, J.R. Circadian polymorphisms associated with affective disorders. *J. Circadian Rhythm.* **2009**, *7*, 2. [CrossRef] [PubMed]

129. Wang, J.; Nuccio, S.R.; Yang, J.Y.; Wu, X.; Bogoni, A.; Willner, A.E. High-speed addition/subtraction/ complement/doubling of quaternary numbers using optical nonlinearities and DQPSK signals. *Opt. Lett.* **2012**, *37*, 1139–1141. [CrossRef] [PubMed]

130. Kovanen, L.; Kaunisto, M.; Donner, K.; Saarikoski, S.T.; Partonen, T. CRY2 genetic variants associate with dysthymia. *PLoS ONE* **2013**, *8*, 71450. [CrossRef] [PubMed]

131. Hua, P.; Liu, W.; Chen, D.; Zhao, Y.; Chen, L.; Zhang, N.; Wang, C.; Guo, S.; Wang, L.; Xiao, H.; et al. *Cry1* and *Tef* gene polymorphisms are associated with major depressive disorder in the Chinese population. *J. Affect. Disord.* **2014**, *157*, 100–103. [CrossRef] [PubMed]

132. Shi, S.Q.; White, M.J.; Borsetti, H.M.; Pendergast, J.S.; Hida, A.; Ciarleglio, C.M.; de Verteuil, P.A.; Cadar, A.G.; Cala, C.; McMahon, D.G.; et al. Molecular analyses of circadian gene variants reveal sex-dependent links between depression and clocks. *Transl. Psychiatry* **2016**, *6*, e748. [CrossRef] [PubMed]

133. Pirovano, A.; Lorenzi, C.; Serretti, A.; Ploia, C.; Landoni, S.; Catalano, M.; Smeraldi, E. Two new rare variants in the circadian "clock" gene may influence sleep pattern. *Genet. Med.* **2005**, *7*, 455–457. [CrossRef] [PubMed]

134. Desan, P.H.; Oren, D.A.; Malison, R.; Price, L.H.; Rosenbaum, J.; Smoller, J.; Charney, D.S.; Gelernter, J. Genetic polymorphism at the *CLOCK* gene locus and major depression. *Am. J. Med. Genet.* **2000**, *96*, 418–421. [CrossRef]

135. Serretti, A.; Gaspar-Barba, E.; Calati, R.; Cruz-Fuentes, C.S.; Gomez-Sanchez, A.; Perez-Molina, A.; de Ronchi, D. 3111T/C clock gene polymorphism is not associated with sleep disturbances in untreated depressed patients. *Chronobiol. Int.* **2010**, *27*, 265–277. [CrossRef] [PubMed]

136. Jones, C.R.; Campbell, S.S.; Zone, S.E.; Cooper, F.; DeSano, A.; Murphy, P.J.; Jones, B.; Czajkowski, L.; Ptácek, L.J. Familial advanced sleep-phase syndrome: A short-period circadian rhythm variant in humans. *Nat. Med.* **1999**, *5*, 1062–1065. [PubMed]

137. Toh, K.L.; Jones, C.R.; He, Y.; Eide, E.J.; Hinz, W.A.; Virshup, D.M.; Ptácek, L.J.; Fu, Y.H. An hPer2 phosphorylation site mutation in familial advanced sleep-phase syndrome. *Science* **2001**, *291*, 1040–1043. [CrossRef] [PubMed]

138. Xu, Y.; Padiath, Q.S.; Shapiro, R.E.; Jones, C.R.; Wu, S.C.; Saigoh, N.; Saigoh, K.; Ptacek, L.J.; Fu, Y.H. Functional consequences of a CKIδ mutation causing familial advanced sleep phase syndrome. *Nature* **2005**, *434*, 640–644. [CrossRef] [PubMed]

139. Meng, Q.-J.; Logunova, L.; Maywood, E.S.; Gallego, M.; Lebiecki, J.; Brown, T.M.; Loudon, A.S.I. Setting clock speed in mammals: The CK1ε *tau* mutation in mice accelerates the circadian pacemaker by selectively destabilizing PERIOD proteins. *Neuron* **2008**, *58*, 78–88. [CrossRef] [PubMed]

140. Xu, Y.; Toh, K.L.; Jones, C.R.; Shin, J.Y.; Fu, Y.H. Modeling of a human circadian mutation yields insights into clock regulation by PER2. *Cell* **2007**, *128*, 59–70. [CrossRef] [PubMed]

141. Hirano, A.; Shi, G.; Jones, C.R.; Lipzen, A.; Pennacchio, L.A.; Xu, Y.; Hallows, W.C.; McMahon, T.; Yamazaki, M.; Ptáček, L.J.; et al. A Cryptochrome 2 mutation yields advanced sleep phase in humans. *Hum. Biol. Med. Neurosci.* **2016**, *10*, 7554. [CrossRef] [PubMed]

142. Wehr, T.A. A circadian signal of change of season in patients with seasonal affective disorder. *Arch. Gen. Psychiatry* **2001**, *58*, 1108–1114. [CrossRef] [PubMed]

143. Lewy, A.J.; Bauer, V.K.; Cutler, N.L.; Sack, R.L.; Ahmed, S.; Thomas, K.H.; Blood, M.L.; Jackson, J.M. Morning versus eveninglight treatment of patients with winter depression. *Arch. Gen. Psychiatry* **1998**, *55*, 890–896. [CrossRef] [PubMed]

144. Johansson, C.; Willeit, M.; Smedh, C.; Ekholm, J.; Paunio, T.; Kieseppa, T.; Lichtermann, D.; Praschak-Rieder, N.; Neumeister, A.; Nilsson, L.G. Circadian clock-related polymorphisms in seasonal affective disorder and their relevance to diurnal preference. *Neuropsychopharmacology* **2003**, *28*, 734–739. [CrossRef] [PubMed]

145. Iwase, T.; Kajimura, N.; Uchiyama, M.; Ebisawa, T.; Yoshimura, K.; Kamei, Y.; Shibui, K.; Kim, K.; Kudo, Y.; Katoh, M.; et al. Mutation screening of the human *Clock* gene in circadian rhythm sleep disorders. *Psychiatry Res.* **2002**, *109*, 121–128. [CrossRef]

146. Takimoto, M.; Hamada, A.; Tomoda, A.; Ohdo, S.; Ohmura, T.; Sakato, H.; Kawatani, J.; Jodoi, T.; Nakagawa, H.; Terazono, H.; et al. Daily expression of clock genes in whole blood cells in healthy subjects and a patient with circadian rhythm sleep disorder. *Am. J. Physiol.* **2005**, *289*, R1273–R1279. [CrossRef] [PubMed]

147. Utge, S.J.; Soronen, P.; Loukola, A.; Kronholm, E.; Ollila, H.M.; Pirkola, S.; Porkka-Heiskanen, T.; Partonen, T.; Paunio, T. Systematic analysis of circadian genes in a population-based sample reveals association of TIMELESS with depression and sleep disturbance. *PLoS ONE* **2010**, *5*, e9259. [CrossRef] [PubMed]

148. Konofal, E.; Lecendreux, M.; Mouren-Siméoni, M.C. Mise au point des études cliniques sur le rapport veille-sommeil dans le trouble déficit de l'attention/hyperactivité de l'enfant. *Ann. Medico-Psychol.* **2002**, *2*, 105–117. [CrossRef]

149. Dickerson Mayes, S.; Calhoun, S.L.; Bixler, E.O.; Vgontzas, A.N.; Mahr, F.; Hillwig-Garcia, J.; Elamir, B.; Edhere-Ekezie, L.; Parvin, M. ADHD Subtypes and Comorbid Anxiety, Depression, and Oppositional-Defiant Disorder: Differences in Sleep Problems. *J. Pediatr. Psychol.* **2009**, *34*, 328–337. [CrossRef] [PubMed]

150. Monti, J.M.; BaHammam, A.S.; Pandi-Perumal, S.R.; Bromundt, V.; Spence, D.W.; Cardinali, D.P.; Brown, G.M. Sleep and circadian rhythm dysregulation in schizophrenia. *Prog. Neuro Psychopharmacol. Biol. Psychiatry* **2013**, *43*, 209–216. [CrossRef] [PubMed]

151. Chouinard, S.; Poulin, J.; Stip, E.; Godbout, R. Sleep in untreated patients with schizophrenia: A meta-analysis. *Schizophr. Bull.* **2004**, *30*, 957–967. [CrossRef] [PubMed]

152. Afonso, P.; Figueira, M.L.; Paiva, T. Sleep-promoting action of the endogenous melatonin in schizophrenia compared to healthy controls. *Int. J. Psychiatry Clin. Pract.* **2011**, *15*, 311–315. [CrossRef] [PubMed]

153. Bromundt, V.; Koster, M.; Georgiev-Kill, A.; Opwis, K.; Wirz-Justice, A.; Stoppe, G.; Cajochen, C. Sleep-wake cycles and cognitive functioning in schizophrenia. *Br. J. Psychiatry J. Ment. Sci.* **2011**, *198*, 269–276. [CrossRef] [PubMed]

154. Morgan, R.; Cheadle, A.J. Circadian body temperature in chronic schizophrenia. *Br. J. Psychiatry J. Ment. Sci.* **1976**, *129*, 350–354. [CrossRef]

155. Rao, M.L.; Gross, G.; Strebel, B.; Halaris, A.; Huber, G.; Braunig, P.; Marler, M. Circadian rhythm of tryptophan, serotonin, melatonin, and pituitary hormones in schizophrenia. *Biol. Psychiatry* **1994**, *35*, 151–163. [CrossRef]

156. Lamont, E.W.; Coutu, D.L.; Cermakian, N.; Boivin, D.B. Circadian rhythms and clock genes in psychotic disorders. *Isr. J. Psychiatry Relat. Sci.* **2010**, *47*, 27–35. [PubMed]

157. Takao, T.; Tachikawa, H.; Kawanishi, Y.; Mizukami, K.; Asada, T. *CLOCK* gene T3111C polymorphism is associated with Japanese schizophrenics: A preliminary study. *Eur. Neuropsychopharmacol.* **2007**, *17*, 273–276. [CrossRef] [PubMed]

158. Rivto, E.R.; Ritvo, R.; Yuwiler, A.; Brothers, A.; Freeman, B.J.; Plotkin, S. Elevated daytime melatonin concentrations in autism: A pilot study. *Eur. Child Adolesc. Psychiatry* **1993**, *2*, 75.

159. Nir, I.; Meir, D.; Zilber, N.; Knobler, H.; Hadjez, J.; Lerner, Y. Brief report: Circadian melatonin, thyroid-stimulating hormone, prolactin, and cortisol levels in serum of young adults with autism. *J. Autism Dev. Disord.* **1995**, *25*, 641. [CrossRef] [PubMed]

160. Kulman, G.; Lissoni, P.; Rovelli, F.; Roselli, M.G.; Brivio, F.; Sequeri, P. Evidence of pineal endocrine hypofunction in autistic children. *Neuroendocrinol. Lett.* **2000**, *21*, 31. [PubMed]

161. Tordjman, S.; Anderson, G.M.; Pichard, N.; Charbuy, H.; Touitou, Y. Nocturnal excretion of 6-sulphatoxymelatonin in children and adolescents with autistic disorder. *Biol. Psychiatry* **2005**, *57*, 134. [CrossRef] [PubMed]

162. Tordjman, S.; Anderson, G.M.; Bellissant, E.; Botbol, M.; Charbuy, H.; Camus, F.; Graignic, R.; Kermarrec, S.; Fougerou, C.; Cohen, D.; et al. Day and nighttime excretion of 6-sulphatoxymelatonin in adolescents and young adults with autistic disorder. *Psychoneuroendocrinology* **2012**, *37*, 1990–1997. [CrossRef] [PubMed]

163. Melke, J.; Goubran-Botros, H.; Chaste, P.; Betancur, C.; Nygren, G.; Anckarsäter, H.; Rastam, M.; Ståhlberg, O.; Gillberg, I.C.; Delorme, R.; et al. Abnormal melatonin synthesis in autism spectrum disorders. *Mol. Psychiatry* **2007**, *13*, 90–98. [CrossRef] [PubMed]

164. Freitag, C.M. The genetics of autistic disorders and its clinical relevance: A review of the literature. *Mol. Psychiatry* **2007**, *12*, 2. [CrossRef] [PubMed]

165. Sarowar, T.; Chhabra, R.; Vilella, A.; Boeckers, T.M.; Zoli, M.; Grabrucker, A.M. Activity and circadian rhythm influence synaptic Shank3 protein levels in mice. *J. Neurochem.* **2016**, *138*, 887–895. [CrossRef] [PubMed]

166. Bourgeron, T. The Possible Interplay of Synaptic and Clock Genes in Autism Spectrum Disorders. *Cold Spring Harb. Symp. Quant. Biol.* **2007**, *72*, 645–654. [CrossRef] [PubMed]
167. Wimpory, D.; Nicholas, B.; Nash, S. Social timing, clock genes and autism: A new hypothesis. *J. Intellect. Disabil. Res.* **2002**, *46*, 352. [CrossRef] [PubMed]
168. Yang, Z.; Matsumoto, A.; Nakayama, K.; Jimbo, E.F.; Kojima, K.; Nagata, K.; Iwamoto, S.; Yamagata, T. Circadian-relevant genes are highly polymorphic in autism spectrum disorder patients. *Brain Dev.* **2016**, *38*, 91–99. [CrossRef] [PubMed]
169. Roubertoux, P.L.; Tordjman, S. The autism spectrum disorders (ASD): From the clinics to the molecular analysis. In *Organism Models of Autism Spectrum Disorders*; Roubertoux, P.L., Ed.; Springer: New York, NY, USA, 2015; pp. 29–66.

International Journal of
Molecular Sciences

MDPI

Review

Dysfunctional mTORC1 Signaling: A Convergent Mechanism between Syndromic and Nonsyndromic Forms of Autism Spectrum Disorder?

Juliana Magdalon [1], Sandra M. Sánchez-Sánchez [1,2], Karina Griesi-Oliveira [1] and Andréa L. Sertié [1,*]

[1] Hospital Israelita Albert Einstein, Centro de Pesquisa Experimental, São Paulo 05652-900, Brazil; juliana.magdalon@einstein.br (J.M.); sandra.mabel@einstein.br (S.M.S.S.); karina.griesi@einstein.br (K.G.O.)
[2] Departamento de Genética e Biologia Evolutiva, Instituto de Biociências, Universidade de São Paulo, São Paulo 05508-090, Brazil
* Correspondence: andrea.sertie@einstein.br; Tel.: +55-11-215-112-265; Fax: +55-11-21510273

Academic Editor: Merlin G. Butler
Received: 20 February 2017; Accepted: 14 March 2017; Published: 18 March 2017

Abstract: Whereas autism spectrum disorder (ASD) exhibits striking heterogeneity in genetics and clinical presentation, dysfunction of mechanistic target of rapamycin complex 1 (mTORC1) signaling pathway has been identified as a molecular feature common to several well-characterized syndromes with high prevalence of ASD. Additionally, recent findings have also implicated mTORC1 signaling abnormalities in a subset of nonsyndromic ASD, suggesting that defective mTORC1 pathway may be a potential converging mechanism in ASD pathology across different etiologies. However, the mechanistic evidence for a causal link between aberrant mTORC1 pathway activity and ASD neurobehavioral features varies depending on the ASD form involved. In this review, we first discuss six monogenic ASD-related syndromes, including both classical and potentially novel mTORopathies, highlighting their contribution to our understanding of the neurobiological mechanisms underlying ASD, and then we discuss existing evidence suggesting that aberrant mTORC1 signaling may also play a role in nonsyndromic ASD.

Keywords: mTORC1 signaling pathway; ASD-related syndromes and nonsyndromic/idiopathic ASD; neuronal cell growth; axonal and dendritic morphogenesis; dendritic spine density and maturation; synaptic plasticity; mTORC1-targeted therapies

1. Introduction

Autism Spectrum Disorder (ASD) is among the most common developmental neuropsychiatric disorders and affects about 1 in 68 individuals [1]. It is characterized by impairments in two core domains: deficits in social communication and restricted, repetitive pattern of behavior or interests [2]. The presentation of symptoms is variable, ranging from mild to severe, and usually coexists with other psychiatric and medical conditions. There is a strong male bias in ASD, especially among individuals less severely affected (~4 males/1 female affected) [3,4].

ASD may be part of the clinical presentation of well-characterized genetic syndromes, hereinafter referred to as ASD-related syndromes, such as tuberous sclerosis complex (TSC) [5], fragile X syndrome (FXS) [6], Rett syndrome (RTT) [7,8], Angelman syndrome (AS) [9,10], phosphatase and tensin homolog (*PTEN*)-related syndromes [11], neurofibromatosis type 1 (NF1) [12], Timothy syndrome [13], 22q13.3 deletion syndrome [14], among others. These ASD-related syndromes, although representing only 5%–10% of all ASD cases, have contributed greatly to our understanding of ASD pathogenesis [15–17].

On the other hand, for most ASD cases, hereinafter called nonsyndromic ASD (NS-ASD)—even if additional phenotypic traits are present—to distinguish it from the well-defined ASD-related syndromes, the underlying causes remain unknown. Several twin and family studies have provided indisputable evidence for a genetic component underlying NS-ASD with heritability estimates ranging from 38% to 90% depending on the study parameters [18–20]. Recent high-throughput genomic techniques accompanied by large well-characterized cohorts of patients have identified a large number of rare and common variants for further characterization in relation to ASD [20–23], but several rare de novo and highly penetrant protein coding mutations appear to be sufficiently pathogenic to cause NS-ASD by themselves [24–33]. To date, a list of >200 ASD-risk genes categorized as "high confidence", "strong candidate" and "suggestive evidence" can be found at the Simons Foundation Autism Research Initiative [34]. However, none of these genes accounts individually for more than 1%–2% of all cases of NS-ASD and, collectively, these forms of NS-ASD caused by highly penetrant mutations represent approximately 10%–20% of all cases, highlighting the enormous genetic heterogeneity of the disorder [35].

However, in spite of the high clinical and genetic heterogeneity of ASD, shared mechanisms between ASD-related syndromes and NS-ASD are being discovered and several mutated genes seem to converge on key biological pathways to give rise to ASD relevant symptoms [32,33,36–38]. One such pathway is the mechanistic target of rapamycin complex 1 (mTORC1) signaling cascade (Box 1, see below), which is a vital regulator of translation that impacts numerous cellular processes in the developing and mature brain [39,40]. Dysfunctional mTORC1 signaling has been described in several monogenic ASD-related syndromes, such as TSC, *PTEN*-associated ASD (PTEN-ASD), NF1, FXS [41,42], RTT [43] and AS [44,45]. Patient-derived cells and brain tissues as well as rodent models for these syndromes have been used to dissect the consequences of aberrant mTORC1 signaling in brain structure and function, as well as in particular aspects of cognition and behavior. In addition, recent findings suggest that *methyl-CpG binding protein 2* (*MECP2*) duplication syndrome [46] and *cyclin-dependent kinase-like 5* (*CDKL5*)-related syndrome [47] may also be associated with defective mTORC1 cascade activity; however, this needs further mechanistic exploration. Finally, more recently, dysregulation of mTORC1-dependent signaling, both upstream and downstream of its kinase activity, has also been observed in patients and animal models of NS-ASD of both known and unknown etiologies [42,48–55]. Nonetheless, our understanding of the mechanisms by which unbalanced mTORC1 signaling leads to NS-ASD is far less explored and many questions require further clarification, such as: (1) Is there enough evidence in the literature to support altered mTORC1 signaling in NS-ASD pathogenesis? (2) How large is the proportion of NS-ASD cases that show altered mTORC1 signaling? (3) Can the mechanistic insights gained from studying ASD-related syndromes be extrapolated to NS-ASD? (4) Will the putative mTORC1-targeted therapies that have been found to be effective for treating some ASD-related syndromes benefit at least a subgroup of NS-ASD patients?

Herein, we review the main mechanistic and therapeutic insights gained from studying six ASD-related syndromes with evidence for aberrant activation of mTORC1 pathway, and then discuss recent findings potentially linking mTORC1 signaling dysfunction to NS-ASD. A comparison of the main neuropathological features found in patients and rodent models of these ASD forms can be found in Table 1. Importantly, those phenotypes that were rescued by targeting mTORC1 pathway at different levels are also depicted in Table 1. All animal models described herein display ASD-relevant behaviors, except the *cytoplasmic FMR1 interacting protein 1* (*Cyfip1*) transgenic mice, in which ASD traits were not analyzed [52].

Table 1. Main brain functional and morphological features found in patients and/or in vitro human pluripotent stem cells models and/or rodent in vivo and in vitro models of autism spectrum disorder (ASD)-related syndromes and nonsyndromic ASD forms with evidence for aberrant mechanistic target of rapamycin complex 1 (mTORC1) signaling pathway (please see the manuscript text for complete details and references). The phenotypes that were rescued and/or prevented by targeting mTORC1 pathway at different levels are indicated.

Phenotypes	ASD-Related Syndromes						Nonsyndromic or Idiopathic ASD		
	TSC	PTEN	NF1	FXS	AS	RTT	Dup15q	eIF4E-ASD	Idiopathic
mTORC1 signaling	↑	↑	↑	↑	↑	↓	↑	↑	↑/↓
Seizures	Present *,1	Present *,1	Present	Present	Present	Present	Present	ND	Present
Brain size	↑*,1	↑*,1	↑	↑	↑	↓	↑	ND	↑
Neuron size	↑*,1	↑*,1,3	ND	↑/↓/normal	ND	↓*,5,6	↑*,1 (Cyfip1)	ND	↓/normal
Neuronal migration	Abnormal *,1	Abnormal *,3	Abnormal	Abnormal	ND	Abnormal	Abnormal	ND	Abnormal
Neurite arborization	↑*,1	↑*,3	ND	↓/normal	↓	↓*,5,6	↑*,1 (Cyfip1)	ND	ND
Neurite length	↑/normal	↑	↓	↓	↓	↓	↑/↓*,1 (Cyfip1)	↑	↑
Spine density	↑*,1/↓*,1/normal	↑*,1	↓	↑/↓/normal	↓*,1	↓	↑*(Cyfip1)	ND	ND
Spine length	↑/↓*,1/normal	↑	ND	↑/↓/normal	↓	ND	ND	ND	ND
Immature spine morphology	↓*,1/↑	↑/↓	ND	↑*,3,4	Abnormal *,1	↑	↓(Cyfip1)	ND	ND
LTP	↓	↑/↓	↓	↑/↓	↓*,1,3	↓	↓(patDp/+)	↑*,7	↓
mGluR-LTD	↓*,1	↑/↓	ND	↑*,3,4	ND	ND	ND	↑*,7	ND
Protein synthesis	↓*,1/↑*,1	↑	↑*,1	↑*,1,2,3,4	ND	↓*,5,6	ND	↑*,7	↑*,8

↑ = increased; ↓ = decreased; * Phenotypes were rescued by: 1, mTORC1 inhibition (rapamycin); 2, phosphatidylinositol 3-kinase (PI3K) inhibition; 3, ribosomal protein S6 kinase 1 (S6K1) depletion; 4, eukaryotic translation initiation factor (eIF) 4E phosphorylation reduction; 5, insulin-like growth factor 1 (IGF-1) and/or brain-derived neurotrophic factor (BDNF); 6, phosphatase and tensin homolog (PTEN) depletion; 7, 4EGI-1, an inhibitor of eIF4E-eIF4G interaction; 8, p110δ inhibition; ND = not determined; patDp/+ = model mice for 15q11-13 duplication; = transgenic mice and/or cultured neuronal cells overexpressing Cyfip1; cytoplasmic FMR1 interacting protein 1 (Cyfip1). TSC = tuberous sclerosis complex; NF1 = neurofibromatosis type I; FXS = fragile X syndrome; AS = Angelman syndrome; RTT = Rett syndrome; LTP = long-term potentiation; mGluR-LTD = metabotropic glutamate receptor-mediated long-term depression.

2. mTORC1 Signaling Pathway in Monogenic Autism Spectrum Disorder-Related Syndromes

2.1. Tuberous Sclerosis Complex (TSC)

TSC (MIM#191100, #613254), a classical mTORopathy, is caused by loss-of-function mutations in the genes encoding TSC1 or TSC2 [56,57], which, together with TBC1D7 [58], form a complex that acts as a guanosine triphosphate (GTP)ase-activating protein for Ras homolog enriched in brain (RHEB) and negatively regulates mTORC1 [59,60] (Box 1, Figure 1). Therefore, constitutively active mTORC1 signaling constitutes the molecular basis of TSC [61,62]. The prevalence of ASD in TSC has been estimated to be ~36% [63]. Brain pathological features in patients include, in addition to epilepsy [64], benign proliferative lesions and focal malformation of cortical architecture termed cortical tubers, characterized by dysregulated mTORC1 activity, disruption of lamination, hyperexcitable synaptic network, giant cells, astrogliosis, reduced myelination, as well as dysplastic neurons with multiple and longer axons [65–67]. Several animal models with TSC downregulation, including constitutive heterozygous mutant mice and conditional knockout (KO) mice with *Tsc1/2*-deficiency in different cell types, have been used to shed light on the mechanisms by which TSC loss of function leads to brain alterations that, ultimately, converge on the neurocognitive impairments observed in TSC. Consistent brain functional and morphological abnormalities observed in these animals include seizures [68–71], larger brains [69,70,72], deficits in neuronal migration and cortical lamination [68–70,72–74], enlarged and dysplastic neurons [73,75–77] astrogliosis [68–70,73], reduced myelination [72,76], multiple and ectopic axons [78,79], enhanced excitatory network [65,80,81], and disrupted synaptic plasticity in the form of impaired hippocampal long-term potentiation (LTP) [82,83] and metabotropic glutamate receptor-mediated long-term depression (mGluR-LTD) [80,84] (Table 1). Notwithstanding, some conflicting results exist regarding normal/increased neurite length [78,79], reduced/normal/increased dendritic spine density and length [51,75,76,80,85,86] and increased spine head width/immature shape [75,85] (Table 1), possibly due to the use of different animal models or experimental conditions. Importantly, while it is still unknown whether increased neurite length and reduced LTP are dependent exclusively on mTORC1 overactivation, the majority of the other brain alterations in rodent models were partially or completely rescued or prevented by the mTORC1 inhibitor rapamycin [51,69,70,72,75,76,78,81,84,86] (Table 1), including astrogliosis, reduced myelination and ASD-relevant behaviors (not described in Table 1), even when treatment is begun in adulthood [77,87,88], demonstrating that they are dependent on mTORC1 overactivation. Notably, it was shown that cell-autonomous and non-cell autonomous mechanisms still not understood drive astrogliosis in *Tsc1*-deficient mice [68–70,73], and that neuron-specific deletion of *Tsc1* impairs oligodentrocyte maturation and myelination through a non-cell autonomously manner [89]. Interestingly, it was suggested that TSC2 deficiency affects neuronal migration through an abnormal crosstalk between mTORC1 and Reelin-Disabled 1 (Dab1) signaling pathways [74], and leads to increased spine density due to diminished postnatal mTORC1-mediated autophagy and spine pruning [51]. Paralleling these findings, it was shown that *Tsc1/2*-deficient neurons present mTORC1-dependent deficits in mitophagy, leading to mitochondria accumulation in cell soma and depletion from axon and pre-synaptic sites [79], which might also contribute to synaptic transmission dysfunction due to lack of adenosine triphosphate (ATP) [90,91]. On the other hand, an in vitro study has shown that *Tsc2*-deficient neurons have increased autophagy through 5′ adenosine monophosphate (AMP)-activated protein kinase (AMPK) stimulation [92], which could potentially explain the controversial results regarding dendritic spine density observed in different studies. In addition, recent reports revealed mTORC1-dependent alterations in hippocampal protein synthesis in TSC mouse models, including decreased expression of synaptic proteins, such as the plasticity-related Arc protein required for mGluR-LTD [84], which could contribute to the impaired synaptic plasticity, but increased expression of stress-responsive proteins and anti-inflammatory cytokines [86], which may suggest an attempt to deal with the oxidative stress and inflammation that are frequently observed in the brain of individuals with ASD [93].

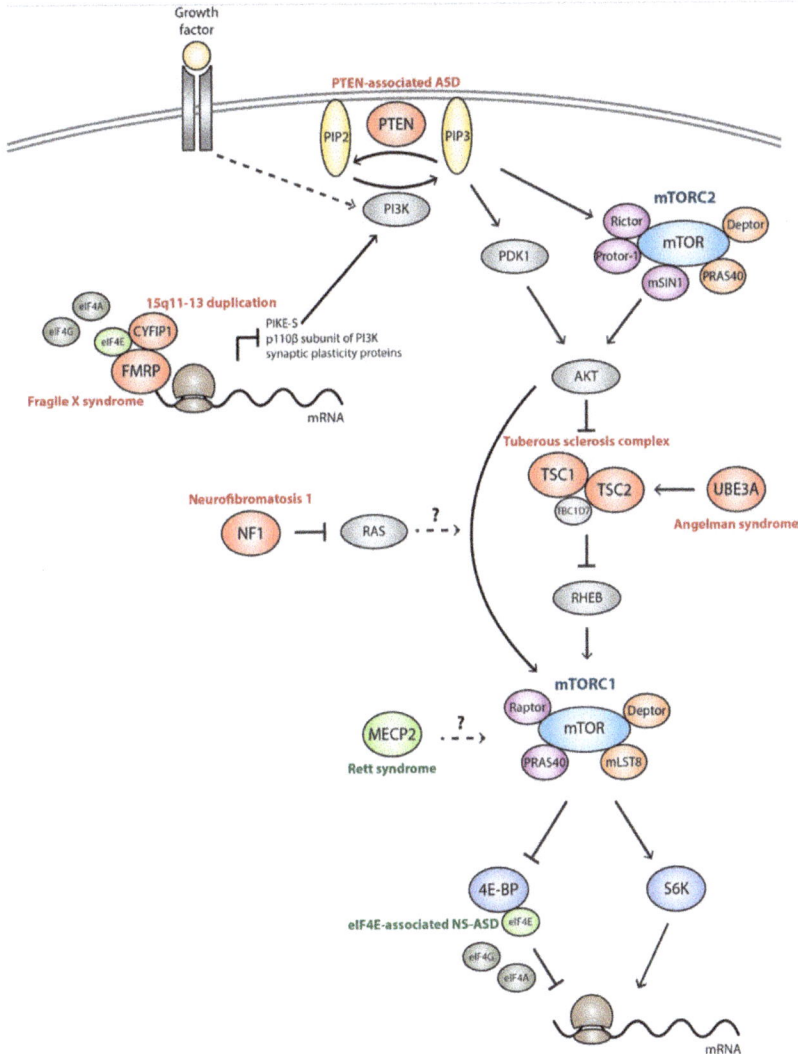

Figure 1. The mechanistic target of rapamycin complex 1 (mTORC1) signaling pathway. mTORC1 signaling components and proteins encoded by genes that inhibit (in red) or enhance (in green) mTORC1 pathway and cause autism spectrum disorder (ASD)-related syndromes and nonsyndromic ASD. Please see Box 1 for further details on mTORC1 signaling pathway. 4E-BP = eIF4E-binding protein; CYFIP1 = cytoplasmic FMR1 interacting protein 1; Deptor = DEP domain-containing mTOR-interacting protein; eIF = eukaryotic initiation factor; FMRP = fragile X mental retardation protein; MECP2 = methyl-CpG binding protein 2; mLST8 = mammalian lethal with SEC13 protein 8; mSIN1 = mammalian stress-activated protein kinase interacting protein 1; NF1 = neurofibromatosis 1; PDK1 = 3-phosphoinositide-dependent protein kinase 1; PI3K = phosphoinositide 3-kinase; PIKE = phosphoinositide 3-kinase enhancer; PIP = phosphatidylinositol; PRAS40 = proline-rich AKT substrate of 40 kDa; Protor-1 = protein observed with Rictor-1; PTEN = phosphatase and tensin homolog; Raptor = regulatory-associated protein of mTOR; RHEB = Ras homolog enriched in brain; Rictor = rapamycin-insensitive companion of mTOR; S6K = S6 kinase; TSC = tuberous sclerosis complex; UBE3A = ubiquitin-protein ligase E3A.

2.2. Phosphatase and Tensin Homolog-Associated ASD (PTEN-ASD)

Germline loss-of-function mutations in *PTEN* have been identified in patients with hamartoma tumor syndromes (PHTS) and in patients with ASD who also display macrocephaly with and without additional developmental features of PHTS (MIM#605309) [11,94,95]. Due to this reason and to the fact that higher lifetime risks for multiple cancers exist in patients with *PTEN* mutations, PTEN-ASD was included herein in the category of monogenic ASD-related syndromes. It has been estimated that the prevalence of *PTEN* mutations in patients with ASD and macrocephaly range from 7% to 27% [95–97]. *PTEN* encodes a lipid and protein phosphatase critical for modulating cellular growth, proliferation and survival [98]. PTEN counteracts the function of phosphoinositide 3-kinase (PI3K) and, similarly to TSC1/2, negatively regulates the mTORC1 pathway [99] (Figure 1). Therefore, *PTEN* deficiency is associated with constitutive activation of downstream AKT/mTORC1 pathways [41,100]. Except for macrocephaly, reports of brain pathological findings in patients are scarce and describe some structural abnormalities [101,102] and seizures in a few patients [103,104]. On the other hand, several heterozygous mice with constitutive *Pten* haploinsufficiency and conditional KO or knockdown mice with *Pten*-deficiency in different subsets of neuronal and glial cells have provided critical insights into the role of PTEN in the central nervous system (CNS), suggesting that it functions largely cell autonomously. Brain pathological features in these animals include seizures, macrocephaly, hypertrophy of both neurons and astrocytes throughout the brain [100,105–110]; enhanced glial cell number [105]; altered neuronal and glial migration [105,111,112]; severe abnormalities in myelination [107]; increased calibers, length and arborization of dendritic and axonal projections; increased dendritic spines density and length [100,106–110] (Table 1); and enhanced excitatory connectivity [108,109]. Importantly, treatment of *Pten*-deficient mice with rapamycin or pharmacological inhibition of S6 kinase 1 (S6K1) during early postnatal life prevented seizures, macrocephaly, aberrant neuronal migration, somatic, dendritic and axonal hypertrophy, increased neurite arborization and spine density (Table 1), enhanced excitatory connectivity, as well as reversed ASD-relevant symptoms [100,110,112], providing a causal link between elevated mTORC1 signaling and these neurobehavioral abnormalities in mouse models of *Pten* deficiency in the CNS (Table 1). Although alterations in glial cell growth, proliferation and migration, as well as in myelin production might be related to disease pathogenesis, further studies are needed to determine whether these abnormalities are linked to mTORC1 overactivation and affect ASD-relevant behaviors. Whereas increased/decreased number of immature-shaped spines [107,109] and reduced/increased LTP [107,113], decreased hippocampal mGluR-LTD [113] and cortical protein synthesis [110] are all neuropathological features also found in PTEN-ASD mouse models (Table 1), which might be in part linked to enhanced mTORC1 signaling activity as observed in TSC, a clear mechanistic link is still lacking and deserves deeper studies.

2.3. Neurofibromatosis Type I (NF1)

NF1 (MIM#162200) is a tumor predisposition syndrome which may also exhibit cognitive impairments and ASD-like symptoms [114]. The prevalence of ASD in NF1 patients has been estimated to be ~18% [63]. NF1 is caused by loss-of-function mutations in the *NF1* gene, which encodes the RAS GTPase-activating protein termed neurofibromin. Consequently, *NF1* defect triggers RAS signaling activation [115,116], a key driver of cancer. The first studies suggesting mTORC1 involvement in NF1 showed that RAS can induce PI3K activation and subsequent TSC2 inhibition by AKT, increasing mTORC1 activity in *Nf1*-null mouse embryonic fibroblasts and astrocytes, as well as in cells derived from NF1 patient tumors [117,118]. Thereafter, however, it was shown that NF1 regulates glial cell proliferation and tumor growth in an AKT/mTORC1 dependent but TSC/RHEB independent manner [119] (Figure 1). Although only less than 10% of NF1 patients report seizures [120–122], several brain pathological features were frequently described in patients, such as macrocephaly [123,124] and reduced myelination [125,126], as well as in mouse models of the disorder, including larger brains [127], structural malformations [128–130], abnormal cerebellar neuronal migration [131,132], increased proliferation and protein synthesis in astrocytes [118,119,133], decreased neurite length [134,135], reduced dendritic spine density [136,137] and impaired LTP [138,139] (Table 1). Among those phenotypic alterations, it has been shown that rapamycin inhibited proliferation and protein synthesis in astrocytes [118,119] (Table 1), indicating that mTORC1 overactivation regulates astrocyte function in NF1 and is probably linked to glioma formation, such that pharmacological inhibition of mTORC1 suppresses tumor growth both in NF1 patients [140,141] and in mouse models [142,143]. Interestingly, not only *Nf1* loss of function in astrocytes may promote astrogliosis [118,119,133], but also neuron-specific *Nf1* deletion induces an increase in astrocyte number via a non-cell autonomous mechanism [144]. Nevertheless, further studies are required to unravel whether a defective glial proliferation affects social and other ASD-associated behaviors. Most of the other neuropathological features found in NF1 has not yet been associated with disrupted mTORC1 activity, and might also be due to mTORC1-independent functions of NF1. In fact, impaired cerebellar neuronal migration and LTP were shown to be dependent on extracellular signal-regulated kinase (ERK) signaling [131,132,139], whereas reduced neurite length is caused by defective cyclic AMP (cAMP) generation independently of RAS signaling [134,135].

2.4. Fragile X Syndrome (FXS)

FXS (MIM#300624) is considered the most commonly inherited cause of intellectual disability and a large percentage of individuals with FXS (~30%) are codiagnosed with ASD [63,145]. FXS is caused by transcriptional silence of the X-linked gene *FMR1* and loss of the protein product, fragile X mental retardation protein (FMRP) [146,147]. FMRP is an RNA-binding protein that negatively regulates the translation, stability and transport of several mRNAs, many of which encode proteins that are essential to synapse maturation, stabilization and elimination and that are well-studied ASD risk genes, such as *SH3 and multiple ankyrin repeat domains 3* (*SHANK3*), *PTEN, TSC2, NF1, CYFIP1, Neuroligin 3* (*NLGN3*) and *Neurexin 1* (*NRXN1*) [147–149]. In addition to high incidence of epilepsy [150,151] and increased head circumference that may be present in FXS patients [152,153], consistent neuronal pathology includes increased dendritic spine density and overabundance of immature-shaped spines on neurons in various brain regions [154,155], which are thought to affect synaptic plasticity and network function (Table 1). In addition, it was shown that in vitro neurons derived from human FXS pluripotent stem cell lines show reduced cell size and neurite length [156–158] (Table 1). Accordingly, *Fmr1*KO mice, a model for human FXS, also exhibit a seizure phenotype [159,160], neurons with shorter neurite length [161] and with the atypical immature feature of FXS spines [160,162–166]. Additionally, it was shown that *Fmr1*KO mice exhibit altered neuronal migration and cortical circuitry [167], cerebellar astrogliosis [168], reduced cerebellar myelination [169], exaggerated hippocampal mGluR-LTD and protein synthesis [160,166,170–172], as well as overall brain hyperexcitability [173,174] (Table 1). Among these latter brain abnormalities, increased mGluR-LTD and protein synthesis were proved to play an important role in the neurological manifestations of FXS. In addition, less consistent results exist regarding decreased/normal neurite arborization in neurons derived from human FXS pluripotent stem cell lines [157,158], as well as normal/increased neuron size [175,176], increased/decreased/normal dendritic spine density [160,162–164,175,177] or length [165,175,178] and decreased/increased LTP [179–181] in *Fmr1*KO mice (Table 1). These discrepancies have been suggested to be due to differences in experimental conditions, brain regions examined, age and genetic background of the animals. It has been well documented that *Fmr1*KO mice exhibit upregulated mTORC1 signaling and elevated translation initiation complex formation in the brain [160,166,172], due at least in part to increased translation of the mRNAs encoding the p110β subunit of PI3K and its upstream activator PI3K enhancer (PIKE)-S, positive regulators of the mTORC1 pathway [172] (Figure 1). These findings suggest that in addition to its RNA-binding activity, FMRP also plays a role in the regulation of PI3K/mTORC1-mediated translation initiation. It is also noteworthy that although it has been suggested that mTORC1 signaling phosphorylates FMRP and inhibits its translation repressor activity [182], this finding was not confirmed in another study [183]. Importantly, pharmacological inhibition of either PI3K or mTORC1 rescues excessive synaptic protein synthesis in neurons from

*Fmr1*KO mice [177]. In addition, genetic deletion of *S6K1* and pharmacological or genetic ablation of eukaryotic translation initiation factor (eIF) 4E phosphorylation, downstream targets of both ERK and mTORC1 pathways (Box 1; Figure 1), in *Fmr1*KO mice prevented dendritic spine morphology defects, synaptic plasticity alterations, exaggerated protein synthesis (Table 1) and ASD-associated behavioral phenotypes [160,166], providing a direct evidence that upregulated mTORC1 signaling and cap-dependent translation play a role in FXS pathophysiology.

2.5. Angelman Syndrome (AS)

Most cases of AS (MIM#105830) are caused by loss of function of the maternally-inherited *ubiquitin protein ligase E3A* (*UBE3A*) allele in neuronal cells [184,185], which encodes a protein that targets other proteins for degradation. This gene is localized on a cluster of imprinted genes on chromosomal region 15q11-13 such that *UBE3A* is paternally imprinted and silenced by a non-coding antisense transcript [186]. A substantial portion of AS patients meets criteria for ASD [187], and the prevalence of ASD in AS has been estimated to be ~34% [63]. Brain pathological features already described in AS patients include epilepsy [188,189], microcephaly [190] and reduced myelination [191,192]. These abnormalities have been also observed in mice with a maternal null mutation in *Ube3a* (AS mice) [193–195], which additionally exhibit abnormal spine morphology, reduced dendritic spine density and length [44,45,196], suggesting that deficient synaptic development may underlie the neurological aspects of AS. Moreover, an overall excitatory network, due to a more severe decrease in inhibitory than excitatory inputs [197], possibly contributes to the increased seizure susceptibility observed in AS patients. In vitro studies using *Ube3a* knockdown or *Ube3a*KO neurons from mice have also shown that *Ube3a* loss of function decreased dendrite arborization, disrupted dendrite polarity and reduced apical dendrite length [198,199]. In addition, disrupted synaptic plasticity in the form of impaired LTP in different brain areas [45,193,200] and enhanced hippocampal mGluR-LTD [201] were also observed in AS mice (Table 1). This is believed to be mainly caused by increased levels of Arc due to its reduced ubiquitination by the UBE3A proteins and, consequently, reduced degradation [202]. Importantly, recent studies have also observed that the increased levels of Arc may also be the result of increased mTORC1 signaling and its downstream target S6K1 activation in the cerebellum and hippocampus of AS mice, triggered by increased inhibitory phosphorylation of TSC2 in the absence of UBE3A [44,45], although the precise mechanism is unknown. In addition to increasing Arc levels and improving LTP deficits, rapamycin or an S6K1 inhibitor also ameliorated dendritic spine density and morphology in Purkinje and pyramidal cells (Table 1), and consequently, motor dysfunction and learning deficits in AS mice [44,45], suggesting that mTORC1 activity may also be affecting synaptic plasticity and function in AS patients. Given the fact that the association between mTORC1 and UBE3A deficiency has been only recently demonstrated, further studies are necessary to test whether the other brain abnormalities found in AS are dependent on mTORC1 overactivation and would benefit from mTORC1-targeted therapies.

2.6. Rett Syndrome (RTT)

RTT (MIM#312750) is a severe progressive neurodevelopmental disorder that manifests mostly in girls during early childhood after a typical perinatal development. Although RTT is no longer considered an ASD in Diagnostic and Statistical Manual of Mental Disorders, fifth edition (DMS-5) [2], children afflicted with RTT often exhibit ASD-like behaviors, and the prevalence of ASD symptoms in RTT has been estimated to be ~61% in female patients [63]. RTT is mainly caused by loss-of-function mutations in the X-linked gene *MECP2* [203,204], which encodes a methyl-CpG binding protein that controls gene expression and chromatin remodeling [205]. In addition to epilepsy [206,207] and reduced brain size [208,209], brain pathology in human patients that are thought to contribute to the neurocognitive deficits in RTT includes reduced neuronal size but increased neuronal cell density in several brain regions [210,211], evidence of cortical astrogliosis [212], decreased dendritic arborization and length [213,214], as well as decreased spine density and maturation in the cortex and hippocampus [215–217] (Table 1). These neuronal abnormalities have been consistently reproduced by several studies using genetically distinct rodent models of RTT [218–225], which additionally exhibit abnormal activity-dependent synaptic plasticity in the form of attenuated LTP and LTD [226,227], and reduced number of excitatory synapses in hippocampal neurons [228] (Table 1). Notably, studies in RTT mouse models have suggested that cell autonomous and non-cell autonomous mechanisms drive neuronal morphology and function [229–231]. In addition, recent in vitro models of RTT using *MECP2*-deficient neurons derived from human pluripotent stem cells have recapitulated many neurological features of RTT [232–236], and have also shown neuronal migration defects [236] (Table 1). Interestingly, in contrast to the majority of the ASD-associated mTORopathies, neurons from $Mecp2^{-/-}$ and $Mecp2^{+/-}$ mice [43], as well as from *MECP2*-deficient human pluripotent stem cells [233], show decreased mTORC1 signaling activity, transcription and protein synthesis rate. These findings suggest that mTORC1 signaling deviations in either direction can adversely affect neuronal connectivity, cognition and social behavior. Although the mechanism by which MECP2 enhances mTORC1 signaling is still unknown (Figure 1), the levels of brain-derived neurotrophic factor (BDNF) are lower in RTT mouse models, possibly resulting in lower activation of PI3K/mTORC1 pathways [43]. Treatment of neurons derived from *MECP2*-deficient human pluripotent stem cells with exogenous growth factors (insulin-like growth factor 1 [IGF-1] or BNDF) or genetically ablation of *PTEN*, promoted protein synthesis via enhancing PI3K/mTORC1 signaling activity and rescued the soma size and neurite complexity deficits [233] (Table 1). These findings suggest that defects in the global control of transcription and PI3K/mTORC1-mediated translation might be the underlying pathomechanisms by which MECP2 dysfunction leads to RTT, although additional mechanistic exploration are needed.

3. mTORC1 Signaling Pathway in Nonsyndromic/Idiopathic Autism Spectrum Disorder

3.1. 15q11-13 Duplication (Dup15q)

Maternally inherited duplications at 15q11-13 (#MIN608636) is one of the most frequent and penetrant copy number variation in ASD, found in ~1%–2% of patients [237,238], suggesting that one or several genes from this region, when duplicated, can lead to ASD. Deletions at this same region give rise to Prader–Willi syndrome (PWS) or AS depending on whether the deletions are paternally or maternally inherited, respectively [239] and, as discussed above, ASD symptoms are usually reported in AS patients. Instability of this region is mediated by the presence of five low copy repeats, termed breakpoint (BP)1 through BP5. The PWS/AS critical region lies on the imprinted region between BP2 and BP3, and evidence from mouse models suggests that increased dosage of the *Ube3a* gene located in this region impairs excitatory synapse transmission and might underlie ASD-relevant behaviors [240]. It is noteworthy that, although duplications of paternal origin show low penetrance, a mouse model with a paternally inherited duplication of the BP2–BP3 interval (patDp/+ mice) displays behaviors associated with ASD [241], increased spine turnover [242] and impairment in cerebellar LTP [243] (Table 1). In addition, there is evidence suggesting that the more proximal non-imprinted region between BP1 and BP2 (15q11.2) is also a hot spot for ASD and that genes located in this region impact neurological and behavioral functions [244–246]. Among the four genes located between BP1 and BP2, *CYFIP1* became a prime candidate for a causal role in ASD [52,245]: it directly interacts with FMRP and with eIF4E and mediates the translational repression activity of FMRP in the brain [247], and also regulates actin polymerization and cytoskeleton remodeling through its interaction with the small GTPase Rac1 [248,249]. It was shown that CYFIP1 levels are increased in lymphoblastoid cells [52,250] and *postmortem* brain tissues (temporal cortex) [52] from ASD subjects with Dup15q, and its overexpression in cultured human and mouse neuronal cells leads to increased neuronal cell size, neurite arborization, as well as increased/decreased neurite length [52,251]. Similar abnormalities in neuronal cell size, neurite outgrowth and branching were also observed in transgenic mice overexpressing *Cyfip1*, which additionally exhibit increased spine density and number of mature spines in cortical neurons [52] (Table 1), defects previously shown to contribute to synaptopathology that drives ASD-related symptoms. However, further research should evaluate whether *Cyfip1* transgenic mice exhibits autistic traits. Importantly, evidence for increased mTORC1 signaling was observed in *postmortem* brains of ASD-Dup15q carriers (n = 3), as well as in embryonic *Cyfip1* transgenic mice and cultured mouse neuronal cells overexpressing *Cyfip1*. Pharmacological treatment of these cultured cells with rapamycin rescued the observed abnormalities in cell size, neurite length and branching [52]. Finally, evidence for hyperfunctional mTORC1 signaling was also observed in cultured stem cells from human exfoliated deciduous teeth (SHED) derived from one ASD patient with Dup15q [53]. Taken together these findings suggest that CYFIP1-mediated mTORC1 signaling overactivation may contribute to disease pathogenesis in ASD-Dup15 patients.

3.2. eIF4E-Associated NS-ASD (eIF4E-NS-ASD)

mTORC1 signaling phosphorylates eIF4E-binding proteins (4E-BPs) and releases them from eIF4E, allowing eIF4E to interact with eIF4G and eIF4A to form the eIF4F complex, a critical step in cap-dependent translation (Box 1; Figure 1). Interestingly, rare mutations in the promoter region of the *eIF4E* gene, which were suspected to enhance promoter activity, were found in few unrelated NS-ASD patients [48]. Later on, it was shown that transgenic mice overexpressing *eif4e* and mice lacking *4e-bp2* display increased dendritic spine density, altered synaptic plasticity, including augmented excitation, enhanced late-phase LTP and mGluR-LTD in the prefrontal cortex and hippocampus, as well as exaggerated brain cap-dependent translation [49,50] (Table 1). Treatment of *eif4e*-transgenic and *4e-bp2*KO mice with 4EGI-1, an inhibitor of eIF4E-eIF4G interaction, rescued protein synthesis and synaptic plasticity abnormalities (Table 1), as well as ASD-relevant behaviors [49,50]. These findings provided a causal link between NS-ASD and excessive cap-dependent translation, and suggest that this is one of the targets by which mTORC1 inhibitors reverse synaptic plasticity deficits and ASD symptoms.

3.3. Idiopathic Autism Spectrum Disorder

Paralleling neuropathological observations from ASD-related syndromes, children with ASD of unknown etiology also frequently display increased risk for developing epilepsy [252,253], macrocephaly in early childhood [254,255], increased neuronal density in several brain regions but reduced number of Purkinje cells [255,256], normal/decreased neuronal size [255,257], altered neuronal migration [258,259], astrogliosis and microglial activation [260], altered myelination [261,262], increased dendritic spine densities on cortical neurons [51,263] and impaired LTP [264]. In addition, increased number of inhibitory synaptic connections was described in cultured neurons derived from patient-induced pluripotent stem cells (iPSCs) [265], and increased spine turnover was found in the BTBR inbred mouse strain that displays the core behavioral deficits of ASD [242]. Although these abnormalities are most probably caused by different etiological origins, recent studies have shown that there is a subgroup of idiopathic ASD with defects in mTORC1 signaling activity. Hyperactivation of mTORC1 pathway was observed in *postmortem* brains from adolescent patients with idiopathic ASD (n = 5) and was shown to impair autophagy and spine pruning during childhood and adolescence, leading to increased basal dendritic spine density and, therefore, enhanced excitatory connectivity [51] (Table 1). As discussed above, similar findings were also observed in TSC mouse models in the same study, suggesting that downregulation of mTORC1 signaling is required for postnatal spine pruning [51]. In addition, mTORC1 pathway upregulation was also observed in non-neuronal cells derived from idiopathic ASD patients, such as in cultured stem cells from SHED derived from 2 out of 12 patients [53], and in lymphoblastoid cell lines (LCLs) derived from 4 out of 58 patients (7% of the patient sample) [55]. Interestingly, in LCLs from one of these patients with elevated mTORC1 pathway activity, it was also observed increased expression of the p110δ subunit of PI3K and enhanced protein synthesis rates, which were corrected by a p110δ-specific inhibitor [55] (Table 1). Curiously, contrary to these findings, a study reported evidence for decreased mTORC1 signaling in *postmortem* brains from patients with idiopathic ASD (n = 11; aged 5–56 years, mean 20.1 years) [54] (Table 1). The high etiological heterogeneity in this group of patients may account for these observed discrepancies in the direction of mTORC1 signaling activation, and additional investigations are needed to provide further mechanistic understanding of the causal link between mTORC1 pathway dysfunction and ASD in this group.

4. Discussion

Research on both ASD-related syndromes and NS-ASD and their corresponding mouse models has shown that abnormalities in brain size and structure, neuronal size, migration and myelination, astrocyte proliferation, neurite and dendritic spine morphology, synapse plasticity, imbalanced synaptic excitation/inhibition, as well as dysregulated brain protein synthesis are all common features of ASD across different etiologies. Although aberrant mTORC1 pathway activation has been suggested as a convergent molecular mechanism in ASD etiopathology, the evidence supporting a causal relationship between abnormal mTORC1 signaling and these brain anatomical and physiological deficits, as well as behavioral alterations found in patients and/or animal models, varies greatly depending on the ASD form involved. Herein, we will discuss only the neurobehavioral abnormalities that were experimentally linked to mTORC1, which include those phenotypes that were rescued or prevented by modulating mTORC1 cascade activity at different levels.

Seizures, enlarged brain and neuron size, neuronal migration abnormalities and increased neurite arborization were mechanistically linked to enhanced mTORC1 signaling in mouse models of ASD-related syndromes caused by mutations in upstream negative regulators of the pathway, such as TSC [69,70,72,75,76,78] and PTEN-ASD [100,112]. Interestingly, overactivation of mTORC1 was also associated with increased neuronal size and neurite arborization in cultured neuronal cells overexpressing *Cyfip1*, located in the 15q11.2 ASD risk locus [52]. On the other hand, diminished mTORC1 pathway seems to be associated with decreased neuronal size and neurite arborization in RTT models [233]. These results suggest that an optimal level of mTORC1 signaling activation is required to maintain proper brain and neuron size, as well as neurite branching patterns, such that abnormalities in these morphological features of neurons may contribute to ASD pathogenesis.

Atypical number and/or length and/or morphology of dendritic spines were mechanistically linked to disinhibited mTORC1 signaling in different mouse models of ASD-related syndromes, including TSC [75,76,86], PTEN-ASD [100], FXS [160,166] and AS [44,45], as well as in NS-ASD models, such as *Cyfip1* transgenic mice [52], eIF4E-NS-ASD mice [49,50] and idiopathic ASD patients [51]. Importantly, a recent study reported that disruption of mTORC1-dependent macroautophagy reduces spine pruning and consequently increases spine density in neurons of individuals with TSC or idiopathic ASD [51]. It would be important to address whether defects in mTORC1-mediated autophagy also play a role in impaired developmental pruning of neuronal connections in other ASD models. Together these findings suggest that mTORC1-mediated abnormalities in dendritic spine number and structure are central in ASD pathogenesis across multiple underlying causes.

Abnormal mTORC1-mediated brain protein synthesis was shown to play a role in the synaptic pathophysiology of TSC [84,86], FXS [160,166,177] and eIF4E-NS-ASD [49,50] mouse models. In addition, different responses for overactivated mTORC1 signaling in synaptic plasticity was observed depending on the ASD mouse model, i.e., mTORC1 may reduce mGluR-LTD in TSC [84] but enhance it in FXS [160,166] and eIF4E-NS-ASD [49,50], as well as decrease LTP in AS [44,45] and increase it in eIF4E-NS-ASD [49,50]. Interestingly, there is a tendency for those ASD forms with heightened mTORC1-dependent translation of synaptic proteins, such as FXS and eIF4E-NS-ASD, to display enhanced mGluR-LTD, whereas TSC, which show decreased mTORC1-dependent translation of synaptic proteins, exhibit impaired mGluR-LTD, suggesting that altered (either enhanced or reduced) mTORC1-mediated protein abundance of synaptic proteins, such as Arc, may influence mGluR-LTD and may be implicated in the synaptic defects and cognitive impairments associated with ASD pathogenesis across different genetic causes. Although PI3K/mTORC1-associated protein synthesis defects have also been observed in *Nf1*-deficient astrocytes [118], in *MECP2*-deficient human pluripotent stem cells (model of RTT) [233] and in lymphoblastoid cell lines from an idiopathic ASD patient [55], further investigation is required in order to unravel a potential link with synaptic plasticity abnormalities and ASD-like symptoms in these different ASD models.

Finally, and perhaps most importantly, a direct role for overactivated mTORC1 signaling in ASD core symptoms is supported by multiple evidence in mouse models of TSC [77,87,88], PTEN-ASD [100], FXS [160,166] and eIF4E-NS-ASD [49,50], showing that pharmacological or genetic inhibition of mTORC1 cascade both upstream and downstream of its kinase activity rescues or attenuates ASD-relevant behaviors, highlighting the potential therapeutic value of drugs targeting this pathway for patients. Indeed, based on evidence for the benefit of rapamycin and similar drugs on neurobehavioral deficits in TSC mouse models, clinical trials aimed to evaluate the effect of mTORC1 inhibitors on neurocognition in TSC patients are currently underway [266].

Therapeutic approaches based on mTORC1 inhibition might benefit NS-ASD patients as well. There are evidences that individuals with Dup15q, which accounts for 1%–2% of all ASD cases, display aberrant mTORC1 pathway activation [52] and, thus, it would be interesting to develop preclinical and clinical studies to evaluate the ability of drugs acting on mTORC1 cascade to ameliorate neurobehavioral function in Dup15q. Lastly, some studies have described mTORC1 signaling defects in a sub-cohort of patients with idiopathic ASD [51,53–55], although further functional studies are required to provide additional and mechanistic proof of a causal link between mTORC1 abnormalities and ASD development in this group of patients. While it is currently difficult to estimate the proportion of patients with ASD of unknown etiology that show aberrant mTORC1 pathway, pilot studies using patient-derived non-neuronal cells [53,55] have opened up the exciting possibility of large-scale screens for mTORC1 signaling defects using more easily accessible patient biological material, which might be used to select those patients who could possibly benefit from treatments targeting mTORC1 pathway. It is noteworthy that mTORC1 signaling abnormalities may be caused by a variety of factors in this group, including genetic, epigenetic and environmental risk factors, which may further complicate clinical studies; however, in spite of these challenges, identifying a subgroup of patients that will benefit from mTORC1-targeted therapies will be of paramount importance.

5. Conclusions and Future Directions

A causal relationship has been established between disturbed activation of mTORC1 signaling pathway and several neurological abnormalities observed in different well-characterized monogenic syndromes with high prevalence of ASD (TSC, PTEN-ASD, FXS, AS and RTT), and preclinical studies have shown that modulation of mTORC1 signaling may provide promising avenues for the treatment of ASD-relevant symptoms. The emerging evidence for aberrant mTORC1 signaling activation in a subgroup of patients with nonsyndromic/idiopathic ASD also provides an exciting possibility for the treatment of behavioral and cognitive deficits in these patients as well. The findings that defective mTORC1 activity can also be detected in non-neuronal more easily accessible cells suggest that mTORC1 cascade components may potentially be used as biomarkers to identify those patients most likely to benefit from mTORC1-targeted therapies. However, given the etiological complexity of ASD in this group, additional studies are required to further explore the mechanistic relevance of mTORC1 pathway alterations to the disease.

Box 1. mTOR signaling biology.

mTOR is a large (predicted molecular weight 280 kD) serine/threonine kinase that can combine with protein binding partners to form one of two functionally distinct mTOR complexes: mTORC1 and mTORC2. In addition to mTOR, mTORC1 consists of regulatory-associated protein of mTOR (Raptor) and mammalian lethal with SEC13 protein 8 (mLST8), which are essential for mTORC1 function, as well as proline-rich AKT substrate of 40 kDa (PRAS40) and DEP domain-containing mTOR-interacting protein (Deptor), inhibitors of mTORC1 activity (Figure 1). Although most of mTORC1 actions are sensitive to rapamycin, mTORC2 is mostly insensitive to rapamycin and contains the core components mTOR, mLST8, Deptor, mammalian stress-activated protein kinase interacting protein 1 (mSIN1), rapamycin-insensitive companion of mTOR (Rictor) and protein observed with Rictor-1 (Protor-1) (Figure 1). Although much less is known about mTORC2 than is known for mTORC1, a growing amount of literature demonstrates a role for mTORC2 in cytoskeletal integrity and neuronal morphology [267,268]. To date, the majority of neurological disorders associated with mTOR signaling have been linked to mTORC1 [39,269]. In the presence of growth factors, such as insulin, the PI3K is activated and stimulates phosphatidylinositol (3,4,5)-trisphosphate (PIP3) production. PIP3 accumulation in the plasma membrane promotes AKT recruitment, phosphorylation and activation by 3-phosphoinositide-dependent protein kinase 1 (PDK1) and mTORC2. When active, AKT phosphorylates and inhibits TSC2 that, together with TSC1 and TBC1D7 [58], is part of the tuberous sclerosis complex (TSC) [270]. TSC functions as a GTPase-activating protein (GAP) toward RAS homolog enriched in brain (RHEB), stimulating the conversion of RHEB-GTP to RHEB-GDP and inactivating this protein. Therefore, the inhibition of TSC by AKT promotes RHEB activation, which then activates mTORC1 in the presence of amino acids [271,272]. Among several processes, mTORC1 inhibits autophagy and stimulates mRNA translation, which is dependent on the phosphorylation and activation of S6 kinase (S6K) and inhibition of eukaryotic translation initiation factor 4E (eIF4E)-binding proteins (4E-BPs), releasing it from eIF4E and enabling interaction with eIF4G and eIF4A to form the eIF4F translation initiation complex, a critical step in cap-dependent translation (Figure 1). In addition to insulin and amino acids, other signals modulate mTORC1 activity, such as levels of ATP, glucose and oxygen. Therefore, mTORC1 is considered a sensor of internal and external cues that maintains cellular homeostasis through modulation of anabolic and catabolic processes [273].

Author Contributions: Juliana Magdalon and Andrea L. Sertie reviewed the literature and wrote the manuscript; Sandra M.S. Sánchez analyzed critically the literature review; Karina Griesi-Oliveira critically revised the manuscript; and Andrea L. Sertie conceived the manuscript. All the authors approved the final version of the manuscript.

Conflicts of Interest: The authors declare no conflict of interest.

References

1. Christensen, D.L.; Baio, J.; van Naarden Braun, K.; Bilder, D.; Charles, J.; Constantino, J.; Daniels, J.; Durkin, M.; Fitzgerald, R.T.; Kurzius-Spencer, M.; et al. Prevalence and Characteristics of Autism Spectrum Disorder among Children Aged 8 Years—Autism and Developmental Disabilities Monitoring Network, 11 Sites, United States, 2012. *MMWR Surveill. Summ.* **2016**, *65*, 1–23. [CrossRef] [PubMed]

2. American Psychiatric Association. *Diagnostic and Statistical Manual of Mental Disorders (5th ed.; DSM-5)*; American Psychiatric Publishing: Arlington, VA, USA, 2013.

3. Newschaffer, C.; Croen, L.A.; Daniels, J.; Giarelli, E.; Grether, J.K.; Levy, S.E.; Mandell, D.S.; Miller, L.A.; Pinto-Martin, J.; Reaven, J.; et al. The Epidemiology of Autism Spectrum Disorders. *Annu. Rev. Public Health* **2007**, *28*, 235–258. [CrossRef] [PubMed]

4. Fombonne, E. Epidemiology of pervasive developmental disorders. *Pediatr. Res.* **2009**, *65*, 591–598. [CrossRef] [PubMed]

5. Jeste, S.S.; Sahin, M.; Bolton, P. Characterization of Autism in Young Children with Tuberous Sclerosis Complex. *J. Child Neurol.* **2008**, *23*, 520–525. [CrossRef] [PubMed]

6. Budimirovic, D.B.; Kaufmann, W.E. What Can We Learn about Autism from Studying Fragile X Syndrome? *Dev. Neurosci.* **2011**, *33*, 379–394. [CrossRef] [PubMed]

7. Chahrour, M.; Zoghbi, H.Y. The Story of Rett Syndrome: From Clinic to Neurobiology. *Neuron* **2007**, *56*, 422–437. [CrossRef] [PubMed]

8. Neul, J.L. The relationship of Rett syndrome and MECP2 disorders to autism. *Dialogues Clin. Neurosci.* **2012**, *14*, 253–262. [PubMed]

9. Williams, C.A.; Beaudet, A.L.; Clayton-Smith, J.; Knoll, J.H.; Kyllerman, M.; Laan, L.A.; Magenis, R.E.; Moncla, A.; Schinzel, A.A.; Summers, J.A.; et al. Angelman Syndrome 2005: Updated Consensus for Diagnostic Criteria. *Am. J. Med. Genet.* **2006**, *140*, 413–418. [CrossRef] [PubMed]

10. Buiting, K. Prader-Willi syndrome and Angelman syndrome. *Am. J. Med. Genet. Part C Semin. Med. Genet.* **2010**, *154*, 365–376. [CrossRef] [PubMed]

11. Butler, M.G.; Dasouki, M.J.; Zhou, X.; Talebizadeh, Z.; Brown, M.; Takahashi, T.N.; Miles, J.H.; Wang, C.H.; Stratton, R.; Pilarski, R.; et al. Subset of individuals with autism spectrum disorders and extreme macrocephaly associated with germline PTEN tumour suppressor gene mutations. *J. Med. Genet.* **2005**, *42*, 318–321. [CrossRef] [PubMed]

12. Ratner, N.; Miller, S.J. A RASopathy gene commonly mutated in cancer: The neurofibromatosis type 1 tumour suppressor. *Nat. Rev. Cancer* **2015**, *15*, 290–301. [CrossRef] [PubMed]

13. Splawski, I.; Timothy, K.W.; Sharpe, L.M.; Decher, N.; Kumar, P.; Bloise, R.; Napolitano, C.; Schwartz, P.J.; Joseph, R.M.; Condouris, K.; et al. CaV1.2 Calcium Channel Dysfunction Causes a Multisystem Disorder Including Arrhythmia and Autism. *Cell* **2004**, *119*, 19–31. [CrossRef] [PubMed]

14. Phelan, K.; McDermid, H.E. The 22q13.3 deletion syndrome (Phelan-McDermid syndrome). *Mol. Syndromol.* **2011**, *2*, 186–201. [CrossRef] [PubMed]

15. Kelleher, R.J.; Bear, M.F. The Autistic Neuron: Troubled Translation? *Cell* **2008**, *135*, 401–406. [CrossRef] [PubMed]

16. Ebrahimi-Fakhari, D.; Sahin, M. Autism and the synapse: Emerging mechanisms and mechanism-based therapies. *Curr. Opin. Neurol.* **2015**, *28*, 91–102. [CrossRef] [PubMed]

17. Sztainberg, Y.; Zoghbi, H.Y. Lessons learned from studying syndromic autism spectrum disorders. *Nat. Neurosci.* **2016**, *19*, 1408–1418. [CrossRef] [PubMed]

18. Ronald, A.; Hoekstra, R.A. Autism spectrum disorders and autistic traits: A decade of new twin studies. *Am. J. Med. Genet. Part B Neuropsychiatr. Genet.* **2011**, *156*, 255–274. [CrossRef] [PubMed]

19. Sandin, S.; Lichtenstein, P.; Larsson, H.; Cm, H.; Reichenberg, A. The familial risk of autism. *JAMA* **2014**, *311*, 1770–1777. [CrossRef] [PubMed]

20. Gaugler, T.; Klei, L.; Sanders, S.J.; Bodea, C.A.; Goldberg, A.P.; Lee, A.B.; Mahajan, M.; Manaa, D.; Pawitan, Y.; Reichert, J.; et al. Most genetic risk for autism resides with common variation. *Nat. Genet.* **2014**, *46*, 881–885. [CrossRef] [PubMed]

21. State, M.W.; Levitt, P. The conundrums of understanding genetic risks for autism spectrum disorders. *Nat. Neurosci.* **2011**, *14*, 1499–1506. [CrossRef] [PubMed]

22. Devlin, B.; Scherer, S.W. Genetic architecture in autism spectrum disorder. *Curr. Opin. Genet. Dev.* **2012**, *22*, 229–237. [CrossRef] [PubMed]

23. Klei, L.; Sanders, S.J.; Murtha, M.T.; Hus, V.; Lowe, J.K.; Willsey, A.J.; Moreno-De-Luca, D.; Yu, T.W.; Fombonne, E.; Geschwind, D.; et al. Common genetic variants, acting additively, are a major source of risk for autism. *Mol. Autism* **2012**, *3*, 9. [CrossRef] [PubMed]

24. Sebat, J.; Lakshmi, B.; Malhotra, D.; Troge, J.; Lese-martin, C.; Walsh, T.; Yamrom, B.; Yoon, S.; Krasnitz, A.; Kendall, J.; et al. Strong Association of De Novo Copy Number Mutations with Autism. *Science* **2007**, *316*, 445–449. [CrossRef] [PubMed]

25. Glessner, J.; Wang, K.; Cai, G.; Korvatska, O. Autism genome-wide copy number variation reveals ubiquitin and neuronal genes. *Nature* **2009**, *459*, 569–573. [CrossRef] [PubMed]

26. Pinto, D.; Pagnamenta, A.T.; Klei, L.; Anney, R.; Merico, D.; Regan, R.; Conroy, J.; Magalhaes, T.R.; Correia, C.; Brett, S.; et al. Functional Impact of Global Rare Copy Number Variation in Autism Spectrum Disorder. *Nature* **2010**, *466*, 368–372. [CrossRef] [PubMed]

27. Betancur, C. Etiological heterogeneity in autism spectrum disorders: More than 100 genetic and genomic disorders and still counting. *Brain Res.* **2011**, *1380*, 42–77. [CrossRef] [PubMed]

28. Iossifov, I.; Ronemus, M.; Levy, D.; Wang, Z.; Hakker, I.; Yamrom, B.; Lee, Y.; Narzisi, G.; Leotta, A.; Grabowska, E.; et al. De novo Gene Disruptions in children on the Autistic Spectrum. *Neuron* **2012**, *74*, 285–299. [CrossRef] [PubMed]

29. Neale, B.; Kou, Y.; Liu, L.; Ma'ayan, A. Patterns and rates of exonic de novo mutations in autism spectrum disorders. *Nature* **2013**, *485*, 242–245. [CrossRef] [PubMed]

30. O'Roak, B.J.; Vives, L.; Girirajan, S.; Karakoc, E.; Krumm, N.; Coe, B.P.; Levy, R.; Ko, A.; Lee, C.; Smith, J.D.; et al. Sporadic autism exomes reveal a highly interconnected protein network of de novo mutations. *Nature* **2012**, *485*, 246–250. [CrossRef] [PubMed]

31. Sanders, S.J.; Murtha, M.T.; Gupta, A.R.; Murdoch, J.D.; Raubeson, M.J.; Willsey, A.J.; Ercan-Sencicek, A.G.; DiLullo, N.M.; Parikshak, N.N.; Stein, J.L.; et al. De novo mutations revealed by whole-exome sequencing are strongly associated with autism. *Nature* **2012**, *485*, 237–241. [CrossRef] [PubMed]

32. De Rubeis, S.; He, X.; Goldberg, A.P.; Poultney, C.S.; Samocha, K.; Ercument Cicek, A.; Kou, Y.; Liu, L.; Fromer, M.; Walker, S.; et al. Synaptic, transcriptional and chromatin genes disrupted in autism. *Nature* **2014**, *515*, 209–215. [CrossRef] [PubMed]

33. Ronemus, M.; Iossifov, I.; Levy, D.; Wigler, M. The role of de novo mutations in the genetics of autism spectrum disorders. *Nat. Rev. Genet.* **2014**, *15*, 133–141. [CrossRef] [PubMed]

34. Simons Foundation Autism Research Initiative (SFARI). Available online: https://gene.sfari.org (accessed on 16 March 2017).

35. Bourgeron, T. From the genetic architecture to synaptic plasticity in autism spectrum disorder. *Nat. Rev. Neurosci.* **2015**, *16*, 551–563. [CrossRef] [PubMed]

36. Gilman, S.R.; Iossifov, I.; Levy, D.; Ronemus, M.; Wigler, M.; Vitkup, D. Rare De Novo Variants Associated with Autism Implicate a Large Functional Network of Genes Involved in Formation and Function of Synapses. *Neuron* **2011**, *70*, 898–907. [CrossRef]

37. Baudouin, S.J.; Gaudias, J.; Gerharz, S.; Hatstatt, L.; Zhou, K.; Punnakkal, P.; Tanaka, K.F.; Spooren, W.; Hen, R.; De Zeeuw, C.I.; et al. Shared Synaptic Pathophysiology in Syndromic and Nonsyndromic Rodent Models of Autism. *Science* **2012**, *338*, 128–132. [CrossRef] [PubMed]

38. Pinto, D.; Delaby, E.; Merico, D.; Barbosa, M.; Merikangas, A.; Klei, L.; Thiruvahindrapuram, B.; Xu, X.; Ziman, R.; Wang, Z.; et al. Convergence of Genes and Cellular Pathways Dysregulated in Autism Spectrum Disorders. *Am. J. Hum. Genet.* **2014**, *94*, 677–694. [CrossRef] [PubMed]

39. Costa-Mattioli, M.; Monteggia, L.M. mTOR complexes in neurodevelopmental and neuropsychiatric disorders. *Nat. Neurosci* **2013**, *16*, 1537–1543. [CrossRef] [PubMed]

40. Lipton, J.O.; Sahin, M. The Neurology of mTOR. *Neuron* **2014**, *84*, 275–291. [CrossRef] [PubMed]

41. Ehninger, D.; Silva, A.J. Rapamycin for treating Tuberous sclerosis and Autism spectrum disorders. *Trends Mol. Med.* **2011**, *17*, 78–87. [CrossRef] [PubMed]

42. Huber, K.M.; Klann, E.; Costa-Mattioli, M.; Zukin, R.S. Dysregulation of Mammalian Target of Rapamycin Signaling in Mouse Models of Autism. *J. Neurosci.* **2015**, *35*, 13836–13842. [CrossRef] [PubMed]

43. Ricciardi, S.; Boggio, E.M.; Grosso, S.; Lonetti, G.; Forlani, G.; Stefanelli, G.; Calcagno, E.; Morello, N.; Landsberger, N.; Biffo, S.; et al. Reduced AKT/mTOR signaling and protein synthesis dysregulation in a Rett syndrome animal model. *Hum. Mol. Genet.* **2011**, *20*, 1182–1196. [CrossRef] [PubMed]

44. Sun, J.; Liu, Y.; Moreno, S.; Baudry, M.; Bi, X. Imbalanced Mechanistic Target of Rapamycin C1 and C2 Activity in the Cerebellum of Angelman Syndrome Mice Impairs Motor Function. *J. Neurosci.* **2015**, *35*, 4706–4718. [CrossRef] [PubMed]

45. Sun, J.; Liu, Y.; Tran, J.; O'Neal, P.; Baudry, M.; Bi, X. mTORC1-S6K1 inhibition or mTORC2 activation improves hippocampal synaptic plasticity and learning in Angelman syndrome mice. *Cell. Mol. Life Sci.* **2016**, *73*, 4303–4314. [CrossRef] [PubMed]

46. Jiang, M.; Ash, R.T.; Baker, S.A.; Suter, B.; Ferguson, A.; Park, J.; Rudy, J.; Torsky, S.P.; Chao, H.-T.; Zoghbi, H.Y.; et al. Dendritic arborization and spine dynamics are abnormal in the mouse model of MECP2 duplication syndrome. *J. Neurosci.* **2013**, *33*, 19518–19533. [CrossRef] [PubMed]

47. Della Sala, G.; Putignano, E.; Chelini, G.; Melani, R.; Calcagno, E.; Michele Ratto, G.; Amendola, E.; Gross, C.T.; Giustetto, M.; Pizzorusso, T. Dendritic Spine Instability in a Mouse Model of CDKL5 Disorder Is Rescued by Insulin-like Growth Factor 1. *Biol. Psychiatry* **2016**, *80*, 302–311. [CrossRef] [PubMed]

48. Neves-Pereira, M.; Müller, B.; Massie, D.; Williams, J.H.G.; O'Brien, P.C.M.; Hughes, A.; Shen, S.-B.; Clair, D.S.; Miedzybrodzka, Z. Deregulation of EIF4E: A novel mechanism for autism. *J. Med. Genet.* **2009**, *46*, 759–765. [CrossRef] [PubMed]

49. Gkogkas, C.G.; Khoutorsky, A.; Ran, I.; Rampakakis, E.; Nevarko, T.; Weatherill, D.B.; Vasuta, C.; Yee, S.; Truitt, M.; Dallaire, P.; et al. Autism-related deficits via dysregulated eIF4E-dependent translational control. *Nature* **2013**, *493*, 371–377. [CrossRef] [PubMed]

50. Santini, E.; Huynh, T.N.; Macaskill, A.F.; Carter, A.G.; Pierre, P.; Ruggero, D.; Kaphzan, H.; Klann, E. Exaggerated Translation Causes Synaptic and Behavioural Aberrations Associated with Autism. *Nature* **2013**, *493*, 411–415. [CrossRef] [PubMed]
51. Tang, G.; Gudsnuk, K.; Kuo, S.H.; Cotrina, M.L.; Rosoklija, G.; Sosunov, A.; Sonders, M.S.; Kanter, E.; Castagna, C.; Yamamoto, A.; et al. Loss of mTOR-Dependent Macroautophagy Causes Autistic-like Synaptic Pruning Deficits. *Neuron* **2014**, *83*, 1131–1143. [CrossRef] [PubMed]
52. Oguro-Ando, A.; Rosensweig, C.; Herman, E.; Nishimura, Y.; Werling, D.; Bill, B.R.; Berg, J.M.; Gao, F.; Coppola, G.; Abrahams, B.S.; et al. Increased CYFIP1 dosage alters cellular and dendritic morphology and dysregulates mTOR. *Mol. Psychiatry* **2015**, *20*, 1069–1078. [CrossRef] [PubMed]
53. Suzuki, A.M.; Griesi-Oliveira, K.; de Oliveira Freitas Machado, C.; Vadasz, E.; Zachi, E.C.; Passos-Bueno, M.R.; Sertie, A.L. Altered mTORC1 signaling in multipotent stem cells from nearly 25% of patients with nonsyndromic autism spectrum disorders. *Mol. Psychiatry* **2015**, *20*, 551–552. [CrossRef] [PubMed]
54. Nicolini, C.; Ahn, Y.; Michalski, B.; Rho, J.M.; Fahnestock, M. Decreased mTOR signaling pathway in human idiopathic autism and in rats exposed to valproic acid. *Acta Neuropathol. Commun.* **2015**, *3*, 3. [CrossRef] [PubMed]
55. Poopal, A.C.; Schroeder, L.M.; Horn, P.S.; Bassell, G.J.; Gross, C. Increased expression of the PI3K catalytic subunit p110δ underlies elevated S6 phosphorylation and protein synthesis in an individual with autism from a multiplex family. *Mol. Autism* **2016**, *7*, 3. [CrossRef] [PubMed]
56. The European Chromosome 16 Tuberous Sclerosis Consortium. Identification and Characterization of the Tuberous Sclerosis Gene on Chromosome 16. *Cell* **1993**, *75*, 1305–1315.
57. Slegtenhorst, V.; de Hoogt, R.; Hermans, C.; Nellist, M.; Janssen, B.; Verhoef, S.; Lindhout, D.; van den Ouweland, AH.D.; Young, J.; Burley, M.; et al. Identification of the Tuberous Sclerosis Gene TSC1 on Chromosome 9q34. *Science* **1997**, *277*, 805–808. [CrossRef] [PubMed]
58. Dibble, C.C.; Elis, W.; Menon, S.; Qin, W.; Klekota, J.; Asara, J.M.; Finan, P.M.; Kwiatkowski, D.J.; Murphy, L.O.; Manning, B.D. TBC1D7 Is a Third Subunit of the TSC1-TSC2 Complex Upstream of mTORC1. *Mol. Cell* **2012**, *47*, 535–546. [CrossRef] [PubMed]
59. Garami, A.; Zwartkruis, F.J.T.; Nobukuni, T.; Joaquin, M.; Roccio, M.; Stocker, H.; Kozma, S.C.; Hafen, E.; Bos, J.L.; Thomas, G. Insulin Activation of Rheb, a Mediator of mTOR/S6K/4E-BP Signaling, Is Inhibited by TSC1 and 2. *Mol. Cell* **2003**, *11*, 1457–1466. [CrossRef]
60. Tee, A.R.; Manning, B.D.; Roux, P.P.; Cantley, L.C.; Blenis1, J. Tuberous Sclerosis Complex Gene Products, Tuberin and Hamartin, Control mTOR Signaling by Acting as a GTPase-Activating Protein Complex toward Rheb. *Curr. Biol.* **2003**, *13*, 1259–1268. [CrossRef]
61. Curatolo, P.; Bombardieri, R.; Jozwiak, S. Tuberous sclerosis. *Lancet* **2008**, *372*, 657–668. [CrossRef]
62. Curatolo, P.; Moavero, R. mTOR Inhibitors in Tuberous Sclerosis Complex. *Curr. Neuropharmacol.* **2012**, *10*, 404–415. [CrossRef] [PubMed]
63. Richards, C.; Jones, C.; Groves, L.; Moss, J.; Oliver, C. Prevalence of autism spectrum disorder phenomenology in genetic disorders: A systematic review and meta-analysis. *Lancet Psychiatry* **2015**, *2*, 909–916. [CrossRef]
64. Chu-Shore, C.J.; Major, P.; Camposano, S.; Muzykewicz, D.; Thiele, E.A. The natural history of epilepsy in tuberous sclerosis complex. *Epilepsia* **2010**, *51*, 1236–1241. [CrossRef] [PubMed]
65. Wang, Y.; Greenwood, J.S.F.; Calcagnotto, M.E.; Kirsch, H.E.; Barbaro, N.M.; Baraban, S.C. Neocortical Hyperexcitability in a Human Case of Tuberous Sclerosis Complex and Mice Lacking Neuronal Expression of TSC1. *Ann. Neurol.* **2007**, *61*, 139–152. [CrossRef] [PubMed]
66. Talos, D.M.; Kwiatkowski, D.J.; Cordero, K.; Black, P.M.; Jensen, F.E. Cell-Specific Alterations of Glutamate Receptor Expression in Tuberous Sclerosis Complex Cortical Tubers. *Ann. Neurol.* **2008**, *63*, 454–465. [CrossRef] [PubMed]
67. Ruppe, V.; Dilsiz, P.; Reiss, C.S.; Carlson, C.; Devinsky, O.; Zagzag, D.; Weiner, H.L.; Talos, D.M. Developmental brain abnormalities in tuberous sclerosis complex: A comparative tissue analysis of cortical tubers and perituberal cortex. *Epilepsia* **2014**, *55*, 539–550. [CrossRef] [PubMed]
68. Uhlmann, E.J.; Wong, M.; Baldwin, R.L.; Bajenaru, M.L.; Onda, H.; Kwiatkowski, D.J.; Yamada, K.; Gutmann, D.H. Astrocyte-Specific TSC1 Conditional Knockout Mice Exhibit Abnormal Neuronal Organization and Seizures. *Ann. Neurol.* **2002**, *52*, 285–296. [CrossRef] [PubMed]
69. Zeng, L.H.; Xu, L.; Gutmann, D.H.; Wong, M. Rapamycin prevents epilepsy in a mouse model of tuberous sclerosis complex. *Ann. Neurol.* **2008**, *63*, 444–453. [CrossRef] [PubMed]

70. Zeng, L.H.; Rensing, N.R.; Zhang, B.; Gutmann, D.H.; Gambello, M.J.; Wong, M. Tsc2 gene inactivation causes a more severe epilepsy phenotype than Tsc1 inactivation in a mouse model of Tuberous Sclerosis Complex. *Hum. Mol. Genet.* **2011**, *20*, 445–454. [CrossRef] [PubMed]

71. Fu, C.; Ess, K.C. Conditional and domain-specific inactivation of the Tsc2 gene in neural progenitor cells. *Genesis* **2013**, *51*, 284–292. [CrossRef] [PubMed]

72. Magri, L.; Cambiaghi, M.; Cominelli, M.; Alfaro-Cervello, C.; Cursi, M.; Pala, M.; Bulfone, A.; Garca-Verdugo, J.M.; Leocani, L.; Minicucci, F.; et al. Sustained Activation of mTOR Pathway in Embryonic Neural Stem Cells Leads to Development of Tuberous Sclerosis Complex-Associated Lesions. *Cell Stem Cell* **2011**, *9*, 447–462. [CrossRef] [PubMed]

73. Crowell, B.; Hwa Lee, G.; Nikolaeva, I.; Dal Pozzo, V.; D'Arcangelo, G. Complex Neurological Phenotype in Mutant Mice Lacking Tsc2 in Excitatory Neurons of the Developing Forebrain. *eNeuro* **2015**, *2*. [CrossRef] [PubMed]

74. Moon, U.Y.; Park, J.Y.; Park, R.; Cho, J.Y.; Hughes, L.J.; McKenna, J.; Goetzl, L.; Cho, S.-H.; Crino, P.B.; Gambello, M.J.; et al. Impaired Reelin-Dab1 Signaling Contributes to Neuronal Migration Deficits of Tuberous Sclerosis Complex. *Cell Rep.* **2015**, *12*, 965–978. [CrossRef] [PubMed]

75. Tavazoie, S.F.; Alvarez, V.A.; Ridenour, D.A.; Kwiatkowski, D.J.; Sabatini, B.L. Regulation of neuronal morphology and function by the tumor suppressors Tsc1 and Tsc2. *Nat. Neurosci.* **2005**, *8*, 1727–1734. [CrossRef] [PubMed]

76. Meikle, L.; Pollizzi, K.; Egnor, A.; Kramvis, I.; Lane, H.; Sahin, M.; Kwiatkowski, D.J. Response of a Neuronal Model of Tuberous Sclerosis to Mammalian Target of Rapamycin (mTOR) Inhibitors: Effects on mTORC1 and Akt Signaling Lead to Improved Survival and Function. *J. Neurosci.* **2008**, *28*, 5422–5432. [CrossRef] [PubMed]

77. Tsai, P.T.; Hull, C.; Chu, Y.; Greene-Colozzi, E.; Sadowski, A.R.; Leech, J.M.; Steinberg, J.; Crawley, J.N.; Regehr, W.G.; Sahin, M. Autistic-like behaviour and cerebellar dysfunction in Purkinje cell Tsc1 mutant mice. *Nature* **2012**, *488*, 647–651. [CrossRef] [PubMed]

78. Choi, Y.; Di Nardo, A.; Kramvis, I.; Meikle, L.; Kwiatkowski, D.J.; Sahin, M.; He, X. Tuberous sclerosis complex proteins control axon formation. *Genes Dev.* **2008**, *22*, 2485–2495. [CrossRef] [PubMed]

79. Ebrahimi-fakhari, D.; Saffari, A.; Wahlster, L.; Di Nardo, A.; Turner, D.; Lewis, T.L., Jr.; Conrad, C.; Rothberg, J.M.; Jonathan, O.; Kölker, S.; et al. Impaired Mitochondrial Dynamics and Mitophagy in Neuronal Models of Tuberous Sclerosis Complex. *Cell Rep.* **2016**, *17*, 1053–1070. [CrossRef] [PubMed]

80. Bateup, H.S.; Takasaki, K.T.; Saulnier, J.L.; Denefrio, C.L.; Sabatini, B.L. Loss of Tsc1 In Vivo Impairs Hippocampal mGluR-LTD and Increases Excitatory Synaptic Function. *J. Neurosci.* **2011**, *31*, 8862–8869. [CrossRef] [PubMed]

81. Bateup, H.S.; Johnson, C.A.; Denefrio, C.L.; Saulnier, J.L.; Kornacker, K.; Sabatini, B.L. Excitatory/Inhibitory Synaptic Imbalance Leads to Hippocampal Hyperexcitability in Mouse Models of Tuberous Sclerosis. *Neuron* **2013**, *78*, 510–522. [CrossRef] [PubMed]

82. Von Der Brelie, C.; Waltereit, R.; Zhang, L.; Beck, H.; Kirschstein, T. Impaired synaptic plasticity in a rat model of tuberous sclerosis. *Eur. J. Neurosci.* **2006**, *23*, 686–692. [CrossRef] [PubMed]

83. Zeng, L.H.; Ouyang, Y.; Gazit, V.; Cirrito, J.R.; Jansen, L.A.; Ess, K.C.; Yamada, K.A.; Wozniak, D.F.; Holtzman, D.M.; Gutmann, D.H.; et al. Abnormal glutamate homeostasis and impaired synaptic plasticity and learning in a mouse model of tuberous sclerosis complex. *Neurobiol. Dis.* **2007**, *28*, 184–196. [CrossRef] [PubMed]

84. Auerbach, B.D.; Osterweil, E.K.; Bear, M.F. Mutations causing syndromic autism define an axis of synaptic pathophysiology. *Nature* **2011**, *480*, 63–68. [CrossRef] [PubMed]

85. Yasuda, S.; Sugiura, H.; Katsurabayashi, S.; Shimada, T.; Tanaka, H.; Takasaki, K.; Iwasaki, K.; Kobayashi, T.; Hino, O.; Yamagata, K. Activation of Rheb, but not of mTORC1, impairs spine synapse morphogenesis in tuberous sclerosis complex. *Sci. Rep.* **2014**, *4*, 5155. [CrossRef] [PubMed]

86. Nie, D.; Chen, Z.; Ebrahimi-Fakhari, D.; Di Nardo, A.; Julich, K.; Robson, V.K.; Cheng, Y.-C.; Woolf, C.J.; Heiman, M.; Sahin, M. The Stress-Induced Atf3-Gelsolin Cascade Underlies Dendritic Spine Deficits in Neuronal Models of Tuberous Sclerosis Complex. *J. Neurosci.* **2015**, *35*, 10762–10772. [CrossRef] [PubMed]

87. Ehninger, D.; Han, S.; Shilyansky, C.; Zhou, Y.; Li, W.; David, J. Reversal of learning deficits in a Tsc2+/− mouse model of tuberous sclerosis. *Nat. Med.* **2008**, *14*, 843–848. [CrossRef] [PubMed]

88. Sato, A.; Kasai, S.; Kobayashi, T.; Takamatsu, Y.; Hino, O.; Ikeda, K.; Mizuguchi, M. Rapamycin reverses impaired social interaction in mouse models of tuberous sclerosis complex. *Nat. Commun.* **2012**, *3*, 1292. [CrossRef] [PubMed]

89. Ercan, E.; Han, J.M.; Di Nardo, A.; Winden, K.; Han, M.-J.; Hoyo, L.; Saffari, A.; Leask, A.; Geschwind, D.H.; Sahin, M. Neuronal CTGF/CCN2 negatively regulates myelination in a mouse model of tuberous sclerosis complex. *J. Exp. Med.* **2017**, *214*, 681–697. [CrossRef] [PubMed]

90. Verstreken, P.; Ly, C.V.; Venken, K.J. T.; Koh, T.W.; Zhou, Y.; Bellen, H.J. Synaptic mitochondria are critical for mobilization of reserve pool vesicles at Drosophila neuromuscular junctions. *Neuron* **2005**, *47*, 365–378. [CrossRef] [PubMed]

91. Ma, H.; Cai, Q.; Lu, W.; Sheng, Z.-H.; Mochida, S. KIF5B Motor Adaptor Syntabulin Maintains Synaptic Transmission in Sympathetic Neurons. *J. Neurosci.* **2009**, *29*, 13019–13029. [CrossRef] [PubMed]

92. Di Nardo, A.; Wertz, M.H.; Kwiatkowski, E.; Tsai, P.T.; Leech, J.D.; Greene-Colozzi, E.; Goto, J.; Dilsiz, P.; Talos, D.M.; Clish, C.B.; et al. Neuronal Tsc1/2 complex controls autophagy through AMPK-dependent regulation of ULK1. *Hum. Mol. Genet.* **2014**, *23*, 3865–3874. [CrossRef] [PubMed]

93. Rossignol, D.A.; Frye, R.E. Evidence linking oxidative stress, mitochondrial dysfunction, and inflammation in the brain of individuals with autism. *Front. Physiol.* **2014**, *5*, 150. [CrossRef] [PubMed]

94. Buxbaum, J.D.; Cai, G.; Chaste, P.; Nygren, G.; Goldsmith, J.; Reichert, J.; Anckarsäter, H.; Rastam, M.; Smith, C.J.; Silverman, J.M.; et al. Mutation Screening of the PTEN Gene in Patients With Autism Spectrum Disorders and Macrocephaly. *Am. J. Med. Genet. Part B Neuropsychiatr. Genet.* **2007**, *144*, 484–491. [CrossRef] [PubMed]

95. McBride, K.L.; Varga, E.A.; Pastore, M.T.; Prior, T.W.; Manickam, K.; Atkin, J.F.; Herman, G.E. Confirmation study of PTEN mutations among individuals with autism or developmental delays/mental retardation and macrocephaly. *Autism Res.* **2010**, *3*, 137–141. [CrossRef] [PubMed]

96. Varga, E.A.; Pastore, M.; Prior, T.; Herman, G.E.; McBride, K.L. The prevalence of PTEN mutations in a clinical pediatric cohort with autism spectrum disorders, developmental delay, and macrocephaly. *Genet. Med.* **2009**, *11*, 111–117. [CrossRef] [PubMed]

97. Hobert, J.A.; Embacher, R.; Mester, J.L.; Frazier, T.W.; Eng, C. Biochemical screening and PTEN mutation analysis in individuals with autism spectrum disorders and macrocephaly. *Eur. J. Hum. Genet.* **2014**, *22*, 273–276. [CrossRef] [PubMed]

98. Song, M.S.; Salmena, L.; Pandolfi, P.P. The functions and regulation of the PTEN tumour suppressor. *Nat. Rev. Mol. Cell. Biol.* **2012**, *13*, 283–296. [CrossRef] [PubMed]

99. Maehama, A.J.; Dixon, J.E. The Tumor Suppressor PTEN/MMAC1, Dephosphorylates the Lipid Second Messenger, Phosphatidylinositol 3,4,5-Trisphosphate. *J. Biol. Chem.* **1998**, *273*, 13375–13379. [CrossRef] [PubMed]

100. Zhou, J.; Blundell, J.; Ogawa, S.; Kwon, C.-H.; Zhang, W.; Sinton, C.; Powell, C.M.; Parada, L.F. Pharmacological inhibition of mTORC1 suppresses anatomical, cellular, and behavioral abnormalities in neural-specific Pten knock-out mice. *J. Neurosci.* **2009**, *29*, 1773–1783. [CrossRef] [PubMed]

101. Vanderver, A.; Tonduti, D.; Kahn, I.; Schmidt, J.; Medne, L.; Vento, J.; Chapman, K.A.; Lanpher, B.; Pearl, P.; Gropman, A.; et al. Characteristic brain magnetic resonance imaging pattern in patients with macrocephaly and PTEN mutations. *Am. J. Med. Genet. Part A* **2014**, *164*, 627–633. [CrossRef] [PubMed]

102. Jansen, L.A.; Mirzaa, G.M.; Ishak, G.E.; O'Roak, B.J.; Hiatt, J.B.; Roden, W.H.; Gunter, S.A.; Christian, S.L.; Collins, S.; Adams, C.; et al. PI3K/AKT pathway mutations cause a spectrum of brain malformations from megalencephaly to focal cortical dysplasia. *Brain* **2015**, *138*, 1613–1628. [CrossRef] [PubMed]

103. Conti, S.; Condo, M.; Posar, A.; Mari, F.; Resta, N.; Renieri, A.; Neri, I.; Patrizi, A.; Parmeggiani, A. Phosphatase and Tensin Homolog (PTEN) Gene Mutations and Autism: Literature Review and a Case Report of a Patient With Cowden Syndrome, Autistic Disorder, and Epilepsy. *J. Child Neurol.* **2012**, *27*, 392–397. [CrossRef] [PubMed]

104. Marchese, M.; Conti, V.; Valvo, G.; Moro, F.; Muratori, F.; Tancredi, R.; Santorelli, F.M.; Guerrini, R.; Sicca, F. Autism-epilepsy phenotype with macrocephaly suggests PTEN, but not GLIALCAM, genetic screening. *BMC Med. Genet.* **2014**, *15*, 26. [CrossRef] [PubMed]

105. Fraser, M.M.; Zhu, X.; Kwon, C.H.; Uhlmann, E.J.; Gutmann, D.H.; Baker, S.J. Pten loss causes hypertrophy and increased proliferation of astrocytes in vivo. *Cancer Res.* **2004**, *64*, 7773–7779. [CrossRef] [PubMed]

106. Kwon, C.H.; Luikart, B.W.; Powell, C.M.; Zhou, J.; Matheny, S.A.; Zhang, W.; Li, Y.; Baker, S.J.; Parada, L.F. Pten Regulates Neuronal Arborization and Social Interaction in Mice. *Neuron* **2006**, *50*, 377–388. [CrossRef] [PubMed]

107. Fraser, M.M.; Bayazitov, I.T.; Zakharenko, S.S.; Baker, S.J. Phosphatase and tensin homolog, deleted on chromosome 10 deficiency in brain causes defects in synaptic structure, transmission and plasticity, and myelination abnormalities. *Neuroscience* **2008**, *151*, 476–488. [CrossRef] [PubMed]

108. Luikart, B.W.; Schnell, E.; Washburn, E.K.; Bensen, A.L.; Tovar, K.R.; Westbrook, G.L. Pten Knockdown In Vivo Increases Excitatory Drive onto Dentate Granule Cells. *J. Neurosci.* **2011**, *31*, 4345–4354. [CrossRef] [PubMed]

109. Williams, M.R.; DeSpenza, T.; Li, M.; Gulledge, A.T.; Luikart, B.W. Hyperactivity of Newborn Pten Knock-out Neurons Results from Increased Excitatory Synaptic Drive. *J. Neurosci.* **2015**, *35*, 943–959. [CrossRef] [PubMed]

110. Huang, W.-C.; Chen, Y.; Page, D.T. Hyperconnectivity of prefrontal cortex to amygdala projections in a mouse model of macrocephaly/autism syndrome. *Nat. Commun.* **2016**, *7*, 13421. [CrossRef] [PubMed]

111. Marino, S.; Krimpenfort, P.; Leung, C.; van der Korput, H.A.G.M.; Trapman, J.; Camenisch, I.; Berns, A.; Brandner, S. PTEN is essential for cell migration but not for fate determination and tumourigenesis in the cerebellum. *Development* **2002**, *129*, 3513–3522. [PubMed]

112. Getz, S.A.; DeSpenza, T.; Li, M.; Luikart, B.W. Rapamycin prevents, but does not reverse, aberrant migration in Pten knockout neurons. *Neurobiol. Dis.* **2016**, *93*, 12–20. [CrossRef] [PubMed]

113. Takeuchi, K.; Gertner, M.J.; Zhou, J.; Parada, L.F.; Bennett, M.V.L.; Zukin, R.S. Dysregulation of synaptic plasticity precedes appearance of morphological defects in a Pten conditional knockout mouse model of autism. *Proc. Natl. Acad. Sci. USA* **2013**, *110*, 4738–4743. [CrossRef] [PubMed]

114. Garg, S.; Green, J.; Leadbitter, K.; Emsley, R.; Lehtonen, A.; Evans, G.; Huson, S.M. Neurofibromatosis type 1 and autism spectrum disorder. *Pediatrics* **2013**, *132*, e1642–e1648. [CrossRef] [PubMed]

115. Xu, G.; Lin, B.; Tanaka, K.; Dunn, D.; Wood, D.; Gesteland, R.; White, R.; Weiss, R.; Tamanoi, F. The catalytic domain of the neurofibromatosis type 1 gene product stimulates ras GTPase and complements ira mutants of *S. cerevisiae*. *Cell* **1990**, *63*, 835–841. [CrossRef]

116. Cawthon, R.M.; Weiss, R.; Xu, G.; Viskochil, D.; Culver, M.; Stevens, J.; Robertson, M.; Dunn, D.; Gesteland, R.; O'Connell, P.; et al. A major segment of the neurofibromatosis type 1 gene: cDNA sequence, genomic structure, and point mutations. *Cell* **1990**, *62*, 193–201. [CrossRef]

117. Johannessen, C.M.; Reczek, E.E.; James, M.F.; Brems, H.; Legius, E.; Cichowski, K. The NF1 tumor suppressor critically regulates TSC2 and mTOR. *Proc. Natl. Acad. Sci. USA* **2005**, *102*, 8573–8578. [CrossRef] [PubMed]

118. Dasgupta, B.; Yi, Y.; Chen, D.Y.; Weber, J.D.; Gutmann, D.H. Proteomic Analysis Reveals Hyperactivation of the Mammalian Target of Rapamycin Pathway in Neurofibromatosis 1—Associated Human and Mouse Brain Tumors. *Cancer Res.* **2005**, *65*, 2755–2760. [CrossRef] [PubMed]

119. Banerjee, S.; Crouse, N.R.; Emnett, R.J.; Gianino, S.M.; Gutmann, D.H. Neurofibromatosis-1 regulates mTOR-mediated astrocyte growth and glioma formation in a TSC/Rheb-independent manner. *Proc. Natl. Acad. Sci. USA* **2011**, *108*, 15996–16001. [CrossRef] [PubMed]

120. Kulkantrakorn, K.; Geller, T.J. Seizures in neurofibromatosis 1. *Pediatr. Neurol.* **1998**, *19*, 347–350. [CrossRef]

121. Vivarelli, R.; Grosso, S.; Calabrese, F.; Farnetani, M.; Di, B.R.; Morgese, G.; Balestri, P. Epilepsy in neurofibromatosis 1. *J. Child Neurol.* **2003**, *18*, 338–342. [CrossRef] [PubMed]

122. Ostendorf, A.P.; Gutmann, D.H.; Weisenberg, J.L.Z. Epilepsy in individuals with neurofibromatosis type 1. *Epilepsia* **2013**, *54*, 1810–1814. [CrossRef] [PubMed]

123. Moore, B.D.; Slopis, J.M.; Jackson, E.F.; De Winter, A.E.; Leeds, N.E. Brain volume in children with neurofibromatosis type 1. *Neurology* **2000**, *54*, 914–920. [CrossRef] [PubMed]

124. Cutting, L.E.; Koth, C.W.; Burnette, C.P.; Abrams, M.T.; Kaufmann, W.E.; Denckla, M.B. Relationship of Cognitive Functioning, Whole Brain Volumes, and T2-Weighted Hyperintensities in Neurofibromatosis-1. *J. Child Neurol.* **2000**, *15*, 157–160. [CrossRef] [PubMed]

125. Margariti, P.N.; Blekas, K.; Katzioti, F.G.; Zikou, A.K.; Tzoufi, M.; Argyropoulou, M.I. Magnetization transfer ratio and volumetric analysis of the brain in macrocephalic patients with neurofibromatosis type 1. *Eur. Radiol.* **2007**, *17*, 433–438. [CrossRef] [PubMed]

126. Karlsgodt, K.H.; Rosser, T.; Lutkenhoff, E.S.; Cannon, T.D.; Silva, A.; Bearden, C.E. Alterations in White Matter Microstructure in Neurofibromatosis-1. *PLoS ONE* **2012**, *7*, e47854. [CrossRef] [PubMed]

127. Petrella, L.I.; Cai, Y.; Sereno, J.V.; Gon??alves, S.I.; Silva, A.J.; Castelo-Branco, M. Brain and behaviour phenotyping of a mouse model of neurofibromatosis type-1: An MRI/DTI study on social cognition. *Genes Brain Behav.* **2016**, *15*, 637–646. [CrossRef] [PubMed]

128. Korf, B.R.; Schneider, G.; Poussaint, T.Y. Structural anomalies revealed by neuroimaging studies in the brains of patients with neurofibromatosis type 1 and large deletions. *Genet. Med.* **1999**, *1*, 136–140. [CrossRef] [PubMed]

129. Balestri, P.; Vivarelli, R.; Grosso, S.; Santori, L.; Farnetani, M.A.; Galluzzi, P.; Vatti, G.P.; Calabrese, F.; Morgese, G. Malformations of cortical development in neurofibromatosis type 1. *Neurology* **2003**, *61*, 1799–1801. [CrossRef] [PubMed]

130. Huijbregts, S.C.; Loitfelder, M.; Rombouts, S.A.; Swaab, H.; Verbist, B.M.; Arkink, E.B.; van Buchem, M.A.; Veer, I.M. Cerebral volumetric abnormalities in Neurofibromatosis type 1: Associations with parent ratings of social and attention problems, executive dysfunction, and autistic mannerisms. *J. Neurodev. Disord.* **2015**, *7*, 32. [CrossRef] [PubMed]

131. Kim, E.; Wang, Y.; Kim, S.-J.; Bornhorst, M.; Jecrois, E.S.; Anthony, T.E.; Wang, C.; Li, Y.E.; Guan, J.-L.; Murphy, G.G.; et al. Transient inhibition of the ERK pathway prevents cerebellar developmental defects and improves long-term motor functions in murine models of neurofibromatosis type 1. *Elife* **2014**, *3*, e05151. [CrossRef] [PubMed]

132. Sanchez-Ortiz, E.; Cho, W.; Nazarenko, I.; Mo, W.; Chen, J.; Parada, L.F. NF1 regulation of RAS/ERK signaling is required for appropriate granule neuron progenitor expansion and migration in cerebellar development. *Genes Dev.* **2014**, *28*, 2407–2420. [CrossRef] [PubMed]

133. Bajenaru, M.L.; Zhu, Y.; Hedrick, N.M.; Donahoe, J.; Parada, L.F.; Gutmann, D.H. Astrocyte-specific inactivation of the neurofibromatosis 1 gene (NF1) is insufficient for astrocytoma formation. *Mol. Cell. Biol.* **2002**, *22*, 5100–5113. [CrossRef] [PubMed]

134. Hegedus, B.; Dasgupta, B.; Shin, J.E.; Emnett, R.J.; Hart-Mahon, E.K.; Elghazi, L.; Bernal-Mizrachi, E.; Gutmann, D.H. Neurofibromatosis-1 Regulates Neuronal and Glial Cell Differentiation from Neuroglial Progenitors In Vivo by Both cAMP- and Ras-Dependent Mechanisms. *Cell Stem Cell* **2007**, *1*, 443–457. [CrossRef] [PubMed]

135. Brown, J.A.; Gianino, S.M.; Gutmann, D.H. Defective cAMP generation underlies the sensitivity of CNS neurons to neurofibromatosis-1 heterozygosity. *J. Neurosci.* **2010**, *30*, 5579–5589. [CrossRef] [PubMed]

136. Lin, Y.L.; Lei, Y.T.; Hong, C.J.; Hsueh, Y.P. Syndecan-2 induces filopodia and dendritic spine formation via the neurofibromin-PKA-Ena/VASP pathway. *J. Cell Biol.* **2007**, *177*, 829–841. [CrossRef] [PubMed]

137. Wang, H.F.; Shih, Y.T.; Chen, C.Y.; Chao, H.W.; Lee, M.J.; Hsueh, Y.P. Valosin-containing protein and neurofibromin interact to regulate dendritic spine density. *J. Clin. Investig.* **2011**, *121*, 4820–4837. [CrossRef] [PubMed]

138. Costa, R.M.; Federov, N.B.; Kogan, J.H.; Murphy, G.G.; Stern, J.; Ohno, M.; Kucherlapati, R.; Jacks, T.; Silva, A.J. Mechanism for the learning deficits in a mouse model of neurofibromatosis type 1. *Nature* **2002**, *415*, 526–530. [CrossRef] [PubMed]

139. Guilding, C.; McNair, K.; Stone, T.W.; Morris, B.J. Restored plasticity in a mouse model of neurofibromatosis type 1 via inhibition of hyperactive ERK and CREB. *Eur. J. Neurosci.* **2007**, *25*, 99–105. [CrossRef] [PubMed]

140. Weiss, B.; Widemann, B.C.; Wolters, P.; Dombi, E.; Vinks, A.; Cantor, A.; Perentesis, J.; Schorry, E.; Ullrich, N.; Gutmann, D.H.; et al. Sirolimus for progressive neurofibromatosis type 1-associated plexiform neurofibromas: A neurofibromatosis clinical trials consortium phase II study. *Neurol. Oncol.* **2015**, *17*, 596–603. [CrossRef] [PubMed]

141. Hua, C.; Zehou, O.; Ducassou, S.; Minard-Colin, V.; Hamel-Teillac, D.; Wolkenstein, P.; Valeyrie-Allanore, L. Sirolimus Improves Pain in NF1 Patients with Severe Plexiform Neurofibromas. *Pediatrics* **2014**, *133*, e1792–e1797. [CrossRef] [PubMed]

142. Johannessen, C.M.; Johnson, B.W.; Williams, S.M.G.; Chan, A.W.; Reczek, E.E.; Lynch, R.C.; Rioth, M.J.; McClatchey, A.; Ryeom, S.; Cichowski, K. TORC1 Is Essential for NF1-Associated Malignancies. *Curr. Biol.* **2008**, *18*, 56–62. [CrossRef] [PubMed]

143. Bhola, P.; Banerjee, S.; Mukherjee, J.; Balasubramanium, A.; Arun, V.; Karim, Z.; Burrell, K.; Croul, S.; Gutmann, D.H.; Guha, A. Preclinical in vivo evaluation of rapamycin in human malignant peripheral nerve sheath explant xenograft. *Int. J. Cancer* **2010**, *126*, 563–571. [CrossRef] [PubMed]

144. Zhu, Y.; Romero, M.I.; Ghosh, P.; Ye, Z.; Charnay, P.; Rushing, E.J.; Marth, J.D.; Parada, L.F. Ablation of NF1 function in neurons induces abnormal development of cerebral cortex and reactive gliosis in the brain. *Genes Dev.* **2001**, *15*, 859–876. [CrossRef] [PubMed]
145. Hagerman, R.; Hoem, G.; Hagerman, P. Fragile X and autism: Intertwined at the molecular level leading to targeted treatments. *Mol. Autism* **2010**, *1*, 12. [CrossRef] [PubMed]
146. Verkerk, A.J.M.H.; Pieretti, M.; Sutcliffe, J.S.; Fu, Y.H.; Kuhl, D.P.A.; Pizzuti, A.; Reiner, O.; Richards, S.; Victoria, M.F.; Zhang, F.; et al. Identification of a gene (FMR-1) containing a CGG repeat coincident with a breakpoint cluster region exhibiting length variation in fragile X syndrome. *Cell* **1991**, *65*, 905–914. [CrossRef]
147. Bagni, C.; Greenough, W.T. From mRNP trafficking to spine dysmorphogenesis: The roots of fragile X syndrome. *Nat. Rev. Neurosci.* **2005**, *6*, 376–387. [CrossRef] [PubMed]
148. Darnell, J.C.; van Driesche, S.J.; Zhang, C.; Hung, K.Y.S.; Mele, A.; Fraser, C.E.; Stone, E.F.; Chen, C.; Fak, J.J.; Chi, S.W.; et al. FMRP stalls ribosomal translocation on mRNAs linked to synaptic function and autism. *Cell* **2011**, *146*, 247–261. [CrossRef] [PubMed]
149. Darnell, J.C.; Klann, E. The translation of translational control by FMRP: Therapeutic targets for FXS. *Nat. Neurosci.* **2013**, *16*, 1530–1536. [CrossRef] [PubMed]
150. Incorpora, G.; Sorge, G.; Sorge, A.; Pavone, L. Epilepsy in fragile X syndrome. *Brain Dev.* **2002**, *24*, 766–769. [CrossRef]
151. Hagerman, P.J.; Stafstrom, C.E. Origins of Epilepsy in Fragile X Syndrome. *Epilepsy Curr.* **2009**, *9*, 108–112. [CrossRef] [PubMed]
152. Butler, M.G.; Brunschwig, A.; Miller, L.K.; Hagerman, R.J. Standards for Selected Anthropometric Measurements in Males With the Fragile X Syndrome. *Pediatrics* **1992**, *89*, 1059–1062. [PubMed]
153. Chiu, S.; Wegelin, J.A.; Blank, J.; Jenkins, M.; Day, J.; Hessl, D.; Tassone, F.; Hagerman, R. Early acceleration of head circumference in children with fragile x syndrome and autism. *J. Dev. Behav. Pediatr.* **2007**, *28*, 31–35. [CrossRef] [PubMed]
154. Hinton, V.J.; Brown, W.T.; Wisniewski, K.; Rudelli, R.D. Analysis of neocortex in three males with the fragile X syndrome. *Am. J. Med. Genet.* **1991**, *41*, 289–294. [CrossRef] [PubMed]
155. Irwin, S.A.; Patel, B.; Idupulapati, M.; Harris, J.B.; Crisostomo, R.A.; Larsen, B.P.; Kooy, F.; Willems, P.J.; Cras, P.; Kozlowski, P.B.; et al. Abnormal Dendritic Spine Characteristics in the Temporal and Visual Cortices of Patients With Fragile-X Syndrome: A Quantitative Examination. *Am. J. Med. Genet.* **2001**, *98*, 161–167. [CrossRef]
156. Sheridan, S.D.; Theriault, K.M.; Reis, S.A.; Zhou, F.; Madison, J.M.; Daheron, L.; Loring, J.F.; Haggarty, S.J. Epigenetic Characterization of the FMR1 Gene and Aberrant Neurodevelopment in Human Induced Pluripotent Stem Cell Models of Fragile X Syndrome. *PLoS ONE* **2011**, *6*, e26203. [CrossRef] [PubMed]
157. Doers, M.E.; Musser, M.T.; Nichol, R.; Berndt, E.R.; Baker, M.; Gomez, T.M.; Zhang, S.-C.; Abbeduto, L.; Bhattacharyya, A. iPSC-Derived Forebrain Neurons from FXS Individuals Show Defects in Initial Neurite Outgrowth. *Stem Cells Dev.* **2014**, *23*, 1777–1787. [CrossRef] [PubMed]
158. Telias, M.; Kuznitsov-Yanovsky, L.; Segal, M.; Ben-Yosef, D. Functional Deficiencies in Fragile X Neurons Derived from Human Embryonic Stem Cells. *J. Neurosci.* **2015**, *35*, 15295–15306. [CrossRef] [PubMed]
159. Musumeci, S.A.; Calabrese, G.; Bonaccorso, C.M.; D'Antoni, S.; Brouwer, J.R.; Bakker, C.E.; Elia, M.; Ferri, R.; Nelson, D.L.; Oostra, B.A.; et al. Audiogenic seizure susceptibility is reduced in fragile X knockout mice after introduction of FMR1 transgenes. *Exp. Neurol.* **2007**, *203*, 233–240. [CrossRef] [PubMed]
160. Bhattacharya, A.; Kaphzan, H.; Alvarez-Dieppa, A.C.; Murphy, J.P.; Pierre, P.; Klann, E. Genetic Removal of p70 S6 Kinase 1 Corrects Molecular, Synaptic, and Behavioral Phenotypes in Fragile X Syndrome Mice. *Neuron* **2012**, *76*, 325–337. [CrossRef] [PubMed]
161. Uutela, M.; Lindholm, J.; Louhivuori, V.; Wei, H.; Louhivuori, L.M.; Pertovaara, A.; Åkerman, K.; Castrén, E.; Castrén, M.L. Reduction of BDNF expression in Fmr1 knockout mice worsens cognitive deficits but improves hyperactivity and sensorimotor deficits. *Genes Brain Behav.* **2012**, *11*, 513–523. [CrossRef] [PubMed]
162. Comery, T.A.; Harris, J.B.; Willems, P.J.; Oostra, B.A.; Irwin, S.A.; Weiler, I.J.; Greenough, W.T. Abnormal dendritic spines in fragile X knockout mice: Maturation and pruning deficits. *Proc. Natl. Acad. Sci. USA* **1997**, *94*, 5401–5404. [CrossRef] [PubMed]
163. Galvez, R.; Greenough, W.T. Sequence of abnormal dendritic spine development in primary somatosensory cortex of a mouse model of the fragile X mental retardation syndrome. *Am. J. Med. Genet.* **2005**, *135A*, 155–160. [CrossRef] [PubMed]

164. Grossman, A.W.; Elisseou, N.M.; McKinney, B.C.; Greenough, W.T. Hippocampal pyramidal cells in adult Fmr1 knockout mice exhibit an immature-appearing profile of dendritic spines. *Brain Res.* **2006**, *1084*, 158–164. [CrossRef] [PubMed]

165. Liu, Z.-H.; Chuang, D.-M.; Smith, C.B. Lithium ameliorates phenotypic deficits in a mouse model of fragile X syndrome. *Int. J. Neuropsychopharmacol.* **2011**, *14*, 618–630. [CrossRef] [PubMed]

166. Gkogkas, C.G.; Khoutorsky, A.; Cao, R.; Jafarnejad, S.M.; Prager-Khoutorsky, M.; Giannakas, N.; Kaminari, A.; Fragkouli, A.; Nader, K.; Price, T.J.; et al. Pharmacogenetic Inhibition of eIF4E-Dependent Mmp9 mRNA Translation Reverses Fragile X Syndrome-like Phenotypes. *Cell Rep.* **2014**, *9*, 1742–1755. [CrossRef] [PubMed]

167. La Fata, G.; Gärtner, A.; Domínguez-Iturza, N.; Dresselaers, T.; Dawitz, J.; Poorthuis, R.B.; Averna, M.; Himmelreich, U.; Meredith, R.M.; Achsel, T.; et al. FMRP regulates multipolar to bipolar transition affecting neuronal migration and cortical circuitry. *Nat. Neurosci.* **2014**, *17*, 1693–1700. [CrossRef] [PubMed]

168. Pacey, L.K.K.; Guan, S.; Tharmalingam, S.; Thomsen, C.; Hampson, D.R. Persistent astrocyte activation in the fragile X mouse cerebellum. *Brain Behav.* **2015**, *5*, e00400. [CrossRef] [PubMed]

169. Pacey, L.K.K.; Xuan, I.C.Y.; Guan, S.; Sussman, D.; Henkelman, R.M.; Chen, Y.; Thomsen, C.; Hampson, D.R. Delayed myelination in a mouse model of fragile X syndrome. *Hum. Mol. Genet.* **2013**, *22*, 3920–3930. [CrossRef] [PubMed]

170. Huber, K.M.; Gallagher, S.M.; Warren, S.T.; Bear, M.F. Altered synaptic plasticity in a mouse model of fragile X mental retardation. *Proc. Natl. Acad. Sci. USA* **2002**, *99*, 7746–7750. [CrossRef] [PubMed]

171. Nosyreva, E.D.; Huber, K.M.; Elena, D.; Metabotropic, K.M. H. Metabotropic Receptor-Dependent Long-Term Depression Persists in the Absence of Protein Synthesis in the Mouse Model of Fragile X Syndrome. *J. Neurophysiol.* **2006**, *95*, 3291–3295. [CrossRef] [PubMed]

172. Sharma, A.; Hoeffer, C.A.; Takayasu, Y.; Miyawaki, T.; McBride, S.M.; Klann, E.; Zukin, R.S. Dysregulation of mTOR signaling in fragile X syndrome. *J. Neurosci.* **2010**, *30*, 694–702. [CrossRef] [PubMed]

173. Gonçalves, J.T.; Anstey, J.E.; Golshani, P.; Portera-Cailliau, C. Circuit level defects in the developing neocortex of Fragile X mice. *Nat. Neurosci.* **2013**, *16*, 903–909. [CrossRef] [PubMed]

174. Deng, P.Y.; Rotman, Z.; Blundon, J.A.; Cho, Y.; Cui, J.; Cavalli, V.; Zakharenko, S.S.; Klyachko, V.A. FMRP Regulates Neurotransmitter Release and Synaptic Information Transmission by Modulating Action Potential Duration via BK Channels. *Neuron* **2013**, *77*, 696–711. [CrossRef] [PubMed]

175. Braun, K.; Segal, M. FMRP involvement in formation of synapses among cultured hippocampal neurons. *Cereb. Cortex* **2000**, *10*, 1045–1052. [CrossRef] [PubMed]

176. Selby, L.; Zhang, C.; Sun, Q. Major Defects in Neocortical GABAergic Inhibitory Circuits in Mice Lacking the Fragile X Mental Retardation Protein. *Neurosci. Lett.* **2007**, *412*, 227–232. [CrossRef] [PubMed]

177. Gross, C.; Nakamoto, M.; Yao, X.; Chan, C.; Yim, S.Y.; Warren, S.T.; Bassell, G.J. Excess PI3K subunit synthesis and activity as a novel therapeutic target in Fragile X Syndrome. *Neuroscience* **2010**, *30*, 10624–10638. [PubMed]

178. Hayashi, M.L.; Rao, B.S.S.; Seo, J.-S.; Choi, H.-S.; Dolan, B.M.; Choi, S.-Y.; Chattarji, S.; Tonegawa, S. Inhibition of p21-activated kinase rescues symptoms of fragile X syndrome in mice. *Proc. Natl. Acad. Sci. USA* **2007**, *104*, 11489–11494. [CrossRef] [PubMed]

179. Zhao, M.-G.; Toyoda, H.; Ko, S.W.; Ding, H.-K.; Wu, L.-J.; Zhou, M. Deficits in Trace Fear Memory and Long-Term Potentiation in a Mouse Model for Fragile X Syndrome. *J. Neurosci.* **2005**, *25*, 7385–7392. [CrossRef] [PubMed]

180. Pilpel, Y.; Kolleker, A.; Berberich, S.; Ginger, M.; Frick, A.; Mientjes, E.; Oostra, B.A.; Seeburg, P.H. Synaptic ionotropic glutamate receptors and plasticity are developmentally altered in the CA1 field of Fmr1 knockout mice. *J. Physiol.* **2009**, *587*, 787–804. [CrossRef] [PubMed]

181. Seese, R.R.; Babayan, A.H.; Katz, A.M.; Cox, C.D.; Lauterborn, J.C.; Lynch, G.; Gall, C.M. LTP induction translocates cortactin at distant synapses in wild-type but not Fmr1 knock-out mice. *J. Neurosci.* **2012**, *32*, 7403–7413. [CrossRef] [PubMed]

182. Narayanan, U.; Nalavadi, V.; Nakamoto, M.; Thomas, G.; Ceman, S.; Bassell, G.J.; Warren, S.T. S6K1 phosphorylates and regulates fragile X mental retardation protein (FMRP) with the neuronal protein synthesis-dependent mammalian target of rapamycin (mTOR) signaling cascade. *J. Biol. Chem.* **2008**, *283*, 18478–18482. [CrossRef] [PubMed]

183. Bartley, C.M.; O'Keefe, R.A.; Bordey, A. FMRP S499 is phosphorylated independent of mTORC1-S6K1 activity. *PLoS ONE* **2014**, *9*, e96956.

184. Vu, T.H.; Hoffman, A.R. Imprinting of the Angelman syndrome gene, UBE3A, is restricted to brain. *Nat. Genet.* **1997**, *17*, 12–13. [CrossRef] [PubMed]

185. Grier, M.D.; Carson, R.P.; Lagrange, A.H. Toward a broader view of Ube3a in a mouse model of Angelman Syndrome: Expression in brain, spinal cord, sciatic nerve, glial cells. *PLoS ONE* **2015**, *10*, 1–14. [CrossRef] [PubMed]

186. Buiting, K.; Williams, C.; Horsthemke, B. Angelman syndrome—Insights into a rare neurogenetic disorder. *Nat. Rev. Neurol.* **2016**, *12*, 584–593. [CrossRef] [PubMed]

187. Bonati, M.T.; Russo, S.; Finelli, P.; Valsecchi, M.R.; Cogliati, F.; Cavalleri, F.; Roberts, W.; Elia, M.; Larizza, L. Evaluation of autism traits in Angelman syndrome: A resource to unfold autism genes. *Neurogenetics* **2007**, *8*, 169–178. [CrossRef] [PubMed]

188. Matsumoto, A.; Kumagai, T.; Miura, K.; Miyazaki, S.; Hayakawa, C.; Yamanaka, T. Epilepsy in Angelman Syndrome Associated with Chromosome 15q Deletion. *Epilepsia* **1992**, *33*, 1083–1090. [CrossRef] [PubMed]

189. Clayton-Smith, J. Clinical research on Angelman syndrome in the United Kingdom: Observations on 82 affected individuals. *Am. J. Med. Genet.* **1993**, *46*, 12–15. [CrossRef] [PubMed]

190. Tan, W.H.; Bacino, C.A.; Skinner, S.A.; Anselm, I.; Barbieri-Welge, R.; Bauer-Carlin, A.; Beaudet, A.L.; Bichell, T.J.; Gentile, J.K.; Glaze, D.G.; et al. Angelman syndrome: Mutations influence features in early childhood. *Am. J. Med. Genet. Part A* **2011**, *155*, 81–90. [CrossRef] [PubMed]

191. Harting, I.; Seitz, A.; Rating, D.; Sartor, K.; Zschocke, J.; Janssen, B.; Ebinger, F.; Wolf, N.I. Abnormal myelination in Angelman syndrome. *Eur. J. Paediatr. Neurol.* **2009**, *13*, 271–276. [CrossRef] [PubMed]

192. Castro-Gago, M.; Gómez-Lado, C.; Eirís-Puñal, J. Abnormal myelination in Angelman syndrome. *Eur. J. Paediatr. Neurol.* **2010**, *14*, 292. [CrossRef] [PubMed]

193. Jiang, Y.-H.; Armstrong, D.; Albrecht, U.; Atkins, C.M.; Noebels, J.L.; Eichele, G.; Sweatt, J.D.; Beaudet, A.L. Mutation of the Angelman ubiquitin ligase in mice causes increased cytoplasmic p53 and deficits of contextual learning and long-term potentiation. *Neuron* **1998**, *21*, 799–811. [CrossRef]

194. Miura, K.; Kishino, T.; Li, E.; Webber, H.; Dikkes, P.; Holmes, G.L.; Wagstaff, J. Neurobehavioral and electroencephalographic abnormalities in Ube3a maternal-deficient mice. *Neurobiol. Dis.* **2002**, *9*, 149–159. [CrossRef] [PubMed]

195. Grier, M.D.; Carson, R.P.; Lagrange, A.H. Of mothers and myelin: Aberrant myelination phenotypes in mouse model of Angelman syndrome are dependent on maternal and dietary influences. *Behav. Brain Res.* **2015**, *291*, 260–267. [CrossRef] [PubMed]

196. Dindot, S.V.; Antalffy, B.A.; Bhattacharjee, M.B.; Beaudet, A.L. The Angelman syndrome ubiquitin ligase localizes to the synapse and nucleus, and maternal deficiency results in abnormal dendritic spine morphology. *Hum. Mol. Genet.* **2008**, *17*, 111–118. [CrossRef] [PubMed]

197. Wallace, M.L.; Burette, A.C.; Weinberg, R.J.; Philpot, B.D. Maternal Loss of Ube3a Produces an Excitatory/Inhibitory Imbalance through Neuron Type-Specific Synaptic Defects. *Neuron* **2012**, *74*, 793–800. [CrossRef] [PubMed]

198. Miao, S.; Chen, R.; Ye, J.; Tan, G.-H.; Li, S.; Zhang, J.; Jiang, Y.-H.; Xiong, Z.-Q. The Angelman Syndrome Protein Ube3a Is Required for Polarized Dendrite Morphogenesis in Pyramidal Neurons. *J. Neurosci.* **2013**, *33*, 327–333. [CrossRef] [PubMed]

199. Tonazzini, I.; Meucci, S.; van Woerden, G.M.; Elgersma, Y.; Cecchini, M. Impaired Neurite Contact Guidance in Ubiquitin Ligase E3a (Ube3a)-Deficient Hippocampal Neurons on Nanostructured Substrates. *Adv. Healthc. Mater.* **2016**, *5*, 850–862. [CrossRef] [PubMed]

200. Yashiro, K.; Riday, T.T.; Condon, K.H.; Roberts, A.C.; Bernardo, D.R.; Prakash, R.; Weinberg, R.J.; Ehlers, M.D.; Philpot, B.D. Ube3a is required for experience-dependent maturation of the neocortex. *Nat. Neurosci.* **2009**, *12*, 777–783. [CrossRef]

201. Pignatelli, M.; Piccinin, S.; Molinaro, G.; Di Menna, L.; Riozzi, B.; Cannella, M.; Motolese, M.; Vetere, G.; Catania, M.V.; Battaglia, G.; et al. Changes in mGlu5 receptor-dependent synaptic plasticity and coupling to homer proteins in the hippocampus of Ube3A hemizygous mice modeling angelman syndrome. *J. Neurosci.* **2014**, *34*, 4558–4566. [CrossRef] [PubMed]

202. Greer, P.L.; Hanayama, R.; Bloodgood, B.L.; Mardinly, A.R.; Lipton, D.M.; Flavell, S.W.; Kim, T.K.; Griffith, E.C.; Waldon, Z.; Maehr, R.; et al. The Angelman Syndrome Protein Ube3A Regulates Synapse Development by Ubiquitinating Arc. *Cell* **2010**, *140*, 704–716. [CrossRef] [PubMed]

203. Amir, R.E.; van den Veyver, I.B.; Wan, M.; Tran, C.Q.; Francke, U.; Zoghbi, H.Y. Rett syndrome is caused by mutations in X-linked MECP2, encoding methyl-CpG-binding protein 2. *Nat. Genet.* **1999**, *23*, 185–188. [PubMed]

204. Bienvenu, T.; Chelly, J. Molecular genetics of Rett syndrome: When DNA methylation goes unrecognized. *Nat. Rev. Genet.* **2006**, *7*, 415–426. [CrossRef] [PubMed]

205. Chahrour, M.; Jung, S.Y.; Shaw, C.; Zhou, X.; Wong, S.T.C.; Qin, J.; Zoghbi, H.Y. MeCP2, a key contributor to neurological disease, activates and represses transcription. *Science* **2008**, *320*, 1224–1229. [CrossRef] [PubMed]

206. Cardoza, B.; Clarke, A.; Wilcox, J.; Gibbon, F.; Smith, P.E.M.; Archer, H.; Hryniewiecka-Jaworska, A.; Kerr, M. Epilepsy in Rett syndrome: Association between phenotype and genotype, and implications for practice. *Seizure* **2011**, *20*, 646–649. [CrossRef] [PubMed]

207. Dolce, A.; Ben-Zeev, B.; Naidu, S.; Kossoff, E.H. Rett syndrome and epilepsy: An update for child neurologists. *Pediatr. Neurol.* **2013**, *48*, 337–345. [CrossRef] [PubMed]

208. Jellinger, K.; Armstrong, D.; Zoghbi, H.Y.; Percy, A.K.; Boltzmann, L.; Wien, A. Neuropathology of Rett syndrome. *Acta Neuropathol.* **1988**, *76*, 142–158. [CrossRef] [PubMed]

209. Hagberg, G.; Stenbom, Y.; Witt Engerström, I. Head growth in Rett syndrome. *Brain Dev.* **2001**, *23*, 227–229. [CrossRef]

210. Bauman, M.L.; Kemper, T.L.; Arin, D.M. Microscopic observations of the brain in Rett syndrome. *Neuropediatrics* **1995**, *26*, 105–108. [CrossRef] [PubMed]

211. Bauman, M.L.; Kemper, T.L.; Arin, D.M. Pervasive neuroanatomic abnormalities of the brain in three cases of Rett's syndrome. *Neurology* **1995**, *45*, 1581–1586. [CrossRef] [PubMed]

212. Lipani, J.D.; Bhattacharjee, M.B.; Corey, D.M.; Lee, D.A. Reduced nerve growth factor in Rett syndrome postmortem brain tissue. *J. Neuropathol. Exp. Neurol.* **2000**, *59*, 889–895. [CrossRef] [PubMed]

213. Armstrong, D.; Dunn, J.K.; Antalffy, B.T.R. Selective dendritic alterations in the cortex of Rett syndrome. *J. Neuropathol. Exp. Neurol.* **1995**, *54*, 195–201. [CrossRef] [PubMed]

214. Armstrong, D.D. Neuropathology of Rett syndrome. *J. Child Neurol.* **2005**, *20*, 747–753. [CrossRef] [PubMed]

215. Belichenko, P.V.; Oldfors, A.; Hagberg, B.; Dahlström, A. Rett syndrome: 3-D confocal microscopy of cortical pyramidal dendrites and afferents. *Neuroreport* **1994**, *5*, 1509–1513. [CrossRef] [PubMed]

216. Kaufmann, W.E.; Taylor, C.; Hohmann, C.; Sanwal, I.; Naidu, S. Abnormalities in neuronal maturation in Rett syndrome neocortex: Preliminary molecular correlates. *Eur. Child Adolesc. Psychiatry* **1997**, *6*, 75–77.

217. Chapleau, C.A.; Larimore, J.L.; Theibert, A.; Pozzo-Miller, L. Modulation of dendritic spine development and plasticity by BDNF and vesicular trafficking: Fundamental roles in neurodevelopmental disorders associated with mental retardation and autism. *J. Neurodev. Disord.* **2009**, *1*, 185–196. [CrossRef] [PubMed]

218. Chen, R.Z.; Akbarian, S.; Tudor, M.; Jaenisch, R. Deficiency of methyl-CpG binding protein-2 in CNS neurons results in a Rett-like phenotype in mice. *Nat. Genet.* **2001**, *27*, 327–331. [CrossRef] [PubMed]

219. Shahbazian, M.D.; Young, J.I.; Yuva-Paylor, L.A.; Spencer, C.M.; Antalffy, B.A.; Noebels, J.L.; Armstrong, D.L.; Paylor, R.; Zoghbi, H.Y. Mice with truncated MeCP2 recapitulate many Rett syndrome features and display hyperacetylation of histone H3. *Neuron* **2002**, *35*, 243–254. [CrossRef]

220. Kishi, N.; Macklis, J.D. MECP2 is progressively expressed in post-migratory neurons and is involved in neuronal maturation rather than cell fate decisions. *Mol. Cell. Neurosci.* **2004**, *27*, 306–321. [CrossRef] [PubMed]

221. Fukuda, T.; Yamashita, Y.; Nagamitsu, S.; Miyamoto, K.; Jin, J.J.; Ohmori, I.; Ohtsuka, Y.; Kuwajima, K.; Endo, S.; Iwai, T.; et al. Methyl-CpG binding protein 2 gene (MECP2) variations in Japanese patients with Rett syndrome: Pathological mutations and polymorphisms. *Brain Dev.* **2005**, *27*, 211–217. [CrossRef] [PubMed]

222. Tropea, D.; Giacometti, E.; Wilson, N.R.; Beard, C.; McCurry, C.; Fu, D.D.; Flannery, R.; Jaenisch, R.; Sur, M. Partial reversal of Rett Syndrome-like symptoms in MeCP2 mutant mice. *Proc. Natl. Acad. Sci. USA* **2009**, *106*, 2029–2034. [CrossRef] [PubMed]

223. Nguyen, M.V.C.; Du, F.; Felice, C.A.; Shan, X.; Nigam, A.; Mandel, G.; Robinson, J.K.; Ballas, N. MeCP2 is critical for maintaining mature neuronal networks and global brain anatomy during late stages of postnatal brain development and in the mature adult brain. *J. Neurosci.* **2012**, *32*, 10021–10034. [CrossRef] [PubMed]

224. Castro, J.; Garcia, R.I.; Kwok, S.; Banerjee, A.; Petravicz, J.; Woodson, J.; Mellios, N.; Tropea, D.; Sur, M. Functional recovery with recombinant human IGF1 treatment in a mouse model of Rett Syndrome. *Proc. Natl. Acad. Sci. USA* **2014**, *111*, 9941–9946. [CrossRef] [PubMed]

225. Baj, G.; Patrizio, A.; Montalbano, A.; Sciancalepore, M.; Tongiorgi, E. Developmental and maintenance defects in Rett syndrome neurons identified by a new mouse staging system in vitro. *Front. Cell. Neurosci.* **2014**, *8*, 18. [CrossRef] [PubMed]

226. Asaka, Y.; Jugloff, D.G.M.; Zhang, L.; Eubanks, J.H.; Fitzsimonds, R.M. Hippocampal synaptic plasticity is impaired in the Mecp2-null mouse model of Rett syndrome. *Neurobiol. Dis.* **2006**, *21*, 217–227. [CrossRef] [PubMed]

227. Moretti, P. Learning and Memory and Synaptic Plasticity Are Impaired in a Mouse Model of Rett Syndrome. *J. Neurosci.* **2006**, *26*, 319–327. [CrossRef] [PubMed]

228. Chao, H.T.; Zoghbi, H.Y.; Rosenmund, C. MeCP2 Controls Excitatory Synaptic Strength by Regulating Glutamatergic Synapse Number. *Neuron* **2007**, *56*, 58–65. [CrossRef] [PubMed]

229. Ballas, N.; Lioy, D.T.; Grunseich, C.; Mandel, G. Non-cell autonomous influence of MeCP2-deficient glia on neuronal dendritic morphology. *Nat. Neurosci.* **2009**, *12*, 311–317. [CrossRef] [PubMed]

230. Maezawa, I.; Swanberg, S.; Harvey, D.; LaSalle, J.M.; Jin, L.-W. Rett Syndrome Astrocytes Are Abnormal and Spread MeCP2 Deficiency through Gap Junctions. *J. Neurosci.* **2009**, *29*, 5051–5061. [CrossRef] [PubMed]

231. Kishi, N.; Macklis, J.D. MeCP2 functions largely cell-autonomously, but also non-cell-autonomously, in neuronal maturation and dendritic arborization of cortical pyramidal neurons. *Exp. Neurol.* **2010**, *222*, 51–58. [CrossRef] [PubMed]

232. Marchetto, M.C.N.; Carromeu, C.; Acab, A.; Yu, D.; Yeo, G.W.; Mu, Y.; Chen, G.; Gage, F.H.; Muotri, A.R. A model for neural development and treatment of rett syndrome using human induced pluripotent stem cells. *Cell* **2010**, *143*, 527–539. [CrossRef] [PubMed]

233. Li, Y.; Wang, H.; Muffat, J.; Cheng, A.W.; Orlando, D.A.; Lovén, J.; Kwok, S.M.; Feldman, D.A.; Bateup, H.S.; Gao, Q.; et al. Global transcriptional and translational repression in human-embryonic-stem-cell-derived rett syndrome neurons. *Cell Stem Cell* **2013**, *13*, 446–458. [CrossRef] [PubMed]

234. Williams, E.C.; Zhong, X.; Mohamed, A.; Li, R.; Liu, Y.; Dong, Q.; Ananiev, G.E.; Choongmok, J.C.; Lin, B.R.; Lu, J.; et al. Mutant astrocytes differentiated from Rett syndrome patients-specific iPSCs have adverse effects on wildtype neurons. *Hum. Mol. Genet.* **2014**, *23*, 2968–2980. [CrossRef] [PubMed]

235. Djuric, U.; Cheung, A.Y.L.; Zhang, W.; Mok, R.S.; Lai, W.; Piekna, A.; Hendry, J.A.; Ross, P.J.; Pasceri, P.; Kim, D.S.; et al. MECP2e1 isoform mutation affects the form and function of neurons derived from Rett syndrome patient iPS cells. *Neurobiol. Dis.* **2015**, *76*, 37–45. [CrossRef] [PubMed]

236. Zhang, Z.-N.; Freitas, B.C.; Qian, H.; Lux, J.; Acab, A.; Trujillo, C.A.; Herai, R.H.; Nguyen Huu, V.A.; Wen, J.H.; Joshi-Barr, S.; et al. Layered hydrogels accelerate iPSC-derived neuronal maturation and reveal migration defects caused by MeCP2 dysfunction. *Proc. Natl. Acad. Sci. USA* **2016**, *113*, 3185–3190. [CrossRef] [PubMed]

237. Cook, E.H.; Lindgren, V.; Leventhal, B.L.; Courchesne, R.; Lincoln, A.; Shulman, C.; Lord, C.; Courchesne, E. Autism or Atypical Autism in Maternally but Not Paternally Derived Proximal 15q Duplication. *Am. J. Hum. Genet.* **1997**, *60*, 928–934. [PubMed]

238. Cook, E.H.; Scherer, S.W. Copy-number variations associated with neuropsychiatric conditions. *Nature* **2008**, *455*, 919–923. [CrossRef] [PubMed]

239. Driscoll, D.; Waters, M.; Williams, C.; Zori, R.; Glenn, C.; Avidano, K.; Nicholls, R. A DNA methylation imprint, determined by the sex of the parent, distinguishes the Angelman and Prader-Willi syndromes. *Genomics* **1992**, *13*, 917–924. [CrossRef]

240. Smith, S.E. P.; Zhou, Y.-D.; Zhang, G.; Jin, Z.; Stoppel, D.C.; Anderson, M.P. Increased gene dosage of Ube3a results in autism traits and decreased glutamate synaptic transmission in mice. *Sci. Transl. Med.* **2011**, *3*, 103ra97. [CrossRef] [PubMed]

241. Nakatani, J.; Tamada, K.; Hatanaka, F.; Ise, S.; Ohta, H.; Inoue, K.; Tomonaga, S.; Watanabe, Y.; Chung, Y.J.; Banerjee, R.; et al. Abnormal Behavior in a Chromosome- Engineered Mouse Model for Human 15q11-13 Duplication Seen in Autism. *Cell* **2009**, *137*, 1235–1246. [CrossRef] [PubMed]

242. Isshiki, M.; Tanaka, S.; Kuriu, T.; Tabuchi, K.; Takumi, T.; Okabe, S. Enhanced synapse remodelling as a common phenotype in mouse models of autism. *Nat. Commun.* **2014**, *5*, 4742. [CrossRef] [PubMed]

243. Piochon, C.; Kloth, A.D.; Grasselli, G.; Titley, H.K.; Nakayama, H.; Hashimoto, K.; Wan, V.; Simmons, D.H.; Eissa, T.; Nakatani, J.; et al. Masanobu Kano8, Samuel S-H Wang2, 3, and C.H. Cerebellar Plasticity and Motor Learning Deficits in a Copy Number Variation Mouse Model of Autism. *Nat. Commun.* **2014**, *5*, 5586. [CrossRef] [PubMed]

244. Doornbos, M.; Sikkema-Raddatz, B.; Ruijvenkamp, C.A.L.; Dijkhuizen, T.; Bijlsma, E.K.; Gijsbers, A.C.J.; Hilhorst-Hofstee, Y.; Hordijk, R.; Verbruggen, K.T.; Kerstjens-Frederikse, W.S.; et al. Nine patients with a microdeletion 15q11.2 between breakpoints 1 and 2 of the Prader-Willi critical region, possibly associated with behavioural disturbances. *Eur. J. Med. Genet.* **2009**, *52*, 108–115. [CrossRef] [PubMed]

245. Van Der Zwaag, B.; Staal, W.G.; Hochstenbach, R.; Poot, M.; Spierenburg, H.; De Jonge, M.V.; Verbeek, N.E.; van 't Slot, R.; van Es, M.A.; Staal, F.J.; et al. A co-segregating microduplication of chromosome 15q11.2 pinpoints two risk genes for autism spectrum disorder. *Am. J. Med. Genet. Part B Neuropsychiatr. Genet.* **2009**, *153*, 960–966. [CrossRef] [PubMed]

246. Burnside, R.D.; Pasion, R.; Mikhail, F.M.; Carroll, A.J.; Robin, N.H.; Youngs, E.L.; Gadi, I.K.; Keitges, E.; Jaswaney, V.L.; Papenhausen, P.R.; et al. Microdeletion/microduplication of proximal 15q11.2 between BP1 and BP2: A susceptibility region for neurological dysfunction including developmental and language delay. *Hum. Genet.* **2011**, *130*, 517–528. [CrossRef] [PubMed]

247. Napoli, I.; Mercaldo, V.; Boyl, P.P.; Eleuteri, B.; Zalfa, F.; De Rubeis, S.; Di Marino, D.; Mohr, E.; Massimi, M.; Falconi, M.; et al. The Fragile X Syndrome Protein Represses Activity-Dependent Translation through CYFIP1, a New 4E-BP. *Cell* **2008**, *134*, 1042–1054. [CrossRef] [PubMed]

248. Kobayashi, K.; Kuroda, S.; Fukata, M.; Nakamura, T.; Nagase, T.; Nomura, N.; Matsuura, Y.; Yoshida-Kubomura, N.; Iwamatsu, A.; Kaibuchi, K. p140Sra-1 (specifically Rac1-associated protein) is a novel specific target for Rac1 small GTPase. *J. Biol. Chem.* **1998**, *273*, 291–295. [CrossRef] [PubMed]

249. DeRubeis, S.; Pasciuto, E.; Li, K.W.; Fernández, E.; DiMarino, D.; Buzzi, A.; Ostroff, L.E.; Klann, E.; Zwartkruis, F.J.T.; Komiyama, N.H.; Grant, S.G.N.; et al. CYFIP1 coordinates mRNA translation and cytoskeleton remodeling to ensure proper dendritic Spine formation. *Neuron* **2013**, *79*, 1169–1182. [CrossRef] [PubMed]

250. Nishimura, Y.; Martin, C.L.; Vazquez-Lopez, A.; Spence, S.J.; Alvarez-Retuerto, A.I.; Sigman, M.; Steindler, C.; Pellegrini, S.; Schanen, N.C.; Warren, S.T.; et al. Genome-wide expression profiling of lymphoblastoid cell lines distinguishes different forms of autism and reveals shared pathways. *Hum. Mol. Genet.* **2007**, *16*, 1682–1698. [CrossRef] [PubMed]

251. Pathania, M.; Davenport, E.C.; Muir, J.; Sheehan, D.F.; López-Doménech, G.; Kittler, J.T. The autism and schizophrenia associated gene CYFIP1 is critical for the maintenance of dendritic complexity and the stabilization of mature spines. *Transl. Psychiatry* **2014**, *4*, e374. [CrossRef] [PubMed]

252. Spence, S.J.; Schneider, M.T. The role of epilepsy and epileptiform EEGs in autism spectrum disorders. *Pediatr. Res.* **2009**, *65*, 599–606. [CrossRef] [PubMed]

253. Tuchman, R.; Alessandri, M.; Cuccaro, M. Autism spectrum disorders and epilepsy: Moving towards a comprehensive approach to treatment. *Brain Dev.* **2010**, *32*, 719–730. [CrossRef] [PubMed]

254. Courchesne, E.; Pierce, K.; Schumann, C.M.; Redcay, E.; Buckwalter, J.A.; Kennedy, D.P.; Morgan, J.T. Mapping early brain development in autism. *Neuron* **2007**, *56*, 399–413. [CrossRef] [PubMed]

255. Courchesne, E.; Campbell, K.; Solso, S. Brain growth across the life span in autism: Age-specific changes in anatomical pathology. *Brain Res.* **2011**, *1380*, 138–145. [CrossRef] [PubMed]

256. Casanova, M.F.; Pickett, J. The Neuropathology of Autism. In *Imaging the Brain in Autism*; Casanova, M., El-Baz, A., Suri, J., Eds.; Springer: New York, NY, USA, 2013; pp. 27–41.

257. Bauman, M.L.; Kemper, T.L. Neuroanatomic observations of the brain in autism: A review and future directions. *Int. J. Dev. Neurosci.* **2005**, *23*, 183–187. [CrossRef] [PubMed]

258. Wegiel, J.; Kuchna, I.; Nowicki, K.; Imaki, H.; Wegiel, J.; Marchi, E.; Ma, S.Y.; Chauhan, A.; Chauhan, V.; Bobrowicz, T.W.; et al. The neuropathology of autism: Defects of neurogenesis and neuronal migration, and dysplastic changes. *Acta Neuropathol.* **2010**, *119*, 755–770. [CrossRef] [PubMed]

259. Stoner, R.; Chow, M.L.; Boyle, M.P.; Sunkin, S.M.; Mouton, P.R.; Roy, S.; Wynshaw-Boris, A.; Colamarino, S.A.; Lein, E.S.; Courchesne, E. Patches of disorganization in the neocortex of children with autism. *N. Engl. J. Med.* **2014**, *370*, 1209–1219. [CrossRef] [PubMed]

260. Laurence, J.A.; Fatemi, S.H. Glial fibrillary acidic protein is elevated in superior frontal, parietal and cerebellar cortices of autistic subjects. *Cerebellum* **2005**, *4*, 206–210. [CrossRef] [PubMed]

261. Gozzi, M.; Nielson, D.M.; Lenroot, R.K.; Ostuni, J.L.; Luckenbaugh, D.A.; Thurm, A.E.; Giedd, J.N.; Swedo, S.E. A magnetization transfer imaging study of corpus callosum myelination in young children with autism. *Biol. Psychiatry* **2012**, *72*, 215–220. [CrossRef] [PubMed]

262. Deoni, S.C. L.; Dean, D.C.; Remer, J.; Dirks, H.; O'Muircheartaigh, J. Cortical maturation and myelination in healthy toddlers and young children. *Neuroimage* **2015**, *115*, 147–161. [CrossRef] [PubMed]

263. Hutsler, J.J.; Zhang, H. Increased dendritic spine densities on cortical projection neurons in autism spectrum disorders. *Brain Res.* **2010**, *1309*, 83–94. [CrossRef] [PubMed]

264. Jung, N.H.; Janzarik, W.G.; Delvendahl, I.; Münchau, A.; Biscaldi, M.; Mainberger, F.; Bäumer, T.; Rauh, R.; Mall, V. Impaired induction of long-term potentiation-like plasticity in patients with high-functioning autism and Asperger syndrome. *Dev. Med. Child Neurol.* **2012**, *55*, 83–89. [CrossRef] [PubMed]

265. Mariani, J.; Coppola, G.; Zhang, P.; Abyzov, A.; Provini, L.; Tomasini, L.; Amenduni, M.; Szekely, A.; Palejev, D.; Wilson, M.; et al. FOXG1-Dependent Dysregulation of GABA/Glutamate Neuron Differentiation in Autism Spectrum Disorders. *Cell* **2015**, *162*, 375–390. [CrossRef] [PubMed]

266. Capal, J.K.; Franz, D.N. Profile of everolimus in the treatment of tuberous sclerosis complex: An evidence-based review of its place in therapy. *Neuropsychiatr. Dis. Treat.* **2016**, *12*, 2165–2172. [PubMed]

267. Angliker, N.; Rüegg, M. In vivo evidence for mTORC2-mediated actin cytoskeleton rearrangement in Neurons. *Bioarchitecture* **2013**, *3*, 113–118. [CrossRef] [PubMed]

268. Thomanetz, V.; Angliker, N.; Cloëtta, D.; Lustenberger, R.M.; Schweighauser, M.; Oliveri, F.; Suzuki, N.; Rüegg, M.A. Ablation of the mTORC2 component rictor in brain or Purkinje cells affects size and neuron morphology. *J. Cell Biol.* **2013**, *201*, 293–308. [CrossRef] [PubMed]

269. Crino, P.B. The mTOR signalling cascade: Paving new roads to cure neurological disease. *Nat. Rev. Neurol.* **2016**, *12*, 379–392. [CrossRef] [PubMed]

270. Sengupta, S.; Peterson, T.R.; Sabatini, D.M. Regulation of the mTOR Complex 1 Pathway by Nutrients, Growth Factors, and Stress. *Mol. Cell* **2010**, *40*, 310–322. [CrossRef] [PubMed]

271. Inoki, K.; Li, Y.; Xu, T.; Guan, K. Rheb GTPase is a direct target of TSC2 GAP activity and regulates mTOR signaling. *Genes Dev.* **2003**, *17*, 1829–1834. [CrossRef] [PubMed]

272. Menon, S.; Dibble, C.C.; Talbott, G.; Hoxhaj, G.; Valvezan, A.J.; Takahashi, H.; Cantley, L.C.; Manning, B.D. Spatial Control of the TSC Complex Integrates Insulin and Nutrient Regulation of mTORC1 at the Lysosome. *Cell* **2014**, *156*, 771–785. [CrossRef] [PubMed]

273. Laplante, M.; Sabatini, D.M. mTOR Signaling in Growth Control and Disease. *Cell* **2012**, *149*, 274–293. [CrossRef] [PubMed]

International Journal of
Molecular Sciences

MDPI

Article

Autism and Intellectual Disability Associated with Mitochondrial Disease and Hyperlactacidemia

José Guevara-Campos [1], Lucía González-Guevara [2] and Omar Cauli [3],*

[1] "Felipe Guevara Rojas" Hospital, Pediatrics Service, University of Oriente, El Tigre-Anzoátegui, 6034 Venezuela, Spain; joguevara90@hotmail.com

[2] "Felipe Guevara Rojas" Hospital, Epilepsy and Encephalography Unit, El Tigre-Anzoátegui, 6034 Venezuela, Spain; proyectoasociacion@hotmail.com

[3] Department of Nursing, University of Valencia, 46010 Valencia, Spain

* Author to whom correspondence should be addressed; omar.cauli@uv.es
Tel.: +34-96-398-32-71; Fax: +34-96-386-43-10.

Academic Editor: Merlin G. Butler

Received: 26 November 2014; Accepted: 4 February 2015; Published: 11 February 2015

Abstract: Autism spectrum disorder (ASD) with intellectual disability (ID) is a life-long debilitating condition, which is characterized by cognitive function impairment and other neurological signs. Children with ASD-ID typically attain motor skills with a significant delay. A sub-group of ASD-IDs has been linked to hyperlactacidemia and alterations in mitochondrial respiratory chain activity. The objective of this report is to describe the clinical features of patients with these comorbidities in order to shed light on difficult diagnostic and therapeutic approaches in such patients. We reported the different clinical features of children with ID associated with hyperlactacidemia and deficiencies in mitochondrial respiratory chain complex II–IV activity whose clinical presentations are commonly associated with the classic spectrum of mitochondrial diseases. We concluded that patients with ASD and ID presenting with persistent hyperlactacidemia should be evaluated for mitochondrial disorders. Administration of carnitine, coenzyme Q10, and folic acid is partially beneficial, although more studies are needed to assess the efficacy of this vitamin/cofactor treatment combination.

Keywords: autism; possible mitochondrial disease; vitamins; intellectual disability; muscular tone

1. Introduction

Intellectual disability (ID) represents a significant social and economic burden; About 3% of the Western population is diagnosed with ID, and these patients require lifelong care [1–3]. An individual is considered to have an ID based on the following three criteria: (1) An intellectual functioning level (IQ) below 70; (2) Significant limitations exist in two or more adaptive skill areas (communication, self-care, home living, social skills, leisure, health and safety, self-direction, functional academics, community use, and work); and (3) the condition manifests itself before the age of 18 [4,5]. ID is sometimes associated with autism spectrum disorders (ASDs), behavioral disorders (hyperactivity, irritability, and self-injurious behavior), epilepsy, and/or other neurological disabilities (ataxia, hypotonia, sensorial alterations), all resulting in psychological, social, and economic burdens [1,6,7]. There are also some genetic causes of ID, which constitute 30%–50% of cases [3,7–9], however in the majority of instances the etiology is unknown. Assessment of molecular defects that result in synaptic deficits as well as defining suitable therapeutic treatments for these deficits remains a challenge [10].

Mitochondrial diseases can develop because of mitochondrial (mtDNA) or nuclear DNA (nDNA) mutations, however in many cases no mutations are found. These mutations are uncommon in the general population and have an estimated prevalence of around 1:8500 [11]. The extremely heterogeneous clinical presentation of mtDNA and nDNA mutations often causes diagnostic difficulties and thus many patients are diagnosed with mitochondrial disease without identifying the mtDNA or nDNA mutation responsible for the ASD-ID [2]. Additionally, the same genetic mutation can give rise to multiple varied phenotypes [12]. Mitochondrial respiratory chain (electron transport chain; ETC) disorders can be difficult to recognize clinically because of their diverse symptoms and clinical presentations and, as we report here, the diagnosis is often reached late in childhood.

Histopathological, biochemical, and genetic findings [13,14] are normally used to aid final diagnosis of possible, probable, or definite mitochondrial disease. The clinical presentation almost always widely varies, making definitive diagnosis of these mitochondrial disorders challenging [15] and although various criteria and checklists have been established these are more reflective of adult disease [13,14,16]. Patients with mitochondrial disease often show signs and symptoms such as heart, pancreas, or liver dysfunction, growth retardation, and fatigability, but sometimes the semiology is different from classical mitochondrial diseases and patients show symptoms associated with neurological deficits such as ID, ASD, abnormal muscle tone, seizures, extrapyramidal movements, and autonomic and ocular dysfunction [17,18].

Here, we report the interesting cases of three patients diagnosed with developmental delay, ID and ASD, and also with a possible mitochondrial disease accompanied by an ETC deficiency accompanied by hyperlactacidemia [14]. These cases have an unusual clinical presentation, which would not normally lead to suspicion of mitochondrial disease. For instance, fewer than three organ systems were involved, and red-flags for neurological involvement *i.e.*, cerebral stroke-like lesions in a nonvascular pattern, basal ganglia disease, recurrent encephalopathy, neurodegeneration, epilepsia partialis continua, myoclonus, ataxia, magnetic resonance imaging (MRI) findings consistent with Leigh disease, *etc.*, were not present [2]. We provide useful clinical evidence that patients presenting only ID and an alteration in muscular tone but in the absence of any other systemic signs may, according to the current classification of these disorders, also have a mitochondrial ETC disorder [14]. Children with ID and ASD who show no systemic signs should therefore be evaluated for comorbid mitochondrial disorders. In addition we describe some beneficial effects of combined pharmacological treatment with antipsychotic drugs and vitamins.

2. Results

2.1. Medical History and Clinical Examination

Patient 1: A 17-month-old boy was referred to the neuropediatrician because of generalized hypotonia and reduced cranial growth perimeter. He was the full-term product of a third pregnancy (preceded by a spontaneous miscarriage) resulting from a non-consanguineous relationship. His birth weight was 4.2 kg (60th percentile) and height was 51 cm (50th percentile), he had normal immunization and had an Apgar score of 9. However, he showed global developmental delay according to the Peabody Developmental Motor Scale 2 (PDMS-2) and the Battelle developmental inventory 2 (BDI-2; see Section 4.1).

Head support was achieved at seven months, with stable seating and autonomous position attained at 12 months; he started to walk autonomously at 30 months, and his first raw words appeared at 12 months. There was no family history of metabolic or neurological disease. Upon physical examination, his weight was 10 kg (50–75th percentile), height was 72 cm (90th percentile), head circumference was 41.5 cm (10th percentile, microcephaly), and his general condition was regular except for generalized hypotonia and loss of strength in the upper part of his lower-limbs but with normal deep tendon I/IV reflexes. No dysmorphic features, skin blemishes, or cardiopulmonary alterations were observed.

Patient 2: A two-year-old girl was referred to the neuropediatrician for psychomotor delay and generalized hypotonia. She was the term product of a first pregnancy resulting from a non-consanguineous relationship. Her birth weight was 2.7 kg (10th percentile), her height was 50 cm (50th percentile), and she had normal immunization. At birth, her Apgar score was 7 because of a delayed birth cry. Head support was achieved at 12 months, with a stable seating position achieved for the first time at 24 months, and autonomous walking at 30 months. Her first raw words appeared at 12 months but she did not produce complete sentences. She exhibited global developmental delay according to the PDMS-2 scale and BDI-2 inventory (see Section 4.1).

There was a family history of diabetes (four aunts and her maternal grandfather). Upon physical examination she had a weight of 13.5 kg (50–75th percentile), a height of 93 cm (90th percentile), head circumference of 43 cm (10th percentile, microcephaly), hypertonia in the left hemibody, and deep tendon hyperreflexia in her left-hand side limbs. There were no dysmorphic features, skin blemishes, or pulmonary alterations, and although her mitral valve was hyperelastic it was competent. Her thorax radiography, karyotype, and the EEG were normal, but an MRI showed ventriculomegaly with a prominent cysterna magna but without signs of displacement or compression (Figure 1A). Her associated clinical features are summarized in Table 1.

Table 1. Summary of the clinical features of the three patients described in this case study.

ID	Patient 1	Patient 2	Patient 3
	Severe	Moderate	Severe
Hypotonia	YES	YES (hypertonia at 1.5 months in the left hemibody)	YES
EEG Alteration	NO	NO	YES
Seizure	NO	NO	NO
Developmental Delay (Before age 5)	YES	YES	YES
Behavioral Problem (Irritability)	YES	NO	YES
PDD-NOS (According to DMS-IV)	YES (fulfilled 2 criteria of ASD diagnosis: Qualitative abnormalities in communication and reciprocal social interaction)	YES (fulfilled 1 criterion of ASD diagnosis: Qualitative abnormalities in communication)	YES (fulfilled 2 criteria of ASD diagnosis: Qualitative abnormalities in communication and reciprocal social interaction)
Microcephaly	YES	YES	NO
MRI Alteration (Ventricle enlargement)	YES	YES	YES

ID: Intellectual disability; EEG: Electroencephalography; PDD-NOS: Pervasive developmental disorder-not otherwise specified; DMS-IV: Diagnostic and statistical manual of mental disorders, fourth edition; and MRI: Magnetic resonance imaging.

Patient 3: A 19-month-old boy was referred to the neuropediatrician for psychomotor delay because he was unable to maintain an autonomous seating position or to autonomously walk, and he made poor visual contact and could not pronounce any words. He was the term product of a first pregnancy resulting from a non-consanguineous relationship, and there was no family history of metabolic or neurological diseases. His birth weight was 2.9 kg (50th percentile), his height was 74 cm (50th percentile), he received normal immunization, and his Apgar score was 9. He started to autonomously stand up from the age of three and there was autonomous ambulation at 40 months. His first raw words appeared at 12 months but he did not communicate intentionally or make any sentences. He exhibited global developmental delay according to the PDMS-2 scale and BDI-2 inventory (see Section 4.1), which was later defined as severe ID.

Upon physical examination his weight was 15 kg (50–75th percentile), height was 83 cm (90th percentile), and he had a normal head circumference of 47 cm. He presented moderate generalized hypotonia accompanied by a reduction of strength in his upper limbs and hyperreflexia (deep tendon II/IV reflexes). There were no dysmorphic features, skin blemishes, visceromegaly, or cardiopulmonary alterations. His thorax radiography and karyotype were normal, but a brain MRI showed cerebral atrophy, prominent cysterna magna, and a moderately increased ventricular volume (Figure 1B). An EEG during natural sleep showed generalized spikes of slow waves (1–2 Hz high voltage), although he never presented seizures. He was diagnosed with severe psychomotor delay, ID, and a possible mitochondrial disease due to decreased mitochondrial respiratory chain complex I activity. His clinical features are summarized in Table 1.

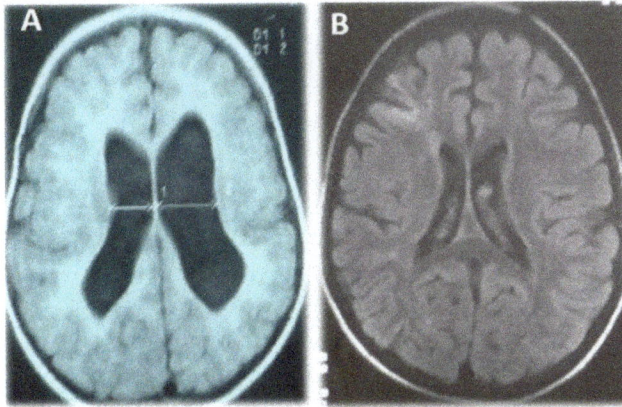

Figure 1. Magnetic resonance image of patient 2 (**A**) and 3 (**B**) showing an increase in lateral cerebral ventricle volume, diffuse enlargement of the cortical spaces, and cortical atrophy.

2.2. Blood and Muscle Laboratory Tests

Blood laboratory test results were normal for hemoglobin, hematocrit, leukocytes, platelets, glucose, urea, creatinine, alkaline phosphatase, alanine and aspartate transaminases (AST and ALT, respectively), total bilirubin concentration, cholesterol (above 145 mg/dL), ammonia, thyroid hormones, and thyrotropin for all three patients. The concentration of organic acids in urine, biotinidase, and cystine/homocysteine, acylcarnitine panel as well as all amino acids (including alanine) were also within the normal range. Due to persistent hyperlactacidemia (analysis was repeated three times on different days) and the clinical phenotypes of all three patients, femoral muscle biopsies were taken to evaluate any metabolic or histological alterations suggestive of a mitochondrial disorders; we measured mitochondrial ETC complex activity at three years for patient 1, 3.5 years for patient 2, and at 2.4 years for patient 3.

Patient 1: Blood gas analysis showed metabolic acidosis with pH 7.20, partial pressure of carbon dioxide (PCO_2) was 50 mmHg, partial pressure of oxygen (PO_2) was 134 mmHg, bicarbonate (HCO_3^-) was 17.6 mEq/L, and the base excess (BE) was −6.8 mEq/L. Serum electrolytes were normal average (sodium, potassium, and chlorine were 139, 4.80 and 105 mEq/L, respectively). Further tests revealed elevated levels of fasting lactate (2.96 mM; normal range 0.95–2.30), postprandial lactate (3.84 mM; normal range 1.95–3.0), fasting pyruvate (0.1 mM; normal range 0.05–0.09), postprandial pyruvate (0.15 mM; normal range 0.08–0.13), fasting lactate pyruvate ratio (29.6), and postprandial lactate pyruvate ratio (25.6). These details are summarized in Table 2.

Analysis of the muscle homogenate revealed a 30%–50% reduction in the enzymatic complexes II, III, and IV, and a slight reduction (12%) in coenzyme Q10 concentration (CoQ10). Mild and non-specific histological alterations in muscle fiber size were observed by optical microscopy. No known mtDNA mutations were found. The final diagnosis was severe ID (35th IQ percentile at age five), a possible mitochondrial disease (based on current diagnostic criteria [13,14]), a pervasive developmental disorder not otherwise specified (according to the DMS-IV criteria and the Autism Diagnostic Interview—Revised), and microcephaly.

Table 2. Metabolic energy study performed on muscle biopsy samples. The enzymatic activity of mitochondrial respiratory chain complexes (I–IV) was corrected for the presence of citrate synthase (CS). Succinate dehydrogenase activity was assessed in the presence of phenazine methosulfate, which acts as part of the electron transfer system. The measurement unit for each enzyme/complex activity was mU/U, and this was normalized to CS activity. Lactate values and the lactate/pyruvate ratio found in the plasma are shown. Values that are underlined represent those that are outside the normal range for children without mitochondrial disease.

Enzyme Activity or Metabolite Concentration	Patient 1	Patient 2	Patient 3	Control Values
NADH: Cit C oxidoreductase (Complex I + III) (mU/U CS)	156	310	253	107–560
Succinate: Cit C oxidoreductase (Complex II + III) (mU/U CS)	49	111	35	75–149
Succinate: DCPIP oxidoreductase (Complex II) (mU/U CS)	29	48	27	33–69
Succinate Dehydrogenase (mU/U CS)	74	119	79	57–239
Decylubiquinone: Cytochrome C oxidoreductase (Complex III) (mU/U CS)	498	597	615	610–1760
Cytochrome C oxidase (Complex IV) (mU/U CS)	287	501	291	590–1300
Coenzyme Q10 (nmol/U CS)	2.3	2.6	2.9	2.6–8.4
Citrate Synthase (nmol/min/mg)	243.7	125.7	350	71–200
Fasting Lactate (mM)	2.96	2.64	4.56	<2.30
Postprandial Lactate (mM)	3.84	2.82	5.15	<3
Fasting lactate/pyruvate ratio	29.6	47.7	14.2	10–15
Postprandial lactate/pyruvate ratio	25.6	42.2	24.5	10–15

Patient 2: Blood gas analysis showed the following: pH 7.35, PCO_2: 26.8 mmHg, PO_2: 98.9 mmHg, HCO_3: 24.3 mEq/L, and BE: −0.3 mEq/L (as shown in Table 2). Sodium, potassium, and chlorine serum electrolytes were 136, 4.93, and 102 mEq/L, respectively. Fasting lactate was elevated at 2.64 mM, while fasting pyruvate and the fasting lactate pyruvate ratio (0.06 mM and 47.67, respectively, were within the normal range). No histological alterations were observed by optical microscopy. Analysis of the muscle homogenate revealed a significant reduction in the activity of the mitochondrial respiratory chain enzymatic complexes III (decylubiquinol cytochrome c oxidoreductase) and IV (cytochrome c oxidase) to within 5% and 15% of the lower limit of their normal ranges, respectively (Table 2). No known mtDNA mutations were found and the final diagnosis was moderate ID (50th IQ percentile at age five), microcephaly, a pervasive developmental disorder not otherwise specified (using DMS-IV and the criteria of Autism Diagnostic Interview—Revised), and based on current diagnostic criteria [13,14], a possible mitochondrial disease.

Patient 3: Blood gas analysis showed metabolic acidosis with pH 7.33, PCO_2 of 25 mmHg, PO_2 of 90 mmHg, HCO_3 at 17.5 mEq/L, and a BE of −5.5 mEq/L. Serum electrolytes were: sodium 144 mEq/L, potassium: 4.7 mEq/L, and chlorine: 101 mEq/L. Other tests revealed significantly elevated fasting lactate (4.56 mM; normal range 0.95–2.30), postprandial lactate (5.15 mM; normal range 1.95–3.0), and fasting pyruvate (0.21 mM; normal range 0.08–0.10). Postprandial pyruvate was slightly above normal (0.15 mM; normal range 0.08–0.13), and fasting lactate pyruvate and postprandial lactate/pyruvate ratios were normal (14.2 and 24.5, respectively). Analysis of the muscle homogenate revealed a significant reduction (by 20%–45%) in the activity of mitochondrial respiratory chain complexes II and IV/I as well as increased CS. These data are summarized in Table 2. Optical microscopy revealed mild and non-specific histological variations in muscular fiber size. No known mtDNA mutations were found. The final diagnosis was moderate ID (55th IQ percentile at age five), a pervasive developmental disorder not otherwise specified (defined by the DMS IV and the criteria of Autism Diagnostic Interview—Revised), and a possible mitochondrial disease based on current diagnostic criteria [13,14].

2.3. Pharmacological Treatment

All three patients were started on the same initial treatment regime of 50 mg/Kg/day L-carnitine, a vitamin B complex (50 mg each of vitamin B1and B2, 15 mg B3, 2 mg B6, and 10 mg B12), and 30 mg CoQ10 divided twice daily, and 5 mg folic acid given once a day.

Patient 1: After one year of pharmacological treatment his intellectual abilities improved and he reached the 45th IQ percentile. This patient is currently seven years old and does not show any motor delay, motor alterations, or hypotonia. He recently displayed a behavioral disorder characterized by hyperactivity and irritability (as defined by the fifth edition of the diagnostic and statistical manual of mental disorders [DSM-5] by the American psychiatric association [19]), which was accompanied by psychotic reactions. These symptoms, as well as a sleep problem, greatly improved after he was started on risperidone (0.5 mg twice a day). It is unlikely that this antipsychotic drug caused the improvement of his ID because these changes were present before risperidone therapy, rather it is more likely that risperidone allowed him to be more responsive to other therapeutic interventions, and gave him a better ability to demonstrate his cognitive status when assessed. His intellectual abilities still remain stable (an additional year after the implementation of initial pharmacological treatment), however, he still uses very few words and is not able to construct complete sentences; he attends a special school to which he has adapted well.

Patient 2: This patient is currently nine years old and she attends a special school; She now walks autonomously with mild ataxia and instability. Once she started pharmacological treatment both signs (ataxia and tremor) disappeared, if CoQ10 is withdrawn she becomes ataxic and a strong hand tremor ensues. She is able to construct simple sentences and to communicate verbally; she understands verbal instructions and presents moderate ID that improved after treatment, reaching the 60th IQ percentile after 10 months of pharmacological treatment. After an additional year of this treatment her ID remains stable and although her hypertonia improved it still remains present at the lower limb level. This patient has never been referred for behavioral or sleep problems.

Patient 3: This patient is currently six years old and he attends a special school, which he has accepted well. His language is very poor and he displays episodes of irritability that improved following administration of 0.75 mg risperidone twice a day. Similar to patient 1, improvement of cognitive status was not due to risperidone treatment since it was started eight months after starting the mitochondrial drug combination treatment. ID also improved by 10 IQ points with this pharmacological treatment compared to before starting treatment (he reached the 60th IQ percentile, compared to the 50th measured a year before treatment started).

3. Discussion

Mitochondrial respiratory chain complex disorders are the most prevalent group of inherited neurometabolic diseases [2,3]. Classically, their clinical manifestation includes central and peripheral neurological manifestations, usually associated with the involvement of other organs including the eye, heart, liver, and kidneys, and may cause diabetes mellitus and/or sensorineural deafness. Patients with ID and a mitochondrial disease generally present several symptoms which are rarely observed in idiopathic ID, which can include tremor, ataxia, pancreatic or liver dysfunction, cardiac or hematological alterations, growth retardation, and/or ophthalmological or auditory signs [16,20–22]. Mitochondrial diseases are caused by abnormalities in the mitochondrial ETC, and may be accompanied by ID and an ASD in some individuals. Hypotonia with a long delay in motor development milestones such as head support, or stable autonomous seating and walking, can also suggest the presence of a mitochondrial disorder.

All three patients described in this study fulfilled the criteria for global developmental delay (under age five) and ID (over age five) according to the consensual terminology [23]. Their cranial MRIs showed a symmetrical increase in lateral cerebral ventricle volume and there was cortical atrophy in case 3. Interestingly, in addition to general hypotonia on the right side of her body, patient 2 also showed additional spasticity on the left side. This clinical finding probably represents a prior insult to the right hemisphere, as shown by asymmetry of lateral ventricles in her MRI image. Non-specific global MRI abnormalities such as abnormal or delayed myelination, atrophy, ventricle enlargement, or specific brain structural changes are common in patients with mitochondrial disorders and/or ID, especially those with clinical central nervous system involvement [16,24,25]. However, specific MRI findings (lesions in specific brain areas) are more likely to be associated with syndromic phenotypes, as in the case of MELAS (mitochondrial encephalomyopathy, lactic acidosis, and stroke-like episodes), Leigh, or Pearson/Kearns-Sayre syndromes [24,26,27]. Nonetheless, it should be stressed that brain MRI results can also appear normal in some patients with mitochondrial diseases.

Microcephaly, which was present in patients 1 and 2, has been described in both mitochondrial disorders and in ID cases [28], but its significance is poorly understood in terms of clinical presentation. Additionally, many patients with ID and mitochondrial disorders also have epilepsy [29] although, as reported here, EEG activity was abnormal in patient 3 and he had no clinical seizures, meaning that the lack of epilepsy does not exclude a mitochondrial disorder diagnosis. Some patients with mitochondrial diseases have hyperlactacidemia, which can be interpreted as a sign of mitochondrial metabolism impairment [30]. However, an increase in plasma lactate concentration is observable in many acute or chronic diseases and lactic acid in blood can be found within the normal range in some patients with mitochondrial disorders [2]. A persistent increase in plasma lactate concentration was found in all three cases and was accompanied by variable changes in pyruvate concentration (a slight increase in case 1, normal range in case 2, and a strong increase in case 3). Persistent hyperlactacidemia and an increased lactate/pyruvate ratio suggested to us that we should evaluate the presence of mitochondrial alterations by analyzing ETC activity in muscle tissue.

Our experience suggests that measuring the ratio of fasting lactate/pyruvate is recommendable in patients with ID with suspected mitochondrial disorders in order to further direct the diagnosis towards more relevant assays (*i.e.*, muscle biopsy, a detailed MRI/MRS study, or more sophisticated physiological measurements). In many cases a mitochondrial protein mutation is likely responsible for the clinical phenotype but in others, as reported here, no known mutations can be identified [31,32]. Based on these difficulties, and given that classical mitochondrial diseases affect only a small number of individuals with ID and ASD, we cannot exclude *de novo* mutations or secondary mitochondrial dysfunction in these patients. When these types of changes occur, they are accompanied by genetic abnormalities or biochemical defects in respiratory chain enzymes. Patient 1 and 3 had a 25%–50% complex II–IV deficiency, patient 1 also had slightly diminished levels of muscular CoQ10 (by 12%), and patient 2 displayed a 5%–15% decrease in complex III and IV activity. These biochemical findings are rare and add clinical relevance to this report since complex I deficiency is the most commonly observed deficit, and it presents either alone or in association with other respiratory complex deficits [33]. These cases adds new information compared to the previous reported cases of ASD and mitochondrial disease since the typical reported features such as gastro-intestinal symptoms (red-flags for, other organ involvement (eyes, liver, hearth, *etc.*) and regression were missing in these patients [21].

Pharmacological treatment for these patients remains limited, and a Cochrane Collaboration systematic review [34] concluded that there is currently no clear evidence supporting the systemic use of such interventions in patients with mitochondrial disorders. Symptomatic therapy can be effective in some patients with mitochondrial disorders [32,35] however a limited, or no, clinical effect is observed in the majority of cases [34]. It is clear that the efficacy of these treatments needs to be confirmed by randomized clinical trials performed in homogeneous study samples (with clinical features similar to the patients described in this report), which have clinically relevant primary endpoints. The limited size of our series limits our ability to recommend the use of any pharmacological treatments to treat patients with similar mitochondrial diseases. However, the treatments we administered to our patients could be described as almost empirical: there is currently very little evidence in the literature for drug combinations, which have a beneficial effect for such cases, and there are no FDA-approved drugs for this patient profile.

The administration of vitamins B1, B2, B3, folic acid (vitamin B9), carnitine, and Coenzyme Q, and their dosage was selected on the basis of the few reports we did find that show some benefit in patients with mitochondrial disease, as reviewed elsewhere [36,37]. Vitamin B6 and B12 were chosen because they have shown some benefit in patients with ASD and mitochondrial dysfunction [32,38]. Encouragingly, we observed a general (although variable) clinical improvement after pharmacological treatment with this combination, and interestingly, this beneficial effect was observed for ID (a condition which is classically considered to be stable over time and rarely scored after pharmacological treatment in this subgroup of patients), ASD, and in muscular tone signs. In contrast, other CNS manifestations such as sleep problems, irritability, and hyperactivity only improved following the administration of the neuroleptic drug risperidone. These results, although limited to case reports, suggest that administration of this mitochondrial "cocktail" as a symptomatic treatment can have some beneficial effects.

Undoubtedly the efficacy of this treatment should be evaluated in a large sample of patients with concomitant ID, ASD, hyperlactacidemia, and possible mitochondrial disorders. It is important to note that mitochondrial energy metabolism disorders may be missed in children with ID-ASD because its symptoms overlap with idiopathic ID, ASD, and other developmental disorders and secondary mitochondrial dysfunctions cannot be completely ruled out in these patients since their clinical phenotype fulfills that of a possible mitochondrial disorder [14]. In conclusion, our report supports the idea that some children with ID, ASD, and hyperlactacidemia may have a mitochondrial disorder with clinical manifestations different from, or less severe than, those patients with the known phenotypes of other well-characterized mitochondrial diseases.

These data offer clinicians a valuable set of benchmark parameters which can be used as a reference for suspected mitochondrial diseases in patients with ID, ASD, hypotonia, and for those with other non-systemic involvement (besides the muscular and nervous system). However it should be pointed out that these patients manifested additional and diverse clinical signs, *i.e.*, acquired microcephaly, hemi-spasticity, EEG alterations without clinical convulsions, and MRI imaging alterations, and non-specific pervasive developmental disorders (according to DMS-IV). The diagnosis of an associated mitochondrial disorder in patients with ID and ASD is sometimes difficult and requires knowledge of several different clinical presentations (as shown in Table 1). Each of these clinical features is shared by a subgroup of patients with idiopathic ID however these features alone are not sufficient to suggest mitochondrial disease unless other signs and symptoms are present. Despite the biochemical, histological, and genetic limitations that determine the final diagnosis of mitochondrial disease, this etiological possibility must be considered for primary or secondary mitochondrial respiratory chain disorders in children with ID and ASD in order to reach an early diagnosis and start prompt treatment and rehabilitation programs. There is a clear need of more advanced genetic testing of mtDNA and nDNA genes using next generation sequencing to help identify new genetic lesions (mutations) seen in these types of patients.

4. Experimental Section

4.1. Clinical Examinations

Each individual was subjected to a general physical, neurological, and psychological examination performed by a neuropediatrician, neurologist, and psychologist. Developmental delay was assessed under the age of 5 years using the standardized and validated Peabody Developmental Motor Scale 2 (PDMS-2) [39]. The Battelle developmental inventory (BDI-2) [40] was used to perform a comprehensive developmental assessment. The diagnosis of ID was made between 4.5–5 years of age using Raven's Colored Progressive Matrices (CPM) [41]. This is a standardized instrument for assessing non-verbal general intelligence in children aged 5 years and over and for adults with ID for the purposes of clinical investigation. The overall score in each scale/inventory was transferred into a centile using the age-appropriate norms given by this scale. Testing was done in all three patients in order to exclude the common genetic causes of ID, such as fragile X syndrome, Prader–Willi syndrome, MECP2, and karyotype alterations (chromosomal deletions and duplications). Genetic testing for fragile X syndrome, Prader–Willi syndrome, MECP2 were done in University of Alabama (Med. Genomics Lab, Dr. L. Messiaen, Birmingham, USA) after extraction of genomic DNA from peripheral blood samples collected in EDTA (1 mM) in a volume of about 5 mL. Fragile X syndrome was analyzed by Southern blot analysis using primary restriction site EcoRI, (5.2 kb) and EagI (internal methyl-sensitive site) digestion, probed with StB12.3. After this analysis, fragile X syndrome was excluded in all three patients (CGG 29; 30; 48 for patient 1, 2 and 3, respectively). Analysis of MECP2 was performed by amplifying the MECP2-coding exons and flanking intronic sequences. The following primer pair was used to amplify exon 1: forward Rett-EX-1F 5'-GCAGCTCAATGGGGGCT-TTCAACTT-3' and reverse Rett-EX-1R 5'-GGCACAGTTA-TGTCTTTAGTCTTTGG-3'. PCR were performed and the products extracted from agarose gel using QIAquick gel extraction kit (Qiagen, Hilden, Germany) according to manufacturer's protocol. Sequencing of the PCR products was performed by ABI prism BigDye Terminator (Applied Biosystem, Foster City, CA, USA) and separated by capillary electrophoresis and detected via laser induced florescence and compared with reference MECP2 sequence. Prader–Willi syndrome (PWS) was performed with the methylation test by Southern blot. For the analysis a double digestion was done with the enzymes Hind III and Hpa II (sensitive to methylation) and subsequent hybridization with the PW71B probe (patients without Prader–Willy or Angelman syndrome displayed a pattern with two bands (6 and 4 kb). Karyotype was performed at Oriente University, Faculty of Medicine in Ciudad Bolivar (Venezuela) by analyzing specimens that were processed using direct methods and unstimulated short-term (24- and 48-h) cultures with G-banding analysis (550–850 band). A minimum of 20 metaphases was required to define a normal karyotype. Analysis of known mitochondrial ASD/ID gene mutations according to patients' phenotype are described in the Section 4.3 Autism spectrum disorder was assessed by Autism Diagnostic Interview—Revised (according to DMS IV criteria).

4.2. Blood Analysis

Blood samples were extracted in both the fasting and postprandial condition for analytical and biochemical assessments. In order to prevent erroneous lactate elevations due to a poor venipuncture technique or because of the use of a tourniquet, an indwelling butterfly needle was placed in order to permit blood sample collection after the patient had settled for 30 min.

4.3. Muscle Tissue Analysis

A quadricep muscle was biopsied and the sample was frozen until analysis with light and electron microscopy, as well as for mtDNA content using real time PCR analysis and for the following mutations: T3271C and A3243G in tRNA-leu (UUR), A8344G and T8356C in tRNA-lys, T8993C and T8993G in subunit 6 of ATPase gene, G3460A and G11778A, mutation at 8993 and common deletion of mitochondrial DNA that causes Kearns-Sayre syndrome (KSS), progressive external ophthalmoplegia (PEO) or Pearson syndrome. MtDNA deletions and depletion were analyzed as previously described [42,43]. The percentage of m.6955G>A transition was analyzed by last-cycle radioactive PCR/RFLP by using primers HmtL6954 (GGATTCATCTTTCTTTTCACCCTAG) and HmtH7114 (TGGCGTAGGTTTGGTCTAGG). HmtL6954 contains a mismatch (G-C) at nucleotide position 6951. The amplicon size is 204 bp and the PCR conditions: 94 °C 2min (94 °C 30 s/55 °C 30 s/72 °C 1 min 30 s) 35 cycles, 72 °C 5 min. 0.5 μL of (α-32P) dCTP at 10 mCi/mL was added to each PCR reaction before the last cycle. The restriction enzyme BlnI (C/CTAGG) cuts the amplicon in two fragments of 183 + 21 bp. The m.6955G>A transition removes the cutting site. The levels of the mutant mtDNA were quantified by using the GelProAnalyzer 4.0 program. The revised human mitochondrial DNA Cambridge reference sequence (GenBank REFSEQ AC_000021.2) was used. All of these mtDNA depletion analyses were negative.

4.4. Determination of Mitochondrial Respiratory Chain Complexes

Small pieces of frozen muscle tissues were homogenized (1/30 weight per volume) in a solution containing 50 mM Tris buffer (pH 7.5), 100 mM potassium chloride, 5 mM $MgCl_2$, and 1 mM ethylenediaminetetraacetic acid using a glass/glass homogenizer. Enzyme activities were assayed at 30 °C using a spectrophotometer and were calibrated against CS activity.

ETC function analysis is presented as enzyme complex activity, namely: Nicotinamide adenine dinucleotide (NADH): Coenzyme Q1-oxidoreductase (complex I), NADH: Cytochrome c-oxidoreductase (complex I + III), succinate: Cytochrome c-oxidoreductase (complex II + III), ubiquinone: Cytochrome c-oxidoreductase (complex III), cytochrome c-oxidase (complex IV), in all cases using CS as a mitochondrial marker enzyme, as previously described [44,45]. The concentration of CoQ10 was determined according to Montero *et al.* [46]. Abnormalities in ETC function were compared to the standard range in age-matched control subjects used in each diagnostic laboratory.

Each patient's parents were requested (and gave) written informed for consent to allow their child's medical information to be anonymously abstracted into a clinical database that contained their medical history, physical examination findings, and the results of neurological, psychological, and metabolic tests. The protocols described in this manuscript follow rules approved by the institute's ethical committee.

5. Conclusions

In conclusion, our experience suggests that children with ID, ASD, and with abnormal neurological or muscular involvement but without other systemic findings should be also evaluated for mitochondrial disorders (blood biochemical screening tests) and these disorders should be included in the differential diagnosis of secondary ID/ASD. Other genetic causes such as single gene disorders with ID or ASD as a feature via candidate gene mutation screening, clinical genetics evaluation and microarray analysis should be performed in patients presenting with these findings. Advanced genetic testing results could impact on clinical care and genetic risk counseling for family members.

Acknowledgments: This work was supported by Grant number UV-INV_PRECOMP13-115500 from The University of Valencia and GV/043 from *Conselleria de Educació-Generalitat Valenciana.*

Author Contributions: José Guevara-Campos, Lucía González-Guevara, and Omar Cauli analyzed the data; Omar Cauli and José Guevara-Campos drafted the manuscript; José Guevara-Campos and Lucía González-Guevara revised the manuscript; Omar Cauli designed and conceived the study and was responsible for obtaining project grant funding for the search for a molecular basis for cognitive impairment in neuropsychiatric disease.

Conflicts of Interest: The authors declare no conflicts of interest.

References

1. Jansen, D.E.; Krol, B.; Groothoff, J.W.; Post, D. People with intellectual disability and their health problems: A review of comparative studies. *J. Intellect. Disabil. Res.* **2004**, *48*, 93–102. [CrossRef] [PubMed]
2. Haas, R.H.; Parikh, S.; Falk, M.J.; Saneto, R.P.; Wolf, N.I.; Darin, N.; Cohen, B.H. Mitochondrial disease: A practical approach for primary care physicians. *Pediatrics* **2007**, *120*, 1326–1333. [CrossRef] [PubMed]
3. Sherr, E.H.; Shevell, M.I. Global developmental delay and mental retardation/intellectual disability. In *Pediatric Neurology: Principles and Practice*, 5th ed.; Swaiman, K.F., Ashwal, S., Ferriero, D.M., Schor, N.F., Eds.; Elsevier Saunders: Philadelphia, PA, USA, 2012; pp. 554–574.
4. American Association of Intellectual and Developmental Disabilities (AAIDD). Definition of Intellectual Disability. Available online: http://aaidd.org/intellectual-disability/definition (accessed on 11 July 2013).
5. Luckasson, R.; Reeve, A. Naming, defining, and classifying mental retardation. *Ment. Retard.* **2001**, *39*, 47–52. [CrossRef] [PubMed]
6. Oeseburg, B.; Jansen, D.E.; Groothoff, J.W.; Dijkstra, G.J.; Reijneveld, S.A. Emotional and behavioural problems in adolescents with intellectual disability with and without chronic diseases. *J. Intellect. Disabil. Res.* **2010**, *54*, 81–89. [CrossRef] [PubMed]
7. Van Karnebeek, C.D.; Stockler, S. Treatable inborn errors of metabolism causing intellectual disability: A systematic literature review. *Mol. Genet. Metab.* **2012**, *105*, 368–381.
8. Bokhoven, H. Genetic and epigenetic networks in intellectual disabilities. *Ann. Rev. Genet.* **2011**, *45*, 81–104. [CrossRef] [PubMed]
9. Sherr, E.H.; Michelson, D.J.; Shevell, M.I.; Moeschler, J.B.; Gropman, A.L.; Ashwal, S. Neurodevelopmental disorders and genetic testing: Current approaches and future advances. *Ann. Neurol.* **2013**, *74*, 164–170. [CrossRef] [PubMed]

10. Baker, K.; Raymond, F.L.; Bass, N. Genetic investigation for adults with intellectual disability: Opportunities and challenges. *Curr. Opin. Neurol.* **2012**, *25*, 150–158. [CrossRef] [PubMed]

11. Chinnery, P.F. Mitochondrial disorders overview. In *GeneReviews®*; Pagon, R.A., Adam, M.P., Ardinger, H.H., Bird, T.D., Dolan, C.R., Fong, C.T., Smith, R.J.H., Stephens, K., Eds.; University of Washington: Seattle, WA, USA; pp. 1993–2014.

12. Friedman, S.D.; Shaw, D.W.; Ishak, G.; Gropman, A.L.; Saneto, R.P. The use of neuroimaging in the diagnosis of mitochondrial disease. *Dev. Disabil. Res. Rev.* **2010**, *16*, 129–135. [CrossRef] [PubMed]

13. Bernier, F.P.; Boneh, A.; Dennett, X.; Chow, C.W.; Cleary, M.A.; Thorburn, D.R. Diagnostic criteria for respiratory chain disorders in adults and children. *Neurology* **2002**, *59*, 1406–1411. [CrossRef] [PubMed]

14. Wolf, N.I.; Smeitink, J.A. Mitochondrial disorders: A proposal for consensus diagnostic criteria in infants and children. *Neurology* **2002**, *59*, 1402–1405. [CrossRef] [PubMed]

15. Thorburn, D.R.; Sugiana, C.; Salemi, R.; Kirby, D.M.; Worgan, L.; Ohtake, A.; Ryan, M.T. Biochemical and molecular diagnosis of mitochondrial respiratory chain disorders. *Biochim. Biophys. Acta.* **2004**, *1659*, 121–128. [CrossRef] [PubMed]

16. Verity, C.M.; Winstone, A.M.; Stellitano, L.; Krishnakumar, D.; Will, R.; McFarland, R. The clinical presentation of mitochondrial diseases in children with progressive intellectual and neurological deterioration: A national, prospective, population-based study. *Dev. Med. Child Neurol.* **2010**, *52*, 434–440. [CrossRef] [PubMed]

17. .McFarland, R.; Taylor, R.W.; Turnbull, D.M. A neurological perspective on mitochondrial disease. *Lancet Neurol.* **2010**, *9*, 829–840. [CrossRef] [PubMed]

18. Nissenkorn, A.; Zeharia, A.; Lev, D.; Watemberg, N.; Fattal-Valevski, A.; Barash, V.; Gutman, A.; Harel, S.; Lerman-Sagie, T. Neurologic presentations of mitochondrial disorders. *J. Child Neurol.* **2000**, *15*, 44–48. [CrossRef] [PubMed]

19. American Psychiatric Association. Intellectual disability. In *Diagnostic and Statistical Manual of Mental Disorders*, 5th ed.; American Psychiatric Publication: Arlington, VA, USA, 2013; p. 33.

20. Di Mauro, S.; Schon, E.A. Mitochondrial respiratory-chain diseases. *N. Engl. J. Med.* **2003**, *348*, 2656–2668. [CrossRef] [PubMed]

21. Weissman, J.R.; Kelley, R.I.; Bauman, M.L.; Cohen, B.H.; Murray, K.F.; Mitchell, R.L.; Kern, R.L.; Natowicz, M.R. Mitochondrial disease in autism spectrum disorder patients: A cohort analysis. *PLoS One* **2008**, *3*, e3815. [CrossRef] [PubMed]

22. Zeviani, M.; Carelli, V. Mitochondrial disorders. *Curr. Opin. Neurol.* **2007**, *20*, 564–571. [CrossRef] [PubMed]

23. Shevell, M. Global developmental delay and mental retardation or intellectual disability: Conceptualization, evaluation, and etiology. *Pediatr. Clin. North Am.* **2008**, *55*, 1071–1084. [CrossRef] [PubMed]

24. Valanne, L.; Ketonen, L.; Majander, A.; Suomalainen, A.; Pihko, H. Neuroradiologic findings in children with mitochondrial disorders. *Am. J. Neuroradiol.* **1998**, *19*, 369–377. [PubMed]

25. Barragan-Campos, H.M.; Valee, J.N.; Lo, D.; Barrera-Ramírez, C.F.; Argote-Greene, M.; Sánchez-Guerrero, J.; Estañol, B.; Guillevin, R.; Chiras, J. Brain magnetic resonance imaging findings in patients with mitochondrial cytopathies. *Arch. Neurol.* **2005**, *62*, 737–742. [CrossRef] [PubMed]

26. Gropman, A.L. Neuroimaging in mitochondrial disorders. *Neurotherapeutics* **2013**, *10*, 273–285. [CrossRef] [PubMed]

27. Saneto, R.P.; Friedman, S.D.; Shaw, D.W. Neuroimaging of mitochondrial disease. *Mitochondrion* **2008**, *8*, 396–413. [CrossRef] [PubMed]

28. Watemberg, N.; Silver, S.; Harel, S.; Lerman-Sagie, T. Significance of microcephaly among children with developmental disabilities. *J. Child. Neurol.* **2002**, *17*, 117–122. [CrossRef] [PubMed]

29. Bindoff, L.A.; Engelsen, B.A. Mitochondrial diseases and epilepsy. *Epilepsia* **2012**, *53*, 92–97. [CrossRef] [PubMed]

30. Koenig, M.K. Presentation and diagnosis of mitochondrial disorders in children. *Pediatr. Neurol.* **2008**, *38*, 305–313. [CrossRef] [PubMed]

31. Anitha, A.; Nakamura, K.; Thanseem, I.; Matsuzaki, H.; Miyachi, T.; Tsujii, M.; Iwata, Y.; Suzuki, K.; Sugiyama, T.; Mori, N. Down-regulation of the expression of mitochondrial electron transport complex genes in autism brains. *Brain Pathol.* **2013**, *23*, 294–302. [CrossRef] [PubMed]

32. Rossignol, D.A.; Frye, R.E. Mitochondrial dysfunction in autism spectrum disorders: A systematic review and meta-analysis. *Mol. Psychiatry* **2011**, *17*, 290–314. [CrossRef] [PubMed]

33. Nissenkorn, A.; Zeharia, A.; Lev, D.; Fatal-Valevskic, A.; Barashd, V.; Gutmand, A.; Harelc, S.; Lerman-Sagiea, T. Multiple presentation of mitochondrial disorders. *Arch Dis. Child.* **1999**, *81*, 209–214. [CrossRef] [PubMed]

34. Pfeffer, G.; Majamaa, K.; Turnbull, D.M.; Thorburn, D.; Chinnery, P.F. Treatment for mitochondrial disorders. *Cochrane Database Syst. Rev.* **2012**, *4*, CD004426. [PubMed]

35. Di Mauro, S.; Mancuso, M. Mitochondrial diseases: Therapeutic approaches. *Biosci. Rep.* **2007**, *27*, 125–137. [CrossRef] [PubMed]

36. Chinnery, P.F.; Turnbull, D.M. Epidemiology and treatment of mitochondrial disorders. *Am. J. Med. Genet.* **2001**, *106*, 94–101. [CrossRef] [PubMed]

37. Avula, S.; Parikh, S.; Demarest, S.; Kurz, J.; Gropman, A. Treatment of mitochondrial disorders. *Curr. Treat. Options Neurol.* **2014**, *16*, 292. [CrossRef] [PubMed]

38. Frye, R.E.; Rossignol, D.A. Treatments for biomedical abnormalities associated with autism spectrum disorder. *Front. Pediatr.* **2014**, *2*, 66. [CrossRef] [PubMed]

39. Connolly, B.H.; McClune, N.O.; Gatlin, R. Concurrent validity of the Bayley-III and the peabody developmental motor scale-2. *Pediatr. Phys. Ther.* **2012**, *24*, 345–352. [CrossRef] [PubMed]

40. Goldin, R.L.; Matson, J.L.; Beighley, J.S.; Jang, J. Autism spectrum disorder severityas a predictor of Battelle Developmental Inventory—Second edition (BDI-2) scores in toddlers. *Dev. Neurorehabil.* **2014**, *17*, 39–43. [CrossRef] [PubMed]

41. Raven, J.; Raven, J.C.; Court, J. *Raven Manual: Standard Colored Progressive Matrices*; Oxford Psychologists Press: London, UK, 1998.

42. Herrero-Martín, M.D.; Pineda, M.; Briones, P.; López-Gallardo, E.; Carreras, M.; Benac, M.; Angel Idoate, M.; Vilaseca, M.A.; Artuch, R.; López-Pérez, M.J.; *et al.* A new pathologic mitochondrial DNA mutation in the cytochrome oxidase subunit I (MT-CO1). *Hum. Mutat.* **2008**, *29*, E112–E122. [PubMed]

43. Marcuello, A.; González-Alonso, J.; Calbet, J.A.; Damsgaard, R.; López-Pérez, M.J.; Díez-Sánchez, C. Skeletal muscle mitochondrial DNA content in exercising humans. *J. Appl. Physiol.* **1985**, *99*, 1372–1377. [CrossRef]

44. Genova, M.L.; Castelluccio, C.; Fato, R. Major changes in complex I activity in mitochondria from aged rats may not be detected by direct assay of NADH:coenzyme Q reductase. *Biochem. J.* **1995**, *311*, 105–109. [PubMed]

45. Trumbeckaite, S.; Opalka, J.R.; Neuhof, C.; Zierz, S.; Gellerich, F.N. Different sensitivity of rabbit heart and skeletal muscle to endotoxin-induced impairment of mitochondrial function. *Eur. J. Biochem.* **2001**, *268*, 1422–1429. [CrossRef] [PubMed]

46. Montero, R.; Artuch, R.; Briones, P.; Nascimento, A.; García-Cazorla, A.; Vilaseca, M.A.; Śnchez-Alćzar, J.A.; Navas, P.; Montoya, J.; Pineda, M. Muscle coenzyme Q10 concentrations in patients with probable and definite diagnosis of respiratory chain disorders. *Biofactors* **2005**, *25*, 109–115. [CrossRef] [PubMed]

International Journal of
Molecular Sciences

MDPI

Short Note

Unbalance between Excitation and Inhibition in Phenylketonuria, a Genetic Metabolic Disease Associated with Autism

Antonella De Jaco [1,*], Dalila Mango [2], Federica De Angelis [1], Flores Lietta Favaloro [1], Diego Andolina [3,4], Robert Nisticò [5], Elena Fiori [2,3,6], Marco Colamartino [3,4] and Tiziana Pascucci [3,4]

[1]   Department of Biology and Biotechnologies "Charles Darwin", Sapienza University of Rome, 00185 Rome, Italy; federica.deangelis@uniroma1.it (F.D.A.); floresl@libero.it (F.L.F.)
[2]   EBRI-European Brain Research Institute, 00143 Rome, Italy; dalilamango@gmail.com (D.M.); elena.fiori@uniroma1.it (E.F.)
[3]   Department of Psychology, "Daniel Bovet", Neurobiology Research Center, Sapienza University of Rome, 00185 Rome, Italy; diego.andolina@uniroma1.it (D.A.); marco.colamartino@uniroma1.it (M.C.); tiziana.pascucci@uniroma1.it (T.P.)
[4]   Foundation Santa Lucia, IRCCS, 00143 Rome, Italy
[5]   Department of Biology, University of Tor Vergata, 00133 Rome, Italy; robert.nistico@uniroma1.it
[6]   Cell Biology and Neurobiology Institute, National Research Council, 00143 Rome, Italy
*   Correspondence: antonella.dejaco@uniroma1.it; Tel.: +39-06-4991-2310; Fax: +39-06-4991-2351

Academic Editor: Merlin G. Butler
Received: 28 February 2017; Accepted: 23 April 2017; Published: 29 April 2017

Abstract: Phenylketonuria (PKU) is the most common genetic metabolic disease with a well-documented association with autism spectrum disorders. It is characterized by the deficiency of the phenylalanine hydroxylase activity, causing plasmatic hyperphenylalaninemia and variable neurological and cognitive impairments. Among the potential pathophysiological mechanisms implicated in autism spectrum disorders is the excitation/inhibition (E/I) imbalance which might result from alterations in excitatory/inhibitory synapse development, synaptic transmission and plasticity, downstream signalling pathways, and intrinsic neuronal excitability. Here, we investigated functional and molecular alterations in the prefrontal cortex (pFC) of BTBR-Pah^{enu2} (ENU2) mice, the animal model of PKU. Our data show higher frequency of inhibitory transmissions and significant reduced frequency of excitatory transmissions in the PKU-affected mice in comparison to wild type. Moreover, in the pFC of ENU2 mice, we reported higher levels of the post-synaptic cell-adhesion proteins neuroligin1 and 2. Altogether, our data point toward an imbalance in the E/I neurotransmission favouring inhibition in the pFC of ENU2 mice, along with alterations of the molecular components involved in the organization of cortical synapse. In addition to being the first evidence of E/I imbalance within cortical areas of a mouse model of PKU, our study provides further evidence of E/I imbalance in animal models of pathology associated with autism spectrum disorders.

Keywords: neurotransmission; excitation and inhibition balance; cognitive delay; prefrontal cortex; neuroligins

---

1. Introduction

Several reports suggest an association between autism and inherited metabolic diseases among which phenylketonuria (PKU), suggesting that autism spectrum disorders might represent the end result of a dysfunction caused by a metabolic block in the brain [1]. PKU is the prototypical human Mendelian disease (OMIM 261600; overall incidence of 1 in 10,000) resulting from impaired activity of phenylalanine hydroxylase (PAH), the enzyme necessary to convert phenylalanine (PHE) to tyrosine. This deficiency causes hyperphenylalaninemia (HPA), which is especially harmful for the brain during the first years of life, resulting in variable neurological and mental impairments [2–4]. Previous evidence from our group demonstrated that the accumulation of PHE in the brain of BTBR-Pah^{enu2} (ENU2) mice impairs protein levels and enzymatic activity of the tryptophan hydroxylase, the rate-limiting enzyme responsible for serotonin biosynthesis [5], and that the serotonin reduction in the brain causes cortical morphological alterations such as a reduction in the dendritic spine density and maturation [6]. Restoring normal levels of brain serotonin in the ENU2 mice, during the third post-natal week, allowed the recovery of some cognitive functions as well as the morphological maturation of pyramidal neuron dendritic spines in the prefrontal cortex (pFC) [6].

We have investigated functional alterations and molecular rearrangements typically associated with neurodevelopmental disorders in an animal model of PKU in order to explore possible common molecular mechanisms in comorbidity with autism. Alterations in excitatory/inhibitory (E/I) ratio in cortical circuitry have been reported in several animal models for neurodevelopmental disorders in association with cognitive delay [7], providing experimental models to define abnormal molecular mechanisms and to identify new therapeutic targets. Since synaptic transmission is regulated by a plethora of molecules where cell-adhesion molecules are emerging as crucial players [8], we have studied the neuroligin/neurexin (NLGN/NRXN) pathway involved in the maturation of the inhibitory and excitatory synapses [9]. Moreover, genes directly involved in the regulation of the ratio between excitation and inhibition represent risk candidate genes [10]. Copy number variations and/or several single point mutations in the NLGN/NRXN synaptic pathway have been detected in association to neurodevelopmental disorders [11] including autism spectrum disorders [12–17].

Here we investigate, for the first time, the functional and molecular features underlying the morphological and biochemical phenotype reported in the pFC of ENU2 mice, the genetic murine model of the most common metabolic inborn error. Our data support the hypothesis that in PKU, unknown mechanisms linked to PHE accumulation lead to a E/I imbalance shifting toward inhibition, accompanied by altered expression levels of specific members of the synaptic family of the Neuroligin proteins, classically linked to autism.

2. Results

2.1. Analysis of Inhibitory and Excitatory Transmission in Layer II/III of ENU2 pFC

ENU2-mutant mice exhibit abnormal behaviors that mimic the intellectual disability symptoms observed in human PKU untreated patients. In order to assess whether immature spine morphology and cognitive impairments described for PKU in the ENU2 mice [6] reflect functionally a different cortical activity in comparison to parental controls, we have measured the spontaneous inhibitory postsynaptic currents (sIPSC) and spontaneous excitatory postsynaptic currents (sEPSC) from layer II/III of brain pFC by using whole-cell patch clamp recordings. We have analyzed synaptic transmission by assessing amplitude and frequency of action potential dependent inhibitory and excitatory spontaneous events from slices obtained by ENU2 and relative control mice at postnatal day 60 (PND 60). We have measured cumulative probability of amplitude and inter-event interval of frequency for sIPSC and sEPSC. As shown in Figure 1a, we have found higher frequency of sIPSC in ENU2 mice compared to wild-type (WT) (K–S test $p < 0.001$, t-test $p = 0.0412$ ENU2 $n = 8$ vs. WT $n = 7$, Figure 1A) and significant reduced frequency of sEPSC in ENU2 mice compared to WT (K–S test $p = 0.0091$, t-test $p = 0.0433$ ENU2 $n = 7$ vs. WT $n = 6$, Figure 1B). Consistently, the E/I ratio was also significantly reduced in ENU2 mice compared to WT (t-test $p = 0.0306$, ENU2 $n = 6$ vs. WT $n = 6$ Figure 1C).

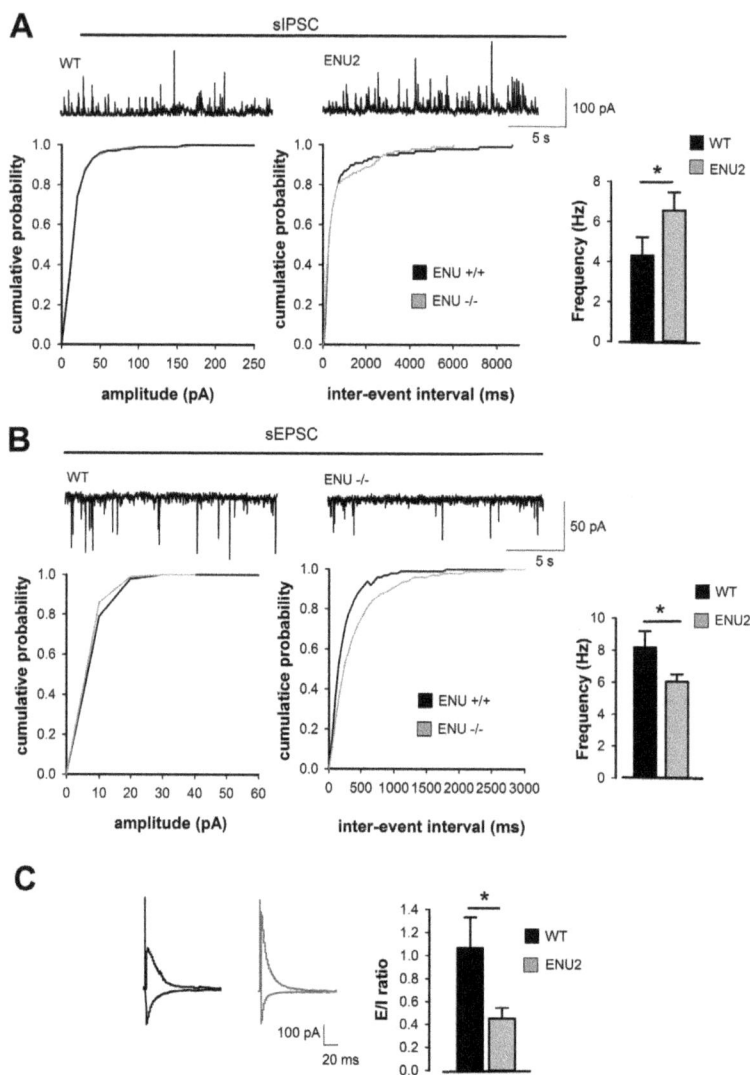

Figure 1. BTBR-Pah[enu2] (ENU2) mice show altered excitatory/inhibitory (E/I) balance. (**A**) Pooled cumulative distributions of spontaneous inhibitory post synaptic currents (sIPSCs) amplitude (**left**; bin size 10 pA) and inter-event interval (**right**; bin size 50 ms) recorded from neurons of wild type (WT, $n = 8$) and ENU2 ($n = 7$) mice. Representative traces are shown on top. (**B**) Pooled cumulative distributions of spontaneous excitatory post synaptic currents (sEPSCs) amplitude (**left**) and inter-event interval (**right**) recorded from neurons of WT ($n = 7$) and ENU2 ($n = 6$) mice. Histograms are averages (mean $\pm$ S.E.M) of the corresponding median values of sEPSCs frequency for the same neurons. Representative traces are shown on top. (**C**) Histograms are averages (mean $\pm$ S.E.M) of E/I ratio recorded from WT ($n = 6$) and ENU2 ($n = 6$). Representative traces are shown on the left. (* $p < 0.05$).

The reported electrophysiological alterations resemble those typically associated with other neurodevelopmental disorders [7], such as autism, where the excitatory and inhibitory balance is functionally impaired and might account for the cognitive phenotype.

2.2. Protein Levels of Synaptic Cell Adhesion Molecules in the pFC of ENU2 Mice

Synaptic cell adhesion molecules operate in concert with neurotransmitter receptors to ensure proper function of synaptic circuits [18]. The NLGN/NRXN pathway is currently one of the most studied trans-synaptic codes acting in the organization of the excitatory and inhibitory synapses. The NLGN family is made of four members (1, 2, 3, 4), encoded by different genes, with NLGN1 being specifically localized to excitatory postsynaptic densities while NLGN2 is found in inhibitory postsynaptic specializations and NLGN3 is present at both [19]. NLGNs play a crucial role in the recruitment of neurotransmitter receptors at the synapse and in the control of the E/I balance in the brain [20].

We have initially quantified the levels of all the NLGNs by western blot using a PAN-antibody. Analysis of NLGNs levels revealed an increasing trend in the ENU2 mice that however did not present significant results in comparison to WT mice (*t*-test $p = 0.2162$, ENU2 $n = 8$ vs. WT $n = 7$ Figure 2A). We have then investigated the levels of each family member by using antibodies specific for each of the NLGNs forms. We found that NLGN1 (*t*-test $p = 0.0142$, ENU2 $n = 14$ vs. WT $n = 18$ Figure 2B) and NLGN2 (*t*-test $p = 0.0266$, ENU2 $n = 14$ vs. WT $n = 13$, Figure 2C) resulted in increased ENU2 in comparison to WT mice. Non-significant differences were observed for NLGN3 between ENU2 and WT mice (*t*-test $p = 0.543$, ENU2 $n = 14$ vs. WT $n = 17$ Figure 2D). These observations suggest that the functional differences in the ENU2 pFC reflect a different regulation of molecular synaptic components.

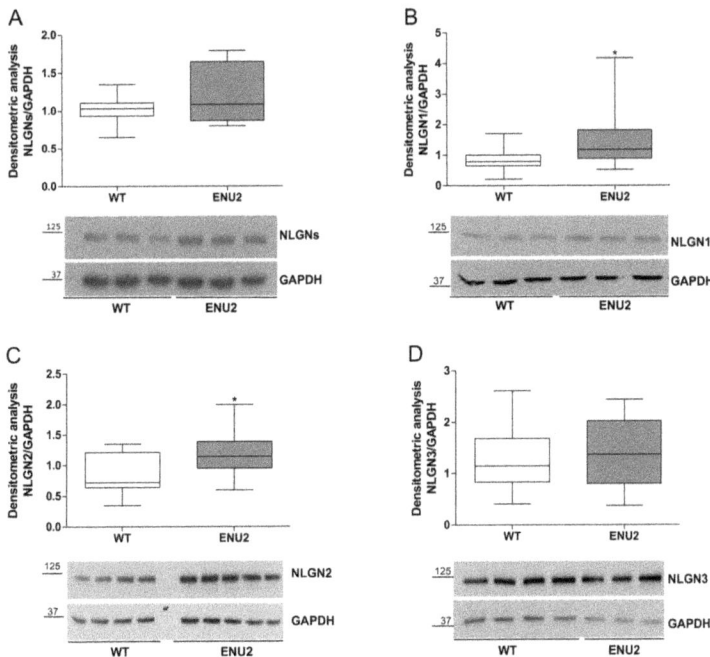

Figure 2. Neuroligins (NLGNs) levels in the pFC of ENU2 mice. Protein levels were quantified by densitometry after western blot analysis for total NLGNs and for NLGN1, 2, and 3 family members. Values were normalized to GAPDH loading control and are represented as a box plot of their distribution (min/max *e* median). (**A**) NLGNs (WT $n = 7$, ENU2 $n = 8$, $p = 0.2162$); (**B**) NLGN1 (WT $n = 18$, ENU2 $n = 13$, $p = 0.0142$); (**C**) NLGN2 (WT $n = 13$, ENU2 $n = 14$, $p = 0.0266$); and (**D**) NLGN3 (WT $n = 17$, ENU2 $n = 14$, $p = 0.5430$). Statistical analysis compared ENU2 values versus WT (* $p < 0.05$). Representative images of western blot analysis are shown. Molecular masses are indicated on the blots in kDa.

3. Discussion

Dysregulation of the excitation/inhibition equilibrium has been postulated to represent a hallmark of neuropsychiatric disorders, including autism and some forms of mental retardation. Several mouse models reproducing behavioral phenotypes common to neurodevelopmental syndromes show alterations of the E/I balance. In particular, a mouse model of Rett syndrome showed a shift favoring inhibition in the pFC [21]. Increased inhibition was observed in the somatosensory cortex of mice expressing the R451C autism-related mutation in NLGN3, and this was associated with impairments in social interaction [22]. In general, while a decrease in inhibition is currently associated with autism spectrum disorders, an excess of inhibition has been described to occur in mental retardation syndromes such as Down [23–25] and Rett syndromes [21]. Interestingly, by using an optogenetic approach to study real-time effects of elevation of cellular E/I balance in vivo, it was shown that elevated E/I balance resulted in impairments on social behavior that are specific for pFC [26].

Although compelling evidence points toward a link between dysregulation of E/I ratio and behavioral phenotypes resembling those observed in neuropsychiatric disorders, the molecular machinery involved in the regulation of this balance remains unclear.

PKU mice, created by chemically induced genetic mutation, display a phenotype that closely resembles untreated human PKU, characterized by reduced PAH activity, PHE plasma levels 10–20 times greater than those of healthy littermates, impaired cerebral protein synthesis, neurochemical reductions in different brain regions, particularly in serotonergic metabolism in prefrontal cortical areas, reduced functional and morphological synaptic plasticity, and cognitive and other behavioral abnormalities [27].

We have postulated that cognitive impairments might be linked to E/I imbalance and found that there was a resultantly higher inhibition and reduced excitation in the layer II/III of ENU2 pFC, suggesting an overall reduced activity in cortical circuits as observed in other animal models of neurodevelopmental disorders where a shift in the balance between excitation and inhibition—favoring inhibition—has been reported [7].

In recent years, several lines of evidence suggest a possible link between the levels of neuroligins and neurotransmission dysfunctions in association to autism spectrum disorders [28]. Gain and loss of function studies in vitro and in vivo have provided experimental support to the hypothesis that the regulation of the levels of the neuroligin proteins might correlate with alteration of the E/I balance. Transgenic mice where NLGN2 expression has been enhanced showed higher frequency of mIPSC in the pFC and an overall reduction in the E/I ratio [29]. NLGN2 function in modulating inhibitory synaptic currents was further highlighted by the selective deletion of NLGN2 in the medial pFC in a conditional knock-out mouse strain. This resulted in chronic changes in E/I balance characterized by a reduction in frequency and amplitude of inhibitory sIPSCs and by cognitive behavioral changes [30]. This has led us to investigate whether the E/I unbalance in the pFC of ENU2 mice might correlate with altered levels of the NLGNs family members. Indeed, in ENU2 mice we have found different levels of neuroligin proteins in comparison to the parental healthy mice. In particular, our data show unchanged levels for NLGN3 along with higher protein levels for NLGN2 and NLGN1. Enhanced NLGN2 protein levels in the pFC of ENU2 mice correlate with the increased inhibitory transmission observed in the layer II/III and strengthen the hypothesis of a shift in the E/I balance favoring inhibition in the ENU2 mice. In fact, NLGN2 is found in inhibitory postsynaptic specializations [31] where it plays a specific role in the regulation of inhibitory synaptic terminals and in the maintenance of E/I balance in the brain [32]. NLGN2 interacts with collybistin and gephyrin in order to recruit and anchoring GABAA receptors to the post-synaptic membrane [33], favoring the maturation of the inhibitory synapses [34]. At this stage, we have not investigated whether the enhancement of sIPSC is due to an increase of the number of inhibitory synapses. However, the increased levels of NLGN2 cannot explain the decrease in the excitatory neurotransmission in ENU2 mice. Recently, it has been shown that selectively deleting NLGN2 from the II/III layer of pFC leads to a decrease in spontaneous mIPSC without affecting mEPSC [30].

The prominent deficit of serotonin in pFC of PKU mice is well documented [5,35] and also a crucial role is played by serotonin in regulating maturational events such as spine morphology through the activation of the serotonin 2A receptor (5HT2A) receptor, expressed in excitatory synapses [36]. Our previous work showed that cortical spine maturation, and consequently cognitive deficits, are affected in ENU2 mice through a serotonin-dependent pathway [6]. Thus, the reduced serotonin release in ENU2 [5] might result in a lower excitatory activity 5HT2A-dependent. This would agree with the data we show in regard to the lower rates of spontaneous EPSCs in the II/III layers of pFC. Our data show a statistically significant increase for NLGN1 in ENU2 mice, classically localized to excitatory synapses [34]. This increase does not agree with the reduced excitatory transmissions found in the layer II/III of the pFC in ENU2 mice. This might be due to the western blot analysis being performed from punchings of the pFC, comprising all of the cortical layers, in contrast to the electrophysiological recordings restricted to layer II/III.

The shift in the ratio from excitation to inhibition is however differentially regulated by the association of the NLGNs with elements critical for synapse formation such as postsynaptic scaffolding proteins, PSD-95 (at the excitatory synapses), and gephryn (at the inhibitory synapses) and to the presynaptic proteins NRXNs [37–39]. Therefore, a further analysis of these components will help clarify the mechanism.

The data presented shows the first evidence of E/I cortical imbalance in a genetic murine model of inherited metabolic disease, PKU. The unbalance toward inhibitory transmission in the pFC of ENU2 mice might impact on the proper development of brain circuits involved in cognitive function. The cascade of events that lead from high blood PHE levels to the E/I cortical imbalance in PKU however is still not understood.

Finally, investigating the molecular and physiological mechanisms underlying cognitive disability in PKU mice can provide insights for autism spectrum disorders, as well as for all syndromes characterized by similar pathogenic mechanisms.

4. Materials and Methods

4.1. Animal Protocols and Housing

All experiments were approved by the ethics committee of the Italian Ministry of Health and conducted under license/approval ID #: 10/2011-B, according with Italian regulations on the use of animals for research (legislation DL 116/92) and the Council Directive 2010/63EU of the European Parliament and the Council of 22 September 2010 on the protection of animals used for scientific purposes. Homozygote $(-/-)$ Pah^{Enu2} (ENU2) and Homozygote $(+/+)$ Pah^{Enu2} (WT) BTBR mice were issued from heterozygous mating. Genetic characterization was performed on DNA prepared from tail tissue using the Easy DNA Kit (Invitrogen, Carlsbad, CA, USA). The ethylnitrosourea (ENU2) mutation was detected after PCR amplification of exon 7 of the Pah gene and digestion with BsmAI restriction enzyme (NEB, USA) as previously described [40]. At PND28, animals (sex matched) were housed 2–4 per standard breeding cage with food and water ad libitum on a 12:12 h dark: light cycle (light on 07.00 a.m.–07.00 p.m. h).

Brain tissue was collected at PND80 from male ENU2 and WT mice. All animals were killed and the brain was removed and stored depending on the experimental procedures. Every effort was made to alleviate animal discomfort and cervical dislocation was applied as the appropriate method of sacrifice.

4.2. Slice Preparation for Electrophysiological Recordings

The brain was rapidly removed from the skull and coronal slices (250 μm thick) were cut with a vibratome (VT 1200S, Leica) in cold (0 °C) artificial cerebrospinal fluid (aCSF) containing (in mM): NaCl 124; KCl 3; MgSO$_4$ 1; CaCl$_2$ 2; NaH$_2$PO$_4$ 1.25; NaHCO$_3$ 26; glucose 10; saturated with 95% O$_2$, 5% CO$_2$ (pH 7.4), and left to recover for 1 h in aCSF at room temperature.

4.3. Whole-Cell Patch Clamp Recordings

Individual slices were placed in a recording chamber, on the stage of an upright microscope (Zeiss, Munich, Germany) and submerged in a continuously flowing (3 mL/min) solution at 30°C (±2 °C). Individual neurons were visualized through a 40× water-immersion objective (Olympus, Tokyo, Japan) connected to infrared video microscopy (Hamamatsu, Hamamatsu City, Japan). Borosilicate glass electrodes (5–7 MΩ), pulled with a PP 83 Narishige puller, were filled with a solution containing the following (in mM): CsCH$_3$SO$_3$ 115; CsCl 10; KCl 10; CaCl$_2$ 0.45; EGTA 1; Hepes 10; QX-314 5; Na$_3$-GTP 0.3; Mg-ATP 4.0; pH adjusted to 7.3 with CsOH.

Whole cell patch-clamp recordings have been performed from layer II/III pyramidal neurons of pFC brain slice of WT and ENU2 mice. To isolate sEPSCs and sIPSCs from the same neurons we recorded in voltage clamp mode while maintaining the membrane potential either at the reversal potential for GABA receptor for EPSCs (−70 mV) or at the reversal potential for ionotropic glutamate receptors for IPSCs (+10 mV). To record evoked responses elicited by monopolar stimulating electrodes placed in layer I of pFC, EPSCs and IPSCs were monitored sequentially in the presence of 50 μm APV at postsynaptic holding voltages of −60 and 0 mV, respectively. The E/I ratio is computed as the ratio of excitatory to inhibitory charges, obtained by the integration of the measured currents from the network response triggered by extracellular stimulations [41]. Kolmogorov–Smirnov test (K–S test) and unpaired Student's *t*-test (*t*-test) have been applied as statistical test with α value set at 0.05, *n* reflected the number of neurons recorded.

4.4. SDS PAGE and Western Blot

For protein analysis, frozen brains were removed and dissected to obtain punches of the pFC from brain slices (coronal sections) not thicker than 300 μm. Stainless steel tubes of 1.0 mm inside diameter were used and the coordinates were measured as previously reported [42].

Samples derived from pFC punching were homogenized by sonication using RIPA buffer (Life Technologies, Monza, Italy) to extract total proteins and total protein concentration was determined by Bradford assay (Biorad, Rome, Italy). Around 70 micrograms of total proteins were loaded for each sample. Immunoblotting used previously optimized standard techniques [43] including 10% *w*/*v* SDS-PAGE (Biorad, Rome, Italy) and immobilon transfer membranes (Millipore, Bedford, MA, USA).

Detection of NLGN proteins employed commercial primary antibodies from Synaptic Systems, used at the 1:1000 dilution: anti-NLGN pan mouse monoclonal antibody (clone 4F9, Cat. No. 129-011); anti-NLGN1 mouse monoclonal (Cat. No. 129-111); anti-NLGN2 polyclonal rabbit (Cat. No. 129-203); anti-NLGN3 polyclonal rabbit (Cat. No. 129-113). The anti-GAPDH polyclonal rabbit antibody (abcam ab37168) was used as a loading control. The anti-mouse-HRP and anti-rabbit-HRP (Sigma-Aldrich, Milan, Italy) secondary antibodies were diluted 1:10,000. The HRP signal was developed using the LiteAblot PLUS and TURBO extra sensitive chemiluminescent substrates (Euroclone, Milan, Italy) and exposed to autoradiographic films (Santa Cruz Biotechnology, through Aurogene, Rome, Italy) or revealed by using the ChemiDoc™ MP System (Biorad, Rome, Italy). Densitometry was performed using the Image-J software (version 1.43, NIH, Bethesda, MD, USA). Punching samples were derived from 7 to 18 animals per group and unpaired Student's *t*-test (*t*-test) statistical analysis was used to compare values from ENU2 and WT mice.

Acknowledgments: This work was supported by the Comitato Telethon Fondazione ONLUS grant (Grant GGP09254) and Scientific Research grants 2010, 2011 and 2015 by Sapienza University of Rome to Tiziana Pascucci and by the Compagnia San Paolo and Pasteur Institute and Cenci Bolognetti Foundation grants and Scientific Research grants 2010, 2011, 2012, 2013 and 2014 to Antonella De Jaco.

Author Contributions: Antonella De Jaco designed the research strategy and wrote the manuscript. Dalila Mango performed the electrophysiological recordings. Federica De Angelis and Flores Lietta Favaloro analyzed protein levels by western blot. Diego Andolina and Elena Fiori conducted mouse colony handling, dissection, and punching for protein analysis. Robert Nisticò contributed to the experimental design and the interpretation of the results. Marco Colamartino performed genetic characterization of ENU2 mice. Tiziana Pascucci conceived the study and contributed to writing the manuscript.

Conflicts of Interest: The authors declare no conflict of interest.

Abbreviations

sEPSC	spontaneous excitatory postsynaptic potential
sIPSC	spontaneous inhibitory postsynaptic potential
NLGNs	neuroligins
HRP	horseradish peroxidase
SDS	sodium dodecyl sulfate
PAGE	polyacrylamide gel electrophoresis
pFC	prefrontal cortex
LTP	long term potentiation
LTD	long term depression

References

1. Ghaziuddin, M.; Al-Owain, M. Autism spectrum disorders and inborn errors of metabolism: An update. *Pediatr. Neurol.* **2013**, *49*, 232–236. [CrossRef] [PubMed]
2. DeRoche, K.; Welsh, M. Twenty-five years of research on neurocognitive outcomes in early-treated phenylketonuria: Intelligence and executive function. *Dev. Neuropsychol.* **2008**, *33*, 474–504. [CrossRef] [PubMed]
3. Stemerdink, B.A.; Kalverboer, A.F.; van der Meere, J.J.; van der Molen, M.W.; Huisman, J.; de Jong, L.W.; Slijper, F.M.; Verkerk, P.H.; van Spronsen, F.J. Behaviour and school achievement in patients with early and continuously treated phenylketonuria. *J. Inherit. Metab. Dis.* **2000**, *23*, 548–562. [CrossRef] [PubMed]
4. Diamond, A.; Prevor, M.B.; Callender, G.; Druin, D.P. Prefrontal cortex cognitive deficits in children treated early and continuously for PKU. *Monogr. Soc. Res. Child Dev.* **1997**, *62*. [CrossRef]
5. Pascucci, T.; Andolina, D.; Mela, I.L.; Conversi, D.; Latagliata, C.; Ventura, R.; Puglisi-Allegra, S.; Cabib, S. 5-Hydroxytryptophan rescues serotonin response to stress in prefrontal cortex of hyperphenylalaninaemic mice. *Int. J. Neuropsychopharmacol.* **2009**, *12*, 1067–1079. [CrossRef] [PubMed]
6. Andolina, D.; Conversi, D.; Cabib, S.; Trabalza, A.; Ventura, R.; Puglisi-Allegra, S.; Pascucci, T. 5-Hydroxytryptophan during critical postnatal period improves cognitive performances and promotes dendritic spine maturation in genetic mouse model of phenylketonuria. *Int. J. Neuropsychopharmacol.* **2011**, *14*, 479–489. [CrossRef] [PubMed]
7. Nelson, S.B.; Valakh, V. Excitatory/Inhibitory Balance and Circuit Homeostasis in Autism Spectrum Disorders. *Neuron* **2015**, *87*, 684–698. [CrossRef] [PubMed]
8. Bemben, M.A.; Shipman, S.L.; Nicoll, R.A.; Roche, K.W. The cellular and molecular landscape of neuroligins. *Trends Neurosci.* **2015**, *38*, 496–505. [CrossRef] [PubMed]
9. Krueger, D.D.; Tuffy, L.P.; Papadopoulos, T.; Brose, N. The role of neurexins and neuroligins in the formation, maturation, and function of vertebrate synapses. *Curr. Opin. Neurobiol.* **2012**, *22*, 412–422. [CrossRef] [PubMed]
10. Lee, E.; Lee, J.; Kim, E. Excitation/Inhibition Imbalance in Animal Models of Autism Spectrum Disorders. *Biol. Psychiatry* **2017**, *81*, 838–847. [CrossRef] [PubMed]
11. De la Torre-Ubieta, L.; Won, H.; Stein, J.L.; Geschwind, D.H. Advancing the understanding of autism disease mechanisms through genetics. *Nat. Med.* **2016**, *22*, 345–361. [CrossRef] [PubMed]
12. Jamain, S.; Quach, H.; Betancur, C.; Råstam, M.; Colineaux, C.; Gillberg, I.C.; Soderstrom, H.; Giros, B.; Leboyer, M.; Gillberg, C.; et al. Paris Autism Research International Sibpair Study Mutations of the X-linked genes encoding neuroligins NLGN3 and NLGN4 are associated with autism. *Nat. Genet.* **2003**, *34*, 27–29. [CrossRef] [PubMed]
13. Laumonnier, F.; Bonnet-Brilhault, F.; Gomot, M.; Blanc, R.; David, A.; Moizard, M.-P.; Raynaud, M.; Ronce, N.; Lemonnier, E.; Calvas, P.; et al. X-linked mental retardation and autism are associated with a mutation in the NLGN4 gene, a member of the neuroligin family. *Am. J. Hum. Genet.* **2004**, *74*, 552–557. [CrossRef] [PubMed]
14. Chih, B.; Afridi, S.K.; Clark, L.; Scheiffele, P. Disorder-associated mutations lead to functional inactivation of neuroligins. *Hum. Mol. Genet.* **2004**, *13*, 1471–1477. [CrossRef] [PubMed]

15. Comoletti, D.; de Jaco, A.; Jennings, L.L.; Flynn, R.E.; Gaietta, G.; Tsigelny, I.; Ellisman, M.H.; Taylor, P. The Arg451Cys-neuroligin-3 mutation associated with autism reveals a defect in protein processing. *J. Neurosci. Off. J. Soc. Neurosci.* **2004**, *24*, 4889–4893. [CrossRef] [PubMed]

16. De Jaco, A.; Lin, M.Z.; Dubi, N.; Comoletti, D.; Miller, M.T.; Camp, S.; Ellisman, M.; Butko, M.T.; Tsien, R.Y.; Taylor, P. Neuroligin trafficking deficiencies arising from mutations in the α/β-hydrolase fold protein family. *J. Biol. Chem.* **2010**, *285*, 28674–28682. [CrossRef] [PubMed]

17. Cao, X.; Tabuchi, K. Functions of synapse adhesion molecules neurexin/neuroligins and neurodevelopmental disorders. *Neurosci. Res.* **2017**, *116*, 3–9. [CrossRef] [PubMed]

18. Siddiqui, T.J.; Craig, A.M. Synaptic organizing complexes. *Curr. Opin. Neurobiol.* **2011**, *21*, 132–143. [CrossRef] [PubMed]

19. Südhof, T.C. Neuroligins and neurexins link synaptic function to cognitive disease. *Nature* **2008**, *455*, 903–911. [CrossRef] [PubMed]

20. Maćkowiak, M.; Mordalska, P.; Wędzony, K. Neuroligins, synapse balance and neuropsychiatric disorders. *Pharmacol. Rep.* **2014**, *66*, 830–835. [CrossRef] [PubMed]

21. Dani, V.S.; Chang, Q.; Maffei, A.; Turrigiano, G.G.; Jaenisch, R.; Nelson, S.B. Reduced cortical activity due to a shift in the balance between excitation and inhibition in a mouse model of Rett syndrome. *Proc. Natl. Acad. Sci. USA* **2005**, *102*, 12560–12565. [CrossRef] [PubMed]

22. Tabuchi, K.; Blundell, J.; Etherton, M.R.; Hammer, R.E.; Liu, X.; Powell, C.M.; Südhof, T.C. A neuroligin-3 mutation implicated in autism increases inhibitory synaptic transmission in mice. *Science* **2007**, *318*, 71–76. [CrossRef] [PubMed]

23. Belichenko, P.V.; Kleschevnikov, A.M.; Masliah, E.; Wu, C.; Takimoto-Kimura, R.; Salehi, A.; Mobley, W.C. Excitatory-inhibitory relationship in the fascia dentata in the Ts65Dn mouse model of Down syndrome. *J. Comp. Neurol.* **2009**, *512*, 453–466. [CrossRef] [PubMed]

24. Fernandez, F.; Garner, C.C. Over-inhibition: A model for developmental intellectual disability. *Trends Neurosci.* **2007**, *30*, 497–503. [CrossRef] [PubMed]

25. Kleschevnikov, A.M.; Belichenko, P.V.; Villar, A.J.; Epstein, C.J.; Malenka, R.C.; Mobley, W.C. Hippocampal long-term potentiation suppressed by increased inhibition in the Ts65Dn mouse, a genetic model of Down syndrome. *J. Neurosci. Off. J. Soc. Neurosci.* **2004**, *24*, 8153–8160. [CrossRef] [PubMed]

26. Yizhar, O.; Fenno, L.E.; Prigge, M.; Schneider, F.; Davidson, T.J.; O'Shea, D.J.; Sohal, V.S.; Goshen, I.; Finkelstein, J.; Paz, J.T.; et al. Neocortical excitation/inhibition balance in information processing and social dysfunction. *Nature* **2011**, *477*, 171–178. [CrossRef] [PubMed]

27. De Groot, M.J.; Hoeksma, M.; Blau, N.; Reijngoud, D.J.; van Spronsen, F.J. Pathogenesis of cognitive dysfunction in phenylketonuria: Review of hypotheses. *Mol. Genet. Metab.* **2010**, *99*, S86–S89. [CrossRef] [PubMed]

28. Levinson, J.N.; El-Husseini, A. Building excitatory and inhibitory synapses: Balancing neuroligin partnerships. *Neuron* **2005**, *48*, 171–174. [CrossRef] [PubMed]

29. Hines, R.M.; Wu, L.; Hines, D.J.; Steenland, H.; Mansour, S.; Dahlhaus, R.; Singaraja, R.R.; Cao, X.; Sammler, E.; Hormuzdi, S.G.; et al. Synaptic imbalance, stereotypies, and impaired social interactions in mice with altered neuroligin 2 expression. *J. Neurosci.* **2008**, *28*, 6055–6067. [CrossRef] [PubMed]

30. Liang, J.; Xu, W.; Hsu, Y.-T.; Yee, A.X.; Chen, L.; Südhof, T.C. Conditional neuroligin-2 knockout in adult medial prefrontal cortex links chronic changes in synaptic inhibition to cognitive impairments. *Mol. Psychiatry* **2015**, *20*, 850–859. [CrossRef] [PubMed]

31. Varoqueaux, F.; Jamain, S.; Brose, N. Neuroligin 2 is exclusively localized to inhibitory synapses. *Eur. J. Cell Biol.* **2004**, *83*, 449–456. [CrossRef] [PubMed]

32. Poulopoulos, A.; Aramuni, G.; Meyer, G.; Soykan, T.; Hoon, M.; Papadopoulos, T.; Zhang, M.; Paarmann, I.; Fuchs, C.; Harvey, K.; et al. Neuroligin 2 drives postsynaptic assembly at perisomatic inhibitory synapses through gephyrin and collybistin. *Neuron* **2009**, *63*, 628–642. [CrossRef] [PubMed]

33. Soykan, T.; Schneeberger, D.; Tria, G.; Buechner, C.; Bader, N.; Svergun, D.; Tessmer, I.; Poulopoulos, A.; Papadopoulos, T.; Varoqueaux, F.; et al. A conformational switch in collybistin determines the differentiation of inhibitory postsynapses. *EMBO J.* **2014**, *33*, 2113–2133. [CrossRef] [PubMed]

34. Chubykin, A.A.; Atasoy, D.; Etherton, M.R.; Brose, N.; Kavalali, E.T.; Gibson, J.R.; Südhof, T.C. Activity-dependent validation of excitatory versus inhibitory synapses by neuroligin-1 versus neuroligin-2. *Neuron* **2007**, *54*, 919–931. [CrossRef] [PubMed]

35. Pascucci, T.; Ventura, R.; Puglisi-Allegra, S.; Cabib, S. Deficits in brain serotonin synthesis in a genetic mouse model of phenylketonuria. *Neuroreport* **2002**, *13*, 2561–2564. [CrossRef] [PubMed]

36. Jones, K.A.; Srivastava, D.P.; Allen, J.A.; Strachan, R.T.; Roth, B.L.; Penzes, P. Rapid modulation of spine morphology by the 5-HT2A serotonin receptor through kalirin-7 signaling. *Proc. Natl. Acad. Sci. USA* **2009**, *106*, 19575–19580. [CrossRef] [PubMed]

37. Prange, O.; Wong, T.P.; Gerrow, K.; Wang, Y.T.; El-Husseini, A. A balance between excitatory and inhibitory synapses is controlled by PSD-95 and neuroligin. *Proc. Natl. Acad. Sci. USA* **2004**, *101*, 13915–13920. [CrossRef] [PubMed]

38. Ko, J.; Choii, G.; Um, J.W. The balancing act of GABAergic synapse organizers. *Trends Mol. Med.* **2015**, *21*, 256–268. [CrossRef] [PubMed]

39. Graf, E.R.; Zhang, X.; Jin, S.-X.; Linhoff, M.W.; Craig, A.M. Neurexins induce differentiation of GABA and glutamate postsynaptic specializations via neuroligins. *Cell* **2004**, *119*, 1013–1026. [CrossRef] [PubMed]

40. Pascucci, T.; Andolina, D.; Ventura, R.; Puglisi-Allegra, S.; Cabib, S. Reduced availability of brain amines during critical phases of postnatal development in a genetic mouse model of cognitive delay. *Brain Res.* **2008**, *1217*, 232–238. [CrossRef] [PubMed]

41. Delattre, V.; La Mendola, D.; Meystre, J.; Markram, H.; Markram, K. NLGN4 knockout induces network hypo-excitability in juvenile mouse somatosensory cortex in vitro. *Sci. Rep.* **2013**, *3*, 2897. [CrossRef] [PubMed]

42. Puglisi-Allegra, S.; Cabib, S.; Pascucci, T.; Ventura, R.; Cali, F.; Romano, V. Dramatic brain aminergic deficit in a genetic mouse model of phenylketonuria. *Neuroreport* **2000**, *11*, 1361–1364. [CrossRef] [PubMed]

43. Ulbrich, L.; Favaloro, F.L.; Trobiani, L.; Marchetti, V.; Patel, V.; Pascucci, T.; Comoletti, D.; Marciniak, S.J.; De Jaco, A. Autism-associated R451C mutation in neuroligin3 leads to activation of the unfolded protein response in a PC12 Tet-On inducible system. *Biochem. J.* **2016**, *473*, 423–434. [CrossRef] [PubMed]

International Journal of
Molecular Sciences

MDPI

Article

Investigating Autism-Related Symptoms in Children with Prader-Willi Syndrome: A Case Study

Jeffrey A. Bennett [1,2], Sandra Hodgetts [3], Michelle L. Mackenzie [1], Andrea M. Haqq [1,*,†]
and Lonnie Zwaigenbaum [1,2,†]

1   Department of Pediatrics, Faculty of Medicine and Dentistry, University of Alberta, 11405 87 Avenue,
    Edmonton, AB T6G1C9, Canada; jabennet@ualberta.ca (J.A.B.); michelle.mackenzie@ualberta.ca (M.L.M.);
    lonnie.zwaigenbaum@albertahealthservices.ca (L.Z.)
2   Autism Research Centre-E209, Glenrose Rehabilitation Hospital, 10230 111 Avenue,
    Edmonton, AB T5G 0B7, Canada
3   Faculty of Rehabilitation Medicine, University of Alberta, 8205 114 Street, Edmonton, AB T6G 2G4, Canada;
    sandra.hodgetts@ualberta.ca
*   Correspondence: haqq@ualberta.ca; Tel.: +1-780-492-0015
†   These authors contributed equally to this work.

Academic Editor: Merlin G. Butler
Received: 21 January 2017; Accepted: 23 February 2017; Published: 28 February 2017

Abstract: Prader-Willi syndrome (PWS), a rare genetic disorder caused by the lack of expression of paternal genes from chromosome 15q11-13, has been investigated for autism spectrum disorder (ASD) symptomatology in various studies. However, previous findings have been variable, and no studies investigating ASD symptomatology in PWS have exclusively studied children. We aimed to characterize social communication functioning and other ASD-related symptoms in children with PWS, and assessed agreement across measures and rates of ASD diagnosis. Measures included the Autism Diagnostic Observation Schedule-2 (ADOS-2), the Social Communication Questionnaire (SCQ), Social Responsiveness Scale-2 (SRS-2), Social Skills Improvement System-Rating Scales (SSIS-RS), and the Vineland Adaptive Behavioral Scales-II (VABS-II). General adaptive and intellectual skills were also assessed. Clinical best estimate (CBE) diagnosis was determined by an experienced developmental pediatrician, based on history and review of all available study measures, and taking into account overall developmental level. Participants included 10 children with PWS, aged 3 to 12 years. Three of the 10 children were male and genetic subtypes were two deletion (DEL) and eight uniparental disomy (UPD) (with a total of 6 female UPD cases). Although 8 of the 10 children exceeded cut-offs on at least one of the ASD assessments, agreement between parent questionnaires (SCQ, SRS-2, SSIS-RS) and observational assessment (ADOS-2) was very poor. None of the children were assigned a CBE diagnosis of ASD, with the caveat that the risk may have been lower because of the predominance of girls in the sample. The lack of agreement between the assessments emphasizes the complexity of interpreting ASD symptom measures in children with PWS.

Keywords: Prader-Willi syndrome; PWS; social communication; Autism Diagnostic Observation Schedule; ADOS; autism spectrum disorder; ASD

1. Introduction

Prader-Willi syndrome (PWS) is a genetic disorder caused by the absence of expression of the paternal contribution of chromosome 15q11-13. The majority of cases are due to deletion (DEL; 65%–75%) or uniparental disomy (UPD; 20%–30%), with a small minority (1%–3%) due to rare imprinting center defects [1]. The PWS phenotype includes hypotonia and failure to thrive during infancy, followed in early childhood by hyperphagia and an insatiable appetite which can lead to morbid obesity if left unchecked [1]. Cognitive disability and problem behaviors are also common and, for some families, represent greater challenges than the food-seeking behaviors [2].

Although variable, the PWS phenotype overlaps to some degree with autism spectrum disorder (ASD), a neurodevelopmental disorder characterized by the presence of symptoms in two core domains: social communication impairment and restricted and repetitive behaviors and interests. Indeed, the prevalence of ASD in PWS has been estimated at 26.7% based on exceeding clinical cut-points on relevant ASD assessments [3]. Also, ASD in the UPD genetic subtype (35.3%; two maternal copies of chromosome 15 present) is almost twice as common as in the DEL subtype (18.3%) [3]. This may be partially due to a genetic finding that overexpression of chromosome 15 is associated with higher rates of ASD [4]. The estimated prevalence of ASD in PWS is significantly higher than the current prevalence estimate of about 1.5% for ASD in the general population [5]. To receive an ASD diagnosis, symptoms in both domains must be present during the early developmental period; however, they may not be fully recognized until a child is older and social demands exceed their capacity [1]. Nevertheless, early social communication impairments (including reduced gazing towards faces and directed vocalizations [6], and reduced response to name being called [7]) in children who later receive a diagnosis of ASD have been reported as early as 12 months. Indeed, ASD diagnosis can be reliably made by 18 to 24 months in a clinical setting. Earlier recognition of ASD is important as studies have demonstrated that early intervention for ASD yields more favorable outcomes [8].

Despite a recent increase of research investigating symptoms of ASD in PWS, no studies have focused on young children with PWS. Eight years of age is the youngest mean age in a study investigating ASD in individuals with PWS [9]. The expression of ASD in PWS may change over development, with two studies reporting more prominent symptoms in adolescents and adults with PWS compared to younger children. Lo et al. [10] reported that none of the 22 children with PWS, ages 7–9, exceeded the cut-off for ASD on the Diagnostic Interview for Social and Communication Disorders (DISCO); however, 24 of 44 of the individuals aged 10–17 years old exceeded the cut-off for ASD. Additionally, Akefeldt and Gillberg [11] reported lower average scores (3.4; SD = 4.4) in toddlers with PWS (mean age = 2.1 years; range = 0.8–3.7 years) compared to the older individuals (19.1; SD = 10.7) with PWS (mean age 18.4 years; range = 4.2–36.3 years) on the Autism Spectrum Screening Questionnaire (ASSQ) [12]. However, this tool was designed to assess children ages 6 to 17 years old with normal intelligence to mild cognitive disability, raising concerns about the validity of the ASSQ in this sample.

Assessments commonly used to investigate ASD symptomatology and adaptive functioning in individuals with PWS include the Social Responsiveness Scale [13,14], Social Communication Questionnaire [15,16], and Vineland Adaptive Behavior Scales [13,17]. A recent systematic review [3] found that few studies investigating ASD in PWS have included the Autism Diagnostic Observation Schedule (ADOS) [18] or the Autism Diagnostic Interview-Revised (ADI-R) [19], considered to be gold standard measures of ASD symptoms based on excellent sensitivity and specificity in differentiating ASD from other developmental disorders [20]. Moreover, rates of ASD reported in PWS are generally not based on ICD or DSM-based clinical diagnoses, but are rather based on clinical cut-offs on proxy measures such as the Social Communication Questionnaire (SCQ) [3]. This is problematic, as no single measure can be used as a proxy for diagnosis.

The primary goal of this study was to characterize ASD symptoms in children ages 3 to 12 with PWS using the ADOS, 2nd edition (ADOS-2) [21], and other standardized ASD assessment tools. As a secondary objective, we investigated other aspects of social-communication development in children with PWS, and compared the agreement between various ASD assessment tools and categorical ASD diagnosis by clinical best estimate (CBE). Based on previous studies, we hypothesized elevated rates of ASD symptomatology in children with PWS, as has been found in adolescents and adults with PWS. We also expected ASD symptoms to be present in some children with PWS who do not have a diagnosis of ASD. Furthermore, we expected a moderate agreement between the various ASD tools used to assess ASD symptomatology in children with PWS.

2. Results

2.1. Demographics

There were a total of 10 participants, all of whom resided in western Canada (Alberta (n = 8); British Columbia (n = 1); Saskatchewan (n = 1)). Nine of the ten patients with PWS in the age range who were contacted directly through a regional specialty clinic agreed to participate in the study. One additional child was recruited through a provincial PWS group. The cohort included seven females and three males, with a mean age of 6.77 years old (SD = 2.85). Participant genetic subtype was confirmed by genetic testing completed previously, or through confirmation of genetic testing by the participants' health-care provider. Only two participants were the DEL subtype; the other eight were the UPD subtype. Each participant completed all of the assessments. Table 1 provides a summary of the study results, including the mean, standard error of the mean, and standard deviation, as well as the minimum and maximum scores attained for all measures. The average full scale IQ (FSIQ) was 63.38 (SD = 17.42), which falls within the range for individuals with PWS reported in a review by Cassidy and colleagues [1]. A negative correlation between age and FSIQ was found (Spearman's $\rho = -0.815$, $p = 0.004$).

Table 1. Results from assessments, given as mean (SE; SD) and range.

	Mean (SE; SD)	Min–Max
Age	6.77 (2.85)	3.42–11.75
# (%) male	3 (30)	–
# (%) UPD	8 (80)	–
FSIQ	64.70 (5.05; 15.96)	40–92
ADOS-2 Severity score	3.00 (0.42; 1.33)	1–5
ADOS-2 SA Severity score	3.50 (0.56; 1.78)	2–7
ADOS-2 RRB Severity score	4.70 (0.80; 2.54)	0–7
SCQ raw score	11.90 (2.20; 6.95)	5–21
SRS-2 Overall T-score	62.70 (3.98; 12.56)	43–85
SRS-2: Social Awareness T-score	61.60 (4.16; 13.16)	43–81
SRS-2: Social Cognition T-score	61.00 (4.83; 15.27)	40–94
SRS-2: Social Communication T-score	61.70 (4.04; 12.78)	43–86
SRS-2: Social Motivation T-score	56.10 (2.47; 7.81)	43–67
SRS-2: RRB T-score	65.80 (4.20; 13.28)	48–88
SSIS-RS: Social skills percentile	23.10 (7.02; 22.18)	2–64
SSIS-RS: Problem behaviors percentile	80.00 (7.06; 22.32)	42–99
SSIS-RS: ASD raw score	19.00 (2.53; 8.01)	8–31
Vineland-II Composite standard score	80.40 (3.57; 11.29)	60–96
Vineland-II: Communication standard score	85.20 (4.60; 14.56)	59–100
Vineland-II: Daily Living Skills standard score	86.50 (3.60; 11.38)	68–101
Vineland-II: Socialization standard score	83.20 (4.79; 15.14)	57–108
Vineland-II: Motor standard score	77.20 (2.59; 8.19)	67–91

n = 10 for all assessments; SE = Standard Error (of the mean); SD = Standard Deviation (of the sample); FSIQ = Full-scale IQ (based on WPPSI/WISC scores); ADOS-2 = Autism Diagnostic Observation Schedule-2; SA = Social Affect; RRB = Restricted and Repetitive Behaviors; SRS-2 = Social Responsiveness Scale-2; SCQ = Social Communication Questionnaire; SSIS-RS = Social Skills Improvement System-Rating Scales.

2.2. Assessments

Table 2 gives detailed results for every participant on each of the ASD assessments, including an indication of which scores exceeded clinical cut-offs. In total, three of 10 children (all 3 female UPD) scored above the cut-off for ASD on the ADOS-2 symptom severity score. However, two of those three children (P8 and P9) scored 0 in the restricted and repetitive behaviors (RRB) domain, which would suggest that ASD is not present, as under DSM-5 observable symptoms are expected in both the social affect (SA) and RRB domains.

Int. J. Mol. Sci. **2017**, *18*, 517

Table 2. Individual participant score profiles.

ID	Sub-Type	Sex	FSIQ	ADOS Module	ADOS SA	ADOS RRB	ADOS Overall	SCQ	SRS AWR	SRS COG	SRS COM	SRS MOT	SRS RRB	SRS Overall	SSIS SS	SSIS PB	SSIS ASD	CBE
1	DEL	F	92	2	2	6	2	6	54	56	55	51	62	56	25	53	11	N
2	UPD	M	63	2	2	5	1	6	43	54	48	57	56	52	50	56	16	N
3	UPD	F	77	2	4	6	4	5	49	40	43	43	48	43	64	42	8	N
4	UPD	M	82	2	2	6	2	14	66	63	57	47	64	59	18	90	21	N
5	UPD	F	69	2	2	6	2	9	57	53	64	54	58	62	14	98	20	N
6	UPD	F	56	3	3	6	3	13	68	77	71	64	82	75	8	97	26	N
7	DEL	M	65	3	4	5	3	20	67	55	65	67	66	65	2	99	31	N
8	UPD	F	47	1	6	0	5	14	81	67	73	64	80	75	4	96	21	N
9	UPD	F	56	3	7	0	5	6	50	51	55	60	54	55	44	71	8	N
10	UPD	F	40	3	3	7	3	26	81	94	86	54	88	85	2	98	23	N

Subtype: DEL = deletion ; UPD = uniparental disomy; Sex: M=Male, F = Female; ADOS-SA = ADOS-2 Social Affect Severity Score; ADOS-RRB = ADOS-2 Restricted or Repetitive Behavior Severity Score; SCQ = SCQ Raw Score; SRS-AWR = SRS-2 Social Awareness T-score; SRS-COG = SRS-2 Social Cognition T-score; SRS-COM = SRS-2 Social Communication T-score; SRS-MOT = SRS-2 Social Motivation T-score; SRS-RRB = SRS-2 Restricted or Repetitive Behaviors T-score; SSIS-SS = SSIS-RS Social Skills Percentile; SSIS-PB = SSIS-RS Problem Behaviors Percentile; SSIS-ASD = SSIS-RS ASD Subscale Raw; CBE = Clinical Best Estimate; ADOS-2: scores of >3 are associated with ASD; SCQ: scores of ≥12 are associated with ASD; SRS-2: scores <60 are not associated with ASD, 60–65 are associated with mild to moderate impairment in social responsiveness and ASD, 66–75 are associated with substantial impairment in social responsiveness and ASD, >75 are associated with severe impairment in social responsiveness and ASD; SSIS-RS: overall score for the following age ranges representts the average amount of ASD behaviors; any score above or below these ranges indicate above or below average ASD symptoms, respectively: age 3–5 years: 4–16; age 5–12 years: 3–14; CBE: N = does not qualify for ASD diagnosis, Y = does qualify for ASD diagnosis; * Scores that exceed cut-off for ASD are *bolded and italicized*. Note: The individuals were ordered from youngest (ID = 1) to oldest (ID = 10) in order to give a sense of participant age (range = 3–12 years); to maintain anonymity (due to the rarity of PWS), participant ages not listed; rather, ordered by age.

2.2.1. ASD Cut-Off Scores

Five children exceeded the cut-off SRS-2 score associated with ASD in the general population, four of whom also exceeded cut-off on the SCQ. With respect to the separate domains, social motivation showed the least impairment (T-score = 56.10), while the RRB domain showed the most impairment (T-score = 65.80). No differences were detected among SRS-2 domain means, based on a non-parametric Kruskal-Wallis Test. It should be noted that, although a modified cut-off score of 12 was used for the SCQ, using the original cut-off score of 15 did not add meaningful clarification to the overall scores or patterns of agreement between assessments. Six of 10 children (1 male DEL, 1 male DEL, 4 female UPD) had above average ASD symptoms on the SSIS-RS ASD subscale. These six children also exceeded cut-off on the SCQ and/or the SRS-2.

2.2.2. Clinical Best Estimate

None of the participants met clinical best estimate (CBE) criteria for ASD based on DSM-5, largely because of inconsistency across measures. Detailed case descriptions of each participant (P) are provided below. To maintain confidentiality, age and gender cannot be reported, although the participants are ordered from youngest (P1) to oldest (P10) to provide a sense of relative age. Each profile describes the unique or extreme findings from each child.

2.2.3. Case Study Descriptions

P1 (female DEL) was the youngest individual in our study. The assessment results from this participant were not indicative of ASD. The Restricted or Repetitive Behaviors and Interests (RRB) sections on both the ADOS-2 and SRS-2 confirmed mild impairment in this symptom domain, but the ADOS-SA domain was low at a severity score of 2. The SSIS-RS percentiles and Vineland-II scores were average, with the exception of the Vineland-II motor skills domain, which was more than one standard deviation below the normative standard.

P2 (male UPD) did not exceed cut-off for ASD on any of the assessments. Again, the SSIS-RS and Vineland-II results were all close to the normative standard, with the exception of motor skills on the Vineland-II.

P3 (female UPD) had parent-reported symptoms from the three ASD questionnaires (SCQ, SRS-2, and SSIS-RS) that were the lowest in the entire sample. Indeed, she had among the most favourable scores on the SSIS-RS and Vineland-II of all participants, scoring above the normative standard on all domains except Vineland-II socialization and Vineland-II motor skills domains, both of which were less than one standard deviation below age-related norms. In contrast, P3 is the only child who exceeded the ASD threshold on both the SA and RRB domains of the ADOS-2, albeit with an overall severity score right at the cut-off associated with ASD.

P4 (male UPD) had SCQ and SSIS-RS ASD subdomain scores that exceeded the cut-off for ASD, while his SRS-2 score was one point below cut-off. The SRS-2 domains that elevated the overall scores included social awareness, social cognition, and RRBs. He also scored below the sample mean on the social skills subdomain of the SSIS-RS (18th percentile), and higher on the problem behaviors subdomain (90th percentile). However, he was assessed as having social-communicative symptoms well below the level associated with ASD (ADOS-SA severity score = 2, and overall severity score = 2), although RRB symptoms were observed. As well, Vineland-II social and communicative domains were rated as close to average.

P5 (female UPD) had similar results as P4: low ADOS-2 scores, despite exceeding cut-off for ASD on two of the three parent-report questionnaires. Critically, the SCQ score was three points below the cut-off associated with ASD and the SRS-2 overall score was only two points above cut-off. She also scored low on the social skills domain of the SSIS-RS (14th percentile), and very high on the problem behaviors domain (98th percentile). Additionally, the Vineland-II scores were all below the sample mean, with motor skills being the most affected.

P6 (female UPD) was assessed as having social-communication symptoms below the level associated with ASD (severity score = 3) on the ADOS, despite RRB symptoms (severity score = 6). However, she was rated as having high levels of symptoms on all three questionnaires. The SRS-2 scores were especially high, with the overall score one point below the severe range. These scores were largely driven by the RRB and the social cognition subdomains, both of which fell in the severe range; comparatively, the social motivation subdomain only showed mild impairment. Additionally, the SSIS-RS social skills (8th percentile) and problem behavior (97th percentile) were both indicative of significant social impairments. Despite the high scores on the questionnaires, P6 scored within one SD of age norms on the communication and daily living skills domains on the Vineland-II, although more than one standard deviation lower on the socialization and motor skills domains. Full scale IQ was 56, so clinically, social impairments were felt to relate, at least in part, to intellectual impairments.

P7 (male DEL) also scored below the clinical threshold on the ADOS-2, with an overall severity score of 3, despite symptoms observed in both domains. He exceeded the cut-off for ASD on all three questionnaires. On the SSIS-RS, this child was rated at the 2nd percentile on the social skills domain, >99th percentile for the problem behaviors domain, and the highest raw score on the ASD subdomain. The participant's SCQ score was also elevated. Interestingly, all of the subdomains on the SRS-2 were extremely similar (scores ranged from 65–67), with the exception of social cognition, which was below cut-off.

P8 (female UPD) had an overall symptom severity rating of 5, above the level associated with ASD, although with no symptoms observed in the RRB domain (severity score = 0). The parent report on the SRS-2 did suggest severe impairments in the RRB subdomain, with overall SRS-2 score at the upper end of the moderate range. SSIS-RS social skills (4th percentile) and problem behaviors (96th percentile) were also indicative of social impairment, and the ASD subscale score was very high as well. She scored over two standard deviations below the normative standard on the communication domain of the Vineland-II, and had slightly higher scores in the other three domains. Other factors taken in consideration in the CBE process included cognitive delay (FSIQ = 47).

P9 (female UPD) had high levels of symptoms rated on the SA domain of the ADOS (severity score = 7), but no observed RRB symptoms (severity score = 0). As well, she had among the lowest scores on the SSIS-RS ASD domain, the SCQ, and the SRS-2, including on the RRB domain. The only SRS-2 subdomain that exceeded cut-off for ASD was social motivation, albeit by one point.

P10 (female UPD) had symptoms rated in both SA (severity score = 3) and RRB. (severity score = 7) of the ADOS, although the overall severity was subthreshold for ASD (severity score = 3). Scores were also elevated on the SCQ and SRS-2. This child's SSIS-RS ASD score was also well above cut-off, and the SSIS-RS social skills (2nd percentile) and problem behaviors (98th percentile) also showed significant impairment. She also had the lowest adaptive behavior composite score on the Vineland-II, with the socialization and communication scores almost three standard deviations below standard. However, the clinician responsible for CBE diagnosis concluded that impairments in socialization and communication were largely attributable to intellectual disability (FSIQ = 40), and that this child did not meet DSM-5 criteria for ASD.

2.3. Agreement among Measures

The three questionnaires (SRS-2, SCQ, and SSIS-RS) all had acceptable agreement with one another (see Table 3). Cohen's κ for the SCQ and SSIS-RS, as well as for the SRS-2 and SSIS-RS, was calculated to be 0.80 (SE = 0.19), and κ for the SCQ and SRS-2 was calculated to be 0.60 (SE = 0.25). Notably, a total of six children exceeded the cut-off on the SSIS-RS; 4 of 6 children exceeded the cut-off on both the SRS-2 and SCQ, a fifth on the SRS-2 but not the SCQ, and the sixth on the SCQ but not the SRS-2. However, agreement between the three questionnaires and the ADOS-2 was poor. Cohen's κ for the ADOS-2 vs. the SCQ, SRS-2, and SSRI-RS were calculated to be −0.20 (SE = 0.28), −0.20 (SE = 0.28), and −0.30 (SE = 0.28), respectively.

Table 3. Agreement among Measures.

Measures	Kohen's κ	Standard Error
SCQ/SRS-2	0.60	0.25
SCQ/SSIS-RS	0.80	0.19
SRS-2/SSIS-RS	0.80	0.19
SCQ/ADOS-2	−0.20	0.28
SRS-2/ADOS-2	−0.20	0.28
SSIS-RS/ADOS-2	−0.30	0.28

3. Discussion

Our main findings were that (1) ASD symptomatology and social competence in children with PWS were highly variable; (2) While there was moderate to high agreement among parent-report measures of ASD symptoms and social behaviors (κ = 0.6–0.8), agreement between these measures and ASD symptom severity as observed on the ADOS-2 was poor (κ < 0.0) (see Table 3). Six of the 10 children (1 male DEL, 1 male DEL, 4 female UPD) exceeded the cut-off for ASD on at least one of the parent-report measures (and 4 of 10 on all three). In contrast, 3 of 10 (all female UPD) had observable symptom severity at the level associated with ASD diagnosis on the ADOS-2, only one of whom had elevated symptoms in both domains. Moreover, only 1 of the 3 children with elevated ASD symptom severity on the ADOS-2 met the scoring cut-off suggestive of ASD on *any* of the parent report measures (not the same child with elevated symptoms on both ADOS-2 domains). Thus, while most (8 of 10) participants had evidence of ASD symptoms on at least one study measure, there was poor agreement between measures and thus, inconsistent findings that complicated clinical diagnostic judgement. Indeed, none of the participants were assigned a clinical diagnosis of ASD. One child (P8) with elevated symptoms on both the ADOS-2 (albeit not both domains) and the parent report measures had an FSIQ of 47, and thus, the clinical presentation was complicated by intellectual disability. This child's inability to perform numerous tasks made the interpretation of social skills deficits more challenging. Although the clinical diagnostic status of this child may be equivocal, the pattern of scoring overall on study measures indicated considerable variability in profile, poor agreement between parent-report measures, and direct behavioral observation (i.e., using the ADOS-2), and raises important questions about the clinical interpretation and even the validity of available ASD symptom measures in this population. Finally, there was a negative correlation between age and FSIQ (Spearman's ρ = −0.815, p = 0.004 in this small sample).

In our study, sample means on ASD assessments in children with PWS were less indicative of ASD than have been previously reported in adolescents and adults with PWS. For example, Zyga et al. [14] and Dimitropoulos et al. [13] used the SRS to assess samples with a mean age over 10 years of age, and found average scores of 82.18 and 76.31, respectively. Both of these scores fall in the severe range on the SRS-2, whereas the average score from our population was 62.70, which falls in the mild range. Dimitropoulos et al. [13] and Milner et al. [17] also used the Vineland-II in their samples of adolescents and adults with PWS, and reported average composite scores of 65.15 and 62.60, respectively. Meanwhile, our sample had an average score of 80.40, over one standard deviation above results from these previously published studies. Lastly, Zyga et al. [14] reported that 8 of 14 adolescents (57.14%) met criteria for ASD on the ADOS, whereas only 3 of 10 (30%) from our cohort met criteria (actually only 1 of 10 (10%), considering that P8 and P9 did not pass cut-off on the RRB domain). These findings indicate interesting differences in ASD symptoms and adaptive functioning between age groups (children vs. adolescents and adults). However, other potential confounders, such as growth hormone treatment or exposure to other interventions, cannot be ruled out.

There were marked discrepancies between direct observational assessment and parent report in the assessment of ASD in PWS in our study. Notably, previous research in non-PWS samples indicates that both parent reports and observational data are essential to valid diagnostic assessment for ASD [22]. In a recent systematic review of ASD in PWS, Bennett et al. [3] reported that most studies investigating ASD in PWS are based on single measures, with parent reports more common than observational assessment (i.e., ADOS-2). Given that parent reports may be less resource-intensive and require less specialized training, they may be more feasible to administer in a clinical program serving a rare population such as PWS. However, given the limited agreement across parent reports and direct observational measures in our sample, further research may be needed to verify which assessments are most accurate and informative in the clinical diagnostic assessment of ASD in PWS.

Another possible explanation for the discrepancy between the results from the ADOS-2 and the parent-report assessments could be the distinction between what a child 'can do' and 'typically does'. The AODS-2 is a 30 to 45-minute observational assessment designed to elicit certain responses from children. Since some children with PWS are more comfortable associating with adults than same-age peers, the ADOS-2 interaction potentially provides a more comfortable situation than their usual social experiences. Parent-report measures, on the other hand, focus on the frequency of the child's social behaviors in their real-life settings, which may be influenced by both ability and opportunities/contexts. Another possible explanation for poor agreement between the SCQ questionnaire and ADOS-2 is that the restrictive repetitive behaviors on the SCQ (which are more prevalent in PWS) are not as heavily weighted on the ADOS-2. However, it would be difficult to know whether any of these possibilities are valid without further research to probe discrepancies between the various ASD assessments employed in this study.

Lack of agreement between ASD assessment measures has been reported in studies investigating ASD in other genetic syndromes, specifically Fragile-X syndrome [23]. These studies indicate poor agreement between the ADOS-2 and SCQ (Cohen's $\kappa = 0.33$ for girls, 0.13 for boys). Additionally, using a combination of the ADI-R, ADOS, and DSM-IV criteria, Harris et al. [24] reported that in a group of 63 males with Fragile X syndrome, 15 participants (24%) met criteria for ASD on all three assessments, while an additional 28 individuals (44%) met criteria on only one or two of the assessments. One potential explanation for these findings is that each genetic syndrome manifests its own set of complex behaviors which are not typical in the general population, some of which may overlap with ASD, even when the full ASD phenotype is not present [25,26]. The overlap in phenotype places individuals with a genetic syndrome closer to ASD cut-off scores on particular measures than typically developing individuals, even in the absence of an ASD diagnosis. Another contributor to higher scores on ASD assessments may be the degree of intellectual disability. Indeed, the SCQ manual [27] mentions that non-ASD individuals with lower IQ (50–69) obtained higher scores on the SCQ (11.40; SD = 5.87), which is quite similar to the results in our study (mean IQ = 63.38; mean SCQ score = 11.90). The SRS-2 manual [28] gives similar caution to its use in individuals with an IQ less than 70. Both overlapping phenotype and intellectual disability may affect the ability for these assessments to reliably detect ASD in PWS.

Limitations to this study include the sample demographics, limited sample size, and cross-sectional design. The limited number of children available for the study resulted in uneven distribution in both gender and genetic subtype. Our study included three boys and seven girls, whereas the gender distribution in the PWS population is roughly equal. Given that ASD is recognized four times more in males than females [5], the high number of females in the study may have reduced the likelihood of ASD diagnosis. The ratio of 7:3 in favor of girls (in our study) is not an extreme variation, which would be expected about 17% of the time based on the binomial distribution. As well, the presence of ASD symptoms on the questionnaire measures and the ADOS did not seem to predominantly characterize males, albeit with only 3 in the sample. ASD in general is less common in females, although sex differences in ASD symptoms are generally not reported in PWS studies [3]. As well, although recruitment involved every known child with PWS in our study age-range, there was an unexpected distribution of eight UPD and two DEL cases in our sample. It has been reported that the UPD genetic subtype is associated with a higher level of ASD symptoms than the DEL subtype [3,29,30]. This sample characteristic might have been expected to lead to a higher level of ASD symptomatology. Notably, 1 of 2 (50%) DEL cases exceeded the cut-off on multiple ASD assessments, whereas 7 of 8 (88%) UPD exceeded the cut-off on at least one of the ASD assessments. It is important to note that it is unlikely that recruitment was biased to include more UPD children, given that 9 of 10 parents initially contacted agreed to participate in the study, without prior knowledge of genetic subtype from those contacting the participants. Notably, within the DEL subtype, there are two types of deletions: the longer type 1, and relatively shorter type 2. Some studies have found differences between type 1 and type 2, with type 1 typically showing greater overall impairment as well as ASD-related impairments [17,31–34]. However, we were unable to distinguish between the 2 types in our study, as the sample population had not received the necessary genotyping. Individuals with truncating mutations on the paternal allele of MAGEL2, a gene within the PWS domain, were recently found to exhibit both features of PWS and autism [35]. The overall implication is that our current limited sample was predominantly female UPD which may have influenced ASD symptom levels, although the two factors (sex and genetic subtype) would have been expected to shift the distribution in opposite directions. Our subsample of 6 girls with UPD is a relatively large grouping of this rarer subtype, and may provide a useful detailed clinical reference for other such patients. Critically, this study was originally intended to make group comparisons between UPD, DEL, and an ASD comparison group, but lacked an available adequate sample size due to lack of children with PWS in the region. Although descriptive studies are able to provide a wealth of information regarding smaller samples in order to generate hypotheses and ideas, they lack the statistical power to compare results to other groups, such as an ASD comparison group. A matched control group could have helped shed further insight into both the ASD symptoms and the relative strengths and weaknesses identified in this study. As well, our clinical best estimate (CBE) was done by an experienced developmental paediatrician, who reviewed all results, watched videotapes of the ADOS-2 assessments and met with individual patients for further discussion. However, there was no second rater that would have been necessary to determine reliability of the CBE in this study. Finally, the negative correlation between age and IQ in children with PWS is of interest, but is difficult to interpret in this small sample, which does not allow us to parse potential confounding effects of sex, genetic subtype, and obesity. A longitudinal design would also be helpful to discriminate possible factors involved in the negative correlation between age and IQ, and to determine if ASD symptoms in PWS are exacerbated with age.

4. Materials and Methods

4.1. Study Design

This cross-sectional study was conducted in Edmonton, AB, Canada. Individuals aged 3 to 12 with PWS were recruited for this study from local and regional care providers, as well as provincial PWS groups. The ratio of UPD to deletion cases of PWS (and girls versus boys) was the result of random chance; those who met study inclusion criteria from the local area were recruited. Parents agreed to receive a report of their child's performance and to be referred for further assessment and treatment if indicated clinically by the findings. Ethical approval was received from the local Health Research Ethics Board Health Panel (Study ID: Pro00051250; Approved 7 January 2015) at the University of Alberta. Parents gave informed consent for their children prior to participation in this study.

4.2. Assessments

All assessments were administered by experienced research psychometrists, who had been trained to reliability on each of the measures (including research reliability on the ADOS-2). Data collection was blinded to genetic subtype to avoid bias. A developmental paediatrician with 20 years experience in ASD diagnosis (LZ) reviewed case files and ADOS videos to determine whether criteria were met for ASD for each participant (hereafter, 'clinical best estimate' diagnosis).

4.2.1. Autism Diagnostic Observation Schedule-2

The Autism Diagnostic Observation Schedule, 2nd edition (ADOS-2) [21] is a semi-structured interactive assessment that provides an opportunity to observe behaviors relevant to ASD diagnosis. Items are scored from 0–3 or 0–2, with '0' representing a continuum of behavior not generally associated with ASD; a code of '1' generally indicates mild impairment of a nature that may be observed in persons with ASD; a code of '2' indicates definite impairment in that area; a code of '3' represents more profound impairment, although for the purpose of the scoring algorithm, scores of '3' are converted to '2'. The ADOS-2 provides a total score for social affect (SA) and restricted and repetitive behaviors (RRB) symptoms, each of which can be translated into a standardized severity score on a scale of 1 to 10. There is also an overall severity score, with ≥4 indicative of ASD. Because the severity of SA symptoms is rated based on ten algorithm items and RRB severity is based on four items, the overall severity score is more heavily weighted towards the SA domain. Additionally, since the RRB domain only consists of 4 algorithm items, if all items are scored '0', the severity is 0, whereas any item scored '1' leads to a severity rating of 5. Thus, the RRB severity scores are highly sensitive to scoring on individual items, and must be interpreted with caution. The ADOS-2 has been validated for ages 12 months to adulthood. One of 5 modules is administered, based on age and verbal fluency; Only modules 1 to 3 were administered in our study. Module 1 is designed for individuals with no speech or single words; Module 2 is for individuals with phrase speech; and Module 3 is for individuals with fluent speech and typically less than 14 years of age.

4.2.2. Social Communication Questionnaire

The Social Communication Questionnaire (SCQ) [27] is a 40-item parent questionnaire used to screen for autistic symptomatology; it is derived from the Autism Diagnostic Interview-Revised (ADI-R), a semi-structured interview used in ASD diagnosis [19]. The SCQ, which uses a yes/no parent response form, was chosen over the ADI-R to be more time-feasible for parents (10 minutes for the SCQ vs. 2 h for the ADI-R). Raw scores of 15 or greater are indicative of ASD. A study released after the publication of the SCQ indicates that a cut-off of 12 greatly improves the sensitivity of the assessment when used in combination with the ADOS, particularly for younger children [36].

4.2.3. Social Responsiveness Scale-2

The Social Responsiveness Scale, 2nd edition (SRS-2) [28] is a questionnaire completed by a primary caregiver and/or a teacher that provides an overall rating of social impairment as well as scores on 5 ASD-specific subdomains: (1) Social Awareness; (2) Social Cognition; (3) Social Communication; (4) Social Motivation; and (5) Restricted or Repetitive Behaviors (RRBs). It is designed to assess individuals from age 2.5 years to adulthood. Total scores on the SRS-2 are standardized as T-scores, based on age and gender, and are further separated into four levels: <60 (Within normal limits; generally not associated with ASD); 60 to 65 (Mild range; indicates deficiencies in reciprocal social behavior that may lead to mild to moderate interference with everyday social interactions); 66 to 75 (Moderate range; indicates deficiencies in reciprocal social interaction that lead to substantial interference with everyday social interaction, and are typical for children with ASD of moderate severity); and >75 (Severe range; indicates deficiencies in reciprocal social interaction that lead to severe interference with everyday social interaction, and are strongly associated with a clinical diagnosis of ASD).

4.2.4. Social Skills Improvement System-Rating Scales

The Social Skills Improvement System-Rating Scales (SSIS-RS) [37] was selected to provide information on a broad range of social behaviours; i.e., not only those specifically impaired in ASD. It is a parent questionnaire that can be administered for children between the ages of 3 and 18. The SSIS-RS provides percentiles for social skills (higher scores indicate more advanced social skills) as well as problem behaviors (higher scores indicate increased problem behaviors), normed by gender and age. There is also a subscale to indicate ASD symptoms (rated as below average, average, or above average). Although the SSIS-RS has not previously been reported in a PWS population, it has been widely used in the ASD population and determined to be a valid research tool for measuring social skills [38]. Furthermore, this questionnaire also assesses non-ASD related social impairments that might be present in PWS.

4.2.5. Vineland Adaptive Behavior Scales-II

The Vineland Adaptive Behavior Scales, 2nd edition (Vineland-II) [39] is a parent interview that quantifies their child's current adaptive behaviors. It provides a composite score, plus sub-scores in the following domains: (1) Daily Living Skills; (2) Socialization; (3) Communication; and (4) Motor Skills. Scores are based on a standardized scale with a mean of 100, and standard deviation of 15. It has been validated for all ages and specifically for individuals with ASD [40].

4.2.6. Wechsler Preschool and Primary Scale of Intelligence-III / Wechsler Intelligence Scale for Children-IV

The Wechsler Preschool and Primary Scale of Intelligence, 3rd edition (WPPSI-III) [41] and Wechsler Intelligence Scale for Children, 4th edition (WISC-IV) [42] are individually administered measures of cognitive abilities that yield full-scale IQ. The WPPSI-III has been validated in children from age 2 to 7 years, and the WICS-IV has been validated in children from age 6 to 16 years. Both the WPSSI-III and WISC-IV have been used previously to assess patients with PWS and ASD [13,14,43].

4.3. Analytic Approach

Descriptive data from the ADOS-2, the 3 questionnaires (SCQ, SRS-2, and SSIS-RS), and the Vineland-II were provided for each child. Clinical best estimate (CBE) ASD diagnosis included a video review of each participant's ADOS-2, review of the questionnaires, and in person meeting with the parent and child. Any test result exceeding previously recommended ASD clinical cut-offs were considered in relation to CBE. Summary statistics were calculated for each assessment, including mean, standard deviation, standard error of the mean, and range. Additional analyses were performed using SPSS version 22 to compare performance among domains within measures. Experiment-wise alpha was set at $p < 0.05$ for all statistical analyses, which were completed using non-parametric tests due to the small sample size and associated non-normal distribution of study measures across our sample.

5. Conclusions

Our data suggest variable ASD symptomatology in children with PWS, with caveats related to the sex and genetic subtype composition of our small sample. Further research is necessary to identify the most appropriate tools to assess ASD in PWS. In particular, the disparity in agreement between direct observational assessment and parent reports warrants investigation into appropriate assessment of ASD in children with PWS.

Acknowledgments: Most importantly, we are grateful for the children with Prader-Willi syndrome and their families for coming, some from great distances, to participate. The research was funded by the Women and Children's Health Research Institute (WCHRI) from the University of Alberta, Department of Pediatrics. Andrea M. Haqq is also supported by the Canadian Institutes of Health Research (CIHR). Lonnie Zwaigenbaum is supported by the Stollery Children's Hospital Foundation Chair in Autism Research, which contributed funding toward Open Access publication.

Author Contributions: Jeffrey A. Bennett, Sandra Hodgetts, Michelle L. Mackenzie, Andrea M. Haqq, and Lonnie Zwaigenbaum designed the research study. Jeffrey A. Bennett performed the research. Jeffrey A. Bennett, Andrea M. Haqq, and Lonnie Zwaigenbaum analyzed the data. Jeffrey A. Bennett wrote the paper. Sandra Hodgetts, Michelle L. Mackenzie, Andrea M. Haqq, and Lonnie Zwaigenbaum all contributed to revisions of the paper. All authors read and approved the final manuscript.

Conflicts of Interest: The authors declare no conflict of interest.

Abbreviations

ADI-R	autism diagnostic interview-revised
ADOS-2	autism diagnostic observation schedule, 2nd edition
ASD	autism spectrum disorder
CBE	clinical best estimate
DEL	deletion (genetic subtype)
PWS:	Prader-Willi syndrome
RRB	restricted or repetitive behaviors
SCQ	Social Communication Questionnaire
SRS-2	Social Responsiveness Scale-2
SSIS-RS	Social Skills Improvement System-Rating Scales
UPD	uniparental disomy (genetic subtype)

References

1. Cassidy, S.B.; Schwartz, S.; Miller, J.L.; Driscoll, D.J. Prader-Willi syndrome. *Genet. Med.* **2012**, *14*, 10–26. [CrossRef] [PubMed]
2. Dykens, E.M.; Maxwell, M.A.; Pantino, E.; Kossler, R.; Roof, E. Assessment of hyperphagia in Prader-Willi syndrome. *Obesity* **2007**, *15*, 1816–1826. [CrossRef] [PubMed]
3. Bennett, J.A.; Germani, T.; Zwaigenbaum, L.; Haqq, A.M. Autism spectrum disorder in Prader-Willi syndrome: A systematic review. *Am. J. Med. Genet. A* **2015**, *167*, 2936–2944. [CrossRef] [PubMed]
4. Vorstman, J.A.; Staal, W.G.; van Daalen, E.; van Engeland, H.; Hochstenbach, P.F.; Franke, L. Identification of novel autism candidate regions through analysis of reported cytogenetic abnormalities associated with autism. *Mol. Psychiatry* **2006**, *11*, 18–28. [CrossRef] [PubMed]
5. Developmental Disabilities Monitoring Network Surveillance Year Principal Investigators; Centers for Disease Control and Prevention. Prevalence of autism spectrum disorder among children aged 8 years—Autism and developmental disabilities monitoring network, 11 sites, United States, 2010. *MMWR Surveill. Summ.* **2014**, *63*, 1–21.
6. Ozonoff, S.; Iosif, A.M.; Baguio, F.; Cook, I.C.; Hill, M.M.; Hutman, T.; Rogers, S.J.; Rozga, A.; Sangha, S.; Sigman, M.; et al. A prospective study of the emergence of early behavioral signs of autism. *J. Am. Acad. Child Adolesc. Psychiatry* **2010**, *49*, 256–266. [CrossRef] [PubMed]
7. Zwaigenbaum, L.; Bryson, S.; Rogers, T.; Roberts, W.; Brian, J.; Szatmari, P. Behavioral manifestations of autism in the first year of life. *Int. J. Dev. Neurosci.* **2005**, *23*, 143–152. [CrossRef] [PubMed]
8. Anagnostou, E.; Zwaigenbaum, L.; Szatmari, P.; Fombonne, E.; Fernandez, B.A.; Woodbury-Smith, M.; Brian, J.; Bryson, S.; Smith, I.M.; Drmic, I. Autism spectrum disorder: Advances in evidence-based practice. *Can. Med. Assoc. J.* **2014**, *186*, 509–519. [CrossRef] [PubMed]
9. Ali, D.H.; Effat, S.; Afifi, H. Prader-Willi syndrome psychobehavioral profile in a clinic based sample. *Eur. Psychiatry* **2013**, *28*, 1. [CrossRef]
10. Lo, S.T.; Siemensma, E.; Collin, P.; Hokken-Koelega, A. Impaired theory of mind and symptoms of autism spectrum disorder in children with Prader-Willi syndrome. *Res. Dev. Disabil.* **2013**, *34*, 2764–2773. [CrossRef] [PubMed]
11. Akefeldt, A.; Gillberg, C. Behavior and personality characteristics of children and young adults with Prader-Willi syndrome: A controlled study. *J. Am. Acad. Child Adolesc. Psychiatry* **1999**, *38*, 761–769. [CrossRef] [PubMed]
12. Ehlers, S.; Gillberg, C.; Wing, L. A screening questionnaire for asperger syndrome and other high-functioning autism spectrum disorders in school age children. *J. Autism Dev. Disord.* **1999**, *29*, 129–141. [CrossRef] [PubMed]
13. Dimitropoulos, A.; Ho, A.; Feldman, B. Social responsiveness and competence in Prader-Willi syndrome: Direct comparison to autism spectrum disorder. *J. Autism Dev. Disord.* **2013**, *43*, 103–113. [CrossRef] [PubMed]
14. Zyga, O.; Russ, S.; Ievers-Landis, C.E.; Dimitropoulos, A. Assessment of pretend play in Prader–Willi syndrome: A direct comparison to autism spectrum disorder. *J. Autism Dev. Disord.* **2014**, *45*, 975–987. [CrossRef] [PubMed]

15. Veltman, M.W.; Thompson, R.J.; Roberts, S.E.; Thomas, N.S.; Whittington, J.; Bolton, P.F. Prader-Willi syndrome—A study comparing deletion and uniparental disomy cases with reference to autism spectrum disorders. *Eur. Child Adolesc. Psychiatry* **2004**, *13*, 42–50. [CrossRef] [PubMed]

16. Moss, J.; Oliver, C.; Arron, K.; Burbidge, C.; Berg, K. The prevalence and phenomenology of repetitive behavior in genetic syndromes. *J. Autism Dev. Disord.* **2009**, *39*, 572–588. [CrossRef] [PubMed]

17. Milner, K.M.; Craig, E.E.; Thompson, R.J.; Veltman, M.W.; Thomas, N.S.; Roberts, S.; Bellamy, M.; Curran, S.R.; Sporikou, C.M.; Bolton, P.F. Prader-Willi syndrome: Intellectual abilities and behavioural features by genetic subtype. *J. Child Psychol. Psychiatry* **2005**, *46*, 1089–1096. [CrossRef] [PubMed]

18. Lord, C.; Rutter, M.; DiLavore, P.C.; Risi, S. *Autism Diagnostic Observation Schedule (ADOS) Manual*, 4th ed.; Western Psychological Services: Los Angeles, CA, USA, 2006.

19. Rutter, M.; le Couteur, A.; Lord, C. *Autism Diagnostic Interview-Revised (ADI-R)*; Western Psychological Services: Los Angeles, CA, USA, 2003.

20. Falkmer, T.; Anderson, K.; Falkmer, M.; Horlin, C. Diagnostic procedures in autism spectrum disorders: A systematic literature review. *Eur. Child Adolesc. Psychiatry* **2013**, *22*, 329–340. [CrossRef] [PubMed]

21. Lord, C.; Rutter, M.; DiLavore, P.C.; Risi, S.; Gotham, K.; Bishop, S. *Autism Diagnostic Observation Schedule, Second Edition (ADOS-2) Manual (Part 1): Modules 1–4*; Western Psychological Services: Torrence, CA, USA, 2012.

22. Risi, S.; Lord, C.; Gotham, K.; Corsello, C.; Chrysler, C.; Szatmari, P.; Cook, E.H., Jr.; Leventhal, B.L.; Pickles, A. Combining information from multiple sources in the diagnosis of autism spectrum disorders. *J. Am. Acad. Child Adolesc. Psychiatry* **2006**, *45*, 1094–1103. [CrossRef] [PubMed]

23. Hall, S.S.; Lightbody, A.A.; Hirt, M.; Rezvani, A.; Reiss, A.L. Autism in fragile X syndrome: A category mistake? *J. Am. Acad. Child Adolesc. Psychiatry* **2010**, *49*, 921–933. [CrossRef] [PubMed]

24. Harris, S.W.; Hessl, D.; Goodlin-Jones, B.; Ferranti, J.; Bacalman, S.; Barbato, I.; Tassone, F.; Hagerman, P.J.; Herman, H.; Hagerman, R.J. Autism profiles of males with fragile X syndrome. *Am. J. Ment. Retard* **2008**, *113*, 427–438. [CrossRef] [PubMed]

25. Cornish, K.; Turk, J.; Hagerman, R. The fragile X continuum: New advances and perspectives. *J. Intell. Disabil. Res.* **2008**, *52*, 469–482. [CrossRef] [PubMed]

26. Cornish, K.; Turk, J.; Levitas, A. Fragile X syndrome and autism: Common developmental pathways? *Curr. Pediatr. Rev.* **2007**, *3*, 61–68. [CrossRef]

27. Rutter, M.; Bailey, A.; Lord, C. *The Social Communication Questionnaire (SCQ)*; Western Psychological Services: Los Angeles, CA, USA, 2003.

28. Constantino, J.N.; Gruber, C.P. *Social Responsiveness Scale-2 (SRS-2) Manual*; Western Psychological Services: Los Angeles, CA, USA, 2012.

29. Dykens, E.M.; Lee, E.; Roof, E. Prader-Willi syndrome and autism spectrum disorders: An evolving story. *J. Neurodev. Disord.* **2011**, *3*, 225–237. [CrossRef] [PubMed]

30. Dimitropoulos, A.; Schultz, R.T. Autistic-like symptomatology in Prader-Willi syndrome: A review of recent findings. *Curr. Psychiatry Rep.* **2007**, *9*, 159–164. [CrossRef] [PubMed]

31. Dykens, E.M.; Roof, E. Behavior in Prader-Willi syndrome: Relationship to genetic subtypes and age. *J. Child Psychol. Psychiatry* **2008**, *49*, 1001–1008. [CrossRef] [PubMed]

32. Butler, M.G.; Bittel, D.C.; Kibiryeva, N.; Talebizadeh, Z.; Thompson, T. Behavioral differences among subjects with Prader-Willi syndrome and type I or type II deletion and maternal disomy. *Pediatrics* **2004**, *113*, 565–573. [CrossRef] [PubMed]

33. Bittel, D.C.; Kibiryeva, N.; Butler, M.G. Expression of 4 genes between chromosome 15 breakpoints 1 and 2 and behavioral outcomes in Prader-Willi syndrome. *Pediatrics* **2006**, *118*, 1276–1283. [CrossRef] [PubMed]

34. Zarcone, J.; Napolitano, D.; Peterson, C.; Breidbord, J.; Ferraioli, S.; Caruso-Anderson, M.; Holsen, L.; Butler, M.G.; Thompson, T. The relationship between compulsive behaviour and academic achievement across the three genetic subtypes of Prader-Willi syndrome. *J. Intellect. Disabil. Res.* **2007**, *51*, 478–487. [CrossRef] [PubMed]

35. Schaaf, C.P.; Gonzalez-Garay, M.L.; Xia, F.; Potocki, L.; Gripp, K.W.; Zhang, B.; Peters, B.A.; McElwain, M.A.; Drmanac, R.; Beaudet, A.L.; et al. Truncating mutations of Magel2 cause Prader-Willi phenotypes and autism. *Nat. Genet.* **2013**, *45*, 1405–1408. [CrossRef] [PubMed]

36. Corsello, C.; Hus, V.; Pickles, A.; Risi, S.; Cook, E.H.; Leventhal, B.L.; Lord, C. Between a roc and a hard place: Decision making and making decisions about using the SCQ. *J. Child Psychol. Psychiatry* **2007**, *48*, 932–940. [CrossRef] [PubMed]

37. Gresham, F.M.; Elliot, S.N. *Social Skills Improvement System: Rating scales manual*; NCS Pearson, Inc.: Minneapolis, MN, USA, 2008.

38. Anagnostou, E.; Jones, N.; Huerta, M.; Halladay, A.K.; Wang, P.; Scahill, L.; Horrigan, J.P.; Kasari, C.; Lord, C.; Choi, D.; et al. Measuring social communication behaviors as a treatment endpoint in individuals with autism spectrum disorder. *Autism* **2014**, *19*, 622–636. [CrossRef] [PubMed]

39. Sparrow, S.; Cicchetti, D.; Balla, D.A. *Vineland Adaptive Behavior Scales: Second Edition (Vineland II), Survey Interview Form*; Pearson Assessments: Livonia, MN, USA, 2005.

40. Carter, A.S.; Volkmar, F.R.; Sparrow, S.S.; Wang, J.-J.; Lord, C.; Dawson, G.; Fombonne, E.; Loveland, K.; Mesibov, G.; Schopler, E. The Vineland Adaptive Behavior Scales: Supplementary norms for individuals with autism. *J. Autism Dev. Disord.* **1998**, *28*, 287–302. [CrossRef] [PubMed]

41. Wechsler, D. *Wechsler Preschool and Primary Scale of Intelligence (WPPSI-III)*, 3rd ed.; Pearson Psychological Corporation: San Antonio, TX, USA, 2002.

42. Wechsler, D. *Wechsler Intelligence Scale for Children (WISC-IV)*, 4th ed.; Pearson Psychological Corporation: San Antonio, TX, USA, 2003.

43. Song, D.K.; Sawada, M.; Yokota, S.; Kuroda, K.; Uenishi, H.; Kanazawa, T.; Ogata, H.; Ihara, H.; Nagai, T.; Shimoda, K. Comparative analysis of autistic traits and behavioral disorders in Prader-Willi syndrome and Asperger disorder. *Am. J. Med. Genet. A* **2015**, *167*, 64–68. [CrossRef] [PubMed]

International Journal of
Molecular Sciences

MDPI

Review

The 15q11.2 BP1–BP2 Microdeletion Syndrome: A Review

Devin M. Cox [†,*] and Merlin G. Butler [†]

Departments of Psychiatry & Behavioral Sciences, University of Kansas Medical Center, 3901 Rainbow Boulevard, MS 4015, Kansas City, KS 66160, USA; mbutler4@kumc.edu
* Author to whom correspondence should be addressed; dcox2@kumc.edu;
  Tel.: +1-913-588-0396; Fax: +1-913-588-1305.
† These authors contributed equally to this work.

Academic Editor: Cesar Borlongan
Received: 5 January 2015; Accepted: 10 February 2015; Published: 13 February 2015

Abstract: Patients with the 15q11.2 BP1–BP2 microdeletion can present with developmental and language delay, neurobehavioral disturbances and psychiatric problems. Autism, seizures, schizophrenia and mild dysmorphic features are less commonly seen. The 15q11.2 BP1–BP2 microdeletion involving four genes (*i.e.*, *TUBGCP5*, *CYFIP1*, *NIPA1*, *NIPA2*) is emerging as a recognized syndrome with a prevalence ranging from 0.57%–1.27% of patients presenting for microarray analysis which is a two to four fold increase compared with controls. Review of clinical features from about 200 individuals were grouped into five categories and included developmental (73%) and speech (67%) delays; dysmorphic ears (46%) and palatal anomalies (46%); writing (60%) and reading (57%) difficulties, memory problems (60%) and verbal IQ scores ≤75 (50%); general behavioral problems, unspecified (55%) and abnormal brain imaging (43%). Other clinical features noted but not considered as common were seizures/epilepsy (26%), autism spectrum disorder (27%), attention deficit disorder (ADD)/attention deficit hyperactivity disorder (ADHD) (35%), schizophrenia/paranoid psychosis (20%) and motor delay (42%). Not all individuals with the deletion are clinically affected, yet the collection of findings appear to share biological pathways and presumed genetic mechanisms. Neuropsychiatric and behavior disturbances and mild dysmorphic features are associated with genomic imbalances of the 15q11.2 BP1–BP2 region, including microdeletions, but with an apparent incomplete penetrance and variable expressivity.

Keywords: 15q11.2 BP1–BP2 microdeletion; Burnside-Butler syndrome; clinical and behavioral phenotype; chromosome breakpoints BP1 and BP2; Prader-Willi and Angelman syndromes; language and motor delays; autism; review

1. Introduction

Chromosome 15 contains five common breakpoint sites along the proximal long arm; they are commonly referred to as BP1–BP5. There is a cluster of low copy DNA repeats located within this chromosome region which can facilitate mis-alignment during meiosis leading to non-allelic homologous recombination [1,2]. These low copy repeat sequences are called duplicons and contain pseudogenes [3]. Duplicons found within breakpoints BP1, BP2 and BP3 have been characterized by the presence of the *HERC2* gene (at BP3) and *HERC2* pseudogenes (at BP1 and BP2) [2].

Prader-Willi syndrome (PWS) and Angelman syndrome are typically caused by a deletion of different parental origin involving the distal breakpoint BP3 and proximally placed breakpoints BP1 or BP2. These cytogenetic deletions of chromosome 15q11–q13 region are classified as typical type I (involving BP1 and BP3) or typical type II (involving BP2 and BP3) (see Figure 1). Type I deletions have an average genomic length of 6.58 Mb while type II deletions have a mean length of 5.33 Mb [4]. Several studies have shown that individuals with the larger typical 15q11–q13 type I deletion which is found in both Prader-Willi and Angelman syndromes are reported to have more severe neurodevelopmental symptoms as compared to those individuals with the smaller typical type II deletion [5–10]. Initially, Butler *et al.* [5] found that several behavioral and intelligence measures were statistically different between the two PWS deletion types (type I and type II). PWS individuals with type I deletions showed more compulsive and self-injurious behaviors and visual perception impairment along with lower intelligence, reading and math scores than in those with type II deletions. Furthermore, Bittel *et al.* [6] reported that the amount of mRNA isolated from lymphoblastoid cell lines established from individuals with PWS for the four genes (*i.e.*, *NIPA1*, *NIPA2*, *CYFIP1*, *TUBGCP5*) found in the genomic area between BP1 and BP2 in the 15q11.2 chromosome band explained between 24% to 99% of the phenotypic variability in behavioral and academic measures. The *NIPA2* gene accounted for the largest number of significant correlations between the mRNA levels and phenotypic features.

In other studies, Varela *et al.* [10] found that individuals with PWS having 15q11–q13 type I deletions acquired speech later than those with type II deletions. Hartley *et al.* [7] also found that individuals with PWS having type I deletions had significantly higher Reiss maladaptive behavior scores for depression (physical signs) than individuals with type II deletions. Similarly, Sahoo *et al.* [9] reported that in individuals with Angelman syndrome and the 15q11–q13 type I deletion had significantly more behavioral and cognitive impairments with lower expressive and total language abilities and a higher likelihood of features for autism spectrum disorder.

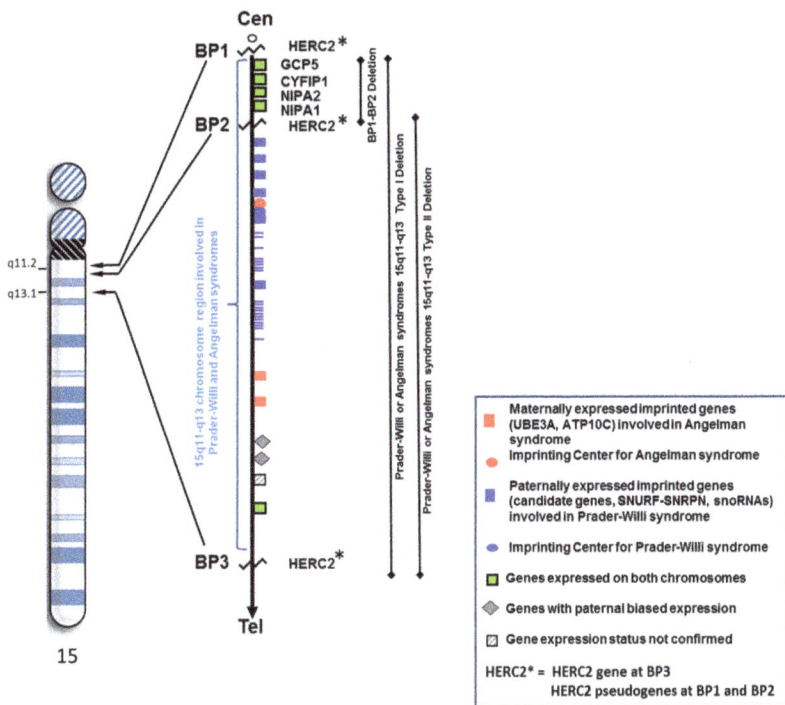

Figure 1. High resolution ideogram representing chromosome 15 showing location of breakpoints BP1 and BP2 (at 15q11.2 band) and BP3 (at 15q13.1 band) involving HERC2 and position of the non-imprinted genes between BP1 and BP2. The three deletion types involving the 15q11–q13 region (*i.e.*, BP1–BP2, typical type I, typical type II) are represented.

Int. J. Mol. Sci. **2015**, *16*, 4068–4082

Valente *et al.* [11] also reported in Angelman syndrome that those with the 15q11–q13 type I deletion had more severe seizures and were refractory to treatment compared with those having the type II deletion. In a separate study Milner *et al.* [8] found significantly higher verbal IQ scores in individuals with PWS and type II deletions *versus* those with type I deletions. They further reported that those with type I deletions performed more poorly on all measures of ability even though they were not significantly different from the patients with type II deletions.

A report by Dykens and Roof [12] examined behaviors in PWS using a mixed cohort of young and old subjects (*n* = 88) and showed a relationship between genetic subtypes and ages. They found negative associations between age and behavior in the 15q11–q13 type I deletion subtype only which implicated non-imprinted genes between breakpoints BP1 and BP2, specifically the *CYFIP1* gene. Disturbed expression of *CYFIP1* is seen in other developmental disabilities including those with 15q disorders without PWS. Although they reported no significant behavioral findings when combining data from subjects at all ages, significant differences were found in those with type I *versus* type II deletions with age. The individuals with the 15q11–q13 type I deletion consistently had lower targeted problem behaviors and adaptive skills and externalizing symptoms with advancing age.

As several studies have shown evidence of disturbed gene expression patterns and behavioral findings in subjects with either PWS or Angelman syndrome with different genetic deletion subtypes implicating genes within the BP1 and BP2 genomic region, Burnside *et al.* [13] summarized the literature and surveyed the first large cohort of patients presenting for genetic testing using high resolution microarrays. They found that 0.86% of the approximate 17,000 individuals had an abnormality (deletion or duplication) of the 15q11.2 BP1–BP2 region. Specifically, 69 subjects were found with the 15q11.2 microdeletion and 77 subjects were found with a microduplication of the same region. They proposed that this genomic area was a susceptibility region for neurological dysfunction, impaired development and characteristic phenotypic features which was initially raised by Butler *et al.* [5] in a study of PWS individuals with the larger 15q11–q13 type I deletion including the four genes in the BP1 and BP2 area and having a more severe behavior phenotype compared with those with the smaller type II deletion. Collectively, they summarized that a deletion involving this genomic region correlated with language or motor delays, behavioral problems, autism, seizures and occasionally mild dysmorphic features. The collection of clinical findings in the microdeletion in a large cohort of patients with the 15q11.2 BP1–BP2 presenting for genetic services supported the original observations by Butler *et al.* [14] of behavioral disturbances seen in PWS patients with the larger 15q11–q13 type I deletion compared with the smaller type II deletion which stimulated interest in additional studies of this chromosome region and, hence, coined the Burnside-Butler syndrome.

Preliminary clinical information about the 15q11.2 microdeletion alone without PWS was first reported by Murthy *et al.* [15] in 2007 in two individuals in a consanguineous family and later by Doornbos *et al.* [16] in 2009 in nine individuals. The vast majority of their combined subjects presented with behavioral or neurological problems. Later, Abdelmoity *et al.* [17] reported a cohort of 1654 consecutive pediatric patients presenting with a range of neurological disorders and found that 21% or 1.27% of the patients carried a 15q11.2 BP1–BP2 deletion. They found that 87.5% of the patients with the deletion had developmental delay or intellectual disability. More recently, Cafferkey *et al.* [18] presented data from 14,605 patients (primarily pediatric) referred for genetic testing using microarray analysis and found 83 (0.57%) with the 15q11.2 BP1–BP2 microdeletion. The majority of their patients presented with some form of behavioral disturbance or developmental/motor delays as summarized by Burnside *et al.* [13]. The area between BP1 and BP2 is approximately 500 kb in size and includes the *NIPA1*, *NIPA2*, *TUBGCP5*, and *CYFIP1* genes and prone to both microdeletions and microduplications. In this review, we summarize information regarding the 15q11.2 BP1–BP2 microdeletion. A review of information regarding the microduplication is beyond the scope of this report.

Chai *et al.* [19] showed that these four genes are highly conserved and biallelically expressed. *NIPA1* or non-imprinted in Prader-Willi/Angelman syndrome 1 gene is the best studied gene within this region and associated with autosomal dominant hereditary spastic paraplegia [20,21]. There has been no case to date where haploinsufficiency of *NIPA1* due to a deletion has led to hereditary spastic paraplegia. *NIPA1* also mediates Mg^{2+} transport and is highly expressed in neuronal tissue [22]. The *NIPA2* or non-imprinted in Prader-Willi/Angelman syndrome 2 gene is used in renal Mg^{2+} transport [22]. Jiang *et al.* [23] examined patients with childhood absence epilepsy and found mutations in *NIPA2* with unknown functional effects. The *TUBGCP5* gene or tubulin gamma complex associated protein 5 gene is involved in neurobehavioral disorders including ADHD and OCD [24]. The final gene within this region is *CYFIP1* or cytoplasmic fragile X mental retardation 1 (FMR1) interacting protein 1 gene. This gene product interacts with FMRP in a ribonucleoprotein complex. FMRP is the product of the *FMR1* gene which is associated with fragile X syndrome, the most common cause of familial intellectual disability that primarily affects males [25]. Both of these gene products play important roles in the regulation of brain mRNAs [24]. Bozdagi *et al.* [26] showed that haploinsufficiency of *CYFIP1* resembles important aspects found in knockout FMR1 mice.

Not all individuals with defects within the 15q11.2 band (*i.e.*, microdeletions or microduplications) share a clinical phenotype or are clinically affected. Therefore, this region contains genetic material showing incomplete penetrance or low penetrance of pathogenicity along with variable expressivity. For example, a review of reported data from control cohorts (*n* = 66,462 subjects) summarized to date show that about 0.25% of controls are found with the 15q11.2 BP1–BP2 microdeletion [18,27–31]. The penetrance of the 15q11.2 BP1–BP2 microdeletion has also been estimated at 10.4% [28] or an approximate two fold increase over the general population risk but may be due to paucity of inheritance data. Other estimates have been lower but results consistently show this region to play a role in autism [32]. The penetrance for the 15q11.2 microdeletion is low compared to other microdeletion syndromes such as the 16p11.2 deletion with a penetrance estimated at 62.4% [28]. A higher penetrance is often seen in copy number variants (CNVs) that have higher *de novo* frequencies while low penetrance estimates may reflect a subclinical presentation or manifestation of features that are recognized as components of disorders such as neuropsychiatric disturbances in parents of affected individuals (e.g., autism in 15q11.2 BP1–BP2 deletion) or in control cohorts. Not only are microarray studies needed on other family members (and parents) of those with the 15q11.2 BP1–BP2 deletion but also neuropsychiatric and behavioral testing as well to appreciate the variability of expression and level of penetrance. The review of literature from six published reports summarized by Cafferkey *et al.* [18] regarding information on the inheritance of the 15q11.2 BP1–BP2 microdeletion indicates that 22/43 (51%) of individuals with the microdeletion, where parental data were available, inherited their deletion from an apparently healthy parent while 10/29 (35%) of individuals inherited their deletion from an abnormal parent [13,15–17,33,34]. The phenotypic information was unavailable or incomplete for all parents found to carry the deletion. The reported *de novo* deletion frequency ranged from 1/21 (5%) [13] to 2/9 (22%) [16] in subjects reviewed by Cafferkey *et al.* [18]. It would be of importance to undertake DNA sequencing analyses of the 15q11.2 BP1–BP2 genomic region to identify subtle deletions or mutations of the non-deleted allele and determine the genetic status of this "normal allele". Other modifying genes outside of the chromosome region may also play a role and will require further investigations.

Recently, individuals have been reported with the 15q11.2 BP1–BP2 microdeletion and additional non-neurological clinical findings. These findings included congenital cataracts [14,35], proximal esophageal atresia and distal tracheoesophageal fistula (type C) [35] and congenital arthrogryposis [36]. These clinical reports lend support for further phenotypic expansion of this susceptibility region impacted by the 15q11.2 BP1–BP2 microdeletion. Our report will focus on the review of the clinical features now recognized in this microdeletion syndrome.

2. Results and Discussion

2.1. Growth and Development

A review of the literature on growth and development in individuals with 15q11.2 BP1–BP2 microdeletions involving a susceptibility region for neurological dysfunction with general developmental and motor delay with speech problems as primary components of this disorder. General developmental delay occurred in 73% of individuals and speech delay occurred in 67% as reported in the literature suggesting characteristic features for this microdeletion syndrome (see Table 1). These features may correlate with the neuronal components of the genes within this deleted region. *NIPA1*, *NIPA2*, and *CYFIP1* are highly expressed in neuronal tissue of the central nervous system and *TUBGCP5* is highly expressed in the subthalamic nuclei [22,24].

Motor delay and microcephaly were considered important features of this microdeletion syndrome after a review of the literature. Motor delay was seen in 42% of individuals and microcephaly was seen in 24%. Occasional features found were intrauterine growth retardation (18%), short stature (10%), and macrocephaly (8%) (see Table 1).

2.2. Dysmorphic Features

A review of the literature on the dysmorphic features for those individuals with the 15q11.2 BP1–BP2 microdeletion or Burnside-Butler syndrome showed that this disorder does not appear to have a clear dysmorphic phenotype (see Table 2). This is supported by studies by Butler *et al.* [5] in examining individuals with Prader-Willi syndrome with 15q11–q13 type I deletion or type II deletion and observing behavioral and cognitive differences but not in growth parameters or other clinical findings indicating that the four genes present within the BP1–BP2 genomic area and when disturbed can lead to neurological or behavioral outcomes.

2.3. Intelligence and Academic Achievement

General, non-specified, dysmorphic features were seen in 55 out of 141 individuals or 39% of those reported within the literature. The only common dysmorphic feature found from a review of the literature was palatal abnormalities with 46% of individuals being reported with these findings. Occasional dysmorphic features included broad forehead (21%), hypertelorism (18%), slender fingers (15%), pectus excavatum (13%), plagiocephaly (10%), dysmorphic nose (8%), dysmorphic teeth (8%), and contractures/arthrogryposis (8%). Reported dysmorphic features that, at present, are uncommon or not associated with this syndrome based on a literature review and include short fingers (5%), long narrow face (3%), small face (3%), and hypotelorism (3%) (see Table 2).

Intelligence and academic achievement for individuals with 15q11.2 BP1–BP2 microdeletions were reviewed (see Table 3). Common reported features included writing difficulties (60%), memory problems (60%), reading difficulties (57%), and a verbal IQ score equal to or below 75 (50%). However, very few individuals were reported within the literature regarding these findings so further studies are needed to determine if they are associated findings. Occasional findings within the literature included intellectual disability (37%) and performance IQ which was equal to or below 75 (33%). There were 43 out of 116 individuals reported within the literature who had intellectual disability; however, there were only 1 out of 3 individuals reportedly with a low performance IQ. Stefansson *et al.* [38] found a modest effect on verbal IQ (effect = 0.38, $p = 0.033$) and performance IQ (effect = 0.43, $p = 0.018$) from the 15q11.2 deletion.

Table 1. Literature Review of Growth and Development for Individuals with Chromosome 15q11.2 BP1–BP2 Microdeletion Syndrome.

Feature	Murthy et al. [15]	Doornbos et al. [16]	Von der Lippe et al. [34]	Burnside et al. [13]	Abdelmoity et al. [17]	Madrigal et al. [33]	Wong et al. [35]	Cafferkey et al. [18]	Usrey et al. [36]	Rudd et al. [37]	Jerkovich & Butler [14]	Total (%)
IUGR [1]	0/1	3/9	0/5	N/A	N/A	N/A	1/2	N/A	0/2	0/2	0/1	4/22 (18)
Short stature	0/1	1/9	1/5	N/A	N/A	N/A	0/2	N/A	0/2	N/A	0/1	2/20 (10)
Microcephaly	0/1	1/9	0/5	N/A	4/16	2/2	0/2	N/A	1/2	N/A	1/1	9/38 (24)
Macrocephaly	0/1	0/9	1/5	N/A	2/16	0/2	0/2	N/A	0/2	N/A	0/1	3/38 (8)
Developmental delay (general)	2/2	7/8	4/7	33/56	13/15*	2/2	0/2	65/77	N/A	0/2	0/1	**126/172 (73)**
Motor delay	1/2	8/9	5/7	20/56	N/A	2/2	1/2	29/77	N/A	0/2	0/1	66/158 (42)
Speech delay	2/2	8/8	5/5	44/49	N/A	2/2	0/2	37/77	N/A	0/2	1/1	**99/148 (67)**

[1] Intrauterine growth retardation; * indicated both Intellectual disability or Global developmental delay; Bold indicate categories that represent 50% or greater incidence within the literature.

Table 2. Literature Review of Dysmorphic Features for Individuals with Chromosome 15q11.2 BP1–BP2 Microdeletion Syndrome.

Feature	Doornbos et al. [16]	Von der Lippe et al. [34]	Burnside et al. [13]	Abdelmoity et al. [17]	Madrigal et al. [33]	Wong et al. [35]	Cafferkey et al. [18]	Usrey et al. [36]	Rudd et al. [37]	Jerkovich & Butler [14]	Total (%)
Dysmorphism, unspecified	N/A	N/A	27/56	N/A	N/A	N/A	28/83	N/A	0/2	N/A	55/141 (39)
Plagiocephaly	4/9	0/7	N/A	0/16	0/2	0/2	N/A	0/2	N/A	0/1	4/39 (10)
Broad forehead	5/9	0/7	N/A	1/16	2/2	0/2	N/A	0/2	N/A	0/1	8/39 (21)
Long narrow face	0/9	1/7	N/A	0/16	0/2	0/2	N/A	0/2	N/A	0/1	1/39 (3)
Small face	0/9	1/7	N/A	1/16	0/2	0/2	N/A	0/2	N/A	0/1	1/39 (3)
Hypertelorism	5/9	1/7	N/A	1/16	0/2	0/2	N/A	0/2	N/A	0/1	7/39 (18)
Hypotelorism	0/9	1/7	N/A	0/16	0/2	0/2	N/A	0/2	N/A	0/1	1/39 (3)
Abnormal nose	0/9	2/7	N/A	1/16	0/2	0/2	N/A	0/2	N/A	0/1	3/39 (8)
Dysmorphic ears	6/9	0/7	N/A	1/16	2/2	0/2	N/A	0/2	N/A	1/1	9/39 (46)
Palatal abnormalities	4/9	0/7	N/A	2/16	2/2	1/2	N/A	0/2	N/A	0/1	9/39 (46)
Abnormal teeth	0/9	2/7	N/A	1/16	0/2	0/2	N/A	0/2	N/A	0/1	3/39 (8)
Pectus excavatum	2/9	0/7	N/A	1/16	0/2	1/2	N/A	0/2	N/A	1/1	5/39 (13)
Contractures/arthrogryposis	1/9	1/7	N/A	0/16	0/2	0/2	N/A	2/2	N/A	0/1	2/39 (5)
Short fingers	0/9	1/7	N/A	1/16	0/2	0/2	N/A	0/2	N/A	0/1	2/39 (5)
Slender fingers	5/9	1/7	N/A	0/16	0/2	0/2	N/A	0/2	N/A	0/1	6/39 (15)

2.4. Behavioral and Psychiatric Problems

Behavioral features including autism spectrum disorder and schizophrenia have been studied as features of the 15q11.2 BP1–BP2 microdeletion [32,33,37–40]. A review of the behavioral and psychiatric features reported within the literature found that 55% of those reported had general behavior problems that were not otherwise specified (see Table 4). Common behavioral features included Attention Deficit Disorder or Attention Deficit Hyperactivity Disorder (35%), Autism Spectrum Disorder (27%), obsessive compulsive disorder (26%), self-injurious behaviors (26%), oppositional defiant disorder (24%), and schizophrenia or paranoid psychosis (20%). Studies have been performed to determine if the 15q11.2 BP1–BP2 microdeletion is a susceptibility locus for schizophrenia [37–40]. Rees *et al.* [40] found a significant association between the microdeletion and schizophrenia with the frequency of individuals having schizophrenia and the microdeletion from their study was 0.64 (44/6882). The 15q11.2 BP1–BP2 microdeletion has also been studied regarding potential susceptibility for autism [32]. Chaste *et al.* [32] found a small risk regarding this microdeletion and autism but a greater effect on the autism phenotype in males and when maternally inherited. They reported an overall modest effect for this copy number variant for autism in studying a sample of 2525 families with autism [32]. They suggested that carriers may need additional copy number variants along with the 15q11.2 microdeletion in order to be diagnosed with autism [32]. Occasionally, reported behavioral features within the literature included an unusually happy expression (12%) and anxiety (6%) [16,34].

2.5. Other Related Medical Concerns

The 15q11.2 BP1–BP2 microdeletion or Burnside-Butler syndrome has been reported with various medical concerns or conditions (see Table 5). A review of the literature indicated that 43% of these individuals had abnormal brain imaging (MRI, EEG, *etc.*) with common clinical features of seizures or epilepsy (26%) and ataxia or balance (coordination) problems (28%). The medical literature surrounding this microdeletion syndrome and seizures indicate that this chromosomal anomaly may be a risk factor for epilepsy [29,41–43]. Occasional features seen within the literature regarding medical conditions included congenital heart defect (9%), genital abnormalities (7%), recurrent infections (7%), cataracts (4%), hearing loss or impairment (4%), tracheoesophageal fistula (2%), and omphalocele (2%). Whether these additional findings represent further phenotypic expansion of the syndrome as raised by others is unknown [14,35,36].

Table 3. Literature Review of Intelligence and Academic Achievement for Individuals with Chromosome 15q11.2 BP1–BP2 Microdeletion Syndrome.

Feature	Murthy et al. [15]	Doornbos et al. [16]	De Kovel et al. [29]	Von der Lippe et al. [34]	Burnside et al. [13]	Abdelmoity et al. [17]	Madrigal et al. [33]	Mullen et al. [43]	Jahn et al. [42]	Usrey et al. [36]	Rudd et al. [37]	Jerkovich & Butler [14]	Total (%)
ID */FSIQ¹ ≤75/special education	2/2	6/12	0/11	5/11	11/49	13/15	2/2	1/6	2/3	1/1	0/3	0/1	43/116 (37)
Verbal IQ ≤75	N/A	1/1	N/A	N/A	N/A	N/A	N/A	N/A	N/A	N/A	1/3	N/A	**2/4 (50)**
Performance IQ ≤75	N/A	N/A	N/A	N/A	N/A	N/A	N/A	N/A	N/A	N/A	1/3	N/A	1/3 (33)
Reading difficulties	N/A	N/A	N/A	4/7	N/A	N/A	N/A	N/A	N/A	N/A	N/A	N/A	**4/7 (57)**
Writing difficulties	N/A	N/A	N/A	3/5	N/A	N/A	N/A	N/A	N/A	N/A	N/A	N/A	**3/5 (60)**
Memory problems	N/A	N/A	N/A	2/4	N/A	N/A	N/A	N/A	N/A	N/A	N/A	1/1	**3/5 (60)**

* Intellectual disability; ¹ Full scale intelligence quotient; Bold indicate categories that represent 50% or greater incidence within the literature.

Table 4. Literature Review of Behavioral and Psychiatric Problems for Individuals with Chromosome 15q11.2 BP1–BP2 Microdeletion Syndrome.

Feature	Murthy et al. [15]	Doornbos et al. [16]	Von der Lippe et al. [34]	Burnside et al. [13]	Abdelmoity et al. [17]	Madrigal et al. [33]	Mullen et al. [43]	Cafferkey et al. [18]	Jahn et al. [42]	Usrey et al. [36]	Rudd et al. [37]	Jerkovich & Butler [14]	Total (%)
General behavior problems, unspecified	2/2	N/A	N/A	35/56	N/A	2/4	N/A	35/73	N/A	N/A	N/A	1/1	**75/136 (55)**
Autism Spectrum Disorder	1/2	4/9	1/7	14/49	2/16	2/4	N/A	19/73	N/A	N/A	N/A	0/1	43/161 (27)
Schizophrenia/paranoid psychosis	N/A	0/9	1/7	N/A	N/A	N/A	N/A	N/A	N/A	3/3	N/A	0/1	4/20 (20)
OCD¹	0/2	2/9	0/7	16/49*	N/A	N/A	N/A	N/A	N/A	N/A	N/A	0/1	18/68 (26)
ODD²	0/2	0/9	0/7	16/49*	N/A	N/A	N/A	N/A	N/A	N/A	N/A	0/1	16/68 (24)
ADD³/ADHD⁴	2/2	2/9	0/7	16/49	7/12	N/A	N/A	N/A	N/A	N/A	N/A	1/1	28/80 (35)
Self-injurious behaviors	0/2	2/9	0/7	16/49*	N/A	N/A	N/A	N/A	N/A	N/A	N/A	0/1	18/68 (26)
Anxiety	N/A	0/9	1/7	N/A	N/A	N/A	N/A	N/A	N/A	N/A	N/A	0/1	1/17 (6)
Happy expression	N/A	2/9	0/7	N/A	N/A	N/A	N/A	N/A	N/A	N/A	N/A	0/1	2/17 (12)

¹ Obsessive compulsive disorder; ² Oppositional defiant disorder; ³ Attention deficit disorder; ⁴ Attention deficit hyperactivity disorder; * indicates obsessive compulsive disorder, oppositional defiant disorder, self-injury, tantrums, etc; Bold indicate categories that represent 50% or greater incidence within the literature.

Table 5. Literature Review of Other Related Medical Concerns for Individuals with Chromosome 15q11.2 BP1–BP2 Microdeletion Syndrome.

Feature	Murthy et al. [15]	Doornbos et al. [16]	De Kovel et al. [29]	Von der Lippe et al. [34]	Burnside et al. [13]	Abdelmoity et al. [17]	Madrigal et al. [33]	Wong et al. [35]	Mullen et al. [43]	Cafferkey et al. [18]	Jahn et al. [42]	Usrey et al. [36]	Rudd et al. [37]	Jerkovich & Butler [14]	Total (%)
Seizures/epilepsy	0/2	2/8	8/23	0/7	14/56	2/16	0/2	0/2	6/6	13/83	4/5	1/2	0/3	0/1	57/216 (26)
Cataracts	0/1	0/9	N/A	0/7	N/A	0/16	0/2	1/2	N/A	N/A	0/3	0/2	0/3	1/1	2/46 (4)
Congenital heart defect	0/1	2/9	N/A	0/7	N/A	0/16	0/2	2/2	N/A	N/A	0/3	0/2	0/3	0/1	4/46 (9)
Genital abnormalities	0/1	2/9	N/A	1/7	N/A	0/16	0/2	0/2	N/A	N/A	0/3	0/2	0/3	0/1	3/46 (7)
Recurrent infections	0/1	2/9	N/A	0/7	N/A	N/A	0/2	0/2	N/A	N/A	0/3	0/2	0/3	0/1	2/30 (7)
Ataxia/balance issues	1/2	4/9	N/A	0/7	15/56	N/A	2/2	0/2	N/A	N/A	N/A	N/A	N/A	N/A	22/78 (28)
TE fistula	0/1	0/9	N/A	1/7	N/A	0/16	0/2	0/2	N/A	N/A	0/3	0/2	0/3	0/1	1/49 (2)
Hearing loss/impairment	0/1	1/9	N/A	0/7	N/A	1/16	0/2	0/2	N/A	N/A	0/3	0/2	0/3	0/1	2/49 (4)
Omphalocele	0/1	1/9	N/A	0/7	N/A	0/16	0/2	0/2	N/A	N/A	0/3	0/2	0/3	0/1	1/49 (2)
Abnormal brain imaging	N/A	0/4	N/A	1/2	20/56*	2/3	N/A	N/A	N/A	5/5	N/A	1/2	3/3	N/A	32/75 (43)

* indicates cases with insomnia, abnormal brain MRI or EEG, etc.

3. Conclusions

The 15q11.2 BP1–BP2 microdeletion (Burnside-Butler) syndrome is now a recognized condition with over 200 individuals identified from the literature using chromosomal microarray analysis. Clinically, neurological dysfunction, developmental and language delay are the most commonly associated findings followed by motor delay, ADD/ADHD and autism spectrum disorder showing incomplete penetrance and variable expressivity. The four non-imprinted biallelically expressed genes (*TUBGCP5, CFYIP1, NIPA1, NIPA2*) in this microdeletion were initially noted to impact severity of clinical presentation and neurological impairment in two classical genomic imprinting disorders (*i.e.*, Prader-Willi and Angelman syndromes) with typical 15q11–q13 deletions depending on the absence or presence of the genomic area between breakpoints BP1 and BP2 containing the four genes leading to studies recognizing this syndrome. The 15q11.2 BP1–BP2 microdeletion syndrome has a reported *de novo* frequency between 5%–22%, with 51% having inherited the microdeletion from an apparently unaffected parent and 35% having inherited the microdeletion from an affected parent. Low penetrance estimates may relate to subclinical manifestations of neuropsychiatric/behavioral problems or incomplete information about the parents of individuals with 15q11.2 BP1–BP2 microdeletion or members of control cohorts [44]. From reported patient cohorts presenting for genetic services and microarray analysis, this microdeletion syndrome can now be recognized as the most common cytogenetic abnormality found in autism spectrum disorder.

Acknowledgments: We thank Carla Meister for preparation of this manuscript and support from NICHD (National Institute of Child Health and Human Development) HD02528 grant.

Author Contributions: Merlin G. Butler conceived the study and wrote and contributed to the manuscript. Devin M. Cox reviewed the literature and wrote and contributed to the content of the manuscript.

Conflicts of Interest: The authors declare no conflict of interest.

References

1. Locke, D.P.; Segraves, R.; Nicholls, R.D.; Schwartz, S.; Pinkel, D.; Alberston, D.G.; Eichler, E.E. BAC microarray analysis of 15q–q13 rearrangements and the impact of segmental duplications. *J. Med. Genet.* **2004**, *41*, 175–182. [CrossRef] [PubMed]
2. Pujana, M.A.; Nadal, M.; Guitart, M.; Armengol, L.; Gratacos, M.; Estivill, X. Human Chromosome 15q11–q14 regions of rearrangements contain clusters of LCR15 duplicons. *Eur. J. Hum. Genet.* **2002**, *10*, 26–35. [CrossRef] [PubMed]
3. Eichler, E.E. Masquerading repeats: Paralogous pitfalls of the human genome. *Genome Res.* **1998**, *8*, 758–762. [PubMed]
4. Butler, M.G.; Fischer, W.; Kibiryeva, N.; Bittel, D.C. Array comparative genomic hybridization (aCGH) analysis in Prader-Willi syndrome. *Am. J. Med. Genet. A* **2008**, *146A*, 854–860. [CrossRef]
5. Butler, M.G.; Bittel, D.C.; Kibiryeva, N.; Talebizadeh, Z.; Thompson, T. Behavioral differences among subjects with Prader-Willi syndrome and type I or type II deletion and maternal disomy. *Pediatrics* **2004**, *113*, 565–573. [CrossRef] [PubMed]
6. Bittel, D.C.; Kibiryeva, N.; Butler, M.G. Expression of 4 genes between chromosome 15 breakpoints 1 and 2 and behavioral outcomes in Prader-Willi syndrome. *Pediatrics* **2006**, *118*, e1276–e1283. [CrossRef] [PubMed]
7. Hartley, S.L.; Maclean, W.E., Jr.; Butler, M.G.; Zarcone, J.; Thompson, T. Maladaptive behaviors and risk factors among the genetic subtypes of Prader-Willi syndrome. *Am. J. Med. Genet. A* **2005**, *136*, 140–145. [CrossRef] [PubMed]
8. Milner, K.M.; Craig, E.E.; Thompson, R.J.; Veltman, M.W.; Thomas, N.S.; Roberts, S.; Bellamy, M.; Curran, S.R.; Sporikou, C.M.; Bolton, P.F. Prader-Willi syndrome: Intellectual abilities and behavioural features by genetic subtype. *J. Child Psychol. Psychiatry* **2005**, *46*, 1089–1096. [CrossRef] [PubMed]
9. Sahoo, T.; Bacino, C.A.; German, J.R.; Shaw, C.A.; Bird, L.M.; Kimonis, V.; Anselm, I.; Waisbren, S.; Beaudet, A.L.; Peters, S.U. Identification of novel deletions of 15q11q13 in Angelman syndrome by array-CGH: Molecular characterization and genotype-phenotype correlations. *Eur. J. Hum. Genet.* **2007**, *15*, 943–949. [CrossRef] [PubMed]

10. Varela, M.C.; Kok, F.; Setian, N.; Kim, C.A.; Koiffmann, C.P. Impact of molecular mechanisms, including deletion size, on Prader-Willi syndrome phenotype: Study of 75 patients. *Clin. Genet.* **2005**, *67*, 47–52. [CrossRef] [PubMed]

11. Valente, K.D.; Varela, M.C.; Koiffmann, C.P.; Andrade, J.Q.; Grossmann, R.; Kok, F.; Marques-Dias, M.J. Angelman syndrome caused by deletion: A genotype-phenotype correlation determined by breakpoint. *Epilepsy Res.* **2013**, *105*, 234–239. [CrossRef] [PubMed]

12. Dykens, E.M.; Roof, E. Behavior in Prader-Willi syndrome: Relationship to genetic subtypes and age. *J. Child. Psychol. Psychiatry* **2008**, *49*, 1001–1008. [CrossRef] [PubMed]

13. Burnside, R.D.; Pasion, R.; Mikhail, F.M.; Carroll, A.J.; Robin, N.H.; Youngs, E.L.; Gadi, I.K.; Keitges, E.; Jaswaney, V.L.; Papenhausen, P.R.; *et al.* Microdeletion/microduplication of proximal 15q11.2 between BP1 and BP2: A susceptibility region for neurological dysfunction including developmental and language delay. *Hum. Genet.* **2011**, *130*, 517–528. [CrossRef] [PubMed]

14. Jerkovich, A.M.; Butler, M.G. Further phenotypic expansion of 15q11.2 BP1–BP2 microdeletion (Burnside-Butler) syndrome. *J. Pediatr. Genet.* **2014**, *3*, 41–44. [PubMed]

15. Murthy, S.K.; Nygren, A.O.H.; El Shakankiry, H.M.; Schouten, J.P.; Al Khayat, A.I.; Righa, A.; Al Ali, M.T. Detection of a novel familial deletion of four genes between BP1 and BP2 of the Prader-Willi/Angelman syndrome critical region by oligo-array CGH in a child with neurological disorder and speech impairment. *Cytogenet. Genome Res.* **2007**, *116*, 135–140. [CrossRef] [PubMed]

16. Doornbos, M.; Sikkema-Raddatz, B.; Ruijvenkamp, C.A.L.; Dijkhuizen, T.; Bijlsma, E.K.; Gijsbers, A.C.J.; Hihorst-Hofstee, Y.; Hordijk, R.; Verbruggen, K.T.; Kerstjens-Frederikse, W.S.M.; *et al.* Nine patients with a microdeletion 15q11.2 between breakpoints 1 and 2 of the Prader-Willi critical region, possibly associated with behavioral disturbances. *Eur. J. Med. Genet.* **2009**, *52*, 108–115. [CrossRef] [PubMed]

17. Abdelmoity, A.T.; LePichon, J.B.; Nyp, S.S.; Soden, S.E.; Daniel, C.A.; Yu, S. 15q11.2 proximal imbalances associated with a diverse array of neuropsychiatric disorders and mild dysmorphic features. *J. Dev. Behav. Pediatr.* **2012**, *33*, 570–576. [CrossRef] [PubMed]

18. Cafferkey, M.; Ahn, J.W.; Flinter, F.; Ogilvie, C. Phenotypic features in patients with 15q11.2 (BP1–BP2) deletion: Further delineation of an emerging syndrome. *Am. J. Med. Genet. Part A* **2013**, *164A*, 1916–1922. [CrossRef]

19. Chai, J.H.; Locke, D.P.; Greally, J.M.; Knoll, J.H.; Ohta, T.; Dunai, J.; Yavor, A.; Eichler, E.E.; Nicholls, R.D. Identification of four highly conserved genes between breakpoint hotspots BP1 and BP2 of the Prader-Willi/Angelman syndromes deletion region that have undergone evolutionary transposition mediated by flanking duplicons. *Am. J. Hum. Genet.* **2003**, *73*, 898–925. [CrossRef] [PubMed]

20. Chen, S.; Song, C.; Guo, H.; Xu, P.; Huang, W.; Zhou, Y.; Sun, J.; Li, C.X.; Du, Y.; Li, X.; *et al.* Distinct novel mutations affecting the same base in the *NIPA1* gene cause autosomal dominant hereditary spastic paraplegia in two Chinese families. *Hum. Mutat.* **2005**, *25*, 135–141. [CrossRef] [PubMed]

21. Rainier, S.; Chai, J.H.; Tokarz, D.; Nicholls, R.D.; Fink, J.K. *NIPA1* gene mutations cause autosomal dominant hereditary spastic paraplegia (SPG6). *Am. J. Hum. Genet.* **2003**, *73*, 967–971. [CrossRef] [PubMed]

22. Goytain, A.; Hines, R.M.; El-Husseini, A.; Quamme, G.A. NIPA1 (SPG6), the basis for autosomal dominant form of hereditary spastic paraplegia, encodes a functional Mg^{2+} transporter. *J. Biol. Chem.* **2007**, *282*, 8060–8068. [CrossRef] [PubMed]

23. Jiang, Y.; Zhang, Y.; Zhang, P.; Sang, T.; Zhang, F.; Ji, T.; Huang, Q.; Xie, H.; Du, R.; Cai, B.; *et al.* NIPA2 located in 15q11.2 is mutated in patients with childhood absence epilepsy. *Hum. Genet.* **2012**, *131*, 1217–1224. [CrossRef] [PubMed]

24. De Wolf, V.; Brison, N.; Devriendt, K.; Peeters, H. Genetic counseling for susceptibility loci and neurodevelopmental disorders: The del15q11.2 as an example. *Am. J. Med. Genet. Part A* **2013**, *161A*, 2846–2854. [CrossRef]

25. Hagerman, R.J.; Hagerman, P.J. *Fragile X Syndrome: Diagnosis, Treatment and Research*, 3rd ed.; The John Hopkins University Press: Baltimore, MD, USA, 2002.

26. Bozdagi, O.; Sakurai, T.; Dorr, N.; Pilorge, M.; Takahashi, N.; Buxbaum, J.D. Haploinsufficiency of Cyfip1 produces fragile X-like phenotypes in mice. *PLoS One* **2012**, *7*, e42422. [CrossRef] [PubMed]

27. Cooper, G.M.; Coe, B.P.; Girirajan, S.; Rosenfeld, J.A.; Vu, T.; Baker, C.; Williams, C.; Stalker, H.; Hamid, R.; Hannig, V.; *et al.* A copy number variation mobidity map of developmental delay. *Nat. Genet.* **2011**, *43*, 838–846. [CrossRef] [PubMed]

28. Rosenfeld, J.A.; Coe, B.P.; Eichler, E.E.; Cuckle, H.; Shaffer, L.G. Estimates of penetrance for recurrent pathogenic copy-number variants. *Genet Med.* **2013**, *15*, 478–481. [CrossRef] [PubMed]

29. De Kovel, C.G.; Trucks, H.; Helbig, I.; Mefford, H.C.; Baker, C.; Leu, C.; Kluck, C.; Muhle, H.; von Spiczak, S.; Ostertag, P.; *et al.* Recurrent microdeletions at 15q11.2 and 16p13.11 predispose to idiopathic generalized epilepsies. *Brain* **2010**, *133*, 23–32. [CrossRef] [PubMed]

30. Itsara, A.; Cooper, G.M.; Baker, C.; Girirajan, S.; Li, J.; Absher, D.; Krauss, R.M.; Myers, R.M.; Ridker, P.M.; Chasman, D.I.; Mefford, H.; Ying, P.; Nickerson, D.A.; Eichler, E.E. Population analysis of large copy number variants and hotspots of human genetic disease. *Am. J. Hum. Genet.* **2009**, *84*, 148–161. [CrossRef] [PubMed]

31. Stefansson, H.; Rujescu, D.; Cichon, S.; Pietilainen, O.P.; Ingason, A.; Steinberg, S.; Fossdal, R.; Sigurdsson, E.; Sigmundsson, T.; Buizer-Voskamp, J.E.; *et al.* Large recurrent microdeletions associated with schizophrenia. *Nature* **2008**, *455*, 232–236. [CrossRef] [PubMed]

32. Chaste, P.; Sanders, S.J.; Mohan, K.N.; Klei, L.; Song, Y.; Murtha, M.T.; Hus, V.; Lowe, J.K.; Willsey, A.J.; Moreno-De-Luca, D.; *et al.* Modest impact on risk for Autism Spectrum Disorder of rare copy number variants at 15q11.2, specifically breakpoints 1 to 2. *Autism Res.* **2014**, *7*, 355–362. [CrossRef] [PubMed]

33. Madrigal, I.; Rodriguez-Revenga, L.; Xuncla, M.; Mila, M. 15q11.2 microdeletion and FMR1 premutation in a family with intellectual disabilities and autism. *Gene* **2012**, *508*, 92–95. [CrossRef] [PubMed]

34. Von der Lippe, C.; Rustad, C.; Heimdal, K.; Rodningen, O.K. 15q11.2 microdeletion—Seven new patients with delayed development and/or behavioral problems. *Eur. J. Med. Genet.* **2011**, *54*, 357–360. [CrossRef] [PubMed]

35. Wong, D.; Johnson, S.M.; Young, D.; Iwamoto, L.; Sood, S.; Slavin, T.P. Expanding the BP1–BP2 15q11.2 microdeletion phenotype: Tracheoesophageal fistula and congenital cataracts. *Case Rep. Genet.* **2013**, *2003*. [CrossRef]

36. Usrey, K.M.; Williams, C.A.; Dasouki, M.; Fairbrother, L.C.; Butler, M.G. Congenital arthrogryposis: An extension of the 15q11.2 BP1–BP2 microdeletion syndrome? *Case Rep. Genet.* **2014**, *2004*. [CrossRef]

37. Rudd, D.S.; Azelsen, M.; Epping, E.A.; Andreasen, N.C.; Wassink, T.H. A genome-wide CNV analysis of schizophrenia reveals a potential role for a multiple-hit model. *Am. J. Med. Genet. Part. B* **2014**, *165B*, 619–626. [CrossRef]

38. Stefansson, H.; Meyer-Lindenberg, A.; Steinberg, S.; Magnusdottir, B.; Morgen, K.; Arnarsdottir, S.; Bjornsdottir, G.; Walters, G.B.; Jonsdottir, G.A.; Doyle, O.M.; *et al.* CNVs conferring risk of autism or schizophrenia affect cognition in controls. *Nature* **2014**, *505*, 361–366. [CrossRef] [PubMed]

39. Kirov, G.; Grozeva, D.; Norton, N.; Ivanov, D.; Mantripragada, K.K.; Holmans, P.; International Schizophrenia Consortium; Wellcome Trust Case Control Consortium; Craddock, N.; Owen, M.J.; *et al.* Support for the involvement of large copy number variants in the pathogenesis of schizophrenia. *Hum. Mol. Genet.* **2009**, *18*, 1497–1503.

40. Rees, E.; Walters, J.T.R.; Georgieva, L.; Isles, A.R.; Chambert, K.D.; Richards, A.L.; Mahoney-Davies, G.; Legge, S.E.; Moran, J.L.; McCarroll, S.A.; *et al.* Analysis of copy number variations at 15 schizophrenia-associated loci. *Br. J. Psychiatry* **2014**, *204*, 108–114. [CrossRef] [PubMed]

41. Helbig, I.; Hartmann, C.; Mefford, H.C. The unexpected role of copy number variations in juvenile myoclonic epilepsy. *Epilepsy Behav.* **2012**, *28*, S66–S68. [CrossRef]

42. Jahn, J.A.; von Spiczak, S.; Muhle, H.; Obermeier, T.; Franke, A.; Mefford, H.C.; Stephani, U.; Helbig, I. Iterative phenotyping of 15q11.2, 15q13.3 and 16p13.11 microdeletion carriers in pediatric epilepsies. *Epilepsy Res.* **2014**, *108*, 109–116. [CrossRef] [PubMed]

43. Mullen, S.A.; Carvill, G.L.; Bellows, S.; Bayly, M.A.; Berkovic, S.F.; Dibbens, L.M.; Scheffer, I.E.; Mefford, H.C. Copy number variants are frequent in genetic generalized epilepsy with intellectual disability. *Neurology* **2013**, *81*, 1507–1514. [CrossRef] [PubMed]

44. Girirajan, S.; Rosenfeld, J.A.; Cooper, G.M.; Antonacci, F.; Siswara, P.; Itsara, A.; Vives, L.; Walsh, T.; McCarthy, S.E.; Baker, C.; *et al.* A recurrent 16p12.1 microdeletion supports a two-hit model for severe developmental delay. *Nat. Genet.* **2010**, *42*, 203–209. [CrossRef] [PubMed]

International Journal of
Molecular Sciences

MDPI

Article

Phenotypic and Molecular Convergence of 2q23.1 Deletion Syndrome with Other Neurodevelopmental Syndromes Associated with Autism Spectrum Disorder

Sureni V. Mullegama [1], Joseph T. Alaimo [1], Li Chen [1,2] and Sarah H. Elsea [1,*]

[1] Department of Molecular and Human Genetics, Baylor College of Medicine, Houston, TX 77030, USA;
      mullegam@bcm.edu (S.V.M.); alaimo@bcm.edu (J.T.A.); li.chen@bcm.edu (L.C.)
[2] Department of Cellular and Genetic Medicine, School of Basic Medical Sciences, Fudan University,
      Shanghai 200032, China
*   Author to whom correspondence should be addressed; elsea@bcm.edu;
      Tel.: +1-713-798-5484; Fax: +1-832-825-1269.

Academic Editor: Merlin G. Butler

Received: 21 January 2015; Accepted: 19 March 2015; Published: 7 April 2015

Abstract: Roughly 20% of autism spectrum disorders (ASD) are syndromic with a well-established genetic cause. Studying the genes involved can provide insight into the molecular and cellular mechanisms of ASD. 2q23.1 deletion syndrome (causative gene, *MBD5*) is a recently identified genetic neurodevelopmental disorder associated with ASD. Mutations in *MBD5* have been found in ASD cohorts. In this study, we provide a phenotypic update on the prevalent features of 2q23.1 deletion syndrome, which include severe intellectual disability, seizures, significant speech impairment, sleep disturbance, and autistic-like behavioral problems. Next, we examined the phenotypic, molecular, and network/pathway relationships between nine neurodevelopmental disorders associated with ASD: 2q23.1 deletion Rett, Angelman, Pitt-Hopkins, 2q23.1 duplication, 5q14.3 deletion, Kleefstra, Kabuki make-up, and Smith-Magenis syndromes. We show phenotypic overlaps consisting of intellectual disability, speech delay, seizures, sleep disturbance, hypotonia, and autistic-like behaviors. Molecularly, MBD5 possibly regulates the expression of *UBE3A*, *TCF4*, *MEF2C*, *EHMT1* and *RAI1*. Network analysis reveals that there could be indirect protein interactions, further implicating function for these genes in common pathways. Further, we show that when *MBD5* and *RAI1* are haploinsufficient, they perturb several common pathways that are linked to neuronal and behavioral development. These findings support further investigations into the molecular and pathway relationships among genes linked to neurodevelopmental disorders and ASD, which will hopefully lead to common points of regulation that may be targeted toward therapeutic intervention.

Keywords: *MBD5*; ASD; networks; overlapping phenotypes; *UBE3A*; *TCF4*; *MEF2C*; *EHMT1*; *RAI1*; transcriptional regulation; pathways; network analysis

---

1. Introduction

Autism spectrum disorder (ASD) is a growing public health concern that affects millions of individuals worldwide [1]. ASD is a broad term encompassing a heterogeneous group of complex, highly heritable neurodevelopmental disorders in which individuals have impairments in social interaction and communication coupled with repetitive and restricted behaviors [2]. Other morbidities such as intellectual disability, epilepsy, neurological disabilities (ataxia and hypotonia), and other behavioral disorders have been associated with ASD [3]. The molecular etiology of ASD involves the interplay of many genes [4]. A contributor to the risk of ASD resides in high-impact rare variants, such as chromosomal abnormalities, copy number variation (CNV), and mutations in genes previously linked to classic monogenenic genetic disorders such as fragile X syndrome (FXS) (MIM 300624), Rett syndrome (RTT) (MIM 312750), Angelman syndrome (AS) (MIM 105830), and Smith-Magenis syndrome (SMS) (MIM 182290) [2,5,6].

With the advancement of genetic diagnostic technologies such as chromosomal microarray analysis, the identification of novel genetic subtypes of ASD and their associated genes have come to light, such as 16p11.2 deletion (MIM 611913, gene unknown), 2q23.1 deletion syndrome (MIM 156200, *MBD5*), Pitt-Hopkins syndrome (PTHS MIM 610954, *TCF4*), 5q14.3 deletion syndrome (MIM 613443, *MEF2C*), and Kleefstra syndrome (MIM 610253, *EHMT1*). Many of these genes are involved in neuronal functions (synaptic transmission and cell-cell interaction), chromatin modification, methylation, and transcriptional regulation [7]. The identification of these disease genes associated with ASD has led to many theories regarding the pathogenesis of ASD. One theory that has garnered support from some in the ASD community proposes that the pathogenesis is due to the disruption of neurodevelopment, which is triggered by genes with global effects on expression (chromatin modification, methylation, and transcriptional regulation) of other genes that are specifically involved in neuronal functions [7]. Further, these genes are thought to be most likely involved in common ASD-associated pathways such as cell adhesion, cadherin signaling, WNT signaling, PTEN signaling, mTOR signaling, PI3K-Akt signaling, and circadian rhythm [4,8].

Corroborating studies have suggested that a disease phenotype is rarely a consequence of an abnormality in a single gene product but a reflection of dysfunction in a variety of genes and pathobiological processes that interact in complex networks and pathways [9–11]. Many neurodevelopmental genetic disorders associated with ASD share common phenotypic features despite their genetic heterogeneity, which further suggests that these genes may act together in complex interconnected pathways that, when perturbed, manifest similar phenotypes.

2q23.1 deletion syndrome (MIM 156200), previously known as "pseudo-Angelman syndrome," was initially identified in one of the first comparative genomic hybridization (CGH) surveys of developmental disorders [12]. Patients are characterized by severe intellectual disability, seizures, significant speech impairment, and autistic-like behavioral problems [13]. 2q23.1 deletion syndrome is caused by deletion in the chromosomal region 2q23.1 or gene specific deletions in methy-CpG-binding domain 5 (MBD5, MIM 611472). Deletions of 2q23.1 range from small deletions of 38 kb to >19 Mb [13]. While some large deletions extend in the chromosomal region 2q22.3, these deletions do not include the zinc finger E box-binding homeobox 2 gene (*ZEB2*, MIM 605802), the causative gene for Mowat-Wilson syndrome (MIM 235730). *MBD5* is part of the methyl-CpG-binding domain (MBD) family, which consists of a well-known ASD-associated gene, *MECP2*, which is the causative gene in Rett syndrome. MBD5 has two known isoforms [14] and is thought to have a role in epigenetic modification [7,14–18]. Further, MBD5/Mbd5 has been shown to regulate expression of genes, suggesting it acts as a transcription factor [15,17,19].

In this study, we conducted a comprehensive phenotypic overview of 2q23.1 deletion syndrome and then briefly examined the phenotypic and molecular relationships between 2q23.1 deletion syndrome and neurodevelopmental disorders associated with ASDs. Finally, we examined the pathways in common between 2q23.1 deletion syndrome and Smith-Magenis syndrome.

2. Results and Discussion

2.1. 2q23.1 Deletion Syndrome Clinical Review

To further update our phenotypic knowledge of 2q23.1 deletion syndrome, we surveyed the most prevalent phenotypic features of all 2q23.1 deletion cases (*MBD5*-specific deletions and 2q23.1 deletions) using the two largest phenotypic studies on 2q23.1 deletion syndrome [13,20] and case reports that were published subsequently after these studies [15,16,19,21–24]. Overall, phenotypic findings are comparable across the 74 cases reported [13,15,16,20–23,25–30], and the frequently reported neurological, neurobehavioral, and craniofacial features associated within syndrome are seen and listed in Figure 1 and Table 1 [13,20]. Neurological and behavioral features are the most prevalent findings in 2q23.1 deletion syndrome patients, while the other abnormalities, such as craniofacial abnormalities and skeletal abnormalities, are not consistently the same among individuals with this deletion (Figure 1).

Developmental delay and motor delay were present in 100% of the 2q23.1 deletion cases studied. Previous reports have identified seizures and severe language impairment as two primary features of 2q23.1 deletion syndrome, and we further show that 94.4% and 84.9% of the 2q23.1 deletion cases exhibit these features, respectively [13,15,16,20–23,25–30]. Infantile hypotonia and feeding difficulties were also present in greater than 85% of reported 2q23.1 deletion cases (Table 1). Autistic-like behaviors and behavioral problems were reported in 98.4%. The most prevalent behaviors were distractibility/short attention span (100%) and sleep disturbances (78.8%). These studies also showed that there is a clear lack of precise documentation by clinicians regarding the exact autistic-like behaviors of individuals with 2q23.1 deletion syndrome. Nonetheless, impairments in communication, social interaction and repetitive behaviors are key behavioral criteria for ASD, which are also present in the majority of 2q23.1 deletion syndrome patients. Overall, these patients should be assessed by ADI-R and ADOS to identify specific ASD phenotypes present in individuals with 2q23.1 deletion syndrome. It is apparent that while craniofacial abnormalities are present in children with 2q23.1 deletion, the features are variable across the population. While mild dysmorphic craniofacial features were present at >70% in reported cases, including broad forehead, arched/thick eyebrows, eye abnormalities, nasal abnormalities, downturned corners of the mouth, open mouth, thin upper lip, tented upper lip, and thick or everted lower lip, a consistent and thorough evaluation, measurement, and documentation of craniofacial and skeletal abnormalities is necessary for an accurate and full understanding of the commonly observed associations with 2q23.1 deletion syndrome (Figure 1).

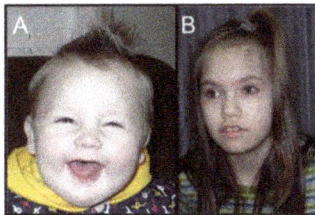

Figure 1. Craniofacial features of 2q23.1 deletion syndrome. Children with 2q23.1 deletion syndrome exhibit broad forehead, open mouth, and tented, thin upper lip. (**A**) 1 year old male. (**B**) 13 year old female. Approvals from parents were obtained to publish these photos.

Table 1. Prevalent features of 2q23.1 deletion syndrome.

2q23.1 Deletion [1]		
Common Features	**Frequency**	**Percentage (%)**
Neurological		
Developmental delay	74/74	100
Motor delay	45/45	100
Language impairment	51/54	94.4
Ataxia	22/32	68.7
Infantile hypotonia	28/30	93.3
Infantile feeding difficulties	17/20	85.0
Seizures	45/53	84.9
Behavioral		
Autistic-like behaviors	60/61	98.4
Behavioral problems	60/61	98.4
Aggression/temper tantrums	13/21	62.9
Distractibility/short attention span	21/21	100
Hyperphagia	8/16	50.0
Self-injurious behaviors	21/33	63.6
Sleep disturbances	41/52	78.8

Table 1. *Cont.*

2q23.1 Deletion [1]		
Common Features	Frequency	Percentage (%)
Neurological		
Growth/Endocrine Abnormalities		
Postnatal growth retardation	25/51	49.0
Obesity	6/17	35.3
Short stature (<5th percentile)	30/43	69.8
Craniofacial Abnormalities		
Cranium		
Brachycephaly	12/36	33.3
Broad forehead	21/30	70.0
Microcephaly	28/46	60.9
Eyes		
Arched/thick eyebrows	19/24	79.2
Myopia/hypermetropia/corrective lenses	8/11	72.7
Synophrys	13/28	46.4
Nose/Ear		
Nasal abnormalities	42/43	97.7
Outer ear abnormalities	22/29	75.9
Mouth/Chin		
Dental abnormalities	18/35	51.4
Downturned corners of the mouth	20/28	71.4
Macroglossia or protruding tongue	8/33	24.2
Micrognathia/retrognathia	16/30	53.3
Open mouth	26/34	76.5
Tented upper lip	19/30	63.3
Thin upper lip	21/28	75.0
Thick or everted lower lip	19/26	73.1
Wide mouth	15/25	60.0
Skeletal Extremity Abnormalities		
Brachydactyly	14/33	42.4
Clinodactyly, 5th finger	24/38	63.2
Sandal gap	12/32	37.5
Short fifth digit	16/37	43.2
Small hands and feet	25/37	67.6

[1] Cases came from [13,15,20–24].

2.2. Overlapping Phenotypes across 2q23.1 Deletion Syndrome and Other Autism Spectrum Disorders

2q23.1 deletion has been aptly termed a "potent masquerader", wherein the clinical features of the disorder were initially thought to be due to other well-known genetic syndromes [20]. Since the neurodevelopmental and behavioral characteristics common to 2q23.1 deletion syndrome are nonspecific and commonly found in multiple other neurodevelopmental disorders associated with ASD [20], many 2q23.1 deletion syndrome patients were initially tested for a variety of disorders, including Rett, Angelman, and Smith-Magenis syndromes [26]. As previously mentioned, Mowat-Wilson syndrome and 2q23.1 deletion syndrome are on neighboring chromosome bands, patients with 2q23.1 deletion do not share key anomalies associated with Mowat-Wilson syndrome, such as Hirschsprung disease, congenital heart defects, genitourinary anomalies, and eye defects; thus, Mowat-Wilson syndrome is not in the differential diagnosis for 2q23.1 deletion syndrome patients [31,32]. Here, we reviewed the neurological and behavioral features of 2q23.1 deletion syndrome and eight other disorders that have been phenotypically linked to or considered in the differential diagnosis for 2q23.1 deletion syndrome [31,32]. The reported phenotypic features of 2q23.1 deletion syndrome, RTT, AS, PTHS, 2q23.1 duplication syndrome, 5q14.3 deletion syndrome, Kleefstra syndrome (KFS), Kabuki syndrome (MIM 147920), and SMS show they share neurological and behavioral co-morbidities coupled with ID, including motor impairments and gaiting abnormalities, hypotonia, language impairments, seizures, sleep disturbance, autistic-like behaviors, and other distinctive behaviors (Table 2). The similar features among many highly penetrant neurodevelopmental disorders associated with ASD allow us to hypothesize several models for

ASD etiology and pathogenesis. First, the ASD-associated genes linked to known neurodevelopmental disorders govern similar cellular functions. Second, the genes play a role in specific cellular functions that converge into common molecular pathways that are associated with specific phenotypes. Third, these genes are involved in entirely different molecular pathways that converge to a common phenotype.

2.3. MBD5 Regulates Disorder-Specific Genes

Due to the high degree of phenotypic similarity between RTT (*MECP2*), AS (*UBE3A*), PTHS (*TCF4*), 2q23.1 duplication syndrome (*MBD5*), 5q14.3 deletion syndrome (*MEF2C*), KFS (*EHMT1*), Kabuki syndrome (*KMT2D* and *KDM6A*), and SMS (*RAI1*) relative to 2q23.1 deletion syndrome (Table 2), we examined the co-expression relationships of these genes to MBD5 toward assessing their involvement in the phenotype of 2q23.1 deletion syndrome. We hypothesized that these genes may be dysregulated in 2q23.1 deletion syndrome and therefore assessed expression of *MECP2*, *UBE3A*, *TCF4*, *MEF2C*, *EHMT1*, *KMT2D*, *KDM6A*, and *RAI1* in 2q23.1 deletion syndrome patient lymphoblastoid cell lines (LCLs) that had various deletions of *MBD5*. As expected, mRNA levels of *MBD5* were significantly down regulated in 2q23.1 deletion syndrome LCLs ($p < 0.0001$), confirming previously reported results [13] (Figure 2). Overall, we observed 5/8 (62.5%) of the tested genes had significantly altered mRNA levels when *MBD5* was haploinsufficient. *TCF4* and *UBE3A* expression levels were elevated to ~1.5-fold (*TCF4*, $p = 0.012$; *UBE3A*, $p < 0.0001$). However, *MEF2C*, *EHMT1*, and *RAI1* expression levels were significantly reduced to ~0.5-fold (*MEF2C*, $p = 0.016$; *EHMT1*, $p = 0.0004$; *RAI1*, $p < 0.0001$). *MECP2*, *KMT2D*, and *KDM6A* had no significant change of expression.

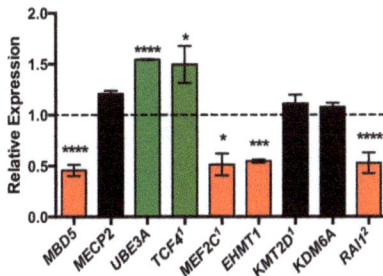

Figure 2. Dysregulation of associated genes in 2q23.1 deletion syndrome. *MBD5*, *MEF2C*, *EHTM1*, and *RAI1* were significantly down regulated (red bars) while *UBE3A* and *TCF4* were significantly up regulated (green bars). *MECP2*, *KMT2D*, and *KDM6A* did not have altered mRNA levels (black bars). Gene expression is shown relative to control set to 1.0 (black line). Graphs represent mean ± SEM (* $p < 0.05$, *** $p < 0.01$, **** $p < 0.0001$). 1 = Manuscript in review, 2 = [20].

Our co-expression data indicate that *MBD5* functions, either directly or indirectly, in the regulation of the expression of other ASD-associated genes (*TCF4*, *MEF2C*, *EHTM1*, *RAI1*, and *UBE3A*), which supports the common pathway ASD pathogenesis hypothesis. *MBD5* and the genes with altered expression possibly converge on common pathways that contribute to the phenotype of 2q23.1 deletion syndrome and genetic neurodevelopmental syndromes associated with ASD. While *MECP2*, *KMT2D*, and *KDM6A*, did not have altered mRNA levels, these genes may share a molecular connection to *MBD5* beyond co-expression, such as physical interactions. Thus, MBD5 and *MECP2*, *KMT2D*, and *KDM6A* could be involved in different genetic pathways that result in a similar phenotypic output, which may indicate the presence of redundant pathways. Alternatively, *MECP2*, *KMT2D*, and *KDM6A* may be in the same or overlapping pathways but may function upstream of MBD5. Our phenotypic and gene expression data (Table 2 and Figure 2) implicate some form of pathway overlap or interaction when *MBD5* is haploinsufficient.

Table 2. Common phenotypes between 2q23.1 deletion syndrome and Rett, Angelman, Pitt-Hopkins, 2q23.1 duplication, 5q14.3 deletion, Kleefstra, Kabuki, and Smith-Magenis syndromes.

DISORDER	2q23.1 del	RTT	AS	PTHS	2q23.1 dup	5q14.3 del	KFS	KMS	SMS
Key References	[13,20]	[33,34]	[31,35]	[36,37]	[38,39]	[40]	[41]	[42,43]	[44]
Causative Gene	*MBD5*	*MECP2*	*UBE3A*	*TCF4*	*MBD5*	*MEF2C*	*EHMT1*	*KMT2D, KDM6A*	*RAI1*
Neurological/Behavioral Characteristics									
Intellectual disability [a]	++	+++	+++	+++	++	+++	+++	++	++
Speech delay [b]	+++	+++	+++	+++	++	+++	++	++	+
Seizures [c]	+++	+++	+++	++	++	++	++	+	+
Sleep disturbance [d]	+++	+	+++	++	++	+	+	+	+++
Delayed walking [e]	++	+++	++	++	+	+++	+	+	+
Hypotonia	+	+	+	+	+	+	+[f]	+	+
Autism-like behaviors	+	+	+	-	+	+	+	+	+
Feeding difficulties	+	+	+	+	+	+	-	+	+
Stereotypic behaviors	+	+	+	+	+	+	+	+	+
Ataxia	+	+	-	+	+	+	-	+	-
Happy disposition (frequent or inappropriate laughing)	+	+	+	+	+	NR	-	+	-
Hyperactivity/short attention span	+	-	+	+	+	-	-	+	-
Self-injurious behavior	+	-	-	+	-	-	+	+	+
Aggressive behavior	-	-	-	+	-	-	+	-	+

2q23.1 del = 2q23.1 deletion syndrome, RTT = Rett syndrome, AS = Angelman syndrome, PTHS = Pitt-Hopkins syndrome, 2q23.1 dup = 2q23.1 duplication syndrome, 5q14.3 del = 5q14.3 deletion syndrome, KFS = Kleefstra syndrome, KMS = Kabuki make-up syndrome, and Smith-Magenis syndrome, [a] + = mild; ++ = moderate; +++ = severe; [b] + = severe; ++ = moderate; +++ = severe; [c] + = absent, [c] + = 0%–40%; ++ = 41%–70%; +++ = 71%–100%, [d] + = 0%–40%; ++ = 40%–70%; +++ = 71%–100%, [e] + = 1–3 y; ++ = 4–6 y; +++ = >6 y or limited mobility, [f] only in childhood, NR = not reported.

2.4. MBD5 Network

Recent studies suggest that genes involved in common endpoints (e.g., phenotypes) show an increased tendency to have protein-protein interactions, similar expression patterns in specific tissues, and exhibit synchronized expression as a group [45,46]. In this study, we observed co-expression relationships between *MBD5* and ASD-associated genes (Figure 2). To comprehensively understand MBD5-dependent genetic pathways relative to these genes, we utilized bioinformatics resources and the network tool, Cognoscente (http://vanburenlab.tamhsc.edu/), to uncover related protein interactions (Figure 3). While data do not exist in the literature to support the key genes (in red boxes) directly interacting with each other, we did observe that some hubs (central network connections) all intervened with one another (Figure 3). This finding could suggest there are intermediary genes that molecularly connect the genes directly involved in known disorders. Interestingly, from this network, several other genes linked to ASD were identified, including *CDKL5*, *HDAC4*, *EP300*, *FMR1*, *SMARCA4*, and *ATRX*, each previously linked to a neurodevelopmental disorder with overlapping or similar phenotypes, as listed in Table 2 (Figure 2). In Mullegama *et al.* (2014), we observed overexpression of *FMR1* in 2q23.1 deletion syndrome cell lines [19]. Additional expression studies of these ASD-associated genes in 2q23.1 deletion syndrome would prove interesting. A limitation using such tools to explore molecular networks, as we demonstrated in Figure 2, is that all programs that generate networks, like Cognoscente, rely on the published literature [47]. This fact revealed that there is clearly a lack of research and understanding of the interactions among the genome and proteome in ASD. Thus, undoubtedly, concerted efforts need to be made to study molecular interactions between these genes collectively to further elucidate the genetic etiology of these and other monogenic neurodevelopmental disorders associated with ASDs. Futures studies such as RNA-seq, methylation sequencing, whole-genome miRNA analysis, and chromatin immunoprecipitation sequencing (ChIP-seq) studies of these disorder-causing genes will allow us to further elucidate the levels of molecular convergence between these disorders.

2.5. Molecular Relationships between MBD5 and RAI1

We propose that functional relationships between genes serve as a strong indicator for involvement in key pathways responsible for phenotypic outcomes. An example of this concept is the relationship between MBD5 and RAI1. We have shown that 2q23.1 deletion syndrome and Smith-Magenis syndrome have overlapping phenotypes (see Table 2). From the data we presented here, it appears that *RAI1* could be regulated directly or indirectly by MBD5. Our previously published work also suggested that *RAI1* is dysregulated when *MBD5* is knocked down through siRNA technology in neuroblastoma SH-SY5Y cell lines [19]. Therefore, we wanted to further examine the functional pathways common to *MBD5* and *RAI1*.

From previously published microarray data [19] where *MBD5* and *RAI1* were knocked down through siRNA technology individually in neuroblastoma cell lines, we conducted Ingenuity Pathway Analysis (IPA) on these microarray data to identify the common pathways that are perturbed when *MBD5* and *RAI1* are haploinsufficient. We propose that these pathways could be key contributors to the overlapping phenotypes we see in both syndromes. IPA was used to determine which biological pathways and function involving genes were differentially expressed in *MBD5* and *RAI1* knockdown SH-SY5Y cell lines. A *p*-value of less than 0.05 for each pathway was determined using Fisher's exact tests to determine the likelihood of those genes assigned by chance. IPA shows that *MBD5* and *RAI1* are implicated in many neurological, cell growth and developmental pathways ($p < 0.05$) (Tables 3 and 4). The top common pathways reveal many interesting and important pathways altered due to *MBD5* and *RAI1* haploinsufficiency. Previously mentioned pathways that are associated with autism and sleep, circadian rhythm signaling, and mTOR signaling pathways are present (Tables 3 and 4) [19]. The CDK5 signaling pathway is crucial to neuronal activity, neuronal migration during development and neurite growth [48]. In previous mouse studies of Mbd5, reduced neurite outgrowth was observed [49]. Further, the *Xenopus laevis* rai1 morphants exhibit aberrant neural crest migration [50]. Thus, the CDK5 signaling pathway could be contributing to the features seen in the above studies. The apoptosis signaling pathway (see Table 4) is involved in the determination of the size and shape of the brain and it regulates the wiring of developing neuronal networks [51]. It is thought that dysregulation of this pathway can lead to neuroanatomic abnormalities and developmental disabilities [51]. There are growing associations between neural cell death and autism [51]. Thus, the involvement of the apoptosis signaling pathway in 2q23.1 deletion syndrome and Smith-Magenis syndrome should be further investigated. Overall, extensive molecular studies on how these shared pathways contribute to the phenotypes present in 2q23.1 deletion syndrome and Smith-Magenis syndrome is necessary. In addition, these results provide further evidence that monogenic neurodevelopmental disorders associated with ASD share common pathways and are crucial to study in regards to phenotypic development.

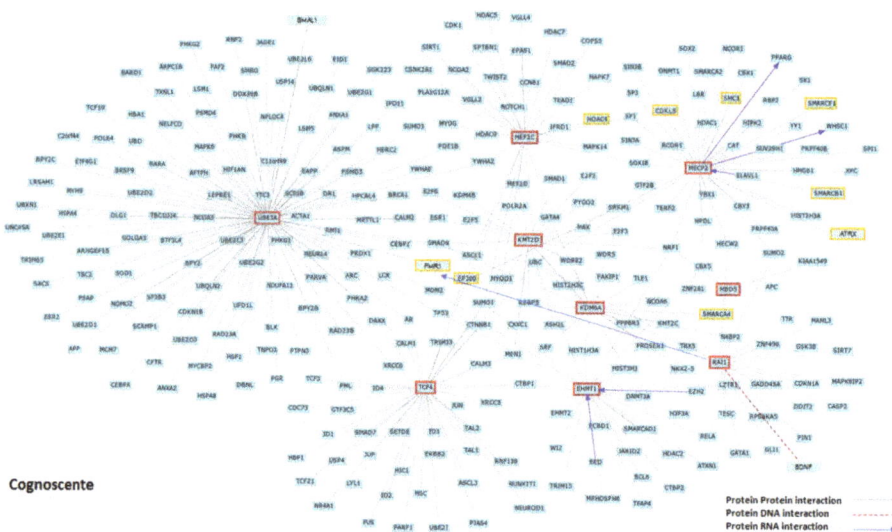

Figure 3. 2q23.1 deletion syndrome gene interaction network. Cognoscente was used to generate direct protein-protein (grey line), protein-DNA (red dash line), and protein-RNA (blue arrow line) interactions of genes previously implicated in ASD: *MBD5*, *MECP2*, *UBE3A*, *TCF4*, *MEF2C*, *EHMT1*, *KMT2D*, *KDM6A* and *RAI1* (red box). The human orthologs of genes of many of the protein products that came up in this network are involved in neuronal functions, chromatin modification, methylation and transcriptional regulation. In addition, other genes associated with ASD also were identified (yellow box).

Table 3. Common neurological pathways in both *MBD5* and *RAI1* knockdown SH-SY5Y cells.

Ingenuity Canonical Pathways *
Acetate conversion to acetyl-CoA
Agrin interactions at neuromuscular junction
Aldosterone signaling in epithelial cells
Amyotrophic lateral sclerosis signaling
Axonal guidance signaling
Cyclin-dependent kinsase 5 signaling
Choline degradation I
Circadian rhythm signaling
Ciliary neurotrophic factor signaling
Docosahexaenoic acid signaling

Table 3. *Cont.*

Ingenuity Canonical Pathways *
Dolichol and dolichyl phosphate biosynthesis
Eukaryotic initiation factor 2 signaling
Ephrin A signaling
Ephrin B signaling
Ephrin receptor signaling
Epidermal growth factor receptor signaling
G protein signaling mediated by Tubby
Gap junction signaling
Glial cell line-derived neurotrophic factors family ligand-receptor interactions
Glioma signaling
Melanoma signaling
Mechanistic target of rapamycin signaling
Netrin signaling
Neuregulin signaling

* All pathways had a *p*-value <0.05 as determined by Fisher Exact Test within Ingenuity Pathway Analysis.

Table 4. Common cell growth and developmental pathways in both *MBD5* and *RAI1* knockdown SH-SY5Y cells.

Ingenuity Canonical Pathways *
14-3-3-mediated signaling
1D-myo-inositol hexakisphosphate biosynthesis II (mammalian)
Acetate conversion to acetyl-CoA
Agrin interactions at neuromuscular junction
Aldosterone signaling in epithelial cells
Angiopoietin signaling
Apoptosis signaling
Aryl hydrocarbon receptor signaling
Assembly of RNA polymerase II complex
Assembly of RNA polymerase III complex
Ataxia telangiectasia mutated signaling
Bone morphogenetic protein signaling
CD40 signaling
Cell division control protein 42 homolog signaling
Cytidine diphosphate diacylglycerol-diacylglycerol biosynthesis I
Cell cycle control of chromosomal replication
Cell Cycle: G1/S checkpoint regulation
Cell Cycle: G2/M DNA damage checkpoint regulation
Ceramide signaling
Cholecystokinin/gastrin-mediated signaling
Choline degradation I
Circadian rhythm signaling
Citrulline-nitric oxide cycle
Clathrin-mediated endocytosis signaling

* All pathways had a *p*-value <0.05 as determined by Fisher Exact Test within Ingenuity Pathway Analysis.

3. Experimental Section

3.1. Patients and Cell Culture Studies

All samples and information were collected after informed consent was obtained and in accordance with Institutional Review Board (IRB) approved protocols from Baylor College of Medicine. Lymphoblastoid cell lines (LCLs) (Epstein–Barr virus-transformed human lymphocytes) were cultured as previously described [25]. LCLs utilized in this study included: 2q23.1 deletion syndrome

(SMS367 [25], SMS185 [25], SMS361 [13], SMS373 [13], SMS375 [13], and SMS368 [13] and five normal control lines).

3.2. Expression Analyses

RNA was isolated from cultured patient and control lymphoblastoid cell lines via TRIzol (Invitrogen, Carlsbad, CA, USA) according to standard protocols. All cell lines were cultured for the same period of time to the same cell density. RNA was quantified using the NanoDrop® ND-100 Spectrophotometer (NanoDrop Technologies, Inc., Wilmington, DE, USA). First-strand cDNA synthesis was carried out using qSCRIPT cDNA SuperMix (Quanta Biosciences, Inc., Gaithersburg, MD, USA) (with 1 μg of RNA) according to the manufacturer's protocol. For quantitative real-time PCR, predesigned Taqman MGB probes from Assays-on-Demand Gene Expression Products (ABI) (Life Technologies Inc., Carlsbad, CA, USA) were used for all genes. A minimum of three unrelated patient samples were run in triplicate in 10 μL reaction volumes. PCR conditions were the default settings of the ABI Prism 7900 HT Sequence Detection System (Life Technologies Inc., Carlsbad, CA, USA). The cycle threshold (C_t) was determined during the geometric phase of the PCR amplification plots as recommended by the manufacturer. Relative differences in transcript levels were quantified with the ΔΔCt method with *GAPDH* (MIM138400) (Hs99999905_m1) mRNA as an endogenous control. All expression values were calculated relative to control levels set at 1.0.

3.3. Network Analysis

Cognoscente (http://vanburenlab.tamhsc.edu/) was used to identify the biomolecular interactions that have been documented in the literature of the genes listed in Table 2. Cognoscente is a highly curated freely available database that identifies interacting components of protein networks and the primary literature to support such interactions [47].

3.4. Ingenuity Pathway Analysis (IPA)

IPA was used to identify biological functions, gene networks and pathways, and likely upstream regulators that were significantly more altered in knockdown cells than in controls. Significant interactions were determined using the Ingenuity Pathway Knowledge Base and a Fisher's exact test to calculate a *p*-value determining the probability that each function network or pathway assigned to that data set is due to chance alone. Statistical significance was determined at a cutoff of $p \leq 0.05$.

3.5. Statistical Analyses

Statistical analysis for gene expression data was performed with Prism 4 version 4.0b (GraphPad Software, Inc., San Diego, CA, USA). Statistical significance was determined at a cutoff of $p \leq 0.05$.

4. Conclusions

In conclusion, we show that 2q23.1 deletion syndrome shares common neurological and behavioral phenotypes with other monogenic neurodevelopmental disorders associated with autism spectrum disorders. Furthermore, haploinsufficiency of *MBD5* impacts expression of several ASD-implicated genes, including *UBE3A*, *TCF4 MEF2C*, *EHMT1*, and *RAI1*; supporting MBD5 as a transcriptional regulator. Further, data suggest that these genes associated with ASD, *MBD5*, *MECP2*, *UBE3A*, *TCF4 MEF2C*, *EHMT1*, *KMT2D* and *KDM6A*, and *RAI1* may be part of a gene network that converges into common or overlapping pathways that results in similar phenotypes when perturbed. As an example, we show that when *MBD5* and *RAI1* are both haploinsufficient they share common perturbed pathways that likely contribute to the overlapping phenotypes exhibited by 2q23.1 deletion syndrome and Smith-Magenis syndrome patients. Overall, future studies identifying and dissecting the genetic and downstream molecular pathways in monogenic neurodevelopmental disorders associated with

ASD may reveal common points of regulation and promote gene candidates for targeted therapeutic and pharmacological intervention.

Acknowledgments: We are deeply indebted to the study participants and their families, as well as support from the Fondation Jérôme Lejeune. This work was supported in part by the resources from Baylor College of Medicine, the Jan and Dan Duncan Neurological Research Institute, and the Texas Children's Hospital. Li Chen was funded by the National Science Foundation-China (NSFC) grant 31200937 and Shanghai Health and Family Planning Commission grant 20144Y0106.

Author Contributions: Sureni V. Mullegama conceived the study, conducted the experiments, analyzed the data and wrote the manuscript. Joseph T. Alaimo and Li Chen helped create figures and critically edited the manuscript. Sarah H. Elsea is the principal investigator and critically edited the manuscript.

Conflicts of Interest: The authors declare no conflicts of interest.

References

1. Spooren, W.; Lindemann, L.; Ghosh, A.; Santarelli, L. Synapse dysfunction in autism: A molecular medicine approach to drug discovery in neurodevelopmental disorders. *Trends Pharmacol. Sci.* **2012**, *33*, 669–684. [CrossRef] [PubMed]

2. Ceroni, F.; Sagar, A.; Simpson, N.H.; Gawthrope, A.J.; Newbury, D.F.; Pinto, D.; Francis, S.M.; Tessman, D.C.; Cook, E.H.; Monaco, A.P.; *et al.* A deletion involving CD38 and BST1 results in a fusion transcript in a patient with autism and asthma. *Autism Res.* **2014**, *7*, 254–263. [CrossRef] [PubMed]

3. Pendergrass, S.; Girirajan, S.; Selleck, S. Uncovering the etiology of autism spectrum disorders: Genomics, bioinformatics, environment, data collection and exploration, and future possibilities. *Pac. Symp. Biocomput.* **2014**, 422–426.

4. De Rubeis, S.; He, X.; Goldberg, A.P.; Poultney, C.S.; Samocha, K.; Cicek, A.E.; Kou, Y.; Liu, L.; Fromer, M.; Walker, S.; *et al.* Synaptic, transcriptional and chromatin genes disrupted in autism. *Nature* **2014**, *515*, 209–215. [CrossRef] [PubMed]

5. Marshall, C.R.; Noor, A.; Vincent, J.B.; Lionel, A.C.; Feuk, L.; Skaug, J.; Shago, M.; Moessner, R.; Pinto, D.; Ren, Y.; *et al.* Structural variation of chromosomes in autism spectrum disorder. *Am. J. Hum. Genet.* **2008**, *82*, 477–488. [CrossRef] [PubMed]

6. Betancur, C. Etiological heterogeneity in autism spectrum disorders: More than 100 genetic and genomic disorders and still counting. *Brain Res.* **2011**, *1380*, 42–77. [CrossRef] [PubMed]

7. Blumenthal, I.; Ragavendran, A.; Erdin, S.; Klei, L.; Sugathan, A.; Guide, J.R.; Manavalan, P.; Zhou, J.Q.; Wheeler, V.C.; Levin, J.Z.; *et al.* Transcriptional consequences of 16p11.2 deletion and duplication in mouse cortex and multiplex autism families. *Am. J. Hum. Genet.* **2014**, *94*, 870–883. [CrossRef] [PubMed]

8. Pasciuto, E.; Bagni, C. Snapshot: Fmrp interacting proteins. *Cell* **2014**, *159*, 218–218.e1. [CrossRef] [PubMed]

9. Van Bokhoven, H. Genetic and epigenetic networks in intellectual disabilities. *Annu. Rev. Genet.* **2011**, *45*, 81–104. [CrossRef] [PubMed]

10. Urdinguio, R.G.; Sanchez-Mut, J.V.; Esteller, M. Epigenetic mechanisms in neurological diseases: Genes, syndromes, and therapies. *Lancet Neurol.* **2009**, *8*, 1056–1072. [CrossRef] [PubMed]

11. Oti, M.; Huynen, M.A.; Brunner, H.G. Phenome connections. *Trends Genet.* **2008**, *24*, 103–106. [CrossRef] [PubMed]

12. Vissers, L.E.; de Vries, B.B.; Osoegawa, K.; Janssen, I.M.; Feuth, T.; Choy, C.O.; Straatman, H.; van der Vliet, W.; Huys, E.H.; van Rijk, A.; *et al.* Array-based comparative genomic hybridization for the genomewide detection of submicroscopic chromosomal abnormalities. *Am. J. Hum. Genet.* **2003**, *73*, 1261–1270. [CrossRef] [PubMed]

13. Talkowski, M.E.; Mullegama, S.V.; Rosenfeld, J.A.; van Bon, B.W.; Shen, Y.; Repnikova, E.A.; Gastier-Foster, J.; Thrush, D.L.; Kathiresan, S.; Ruderfer, D.M.; *et al.* Assessment of 2q23.1 microdeletion syndrome implicates MBD5 as a single causal locus of intellectual disability, epilepsy, and autism spectrum disorder. *Am. J. Hum. Genet.* **2011**, *89*, 551–563. [CrossRef] [PubMed]

14. Laget, S.; Joulie, M.; le Masson, F.; Sasai, N.; Christians, E.; Pradhan, S.; Roberts, R.J.; Defossez, P.A. The human proteins MBD5 and MBD6 associate with heterochromatin but they do not bind methylated DNA. *PLoS ONE* **2010**, *5*, e11982. [CrossRef] [PubMed]

15. Bonnet, C.; Ali Khan, A.; Bresso, E.; Vigouroux, C.; Beri, M.; Lejczak, S.; Deemer, B.; Andrieux, J.; Philippe, C.; Moncla, A.; *et al.* Extended spectrum of mbd5 mutations in neurodevelopmental disorders. *Eur. J. Hum. Genet.* **2013**, *21*, 1457–1461. [CrossRef] [PubMed]

16. Cukier, H.N.; Lee, J.M.; Ma, D.; Young, J.I.; Mayo, V.; Butler, B.L.; Ramsook, S.S.; Rantus, J.A.; Abrams, A.J.; Whitehead, P.L.; *et al.* The expanding role of mbd genes in autism: Identification of a mecp2 duplication and novel alterations in mbd5, mbd6, and setdb1. *Autism Res.* **2012**, *5*, 385–397. [CrossRef] [PubMed]

17. Du, Y.; Liu, B.; Guo, F.; Xu, G.; Ding, Y.; Liu, Y.; Sun, X.; Xu, G. The essential role of MBD5 in the regulation of somatic growth and glucose homeostasis in mice. *PLoS ONE* **2012**, *7*, e47358. [CrossRef] [PubMed]

18. Ladha, S. Getting to the bottom of autism spectrum and related disorders: MBD5 as a key contributor. *Clin. Genet.* **2012**, *81*, 363–364. [CrossRef] [PubMed]

19. Mullegama, S.V.; Pugliesi, L.; Burns, B.; Shah, Z.; Tahir, R.; Gu, Y.; Nelson, D.L.; Elsea, S.H. MBD5 haploinsufficiency is associated with slee disturbance and disrupts circadian pathways common to smith-magenis and fragile X syndromes. *Eur J. Hum. Genet.* **2014**. [CrossRef]

20. Hodge, J.C.; Mitchell, E.; Pillalamarri, V.; Toler, T.L.; Bartel, F.; Kearney, H.M.; Zou, Y.S.; Tan, W.H.; Hanscom, C.; Kirmani, S.; *et al.* Disruption of mbd5 contributes to a spectrum of psychopathology and neurodevelopmental abnormalities. *Mol. Psychiatry* **2014**, *19*, 368–379. [CrossRef] [PubMed]

21. Du, X.; An, Y.; Yu, L.; Liu, R.; Qin, Y.; Guo, X.; Sun, D.; Zhou, S.; Wu, B.; Jiang, Y.H.; *et al.* A genomic copy number variant analysis implicates the mbd5 and hnrnpu genes in chinese children with infantile spasms and expands the clinical spectrum of 2q23.1 deletion. *BMC Med. Genet.* **2014**, *15*, 62. [CrossRef] [PubMed]

22. Lund, C.; Brodtkorb, E.; Rosby, O.; Rodningen, O.K.; Selmer, K.K. Copy number variants in adult patients with lennox-gastaut syndrome features. *Epilepsy Res.* **2013**, *105*, 110–117. [CrossRef] [PubMed]

23. Shichiji, M.; Ito, Y.; Shimojima, K.; Nakamu, H.; Oguni, H.; Osawa, M.; Yamamoto, T. A cryptic microdeletion including MBD5 occurring within the breakpoint of a reciprocal translocation between chromosomes 2 and 5 in a patient with developmental delay and obesity. *Am. J. Med. Genet. A* **2013**, *161*, 850–855. [CrossRef]

24. Girirajan, S.; Dennis, M.Y.; Baker, C.; Malig, M.; Coe, B.P.; Campbell, C.D.; Mark, K.; Vu, T.H.; Alkan, C.; Cheng, Z.; *et al.* Refinement and discovery of new hotspots of copy-number variation associated with autism spectrum disorder. *Am. J. Hum. Genet.* **2013**, *92*, 221–237. [CrossRef] [PubMed]

25. Williams, S.R.; Mullegama, S.V.; Rosenfeld, J.A.; Dagli, A.I.; Hatchwell, E.; Allen, W.P.; Williams, C.A.; Elsea, S.H. Haploinsufficiency of mbd5 associated with a syndrome involving microcephaly, intellectual disabilities, severe speech impairment, and seizures. *Eur. J. Hum. Genet.* **2010**, *18*, 436–441. [CrossRef]

26. Van Bon, B.W.; Koolen, D.A.; Brueton, L.; McMullan, D.; Lichtenbelt, K.D.; Ades, L.C.; Peters, G.; Gibson, K.; Moloney, S.; Novara, F.; *et al.* The 2q23.1 microdeletion syndrome: Clinical and behavioural phenotype. *Eur J. Hum. Genet.* **2010**, *18*, 163–170.

27. Noh, G.J.; Graham, J.M., Jr. 2q23.1 microdeletion of the MBD5 gene in a female with seizures, developmental delay and distinct dysmorphic features. *Eur. J. Med. Genet.* **2012**, *55*, 59–62. [CrossRef] [PubMed]

28. Motobayashi, M.; Nishimura-Tadaki, A.; Inaba, Y.; Kosho, T.; Miyatake, S.; Niimi, T.; Nishimura, T.; Wakui, K.; Fukushima, Y.; Matsumoto, N.; *et al.* Neurodevelopmental features in 2q23.1 microdeletion syndrome: Report of a new patient with intractable seizures and review of literature. *Am. J. Med. Genet. A* **2012**, *158*, 861–868. [CrossRef]

29. Chung, B.H.; Stavropoulos, J.; Marshall, C.R.; Weksberg, R.; Scherer, S.W.; Yoon, G. 2q23 de novo microdeletion involving the mbd5 gene in a patient with developmental delay, postnatal microcephaly and distinct facial features. *Am. J. Med. Genet. A* **2011**, *155*, 424–429. [CrossRef]

30. Wagenstaller, J.; Spranger, S.; Lorenz-Depiereux, B.; Kazmierczak, B.; Nathrath, M.; Wahl, D.; Heye, B.; Glaser, D.; Liebscher, V.; Meitinger, T.; *et al.* Copy-number variations measured by single-nucleotide-polymorphism oligonucleotide arrays in patients with mental retardation. *Am. J. Hum. Genet.* **2007**, *81*, 768–779. [CrossRef] [PubMed]

31. Tan, W.H.; Bird, L.M.; Thibert, R.L.; Williams, C.A. If not angelman, what is it? A review of angelman-like syndromes. *Am. J. Med. Genet. A* **2014**, *164*, 975–992. [CrossRef]

32. Kleefstra, T.; Kramer, J.M.; Neveling, K.; Willemsen, M.H.; Koemans, T.S.; Vissers, L.E.; Wissink-Lindhout, W.; Fenckova, M.; van den Akker, W.M.; Kasri, N.N.; *et al.* Disruption of an ehmt1-associated chromatin-modification module causes intellectual disability. *Am. J. Hum. Genet.* **2012**, *91*, 73–82. [CrossRef] [PubMed]

33. Smeets, E.E.; Pelc, K.; Dan, B. Rett syndrome. *Mol. Syndromol.* **2012**, *2*, 113–127. [CrossRef] [PubMed]

34. Neul, J.L. The relationship of Rett syndrome and MECP2 disorders to autism. *Dialogues Clin. Neurosci.* **2012**, *14*, 253–262. [PubMed]

35. Kyllerman, M. Angelman syndrome. *Handb. Clin. Neurol.* **2013**, *111*, 287–290. [PubMed]

36. Whalen, S.; Heron, D.; Gaillon, T.; Moldovan, O.; Rossi, M.; Devillard, F.; Giuliano, F.; Soares, G.; Mathieu-Dramard, M.; Afenjar, A.; *et al.* Novel comprehensive diagnostic strategy in Pitt-hopkins syndrome: Clinical score and further delineation of the TCF4 mutational spectrum. *Hum. Mutat.* **2012**, *33*, 64–72. [CrossRef] [PubMed]

37. Marangi, G.; Ricciardi, S.; Orteschi, D.; Tenconi, R.; Monica, M.D.; Scarano, G.; Battaglia, D.; Lettori, D.; Vasco, G.; Zollino, M. Proposal of a clinical score for the molecular test for Pitt-hopkins syndrome. *Am. J. Med. Genet. A* **2012**, *158*, 1604–1611. [CrossRef]

38. Chung, B.H.; Mullegama, S.; Marshall, C.R.; Lionel, A.C.; Weksberg, R.; Dupuis, L.; Brick, L.; Li, C.; Scherer, S.W.; Aradhya, S.; *et al.* Severe intellectual disability and autistic features associated with microduplication 2q23.1. *Eur. J. Hum. Genet.* **2012**, *20*, 398–403. [CrossRef] [PubMed]

39. Mullegama, S.V.; Rosenfeld, J.A.; Orellana, C.; van Bon, B.W.; Halbach, S.; Repnikova, E.A.; Brick, L.; Li, C.; Dupuis, L.; Rosello, M.; *et al.* Reciprocal deletion and duplication at 2q23.1 indicates a role for MBD5 in autism spectrum disorder. *Eur. J. Hum. Genet.* **2014**, *22*, 57–63. [CrossRef] [PubMed]

40. Novara, F.; Rizzo, A.; Bedini, G.; Girgenti, V.; Esposito, S.; Pantaleoni, C.; Ciccone, R.; Sciacca, F.L.; Achille, V.; Della Mina, E.; *et al.* Mef2c deletions and mutations versus duplications: A clinical comparison. *Eur. J. Med. Genet.* **2013**, *56*, 260–265. [CrossRef] [PubMed]

41. Willemsen, M.H.; Vulto-van Silfhout, A.T.; Nillesen, W.M.; Wissink-Lindhout, W.M.; van Bokhoven, H.; Philip, N.; Berry-Kravis, E.M.; Kini, U.; van Ravenswaaij-Arts, C.M.; Delle Chiaie, B.; *et al.* Update on Kleefstra syndrome. *Mol. Syndromol.* **2012**, *2*, 202–212. [PubMed]

42. Banka, S.; Lederer, D.; Benoit, V.; Jenkins, E.; Howard, E.; Bunstone, S.; Kerr, B.; McKee, S.; Lloyd, I.C.; Shears, D.; *et al.* Novel KDM6A (UTX) mutations and a clinical and molecular review of the X-linked Kabuki syndrome (KS2). *Clin. Genet.* **2015**, *87*, 252–258. [CrossRef] [PubMed]

43. Bogershausen, N.; Wollnik, B. Unmasking Kabuki syndrome. *Clin. Genet.* **2013**, *83*, 201–211. [CrossRef] [PubMed]

44. Elsea, S.H.; Girirajan, S. Smith-magenis syndrome. *Eur. J. Hum. Genet.* **2008**, *16*, 412–421. [CrossRef] [PubMed]

45. Goh, K.I.; Cusick, M.E.; Valle, D.; Childs, B.; Vidal, M.; Barabasi, A.L. The human disease network. *Proc. Natl. Acad. Sci. USA* **2007**, *104*, 8685–8690. [CrossRef] [PubMed]

46. Goh, K.I.; Choi, I.G. Exploring the human diseasome: The human disease network. *Brief. Funct. Genomics* **2012**, *11*, 533–542. [CrossRef] [PubMed]

47. Jupiter, D.C.; VanBuren, V. A visual data mining tool that facilitates reconstruction of transcription regulatory networks. *PLoS ONE* **2008**, *3*, e1717. [CrossRef] [PubMed]

48. Duhr, F.; Deleris, P.; Raynaud, F.; Seveno, M.; Morisset-Lopez, S.; Mannoury la Cour, C.; Millan, M.J.; Bockaert, J.; Marin, P.; Chaumont-Dubel, S. Cdk5 induces constitutive activation of 5-HT$_6$ receptors to promote neurite growth. *Nat. Chem. Biol.* **2014**, *10*, 590–597. [CrossRef] [PubMed]

49. Camarena, V.; Cao, L.; Abad, C.; Abrams, A.; Toledo, Y.; Araki, K.; Araki, M.; Walz, K.; Young, J.I. Disruption of mbd5 in mice causes neuronal functional deficits and neurobehavioral abnormalities consistent with 2q23.1 microdeletion syndrome. *EMBO Mol. Med.* **2014**, *6*, 1003–1015. [CrossRef] [PubMed]

50. Tahir, R.; Kennedy, A.; Elsea, S.H.; Dickinson, A.J. Retinoic acid induced-1 (Rai1) regulates craniofacial and brain development in *Xenopus*. *Mech. Dev.* **2014**, *133*, 91–104. [CrossRef] [PubMed]

51. Wei, H.; Alberts, I.; Li, X. The apoptotic perspective of autism. *Int. J. Dev. Neurosci.* **2014**, *36*, 13–18. [CrossRef] [PubMed]

International Journal of
Molecular Sciences

MDPI

Article

Examining the Overlap between Autism Spectrum Disorder and 22q11.2 Deletion Syndrome

Opal Ousley [1,*], A. Nichole Evans [2], Samuel Fernandez-Carriba [2,3], Erica L. Smearman [4], Kimberly Rockers [1,5], Michael J. Morrier [1], David W. Evans [6], Karlene Coleman [2] and Joseph Cubells [1,5]

[1] Emory Autism Center, Department of Psychiatry and Behavioral Sciences, Emory University School of Medicine, 1551 Shoup Court, Atlanta, GA 30322, USA; kimberly.rockers@gmail.com (K.R.); mmorrier@emory.edu (M.J.M.); jcubell@emory.edu (J.C.)
[2] Marcus Autism Center, Children's Healthcare of Atlanta, 1920 Briarcliff Road, Atlanta, GA 30329, USA; andrea.evans@choa.org (A.N.E.); samuel.fernandez-carriba@emory.edu (S.F.-C.); karlene.coleman@choa.org (K.C.)
[3] Marcus Autism Center, Department of Pediatrics, Emory University School of Medicine, 1920 Briarcliff Road, Atlanta, GA 30329, USA
[4] Department of Psychology, Emory University, Psychology and Interdisciplinary Studies (PAIS) Building, 36 Eagle Row, Atlanta, GA 30322, USA; esmearm@emory.edu
[5] Department of Human Genetics, Emory University School of Medicine, 615 Michael Street, Whitehead Biomedical Research Building, Suite 301, Atlanta, GA 30322, USA
[6] Department of Psychology, Bucknell University, 1 Dent Drive, Lewisburg, PA 17837, USA; dwevans@bucknell.edu
* Correspondence: oousley@emory.edu; Tel.: +1-404-727-8350

Academic Editor: Merlin G. Butler
Received: 1 March 2017; Accepted: 5 May 2017; Published: 18 May 2017

Abstract: 22q11.2 deletion syndrome (22q11.2DS) is a genomic disorder reported to associate with autism spectrum disorders (ASDs) in 15–50% of cases; however, others suggest that individuals with 22q11.2DS present psychiatric or behavioral features associated with ASDs, but do not meet full criteria for ASD diagnoses. Such wide variability in findings may arise in part due to methodological differences across studies. Our study sought to determine whether individuals with 22q11.2DS meet strict ASD diagnostic criteria using research-based guidelines from the Collaborative Programs of Excellence in Autism (CPEA), which required a gathering of information from three sources: the Autism Diagnostic Interview-Revised (ADI-R), the Autism Diagnostic Observational Schedule (ADOS), and a clinician's best-estimate diagnosis. Our study examined a cohort of children, adolescents, and young adults (*n* = 56) with 22q11.2DS, who were ascertained irrespective of parents' behavioral or developmental concerns, and found that 17.9% (*n* = 10) of the participants met CPEA criteria for an ASD diagnosis, and that a majority showed some level of social-communication impairment or the presence of repetitive behaviors. We conclude that strictly defined ASDs occur in a substantial proportion of individuals with 22q11.2DS, and recommend that all individuals with 22q11.2DS be screened for ASDs during early childhood.

Keywords: 22q11.2 deletion; autism; autism spectrum; diagnosis; copy number variation; CNV; Research Domain Criteria; RDoC

1. Introduction

22q11.2 deletion syndrome (22q11.2DS) is among a growing number of genomic disorders that associate with autism spectrum disorders (ASDs) [1]. 22q11.2DS, also referred to as DiGeorge syndrome, or velo-cardio-facial syndrome (VCFS), occurs in approximately 1/4000 live births [2–5], making it the most common recurrent copy-number variant (CNV) associated with developmental disorders described to date. 22q11.2DS arises from an interstitial deletion of up to 3 Mb of DNA on chromosome 22q11.2 and produces a diverse range of physical, behavioral, social, and neurocognitive impairments [2,5–7], which overlap with the social-communication impairments found among patients with ASD [7–9].

While previous studies have described both the presence of ASDs among individuals with 22q11.2DS and the presence of 22q11.2DS among individuals diagnosed with ASD, there is no consensus on the rate of co-occurrence [10]. Previous studies have reported that 15–50% of those with 22q11.2DS also have an ASD [8,9,11], and that between 0.3% and 1% of individuals with an ASD also have 22q11.2DS [12,13]—an approximately 10–40-fold increase relative to the general population. The substantial discrepancies in reported rates of ASDs in 22q11.2DS could reflect methodological differences across studies, including the use of varied subject ascertainment methods and diagnostic assessment approaches [8]. However, some researchers argue that, although ASD-like symptoms are present in individuals with 22q11.2DS, these individuals do not truly meet criteria for an ASD, and may represent misclassified persons with, for example, prodromal symptoms of schizophrenia [14,15].

This issue resembles the ongoing debate faced by Fragile X syndrome (FXS) researchers, on the association between ASD and FXS [16–18]. Although debate over the overlap between FXS and autism continues, the clinical evaluation of individuals with ASD now regularly includes molecular testing for FXS. Furthermore, basic research on FXS has led to the testing of novel psychopharmacological interventions for FXS that appear to impact ASD-related symptoms [19,20]. Thus, the value of understanding relationships among molecularly defined genetic and genomic disorders and ASD is abundantly clear.

Our study aims to clarify the association between 22q11.2DS and ASD by applying rigorous, research-based diagnostic methods applied to a cohort of individuals ascertained solely on the basis of a molecular diagnosis of 22q11.2DS. We sought to determine the proportion of individuals with 22q11.2DS who meet research criteria for an ASD diagnosis according to the guidelines set forth by the Collaborative Programs of Excellence in Autism (CPEA) Simons Simplex Collection study (Simons Foundation Autism Research Initiative) [21]. The CPEA guidelines require individuals to meet cutoff scores on gold standard observational and parent report measures (i.e., the Autism Diagnostic Interview-Revised (ADI-R) [22] and Autism Diagnostic Observation Schedule (ADOS) [23]) in conjunction with clinicians' best-estimate consensus diagnosis in order to receive a research-based diagnosis of ASD. We assessed 56 children, adolescents, and young adults with fluorescent in situ hybridization (FISH) confirmed 22q11.2DS, ascertained from a registry of all cases of 22q11.2DS diagnosed at Children's Healthcare of Atlanta since the mid-1990s. We hypothesized that we would identify strictly defined ASDs in a substantial proportion of patients with 22q11.2DS.

2. Results

We found that a substantial number of individuals with 22q11.2DS within our sample met strictly defined ASD (17.9%, $n = 10$; 2.3:1 male to female ratio), using rigorous research diagnostic criteria, and that a significant level of ASD-related symptoms occurred for a portion of our sample, even in the absence of an ASD diagnosis. An additional subset of individuals did not meet overall autism or ASD cutoffs on the standardized assessments. Details of these findings are outlined below and depicted in Figures 1–4.

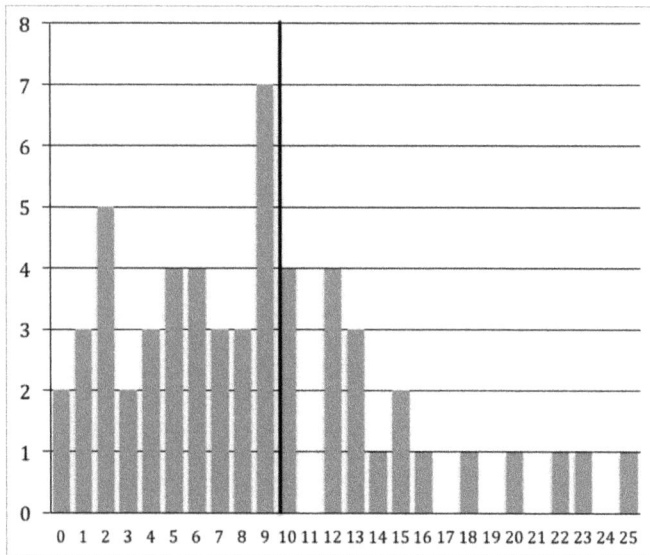

Figure 1. Figure 1 shows the broad range of Autism Diagnostic Interview-Revised (ADI-R) Reciprocal Social Interaction algorithm total scores (0–25; the autism spectrum disorder (ASD) cutoff score $\geq$ 10 is indicated by the black vertical line). The x-axis shows the algorithm score, and the y-axis indicates the number of individuals with each score.

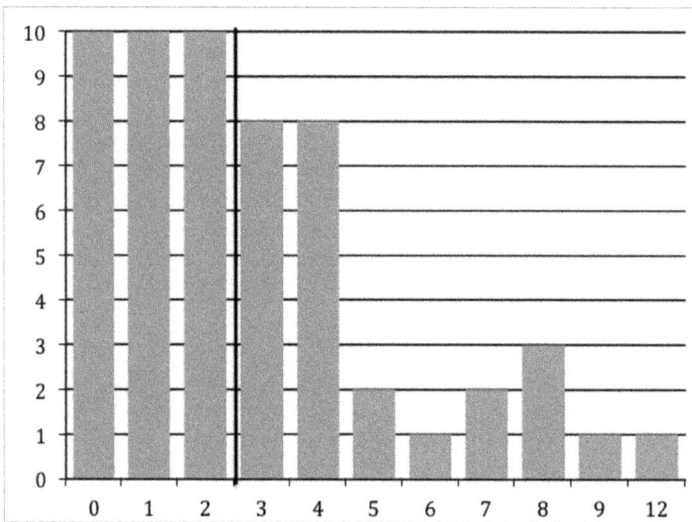

Figure 2. Figure 2 shows the broad range of ADI-R Restricted, Repetitive, and Stereotyped Behavior algorithm total scores (0–12; the ASD cutoff score $\geq$3 is indicated by the black vertical line). The x-axis shows the algorithm score, and the y-axis indicates the number of individuals with each score.

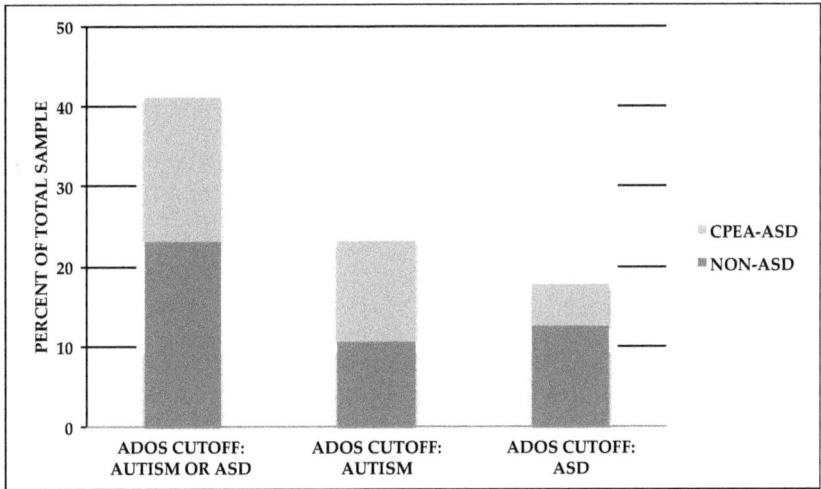

Figure 3. Figure 3 shows the percentages of individuals meeting the Autism Diagnostic Observational Schedule (ADOS) cutoff criteria. Within each cutoff category, diagnostic groups are designated as those who either met the Collaborative Programs of Excellence in Autism (CPEA) criteria for an ASD diagnosis (CPEA-ASD), or those who did not meet these criteria (non-ASD).

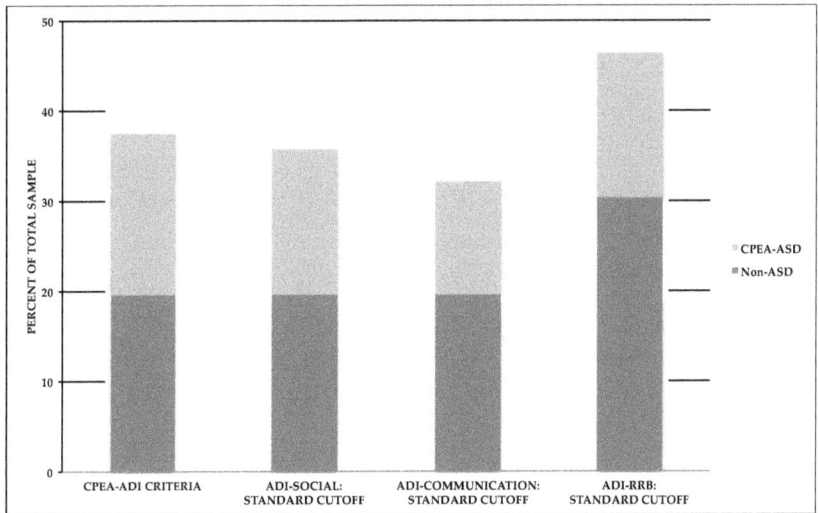

Figure 4. Figure 4 shows the percentages of individuals meeting ADI-R criteria set forth by the CPEA (CPEA-ADI) or standard ADI-R domain cutoff scores. Within each cutoff category, diagnostic groups are designated as those who met CPEA-ASD criteria or those who did not meet these criteria (non-ASD).

As described in the Materials and Methods section, we examined various cutoff scores for the Autism Diagnostic Observation Schedule (ADOS) and the Autism Diagnostic Interview-Revised (ADI-R), and determined who met CPEA criteria for an ASD diagnosis. A majority of individuals (*n* = 53) completed either ADOS Modules 3 or 4, which is given to children or adults, respectively, with language fluency. Three individuals completed either Module 1 (*n* = 2) or Module 2 (*n* = 1), which is given to individuals with no speech/single words or phrase speech, respectively. Examination of the

ADOS scores revealed that 23.2% (*n* = 13) of the participants met the autism cutoff score and 17.9% (*n* = 10) met the ASD cutoff score, for a total of 41.1% (*n* = 23). We also found that more than a third of the study participants (37.5%, *n* = 21) met ADI-R criteria as specified by the CPEA criteria, which follows either the standard ADI-R cutoff scoring parameters for autism (17.9%, *n* = 10) or a more flexible set of parameters (19.6%, *n* = 11; following Risi et al.; see Methods section below) [24]. An examination of each ADI-R domain score revealed that 35.7% (*n* = 20) met the ADI-R Reciprocal Social Interaction domain cutoff score, 32.1% (*n* = 18) met the ADI-R Communication domain cutoff score, and 46.4% (*n* = 26) met the Restricted, Repetitive, and Stereotyped Behavior domain cutoff score. Moreover, 66.1% (*n* = 37) of our total sample met cutoff scores for at least one of the social, communication, or repetitive behavior domains on the ADI-R. A wide range of algorithm scores was obtained for the ADI-R: Reciprocal Social Interaction range = 0–25, Communication range (verbal algorithm) = 0–18 (*n* = 53) and Communication range (nonverbal algorithm) = 0–12 (*n* = 3), and Restricted, Repetitive, and Stereotyped Behavior range = 0–12. The range of scores for the Reciprocal Social Interaction algorithm is shown below in Figure 1, whereas the range for the Restricted, Repetitive, and Stereotyped Behavior algorithm score is shown in Figure 2.

We also examined the presence of a strictly defined ASD using the CPEA guidelines, which required us to examine cutoff scores on the both the ADI-R and ADOS, in conjunction with best-estimate Diagnostic and Statistical Manual of Mental Disorders, Fourth Edition (DSM-IV) diagnosis [21,25]. We found that, of 56 participants, 17.9% (*n* = 10) met CPEA research-based criteria for a strictly defined ASD. That is, each participant met cutoff criteria for both the ADI-R and ADOS, and received a best-estimate diagnosis of a DSM-IV ASD. Our participants meeting CPEA criteria included individuals with a best-estimate diagnosis of DSM-IV Autistic Disorder or PDD-NOS. Although DSM-IV Asperger's Disorder would have qualified as a CPEA diagnosis, no participants in this sample met Asperger's diagnostic criteria. For those individuals who did not meet CPEA diagnostic criteria, two subsets emerged. In one subset (39.3%, *n* = 22), neither the ADI-R nor the ADOS overall cutoff scores were met; in the second subset (42.9%, *n* = 24), overall cutoff scores were met for at least one of the standardized instruments. Thus, even in the absence of a CPEA-ASD diagnosis, a significant minority of individuals met ADOS cutoff scores, or ADI-R cutoff scores, depicted in Figures 3 and 4.

3. Discussion

Debate continues regarding the association between 22q11.2DS and ASDs. While multiple studies show that individuals with 22q11.2DS present with ASD-like symptomatology, not all studies find that such individuals qualify for a formal diagnosis of an ASD [26]. Our study attempts to resolve these discrepant findings by considering ascertainment bias and drawing from a strict and thorough research-based definition of ASD, as outlined by CPEA guidelines, which require individuals to attain cutoff scores for an ASD from both the ADI-R and ADOS as well as a clinician's best estimate diagnosis of DSM-IV Autism, Asperger's Disorder, or PDD-NOS. Our findings showed that a substantial proportion of study participants with 22q11.2DS met research-based CPEA criteria for an ASD, supporting an association between 22q11.2DS and ASD. We also found in our sample that significant social and communication impairments, as well as restricted and repetitive behaviors, occurred in the absence of an ASD diagnosis.

Our finding that 17.9% of study participants with 22q11.2DS met research CPEA criteria for an ASD diagnosis falls within the range of prior studies reporting that 15–50% of individuals meet diagnostic criteria for an ASD [8,9,11]. To obtain a conservative estimate of ASD in individuals with 22q11.2DS, we chose (1) to use the CPEA research criteria; and (2) to recruit from a hospital and medical clinic database of individuals with 22q11.2DS, with a molecular diagnosis of the disorder as the only criterion for participation. We thus avoided the potential ascertainment bias inherent in studying subjects independently seeking behavioral, mental health, or neurodevelopmental evaluations. Importantly, while this is not an epidemiologic study and individuals in this sample may not be representative of all individuals with 22q11.2DS, a substantial portion of participants met

ASD criteria, suggesting the presence of an association. We also found that almost 40% of 22q11.2DS participants met neither the ADI-R nor the ADOS overall cutoffs for autism or ASD, but that more than 60% met at least one domain cutoff on the ADI-R.

These findings should be considered in light of the RDoC (Research Domain Criteria) initiative of the National Institutes of Mental Health [27]. Here, we applied a conservative threshold for ASD diagnosis. We recognize, however, that neurodevelopmental disorders are characterized by deviations and delays on dimensional traits that extend well into the general population [28,29]. Thus, understanding the morbidity of CNV syndromes requires an appreciation for the more subtle deviations and delays that may arise throughout development, reflecting, for example, a psychiatric prodrome. In some instances, even probands with de novo mutations that appear clinically unaffected nonetheless exhibit a "shift" in symptom expression relative to non-carrier first-degree relatives [30,31]. It is important therefore to bear in mind the impact of a CNV as it increases risk—both in terms of traditional clinical thresholds and categories, as well as dimensions of cognitive, social, motor, and behavioral phenotypes.

Our study's limitations include its cross-sectional design, which precludes examination of behavioral trajectories or emergence of ASD-related symptoms. Additionally, although the presence of co-morbid diagnoses was not the focus of this paper, individuals with 22q11.2DS are at risk for symptoms that extend beyond an ASD diagnosis, including schizophrenia spectrum disorder, depressive disorders, and anxiety disorders [21,32]. We do not yet know precisely how the presence of an ASD diagnosis or ASD symptoms in the absence of an ASD diagnosis affects the emergence of these additional disorders, and only longitudinal studies will allow us to understand these associations. Further, our study relied on the DSM-IV diagnostic system, which closely mirrors the International Classification of Diseases (ICD-10) published by the World Health Organization; however, we did not re-code our autism diagnoses based on the revised specifications for "autism spectrum disorder" outlined in the Diagnostic and Statistical Manual, Fifth Edition (DSM-5). We would predict that our findings would be similar using the DSM-5 criteria, which were developed in part based on prior research from the ADOS, ADI-R, and CPEA [21,23,24]. Future studies examining the various diagnostic classification systems are needed, however, to verify the stability of our findings.

Findings from this study have both clinical and scientific relevance. Because this study supports an association between 22q11.2DS and ASD, we recommend that individuals with 22q11.2DS (1) receive earlier evaluations for ASDs and (2) receive targeted interventions and therapies, such as pharmacological and behavioral approaches, which, when started earlier, may lead to more positive long-term outcomes [9,19]. Furthermore, clinicians evaluating or treating patients with ASDs should consider diagnostic testing for 22q11.2DS (and other CNV disorders), particularly when other associated features such as congenital heart defects, velo-palatal pathology, and immune-related difficulties are present. Awareness of the association between 22q11.2DS and ASDs may also lead to specific hypotheses regarding the etiology and final common pathways that increase risk for childhood social disability and ASDs [33,34].

4. Materials and Methods

4.1. Participants

4.1.1. Overview

This study evaluated 56 participants between the ages of 6–29 who were diagnosed with 22q11.2DS and who had participated in a larger study investigating neuropsychological and behavioral outcomes in 22q11.2DS [21,35]. All individuals in the study underwent neurocognitive screening and diagnostic assessments for ASD, which included the Autism Diagnostic Interview- Revised (ADI-R) [22] and the Autism Diagnostic Observation Schedule (ADOS) [23]. Data collection procedures were approved by the Emory University Institutional Review Board (#IRB00024756).

4.1.2. Recruitment and Eligibility

We recruited study participants from a children's 22q11.2DS medical clinic and hospital case registry maintained at the Children's Healthcare of Atlanta, which also includes adult subjects who were first diagnosed with 22q11.2DS in a congenital heart defects follow-up clinic. These clinics serve all of metro Atlanta and pull from a large regional catchment area in the Southeastern United States. For all study participants, the diagnosis of 22q11.2DS was identified by FISH using the standard *TUPLE* probe (LSI TUPLE1(22q11.2)/ARSA(22q13.3); Abbott Molecular #05J21-028; Abbott Park, IL, USA). This probe detects the common 3 Mb and 1.5 Mb deletions, which account for approximately 90% of the deletions observed in the clinically ascertained in the DiGeorge and velo-cardio-facial syndrome populations, but does not distinguish between these deletions.

Having a diagnosis of 22q11.2DS was the primary eligibility requirement for each study, and recruitment occurred without regard to the level of behavioral, developmental, or medical difficulties. The adolescent/adult participants were recruited for a longitudinal study of psychiatric symptoms in 22q11.2DS, and the child participants were recruited for a longitudinal study investigating language development. A majority of subjects ($n = 51$) were ascertained directly from this case registry, based on age eligibility. Five other participants were referred directly to the study from the Children's Healthcare of Atlanta 22q11.2DS multidisciplinary medical clinic or an adult congenital heart clinic, which follows children formerly served at Children's Healthcare of Atlanta. Additionally, no subjects in the study were from the same family.

4.1.3. Demographic Characteristics

Our 56 participants include an adolescent/adult group ($n = 32$; mean age = 19.21, SD = 4.12; range = 14–29) and a child group ($n = 24$; mean age = 9.58, SD = 1.91; range = 6–11). For the adult group, females comprise 53.13% ($n = 17$) of the group, while among children, females comprise 45.83% ($n = 11$). The majority of our adult participants are Caucasian (78.12%, $n = 25$) with three additional ethnicities represented: African-American (12.50%, $n = 4$), Hispanic (6.25%, $n = 2$), and Asian (3.12%, $n = 1$). Similarly among our child participants, Caucasian (70.83%, $n = 17$) is the majority, with four additional ethnicities represented: Hispanic (12.50%, $n = 3$), African-American (8.33%, $n = 2$), Asian (4.16%, $n = 1$), and bi-racial: African-American and Caucasian (4.16%, $n = 1$). Parent-reported medical, developmental and family history is provided in Table 1, which indicates a high rate of cardiac, immune, and palatal problems in the study participants. Parents also commonly reported a history of speech delay and ADHD in the study participants, although a history of known autism spectrum disorder was reported for only one participant. Parents reported ADHD as the most common disorder among family members (i.e., parents or siblings); only one family member was known to have 22q11.2DS.

For the adolescent/adult group, the estimated Verbal IQ (VIQ) was based on the Vocabulary and Similarities subtests from either the Wechsler Adult Intelligence Scale–Third Edition (WAIS-III) or the Wechsler Intelligence Scale for Children-Third Edition (WISC-III), depending on chronological age. To calculate the estimated VIQ, we converted the T-scores from these subtests to standard scores, and calculated their average (i.e., mean VIQ = 84.61, SD = 16.81; range = 57–120; $n = 32$). We also converted the WAIS-III or WISC-III Block Design T-score to a standard score to determine an estimated nonverbal IQ (NVIQ) (i.e., mean NVIQ = 73.29, SD = 14.96; range = 55–100; $n = 32$). For the child group, we used the Differential Ability Scales–Second Edition (DAS-II) Verbal score as the indicator of VIQ (i.e., mean VIQ = 72.87; SD = 17.60; range = 3–98; $n = 23$) and used the DAS-II Pattern Construction score to estimate NVIQ (i.e., mean NVIQ = 79.48, SD = 15.69; range = 43–102; $n = 22$). The sample sizes used to calculate the children's VIQ and NVIQ mean scores vary by one and two data points, respectively, due to missing data.

Table 1. Parent report of known medical, developmental, and family history.

	Child	Adult	Total	Family
	n = 24	*n* = 32	*n* = 56	
Mean age (Standard deviation)	9.58 (1.91)	19.21 (4.12)	15.1 (5.85)	-
Gender				
Female, n (percent, %)	11 (45.8)	17 (53.1)	28 (50.0)	-
Male, n (%)	13 (54.2)	15 (46.9)	28 (50.0)	-
Congenital heart defect				
n (%)	16 (66.7)	17 (53.1)	33 (58.9)	1 (1.8)
Immune deficiency				
n (%)	15 (62.5)	14 (43.8)	29 (51.8)	0 (0)
Palatal anomalies				
n (%)	11 (45.8)	19 (59.4)	30 (53.6)	0 (0)
Speech-language delay				
n (%)	17 (70.8)	15 (46.9)	32 (57.1)	8 (14.3)
Attention Deficit/Hyperactivity Disorder (ADHD)				
n (%)	11 (45.8)	7 (21.9)	18 (32.1)	12 (21.4)
Autism spectrum diagnosis (known prior to study entry)				
n (%)	0 (0)	1 (3.1)	1 (1.8)	2 (3.6)
22q11.2 deletion				
n (%)	24 (100)	32(100)	56 (100)	1 (1.8)

4.2. Diagnostic Procedures

We applied a strict, research-based definition of ASD, as designated by CPEA guidelines and outlined by the Simons Simplex Collection study (Simons Foundation Autism Research Initiative), previously described [21]. Essentially, these criteria require participants to attain cutoff scores for an autism spectrum disorder from all of the following sources: (1) the Autism Diagnostic Interview-Revised (ADI-R); (2) the Autism Diagnostic Observation Schedule (ADOS); and (3) the clinician's best estimate DSM-IV diagnosis, which considers a hierarchical classification strategy for identifying autism, Asperger's Disorder, or PDD-NOS [25]. To achieve the clinician's best estimate diagnosis, two clinicians conducted a case conference during which they reviewed all diagnostic and neurocognitive assessments available.

4.3. Diagnostic Assessments

The diagnostic assessments included the Autism Diagnostic Interview-Revised (ADI-R) and the Autism Diagnostic Observation Schedule (ADOS) [22,23]. The ADI-R is a semi-structured interview used to evaluate autistic symptomatology and, in particular, to differentiate between autism and other developmental disorders [22]. It is used for a wide range of ages, with a minimum mental age of 18 months. As such, both adults and children were assessed with the ADI-R. This interview evaluates three functional domains: Reciprocal Social Interaction, Communication skills, and Restricted, Repetitive, and Stereotyped behaviors. We classified participants as meeting CPEA-ADI-R cutoff criteria if they met either the standard ADI-R cutoff scores, or an additional set of cutoff scores suggested by Risi et al. [24]. The standard cutoff scores included (1) the total Social algorithm score $\geq$10; (2) the total (verbal) Communication algorithm score $\geq$8 or the total (nonverbal) Communication algorithm score $\geq$7; and (3) the total Restricted, Repetitive and Stereotyped behaviors algorithm score $\geq$3 [22]. ADI-R cutoff rule, as designated by the Collaborative Programs of Excellence in Autism (CPEA), required that one of the following conditions be met: (1) meeting standard cutoffs for both Social and Communication domains; (2) meeting the standard cutoff for the Social domain and within 2 points of the standard Communication domain cutoff; (3) meeting the standard cutoff for Communication domain and within

2 points of standard Social domain cutoff; or (4) meeting the standard cutoff within 1 point for both the Social and Communication domains [24].

The Autism Diagnostic Observation Schedule (ADOS) is a semi-structured assessment that uses tasks and play activities to elicit social communication and interaction [23,36]. Observations are made concerning social and communication behaviors associated with ASD. Its structure allows for diagnoses across all ages, developmental stages, and language abilities and includes four different modules (1–4) dependent on age, development, and language ability [36]. We used ADOS for both adults and children. For this study, we used the revised algorithms for Modules 1, 2, and 3 were used based on Gotham et al. [23], whereas the standard algorithm for Module 4 was used. Please note, although we used the first edition of the ADOS in combination with the revised algorithms, these revised algorithms are parallel to the cutoffs used for the second revision of the ADOS (i.e., ADOS-2).

Modules 1–3 consist of two subdomains, Social Affect and Restricted, Repetitive, and Stereotyped Behavior, which yield one collapsed score, with two cutoffs provided: an autism cutoff score and an ASD cutoff. For Module 4, scores are obtained for two subdomains, Communication and Social, in addition to a combined subdomain score, Communication plus Social. Two cutoff scores are provided for each domain: autism and ASD. Individuals must meet all three cutoff scores in order to receive an ADOS classification of either autism or ASD.

These cutoffs differ according to which module is used. Modules 1, 2, and 3 derive a Total raw score from the Social Affect algorithm score and the Restricted, Repetitive, and Stereotyped Behaviors algorithm score. Module 1 ($n = 2$) is for those without no speech or single words, who do not use phrase speech consistently (total raw score ≥ 16 = autism, ≥ 11 = ASD). Module 2 ($n = 1$) is for those who use phrase speech but are not verbally fluent (Total raw score ≥ 9 = autism, ≥ 8 = ASD). Module 3 ($n = 23$) is for verbally fluent children (Total raw score ≥ 9 = autism, ≥ 7 = ASD). Module 4 ($n = 30$) is for verbally fluent adults and adolescents. Scoring parameters for Module 4 include the following: Communication total raw score ≥ 3 = autism, ≥ 2 = ASD; Social total raw score ≥ 6 = autism, ≥ 4 = ASD; Communication and Social total raw score ≥ 10 = autism, ≥ 7 = ASD [23].

Acknowledgments: This work was supported in part by the Robert W. Woodruff Foundation, Predictive Medicine Grant, awarded to Cubells and Ousley; a Simons Foundation Junior Investigator Award and a NARSAD award to Ousley. The NARSAD award provided some funds for open access publishing. We would like to thank Karen Wallace (Emory Autism Center, Department of Psychiatry and Behavioral Sciences, Emory School of Medicine, Atlanta, GA, USA) for her assistance in recruiting families and organizing the study. In addition, we would like to express our gratitude to all children, adults, and families who participated in the study.

Author Contributions: Opal Ousley and Joseph Cubells conceived and designed the experiments; Kimberly Rockers, Samuel Fernandez-Carriba, and Michael J. Morrier performed the experiments; Opal Ousley, A. Nichole Evans, Samuel Fernandez-Carriba, Erica L. Smearman, Kimberly Rockers and Karlene Coleman contributed to the data analysis; Opal Ousley, A. Nichole Evans, David W. Evans and Joseph Cubells contributed to the interpretation of the data analysis; all authors contributed to writing and editing the paper.

Conflicts of Interest: The authors declare no conflict of interest. The founding sponsors had no role in the design of the study; in the collection, analyses, or interpretation of data; in the writing of the manuscript; or in the decision to publish the results.

Abbreviations

22q11.2DS	22q11.2 Deletion Syndrome
FISH	Fluorescent In Situ Hybridization
ASD	Autism Spectrum Disorder
ADOS	Autism Diagnostic Observation Schedule
ADI-R	Autism Diagnostic Interview-Revised
CNV	Copy Number Variation
PDD-NOS	Pervasive Developmental Disorder-Not Otherwise Specified
RDoC	Research Domain Criteria
DSM-IV	Diagnostic and Statistical Manual of Mental Disorders, Fourth edition

References

1. Artigas-Pallares, J.; Gabau-Vila, E.; Guitart-Feliubadalo, M. Syndromic autism: II. Genetic syndromes associated with autism. *Rev. Neurol.* **2005**, *40*, S151–S162. [PubMed]
2. Bassett, A.S.; McDonald-McGinn, D.M.; Devriendt, K.; Digilio, M.C.; Goldenberg, P.; Habel, A.; Vorstman, J. Practical guidelines for managing patients with 22q11.2 deletion syndrome. *J. Pediatr.* **2011**, *159*, 332–339. [CrossRef] [PubMed]
3. Botto, L.D.; May, K.; Fernhoff, P.M.; Correa, A.; Coleman, K.; Rasmussen, S.A.; Merritt, R.K.; O'Leary, L.A.; Wong, L.Y.; Elixson, E.M.; et al. A population-based study of the 22q11.2 deletion: Phenotype, incidence, and contribution to major birth defects in the population. *Pediatrics* **2003**, *112*, 101–107. [CrossRef] [PubMed]
4. Oskarsdottir, S.; Vujic, M.; Fasth, A. Incidence and prevalence of the 22q11 deletion syndrome: A population-based study in Western Sweden. *Arch. Dis. Child.* **2004**, *89*, 148–151. [CrossRef] [PubMed]
5. Tezenas Du Montcel, S.; Mendizabai, H.; Ayme, S.; Levy, A.; Philip, N. Prevalence of 22q11 microdeletion. *J. Med. Genet.* **1996**, *33*, 719. [CrossRef] [PubMed]
6. Gothelf, D.; Frisch, A.; Michaelovsky, E.; Weizman, A.; Shprintzen, R.J. Velo-Cardio-Facial Syndrome. *J. Ment. Health Res. Intellect. Disabil.* **2009**, *2*, 149–167. [CrossRef] [PubMed]
7. Ousley, O.; Rockers, K.; Dell, M.L.; Coleman, K.; Cubells, J.F. A review of neurocognitive and behavioral profiles associated with 22q11 deletion syndrome: Implications for clinical evaluation and treatment. *Curr. Psychiatry Rep.* **2007**, *9*, 148–158. [CrossRef] [PubMed]
8. Fine, S.E.; Weissman, A.; Gerdes, M.; Pinto-Martin, J.; Zackai, E.H.; McDonald-McGinn, D.M.; Emanuel, B.S. Autism spectrum disorders and symptoms in children with molecularly confirmed 22q11.2 deletion syndrome. *J. Autism Dev. Disord.* **2005**, *35*, 461–470. [CrossRef] [PubMed]
9. Vorstman, J.A.; Morcus, M.E.; Duijff, S.N.; Klaassen, P.W.; Heineman-de Boer, J.A.; Beemer, F.A.; van Engeland, H. The 22q11.2 deletion in children: High rate of autistic disorders and early onset of psychotic symptoms. *J. Am. Acad. Child Adolesc. Psychiatry* **2006**, *45*, 1104–1113. [CrossRef] [PubMed]
10. Bruining, H.; de Sonneville, L.; Swaab, H.; de Jonge, M.; Kas, M.; van Engeland, H.; Vorstman, J. Dissecting the clinical heterogeneity of autism spectrum disorders through defined genotypes. *PLoS ONE* **2010**, *5*, e10887. [CrossRef] [PubMed]
11. Antshel, K.M.; Aneja, A.; Strunge, L.; Peebles, J.; Fremont, W.P.; Stallone, K.; Kates, W.R. Autistic spectrum disorders in velo-cardio facial syndrome (22q11.2 deletion). *J. Autism Dev. Disord.* **2007**, *37*, 1776–1786. [CrossRef] [PubMed]
12. Bassett, A.S.; Costain, G.; Fung, W.L.; Russell, K.J.; Pierce, L.; Kapadia, R.; Carter, R.F.; Chow, E.W.; Forsythe, P.J. Clinically detectable copy number variations in a Canadian catchment population of schizophrenia. *J. Psychiatr. Res.* **2010**, *44*, 1005–1009. [CrossRef] [PubMed]
13. Levinson, D.F.; Duan, J.; Oh, S.; Wang, K.; Sanders, A.R.; Shi, J.; Zhang, N.; Mowry, B.J.; Olincy, A.; Amin, F.; et al. Copy number variants in schizophrenia: Confirmation of five previous findings and new evidence for 3q29 microdeletions and VIPR2 duplications. *Am. J. Psychiatry* **2011**, *168*, 302–316. [CrossRef] [PubMed]
14. Karayiorgou, M.; Simon, T.J.; Gogos, J.A. 22q11.2 microdeletions: Linking DNA structural variation to brain dysfunction and schizophrenia. *Nat. Rev. Neurosci.* **2010**, *11*, 402–416. [CrossRef] [PubMed]
15. Ogilvie, C.M.; Moore, J.; Daker, M.; Palferman, S.; Docherty, Z. Chromosome 22q11 deletions are not found in autistic patients identified using strict diagnostic criteria. *Am. J. Med. Genet.* **2000**, *96*, 15–17. [CrossRef]
16. Einfeld, S.; Molony, H.; Hall, W. Autism is not associated with the fragile X syndrome. *Am. J. Med. Genet.* **1989**, *34*, 187–193. [CrossRef] [PubMed]
17. Fisch, G.S. Is autism associated with the fragile X syndrome? *Am. J. Med. Genet.* **1992**, *43*, 47–55. [CrossRef] [PubMed]
18. Hall, S.S.; Lightbody, A.A.; Hirt, M.; Rezvani, A.; Reiss, A.L. Autism in fragile X syndrome: A category mistake? *J. Am. Acad. Child Adolesc. Psychiatry* **2010**, *49*, 921–933. [CrossRef] [PubMed]
19. McCary, L.M.; Roberts, J.E. Early identification of autism in fragile X syndrome: A review. *J. Intellect. Disabil. Res.* **2012**, *57*, 803–814. [CrossRef] [PubMed]
20. Schaefer, G.B.; Mendelsohn, N.J. Genetics evaluation for the etiologic diagnosis of autism spectrum disorders. *Genet. Med.* **2008**, *10*, 4–12. [CrossRef] [PubMed]

21. Ousley, O.Y.; Smearman, E.; Fernandez-Carriba, S.; Rockers, K.A.; Coleman, K.; Walker, E.F.; Cubells, J.F. Axis I psychiatric diagnoses in adolescents and young adults with 22q11 deletion syndrome. *Eur. Psychiatry* **2013**, *28*, 417–422. [CrossRef] [PubMed]

22. Lord, C.; Rutter, M.; Le Couteur, A. Autism diagnostic interview-revised: A revised version of a diagnostic interview for caregivers of individuals with possible pervasive developmental disorders. *J. Autism Dev. Disord.* **1994**, *24*, 659–685. [CrossRef] [PubMed]

23. Gotham, K.; Risi, S.; Pickles, A.; Lord, C. The autism diagnostic observation schedule: Revised algorithms for improved diagnostic validity. *J. Autism Dev. Disord.* **2007**, *7*, 613–627. [CrossRef] [PubMed]

24. Risi, S.; Lord, C.; Gotham, K.; Corsello, C.; Chrysler, C.; Szatmari, P.; Cook, E.H.; Leventhal, B.L.; Pickles, A. Combining information from multiple sources in the diagnosis of autism spectrum disorders. *J. Am. Acad. Child Adolesc. Psychiatry* **2006**, *45*, 1094–1103. [CrossRef] [PubMed]

25. American Psychiatric Association. *DSM-IV: Diagnostic and Statistical Manual*; American Psychiatric Association: Washington, DC, USA, 1994.

26. Wang, P.P.; Woodin, M.F.; Kreps-Falk, R.; Moss, E.M. Research on behavioral phenotypes: Velocardiofacial syndrome (deletion 22q11.2). *Dev. Med. Child Neurol.* **2000**, *42*, 422–427. [CrossRef] [PubMed]

27. Cuthbert, B.N.; Insel, T.R. Toward the future of psychiatric diagnosis: The seven pillars of RDoC. *BMC Med.* **2013**, *11*, 126. [CrossRef] [PubMed]

28. Constantino, J.N.; Todd, R.D. Autistic traits in the general population: A twin study. *Arch. Gen. Psychiatry* **2003**, *60*, 524–530. [CrossRef] [PubMed]

29. Evans, D.W.; Uljarevic, M.; Lusk, L.G.; Loth, E.; Frazier, T. Development of two dimensional measures of restricted and repetitive behaviors in parents and children. *J. Am. Acad. Child Adolesc. Psychiatry* **2017**, *56*, 51–58. [CrossRef] [PubMed]

30. Moreno-De-Luca, A.; Myers, S.M.; Challman, T.D.; Moreno-De-Luca, D.; Evans, D.W.; Ledbetter, D.H. Developmental brain dysfunction (DBD): Revival and expansion of an old concept based on new genetic evidence. *Lancet Neurol.* **2013**, *12*, 406–414. [CrossRef]

31. Moreno-De-Luca, A.; Evans, D.W.; Boomer, K.; Hanson, E.; Bernier, R.; Goin-Kochel, R.; Myers, S.M.; Challman, T.D.; Moreno-de-Luca, D.; Spiro, J.; et al. Parental cognitive, behavioral and motor profiles impact the neurodevelopmental profile of individuals with de novo mutations. *JAMA Psychiatry* **2015**, *72*, 119–126. [CrossRef] [PubMed]

32. Jonas, R.K.; Montojo, C.A.; Bearden, C.E. The 22q11.2 deletion syndrome as a window into complex neuropsychiatric disorders over the lifespan. *Biol. Psychiatry* **2014**, *75*, 351–360. [CrossRef] [PubMed]

33. Grayton, H.M.; Fernandes, C.; Rujescu, D.; Collier, D.A. Copy number variations in neurodevelopmental disorders. *Prog. Neurobiol.* **2012**, *99*, 81–91. [CrossRef] [PubMed]

34. Karam, C.S.; Ballon, J.S.; Bivens, N.M.; Freyberg, Z.; Girgis, R.R.; Lizardi-Ortiz, J.E.; Javitch, J.A. Signaling pathways in schizophrenia: Emerging targets and therapeutic strategies. *Trends Pharmacol. Sci.* **2010**, *31*, 381–390. [CrossRef] [PubMed]

35. Rockers, K.; Ousley, O.; Sutton, T.; Schoenberg, E.; Coleman, K.; Walker, E.; Cubells, J.F. Performance on the modified card sorting test and its relation to psychopathology in adolescents and young adults with 22q11.2 deletion syndrome. *J. Intellect. Disabil. Res.* **2009**, *53*, 665–676. [CrossRef] [PubMed]

36. Lord, C; Rutter, M.; DiLavore, P.; Risi, S. *Autism Diagnostic Observation Schedule Manual*; Western Psychological Services: Los Angeles, CA, USA, 1999.

Chapter II:
Genetics

International Journal of
Molecular Sciences

MDPI

Article

High-Resolution Chromosome Ideogram Representation of Currently Recognized Genes for Autism Spectrum Disorders

Merlin G. Butler *, Syed K. Rafi † and Ann M. Manzardo †

Departments of Psychiatry & Behavioral Sciences and Pediatrics, University of Kansas Medical Center, Kansas City, KS 66160, USA; rafigene@yahoo.com (S.K.R.); amanzardo@kumc.edu (A.M.M.)
* Author to whom correspondence should be addressed; mbutler4@kumc.edu;
    Tel.: +1-913-588-1873; Fax: +1-913-588-1305.
† These authors contributed to this work equally.

Academic Editor: William Chi-shing Cho
Received: 23 January 2015; Accepted: 16 March 2015; Published: 20 March 2015

Abstract: Recently, autism-related research has focused on the identification of various genes and disturbed pathways causing the genetically heterogeneous group of autism spectrum disorders (ASD). The list of autism-related genes has significantly increased due to better awareness with advances in genetic technology and expanding searchable genomic databases. We compiled a master list of known and clinically relevant autism spectrum disorder genes identified with supporting evidence from peer-reviewed medical literature sources by searching key words related to autism and genetics and from authoritative autism-related public access websites, such as the Simons Foundation Autism Research Institute autism genomic database dedicated to gene discovery and characterization. Our list consists of 792 genes arranged in alphabetical order in tabular form with gene symbols placed on high-resolution human chromosome ideograms, thereby enabling clinical and laboratory geneticists and genetic counsellors to access convenient visual images of the location and distribution of ASD genes. Meaningful correlations of the observed phenotype in patients with suspected/confirmed ASD gene(s) at the chromosome region or breakpoint band site can be made to inform diagnosis and gene-based personalized care and provide genetic counselling for families.

Keywords: high-resolution chromosome ideograms; autism; genetic evidence; autism spectrum disorders (ASD); ASD genes

---

1. Introduction

Classical autism or autistic disorder is common, with developmental difficulties noted by three years of age. It belongs to a group of heterogeneous conditions known as autism spectrum disorders (ASDs) with significant impairments in verbal and non-verbal communication and social interactions with restricted repetitive behaviors, specifically in movements and interests [1–3]. Other symptoms include lack of eye contact or focus, sleep disturbances and tactile defensiveness beginning at an early age. Several validated rating scales are used at a young age to help establish the diagnosis, including the autism diagnostic observation schedule (ADOS) and the autism diagnostic interview-revised (ADI-R) supported by pertinent medical history and clinical findings [4–6]. ASD affects about 1% of children in the general U.S. population with a 4:1 male to female ratio, usually without congenital anomalies or growth retardation [7,8].

Autism was first used as a term by Kanner in 1943 when describing a group of children lacking the ability to establish interpersonal contact and communication [9]. About one-fourth of children with autism are diagnosed by 2–3 years of age and show regression of skills in about 30% of cases. About 60% of ASD subjects show intellectual disabilities at a young age [10,11]. When comparing the prevalence of health disorders involving the central nervous system, autism ranks higher than epilepsy (6.5 cases per 1000), brain paralysis or dementia (2.5 cases/1000 for each) and Parkinson disease (two cases per 1000); genetic factors are related to many of these disorders [12,13]. Autism also occurs more commonly than congenital malformations in the general population, but dysmorphic findings are present in about 25% of children with autism. Microcephaly is seen in about 10% of cases, but macrocephaly is documented with larger frontal and smaller occipital lobes in about 20% of children with autism. Those with autism and extreme macrocephaly are at a greater risk to have *PTEN* tumor suppressor gene mutations [14], while another autism-related gene (*CHD8*) can also lead to macrocephaly and autism [15].

Autism is due to a wide range of genetic abnormalities, as well as non-genetic causes, including the environment, environmental and gene interaction (epigenetics) and metabolic disturbances (e.g., mitochondrial dysfunction), with the recurrence risk dependent on the family history and presence or absence of dysmorphic features. Candidate genes for ASD are identified by different means, including cytogenetic abnormalities (*i.e.*, translocations at chromosome breakpoints or deletions (e.g., the 22q11.2 deletion) indicating the location or loss of specific genes) in individuals with ASD along with overlapping linkage and functional data related to the clinical presentation, with certain chromosome regions identified by genetic linkage using DNA markers that co-inherit with the specific phenotype [16,17]. A representative example for such an occurrence is the proto-oncogene (MET) involved in pathways related to neuronal development [18] and found to be linked to the chromosome 7q31 band, where this gene is located. Decreased activity of the gene promoter was recognized when specific single nucleotide polymorphisms (SNPs) were present in this region by linkage studies. However, genetic linkage studies have received only limited success in the study of the genetics of autism. On the other hand, chromosomal microarray analysis using DNA probes disturbed across the genome can be used to detect chromosomal abnormalities at >100-times smaller than seen in high-resolution chromosome studies. Microarray studies have also become the first tier of genetic testing for this patient population and are recommended for all ASD patients [19]. Greater than 20% of studied patients with microarray analysis are found to have submicroscopic deletions or duplications in the genome containing genes that play a role in causing autism [20,21]. Identification of causative mutations is important to guide treatment selection and to manage medical co-morbidities, such as risks for seizures, developmental regression or for cancer (e.g., the *PTEN* gene).

Routine cytogenetic studies have shown abnormalities of chromosomes 2, 3, 4, 5, 7, 8, 11, 13, 15, 16, 17, 19, 22 and X, including deletions, duplications, translocations and inversions involving specific chromosome regions where known or candidate genes for ASD are located [22]. These studies further support the role of genetic factors in the causation of this common neurodevelopment disorder. Specifically, cytogenetic abnormalities involving the 15q11–q13 region are found in at least 1% of individuals with ASD and include *CYFIP1*, *GABRB3* and *UBE3A* genes in this chromosome region [23] and most recently the 15q11.2 BP1-BP2 microdeletion (Burnside-Butler) syndrome [24]. DNA copy number changes have also shown recurrent small deletions or duplications of the chromosome 16p11.2 band using microarray analysis [25,26] and the chromosome 15q13.2–q13.3 region [27], whereas copy number changes are noted throughout the genome in individuals with ASD, indicating the presence of multiple candidate genes on every human chromosome. These copy number changes are more often of the deletion type.

For idiopathic or non-syndromic autism, the empirical risk for siblings to be similarly affected is between 2% and 8% with an average of 4% [28]. In multiplex families having two or more affected children with autism, the recurrence risk may be as high as 25%, but generally ranges from 13% [29] to 19% [30] if due to single-gene disturbances as the cause, a major focus of this illustrative review. Advances in genetic technology beyond linkage or cytogenetic analysis of affected families with ASD or other complex disorders have led to genome-wide association studies (GWAS) involving hundreds of affected and control individuals by analyzing the distribution and clustering of hundreds and thousands of SNPs that have successfully been searched for candidate genes. The first GWAS for ASD was undertaken by Lauritsen *et al.* in 2006 [31] using 600 DNA markers in an isolated population of affected individuals from the Faroe Islands. They found an association of the chromosome 3p25.3 band, and later, other investigators studied more subjects with larger collections of genotyped markers and found several chromosome bands and regions ascertained when specific SNPs were over-represented in the ASD subjects, including 5p14.1, 5p15 and 16p13–p21 [32–37]. The studies implicated several gene families, including the cadherin family, encoding proteins for neuronal cell adhesion, while other genes (e.g., *SEMA5A*) were implicated in axonal guidance with lower gene expression levels in brain specimens from individuals with ASD [33], reviewed by Holt and Monaco [17]. Since that time, several additional studies searching for clinically relevant and known genes for ASD have identified a new collection of ASD genes [38–53].

The ability to identify an increased number of SNPs with advanced genetic platforms and extensive approaches using bioinformatics have led to improved access and a more thorough analysis. This has led to comparing genotyping data from GWAS and DNA copy number variants (CNVs) with the identification of structural genetic defects, such as submicroscopic deletions or duplications of the genome, which was not possible a few years ago. Separate studies using array comparative genomic hybridization or microarray analysis to investigate those individuals with ASD continue to yield useful data in identifying candidate genes for ASD in affected individuals [20,21,54]. The yield for microarray analysis is reported to be approximately 20% for identifying deletions or duplications at sites where known or candidate ASD genes are present. The use of more advanced technology, such as next-generation sequencing (whole genome or exome) will yield additional valuable information on the location and description of lesions of genes contributing to ASD with increasing evidence for specific and recurring mutations of single genes involved with neurodevelopment and function, leading to potential therapeutic discoveries and interventions.

Autism is frequent in single-gene conditions, such as fragile X syndrome, tuberous sclerosis, Rett syndrome or neurofibromatosis, but single-gene disorders as a whole account for less than 20% of all cases; therefore, most individuals with ASD are non-syndromic. The heritability of ASD, which takes into consideration the extent of genetic factors contributing to autism, is estimated to be as high as 90% [55]; hence the relevance and continued importance of investigating the role of genetics in the causation of ASD and expanded diagnostic testing to inform and guide treatment for individuals with identifiable genetic disturbances.

A current list of clinically relevant and known candidate genes for ASD is needed for diagnostic testing and genetic counselling purposes in the clinical setting. Historically, a previous list of known or candidate genes showing an association with ASD was reported in 2011 by Holt and Monaco [17] with the placement of 175 genes on chromosome ideograms. A much greater number of validated genes are now recognized as playing a pivotal role in ASD, warranting an updated, revised summary. We will utilize high-resolution chromosome ideograms (850 band level) to plot the location of genes now recognized by searching the literature and website information as playing a documented role in ASD. In tabular form, we will list the individual gene symbol, expanded name or description and chromosome location.

2. Results and Discussion

The diagnostic approach for an individual with ASD should include a clinical genetics evaluation with interviews of parents and health caregivers for the collection and overview of historical problems, a three-generation family pedigree, recording of developmental milestones and description of atypical behaviors along with medical and surgical procedures and a current list of medications and ongoing treatments. Laboratory tests should include lead, thyroid function, lactate and pyruvate levels in order to assess metabolic and mitochondrial functions that may be impacted by an underlying genetic disturbance along with cholesterol and urine collection for organic acid levels. Brain imaging and electroencephalogram patterns should be reviewed, if available. In addition, the ADI-R and ADOS instruments are used to test the diagnosis of ASD.

To further increase the diagnostic yield in individuals with ASD presenting for genetic service, Schaefer *et al.* [19] proposed and utilized a three-tier approach to include a genetic work-up by a clinical geneticist with expertise in dysmorphology to identify known syndromes with or without dysmorphic features (e.g., birth marks), growth anomalies (e.g., microcephaly, macrocephaly and short stature), viral titers (e.g., rubella) and metabolic screening (urine for organic acids and mucopolysaccharides, plasma lactate and amino acid levels). DNA testing for fragile X syndrome and Rett syndrome in females and males is also available, along with chromosomal and DNA microarrays to examine structural DNA lesions in those with a sporadic form of autism and the use of SNP arrays to examine for regions of homozygosity or uniparental disomy, whereby both members of a chromosome pair come from one parent [56]. Exome sequencing is now available particularly to those affected subjects with a positive family history of autism (multiplex families), if other diagnostic tests are uninformative. *PTEN* gene mutation screening would be indicated in those patients with extreme macrocephaly (head size > 2 SD) [14], if not previously done, and a review of brain MRI results. Serum and urine uric acid levels and assays for adenylate succinase deficiency should be done to include biochemical genetic studies and mitochondrial genome screening and function [57] if the above testing protocols are not diagnostic. Up to one in five children with ASD show findings of mitochondrial dysfunction [57], and a detailed genetic work-up will significantly increase the yield for the diagnosis of ASD, leading to a better understanding of causation, treatment and more accurate genetic counselling for those presenting for genetic services [20,21,54].

Advances made in genetic technology and bioinformatics have led to vastly improved genetic testing options for application in the clinical setting in patients presenting for genetic services [54]. Significant discoveries have been made with the recognition of genetic defects in the causation of ASD using microarray technology and, now, next generation sequencing. This technology has flourished with a combination of DNA probes used for both copy number variation and SNPs being required to identify segmental deletions and duplications in the genome and regions of homozygosity for the determination of identical by descent for the calculation of inbreeding coefficients or consanguinity status along with uniparental disomy of individual chromosomes [56].

Next generation exome DNA sequencing and RNA sequencing allows for discoveries of disease-causing genes and regulatory sequences required for normal function. Identifying and characterizing molecular signatures for novel or disturbed gene or exon expression and disease-specific profiles and patterns with expression heat maps have led to the recognition of interconnected disturbed gene pathways in many diseases, including a growing body of genetic evidence for autism and other psychiatric or aberrant behavioral disorders [54].

The position for each known or candidate gene for ASD susceptibility is plotted on high-resolution chromosome ideograms (850 band level), as shown in Figure 1 below. We have included gene symbols and expanded names along with the chromosome band location in Table 1 for the 792 genes recognized as playing a role in ASD.

Figure 1. *Cont.*

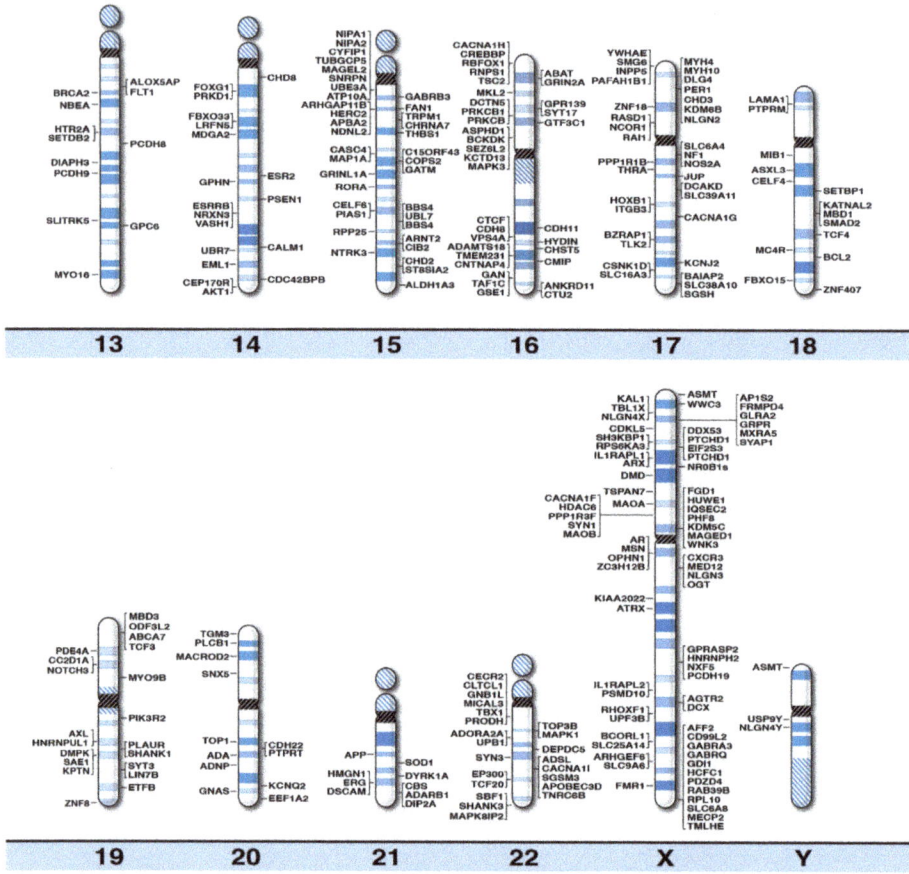

Figure 1. High-resolution human chromosome ideograms (850 band level) with the ASD gene symbol placed at the chromosomal band location. The centromere area, highlighted in black, separates the upper short "p" arm and lower long "q" arm for each chromosome. The gene symbols are arranged in alphabetical order with the expanded name and chromosome band position listed in Table 1.

Table 1. Recognized genes for autism spectrum disorders (ASD) and their chromosome locations.

Gene Symbol	Gene Name	Location
ABAT	4-aminobutyrate aminotransferase	16p13.2
ABCA7	ATP-binding cassette, sub-family A (ABC1), member 7	19p13.3
ABI1	Abl-interactor 1	10p12.1
ABI2	Abl-interactor 2	2q33.2
ABL1	C-Abl oncogene 1, non-receptor tyrosine kinase	9q34.12
ACY1	Aminoacylase 1	3p21.2
ADA	Adenosine deaminase	20q13.12
ADAMTS18	A disintegrin-like and metalloproteinase with thrombospondin type 1 motif, 18	16q23.1
ADARB1	Adenosine deaminase, RNA-specific, B1	21q22.3
ADCY5	Adenylate cyclase 5	3q21.1
ADK	Adenosine kinase	10q22.2
ADNP	Activity-dependent neuroprotector homeobox	20q13.13

Table 1. *Cont.*

Gene Symbol	Gene Name	Location
ADORA2A	Adenosine A2A receptor	22q11.23
ADORA3	Adenosine A3 receptor	1p13.2
ADRB2	Adrenergic, β 2 receptor	5q32
ADSL	Adenylosuccinate lyase	22q13.1
AFF2	AF4/fragile X mental retardation 2 (FMR2) family, member 2	Xq28
AFF4	AF4/fragile X mental retardation 2 (FMR2) family, member 4	5q31.1
AGBL4	ATP/GTP binding protein-like 4	1p33
AGMO	Alkylglycerol monooxygenase	7p21.1
AGTR2	Angiotensin II receptor, type 2	Xq23
AHI1	Abelson helper integration site 1	6q23.3
AHRR	Aryl hydrocarbon receptor repressor	5p15.33
AKT1	v-Akt murine thymoma viral oncogene homolog 1	14q32.33
ALDH1A3	Aldehyde dehydrogenase 1 family, member A3	15q26.3
ALDH5A1	Aldehyde dehydrogenase 5 family, member A1	6p22.3
ALOX5AP	Arachidonate 5-lipoxygenase-activating protein	13q12.3
AMPD1	Adenosine monophosphate deaminase 1	1p13.2
AMT	Aminomethyltransferase	3p21.31
ANK2	Ankyrin 2	4q25
ANK3	Ankyrin 3	10q21.2
ANKRD11	Ankyrin repeat domain 11	16q24.3
ANXA1	Annexin A1	9q21.13
AP1S2	Adaptor-related protein complex 1, sigma 2 subunit	Xp22.2
APBA2	Amyloid β precursor protein-binding, family A, member 2	15q13.1
APC	Adenomatosis polyposis coli	5q22.2
APH1A	APH1A γ secretase subunit	1q21.2
APOBEC3D	Apolipoprotein B mRNA editing enzyme, catalytic polypeptide-like 3D	22q13.1
APP	Amyloid β precursor protein	21q21.3
AR	Androgen receptor	Xq12
ARHGAP11B	Rho GTPase activating protein 11B	15q13.2
ARHGAP15	Rho GTPase activating protein 15	2q22.2
ARHGAP24	Rho GTPase activating protein 24	4q22.1
ARHGEF6	RAC/CDC42 guanine nucleotide exchange factor (GEF) 6	Xq26.3
ARID1B	AT rich interactive domain 1B (SWI1-like)	6q25.3
ARID5A	AT rich interactive domain 5A (MRF1-like)	2q11.2
ARL6IP6	ADP-ribosylation-like factor 6 interacting protein 6	2q23.3
ARNT2	Aryl-hydrocarbon receptor nuclear translocator 2	15q25.1
ARX	Aristaless related homeobox	Xp21.3
ASH1L	Ash1 (absent, small, or homeotic)-like (Drosophila)	1q22
ASMT	Acetylserotonin *O*-methyltransferase, X-chromosomal	Xp22.33
ASMT	Acetylserotonin *O*-methyltransferase, Y-chromosomal	Yp11.32
ASPHD1	Aspartate β-hydroxylase domain containing 1	16p11.2
ASPM	Asp (abnormal spindle) homolog, microcephaly associated	1q31.3
ASS1	Argininosuccinate synthetase	9q34.1
ASTN2	Astrotactin 2	9q33.1
ASXL3	Additional sex combs-like 3	18q12.1
ATG7	Autophagy related 7	3p25.3
ATP10A	ATPase, Class V, type 10A	15q11.2
ATP2B2	ATPase, Ca++ transporting, plasma membrane 2	3p25.3
ATRNL1	Attractin-like 1	10q25.3
ATRX	α thalassemia/mental retardation syndrome X-linked	Xq21.1
ATXN7	Ataxin 7	3p14.1
AUTS2	Autism susceptibility candidate 2	7q11.22
AVPR1A	Arginine vasopressin receptor 1A	12q14.2
AXL	AXL receptor tyrosine kinase	19q13.2
BAIAP2	BAI1-associated protein 2	17q25.3
BBS4	Bardet-Biedl syndrome 4	15q24.1
BCKDK	Branched chain ketoacid dehydrogenase kinase	16p11.2
BCL11A	B-Cell CLL/lymphoma 11A (zinc finger protein)	2p16.1
BCL2	B-cell CLL/lymphoma 2	18q21.33

Table 1. *Cont.*

Gene Symbol	Gene Name	Location
BCORL1	Bc16 co-repressor-like 1	Xq26.1
BDNF	Brain-derived neurotrophic factor	11p14.1
BIN1	Bridging integrator 1	2q14.3
BIRC6	Baculoviral IAP repeat containing 6	2p22.3
BRAF	v-Raf murine sarcoma viral oncogene homolog B	7q34
BRCA2	Breast cancer 2, early onset	13q13.1
BTAF1	RNA polymerase II, B-TFIID transcription factor-associated, 170 kDa (Mot1 homolog, *S. cerevisiae*)	10q23.32
BZRAP1	Benzodiazepine receptor (peripheral) associated protein 1	17q23.2
C11ORF30	Chromosome 11 open reading frame 30	11q13.5
C12ORF57	Chromosome 12 open reading frame 57	12p13.31
C15ORF43	Chromosome 15 open reading frame 43	15q21.1
C3ORF58	Chromosome 3 open reading frame 58	3q24
C4B	Complement component 4B	6p21.33
CA6	Carbonic anhydrase VI	1p36.2
CACNA1B	Calcium channel, voltage-dependent, N type, α 1B subunit	9q34.3
CACNA1C	Calcium channel, voltage-dependent, L type, α 1C subunit	12p13.33
CACNA1D	Calcium channel, voltage-dependent, L type, α 1D subunit	3p14.3
CACNA1F	Calcium channel, voltage-dependent, α 1F subunit	Xp11.23
CACNA1G	Calcium channel, voltage-dependent, T type, α 1G subunit	17q21.33
CACNA1H	Calcium channel, voltage-dependent, α 1H subunit	16p13.3
CACNA1I	Calcium channel, voltage-dependent, T type, α 1I subunit	22q13.1
CACNA2D3	Calcium channel, voltage-dependent, α 2/δ subunit 3	3p21.1
CACNB2	Calcium channel, voltage-dependent, β 2 subunit	10p12.33
CADM1	Cell adhesion molecule 1	11q23.3
CADPS2	Ca+++-dependent activator protein for secretion 2	7q31.32
CALM1	Calmodulin 1 (phosphorylase kinase, δ)	14q32.11
CAMK4	Calcium/calmodulin-dependent protein kinase	5q22.1
CAMSAP2	Calmodulin regulated spectrin-associated protein family, member 2	1q32.1
CAMTA1	Calmodulin binding transcription activator 1	1p36.31
CAPRIN1	Cell cycle associated protein 1	11p13
CASC4	Cancer susceptibility candidate 4	15q15.3
CBS	Cystathionine β-synthase	21q22.3
CCAR2	Cell cycle and apoptosis regulator 2	8p21.3
CC2D1A	Coiled-coil and C2 domain-containing 1A	19p13.12
CCDC19	Coiled-coil domain-containing protein 19	1q23.2
CCDC64	Coiled-coil domain-containing 64	12q24.23
CD38	CD38 molecule	4p15.32
CD44	CD44 molecule	11p13
CD163L1	CD163 molecule-like 1	12p13.31
CD99L2	CD99 molecule-like 2	Xq28
CDC42BPB	CDC42 binding protein kinase β (DMPK-like)	14q32.32
CDH10	Cadherin 10, type 2	5p14.2
CDH22	Cadherin-like 22	20q13.1
CDH8	Cadherin 8, type 2	16q22.1
CDH9	Cadherin 9, type 2	5p14.1
CDH11	Cadherin 11, type 2	16q21
CDKL5	Cyclin-dependent kinase-like 5	Xp22.13
CDKN1B	Cyclin-dependent kinase inhibitor 1B	12p13.1
CECR2	Cat eye syndrome chromosome region, candidate 2	22q11.21
CELF4	CUGBP, Elav-like family, member 4	18q12.2
CELF6	CUGBP, Elav-like family, member 6	15q23
CENTG2	Centaurin γ-2	2q37.2
CEP170R	Centrosomal protein 170B	14q32.33
CEP290	Centrosomal protein 290 kDa	12q21.32
CEP41	Centrosomal protein 41 kDa	7q32.2
CHD1	Chromodomain helicase DNA binding protein 1	5q21.1
CHD2	Chromodomain helicase DNA binding protein 2	15q26.1

Table 1. *Cont.*

Gene Symbol	Gene Name	Location
CHD3	Chromodomain helicase DNA binding protein 3	17p13.1
CHD7	Chromodomain helicase DNA binding protein 7	8q12.2
CHD8	Chromodomain helicase DNA binding protein 8	14q11.2
CHRM3	Cholinergic receptor, muscarinic 3	1q43
CHRNA7	Cholinergic receptor, neuronal nicotinic, α 7	15q13.3
CHRNB3	Cholinergic receptor, neuronal nicotinic, β 3	8p11.21
CHST5	Carbohydrate sulfotransferase 5	16q22.3
CIB2	Calcium and integrin binding family member 2	15q25.1
CKAP5	Cytoskeleton associated protein 5	11p11.2
CLCNKB	Chloride channel voltage-sensitive kidney, B	1p36.13
CLSTN3	Calsyntenin 3	12p13.31
CLTCL1	Clathrin, heavy chain-like 1	22q11.21
CMIP	c-MAF inducing protein	16q23.2
CNR1	Cannabinoid receptor 1	6q15
CNR2	Cannabinoid receptor 2	1p36.11
CNTN3	Contactin 3	3p12.3
CNTN4	Contactin 4	3p26.3
CNTN5	Contactin 5	11q22.1
CNTN6	Contactin 6	3p26.3
CNTNAP2	Contactin associated protein-like 2	7q35
CNTNAP3	Contactin associated protein-like 3	9p13.1
CNTNAP4	Contactin associated protein-like 4	16q23.1
CNTNAP5	Contactin associated protein-like 5	2q14.3
COL7A1	Collagen, type VII, α 1	3p21.31
COPS2	Thyroid hormone receptor interactor 15	15q21.1
CREBBP	CREB binding protein	16p13.3
CSMD1	Cytoskeleton associated protein 5	11p11.2
CSNK1D	Casein kinase 1, δ	17q25
CSTF2T	Cleavage stimulation factor, 3' pre-RNA, subunit 2, 64 kDa, tau	10q21.1
CTCF	CCCTC-binding factor	16q22.1
CTNNA3	Catenin (cadherin-associated protein), α 3	10q21.3
CTNNB1	Catenin (cadherin-associated protein), β 1, 88 kDa	3p22.1
CTSB	Cathepsin B	8p23.1
CTTNBP2	Cortactin binding protein 2	7q31.31
CTU2	Cytosolic thiouridylase subunit 2 homolog (S. pombe)	16q24.3
CUEDC2	CUE domain containing 2	10q24.32
CUL5	Cullin 5	11q22.3
CUL3	Cullin 3	2q36.2
CX3CR1	Chemokine (C-X3-C motif) receptor 1	3p22.2
CXCR3	Chemokine, CXC motif, receptor 3	Xq13.1
CYFIP1	Cytoplasmic FMRP interacting protein 1	15q11.2
CYP11B1	Cytochrome P450, subfamily XIB, polypeptide 1	8q24.3
DAB1	Disabled homolog 1	1p32.2
DAG1	Dystroglycan 1 (dystrophin-associated glycoprotein 1)	3p21.31
DAGLA	Diacylglycerol lipase, α	11q12.2
DAPK1	Death-associated protein kinase 1	9q21.33
DAPP1	Dual adaptor of phosphotyrosine and 3-phosphoinositides 1	4q23
DCAF13	DDB1 and CUL4 associated factor 13	8q22.3
DCAKD	Dephospho-CoA kinase domain-containing protein	17q21.31
DCTN5	Dynactin 5	16p12.2
DCUN1D1	DCN1, domain containing protein 1	3q27.1
DCX	Doublecortin	Xq23
DDC	DOPA decarboxylase	7p12.1
DDX11	DEAD (Asp-Glu-Ala-Asp)/H box 11	12p11.21
DDX53	DEAD (Asp-Glu-Ala-Asp) box polypeptide 53	Xp22.11
DEAF1	DEAF1 transcription factor	11p15.5
DEPDC5	DEP domain containing 3 protein 5	22q12.2
DHCR7	7-dehydrocholesterol reductase	11q13.4

Table 1. *Cont.*

Gene Symbol	Gene Name	Location
DHX9	DEAH (Asp-Glu-Ala-His) box helicase 9	1q25.3
DIAPH3	Diaphanous, Drosophila, homolog 3	13q21.2
DIP2A	DIP2 disco-interacting protein 2 homolog A (Drosophila)	21q22.3
DISC1	Disrupted in schizophrenia 1	1q42.2
DLG4	Discs, large, Drosophila, homolog 4	17p13.1
DLGAP2	Discs, large- associated protein 2	8p23.3
DLGAP3	Discs, large- associated protein 3	1p34.3
DLL1	δ-like 1 (Drosophila)	6q27
DLX1	Distal-less homeobox 1	2q31.1
DLX2	Distal-less homeobox 2	2q31.1
DLX6	Distal-less homeobox 6	7q21.3
DMD	Dystrophin	Xp21.1
DMPK	Dystrophia myotonica-protein kinase	19q13.32
DNAJC19	DNAJ Hsp40 homolog, subfamily C, member 19	3q26.33
DNER	δ- and notch-like epidermal growth factor-related receptor	2q36.3
DNM1L	Dynamin 1-like	12p11.21
DNMT3A	DNA (cytosine-5)-methyltransferase 3 α	2p23.3
DOCK4	Dedicator of cytokinesis 4	7q31.1
DOCK10	Dedicator of cytokinesis 10	2q36.2
DOLK	Dolichol kinase	9q34.1
DPP10	Dipeptidyl peptidase 10	2q14.1
DPP6	Dipeptidyl peptidase 6	7q36.2
DPYD	Dihydropyrimidine dehydrogenase	1p21.3
DRD1	Dopamine receptor D1	5q35.2
DRD2	Dopamine receptor D2	11q23.2
DRD3	Dopamine receptor D3	3q13.31
DSCAM	Down syndrome cell adhesion molecule	21q22.2
DST	Dystonin	6p12.1
DUSP22	Dual specificity phosphatase 22	6p25.3
DYDC1	DPY30 domain containing 1	10q23.1
DYDC2	DPY30 domain containing 2	10q23.1
DYRK1A	Dual-specificity tyrosine-phosphorylation-regulated kinase 1A	21q22.13
EEF1A2	Eukaryotic translation elongation factor 1 α 2	20q13.33
EFR3A	EFR3 homolog A (*S. cerevisiae*)	8q24.22
EGR2	Early growth response 2	10q21.3
EHMT1	Euchromatic histone methyltransferase 1	9q34.3
EIF2S3	Eukaryotic translation initiation factor 2, subunit 3 γ	Xp22.11
EIF4E	Eukaryotic translation initiation factor 4E	4q23
EIF4EBP2	Eukaryotic translation initiation factor 4E binding protein 2	10q22.1
EML1	Echinoderm microtubule associated protein like 1	14q32.2
EN2	Engrailed 2	7q36.3
EP300	E1A binding protein p300	22q13.2
EP400	E1A binding protein p400	12q24.33
EPC2	Enhancer of polycomb, Drosophila homolog of 2	2q23.1
EPHA6	Ephrin receptor A6	3q11.2
EPHB2	Ephrin receptor B2	1p36.12
EPHB6	Ephrin receptor B6	7q34
EPS8	Epidermal growth factor receptor pathway substrate 8	12p12.3
ERBB4	v-ERB-A avian erythroblastic leukemia viral oncogene homolog 4	2q34
ERG	v-ETS avian erythroblastosis virus E26 oncogene homolog	21q22.2
ESR1	Estrogen receptor 1	6q25.1
ESR2	Estrogen receptor 2	14q23.2
ESRRB	Estrogen-related receptor β	14q24.3
ETFB	Electron-transfer-flavoprotein, β polypeptide	19q13.41
ETV1	Ets variant 1	7p21.2
EXOC6B	Exocyst complex component 6B	2p13.2
EXT1	Exostosin 1	8q24.11
F13A1	Factor XIII, A1 subunit	6p25.1

Table 1. *Cont.*

Gene Symbol	Gene Name	Location
FABP3	Fatty acid binding protein 3, muscle and heart (mammary-derived growth inhibitor)	1p35.2
FABP5	Fatty acid binding protein 5	8q21.13
FABP7	Fatty acid binding protein 7	6q22.31
FAM135B	Family with sequence similarity 135, member B	8q24.23
FAN1	FANCD2/FANCI-associated nuclease 1	15q13.2
FAT1	FAT tumor suppressor, Drosophila homolog of, 1	4q35.2
FAT3	FAT tumor suppressor, Drosophila homolog of , 3	11q14.3
FBXO15	F-box protein 15	18q22.3
FBXO33	F-box protein 33	14q21.1
FBXO40	F-box protein 40	3q13.33
FBXW7	F-box and WD repeat domain containing 7, E3 ubiquitin protein	4q31.3
FER	FPS/FES related tyrosine kinase	5q21.3
FEZF2	FEZ family zinc finger 2	3p14.2
FGA	Fibrinogen, A α polypeptide	4q31.3
FGD1	FYVE, Rho GEF and PH domain containing 1	Xp11.22
FGFBP3	Fibroblast growth factor binding protein 3	10q23.32
FHIT	Fragile histidine triad	3p14.2
FLT1	c-FMS-related tyrosine kinase 1	13q12.3
FMR1	Fragile X mental retardation 1 (FMR1)	Xq27.3
FOLH1	Folate hydrolase 1	11p11.2
FOXG1	Forkhead box G1	14q12
FOXP1	Forkhead box P1	3p13
FOXP2	Forkhead box P2	7q31.1
FRK	FYN-related kinase	6q22.1
FRMPD4	FERM and PDZ domain containing protein 4	Xp22.2
GABRA1	γ-aminobutyric acid A receptor, α 1	5q34
GABRA3	γ-aminobutyric acid receptor, α 3	Xq28
GABRA4	γ-aminobutyric acid receptor, α 4	4p12
GABRB1	γ-aminobutyric acid receptor, β 1	4p12
GABRB3	γ-aminobutyric acid receptor, β 3	15q12
GABRQ	γ-aminobutyric acid receptor, θ	Xq28
GAD1	Glutamate decarboxylase 1 (brain, 67 kDa)	2q31.1
GALNT13	UDP-N-acetyl-α-D-galactosamine:polypeptide N-acetylgalactosaminyl-transferase 13	2q23.3
GALNT14	UDP-N-acetyl-α-D-galactosamine:polypeptide N-acetylgalactosaminyl-transferase 14	2p23.1
GAN	Gigaxonin	16q24.1
GAP43	Growth associated protein 43	3q13.31
GAS2	Growth arrest-specific 2	11p14.3
GATM	Glycine amidinotransferase (L-arginine:glycine amidinotransferase)	15q21.1
GDI1	GDP dissociation inhibitor 1	Xq28
GIGYF1	GRB10 interacting GYF protein 1	7q22.1
GLO1	Glyoxalase I	6p21.2
GLRA2	Glycine receptor, α 2 subunit	Xp22.2
GNA14	Guanine nucleotide-binding protein, α 14	9q21.2
GNAS	Guanine nucleotide-binding protein, α-stimulating activity polypeptide I complex locus	20q13.32
GNB1L	Guanine nucleotide-binding protein, β 1-like	22q11.21
GPC6	Glypican 6	13q31.3
GPD2	Glycerol-3-phosphate dehydrogenase 2	2q24.1
GPHN	Gephyrin	14q23.3
GPR139	G protein-coupled receptor 139	16p12.3
GPR37	G protein-coupled receptor 37	7q31.33
GPRASP2	G protein-coupled receptor associated sorting protein 2	Xq22.1
GPX1	Glutathione peroxidase 1	3p21.31
GRID1	Glutamate receptor, ionotropic, δ 1	10q23.2
GRID2	Glutamate receptor, ionotropic, δ 2	4q22.1

Table 1. *Cont.*

Gene Symbol	Gene Name	Location
GRIK2	Glutamate receptor, ionotropic, kainate 2	6q16.3
GRIN1	Glutamate receptor, ionotropic, N-methyl D-aspartate 1	9q34.3
GRIN2A	Glutamate receptor, ionotropic, N-methyl D-aspartate 2A	16p13.2
GRIN2B	Glutamate receptor, ionotropic, N-methyl D-aspartate 2B	12p13.1
GRINL1A	GRINL1A complex locus 1	15q21.3
GRIP1	Glutamate receptor interacting protein 1	12q14.3
GRM1	Glutamate receptor, metabotropic 1	6q24.3
GRM4	Glutamate receptor, metabotropic 4	6p21.31
GRM5	Glutamate receptor, metabotropic 5	11q14.3
GRM8	Glutamate receptor, metabotropic 8	7q31.33
GRPR	Gastrin-releasing peptide receptor	Xp22.2
GSE1	Gse1 coiled-coil protein	16q24.1
GSK3B	Glycogen synthase kinase 3 β	3q13.33
GSN	Gelsolin	9q33.2
GSTM1	Glutathione S-transferase M1	1p13.3
GTF2I	General transcription factor III	7q11.23
GTF2IRD1	GTF2I repeat domain containing 1	7q11.23
GTF3C1	General transcription factor IIIC, polypeptide 1, α	16p12.1
GUCY1A2	Guanylate cyclase 1, soluble, α 2	11q22.3
HCAR1	Hydroxycarboxylic acid receptor 1/G protein-coupled receptor 81	12q24.31
HCFC1	Host cell factor C1	Xq28
HCN1	Hyperpolarization activated cyclic nucleotide-gated potassium channel 1	5p12
HDAC4	Histone deacetylase 4	2q37.3
HDAC6	Histone deacetylase 6	Xp11.23
HDAC9	Histone deacetylase 9	7p21.1
HDLBP	High density lipoprotein binding protein	2q37.3
HEPACAM	Hepatic and glial cell adhesion molecule	11q24.2
HERC2	HECT domain and RCC1-like domain 2	15q13.1
HLA-A	Major histocompatibility complex, class I, A	6p22.1
HLA-DRB1	Major histocompatibility complex, class II, DR β 1	6p21.32
HMGN1	High mobility group nucleosome binding domain 1	21q22.2
HNRNPF	Heterogeneous nuclear ribonucleoprotein F	10q11.21
HNRNPH2	Heterogeneous nuclear ribonucleoprotein H2	Xq22.1
HNRNPUL1	Heterogeneous nuclear ribonucleoprotein U-like 1	19q13.2
HOMER1	Homer, Drosophila, homolog 1 of 1	5q14.1
HOXA1	Homeobox A1	7p15.3
HOXB1	Homeobox B1	17q21.32
HRAS	v-HA-RAS Harvey rat sarcoma viral oncogene homolog	11p15.5
HS3ST5	Heparan sulfate 3-O-sulfotransferase 5	6q22.31
HSD11B1	11-β-hydroxysteroid dehydrogenase type 1	1q32.2
HSPA4	Heat shock 70 kDa protein 4	5q31.1
HTR1B	5-hydroxytryptamine receptor 1B	6q14.1
HTR2A	5-hydroxytryptamine receptor 2A	13q14.2
HTR3A	5-hydroxytryptamine receptor 3A	11q23.2
HTR3C	5-hydroxytryptamine receptor 3, family member C	3q27.1
HTR7	5-hydroxytryptamine receptor 7	10q23.31
HUWE1	HECT, UBA and WWE domain containing 1, E3 ubiquitin protein ligase	Xp11.22
HYDIN	Hydrocephalus-inducing, mouse, homolog of	16q22.2
ICA1	Islet cell autoantigen 1	7p21.3
IL1R2	Interleukin 1 receptor, type II	2q11.2
IL1RAPL1	Interleukin 1 receptor accessory protein-like 1	Xp21.3
IL1RAPL2	Interleukin 1 receptor accessory protein-like 2	Xq22.3
IMMP2L	Inner mitochondrial membrane peptidase, subunit 2, *S. cerevisiae*, homolog of	7q31.1
IMPDH2	Inosine-5-prime monophosphate dehydrogenase 2	3p21.31
INADL	Inactivation no after-potential D-like	1p31.3
INPP1	Inositol polyphosphate-1-phosphatase	2q32.2
INPP5	Inositol polyphosphate-5-phosphatase	17p13.3
IQSEC2	IQ motif and Sec7 domain 2	Xp11.22
ITGA4	Integrin, α 4	2q31.3

Table 1. Cont.

Gene Symbol	Gene Name	Location
ITGB3	Integrin, β 3	17q21.32
ITGB7	Integrin, β 7	12q13.13
ITK	IL20 inducible t-cell kinase	5q33.3
JARID2	Jumonji, AT rich interactive domain 2	6p22.3
JMJD1C	Jumonji domain containing 1C	10q21.3
JUP	Junction plakoglobin	17q21.2
KAL1	Kallmann syndrome interval 1	Xp22.31
KANK1	KN motif and ankyrin repeat domains 1	9p24.3
KATNAL2	Katanin p60 subunit A-like 2	18q21.1
KCND2	Potassium voltage-gated channel, Shal-related subfamily, member 2	7q31.31
KCNJ2	Potassium inwardly-rectifying channel, subfamily J, member 2	17q24.3
KCNJ10	Potassium inwardly-rectifying channel, subfamily J, member 10	1q23.2
KCNMA1	Potassium large conductance calcium-activated channel, subfamily M, α member 1	10q22.3
KCNQ2	Potassium voltage-gated channel, KQT-like subfamily, member 2	20q13.3
KCNQ3	Potassium voltage-gated channel, KQT-like subfamily, member 3	8q24.22
KCNT1	Potassium channel, subfamily T, member 1	9q34.3
KCTD13	Potassium channel tetramerization domain containing protein 13	16p11.2
KDM5A	Lysine (K)-specific demethylase 5A	12p13.33
KDM5B	Lysine (K)-specific demethylase 5B	1q32.1
KDM5C	Lysine (K)-specific demethylase 5C	Xp11.22
KDM6B	Lysine (K)-specific demethylase 6B	17p13.1
KHDRBS2	KH domain containing, RNA binding, signal transduction associated protein 2	6q11.1
KIAA1217	Sickle tail protein homolog	10p12.31
KIAA1586	KIAA1586	6p12.1
KIAA2022	KIAA2022	Xq13.3
KIF5C	Kinesin family member 5C	2q23.1
KIRREL3	Kin of IRRE like 3	11q24.2
KIT	v-KIT Hardy-Zuckerman 4 feline sarcoma viral oncogene homolog	4q12
KLC2	Kinesin light chain 2	11q13.2
KMO	Kynurenine 3-monooxygenase	1q43
KMT2A	Lysine (K)-specific methyltransferase 2A	11q23.3
KMT2C	Lysine (K)-specific methyltransferase 2C	7q36.1
KMT2E	Lysine (K)-specific methyltransferase 2E	7q22.3
KPTN	Kaptin (actin binding protein)	19q13.32
LAMA1	Laminin, α 1	18p11.23
LAMB1	Laminin, β 1	7q31.1
LAMC3	Laminin, γ 3	9q34.1
LEP	Leptin	7q32.1
LIN7B	Lin-7 homolog B (*C. elegans*)	19q13.33
LMNA	Lamin A/C	1q22
LMX1B	LIM homeobox transcription factor 1, β	9q33.3
LRFN5	Leucine-rich repeats and fibronectin type III domain containing 5	14q21.1
LRGUK	Leucine-rich repeats and guanylate kinase domain containing	7q33
LRP2	Low density lipoprotein receptor-related protein 2	2q31.1
LRPPRC	Leucine-rich PPR motif containing protein	2p21
LRRC1	Leucine-rich repeat-containing protein 1	6p12.1
LRRC4	Leucine-rich repeat-containing protein 4	7q32.1
LRRC7	Leucine-rich repeat-containing protein 7	1p31.1
LZTS2	Leucine zipper, putative tumor suppressor 2	10q24.31
MACROD2	Macro domain containing 2	20p12.1
MAGED1	Melanoma antigen family D, 1	Xp11.22
MAGEL2	MAGE-like 2	15q11.2
MAOA	Monoamine oxidase A	Xp11.3
MAOB	Monoamine oxidase B	Xp11.23
MAP1A	Microtubule-associated protein 1A	15q15.3
MAP2	Microtubule-associated protein (MAP) 2	2q34
MAP4	Microtubule-associated protein (MAP) 4	3p21.31

Table 1. *Cont.*

Gene Symbol	Gene Name	Location
MAPK1	Mitogen-activated protein kinase 1	22q11.22
MAPK3	Mitogen-activated protein kinase 3	16p11.2
MAPK8IP2	Mitogen-activated protein kinase 8 interacting protein 2	22q13.33
MARK1	MAP/microtubule affinity-regulating kinase 1	1q41
MBD1	Methyl-CpG binding domain protein 1	18q21.1
MBD3	Methyl-CpG binding domain protein 3	19p13.3
MBD4	Methyl-CpG binding domain protein 4	3q21.3
MBD5	Methyl-CpG binding domain protein 5	2q23.1
MBD6	Methyl-CpG binding domain protein 6	12q13.2
MC4R	Melanocortin 4 receptor	18q21.32
MCC	Mutated in colorectal cancers	5q22.2
MCPH1	Microcephalin 1	8p23.1
MDGA2	Mephrin, A5 antigen, protein tyrosine phosphatase mu (MAM) domain containing glycosylphosphatidylinositol anchor 2	14q21.3
MDM2	MDM2 oncogene, E3 ubiquitin protein ligase	12q15
MECP2	Methyl CpG binding protein 2	Xq28
MED12	Mediator complex subunit 12	Xq13.1
MED13L	Mediator complex subunit 13-like	12q24.21
MEF2C	MADS box transcription myocyte enhancer factor 2, polypeptide C	5q14.3
MET	Met proto-oncogene	7q31.2
MIB1	Mind bomb E3 ubiquitin protein ligase 1	18q11.2
MICAL3	Microtubule-associated monooxygenase, calponin and lim domains-containing, 3	22q11.21
MICALCL	MICAL C-terminus-like protein	11p15.3
MKL2	Myocardin-like 2	16p13.12
MOV10	Moloney leukemia virus 10, mouse, homolog of	1p13.2
MSN	Moesin	Xq12
MSNP1AS	Moesin pseudogene 1 antisense	5p14.1
MSR1	Macrophage scavenger receptor	8p22
MTF1	Metal-regulatory transcription factor 1	1p34.3
MTHFR	5-10-methylene-tetrahydrofolate reductase	1p36.22
MTR	5-methyltetrahydrofolate-homocysteine *S*-methyltransferase	1q43
MTX2	Metaxin 2	2q31.1
MXRA5	Matrix-remodelling associated 5	Xp22.2
MYH4	Myosin, heavy chain 4, skeletal muscle	17p13.1
MYH10	Myosin, heavy chain 10, non-muscle	17p13.1
MYO16	Myosin XVI	13q33.3
MYO1A	Myosin IA	12q13.3
MYO9B	Myosin IXB	19p13.11
MYT1L	Myelin transcription factor 1-like	2p25.3
NAA15	$N(\alpha)$-acetyltransferase 15, NatA auxiliary subunit	4q31.1
NASP	Nuclear autoantigenic sperm protein (histone-binding)	1p34.1
NAV1	Neuron navigator 1	1q32.1
NBEA	Neurobeachin	13q13.3
NCKAP1	NCK-associated protein 1	2q32.1
NCKAP5	NCK-associated protein 5	2q21.2
NCKAP5L	NCK-associated protein 5-like	12q13.12
NCOR1	Nuclear receptor corepressor 1	17p11.2
NDNL2	Necdin-like gene 2	15q13.1
NDUFA5	NADH-ubiquinone oxidoreductase 1 α subcomplex, 5	7q31.32
NEFL	Neurofilament protein, light polypeptide	8p21.2
NELL1	NEL-like 1	11p15.1
NF1	Neurofibromin 1	17q11.2
NFIA	Nuclear factor I/A	1p31.3
NIPA1	Non imprinted gene in Prader-Willi/Angelman syndrome chromosomal region 1	15q11.2
NIPA2	Non imprinted gene in Prader-Willi/Angelman syndrome chromosomal region 2	15q11.2
NIPBL	Nipped-B-like	5p13.2
NLGN1	Neuroligin 1	3q26.31
NLGN2	Neuroligin 2	17p13.1

Table 1. *Cont.*

Gene Symbol	Gene Name	Location
NLGN3	Neuroligin 3	Xq13.1
NLGN4X	Neuroligin 4, X-linked	Xp22.31
NLGN4Y	Neuroligin 4, Y-linked	Yq11.221
NOS1AP	Nitric oxide synthase 1 (neuronal) adaptor protein	1q23.3
NOS2A	Nitric oxide synthase 2A	17q11.2
NOTCH3	Notch 3	19p13.12
NPAS2	Neuronal PAS domain protein 2	2q11.2
NR0B1	Nuclear receptor subfamily 0, group B, member 1	Xp21.2
NR3C2	Nuclear receptor subfamily 3, group C, member 2	4q31.23
NR4A1	Nuclear receptor subfamily 4, group A, member 1	12q13.13
NRCAM	Neuronal cell adhesion molecule	7q31.1
NRG1	Neuregulin 1	8p12
NRP2	Neuropilin 2	2q33.3
NRXN1	Neurexin I	2p16.3
NRXN2	Neurexin II	11q13.1
NRXN3	Neurexin III	14q24.3
NSD1	Nuclear receptor-binding Sa-var, enhancer of zeste, and trithorax domain protein 1	5q35.3
NTNG1	Netrin G1	1p13.3
NTRK1	Neurotrophic tyrosine kinase, receptor, type 1	1q23.1
NTRK3	Neurotrophic tyrosine kinase, receptor, type 3	15q25.3
NXF5	Nuclear RNA export factor 5	Xq22.1
NXPH1	Neurexophilin 1	7p21.3
ODF3L2	Outer dense fiber of sperm tails 3-like 2	19p13.3
OGT	O-linked *N*-acetylglucosamine transferase	Xq13.1
OPHN1	Oligophrenin 1	Xq12
OPRM1	Opioid receptor, mu 1	6q25.2
OR1C1	Olfactory receptor, family 1, subfamily C, member 1	1q44
OTX1	Orthodenticle Drosophila, homolog of	2p15
OXTR	Oxytocin receptor	3p25.3
P2RX4	Purinergic receptor P2X, ligand-gated ion channel, 4	12q24.31
PAFAH1B1	Platelet-activating factor acetylhydrolase 1B, regulatory subunit 1	17p13.3
PAH	Phenylalanine hydroxylase	12q23.2
PARD3B	PAR-3 family cell polarity regulator β	2q33.3
PARK2	Parkin	6q26
PAX5	Paired box 5	9p13.2
PBRM1	Polybromo 1	3p21.1
PCDH10	Protocadherin 10	4q28.3
PCDH15	Protocadherin 15	10q21.1
PCDH19	Protocadherin 19	Xq22.1
PCDH8	Protocadherin 8	13q14.3
PCDH9	Protocadherin 9	13q21.32
PCDHA1	Protocadherin α 1	5q31.3
PCDHA10	Protocadherin α 10	5q31.3
PCDHA11	Protocadherin α 11	5q31.3
PCDHA12	Protocadherin α 12	5q31.3
PCDHA13	Protocadherin α 13	5q31.3
PCDHA2	Protocadherin α 2	5q31.3
PCDHA3	Protocadherin α 3	5q31.3
PCDHA4	Protocadherin α 4	5q31.3
PCDHA5	Protocadherin α 5	5q31.3
PCDHA6	Protocadherin α 6	5q31.3
PCDHA7	Protocadherin α 7	5q31.3
PCDHA8	Protocadherin α 8	5q31.3
PCDHA9	Protocadherin α 9	5q31.3
PCDHAC1	Protocadherin α subfamily C, member 1	5q31.3
PCDHAC2	Protocadherin α subfamily C, member 2	5q31.3
PCDHGA11	Protocadherin γ subfamily A, member 11	5q31.3
PDE1C	Phosphodiesterase 1C	7p14.3

Table 1. *Cont.*

Gene Symbol	Gene Name	Location
PDE4A	Phosphodiesterase 4A, cAMP-specific	19p13.2
PDE4B	Phosphodiesterase 4B, cAMP-specific	1p31.3
PDZD4	PDZ domain containing 4	Xq28
PECR	Peroxisomal trans-2-enoyl-CoA reductase	2q35
PER1	Period, Drosophila, homolog of	17p13.1
PEX7	Peroxisomal biogenesis factor 7	6q23.3
PGD	Phosphogluconate dehydrogenase	1p36.22
PHF2	PHD finger protein 2	9q22.31
PHF8	PHD finger protein 8	Xp11.22
PIAS1	Protein inhibitor of activated STAT, 1	15q23
PIK3CG	Phosphatidylinositol-3-kinase, catalytic, γ	7q22.3
PIK3R2	Phosphatidylinositol-3-kinase, regulatory subunit 2	19q13.11
PINX1	PIN2 interacting protein 1	8p23.1
PITX1	Paired-like homeodomain transcription factor 1	5q31.1
PLAUR	Plasminogen activator receptor, urokinase-type	19q13.31
PLCB1	Phospholipase C, β 1	20p12.3
PLCD1	Phospholipase C, δ 1	3p22.2
PLN	Phospholamban	6q22.31
PLXNA4	Plexin A4	7q32.3
POGZ	POGO transposable element with ZNF domain	1q21.3
POLR2L	Polymerase (RNA) II (DNA directed) polypeptide L, 7.6 kDa	11p15.5
POMGNT1	Protein *O*-mannose β-1, 2-*N*-acetylglucosaminyl-transferase	1p34.1
PON1	Paraoxonase 1	7q21.3
POT1	Protection of telomeres 1	7q31.33
PPFIA1	Protein tyrosine phosphatase, receptor type, F polypeptide, interacting protein, α 1	11q13.3
PPP1CB	Protein phosphatase 1, catalytic subunit, β isozyme	2p23.2
PPP1R1B	Protein phosphatase 1, regulatory (inhibitor) subunit 1B	17q12
PPP1R3F	Protein phosphatase 1, regulatory (inhibitor) subunit 3F	Xp11.23
PRODH	Proline dehydrogenase (oxidase) 1	22q11.21
PRICKLE1	Prickle, Drosophila, homolog of, 1	12q12
PRICKLE2	Prickle, Drosophila, homolog of, 2	3p14.1
PRKCB	Protein kinase C, β	16p12.2
PRKCB1	Protein kinase C, β-1	16p12.2
PRKD1	Protein kinase D1	14q12
PRDX1	Peroxiredoxin 1	1p34.1
PRSS38	Protease, serine, 38	1q42.13
PRUNE2	Prune, Drosophila, homolog of, 2	9q21.2
PSD3	Pleckstrin and Sec7 domains-containing protein 3	8p22
PSEN1	Presenilin 1	14q24.2
PSMD10	Proteasome 26S subunit, non-ATPase, 10	Xq22.3
PTCHD1	Patched domain containing protein 1	Xp22.11
PTEN	Phosphatase and tensin homolog	10q23.31
PTGER3	Prostaglandin E receptor 3, EP3 subtype	1p31.1
PTGS2	Prostaglandin-endoperoxide synthase 2	1q31.1
PTPN11	Protein tyrosine phosphatase, non-receptor type 11	12q24.13
PTPRB	Protein tyrosine phosphatase, receptor type, B	12q15
PTPRC	Protein tyrosine phosphatase, receptor type, C	1q31.3
PTPRM	Protein tyrosine phosphatase, receptor type, M	18p11.23
PTPRT	Protein tyrosine phosphatase, receptor type, T	20q13.11
PXDN	Peroxidasin, Drosophila homolog of	2p25.3
RAB11FIP5	RAB11 family-interacting protein 5	2p13.2
RAB19	RAB19, member RAS oncogene family	7q34
RAB39B	RAS-associated protein RAB39B	Xq28
RAI1	Retinoic acid induced gene 1	17p11.2
RAPGEF4	Rap guanine nucleotide exchange factor	2q31.1
RASD1	RAS protein, dexamethasone-induced, 1	17p11.2
RASSF1	RAS association (ralGDS/AF-6) domain family member 1	3p21.31
RASSF5	RAS association domain family protein 5	1q32.1

Table 1. *Cont.*

Gene Symbol	Gene Name	Location
RB1CC1	RB1-inducible coiled-coil 1	8q11.23
RBFOX1	RNA binding protein FOX-1, *C. elegans*, homolog of, 1	16p13.3
RBM8A	RNA binding motif protein 8A	1q21.1
RBMS3	RNA binding motif protein, single stranded interacting, 3	3p24.1
REEP3	Receptor expression-enhancing protein 3	10q21.3
RELN	Reelin	7q22.1
RERE	RE-repeats encoding gene	1p36.23
RFWD2	Ring finger and WD repeat domains-containing protein 2	1q25.2
RGS7	Regulator of G protein signaling 7	1q43
RHOXF1	RHOX homeobox family, member 1	Xq24
RIC8A	RIC8 guanine nucleotide exchange factor A	11p15.5
RIMS1	Regulating synaptic membrane exocytosis 1	6q13
RIMS3	Protein regulating synaptic membrane exocytosis 3	1p34.2
RNPS1	RNA binding protein S1	16p13.3
ROBO1	Roundabout, Drosophila, homolog of, 1	3p12.2
ROBO2	Roundabout, Drosophila, homolog of, 2	3p12.3
RORA	RAR-related orphan receptor A	15q22.2
RPL10	Ribosomal protein L10	Xq28
RPP25	Ribonuclease P/MRP 25 kDa subunit	15q24.2
RPS6KA1	Ribosomal protein S6 kinase, 90 kDa, polypeptide 1	1p36.11
RPS6KA2	Ribosomal protein S6 kinase, 90 kDa, polypeptide 2	6q27
RPS6KA3	Ribosomal protein S6 kinase, 90 kDa, polypeptide 3	Xp22.12
RUVBL1	RuvB-E. coli, homolog-like 1	3q21.3
SAE1	SUMO1 activating enzyme, subunit 1	19q13.32
SATB2	Special AT-rich sequence-binding protein 2	2q33.1
SBF1	SET binding factor 1	22q13.33
SCFD2	Sec1 family domain containing 2	4q12
SCN1A	Sodium channel, neuronal, type I, α subunit	2q24.3
SCN2A	Sodium channel, voltage-gated, type II, α subunit	2q24.3
SCN7A	Sodium channel, voltage-gated, type VII, α subunit	2q24.3
SCN8A	Sodium channel, voltage-gated, type VIII, α subunit	12q13.13
SDC2	Syndecan 2	8q22.1
SDK1	Sidekick cell adhesion molecule 1	7p22.2
SEMA3F	Sema domain, immunoglobulin domain (Ig), short basic domain, secreted, (semaphorin) 3F	3p21.31
SEMA5A	Semaphorin 5A	5p15.31
SERPINE1	Serpin peptidase inhibitor, clade E (nexin, plasminogen activator inhibitor type 1), member 1	7q22.1
SETBP1	SET binding protein 1	18q12.3
SETD2	SET domain containing protein 2	3p21.31
SETD5	SET domain containing protein 5	3p25.3
SETDB1	SET domain, bifurcated, 1	1q21.3
SETDB2	SET domain, bifurcated, 2	13q14.2
SEZ6L2	Seizure related 6 homolog (mouse)-like 2	16p11.2
SF1	Splicing factor 1	11q13.1
SFPQ	Splicing factor proline/glutamine-rich	1p34.3
SFTPD	Surfactant, pulmonary-associated protein D	10q22.3
SGSH	N-sulfoglucosamine sulfohydrolase	17q25.3
SGSM3	Small G protein signaling modulator 3	22q13.1
SH3KBP1	SH3-domain kinase binding protein 1	Xp22.12
SHANK1	SH3 and multiple ankyrin repeat domains 1	19q13.3
SHANK2	SH3 and multiple ankyrin repeat domains 2	11q13.4
SHANK3	SH3 and multiple ankyrin repeat domains 3	22q13.33
SLC16A3	Solute carrier family 16 (monocarboxylic acid transporter), member 3	17q25
SLC16A7	Solute carrier family 16 (monocarboxylic acid transporter), member 7	12q14.1
SLC1A1	Solute carrier family 1 (neuronal/epithelial high affinity glutamate transporter), member 1	9p24.2
SLC22A15	Solute carrier family 22, (organic cation transporter), member 15	1p13.1

Table 1. *Cont.*

Gene Symbol	Gene Name	Location
SLC24A2	Solute carrier family 24 (sodium/potassium/calcium exchanger), member 2	9p22.1
SLC25A12	Solute carrier family 25 (mitochondrial carrier, Aralar), member 12	2q31.1
SLC25A14	Solute carrier family 25 (mitochondrial carrier, brain), member 14	Xq26.1
SLC25A24	Solute carrier family 25 (mitochondrial carrier, phosphate carrier), member 24	1p13.3
SLC25A27	Solute carrier family 25, member 27	6p12.3
SLC29A4	Solute carrier family 29 (equilibrative nucleoside transporter), member 4	7p22.1
SLC30A5	Solute carrier family 30 (zinc transporter), member 5	5q13.1
SLC35A3	Solute carrier family 35 (UDP-N-acetylglucosamine transporter), member 3	1p21.2
SLC38A10	Solute carrier family 38, member 10	17q25.3
SLC39A11	Solute carrier family 39 (metal ion transporter), member 11	17q21.31
SLC4A10	Solute carrier family 4 (sodium bicarbonate transporter-like), member 10	2q24.2
SLC6A1	Solute carrier family 6 (neurotransmitter transporter), member 1	3p25.3
SLC6A3	Solute carrier family 6 (neurotransmitter transporter, dopamine), member 3	5p15.33
SLC6A4	Solute carrier family 6 (neurotransmitter transporter, serotonin), member 4	17q11.2
SLC6A8	Solute carrier family 6 (neurotransmitter transporter, creatine), member 8	Xq28
SLC9A6	Solute carrier family 9 (sodium/hydrogen exchanger), member 6	Xq26.3
SLC9A9	Solute carrier family 9 (sodium/hydrogen exchanger), member 9	3q24
SLCO1B1	Solute carrier organic anion transporter family, member 1B1	12p12.2
SLCO1B3	Solute carrier organic anion transporter family, member 1B3	12p12.2
SLIT3	Slit, Drosophila, homolog of, 3	5q35.1
SLITRK5	SLIT and NTRK-like family, member 5	13q31.2
SLK	STE20-like kinase	10q24.33
SMAD2	SMAD family member 2	18q21.1
SMARCC2	SWI/SNF related, matrix associated, actin dependent regulator of chromatin, subfamily C, member 2	12q13.2
SMG6	SMG 6, *C. elegans*, homolog of	17p13.3
SND1	EBNA2 coactivator p100	7q32.1
SNRPN	Small nuclear ribonucleoprotein polypeptide N	15q11.2
SNTG2	Syntrophin, γ 2	2p25.3
SNX19	Sorting nexin 19	11q25
SNX5	Sorting nexin 5	20p11.23
SOD1	Superoxide dismutase 1, soluble	21q22.11
SOS1	Son of sevenless (SOS), Drosophila, homolog 1	2p22.1
SOX5	SRY (sex determining region Y)-box 5	12p12.1
SOX7	SRY (sex determining region Y)-box 7	8p23.1
SPAST	Spastin	2p22.3
SRD5A2	Steroid-5-α-reductase, 2	2p23.1
ST7	Suppressor of tumorigenicity 7	7q31.2
ST8SIA2	ST8 α-N-acetyl-neuraminide α-2,8-sialyltransferase 2	15q26.1
STK39	Serine/threonine protein kinase 39	2q24.3
STX6	Syntaxin 6	1q25.3
STX1A	Syntaxin 1A	7q11.23
STXBP1	Syntaxin-binding protein 1	9q34.1
STXBP5	Syntaxin-binding protein 5	6q24.3
STXBP5L	Syntaxin-binding protein 5-like	3q13.33
SUCLG2	Succinate-CoA ligase, GDP-forming, β subunit	3p14.1
SUV420H1	Suppressor of variegation 4–20, Drosophila, homolog of, 1	11q13.2
SYAP1	Synapse associated protein 1	Xp22.2
SYN1	Synapsin 1	Xp11.23
SYN2	Synapsin II	3p25.2
SYN3	Synapsin III	22q12.3
SYNE1	Spectrin repeat containing nuclear envelope 1	6q25.2
SYNGAP1	Synaptic RAS-GTPase-activating protein 1	6p21.32
SYT17	Synaptotagmin XVII	16p12.3
SYT3	Synaptotagmin III	19q13.33
TAF1C	TATA box-binding protein-associated factor 1C	16q24.1
TAF1L	TATA box-binding protein-associated factor 1-like	9p21.1
TAS2R1	Taste receptor, type 2, member 1	5p15.31

Table 1. *Cont.*

Gene Symbol	Gene Name	Location
TBC1D30	TBC1 domain family, member 30	12q14.3
TBC1D5	TBC1 domain family, member 5	3p24.3
TBC1D7	TBC1 domain family, member 7	6p24
TBL1X	Transducin-β-like 1, X-linked	Xp22.31
TBL1XR1	Transducin-β-like 1 receptor 1	3q26.32
TBR1	T-box, brain, 1	2q24.2
TBX1	T-box 1	22q11.21
TCF3	Transcription factor 3	19p13.3
TCF4	Transcription factor 4	18q21.2
TCF20	Transcription factor 20 (AR1)	22q13.2
TCF7L2	Transcription factor 7-like 2 (t-cell specific, HMG-box)	10q25.2
TDO2	Tryptophan 2,3-dioxygenase	4q32.1
TGM3	Transglutaminase 3	20p13
TH	Tyrosine hydroxylase	11p15.5
THBS1	Thrombospondin 1	15q14
THRA	Thyroid hormone receptor, α-1	17q21.1
TLK2	Tousled-like kinase 2	17q23.2
TLX1	T-cell leukemia homeobox 1	10q24.31
TM4SF20	Transmembrane 4 L6 family, member 20	2q36.3
TMEM231	Transmembrane protein 231	16q23.1
TMLHE	Epsilon-trimethyllysine hydroxylase	Xq28
TNIP2	TNFAIP3 interacting protein 2	4p16.3
TNRC6B	Trinucleotide repeat containing 6B	22q13.1
TOMM20	MAS20P, *S. cerevisiae*, homolog of	1q42.3
TOP1	Topoisomerase, DNA, I	20q12
TOP3B	Topoisomerase, DNA, III, β	22q11.22
TOPBP1	Topoisomerase (DNA) II-binding protein 1	3q22.1
TOPORS	Topoisomerase I-binding, arginine/serine-rich, E3 ubiquitin protein ligase	9p21.1
TPH2	Tryptophan hydroxylase 2	12q21.1
TPO	Thyroid peroxidase	2p25.3
TRIM33	Tripartite motif containing protein 33	1p13.2
TRIO	Trio Rho guanine nucleotide exchange factor	5p15.2
TRIP12	Thyroid hormone receptor interactor 12	2q36.3
TRPC6	Transient receptor potential cation channel, subfamily C, member 6	11q22.1
TRPM1	Transient receptor potential cation channel, subfamily M, member 1	15q13.3
TSC1	Tuberous sclerosis 1	9q34.1
TSC2	Tuberous sclerosis 2	16p13.3
TSN	Translin	2q14.3
TSPAN7	Tetraspanin 7	Xp11.4
TTI2	TELO2-interacting protein 2	8p12
TTN	Titin	2q31.2
TUBA1A	Tubulin, α-1A	12q13.12
TUBGCP5	Tubulin-γ complex-associated protein 5	15q11.2
TYR	Tyrosinase	11q14.3
UBE1L2	Ubiquitin-activating enzyme, E1-like 2	4q13.2
UBE2H	Ubiquitin-conjugating enzyme E2H	7q32.2
UBE3A	Ubiquitin protein ligase E3A	15q11.2
UBE3B	Ubiquitin protein ligase E3B	12q24.11
UBE3C	Ubiquitin protein ligase E3C	7q36.3
UBL7	Ubiquitin-like 7	15q24.1
UBR5	Ubiquitin protein ligase E3 component *N*-recognin 5	8q22.3
UBR7	Ubiquitin protein ligase E3 component *N*-recognin 7	14q32.12
UIMC1	Ubiquitin interaction motif containing 1	5q35.2
UPB1	Ureidopropionase, β 1	22q11.23
UPF2	UPF2, yeast, homolog of	10p14
UPF3B	UPF3, yeast, homolog of, B	Xq24
USP54	Ubiquitin specific peptidase 54	10q22.2
USP9Y	Ubiquitin specific protease 9, Y-chromosome	Yq11.21

Table 1. *Cont.*

Gene Symbol	Gene Name	Location
VASH1	Vasohibin 1	14q24.3
VCP	Valosin containing protein	9p13.3
VIL1	Villin 1	2q35
VIP	Vasoactive intestinal peptide (VIP)	6q25.2
VPS13B	Vacuolar protein sorting 13, yeast, homolog of, B	8q22.2
VPS4A	Vacuolar protein sorting 4 homolog A (*S. cerevisiae*)	16q22.1
WAC	WW domain containing adaptor with coiled-coil	10p12.1
WDFY3	WD repeat and FYVE domain containing 3	4q21.23
WHSC1	Wolf-Hirschhorn syndrome candidate 1	4p16.3
WNK3	Protein kinase lysine deficient 3	Xp11.22
WNT1	Wingless-type MMTV integration site family, member 1	12q13.12
WNT2	Wingless-type MMTV integration site family, member 2	7q31.2
WWC3	WWC family member 3	Xp22.32
XIRP1	Cardiomyopathy-associated protein 1	3p22.2
XPC	Xeroderma pigmentosum complementation group C	3p25.1
XPO1	Exportin 1	2p15
XPO5	Exportin 5	6p21.1
YEATS2	YEATS domain containing 2	3q27.1
YTHDC2	YTH domain containing 2	5q22.2
YWHAE	Tyrosine 3-monooxygenase, tryptophan 5-monooxygenase activation protein, epsilon isoform	17p13.3
ZBTB16	Zinc finger- and BTB domain-containing protein 16	11q23.1
ZBTB20	Zinc finger- and BTB domain-containing protein 20	3q13.31
ZC3H12B	Zinc finger CCCH domain-containing protein 12B	Xq12
ZFPL1	Zinc finger protein-like 1	11q13.1
ZMYND11	Zinc finger, MYND-type containing 11	10p15.3
ZNF18	Zinc finger protein 18	17p12
ZNF365	Zinc finger protein 365	10q21.2
ZNF385B	Zinc finger protein 385B	2q31.3
ZNF407	Zinc finger protein 407	18q23
ZNF517	Zinc finger protein 517	8q24.3
ZNF8	Zinc finger protein 8	19q13.43
ZNF713	Zinc finger protein 713	7p11.2
ZNF804A	Zinc finger protein 804A	2q32.1
ZNF827	Zinc finger protein 827	4q31.22
ZSWIM5	Zinc finger, SWIM-type containing 5	1p34.1

3. Experimental Section

We used computer-based internet websites and PubMed (https://www.ncbi.nlm.nih.gov/pubmed) to search key words for genetics and autism. This included the integrated catalogue of human genetic studies related to autism found at the Simons Foundation Autism Research Initiative (SFARI) website (https://gene.sfari.org), which currently lists 667 genes reported as of 25 February 2015. This public access initiative is an ongoing curated collection of clinically proven ASD genes supported by clinical and autism experts, medical geneticists and laboratory specialists in the study of autism. This site includes gene description and evidence of support for causation with cited literature reports. We examined peer-reviewed articles found in the medical literature following our search for genetic evidence (*i.e.*, gene variants, mutations or disturbed gene function) and the involvement of genetics playing a role in autism. Sources included whole-genome sequencing of ASD families randomly selected with at least one unaffected sibling [40] or gene expression profiles in ASD [39] along with other informative websites (e.g., Online Mendelian Inheritance in Man, www.OMIM.org). We then compiled the list of genes from these major sources for a total of 792 genes, whereby at least one mechanism was involved for each gene that could lead to ASD, a heterogeneous condition involving many genes; as our report is focused on the compilation of ASD genes from peer-reviewed research articles and authoritative computer website genomic databases for autism and not necessarily related to causal relationships between the individual gene and ASD. Those genes recognized, to date, as playing a role in ASD susceptibility and causation generally appear to impact chromatin remodeling, metabolism, mRNA translation, cell adhesion and synaptic function [39].

SFARI is a publicly available manually curated web-based searchable site of human genes with links to ASD and includes genes in catalogue form based on five categories—genetic association, syndromic, rare single-gene variant and functional and multi-genetic copy number variation—supported by cited research publications for each. Additional literature sources in our study consisted of both primary research articles and reviews summarizing genetic evidence. Many of the listed genes were identified in multiple research studies and widely reported in literature reviews, data repositories and/or computer genomic-based websites for autism (e.g., SFARI). A large number of genes showed a varied relationship to autism and neurodevelopment, but the mass of the literature surveyed limits the reliability of our relative strength estimates for the ASD and gene associations. The gene would be included if cited and recognized in peer-reviewed publications (e.g., PubMed) with supportive genetic evidence (e.g., genetic linkage, GWAS, functional gene expression patterns, informative SNPs, CNVs or identified gene mutations). Other supporting genetic evidence can be found at Simons Foundation Autism Research Initiative (SFARI) at https://sfari.org/sfari-initiatives/simons-simplex-collection, the National Institutes of Health (NIH) at https://www.ncbi.nlm.nih.gov/gap, the Online Inheritance in Man (OMIM) at www.omim.org or Genecards at https://www.genecards.org.

4. Conclusions

Readily available tissue sources, such as peripheral blood, established lymphoblastoid cell lines and saliva, hold promise for more advances in ASD by enabling the identification of new genes and a better understanding of the causation and disease mechanisms to further stimulate research with the hope to discover new treatment modalities impacted by the recognition of known disease-causing or candidate genes for ASD. We illustrated the master list of clinically relevant and known ASD genes in our summary by plotting individual genes on high-resolution chromosome ideograms and generated a tabular form to increase the awareness required for genetic testing and counselling purposes for family members presenting for genetic services. Creating a master list of genes related to ASD is a complicated process; new genes are continually identified, but not all genes are equally important or certain to be causative. Additional research is needed to further investigate the causal relationships between the specific gene and ASD. The authors encourage the use of this collection of known and clinically relevant candidate genes for ASD in their evaluation of patients and families presenting for genetic testing options and for accurate genetic counselling.

Acknowledgments: We thank Carla Meister for expert preparation of the manuscript and Lorie Gavulic for excellent artistic design and preparation of chromosome ideograms. We acknowledge support from National Institute of Child Health and Human Development (NICHD) HD02528.

Author Contributions: Merlin G. Butler conceived of the study, reviewed data from ASD gene literature reports and wrote the manuscript; Syed K. Rafi obtained and reviewed articles pertaining to ASD genes and summarized the master gene list; and Ann M. Manzardo contributed to gene data review and interpretation, contributed to the content of the manuscript and reviewed the literature.

Conflicts of Interest: The authors declare no conflict of interest.

References

1. American Psychiatric Association. *Diagnostic and Statistical Manual of Mental Disorders*, 4th ed.; American Psychiatric Association: Washington, DC, USA, 2000.
2. Johnson, C.P.; Myers, S.M.; American Academy of Pediatrics Counsel on Children with Disablities. Identifiction and evaluation of children with autism spectrum disorders. *Pediatrics* **2007**, *120*, 1183–1215. [CrossRef]
3. Hughes, J.R. Update on autism: A review of 1300 reports published in 2008. *Epilepsy Behav.* **2009**, *16*, 569–589. [CrossRef] [PubMed]

4. Lord, C.; Risi, S.; Lambrecht, L.; Cook, E.H., Jr.; Leventhal, B.L.; DiLavore, P.C.; Pickles, A.; Rutter, M. The autism diagnostic observation schedule-generic: A standard measure of social and communication deficits associated with the spectrum of autism. *J. Autism Dev. Disord.* **2000**, *30*, 205–223. [CrossRef] [PubMed]
5. Le Couteur, A.; Lord, C.; Rutter, M. *Autism Diagnostic Interview-Reviewed (ADI-R)*; Western Psychological Services: Los Angeles, CA, USA, 2003.
6. Constantino, J.N.; Davis, S.A.; Todd, R.D.; Schindler, M.K.; Gross, M.M.; Brophy, S.L.; Metzger, L.M.; Shoushtari, C.S.; Splinter, R.; Reich, W. Validation of a brief quantitative measure of autistic traits: Comparison of the social responsiveness scale with the autism diagnostic interview-revised. *J. Autism Dev. Disord.* **2003**, *33*, 427–433. [CrossRef] [PubMed]
7. Rice, C. Prevalence of autism spectrum disorders-autism and developmental disabilities monitoring network, United States, 2006. *MMWR Surveill. Summ.* **2009**, *58*, 3–8.
8. Fombonne, E. Epidemiology of autistic disorder and other pervasive developmental disorders. *J. Clin. Psychiatry* **2005**, *66* (Suppl. 10), 3–8. [PubMed]
9. Kanner, L. Autistic psychopathy in childhood. *Nerv. Child.* **1943**, *2*, 217–250.
10. Rapin, I. Autistic regression and disintegrative disorder: How important the role of epilepsy? *Semin. Pediatr. Neurol.* **1995**, *2*, 278–285. [CrossRef] [PubMed]
11. Geschwind, D.H.; Levitt, P. Autism spectrum disorders: Developmental disconnection syndromes. *Curr. Opin. Neurobiol.* **2007**, *17*, 103–111. [CrossRef] [PubMed]
12. Kurtzke, J. Neuroepidemiology. In *Neurology in Clinical Practice*; Bradley, W., Daroff, R., Fenichel, G., Marsden, C., Eds.; Butterworth-Heinemann: Stoneham, MA, USA, 1991; pp. 545–560.
13. Gadia, C.A.; Tuchman, R.; Rotta, N.T. Autism and pervasive developmental disorders. *J. Pediatr.* **2004**, *80*, S83–S94.
14. Butler, M.G.; Dasouki, M.J.; Zhou, X.P.; Talebizadeh, Z.; Brown, M.; Takahashi, T.N.; Miles, J.H.; Wang, C.H.; Stratton, R.; Pilarski, R.; *et al.* Subset of individuals with autism spectrum disorders and extreme macrocephaly associated with germline PTEN tumour suppressor gene mutations. *J. Med. Genet.* **2005**, *42*, 318–321. [CrossRef]
15. Prontera, P.; Ottaviani, V.; Toccaceli, D.; Rogaia, D.; Ardisia, C.; Romani, R.; Stangoni, G.; Pierini, A.; Donti, E. Recurrent approximately 100 kb microdeletion in the chromosomal region 14q11.2, involving *CHD8* gene, is associated with autism and macrocephaly. *Am. J. Med. Genet. A* **2014**, *164*, 3137–3141. [CrossRef]
16. Benvenuto, A.; Moavero, R.; Alessandrelli, R.; Manzi, B.; Curatolo, P. Syndromic autism: Causes and pathogenetic pathways. *World J. Pediatr.* **2009**, *5*, 169–176. [CrossRef] [PubMed]
17. Holt, R.; Monaco, A.P. Links between genetics and pathophysiology in the autism spectrum disorders. *EMBO Mol. Med.* **2011**, *3*, 438–450. [CrossRef] [PubMed]
18. Campbell, D.B.; Sutcliffe, J.S.; Ebert, P.J.; Militerni, R.; Bravaccio, C.; Trillo, S.; Elia, M.; Schneider, C.; Melmed, R.; Sacco, R.; *et al.* A genetic variant that disrupts MET transcription is associated with autism. *Proc. Natl. Acad. Sci. USA* **2006**, *103*, 16834–16839. [CrossRef]
19. Schaefer, G.B.; Starr, L.; Pickering, D.; Skar, G.; Dehaai, K.; Sanger, W.G. Array comparative genomic hybridization findings in a cohort referred for an autism evaluation. *J. Child Neurol.* **2010**, *25*, 1498–1503. [CrossRef] [PubMed]
20. Roberts, J.L.; Hovanes, K.; Dasouki, M.; Manzardo, A.M.; Butler, M.G. Chromosomal microarray analysis of consecutive individuals with autism spectrum disorders or learning disability presenting for genetic services. *Gene* **2014**, *535*, 70–78. [CrossRef] [PubMed]
21. Butler, M.G.; Usrey, K.; Roberts, J.L.; Schroeder, S.R.; Manzardo, A.M. Clinical presentation and microarray analysis of Peruvian children with atypical development and/or aberrant behavior. *Genet. Res. Int.* **2014**, *2014*, 408516. [PubMed]
22. Miles, J.H. Autism spectrum disorders—A genetics review. *Genet. Med.* **2011**, *13*, 278–294. [CrossRef] [PubMed]
23. Liu, X.; Takumi, T. Genomic and genetic aspects of autism spectrum disorder. *Biochem. Biophys. Res. Commun.* **2014**, *452*, 244–253. [CrossRef] [PubMed]
24. Cox, D.M.; Butler, M.G. The 15q11.2 BP1-BP2 microdeletion syndrome: A review. *Int. J. Mol. Sci.* **2015**, *16*, 4068–4082. [CrossRef] [PubMed]

25. Hempel, M.; Rivera Brugues, N.; Wagenstaller, J.; Lederer, G.; Weitensteiner, A.; Seidel, H.; Meitinger, T.; Strom, T.M. Microdeletion syndrome 16p11.2-p12.2: Clinical and molecular characterization. *Am. J. Med. Genet. A* **2009**, *149*, 2106–2112. [CrossRef]

26. Fernandez, B.A.; Roberts, W.; Chung, B.; Weksberg, R.; Meyn, S.; Szatmari, P.; Joseph-George, A.M.; Mackay, S.; Whitten, K.; Noble, B.; *et al.* Phenotypic spectrum associated with de novo and inherited deletions and duplications at 16p11.2 in individuals ascertained for diagnosis of autism spectrum disorder. *J. Med. Genet.* **2010**, *47*, 195–203. [CrossRef]

27. Miller, D.T.; Shen, Y.; Weiss, L.A.; Korn, J.; Anselm, I.; Bridgemohan, C.; Cox, G.F.; Dickinson, H.; Gentile, J.; Harris, D.J.; *et al.* Microdeletion/duplication at 15q13.2q13.3 among individuals with features of autism and other neuropsychiatric disorders. *J. Med. Genet.* **2009**, *46*, 242–248. [CrossRef]

28. Ritvo, E.R.; Jorde, L.B.; Mason-Brothers, A.; Freeman, B.J.; Pingree, C.; Jones, M.B.; McMahon, W.M.; Petersen, P.B.; Jenson, W.R.; Mo, A. The UCLA-university of Utah epidemiologic survey of autism: Recurrence risk estimates and genetic counseling. *Am. J. Psychiatry* **1989**, *146*, 1032–1036. [CrossRef] [PubMed]

29. Sandin, S.; Lichtenstein, P.; Kuja-Halkola, R.; Larsson, H.; Hultman, C.M.; Reichenberg, A. The familial risk of autism. *JAMA* **2014**, *311*, 1770–1777. [CrossRef] [PubMed]

30. Sebat, J.; Lakshmi, B.; Malhotra, D.; Troge, J.; Lese-Martin, C.; Walsh, T.; Yamrom, B.; Yoon, S.; Krasnitz, A.; Kendall, J.; *et al.* Strong association of de novo copy number mutations with autism. *Science* **2007**, *316*, 445–449. [CrossRef] [PubMed]

31. Lauritsen, M.B.; Als, T.D.; Dahl, H.A.; Flint, T.J.; Wang, A.G.; Vang, M.; Kruse, T.A.; Ewald, H.; Mors, O. A genome-wide search for alleles and haplotypes associated with autism and related pervasive developmental disorders on the Faroe islands. *Mol. Psychiatry* **2006**, *11*, 37–46. [CrossRef] [PubMed]

32. Ma, D.; Salyakina, D.; Jaworski, J.M.; Konidari, I.; Whitehead, P.L.; Andersen, A.N.; Hoffman, J.D.; Slifer, S.H.; Hedges, D.J.; Cukier, H.N.; *et al.* A genome-wide association study of autism reveals a common novel risk locus at 5p14.1. *Ann. Hum. Genet.* **2009**, *73*, 263–273. [CrossRef]

33. Weiss, L.A.; Arking, D.E.; Gene Discovery Project of Johns Hopkins & the Autism Consortium; Daly, M.J.; Chakravarti, A. A genome-wide linkage and association scan reveals novel loci for autism. *Nature* **2009**, *461*, 802–808. [CrossRef] [PubMed]

34. Anney, R.; Klei, L.; Pinto, D.; Regan, R.; Conroy, J.; Magalhaes, T.R.; Correia, C.; Abrahams, B.S.; Sykes, N.; Pagnamenta, A.T.; *et al.* A genome-wide scan for common alleles affecting risk for autism. *Hum. Mol. Genet.* **2010**, *19*, 4072–4082. [CrossRef]

35. Wang, K.; Zhang, H.; Ma, D.; Bucan, M.; Glessner, J.T.; Abrahams, B.S.; Salyakina, D.; Imielinski, M.; Bradfield, J.P.; Sleiman, P.M.; *et al.* Common genetic variants on 5p14.1 associate with autism spectrum disorders. *Nature* **2009**, *459*, 528–533. [CrossRef]

36. Pinto, D.; Pagnamenta, A.T.; Klei, L.; Anney, R.; Merico, D.; Regan, R.; Conroy, J.; Magalhaes, T.R.; Correia, C.; Abrahams, B.S.; *et al.* Functional impact of global rare copy number variation in autism spectrum disorders. *Nature* **2010**, *466*, 368–372. [CrossRef]

37. Pagnamenta, A.T.; Khan, H.; Walker, S.; Gerrelli, D.; Wing, K.; Bonaglia, M.C.; Giorda, R.; Berney, T.; Mani, E.; Molteni, M.; *et al.* Rare familial 16q21 microdeletions under a linkage peak implicate cadherin 8 (CDH8) in susceptibility to autism and learning disability. *J. Med. Genet.* **2011**, *48*, 48–54. [CrossRef]

38. Klei, L.; Sanders, S.J.; Murtha, M.T.; Hus, V.; Lowe, J.K.; Willsey, A.J.; Moreno-De-Luca, D.; Yu, T.W.; Fombonne, E.; Geschwind, D.; *et al.* Common genetic variants, acting additively, are a major source of risk for autism. *Mol. Autism* **2012**, *3*, 9. [CrossRef]

39. Campbell, M.G.; Kohane, I.S.; Kong, S.W. Pathway-based outlier method reveals heterogeneous genomic structure of autism in blood transcriptome. *BMC Med. Genomics* **2013**, *6*, 34. [CrossRef] [PubMed]

40. Jiang, Y.H.; Yuen, R.K.; Jin, X.; Wang, M.; Chen, N.; Wu, X.; Ju, J.; Mei, J.; Shi, Y.; He, M.; *et al.* Detection of clinically relevant genetic variants in autism spectrum disorder by whole-genome sequencing. *Am. J. Hum. Genet.* **2013**, *93*, 249–263. [CrossRef] [PubMed]

41. Huguet, G.; Ey, E.; Bourgeron, T. The genetic landscapes of autism spectrum disorders. *Ann. Rev. Genomics Hum. Genet.* **2013**, *14*, 191–213. [CrossRef]

42. Correia, C.; Oliveira, G.; Vicente, A.M. Protein interaction networks reveal novel autism risk genes within gwas statistical noise. *PLoS One* **2014**, *9*, e112399. [CrossRef] [PubMed]

43. Merikangas, A.K.; Segurado, R.; Heron, E.A.; Anney, R.J.; Paterson, A.D.; Cook, E.H.; Pinto, D.; Scherer, S.W.; Szatmari, P.; Gill, M.; *et al.* The phenotypic manifestations of rare genic CNVs in autism spectrum disorder. *Mol. Psychiatry* **2014**. [CrossRef]

44. Lossifov, I.; O'Roak, B.J.; Sanders, S.J.; Ronemus, M.; Krumm, N.; Levy, D.; Stessman, H.A.; Witherspoon, K.T.; Vives, L.; Patterson, K.E.; *et al.* The contribution of de novo coding mutations to autism spectrum disorder. *Nature* **2014**, *515*, 216–221. [CrossRef] [PubMed]

45. Goldani, A.A.S.; Downs, S.R.; Widjaja, F.; Lawton, B.; Hendren, R.L. Biomarkers in autism. *Front. Psychiatry* **2014**, *5*, 100. [CrossRef] [PubMed]

46. Ch'ng, C.; Kwok, W.; Rogic, S.; Pavlidis, P. Meta-analysis of gene expression in autism spectrum disorder. **2015**. [CrossRef]

47. Uddin, M.; Tammimies, K.; Pellecchia, G.; Alipanahi, B.; Hu, P.; Wang, Z.; Pinto, D.; Lau, L.; Nalpathamkalam, T.; Marshall, C.R.; *et al.* Brain-expressed exons under purifying selection are enriched for de novo mutations in autism spectrum disorder. *Nat. Genet.* **2014**, *46*, 742–747. [CrossRef] [PubMed]

48. De Rubeis, S.; He, X.; Goldberg, A.P.; Poultney, C.S.; Samocha, K.; Cicek, A.E.; Kou, Y.; Liu, L.; Fromer, M.; Walker, S.; *et al.* Synaptic, transcriptional and chromatin genes disrupted in autism. *Nature* **2014**, *515*, 209–215. [CrossRef] [PubMed]

49. Hadley, D.; Wu, Z.L.; Kao, C.; Kini, A.; Mohamed-Hadley, A.; Thomas, K.; Vazquez, L.; Qiu, H.; Mentch, F.; Pellegrino, R.; *et al.* The impact of the metabotropic glutamate receptor and other gene family interaction networks on autism. *Nat. Commun.* **2014**. [CrossRef]

50. Butler, M.G.; Rafi, S.K.; Hossain, W.; Stephan, D.A.; Manzardo, A.M. Whole exome sequencing in females with autism implicates novel and candidate genes. *Int. J. Mol. Sci.* **2015**, *16*, 1312–1335. [CrossRef] [PubMed]

51. Yuen, R.K.; Thiruvahindrapuram, B.; Merico, D.; Walker, S.; Tammimies, K.; Hoang, N.; Chrysler, C.; Nalpathamkalam, T.; Pellecchia, G.; Liu, Y.; *et al.* Whole-genome sequencing of quartet families with autism spectrum disorder. *Nat. Med.* **2015**, *21*, 185–191. [CrossRef] [PubMed]

52. Willsey, J.A.; State, M.W. Autism spectrum disorders: From genes to neurobiology. *Curr. Opin. Neurobiol.* **2015**, *30*, 92–99. [CrossRef] [PubMed]

53. Hormozdiari, F.; Penn, O.; Borenstein, E.; Eichler, E.E. The discovery of integrated gene networks for autism and related disorders. *Genome Res.* **2015**, *25*, 142–154. [CrossRef] [PubMed]

54. Butler, M.G.; Youngs, E.L.; Roberts, J.L.; Hellings, J.A. Assessment and treatment in autism spectrum disorders: A focus on genetics and psychiatry. *Autism Res. Treat.* **2012**. [CrossRef]

55. Herman, G.E.; Henninger, N.; Ratliff-Schaub, K.; Pastore, M.; Fitzgerald, S.; McBride, K.L. Genetic testing in autism: How much is enough? *Genet. Med.* **2007**, *9*, 268–274. [CrossRef] [PubMed]

56. Butler, M.G. Prader-willi syndrome: Obesity due to genomic imprinting. *Curr. Genomics* **2011**, *12*, 204–215. [CrossRef] [PubMed]

57. Dhillon, S.; Hellings, J.A.; Butler, M.G. Genetics and mitochondrial abnormalities in autism spectrum disorders: A review. *Curr. Genomics* **2011**, *12*, 322–332. [CrossRef] [PubMed]

International Journal of
Molecular Sciences

MDPI

Article

Morphometric Analysis of Recognized Genes for Autism Spectrum Disorders and Obesity in Relationship to the Distribution of Protein-Coding Genes on Human Chromosomes

Austen B. McGuire [†], Syed K. Rafi [†], Ann M. Manzardo and Merlin G. Butler *

Departments of Psychiatry & Behavioral Sciences and Pediatrics, University of Kansas Medical Center, Kansas City, KS 66160, USA; austenmcguire@gmail.com (A.B.M.); rafigene@yahoo.com (S.K.R.); amanzardo@kumc.edu (A.M.M.)
* Correspondence: mbutler4@kumc.edu; Tel.: +1-913-588-1873
† These authors contributed equally to this work.

Academic Editor: Nicholas Delihas
Received: 10 February 2016; Accepted: 28 April 2016; Published: 5 May 2016

Abstract: Mammalian chromosomes are comprised of complex chromatin architecture with the specific assembly and configuration of each chromosome influencing gene expression and function in yet undefined ways by varying degrees of heterochromatinization that result in Giemsa (G) negative euchromatic (light) bands and G-positive heterochromatic (dark) bands. We carried out morphometric measurements of high-resolution chromosome ideograms for the first time to characterize the total euchromatic and heterochromatic chromosome band length, distribution and localization of 20,145 known protein-coding genes, 790 recognized autism spectrum disorder (ASD) genes and 365 obesity genes. The individual lengths of G-negative euchromatin and G-positive heterochromatin chromosome bands were measured in millimeters and recorded from scaled and stacked digital images of 850-band high-resolution ideograms supplied by the International Society of Chromosome Nomenclature (ISCN) 2013. Our overall measurements followed established banding patterns based on chromosome size. G-negative euchromatic band regions contained 60% of protein-coding genes while the remaining 40% were distributed across the four heterochromatic dark band sub-types. ASD genes were disproportionately overrepresented in the darker heterochromatic sub-bands, while the obesity gene distribution pattern did not significantly differ from protein-coding genes. Our study supports recent trends implicating genes located in heterochromatin regions playing a role in biological processes including neurodevelopment and function, specifically genes associated with ASD.

Keywords: G-negative euchromatin; G-positive heterochromatin; chromosome organization; high-resolution chromosome ideograms; protein-coding genes; obesity genes; autism spectrum disorder (ASD) genes

1. Introduction

Over the course of evolution, the architecture of chromosome structure has become substantially complex with the specific assembly and configuration of each chromosome influencing gene expression and function [1]. There is a need within the field of genetics to better understand chromosome structure and organization, including factors that influence chromosome function and gene location [2]. One aspect of chromosomal organizational research has sought to understand chromatin architecture, the combination of DNA and proteins within a nucleosome with condensation of the expansively long strands of DNA to help store and maintain DNA, the building blocks of genes (e.g., [3,4]).

The chromatin of eukaryotic chromosomes is divided into two main categories: euchromatin and heterochromatin. These regions in each chromosome contain different histone modifications that impact gene expression and DNA packaging styles [5].

Euchromatin chromosome regions tend to be less compact than the more tightly packed heterochromatin regions [5]. At most times during the cell cycle, the euchromatin region decondenses during interphase [6], and the less condensed packaging style is associated with greater gene transcription and activity with early replication [7,8]. Unlike euchromatin, heterochromatin regions are thought to contain less transcription and gene activity with later replication. Heterochromatin is further divided into two subtypes with distinct gene activity (expression) profiles: constitutive and facultative [9–11]. Constitutive heterochromatin is thought to be inactive (not expressed) while facultative heterochromatin may be either active or inactive [12] and less studied compared to the euchromatin regions in the genome [5,13,14]. The remodeling of facultative heterochromatin is known to silence euchromatin-based functional genes due to the gradual gain of lysine methylation in heterochromatic regions, but DNA sequencing of the heterochromatin regions is challenging due to the length and repetitive nature of the code [14]. Recent evidence suggests that heterochromatin may play an important role in development [13,15,16], such as the role of facultative heterochromatin in transcription-associated chromatin remodeling complexes. Heterochromatin regions were once thought to be largely inactive and only important in gene-expression silencing, but facultative heterochromatin is known to change and decondense, thereby allowing for transcription [10,17]. Chromatin remodeling is therefore reversible and dynamic but required for timely activation of functional genes during development [13]. More research is needed to better understand and elucidate differences and similarities between euchromatin and heterochromatin regions and their influence on gene position and activity.

Chromosome-banding methods have been used to visually delineate euchromatin from heterochromatin regions in the study of human chromosomes for more than 45 years. Chromosome banding is applied in the clinical setting for identification of structural and numerical anomalies with an average band accounting for about five megabases of DNA. Chromosome-banding techniques assimilate the DNA nucleotide sequence, associated proteins and functional organization of the chromosome through the use of DNA-staining procedures, most commonly Giesma stain. Giemsa preferentially stains transcriptionally less active, AT-nucleotide-rich sequences associated with heterochromatin regions, referred to as G-dark or Giemsa-positive, while the relatively active GC-nucleotide-rich euchromatin regions stain less intensely, leading these locations to be referred to as G-light or Giemsa-negative [18]. Earlier studies report that GC-nucleotide-poor regions constitute 63% of the human genome, whereas GC-nucleotide-rich regions make up 37% [19].

Protein-coding genes are responsible for the production of proteins to support cellular growth and functioning. The number of protein-coding genes has been estimated at 21,000 (e.g., [14]) and are distributed unevenly among the 24 different chromosomes as represented on high-resolution Ensembl ideograms (Ensembl, available at: http://uswest.ensembl.org/Homo_sapiens/Location/Genome?redirect=no) [20]. The greatest number of protein-coding genes are located on chromosomes 1, 19, 11, and 17, respectively, with the least number of protein-coding genes located on chromosomes Y, 21, 13, and 18 [21], not reflecting the size of the chromosome. However, chromosome 1 is considered the largest chromosome, and chromosome 19 is one of the smallest [21]. The distribution of genome-wide protein-coding genes among the chromosomes is reported by numerous credited sources, such as Ensembl [21] and the Genetics Home Reference provided through the National Institutes of Health (available at: http://ghr.nlm.nih.gov/chromosomes); however, less is known about the distribution of protein-coding genes among euchromatin and heterochromatin regions of the chromosomes. Distribution patterns associated with chromosome banding and proximity to fragile sites (regions susceptible to changes, disturbances and instability) could affect gene function or their location in relationship to subsets of protein-coding genes impacting different organ systems' development and function.

The current study was set forth to descriptively characterize and advance knowledge on the distribution of known genome-wide protein-coding genes in relationship to the G-negative euchromatin, G-positive heterochromatin banding regions and fragile sites, chromosome band locations and chromosome size and disease states for genes recognized in autism spectrum disorders (ASD) and obesity [22,23]. Analyzing the distribution and relationship of genome-wide protein-coding genes in comparison with clinically relevant and associated genes for neurodevelopment and brain function (e.g., ASD) and those involved in the peripheral system (e.g., obesity) could increase our understanding of gene location and function influenced by chromosomal euchromatin and heterochromatin regions in development and disease. Genome-wide protein-coding genes associated with these two gene disorder groups will encompass the full range of the human genome with a large number of genes distributed across the 24 chromosomes. ASD is a spectrum of neurological disorders with known genetic influences and estimated heritability as high as 90% [24], while obesity is a systemic-based energy imbalanced disorder with an average heritability of approximately 50% [25]. The number of recognized genes and their location in both disease state gene sets have recently been summarized [26,27]. The functional status of known and candidate ASD and obesity genes could be implied by their locations within the predominately active G-negative euchromatin or inactive G-positive heterochromatin bands on high-resolution chromosome ideograms. Hence, the purpose of the current study is to compile and examine the distribution of all known genome-wide protein-coding genes from published ideograms among the individual human chromosomes and their location in euchromatin and heterochromatin regions at each band level. Additionally, we will compare for the first time the distribution of protein-coding genes with clinically relevant and known ASD and obesity disease-causing genes to characterize and compare their relationship at the chromosome level and to recognize any deviation from the total protein-coding gene distribution patterns.

2. Results and Discussion

We conducted a morphometric analysis of published high-resolution chromosome ideograms to descriptively characterize the distribution and location of currently recognized relevant candidate or known genes for ASD, obesity and the total number of genome-wide protein-coding genes (see Figure 1). We measured and recorded the physical size (length in millimeters) of each chromosome as rendered ideogram representation of metaphase chromosomes along with their gene group status (*i.e.*, protein-coding [20], ASD [27], or obesity [26]). The location for each gene and distribution were determined from data collected from authoritative websites or published peer-reviewed sources. We then determined the location of these genes either on G-negative euchromatin bands or G-positive heterochromatin bands (and sub-bands) across each chromosome-based ideogram (see Figure 2). We determined their regional and chromosomal distribution patterns, and studied their position in relation to the physical size of the G-negative and G-positive chromatin as seen at the 850 high-resolution band level on the ISCN (2013) chromosome ideograms developed and scaled from cytological data [18] which are mostly equivalent to the high-resolution Ensembl ideograms, or in relation to the known genome-wide protein-coding, ASD, and obesity gene locations across the chromosomes. The Ensembl ideograms are based on chromosomes in the uncondensed state and may not necessarily reflect differential chromatin condensation in the structure of a metaphase chromosome whereby the heterochromatin may occupy different linear space than the euchromatin. However, a large positive correlation was found between the number of protein-coding and ASD genes ($r = 0.65$) per chromosome and between protein-coding and obesity genes ($r = 0.85$) per chromosome.

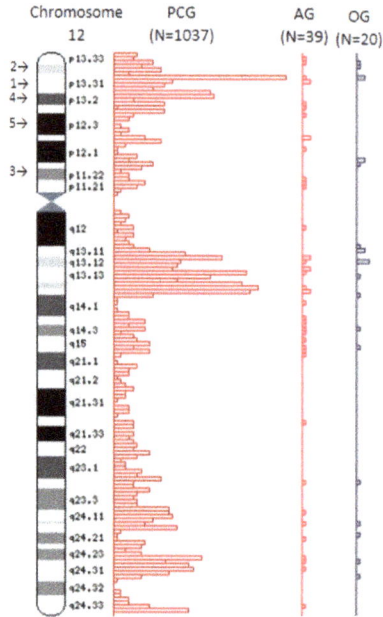

Figure 1. Sample chromosome ideogram with protein-coding, autism and obesity gene frequency distributions by Giemsa band. Ideogram representation of chromosome 12 taken from Genome Reference Consortium Ensembl website (http://uswest.ensembl.org/Homo_sapiens/Location/Genome) [21]. PCG = Protein-coding gene distribution, AG = Autism gene distribution, OG = Obesity gene distribution. 1 = Example of color 1 (euchromatin), 2 = Example of color 2 (heterochromatin), 3 = Example of color 3 (heterochromatin), 4 = Example of color 4 (heterochromatin), 5 = Example of color 5 (heterochromatin).

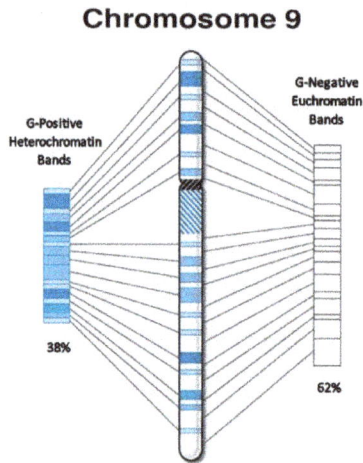

Figure 2. Sample stacked chromosome ideogram.

2.1. Gene Distributions

Our overall measurements followed known and established banding patterns based upon chromosome size (see Table 1 and Figure 3). Chromosomes were arranged in size from number 1 to chromosome Y, except for the X chromosome and when the removal of the qh, centromeric, and acrocentric chromosome p arm regions altered this pattern. For example, chromosome 1 is the longest chromosome, but after removal of the qh region, it became shorter in size than chromosome 2. Altogether, G-negative euchromatin regions encompassed 56.4% of the genome (see Table 1), which contrasts earlier reports of GC-nucleotide-rich regions which only constitute approximately 37% of the human genome [19]. Each chromosome appears to follow this same distribution of approximately 60% for G-negative euchromatin DNA and 40% for G-positive heterochromatin DNA. However, chromosomes 16, 17 and 22 deviated from this pattern with >70% G-negative euchromatin DNA. These chromosomes had at least a threefold difference in euchromatin *vs.* heterochromatin DNA with chromosome 22 having the highest G-negative euchromatin/G-positive heterochromatin ratio of 4.28. This observation deviated considerably from G-negative euchromatin/G-positive heterochromatin expected ratios based on similarly sized chromosomes (*i.e.*, chromosome 21 (1.80 ratio) and chromosome Y (2.03 ratio)).

Protein-coding, ASD, and obesity genes followed a similar distributional pattern across the genome (all 24 chromosomes) based upon the relative length of the individual chromosome (see Table 2) with the number of genes located on a chromosome proportional to the length of the chromosome. As chromosome size decreases, the percentage of total genes per chromosome also decreases. For example, chromosome 1 is the largest chromosome, prior to removal of the qh region, and harbors one of the greatest numbers of protein-coding, ASD, and obesity genes, whereas chromosome 22 is one of the smallest chromosomes and contains one of the smallest numbers of protein-coding, ASD, and obesity genes. Deviations from normal patterns were found, e.g., for chromosome 19 which encompassed 2.3% of the total genome size but contained 7.2% of the total number of protein-coding genes. Chromosome 11 makes up 4.6% of the genome length but possessed 7.4% of obesity genes. Chromosome X makes up 5.6% of the genome length but contained 9.1% of the ASD genes reflecting the established gender disparity with male preponderance seen in ASD (4:1 male:female) [28]. The influence of the observed deviations relative to the typical distribution pattern is unclear and will require further research.

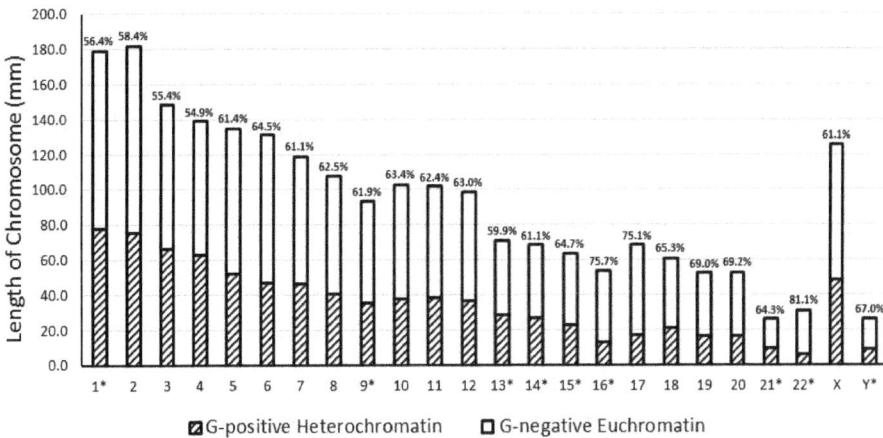

Figure 3. Giemsa band distributions as a proportion of chromosome length. * = Length of qh, centromeric and/or acrocentric chromosome p arm regions were excluded. Percentage above bar for each chromosome represents the proportion of G-negative euchromatin per chromosome.

Int. J. Mol. Sci. **2016**, *17*, 673

Table 1. High-resolution chromosome ideogram measurements and Giemsa banding patterns.

Chromosome	Number of Bands	Total Length (mm)	% of Total Length of all Chromosomes	Total Euchromatin Length (mm)	% Euchromatin	Total Heterochromatin Length (mm)	% Heterochromatin	Ratio of Euchromatin/Heterochromatin
1 *	62	178.9	8.0	101.0	56.4	78.0	43.6	1.30
2	62	182.0	8.1	106.3	58.4	75.7	41.6	1.40
3	59	148.8	6.6	82.4	55.4	66.4	44.6	1.24
4	45	139.5	6.2	76.5	54.9	62.9	45.1	1.22
5	45	134.9	6.0	82.8	61.4	52.1	38.6	1.59
6	48	131.2	5.9	84.6	64.5	46.6	35.5	1.81
7	42	118.9	5.3	72.6	61.1	46.2	38.9	1.57
8	38	107.9	4.8	67.5	62.5	40.4	37.5	1.67
9 *	38	93.6	4.2	58.0	61.9	35.6	38.1	1.63
10	40	102.9	4.6	65.2	63.4	37.7	36.6	1.73
11	34	102.1	4.6	63.7	62.4	38.4	37.6	1.66
12	39	98.8	4.4	62.3	63.0	36.6	37.0	1.70
13 *	31	71.1	3.2	42.6	59.9	28.6	40.1	1.49
14 *	27	68.6	3.1	41.9	61.1	26.7	38.9	1.57
15 *	27	63.7	2.8	41.2	64.7	22.5	35.3	1.83
16 *	22	54.0	2.4	40.9	75.7	13.1	24.3	3.11
17	22	68.7	3.1	51.6	75.1	17.1	24.9	3.02
18	18	60.5	2.7	39.5	65.3	21.0	34.7	1.88
19	15	52.4	2.3	36.2	69.0	16.3	31.1	2.22
20	18	52.8	2.4	36.6	69.2	16.3	30.8	2.25
21 *	9	25.9	1.2	16.7	64.3	9.3	35.7	1.80
22 *	11	30.8	1.4	25.0	81.1	5.8	18.9	4.28
X	38	125.3	5.6	76.6	61.1	48.7	38.9	1.57
Y *	8	26.4	1.2	17.7	67.0	8.7	33.0	2.03
Total / Average	798	2239.7	100.0	1389.1	56.4	850.7	43.6	1.63

Chromosome bands and lengths were measured from ISCN (2013) high-resolution ideograms magnified ×125%. Measurements do not reflect the actual size of human mitotic metaphase chromosomes. * = Length of qh, centromeric and/or acrocentric chromosome p arm regions were excluded.

Table 2. Protein coding gene distribution among G-negative euchromatin and G-positive heterochromatin chromosome regions and relationship to autism and obesity genes.

Chromosome	PCG Sum	% of Total PCG	PCG in Eu	% of PCG in Eu	PCG in Het	% of PCG in Het	AG Sum	% of Total AG	AG in Eu	% of AG in Eu	AG in Het	% of AG in Het	OG Sum	% of Total OG	OG in Eu	% of OG in Eu	OG in Het	% of OG in Het
1 *	2056	10.1	1259	61.2	797	38.8	67	8.5	32	47.8	35	52.2	36	9.9	19	52.8	17	47.2
2	1255	6.3	867	69.1	388	30.9	68	8.6	40	58.8	28	41.2	31	8.5	19	61.3	12	38.7
3	1069	5.3	623	58.3	446	41.7	55	7.1	28	50.9	27	49.1	18	4.9	11	61.1	7	38.9
4	763	3.8	467	61.2	296	38.8	23	2.9	16	69.6	7	30.4	18	4.9	12	66.7	6	33.3
5	864	4.3	513	59.4	351	40.6	44	5.6	29	65.9	15	34.1	16	4.4	12	75.0	4	25.0
6	1041	5.1	520	50.0	521	50.0	36	4.7	13	36.1	23	63.9	25	6.8	13	52.0	12	48.0
7	962	4.7	655	68.1	307	31.9	52	6.6	27	51.9	25	48.1	17	4.7	10	58.8	7	41.2
8	662	3.3	409	61.8	253	38.2	25	3.2	17	68.0	8	32.0	14	3.8	7	50.0	7	50.0
9 *	769	3.8	529	68.8	240	31.2	25	3.2	17	68.0	8	32.0	7	1.9	5	71.4	2	28.6
10	737	3.6	401	54.4	336	45.6	33	4.2	15	45.5	18	54.5	12	3.3	5	41.7	7	58.3
11	1284	6.4	590	46.0	694	54.0	39	4.9	24	61.5	15	38.5	27	7.4	16	59.3	11	40.7
12	1037	5.1	610	58.8	427	41.2	39	4.9	25	64.1	14	35.9	20	5.5	14	70.0	6	30.0
13 *	311	1.5	181	58.2	130	41.8	12	1.5	6	50.0	6	50.0	6	1.6	2	33.3	4	66.7
14 *	807	4.0	550	68.2	257	31.8	17	2.2	8	47.1	9	52.9	6	1.6	4	66.7	2	33.3
15 *	604	3.0	340	56.3	264	43.7	35	4.4	20	57.1	15	42.9	18	4.9	12	66.7	6	33.3
16 *	852	4.2	622	73.0	230	27.0	34	4.3	23	67.6	11	32.4	22	6.0	12	54.5	10	45.5
17	1176	5.8	696	59.2	480	40.8	34	4.3	26	76.5	8	23.5	15	4.1	13	86.7	2	13.3
18	278	1.4	170	61.2	108	38.8	14	1.8	9	64.3	5	35.7	5	1.4	4	80.0	1	20.0
19	1422	7.2	708	49.8	714	50.2	20	2.5	10	50.0	10	50.0	17	4.7	6	35.3	11	64.7
20	539	2.7	337	62.5	202	37.5	12	1.5	6	50.0	6	50.0	9	2.5	5	55.6	4	44.4
21 *	221	1.2	157	71.0	64	29.0	9	1.1	5	55.6	4	44.4	0	0.0	0	0.0	0	0.0
22 *	487	2.5	301	61.8	186	38.2	22	2.8	17	77.3	5	22.7	7	1.9	2	28.6	5	71.4
X	811	4.0	503	62.0	308	38.0	72	9.1	40	55.6	32	44.4	19	5.2	8	42.1	11	57.9
Y *	138	0.7	75	54.3	63	45.7	3	0.4	2	66.7	1	33.3	0	0.0	0	0.0	0	0.0
Total/Average	20,145	100.0	12,083	60.0	8062	40.0	790	100.0	455	57.6	335	42.4	365	100.0	211	57.8	154	42.2

* = Length and genes of qh, centromeric and/or acrocentric chromosome p arm regions were excluded. PCG = Protein Coding Genes, AG = Autism Genes, OG = Obesity Genes, Eu = Euchromatin, Het = Heterochromatin.

This review of morphometric and Giemsa banding chromosome characteristics with respect to the distribution of selected gene groups per length of chromosome did find that many chromosomes possessed a higher proportion of protein-coding genes (e.g., 11, 14, 16, 17 and 22), ASD genes (e.g., 15, 16, 17 and 22), and/or obesity genes (e.g., 11, 15, 16, 17 and 22) than predicted based on their chromosome length. However, eight chromosomes (4, 5, 8, 9, 10, 13, 18 and Y) contained proportionally fewer genes representing the three gene groups than expected based upon their size. In addition, 13 chromosomes (2, 3, 4, 5, 6, 7, 8, 9, 10, 13, 18, X and Y) contained fewer protein-coding genes, 11 chromosomes (4, 5, 6, 8, 9, 10, 13, 14, 18, 20 and Y) had fewer ASD genes, and 10 chromosomes (3, 4, 5, 7, 8, 9, 10, 13, 14 and 18) had fewer obesity genes than expected based on size. The X chromosome had the greatest proportion of ASD genes above the expected level, chromosome 19 had the greatest proportion of protein-coding genes above expected, and chromosome 16 had the greatest proportion of obesity genes, again above expected based upon the length of the chromosome.

2.2. Chromatin Subtyping by Giemsa Band Intensity and Fragile Sites

2.2.1. Chromatin Subtyping

Analysis of chromatin subtype considered both genome-wide and chromosome-level gene distributions, and, as may be anticipated, the greatest number of genome-wide protein-coding (60%), ASD (57.6%) and obesity (57.8%) genes were located in the G-negative euchromatin band type (see Table 3). No significant differences were found between the proportion of ASD, obesity or protein-coding genes for euchromatin *vs.* overall heterochromatin regions ($\chi^2 = 2.4$, df = 2, $p = 0.29$). However, an asymmetric distribution pattern of ASD, obesity and protein-coding genes over the range of G-band intensity levels was observed across the genome. Figure 4 shows the proportion of genes per group by chromatin G-band intensity with light to dark banding scaled numerically by color from 1 to 5. As shown, the proportion of protein-coding genes progressively decreased from 12.6% to 6.8% as the banding color intensity increased (became darker representing colors 2 through 5) for the heterochromatin regions. However, the ASD and obesity genes appear to cluster more in the G-positive heterochromatin (colors 2–5) bands, particularly in medium grey (color 3) and dark grey (colors 4 and 5) as compared to light grey (color 2) and G-negative euchromatin (white, color 1) bands. The lowest number of ASD genes (*i.e.*, 69) were found in the G-positive heterochromatin band color 2 and the lowest number of obesity genes were found in the G-positive heterochromatin band color 5 (see Table 3). This overall difference was statistically significant ($\chi^2 = 31.6$, df = 8, $p < 0.0001$). Examination of standard residuals for ASD genes showed z = +2.21 (G-positive band color 4) and z = +3.14 (G-positive band color 5) relative to obesity and protein-coding genes and z = −2.97 for G-band color 2 (Table 4).

Table 3. Summary data of chromosome bands and genes by group for each chromatin type.

Chromatin Type	Number of Bands	PCG Sum	% of Total PCG	AG Sum	% of Total AG	OG Sum	% of Total OG
G-negative Euchromatin (Color 1)	417	12,083	60.0	455	57.6	211	57.8
G-positive Heterochromatin (Colors 2–5)	381	8062	40.0	335	42.4	154	42.2
Heterochromatin-Color 2	89	2547	12.6	69	8.7	40	11.0
Heterochromatin-Color 3	123	2499	12.4	105	13.3	57	15.6
Heterochromatin-Color 4	88	1638	8.1	83	10.5	36	9.9
Heterochromatin-Color 5	81	1378	6.8	78	9.9	21	5.8
Total	798	20,145	100.0	790	100.0	365	100.0

PCG = Protein-coding Genes, AG = Autism Genes, OG = Obesity Genes. Number of bands and genes were calculated after removal of qh, centromeric and acrocentric chromosome p arm regions.

Table 4. Summary of the percentage deviation with standardized residuals (z-scores) for chromosome bands and genes by group and chromatin type.

| Group | G-Negative Euchromatin Band | | | | G-Positive Heterochromatin Bands | | | | | |
| | Color 1 | | Color 2 | | Color 3 | | Color 4 | | Color 5 | |
	Number of Genes	Percentage (z-Score)	Number of Genes	Percentage (z-Score)	Number of Genes	Percentage (z-Score)	Number of Genes	Percentage (z-Score)	Number of Genes	Percentage (z-Score)
ASD	455	−3.8 (−0.82)	69	−30 (−2.97) *	105	+6.4 (+0.63)	83	+27.4 (+2.21) *	78	+42.4 (+3.14) *
Obesity	211	−3.4 (−0.51)	40	−12.1 (−0.82)	57	+25 (+1.69)	36	+19.6 (+1.07)	21	−17 (−0.86)
PCG	12,083	+0.2 (+0.23)	2547	+1.4 (+0.7)	2499	−0.7 (−0.35)	1638	−1.4 (−0.58)	1378	−1.4 (−0.51)

Chi-Square test percentage deviation and standardized residuals for each cell. ASD = Autism spectrum disorder, N = 790 genes; Obesity, N = 365 genes; PCG = Protein-coding genes, N = 20,145 genes. * Greater than or less than 2 standard deviational z-scores.

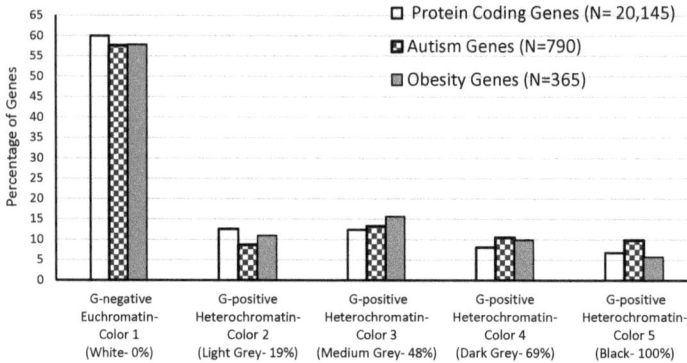

Figure 4. Genome-wide distribution of protein-coding, autism and obesity genes by Giemsa band intensity rating. Calculations excluded genes located in the qh, centromeric and acrocentric chromosome p arm regions.

Further analysis at the chromosome level (see Figure 5) showed that the X chromosome harbored a 2.28-fold higher proportion of ASD genes in relation to protein-coding genes, followed by chromosome 15 with a 1.5-fold higher portion of ASD genes; chromosome 7 (1.4-fold); chromosomes 2 (1.4-fold); chromosome 3 (1.3-fold); and chromosome 5 (1.3-fold). Examination of gene distributions for ASD, obesity and protein-coding genes on the X chromosome did not achieve statistical significance ($\chi^2 = 4.1$, $df = 2$, $p = 0.13$). The chromosomes with the highest proportion of ASD genes also had known ASD gene hotspots, such as: 15q11-q13, 15q13.3, 7q11.23, 7q32.2, Xq28, 22q13 [29], indicating these gene hotspots are dispersed throughout the genome. Chromosome 19 contained the lowest portion of ASD genes in relation to the number of protein-coding genes (0.4-fold differences). Unlike the X chromosome with the highest proportion of ASD genes (*i.e.*, 2.3-fold) in relation to protein-coding genes, the Y chromosome contained nearly four-fold fewer ASD genes (*i.e.*, 0.6-fold) in relation to protein-coding genes. Chromosome 15 had the greatest disparity between protein-coding and obesity genes (2.5-fold) followed by chromosome 9 with a twofold difference and chromosome 19 with 1.53-fold. The least difference in the proportion of obesity genes relative to protein-coding genes was seen for chromosome 15 (0.6-fold), while excluding chromosome 21 and Y as they did not contain recognized obesity genes.

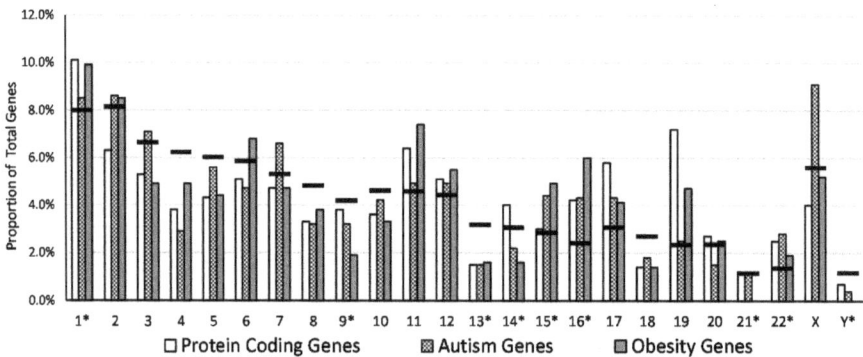

Figure 5. Distribution and proportion of protein-coding, autism and obesity genes by chromosome. * = Length and genes of qh, centromeric and/or acrocentric chromosome p arm regions were excluded. Horizontal black line (—) for each chromosome represents the expected proportion of genes based on chromosome size relative to the total length of all chromosomes summed together.

To further understand and conceptualize the different gene group distributions across chromosomes, chromosomes were ranked based on length, number of protein-coding, ASD, and obesity genes from the longest length or with the most number of genes to the shortest length or with the least number of genes per each gene group (see Table 5). As previously described, the gene number is generally positively correlated with the chromosome length which is reflected in the rank designation for the different gene groups. Deviations from this pattern were found for chromosome 4 with a low ranking for the number of protein-coding and autism genes compared with a relatively high rank based on length.

Table 5. Chromosome rank order by length and gene frequency distribution for protein coding, autism and obesity genes.

Chromosome	Length Rank	Ratio of Euchromatin/ Heterochromatin	PCG Number Rank	AG Number Rank	OG Number Rank
1 *	2	1.30	1	3	1
2	1	1.40	4	2	2
3	3	1.24	6	4	8
4	4	1.22	15	16	9
5	5	1.59	10	6	13
6	6	1.81	7	9	4
7	8	1.57	9	5	11
8	9	1.67	12	14	15
9 *	13	1.63	14	15	18
10	10	1.73	16	13	16
11	11	1.66	3	7	3
12	12	1.70	8	8	6
13 *	14	1.49	21	21	20
14 *	15	1.57	13	19	21
15 *	17	1.83	18	10	10
16 *	19	3.11	11	11	5
17	16	3.02	5	12	14
18	18	1.88	22	20	22
19	21	2.22	2	18	12
20	20	2.25	19	22	17
21 *	23	1.80	23	23	23
22 *	22	4.28	20	17	19
X	7	1.57	12	1	7
Y *	24	2.03	24	24	24

Chromosomes were ranked from 1 (greatest) to 24 (least) based on either length or number of genes per chromosome. PCG = Protein Coding Genes, AG = Autism Genes, OG = Obesity Genes. * = Length and genes of qh, centromeric and/or acrocentric chromosome p arm regions were excluded.

Heterochromatin regions are historically associated with gene silencing or inactivity and areas of less genomic activity [16], but the distribution of euchromatin and heterochromatin regions is not uniform across the chromosomes. We found that heterochromatin DNA regions did contribute genetically with 42.4% of the ASD genes and 42.2% of obesity genes located within the G-positive heterochromatin regions (colors 2–5). This observation was similar to the distribution patterns seen in genome-wide protein-coding genes. Furthermore, our results when analyzing the distribution of genes among G-negative euchromatin and G-positive heterochromatin bands at the chromosomal level supported that heterochromatin regions are also places of active gene expression and not silenced, with an average distribution of the three gene groups across each chromosome with 60% for euchromatin bands and 40% for heterochromatin bands (see Table 2). However, the majority of genes were located within the heterochromatin DNA regions in several select chromosomes with chromosomes 6, 11 and 19 containing ⩾50% of the protein-coding genes as opposed to euchromatin DNA regions. Similarly, chromosomes 1, 6, 10 and 14 contained ⩾50% of the recognized ASD genes in the heterochromatin *vs.* euchromatin regions while chromosomes 8, 10, 13, 19, 22 and X contained ⩾50% of the recognized obesity genes in the heterochromatin *vs.* euchromatin regions.

The G-negative *vs.* G-positive chromosome banding is based on Giemsa staining patterns. Giemsa binds to phosphate groups along the chromosome and perhaps more intensely to those phosphate groups at regions of DNA where there are high amounts of adenine-thymine bonding and that are relatively gene poor. In contrast, less condensed chromatin which tends to be rich in guanine and cytosine (GC-rich) and more transcriptionally active incorporates less Giemsa stain. These regions appear as light bands of varying intensities depending on the degree of AT/GC distribution pattern and transcriptional activity within the light banded regions [30–32]. The Giesma banding and splitting of bands are evident when reviewing the ISCN (2013) ideograms in progression from the 300, 400, 550, 700, and 850-band levels [18]. Giemsa negative and light band(s) appear *de novo* within Giemsa-positive (dark), and rarely, Giemsa-positive (dark) finer bands have been depicted as appearing from larger Giemsa-negative band regions [33]. Hence, the heterochromatic G-positive bands have been shown to contain long stretches of euchromatin DNA which may become G-negative upon extension of the chromosome length, as observed in prometaphase or late prophase banding patterns which far exceed the high-resolution 850-band level [34]. Conversely, as chromosomes condense during early mitosis, their sub-bands fuse in a highly coordinated fashion [35]. Sub-band fusion occurs when two large sub-bands flanking one minor sub-band come together to form one band, which takes on the cytological characteristics of the original flanking sub-bands, when studied from prophase (>1250 bands per haploid set) to late metaphase (~300 bands). Transcriptionally active autism or obesity-susceptibility genes located within the dark Giemsa bands at the 850-band level may be explained if the chromatin is stretched farther (e.g., >2000 band level at the early prophase stage) as narrow, unrecognized G-negative euchromatin regions are embedded within the larger heterochromatin DNA region [36,37]. Overrepresentation of ASD genes in darker heterochromatin bands may reflect the sequestration of transcriptionally active neurodevelopmental genes to inactive chromosome regions following their phase of functional developmental activity.

2.2.2. Fragile Sites

Chromosome fragile sites are also of interest when examining the location or distribution of protein-coding or disease-causing genes located on chromosome bands within the genome. Fragile sites are highly susceptible to changes, disturbances, and instability and thus might select against the location of important protein-coding genes. Analysis of chromosomal fragile sites, specifically aphidicolin-induced CFSs (aCFSs), has revealed that chromatin band-type coverage was the greatest predictor of genome-wide chromosomal fragility and that the majority of aCFSs were within euchromatin regions [38]. Furthermore, Butler [39] also reported on folate-sensitive fragile sites or lesions located at the 350 (mid-metaphase) chromosome band level from peripheral blood cells in a cohort of 117 males with intellectual disability. All but three chromosomes (*i.e.*, 19, 21 and Y) contained fragile sites in cells grown in folate deficient-culture conditions in Medium 199 [39]. Although chromosome 19 did not show fragile sites, it contained the second highest number of protein-coding genes in our morphometric study. Recent studies have reported fragile sites on chromosome 19, but at a lower number in relationship to other chromosomes with a high number of protein-coding genes [40]. One could speculate that a selection against fragile sites on chromosome 19 could exist and is less likely to be susceptible to chromosome breakage or damage resulting from fragile sites due to the quality and/or quantity of specific genes (e.g., housekeeping) that are important for survival. Links have also been reported between autism and fragile sites (e.g., fragile X syndrome) with fragile site stability involving autism-susceptibility genes impacted by folate levels and metabolism [41]. Folate has a key role in the synthesis of DNA and control of DNA methylation [42].

Additionally, the study by Butler [39] reported that spontaneous fragile sites were more concentrated within G-positive heterochromatin bands as compared to G-negative euchromatin bands (e.g., 157 fragile sites distributed over 171 G-negative euchromatin bands *vs.* 144 fragile sites distributed over 127 G-positive heterochromatin bands excluding centromeric, qh and acrocentric chromosome p arm regions) from 6009 cells grown in folate deficient-culture conditions using Medium 199. There was a lower ratio for the number of euchromatin fragile sites and chromosome bands (*i.e.*, 0.92) as compared to the number of heterochromatin fragile sites and chromosome bands (*i.e.*, 1.13). This difference may suggest that fragile sites tend to appear in areas that are less gene-rich and thus less likely to impact genomic function. The ratio of 1.1 was lower for G-negative euchromatin fragile sites ($N = 157$) and G-positive heterochromatin fragile sites ($N = 144$) compared with the ratio of protein-coding genes at 1.5, ASD genes at 1.4 and obesity genes at 1.4 for G-negative euchromatin and G-positive heterochromatin bands reported in our morphometric study. A better understanding of the effects of autism and obesity-susceptibility genes in relation to location of chromatin type and fragile sites could help researchers in understanding the etiology of autism and obesity, and future studies could analyze the connection between fragile site and gene location and chromatin type for specific gene disorder groups beyond autism and obesity. Furthermore, one could determine if chromatin type has an effect on cancer-susceptibility genes and fragile site location, given the well-documented connection between cancer and fragile sites [43].

The molecular mechanism initiated to silence or activate heterochromatic genes appears to result from a balance between negative factors that promote formation of condensed higher-order chromatin structure and positively acting transcription factors that bind to regulatory sequences which activate gene expression [44]. In general, the acetylation of histones is linked to transcriptional activation with histone acetylation decreasing inter-nucleosome interaction, thereby allowing greater accessibility for gene regulation. Histone methylation of both histones and the DNA molecule further directs gene control implicated in disease which underscores the importance of the functional relationships between histone and DNA methylation in maintaining epigenetic traits. Those ASD and obesity genes that are found to be present in the Giemsa-positive dark regions that are of various shades at the 850-band level are expected to be relatively GC-rich regions in defined euchromatin regions embedded within the current dark Giemsa bands, and transcriptionally active with H3K79me1-active histone modifications, and perhaps, acetylation with H3K27ac [45]. At the fiber FISH chromatin level which is greater than 15-fold magnifications to 850-band ideograms, the so-called heterochromatic—dark band domains contain approximately 17% of active gene expression [37]. Even during the cell division at the metaphase stage, one can expect them to contain brief H3K79me1-rich stretches of nucleosomes/chromatin fiber. Additionally, there are at least 39 histone modifications that are classified into active histone modifications and repressive histone modifications for use in chromatin domain prediction. Active modifications are positively correlated with gene expression levels and are known to mark euchromatin genomic regions, whereas repressive modifications are negatively correlated with expression levels and marking heterochromatic domains. Given the fact that the functionality of protein-coding genes is dynamic (euchromatin to facultative heterochromatin status), and the fact that ASD and obesity-causing genes code for functional proteins—either structural or regulatory proteins—their apparent cytogenetic location at the Giemsa-facultative heterochromatic-dark banded regions of varying intensities, cannot necessarily be construed as entirely indicative of their functional inactivation. Hence, the importance of studying histone modifications is emphasized, as mutations in this process may affect most gene structure and biological processes [46,47].

The current study at the 850-band level shows a threefold decrease in the number of protein-coding genes as well as the ASD and obesity genes with an overrepresentation of ASD genes in the facultative G-positive heterochromatic dark band regions. Our examination of the distribution of the protein-coding genes, autism and obesity genes per chromosome and assessment of the disease gene frequency in relation to the chromosome length and G-band characterization was undertaken to examine for bias or skewness in the distribution of disease genes. It is established that a subset of current human chromosome arms or segments were derived from acrocentric chromosomes of ancestral origin including chromosomes 2 and 4 (with relatively recent changes) [48,49]. In addition, the Y chromosome was recently derived or evolved from the X chromosome through shedding of duplicated genes and by retaining and amplifying male-specific genes to compensate for the loss of recombination in order to maintain the integrity of those genes in the absence of recombination with the X chromosome [50].

3. Experimental Section

The individual length of each G-negative euchromatin and G-positive heterochromatin chromosome bands was measured in millimeters and recorded from the 850-band high-resolution ideograms supplied by the International Society of Chromosome Nomenclature (ISCN) 2013 based on scaled cytological data [18] then utilized to calculate the ratio of the two band types per chromosome and chromosome arm. Digital representations were prepared for each chromosome with scaled and stacked images that summarized euchromatin and heterochromatin band distributions over the length of each chromosome (see Figure 2). The images were devoid of centromeric regions, constitutive heterochromatic regions at 1qh, 9qh, 16qh, and Yqh, and acrocentric short (p) arms for chromosomes 13, 14, 15, 21 and 22. To increase size and improve resolution for measurement purposes, each ideogram was uniformly magnified ($\times 125\%$) from the original source [18]. Each scaled image of the summarized euchromatin and heterochromatin chromosome regions was carefully measured using a battery-operated Pittsburgh 6-inch digital caliper (Harbor Freight Tools, Camarillo, CA, USA) and recorded to the one-hundredth of a millimeter. In addition, the total length of G-negative euchromatin and G-positive heterochromatin bands was measured and recorded for each individual chromosome and summarized over the entire genome. The total length of euchromatin and heterochromatin regions per chromosome was then used to calculate the percent length for each band type by dividing the length of each chromatin region for a given chromosome by the overall length of the whole genome.

The location of known genome-wide protein-coding genes was displayed on electronic high-resolution chromosome ideograms supplied by the Genome Reference Consortium at the public access authoritative Ensembl website (available at: http://uswest.ensembl.org/ Homo_sapiens/ Location/Genome) via whole-genome location-based displays [21]. The ideograms were last accessed from the website on 7 December 2014 and updated in August of 2014 using Gencode version GENCODE 21. The total number of genome-wide protein-coding genes for each band was estimated based upon the length in millimeters of each histogram bar illustrating the location of protein-coding genes on the images and arranged perpendicularly to the axis of the high-resolution G-banded represented chromosome ideograms [21]. Figure 1 provides an example of the images used and protein-coding gene distribution, along with the distributions and numbers of recognized ASD and obesity gene sets at the chromosome band level. The total length of the measured histogram bars representing the number of protein-coding genes was then summarized for each chromosome. This sum was divided by the number of protein-coding genes for that specific chromosome. The resulting quotient was used to derive the number of protein-coding genes in each individual histogram bar unit representing these genes in humans. Protein-coding genes were then counted by rounding to the nearest number representing a gene. Each horizontal bar was matched with its respective specific band on the chromosome, showing the distribution and location of the genes. If a band had multiple protein-coding gene histogram bars, the sum of all the bars for that band was then calculated to identify the number of genes per high-resolution chromosome band. Because we focused on euchromatin and heterochromatin chromosome regions, the negligible number of protein-coding genes located at the centromeric, qh and acrocentric chromosome short (p) arm regions were excluded from data analysis. The total number of genome-wide protein-coding genes calculated equaled 20,145, in agreement with the total gene count information from the Ensembl website.

The Ensembl 2014 chromosome ideograms matched the ISCN 2013 chromosome ideograms [18,21], except for seven locations. In each of these instances, the Ensembl ideogram did not contain sub-bands as noted in the ISCN ideograms (e.g., the Ensembl 2014 ideogram showed one band at 1q32.1, whereas the ISCN 2013 ideogram showed three sub-bands at 1q32.11, 1q32.12, and 1q32.13). In these instances, the total number of protein-coding genes for the band on the Ensembl 2014 ideogram was divided by three and evenly distributed across the three more specific sub-bands found in the ISCN 2013 ideogram. The fractional number of genes were rounded to the nearest whole number.

The comparison of protein-coding genes was undertaken in the current study with the 792 neurodevelopmental or functional genes currently recognized as playing a role in ASD and their known chromosome locations [27]. Two ASD genes were excluded from analysis because of their location in a qh, centromeric, or acrocentric chromosome p arm region. Locations for the remaining 790 genes were further refined based on their promotor-molecular locations on the chromosome using website sources such as the Online Inheritance of Man (OMIM) (available at: www.omim.org) and GeneCards (available at: https://www.genecards.org). Additionally, the recognized genes for ASD were then identified as either located on the G-negative (light) euchromatin or G-positive (dark) heterochromatin bands represented in the 850-band chromosome ideograms supplied by ISCN. The distribution of genes from a second gene group representing the obesity-related genes with metabolic or systemic function were also evaluated in a similar manner. A list of 365 clinically relevant and candidate genes for obesity were analyzed (five genes were excluded from the master list of 370 reported obesity genes [26]) based on their location in the qh, centromeric, or acrocentric chromosome p arm regions, and their locations were further refined as stated above before being placed on G-negative euchromatin or G-positive heterochromatin bands on each chromosome.

We further investigated differences among the varying levels of G-positive banding intensity (coloring) within specific chromosome regions compared with the single level (white color) for G-negative bands. Adobe Photoshop (2015) (Adobe Systems Incorporated, San Jose, CA, USA) was used to determine the levels of the G-positive band shading intensity (scaled numerically from 2 to 5 for lightest to darkest color) patterns within the heterochromatin regions on the high-resolution Ensembl chromosome ideograms. Each distinct band on the chromosome was scanned and examined using the Color Picker Tool in Adobe Photoshop to determine the degree of color intensity or darkness. Briefly, the tool pointer was hovered over the band and the color recorded using a greyscale format from 0% (white) to 100% (black). There were a total of five different greyscale grades, one for G-negative and four for G-positive bands. White (color 1) represented the G-negative euchromatin band regions, while 19% were light grey (color 2), 48% medium grey (color 3), 69% dark grey (color 4), or 100% black (color 5) representing the G-positive heterochromatin band regions. The short (p) arm of the acrocentric chromosomes (*i.e.*, 13, 14, 15, 21, and 22) and qh regions (*i.e.*, 1, 9, 16, and Y) which lack protein-coding genes were excluded from the analysis. The Chi-Square test was used to compare the distribution of ASD, obesity and protein-coding genes among euchromatin *vs.* heterochromatin regions genome-wide. Due to the known male prevalence of ASD, *ad hoc* analyses also considered the relative distribution of ASD, obesity and protein-coding genes for euchromatin *vs.* heterochromatin regions of the X chromosome alone.

In review of the literature and our research to address gene-chromosome band relationships (location and type), we reviewed published resources pertaining to the chromosome distribution and signal patterns associated with DNA methylation. We previously reported global DNA promoter methylation patterns from the frontal cortex of alcoholics and controls and found the methylation density patterns targeting CpG islands of the promoters of genes correlated with recognized chromosome banding patterns [51]. Higher CpG methylation peaks or intensity readings at genes were found in G-negative (more genes) chromosome bands and decreased size of peaks in the G-positive (fewer genes) bands in alcoholic and control subjects. For example, we found that 16 of the 20 highest methylation peaks representing CpG islands at gene promoters on chromosome 6 were located on G-negative bands when superimposed over the human chromosome 6 ideogram (data not shown). Thus, the results of our methylation signal data based on global DNA promoter methylation found in high-resolution methylation-specific microarrays and characterization in alcoholics were similar to the visual chromosome G-positive and G-negative bands associated with the distribution of protein-coding genes in ideograms.

4. Conclusions

Our study supports recent trends implicating genes located in heterochromatin regions as playing a role in biological processes including neurodevelopment and function, specifically genes associated with autism spectrum disorder (ASD). For example, almost one-half of the genome-wide protein-coding genes and genes associated with ASD and obesity were located in the G-positive heterochromatin regions. We found a significant overrepresentation of genes contributing to neurological function or development (*i.e.*, ASD) in darker G-positive heterochromatin bands relative to protein-coding genes and those with a systemic basis of function or disease (*i.e.*, obesity). Some genes were overly represented in specific chromosomes (e.g., X chromosome and ASD genes). One could propose analyzing these cytogenetic regions (individually and collectively) in the future by examining the ratios between the protein-coding and ASD genes to further identify ASD gene congregation (if any) in these known ASD-critical regions (e.g., 15q11-q13, 7q11.23, *etc.*) in the chromosomes represented in ideograms, and to simultaneously check for protein-coding gene status at possibly unstable and highly recombinant chromatin locations. Similar questions could be raised regarding the obesity-related genes and they could be examined for obesity gene congregation on chromosome ideograms. Our observations may stimulate future research to analyze the distribution of other gene groups in relationship to chromatin regions and bands including the examination of epigenetically and bioinformatically defined methylation domains in chromatin from different tissues (e.g., Schroeder *et al.*, 2011 [52]). In addition, of interest to genetic researchers would be to investigate genes found in different cell sources with distinct functions, such as ASD genes expressed in neuron cells and obesity genes in hepatic cells, and their relationship, if any, between the location and position of genes having different functions (*i.e.*, ASD genes on behavior/cognition expressed in the central system or brain and obesity-related genes expressed systemically or in peripheral systems). The study of specific G-band (positive or negative) patterns and respective histone maps may correlate with different genome-wide expression, and accessibility could utilize the data from the recently published Epigenome Roadmaps project (available at: http://www.roadmapepigenomics.org/) and yield new information about clustering of specific groups of genes at the tissue or organ (brain, liver, blood, adipose) level or disease (ASD, obesity) state [53]. The above in-depth analysis is beyond the scope of our descriptive approach of examining the location and interaction of protein-coding, ASD and obesity genes at the chromosome or chromosome ideogram or band level. Our study may help researchers gain a better understanding of the foundation of gene clustering and distributions in relationship to chromosome size and proportion of chromosome banding type, as well as specific gene group distribution with similar or dissimilar function as a hierarchical arrangement of gene function and dynamics.

Acknowledgments: We thank Lorie Gavulic for excellent artistic design and preparation of chromosome ideograms and support from the National Institute of Child Health and Human Development (NICHD) grant HD02528.

Author Contributions: Austen B. McGuire, Syed K. Rafi, Merlin G. Butler, and Ann M. Manzardo conceived and designed the experiments; Austen B. McGuire and Syed K. Rafi performed the experiments; Austen B. McGuire, Syed K. Rafi, Ann M. Manzardo, and Merlin G. Butler analyzed the data; Merlin G. Butler contributed reagents/materials/analysis tools; Austen B. McGuire, Syed K. Rafi, Ann M. Manzardo, and Merlin G. Butler wrote the paper.

Conflicts of Interest: The authors declare no conflict of interest.

Abbreviations

ASD	Autism Spectrum Disorder
AT	Adenine-Thymine
CFS	Chromosome Fragile Sites
df	Degrees of Freedom
DNA	Deoxyribonucleic Acid
GC	Guanine-Cytosine
ISCN	International Society of Chromosome Nomenclature
OMIM	Online Inheritance of Man

References

1. Oberdoerffer, P.; Sinclair, D.A. The role of nuclear architecture in genomic instability and ageing. *Nat. Rev. Mol. Cell Biol.* **2007**, *8*, 692–702. [CrossRef] [PubMed]
2. Zhang, Y.; Máté, G.; Müller, P.; Hillebrandt, S.; Krufczik, M.; Bach, M.; Kaufmann, R.; Hausmann, M.; Heermann, D.W. Radiation induced chromatin conformation changes analysed by fluorescent localization microscopy, statistical physics, and graph theory. *PLoS ONE* **2015**, *10*, e0128555. [CrossRef] [PubMed]
3. Annunziato, A. DNA packaging: Nucleosomes and chromatin. *Nat. Educ.* **2008**, *1*, 26.
4. Mello, M.L.S. Cytochemical properties of euchromatin and heterochromatin. *Histochem. J.* **1983**, *15*, 739–751. [CrossRef] [PubMed]
5. Murakami, Y. Heterochromatin and Euchromatin. In *Encyclopedia of Systems Biology*; Dubitzky, W., Wolkenhauer, O., Yokota, H., Cho, K., Yokota, H., Eds.; Springer: New York, NY, USA, 2013; pp. 881–884.
6. Hsieh, T.; Fischer, R.L. Biology of chromatin dynamics. *Annu. Rev. Plant Biol.* **2005**, *56*, 327–351. [CrossRef] [PubMed]
7. Clamp, M.; Fry, B.; Kamal, M.; Xie, X.; Cuff, J.; Lin, M.F.; Kellis, M.; Lindblad-Toh, K.; Lander, E.S. Distinguishing protein-coding and noncoding genes in the human genome. *Proc. Natl. Acad. Sci. USA* **2007**, *104*, 19428–19433. [CrossRef] [PubMed]
8. Harrow, J.; Nagy, A.; Reymond, A.; Alioto, T.; Patthy, L.; Antonarakis, S.E.; Guigó, R. Identifying protein-coding genes in genomic sequences. *Genome Biol.* **2009**, *10*. [CrossRef] [PubMed]
9. Brown, T.; Robertson, F.W.; Dawson, B.M.; Hanlin, S.J.; Page, B.M. Individual variation of centric heterochromatin in man. *Hum. Genet.* **1980**, *55*, 367–373. [CrossRef] [PubMed]
10. Kwon, S.H.; Workman, J.L. The changing faces of HP1: From heterochromatin formation and gene silencing to euchromatic gene expression. *Bioessays* **2011**, *33*, 280–289. [CrossRef] [PubMed]
11. Belmont, A.S.; Dietzel, S.; Nye, A.C.; Strukov, Y.G.; Tumbar, T. Large scale chromatin structure and function. *Curr. Opin. Cell Biol.* **1999**, *11*, 307–311. [CrossRef]
12. Fraser, F.C.; Nora, J. *Genetics of Man*; Lea and Febiger: Philadelphia, PA, USA, 1986.
13. Dimitri, P.; Caizzi, R.; Giordano, E.; Carmela Accardo, M.; Lattanzi, G.; Biamonti, G. Constitutive heterochromatin: A surprising variety of expressed sequences. *Chromosoma* **2009**, *118*, 419–435. [CrossRef] [PubMed]
14. Strachan, T.; Read, A. Human Molecular Genetics. Organization of the Human Genome. In *Human Molecular Genetics*, 4th ed.; Owen, E., Ed.; Garland Science: New York, NY, USA, 2010; pp. 255–295.
15. Grewal, S.; Jia, S. Heterochromatin revisited. *Nat. Rev. Genet.* **2007**, *8*, 35–46. [CrossRef] [PubMed]
16. Yasuhara, J.C.; Wakimoto, B.T. Oxymoron no more: The expanding world of heterochromatic genes. *Trends Genet.* **2006**, *22*, 330–338. [CrossRef] [PubMed]
17. Trojer, P.; Reinberg, D. Facultative heterochromatin: Is there a distinctive molecular signature? *Mol. Cell* **2007**, *28*, 1–13. [CrossRef] [PubMed]
18. Shaffer, L.G.; McGowan-Jordan, J.; Schmid, M. *ISCN (2013): An International System for Human Cytogenetic Nomenclature*; S Karger: Basel, Switzerland, 2013.
19. Saccone, S.; Federico, C.; Solovei, I.; Croquette, M.F.; della Valle, G.; Bernardi, G. Identification of the gene-richest bands in human prometaphase chromosomes. *Chromosome Res.* **1999**, *7*, 379–386. [CrossRef] [PubMed]
20. Whole Genome. Ensembl. Available online: http://uswest.ensembl.org/Homo_sapiens/Location/Genome?redirect=no (accessed on 20 December 2014).
21. Flicek, P.; Amode, M.R.; Barrell, D.; Beal, K.; Billis, K.; Brent, S.; Carvalho-Silva, D.; Clapham, P.; Coates, G.; Fitzgerald, S.; *et al.* Ensembl 2014. *Nucleic Acids Res.* **2014**, *42*, D749–D755. [CrossRef] [PubMed]
22. Ogden, C.L.; Carroll, M.D.; Curtin, L.R.; McDowell, M.A.; Tabak, C.J.; Flegal, K.M. Prevalence of overweight and obesity in the United States, 1999–2004. *JAMA* **2006**, *295*, 1549–1555. [CrossRef] [PubMed]
23. Wingate, M.; Kirby, R.S.; Pettygrove, S.; Cunniff, C.; Schulz, E.; Ghosh, T.; Yeargin-Allsopp, M. Prevalence of autism spectrum disorder among children aged 8 years-autism and developmental disabilities monitoring network, 11 sites, United States, 2010. *MMWR Surveill. Summ.* **2014**, *63*, 1–21.
24. Herman, G.E.; Henninger, N.; Ratliff-Schaub, K.; Pastore, M.; Fitzgerald, S.; McBride, K.L. Genetic testing in autism: How much is enough? *Genet. Med.* **2007**, *9*, 268–274. [CrossRef] [PubMed]
25. Fagnani, C.; Silventoinen, K.; McGue, M.; Korkeila, M.; Christensen, K.; Rissanen, A.; Kaprio, J. Genetic influences on growth traits of BMI: A longitudinal study of adult twins. *Obesity* **2008**, *16*, 847–852.

26. Butler, M.G.; McGuire, A.M.; Manzardo, A.M. Clinically relevant known and candidate genes for obesity and their overlap with human infertility and reproduction. *J. Assist. Reprod. Genet.* **2015**, *32*, 495–508. [CrossRef] [PubMed]

27. Butler, M.G.; Rafi, S.K.; Manzardo, A.M. High-resolution chromosome ideogram representation of currently recognized genes for autism spectrum disorders. *Int. J. Mol. Sci.* **2015**, *16*, 6464–6495. [CrossRef] [PubMed]

28. Rice, C. Prevalence of autism spectrum disorders-autism and developmental disabilities monitoring network, United States, 2006. *MMWR Surveill. Summ.* **2009**, *58*, 3–8.

29. Marshall, C.R.; Noor, A.; Vincent, J.B.; Lionel, A.C.; Feuk, L.; Skaug, J.; Shago, M.; Moessner, R.; Pinto, D.; Ren, Y.; *et al.* Structural variation of chromosomes in autism spectrum disorder. *Am. J. Hum. Genet.* **2008**, *82*, 477–488. [CrossRef] [PubMed]

30. Comings, D.E. Mechanism of chromosome banding and implications for chromosome structure. *Annu. Rev. Genet.* **1978**, *12*, 25–46. [CrossRef] [PubMed]

31. Sumner, A.T. The nature and mechanisms of chromosome-banding. *Cancer Genet. Cytogenet.* **1982**, *6*, 59–87. [CrossRef]

32. Wittekind, D.H.; Gehring, T. On the nature of Romanowsky-Giemsa staining and the Romanowsky-Giemsa effect. I. Model experiments on the specificity of Azure B-Eosin Y stain as compared with other thiazine dye-Eosin Y combinations. *Histochem. J.* **1985**, *17*, 263–289. [CrossRef] [PubMed]

33. Bickmore, W.A. Karyotype Analysis and chromosome banding. *eLS* **2001**. [CrossRef]

34. Holmquist, G.P. Chromosome bands, their chromatin flavors, and their functional features. *Am. J. Hum. Genet.* **1992**, *51*, 17–37. [PubMed]

35. Drouin, R.; Lemieux, N.; Richer, C.L. Chromosome condensation from prophase to late metaphase: Relationship to chromosome bands and their replication time. *Cytogenet. Cell Genet.* **1991**, *57*, 91–99. [CrossRef] [PubMed]

36. Kosyakova, N.; Wiese, A.; Mrasek, K.; Claussen, U.; Leihr, T.; Nelle, H. The hierarchically organized splitting of chromosomal bands for all human chromosomes. *Mol. Cytogenet.* **2009**, *2*. [CrossRef] [PubMed]

37. Wang, J.; Lunyak, V.V.; Jordan, I.K. Genome-wide prediction and analysis of human chromatin boundary elements. *Nucleic Acid Res.* **2012**, *40*, 511–529. [CrossRef] [PubMed]

38. Fungtammasan, A.; Walsh, E.; Chiaromonte, F.; Eckert, K.A.; Makova, K.D. A genome-wide analysis of common fragile sites: What features determine chromosomal instability in the human genome? *Genome Res.* **2012**, *22*, 993–1005. [CrossRef] [PubMed]

39. Butler, M.G. Frequency and distribution of chromosome fragile sites or lesions in males with mental retardation: A descriptive study. *J. Tenn. Acad. Sci.* **1998**, *73*, 87–99.

40. Mrasek, K.; Schoder, C.; Teichmann, A.C.; Behr, K.; Franze, B.; Wilhelm, K.; Blaurock, N.; Claussen, U.; Liehr, T.; Weise, A. Global screening and extended nomenclature for 230 aphidicolin-inducible fragile sites, including 61 yet unreported ones. *Int. J. Oncol.* **2010**, *36*, 929–940. [PubMed]

41. Smith, C.L.; Bolton, A.; Nguyen, G. Genomic and epigenomic instability, fragile sites, schizophrenia and autism. *Curr. Genom.* **2010**, *11*, 447–469. [CrossRef] [PubMed]

42. Nazki, F.H.; Sameer, A.S.; Ganaie, B.A. Folate: Metabolism, genes, polymorphisms and the associated diseases. *Gene* **2014**, *533*, 11–20. [CrossRef] [PubMed]

43. Thys, R.G.; Lehman, C.E.; Pierce, L.C.; Wang, Y.H. DNA secondary structure at chromosomal fragile sites in human disease. *Curr. Genom.* **2015**, *16*, 60–70. [CrossRef] [PubMed]

44. Dillon, N.; Festenstein, R. Unravelling heterochromatin: Competition between positive and negative factors regulates accessibility. *Trends Genet.* **2002**, *18*, 252–258. [CrossRef]

45. Rougeulle, C.; Chaumeil, J.; Sarma, K.; Allis, C.D.; Reinberg, D.; Avner, P.; Heard, E. Differential histone H3 lys-9 and lys-27 methylation profiles on the X chromosome. *Mol. Cell. Biol.* **2004**, *24*, 5475–5484. [CrossRef] [PubMed]

46. Lan, L.; Nakajima, S.; Wei, L.; Sun, L.; Hsieh, C.L.; Sobol, R.W.; Bruchez, M.; van Houten, B.; Yasui, A.; Levine, A.S. Novel method for site-specific induction of oxidative DNA damage reveals differences in recruitment of repair proteins to heterochromatin and euchromatin. *Nucleic Acids Res.* **2013**, *42*, 2330–2345. [CrossRef] [PubMed]

47. Sims, R.J.; Nishioka, K.; Reinberg, D. Histone lysine methylation: A signature for chromatin function. *Trends Genet.* **2003**, *19*, 629–639. [CrossRef] [PubMed]

48. Ijdo, J.W.; Baldini, A.; Ward, D.C.; Reeders, S.T.; Wells, R.A. Origin of human chromosome 2: An ancestral telomere-telomere fusion. *Proc. Natl. Acad. Sci. USA* **1991**, *88*, 9051–9055. [CrossRef] [PubMed]

49. Yunis, J.J.; Prakash, O. The origin of man: A chromosomal pictorial legacy. *Science* **1982**, *215*, 1525–1530. [CrossRef] [PubMed]

50. Skawiński, W.; Parcheta, B. Polymorphism of the human Y chromosome: The evaluation of the correlation between the DNA content and the size of the heterochromatin and euchromatin. *Clin. Genet.* **1984**, *25*, 125–130. [CrossRef] [PubMed]

51. Manzardo, A.M.; Henkhaus, R.S.; Butler, M.G. Global DNA promoter methylation in frontal cortex of alcoholics and controls. *Gene* **2012**, *498*, 5–12. [CrossRef] [PubMed]

52. Schroeder, D.I.; Lott, P.; Korf, I.; LaSalle, J.M. Large-scale methylation domains mark a functional subset of neuronally expressed genes. *Genome Res.* **2011**, *21*, 1583–1591. [CrossRef] [PubMed]

53. Roadmap Epigenomics Consortium. Integrative analysis of 111 reference human epigenomes. *Nature* **2015**, *518*, 317–330.

International Journal of
Molecular Sciences

MDPI

Article

Chromosomal Microarray Analysis of Consecutive Individuals with Autism Spectrum Disorders Using an Ultra-High Resolution Chromosomal Microarray Optimized for Neurodevelopmental Disorders

Karen S. Ho [1,2,*], E. Robert Wassman [1], Adrianne L. Baxter [1], Charles H. Hensel [1], Megan M. Martin [1], Aparna Prasad [1], Hope Twede [1], Rena J. Vanzo [1] and Merlin G. Butler [3]

[1] Lineagen, Inc., Salt Lake City, UT 84109, USA; bwassman@lineagen.com (E.R.W.); abaxter@lineagen.com (A.L.B.); chensel@lineagen.com (C.H.H.); mmartin@lineagen.com (M.M.M.); aprasad@lineagen.com (A.P.); htwede@lineagen.com (H.T.); rvanzo@lineagen.com (R.J.V.)

[2] Department of Pediatrics, University of Utah, Salt Lake City, UT 84132, USA

[3] Departments of Psychiatry, Behavioral Sciences and Pediatrics, University of Kansas Medical Center, Kansas City, UT 66160, USA; mbutler4@kumc.edu

* Correspondence: kho@lineagen.com; Tel.: +1-801-931-6200; Fax: +1-801-931-6201

Academic Editor: Michele Fornaro

Received: 15 October 2016; Accepted: 4 December 2016; Published: 9 December 2016

Abstract: Copy number variants (CNVs) detected by chromosomal microarray analysis (CMA) significantly contribute to understanding the etiology of autism spectrum disorder (ASD) and other related conditions. In recognition of the value of CMA testing and its impact on medical management, CMA is in medical guidelines as a first-tier test in the evaluation of children with these disorders. As CMA becomes adopted into routine care for these patients, it becomes increasingly important to report these clinical findings. This study summarizes the results of over 4 years of CMA testing by a CLIA-certified clinical testing laboratory. Using a 2.8 million probe microarray optimized for the detection of CNVs associated with neurodevelopmental disorders, we report an overall CNV detection rate of 28.1% in 10,351 consecutive patients, which rises to nearly 33% in cases without ASD, with only developmental delay/intellectual disability (DD/ID) and/or multiple congenital anomalies (MCA). The overall detection rate for individuals with ASD is also significant at 24.4%. The detection rate and pathogenic yield of CMA vary significantly with the indications for testing, age, and gender, as well as the specialty of the ordering doctor. We note discrete differences in the most common recurrent CNVs found in individuals with or without a diagnosis of ASD.

Keywords: chromosomal microarray; copy number variants; neurodevelopmental disorders; autism spectrum disorder; variants of unknown significance; FirstStepDx PLUS

1. Introduction

Neurodevelopmental disabilities, including developmental delay (DD), intellectual disability (ID), and autism spectrum disorder (ASD) affect up to 15% of children [1]. However, in the vast majority of cases, a child's clinical presentation does not allow for a definitive etiological diagnosis.

Autism spectrum disorder is characterized by impairment in three domains with onset of one or more of these before the age of 3 years: social interaction; communication skills; and restricted, repetitive, and stereotyped patterns of behavior, interests, and activities. About 40% of individuals with ASD also have a learning disability, and roughly 30% have other co-morbidities such as seizures [1–3].

The etiology of ASD is complex and prominently involves genetic factors, including single gene changes, large genomic structural changes (i.e., deletions or duplications) known as copy number variants (CNV), and other polygenic conditions often influenced by the environment and epigenetic changes [2,3]. Genetic testing to pinpoint the underlying cause of ASD is critical to an individual's clinical management. Further, chromosomal microarray analysis (CMA) has demonstrated the highest diagnostic yield for individuals with ASD as compared to other genetic tests. Therefore, along with previously recognized indications of DD, ID and multiple congenital anomalies (MCA), children and adults presenting with ASD should be offered CMA as a first tier genetic evaluation based on the clinical guidelines from multiple professional societies [2–10].

Microarrays that employ a variety of designs and range of coverage for certain genomic regions have been applied to the clinical testing of individuals with these conditions. Diagnostic yield has increased over time as such arrays have evolved to include better coverage [7–21]. In 2011 the ACMG issued a guideline regarding optimal microarray design and recommended inclusion of additional probe content in areas of known clinical relevance [22]. The following data summarizes our experience with real-world clinical CMA testing of individuals with a diagnosis of ASD in a CLIA-certified laboratory over a period of 4.2 years. The microarray platform utilized in this study was specifically designed to increase detection of CNVs in genomic regions of demonstrated relevance to DD/ID/ASD. We also compare our experience to a non-ASD population clinically tested in parallel in the same laboratory and on the same platform.

2. Results

A total of 10,351 custom, ultra-high resolution CMAs optimized for the detection of neurodevelopmental disorders (FirstStepDx PLUS® (FSDX PLUS®)) were performed over a period of four years. This testing population had a M:F ratio of 2.5:1 and a mean age of 7.0 years. Based on ICD-9 and ICD-10 codes at the time of referral, 55% of cases represented patients with a diagnosis of ASD with or without other features (ASD+ and ASD only, respectively). Tables 1 and 2 show summary data of our neurodevelopmental patient cohort.

Table 1. Summary data of our neurodevelopmental patient cohort.

Cohort Characteristics	Total
Number of Samples	10,351
Number of Males/Females (2.5:1)	7422/2929
Non-ASD *	4657
Any ASD [†]	5694
ASD+ [‡]	2844 (27.4%)
ASD only [§]	2850 (27.5%)

* "Non-ASD" represents that portion of the cohort with no testing indication of ASD; [†] "Any ASD" refers to the portion of the cohort that has ASD as a sole testing indication, or in combination with any other testing indications, thus it represents both "ASD only" and "ASD+" cohorts combined; [‡] "ASD+" refers to the portion of the cohort with an indication of ASD as well as another testing indication, such as MCA, seizures, DD, and/or ID; [§] "ASD only" refers to the portion of the cohort with ASD as the only testing indication.

The mean age of testing was younger for the non-ASD group versus the ASD only group or the ASD+ group (Table 2). Overall, neurologists were the most common referring physicians (36%), followed by developmental pediatricians (31%), pediatricians (16%), and medical geneticists (14%). Although psychiatrists referred only 2% of total cases, they had the highest percentage of their referrals for an indication of ASD (72%) with or without other features, while only 29% of the cases referred by geneticists had an indication including ASD. Of the total caseload, 74% of the ASD cases were referred by pediatric neurologists and developmental/behavioral pediatricians.

Table 2. Mean age at chromosomal microarray analysis (CMA) testing, grouped by diagnostic referral codes.

Population	Mean Age at Testing (Years)	Standard Deviation (Years)
All	7.0	5.6
Non ASD *	6.5	6.0
ASD only [†]	7.3	5.0
ASD+ [‡]	7.5	5.1

* Non-ASD" represents that portion of the cohort with no testing indication of ASD; [†] "ASD only" refers to the portion of the cohort with ASD as the only testing indication; [‡] "ASD+" refers to the portion of the cohort with an indication of ASD as well as another testing indication, such as MCA, seizures, DD, and/or ID.

Overall, we observe a 28.0% diagnostic yield for potentially abnormal CNVs (Table 3), with an average of 1.2 reportable CNVs detected per individual. Interestingly, the rate of pathogenic findings is significantly lower (4.4%) when the diagnostic indication is ASD only compared to diagnostic indication of DD/ID/MCA without a reported diagnosis of ASD (non-ASD cohort) (12.5%) ($p < 0.001$). The pathogenic rate is only slightly higher in the ASD+ group (6.7%). However, VOUS rates are similar across the ASD only, ASD+, and non-ASD cohorts. The observation of lower rates of reportable CNVs for the ASD only cohort as compared to the ASD+ cohort, and the non-ASD cohort is maintained when looking at the overall yields for each group.

Table 3. Diagnostic yields of genetic testing in 10,351 consecutive children with neurodevelopmental disorders by diagnostic referral codes.

Result	All	Non-ASD *	Any ASD [†]	ASD+ [‡]	ASD Only [§]
Pathogenic	8.6%	12.5%	5.4%	6.5%	4.4%
VOUS	19.4%	20.1%	19.0%	19.4%	18.5%
Overall Yield	28.1%	32.6%	24.4%	25.9%	22.9%

* "Non-ASD" represents that portion of the cohort with no testing indication of ASD; [†] "Any ASD" refers to the portion of the cohort that has ASD as a sole testing indication, or in combination with any other testing indications, thus it represents both "ASD only" and "ASD +" cohorts combined; [‡] "ASD+" refers to the portion of the cohort with an indication of ASD as well as another testing indication, such as MCA, seizures, DD, and/or ID; [§] "ASD only" refers to the portion of the cohort with ASD as the only testing indication.

We stratified our cohort by age to determine whether there were differences in the rate of pathogenic findings when data were viewed this way, and found that rates were highest in the non-ASD cohort in the first year of life (18.9%), which then dropped to 10.7%–12.4% during childhood and adolescence (Table 4). In the ASD cohort, the overall pathogenic rate was slightly higher for individuals with ASD+ as compared to the overall pathogenic rate for individuals with ASD only. The pathogenic rate in the ASD+ cohort started at 4.1% in the youngest group and rose to 8.5% in the 5.5–10 years range (Table 5). The pathogenic rate in the ASD only cohort rose gradually with age from 3.4% in the youngest cohort (0–3.4 years) to a peak at 7.0% in adolescence (Table 6).

While largely targeting a pediatric population, a subset of 383 patients comprised adults over 18 years of age at the time of testing (parental and sibling studies excluded). Interestingly, in the non-ASD cohort the pathogenic rate in adults tested was 18.1%, the highest for any age cohort in this population after the first year of life (Table 4). While the percentage of pathogenic findings in adults with an indication of ASD (9.8% for ASD+; 5.6% for ASD only) was much lower than in the non-ASD population. It is worth noting that older age cohorts maintain high diagnostic yields with or without indications of ASD (Tables 4–6).

Table 4. Diagnostic yield by age in patients without autism spectrum disorder (ASD) (1750 (37.6%) females, 2907 (62.4%) males, Total n = 4657).

Age in Years	Number of Tests	Pathogenic (% Yield)	VOUS (% Yield)	Normal (% Yield)
0–1.0	439	83 (18.9%)	84 (19.1%)	272 (62.0%)
1.0–3.5	1407	151 (10.7%)	275 (19.5%)	981 (69.7%)
3.5–5.4	688	81 (11.8%)	147 (21.4%)	460 (66.8%)
5.5–10	1107	132 (11.9%)	244 (22.0%)	731 (66.1%)
10.1–18	834	103 (12.4%)	155 (18.6%)	576 (69.0%)
18+	182	33 (18.1%)	29 (15.9%)	120 (66.0%)
Total	4657	583 (12.5%)	934 (20.1%)	3140 (67.4%)

Table 5. Diagnostic yield by age in patients with ASD and other indications (610 (21.4%) females, 2234 (78.6%) males, Total n = 2844).

Age in Years	Number of Tests	Pathogenic (% Yield)	VOUS (% Yield)	Normal (% Yield)
0–3.4 *	735	30 (4.1%)	156 (21.2%)	549 (74.7%)
3.5–5.4	630	34 (5.4%)	114 (18.1%)	482 (76.5%)
5.5–10	710	60 (8.5%)	132 (18.6%)	518 (72.9%)
10.1–18	657	49 (7.5%)	126 (19.2%)	482 (73.3%)
18+	112	11 (9.8%)	24 (21.4%)	77 (68.8%)
Total	2844	184 (6.5%)	552 (19.4%)	2108 (74.1%)

* Due to the typical age of clinical recognition and diagnoses for ASD, the range of 0–3.4 years was used for the youngest grouping in this Table.

Table 6. Diagnostic yield by age in patients with only ASD indicated (569 (20.0%) females, 2281 (80.0%) males, Total n = 2850).

Age in Years	Number of Tests	Pathogenic (% Yield)	VOUS (% Yield)	Normal (% Yield)
0–3.4 *	701	24 (3.4%)	126 (18.0%)	551 (78.6%)
3.5–5.4	661	21 (3.2%)	119 (18.0%)	521 (78.8%)
5.5–10	768	32 (4.2%)	160 (20.8%)	576 (75.0%)
10.1–18	631	44 (7.0%)	110 (17.4)	477 (75.6%)
18+	89	5 (5.6%)	14 (15.7%)	70 (78.7%)
Total	2850	126 (4.4%)	529 (18.6%)	2195 (77.0%)

* Due to the typical age of clinical recognition and diagnoses for ASD, the range of 0–3.4 years was used for the youngest grouping in this Table.

The most common pathogenic findings detected in this series of individuals evaluated by CMA are shown in Figures 1–3. We observed, in some cases, striking differences in the frequencies of pathogenic findings when patients are grouped by testing indications and/or gender. For example, the 22q11.2 deletion and, to a lesser extent, the proximal 16p11.2 deletion, were far more prevalent in the non-ASD group than in the combined ASD group, suggesting that indications other than ASD in these patients are common or that ASD is less readily diagnosed in these subgroups (Figure 1). However, the 15q11.2 BP1–BP2 deletion (also known as the Burnside-Butler susceptibility locus) and the proximal 16p11.2 duplication were equally likely to be detected if ASD was indicated or not (Figure 1). In contrast, *NRXN1* gene deletions were much more common when ASD was indicated in comparison to when it was not, by a factor of nearly 4-fold (Figure 1). Figure 2 displays some similarities and differences between the ASD+ and ASD only populations; for example the 15q11.2 BP1–BP2 deletion is the most common finding for both cohorts, but detection frequencies vary significantly for other findings, with 47,XXY being the next most common for the ASD-only cohort while the proximal 16p11.2 deletions and duplications are the next most frequent finding in the ASD+ cohort.

Males outnumbered females in our study population, and the rate of abnormality differed significantly with females having higher rates of pathogenic findings across all diagnostic groupings

($p < 0.001$ for non-ASD and ASD+ groups as well as overall); ASD only detection rates in females vs. males were not statistically different ($p = 0.22$) (Table 7). In the combined ASD group, in addition to specific sex-limited diagnoses like 47,XXY and 47,XYY, there were excesses by gender in the prevalence of several common abnormalities, notably with the 15q11.2 BP1–BP2 deletion, 15q duplication, 15q13.3 deletion, proximal 16p11.2 duplication, 16p12.2 deletion, and 22q11.2 deletion all skewed toward a female preponderance by up to 2–3-fold (Figure 3).

Table 7. Rates of diagnostic findings on CMA by gender and diagnostic referral codes grouping.

CMA Result Type Rates by Gender	Non-ASD	ASD+	ASD only	All
Female				
Pathogenic	14.4%	8.5%	5.3%	11.4%
VOUS	20.9%	19.5%	17.2%	19.9%
Normal	64.7%	72.0%	77.5%	68.7%
Male				
Pathogenic	11.4%	5.9%	4.2%	7.5%
VOUS	19.6%	19.4%	18.9%	19.3%
Normal	69.0%	74.7%	76.9%	73.2%

3. Discussion

CMA is the guideline-recognized first-tier test in the evaluation of individuals with DD/ID, MCA, and most recently ASD [4,6–10,15]. CMA yields significant rates of pathogenic or potentially pathogenic (VOUS) results [2,11–20], which have clinical utility for the case-by-case clinical management of individuals with these individually rare disorders [23–34].

Since the introduction of CMA technology, the total genomic content with probe coverage has progressively increased leading to higher diagnostic yields and better resolution of chromosomal abnormalities. Collectively this trend has resulted in corresponding increases in the clinical value of CMA testing [11–21,24–34]. In addition to guidelines on the clinical indications for CMA, the American College of Medical Genetics and Genomics (ACMG) has issued guidance on the appropriate content and design of such arrays and specifically opined that, "It is desirable to have enrichment of probes targeting dosage-sensitive genes known to result in phenotypes consistent with common indications for a genomic screen (e.g., intellectual disability, developmental delays, autism, and congenital anomalies)" [22]. We report here on over four years of clinical experience with a real-world referral base for testing on an ultra-high resolution chromosomal microarray specifically designed to extend the scope of detection for individuals with ASD and other neurodevelopmental disorders. This microarray was optimized through the addition of probes targeting genomic regions more recently identified to have pathogenic relevance for DD, ID, and ASD [21,35,36].

The overall detection rate in this series for clinically established pathogenic CNVs of 8.6% is comparable to other reported series/platforms [2,11–20], despite the inherent bias toward lower rates based on the real-world referral base and higher percentage of individuals with ASD in this population. When ASD is not among the testing indications, the rate of pathogenic findings is 12.5% and the overall diagnostic yield is 32.6%, both of which are at the upper end of reported diagnostic rates. We have previously shown that diagnostic yield varies significantly on a multivariate basis including but not limited to: referring physician specialty, age of patient at testing, patient gender, and referring indication or combination of indications for testing [20]. When we look in this study at the influence of an ASD diagnosis on pathogenic diagnosis rates by practitioners, geneticists have the highest detection rate when there is no indication of ASD for testing, but also the lowest rate for individuals with ASD only.

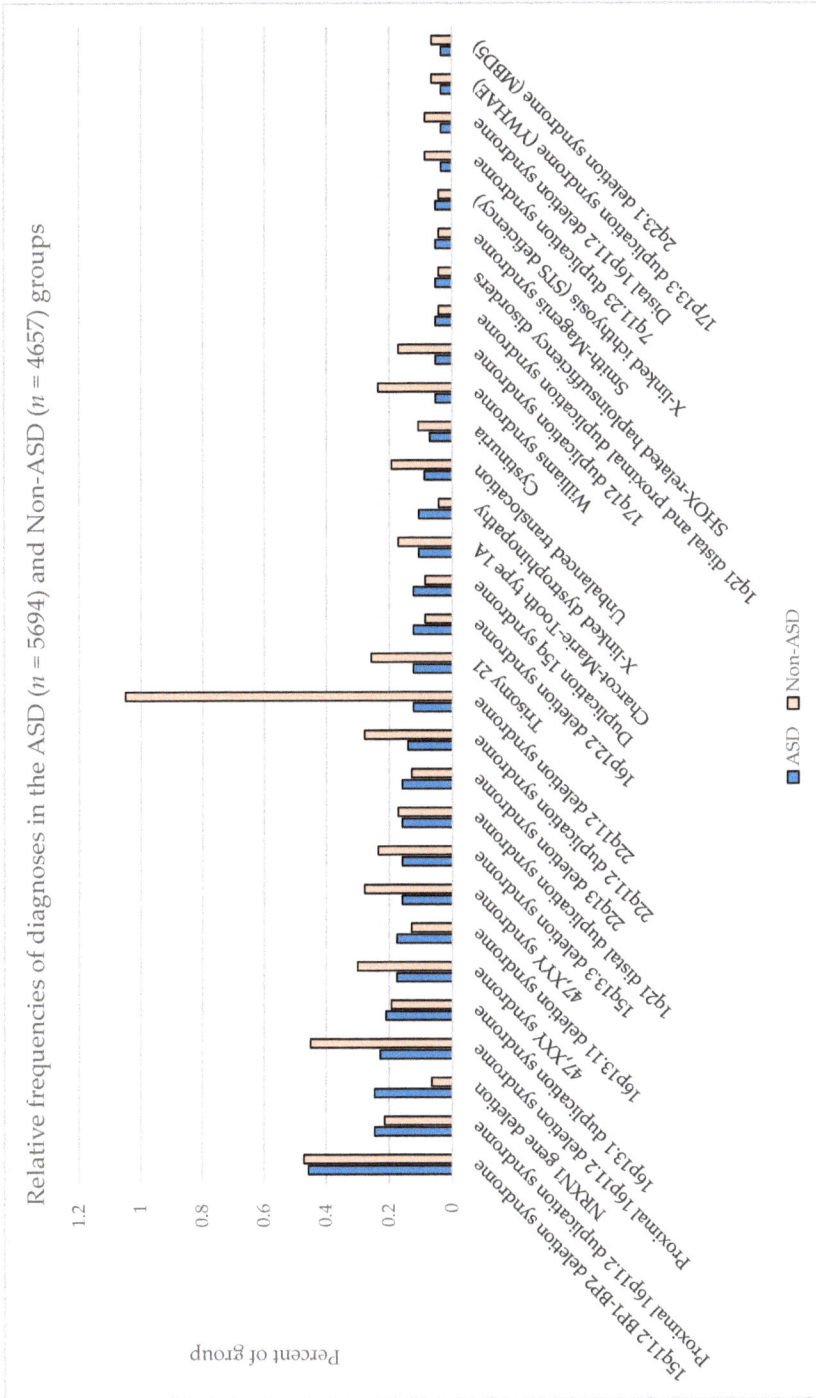

Figure 1. Relative frequencies of diagnoses in the combined ASD (*n* = 5694) and Non-ASD (*n* = 4657) groups.

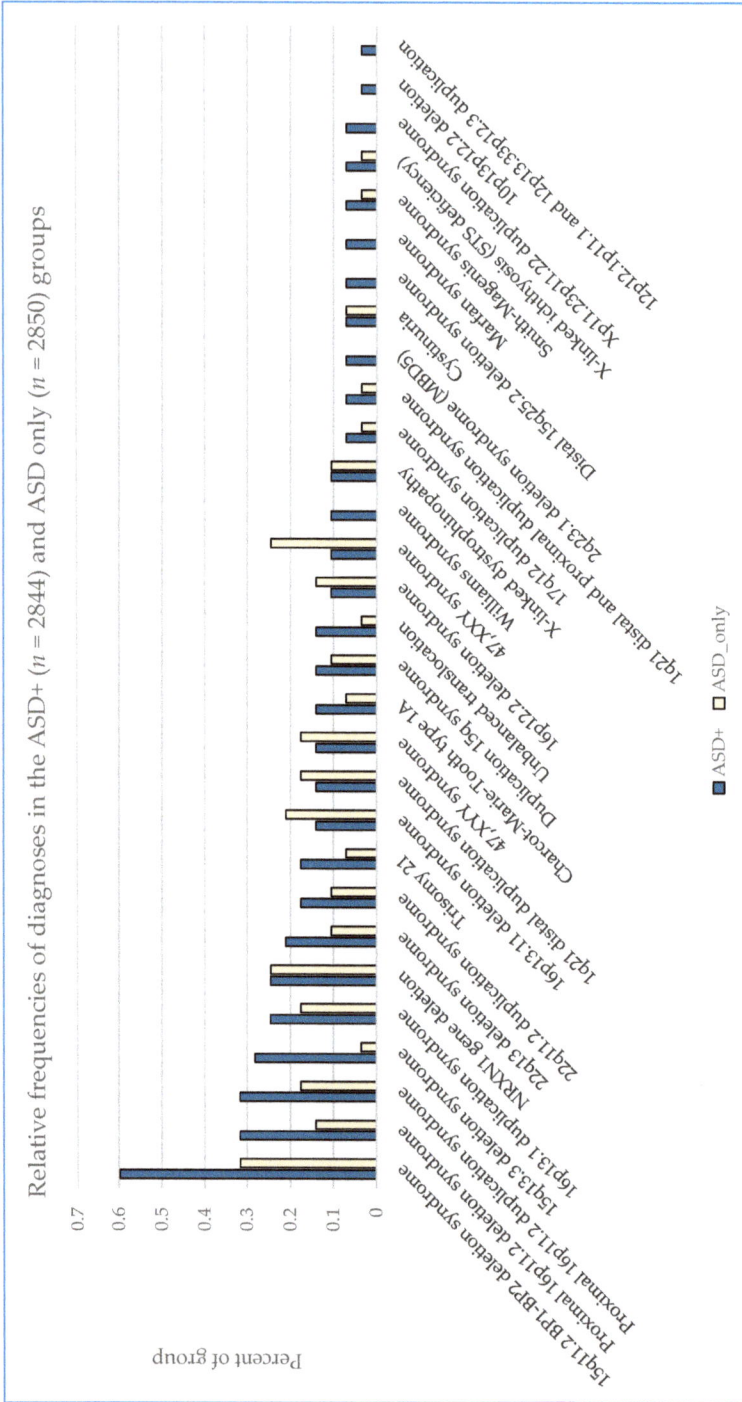

Figure 2. Relative frequencies of diagnoses in the ASD+ (*n* = 2844) and ASD only (*n* = 2850) groups.

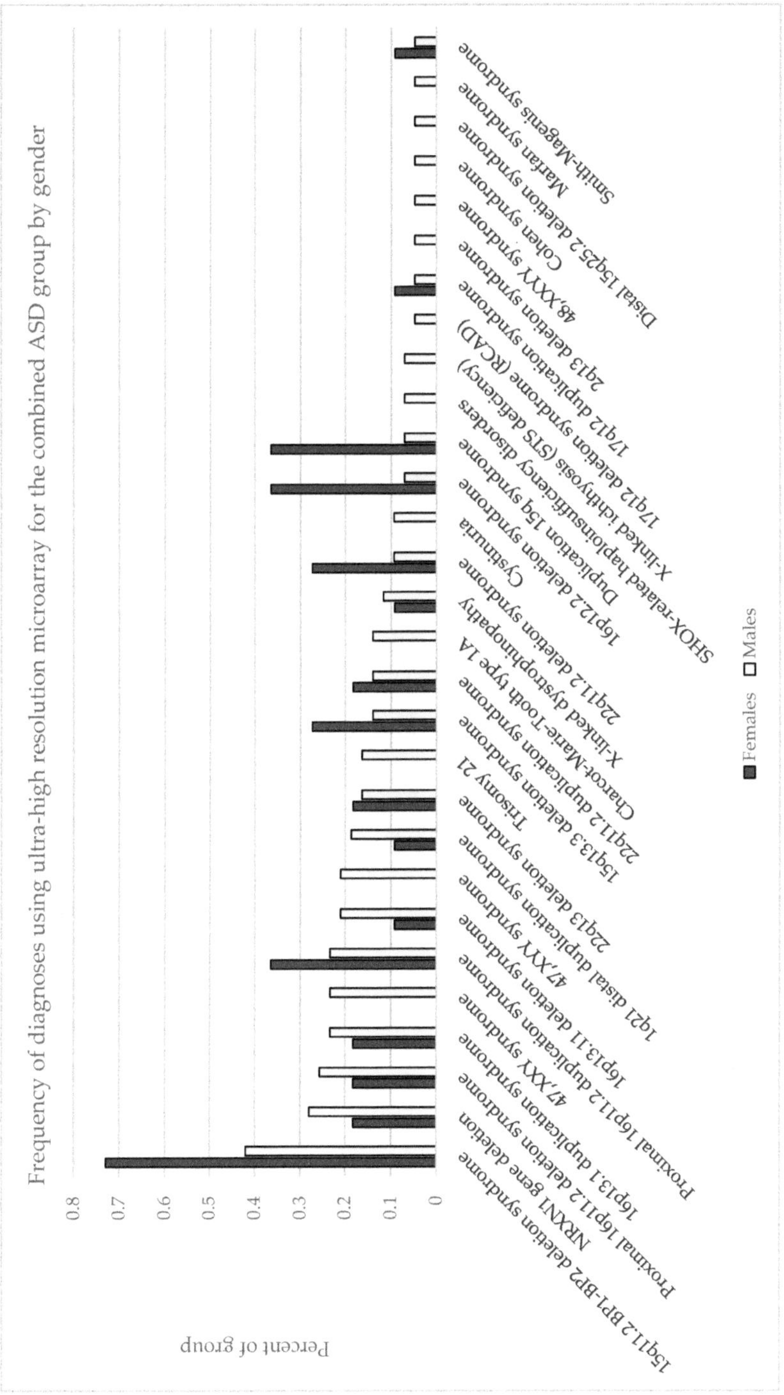

Figure 3. Frequency of diagnoses for combined ASD group by gender. Total females = 2929; total males = 7422.

The overall detection rate in this series for clinically established pathogenic CNVs of 8.6% is comparable to other reported series/platforms [2,11–20], despite the inherent bias toward lower rates based on the real-world referral base and higher percentage of individuals with ASD in this population. When ASD is not among the testing indications, the rate of pathogenic findings is 12.5% and the overall diagnostic yield is 32.6%, both of which are at the upper end of reported diagnostic rates. We have previously shown that diagnostic yield varies significantly on a multivariate basis including but not limited to: referring physician specialty, age of patient at testing, patient gender, and referring indication or combination of indications for testing [20]. When we look in this study at the influence of an ASD diagnosis on pathogenic diagnosis rates by practitioners, geneticists have the highest detection rate when there is no indication of ASD for testing, but also the lowest rate for individuals with ASD only.

While significantly lower than both the overall population and the ASD-excluded sub-population ($p < 0.0001$), the diagnostic yield in all cases including ASD (ASD only and ASD+) are 5.4% pathogenic, 19.0% VOUS, and 24.4% overall. This overall rate exceeds those previously reported [3,14–17] and this supports the value of incremental targeted content for CMA design.

We detected VOUS at an overall rate of 19.4%. Although earlier literature did not typically consider VOUS in the diagnostic yield, this was due to inconsistent criteria for reporting, lack of established databases of normal population variants, and limited sharing of data [10]. Today, it is common and reasonable to consider VOUS in the overall diagnostic yield [3,14,15]. Many VOUS results will evolve into clearly pathogenic findings based on emerging clinical evidence [37]. The excess of males to females in our overall and combined ASD cohorts are consistent with previous reports. In addition, we confirmed the higher rates of abnormalities in the tested female populations in comparison to males, which has been previously observed [3,14].

A recent report by one of us (MGB) on CMA use in individuals with DD and ASD at a single midwest genetics center using relatively low resolution (<180 K) oligonucleotide arrays found 6 of 65 patients with ASD (9%) to have a pathogenic finding and 20% overall (13/65) had a reportable CNV. This is higher than the rate of pathogenic findings in our ASD population (6.7% ASD+; 4.4% ASD only); however, this was a much smaller total population and in addition was a closely-studied cohort within an academic medical center where all patients had complete clinical genetics evaluations [3]. The overall diagnostic yield of 20% in that study was similar to rates reported by Shen et al. [14] (18.2% in 932 patients with ASD) and Schaefer et al. [15] (21% or 14 patients 68 with ASD) but is lower than the 24.4% observed with the optimized array in this series.

Numerous studies have now demonstrated the clinical actionability and utility of CMA testing [23–28,38,39], and increased yield as described here will extend the range and scope of this utility. The increased rate of CNVs classified as VOUS is therefore of potential clinical importance in this setting and consideration of location, gene content, and other factors may help clinicians with such patients despite the complexities of interpreting them and counseling families as to their potential significance. Of critical importance is the ongoing evaluation of novel methods to assess the potential role of VOUS in the underlying pathology of individual patients. This process will allow us as a community to realize the maximum benefit of the increased detection rate achieved through array and interpretation optimization. Furthermore, VOUS results have been clearly demonstrated to be of great importance to parents of patients with DD/ID/ASD [34,40–43].

It is estimated that at least 20% of individuals with ASD have an underlying genetic syndrome, but a survey of a large autism center showed that less than 10% of their population had received any form of genetic evaluation [44]. The evidence here also supports that patients with a diagnosis of ASD remain under-tested overall. The fact that the age of CMA diagnosis in those with ASD is a full standard deviation greater than the potential age of clinical diagnosis speaks to the delay or reticence in taking critical steps to better medically manage these patients. The direct correlation between higher rate of detected abnormalities and age in the ASD cohort suggests that earlier use of CMA and perhaps other genetic testing methods may be important for early intervention.

While still a relatively small sub-cohort, it is remarkable that adults (>18 years old) tested also have the highest pathogenic CNV rate of all diagnostic groups examined. This could be reflective of severity in that particular age group. For example, clinicians/families might believe that testing isn't as valuable for adults but perform it anyway when the individual is considered to be relatively severely impaired. In addition, this may also reflect the desire for adults (or adult siblings of the individual with clinical features) to define recurrence risk to their potential offspring.

Even before prior CMA was introduced into clinical use, the most common chromosome abnormalities associated with apparently isolated ASD were duplication of the 15q11–q13 region (typically of maternal origin) [45] and large microdeletions in the chromosome 16p11.2 and 22q regions reportedly accounting for as many as 1%–5% of ASD related abnormalities each [3,23]. Although these well-described recurrent abnormalities were prominent and relatively more abundant in our ASD cohorts, their prevalence was not as high as predicted by the literature; however, some of the reports suggesting high rates may have been biased by multiplex families.

Partial deletions involving the *NRXN1* gene are now well-described abnormalities, and impairment of the function of the synaptic adhesion protein it encodes, leading to a potential loss of synaptic integrity, is thought to be central to the pathogenesis of ASD [46,47]. *NRXN1* gene deletions were significantly over-represented in both our ASD+ and ASD only groups but were also observed at least occasionally in the non-ASD group. The latter may be due to either early testing for some other clinical feature prior to the formal recognition of ASD-related features, the indubitable inadequacies of relying on physician coding on test requisitions for phenotypic data, or true overlap into other neurodevelopmental conditions without ASD.

The observational study of a large consecutive series of genetic testing for neurodevelopmental disorders, over half of which had feature of ASD highlights the significant value of CMA in defining not merely the underlying etiology but in directing future research into the underlying pathophysiology for improved and ultimately targeted treatments. While it is not ideal to rely on ICD-9/10 coding on test requisitions to define the phenotypic sub-groups of this population and parse the results relevant to ASD, the comparability and trends of this large data set suggest that the conclusions are neither random nor merely directional, and likely reflects a reasonable picture of the scope of abnormalities in these populations. Improved diagnostic tools will lead to increased clinical utility and in the end better clinical management.

4. Materials and Methods

4.1. Patient Ascertainment

A consecutive series of 10,351 real-world samples referred for CMA to a CLIA-licensed clinical laboratory for etiological diagnosis of DD/ID/ASD and MCAs between July 2012 and September 2016 was reviewed for clinical characteristics and related diagnoses. The overwhelming majority of samples were buccal swabs, however 1037/10351 (~10%) cases were conducted on blood specimens. Testing indications were delineated based on International Classification of Diseases, Clinical Modification, Revisions 9 or 10, (ICD-9, ICD-10) (Centers for Medicare & Medicaid Services (cms.gov)).

4.2. Microarray Design

The FSDX PLUS® microarray utilized in this study, and its analytical and clinical validation, have been described in detail elsewhere [21]. It is an expanded whole genome chromosome microarray (CMA) built upon the ultra-high resolution Affymetrix CytoscanHD® platform (Santa Clara, CA, USA) plus 88,435 custom probes targeting genomic regions strongly associated with DD/ID/ASD [14–24]. Both copy number (CNV) and single nucleotide polymorphic (SNP) probes are included in the array, which is consistent with the ACMG guideline for CMA design, as is the "enrichment of probes targeting dosage-sensitive genes known to result in phenotypes consistent with common indications for a genomic screen" [22]. Such critical regions that did not contain >1 probe/1000 bp on the baseline array were supplemented with additional probe content to provide improved detection of smaller deletions and duplications. Additional probe enrichment was of genomic regions identified by our prior studies and elsewhere in the medical literature of published copy number variants and individual genes associated with DD/ID/ASD [23–36]. The increase in analytical sensitivity resulting from this additional 3.3% probe content has been calculated to be 2.6% [21].

4.3. CMA Performance and Interpretation

CMA was routinely performed on DNA extracted by standard methodologies from buccal swab samples (ORAcollect®) in a CLIA-certified laboratory. CMA reagents and equipment were as specified by Affymetrix. Established cytogenetic criteria for interpretation were routinely applied with a minimum of 25-consecutive impacted probes as the baseline determinant for deletions and 50 probes for duplications [35]. Rare CNVs (<1% overall population frequency) were determined to be "pathogenic" if there is sufficient published clinical evidence (at least two independent publications) to indicate that haploinsufficiency or triplosensitivity of the region or gene(s) involved is causative of clinical features. If only preliminary evidence for a causative role for the region or gene(s) therein was found, they were classified as variants of unknown significance (VOUS) as were areas of absence of heterozygosity (AOH) which may increase the risk for conditions with autosomal recessive inheritance or conditions with parent-of-origin/imprinting effects. Cases with only CNVs contained in databases such as the Database of Genomic Variants (DGV) [39] that document presumptively benign CNVs were reported as normal.

4.4. Statistical Methods

Chi-square tests for independence and 2-sided *t*-tests for normal distribution were applied to the data for determination of significance of findings.

5. Conclusions

Ultra-high resolution CMA has demonstrated great value in the clinical assessment of neurodevelopmental disorders. The diagnostic yield of the optimized CMA platform described here is dependent on many factors, including patient gender, age at testing, clinical presentation, and specialty of the ordering physician. Pathogenic findings give insights into the etiology of patients' neurodevelopmental conditions, and in many cases positively impact medical management decisions. The development of novel and accurate methods to interpret the potential pathenogenicity of VOUS will further enable patients and their physicians to realize the maximum benefits of genetic testing for clinical care.

Acknowledgments: We thank the participating providers and their patients for their dedication and support, Suresh Venkatasubramanian for developing custom code to aid in data analysis, Sean Dixon, Kenny Lentz, and Jon Thornton for database management and technical support. This study was made possible by the entire Lineagen team's remarkable energy and dedication to those with neurodevelopmental disabilities. The study was funded by Lineagen, Inc.

Author Contributions: E. Robert Wassman, Merlin G. Butler and Karen S. Ho conceived and designed the experiments; Hope Twede and Karen S. Ho performed the experiments; E. Robert Wassman, Hope Twede, and Karen S. Ho analyzed the data; Adrianne L. Baxter, Charles H. Hensel, Megan M. Martin, Aparna Prasad, and Rena Vanzo contributed to the analysis; E. Robert Wassman wrote the paper.

Conflicts of Interest: Merlin G. Butler has no conflicts. E. Robert Wassman, Karen S. Ho, Hope Twede, Adrianne L. Baxter, Rena Vanzo, Charles H. Hensel, Megan M. Martin, and Aparna Prasad are employees of Lineagen, Inc. which is a clinical reference laboratory performing genetic testing for individuals with neurodevelopmental disorders.

Abbreviations

CNV	copy number variant
CMA	chromosomal microarray
ASD	autism spectrum disorder
ID	intellectual disability
DD	developmental delay
MCA	multiple congenital anomalies

References

1. Boyle, C.A.; Boulet, S.; Schieve, L.A.; Cohen, R.A.; Blumberg, S.J.; Yeargin-Allsopp, M.; Visser, S.; Kogan, M.D. Trends in the prevalence of developmental disabilities in US children, 1997–2008. *Pediatrics* **2011**, *127*, 1034–1042. [CrossRef] [PubMed]

2. Heil, K.M.; Schaaf, C.P. The genetics of autism spectrum disorders—A guide for clinicians. *Curr. Psychiatry Rep.* **2013**, *15*, 334. [CrossRef] [PubMed]

3. Roberts, J.L.; Hovanes, K.; Dasouki, M.; Manzardo, A.M.; Butler, M.G. Chromosomal microarray analysis of consecutive individuals with autism spectrum disorders or learning disability presenting for genetic services. *Gene* **2014**, *535*, 70–78. [CrossRef] [PubMed]

4. Manning, M.; Hudgins, L. Array-based technology and recommendations for utilization in medical genetics practice for detection of chromosomal abnormalities. *Genet. Med.* **2010**, *12*, 742–745. [CrossRef] [PubMed]

5. Schaefer, G.B.; Mendelsohn, N.J. Professional practice and guidelines committee clinical genetics evaluation in identifying the etiology of autism spectrum disorders: 2013 Guideline revisions. *Genet. Med.* **2013**, *15*, 399–407. [CrossRef] [PubMed]

6. Volkmar, F.; Siegel, M.; Woodbury-Smith, M.; King, B.; McCracken, J.; State, M. American academy of child and adolescent psychiatry (AACAP) committee on quality issues (CQI) practice parameter for the assessment and treatment of children and adolescents with autism spectrum disorder. *J. Am. Acad. Child Adolesc. Psychiatry* **2014**, *53*, 237–257. [CrossRef] [PubMed]

7. Moeschler, J.B.; Shevell, M. Committee on genetics comprehensive evaluation of the child with intellectual disability or global developmental delays. *Pediatrics* **2014**, *134*, e903–e918. [CrossRef] [PubMed]

8. Michelson, D.J.; Shevell, M.I.; Sherr, E.H.; Moeschler, J.B.; Gropman, A.L.; Ashwal, S. Evidence report: Genetic and metabolic testing on children with global developmental delay: Report of the quality standards subcommittee of the american academy of neurology and the practice committee of the child neurology society. *Neurology* **2011**, *77*, 1629–1635. [CrossRef] [PubMed]

9. Battaglia, A.; Doccini, V.; Bernardini, L.; Novelli, A.; Loddo, S.; Capalbo, A.; Filippi, T.; Carey, J.C. Confirmation of chromosomal microarray as a first-tier clinical diagnostic test for individuals with developmental delay, intellectual disability, autism spectrum disorders and dysmorphic features. *Eur. J. Paediatr. Neurol.* **2013**, *17*, 589–599. [CrossRef] [PubMed]

10. Miller, D.T.; Adam, M.P.; Aradhya, S.; Biesecker, L.G.; Brothman, A.R.; Carter, N.P.; Church, D.M.; Crolla, J.A.; Eichler, E.E.; Epstein, C.J.; et al. Consensus statement: Chromosomal microarray is a first-tier clinical diagnostic test for individuals with developmental disabilities or congenital anomalies. *Am. J. Hum. Genet.* **2010**, *86*, 749–764. [CrossRef] [PubMed]

11. Bernardini, L.; Alesi, V.; Loddo, S.; Novelli, A.; Bottillo, I.; Battaglia, A.; Digilio, M.C.; Zampino, G.; Ertel, A.; Fortina, P.; et al. High-resolution SNP arrays in mental retardation diagnostics: How much do we gain? *Eur. J. Hum. Genet.* **2010**, *18*, 178–185. [CrossRef] [PubMed]

12. McGrew, S.G.; Peters, B.R.; Crittendon, J.A.; Veenstra-Vanderweele, J. Diagnostic yield of chromosomal microarray analysis in an autism primary care practice: Which guidelines to implement? *J. Autism Dev. Disord.* **2012**, *42*, 1582–1591. [CrossRef] [PubMed]

13. Howell, K.B.; Kornberg, A.J.; Harvey, A.S.; Ryan, M.M.; Mackay, M.T.; Freeman, J.L.; Rodriguez Casero, M.V.; Collins, K.J.; Hayman, M.; Mohamed, A.; et al. High resolution chromosomal microarray in undiagnosed neurological disorders. *J. Paediatr. Child Health* **2013**, *49*, 716–724. [CrossRef] [PubMed]

14. Shen, Y.; Dies, K.A.; Holm, I.A.; Bridgemohan, C.; Sobeih, M.M.; Caronna, E.B.; Miller, K.J.; Frazier, J.A.; Silverstein, I.; Picker, J.; et al. Autism consortium clinical genetics/DNA diagnostics collaboration clinical genetic testing for patients with autism spectrum disorders. *Pediatrics* **2010**, *125*, e727–e735. [CrossRef] [PubMed]

15. Schaefer, G.B.; Starr, L.; Pickering, D.; Skar, G.; Dehaai, K.; Sanger, W.G. Array comparative genomic hybridization findings in a cohort referred for an autism evaluation. *J. Child Neurol.* **2010**, *25*, 1498–1503. [CrossRef] [PubMed]

16. Edelmann, L.; Hirschhorn, K. Clinical utility of array CGH for the detection of chromosomal imbalances associated with mental retardation and multiple congenital anomalies. *Ann. N. Y. Acad. Sci.* **2009**, *1151*, 157–166. [CrossRef] [PubMed]

17. Beaudet, A.L. The utility of chromosomal microarray analysis in developmental and behavioral pediatrics. *Child Dev.* **2013**, *84*, 121–132. [CrossRef] [PubMed]

18. Mason-Suares, H.; Kim, W.; Grimmett, L.; Williams, E.S.; Horner, V.L.; Kunig, D.; Goldlust, I.S.; Wu, B.-L.; Shen, Y.; Miller, D.T.; et al. Density matters: Comparison of array platforms for detection of copy-number variation and copy-neutral abnormalities. *Genet. Med.* **2013**, *15*, 706–712. [CrossRef] [PubMed]

19. Pfundt, R.; Kwiatkowski, K.; Roter, A.; Shukla, A.; Thorland, E.; Hockett, R.; DuPont, B.; Fung, E.T.; Chaubey, A. Clinical performance of the CytoScan Dx Assay in diagnosing developmental delay/intellectual disability. *Genet. Med.* **2016**, *18*, 168–173. [CrossRef] [PubMed]

20. Ho, K.S.; Twede, H.; Vanzo, R.; Harward, E.; Hensel, C.H.; Martin, M.M.; Page, S.; Peiffer, A.; Mowery-Rushton, P.; Serrano, M.; et al. Clinical performance of an ultra-high resolution chromosomal microarray optimized for neurodevelopmental disorders. *BioMed Res. Int.* **2016**. [CrossRef]

21. Hensel, C.H.; Vanzo, R.; Martin, M.; Dixon, S.; Lambert, C.G.; Levy, B.; Nelson, L.; Peiffer, A.; Ho, K.S.; Serrano, M.; et al. Analytical and clinical validity study of FirstStepDx PLUS: A chromosomal microarray optimized for patients with neurodevelopmental conditions. *bioRxiv* **2016**. [CrossRef]

22. Kearney, H.M.; South, S.T.; Wolff, D.J.; Lamb, A.; Hamosh, A.; Rao, K.W. Working group of the American college of medical genetics american college of medical genetics recommendations for the design and performance expectations for clinical genomic copy number microarrays intended for use in the postnatal setting for detection of constitutional abnormalities. *Genet. Med.* **2011**, *13*, 676–679. [PubMed]

23. Weiss, L.A.; Shen, Y.; Korn, J.M.; Arking, D.E.; Miller, D.T.; Fossdal, R.; Saemundsen, E.; Stefansson, H.; Ferreira, M.A.R.; Green, T.; et al. Association between microdeletion and microduplication at 16p11.2 and autism. *N. Engl. J. Med.* **2008**, *358*, 667–675. [CrossRef] [PubMed]

24. Jacquemont, M.-L.; Sanlaville, D.; Redon, R.; Raoul, O.; Cormier-Daire, V.; Lyonnet, S.; Amiel, J.; Le Merrer, M.; Heron, D.; de Blois, M.-C.; et al. Array-based comparative genomic hybridisation identifies high frequency of cryptic chromosomal rearrangements in patients with syndromic autism spectrum disorders. *J. Med. Genet.* **2006**, *43*, 843–849. [CrossRef] [PubMed]

25. Saam, J.; Gudgeon, J.; Aston, E.; Brothman, A.R. How physicians use array comparative genomic hybridization results to guide patient management in children with developmental delay. *Genet. Med.* **2008**, *10*, 181–186. [CrossRef] [PubMed]

26. Coulter, M.E.; Miller, D.T.; Harris, D.J.; Hawley, P.; Picker, J.; Roberts, A.E.; Sobeih, M.M.; Irons, M. Chromosomal microarray testing influences medical management. *Genet. Med.* **2011**, *13*, 770–776. [CrossRef] [PubMed]

27. Ellison, J.W.; Ravnan, J.B.; Rosenfeld, J.A.; Morton, S.A.; Neill, N.J.; Williams, M.S.; Lewis, J.; Torchia, B.S.; Walker, C.; Traylor, R.N.; et al. Clinical utility of chromosomal microarray analysis. *Pediatrics* **2012**, *130*, e1085–e1095. [CrossRef] [PubMed]

28. Riggs, E.R.; Wain, K.E.; Riethmaier, D.; Smith-Packard, B.; Faucett, W.A.; Hoppman, N.; Thorland, E.C.; Patel, V.C.; Miller, D.T. Chromosomal microarray impacts clinical management. *Clin. Genet.* **2014**, *85*, 147–153. [CrossRef] [PubMed]

29. Henderson, L.B.; Applegate, C.D.; Wohler, E.; Sheridan, M.B.; Hoover-Fong, J.; Batista, D.A.S. The impact of chromosomal microarray on clinical management: A retrospective analysis. *Genet. Med.* **2014**, *16*, 657–664. [CrossRef] [PubMed]

30. Tao, V.Q.; Chan, K.Y.K.; Chu, Y.W.Y.; Mok, G.T.K.; Tan, T.Y.; Yang, W.; Lee, S.L.; Tang, W.F.; Tso, W.W.Y.; Lau, E.T.; et al. The clinical impact of chromosomal microarray on paediatric care in Hong Kong. *PLoS ONE* **2014**, *9*, e109629. [CrossRef] [PubMed]

31. Sdano, M.R.; Vanzo, R.J.; Martin, M.M.; Baldwin, E.E.; South, S.T.; Rope, A.F.; Allen, W.P.; Kearney, H. Clinical utility of chromosomal microarray analysis of DNA from buccal cells: Detection of mosaicism in three patients. *J. Genet. Counsel.* **2014**, *23*, 922–927. [CrossRef] [PubMed]

32. Martin, M.M.; Vanzo, R.J.; Sdano, M.R.; Baxter, A.L.; South, S.T. Mosaic deletion of 20pter due to rescue by somatic recombination. *Am. J. Med. Genet. A* **2016**, *170*, 243–248. [CrossRef] [PubMed]

33. Ho, K.S.; South, S.T.; Lortz, A.; Hensel, C.H.; Sdano, M.R.; Vanzo, R.J.; Martin, M.M.; Peiffer, A.; Lambert, C.G.; Calhoun, A.; et al. Chromosomal microarray testing identifies a 4p terminal region associated with seizures in Wolf-Hirschhorn syndrome. *J. Med. Genet.* **2016**, *53*, 256–263. [CrossRef] [PubMed]

34. Gurrieri, F. Working up autism: The practical role of medical genetics. *Am. J. Med. Genet. C Semin. Med. Genet.* **2012**, *160*, 104–110. [CrossRef] [PubMed]

35. Matsunami, N.; Hensel, C.H.; Baird, L.; Stevens, J.; Otterud, B.; Leppert, T.; Varvil, T.; Hadley, D.; Glessner, J.T.; Pellegrino, R.; et al. Identification of rare DNA sequence variants in high-risk autism families and their prevalence in a large case/control population. *Mol. Autism* **2014**, *5*, 5. [CrossRef] [PubMed]

36. Matsunami, N.; Hadley, D.; Hensel, C.H.; Christensen, G.B.; Kim, C.; Frackelton, E.; Thomas, K.; da Silva, R.P.; Stevens, J.; Baird, L.; et al. Identification of rare recurrent copy number variants in high-risk autism families and their prevalence in a large ASD population. *PLoS ONE* **2013**, *8*, e52239. [CrossRef] [PubMed]

37. Palmer, E.; Speirs, H.; Taylor, P.J.; Mullan, G.; Turner, G.; Einfeld, S.; Tonge, B.; Mowat, D. Changing interpretation of chromosomal microarray over time in a community cohort with intellectual disability. *Am. J. Med. Genet. A* **2014**, *164*, 377–385. [CrossRef] [PubMed]

38. South, S.T.; Lee, C.; Lamb, A.N.; Higgins, A.W.; Kearney, H.M. Working group for the American college of medical genetics and genomics laboratory quality assurance committee ACMG standards and guidelines for constitutional cytogenomic microarray analysis, including postnatal and prenatal applications: Revision 2013. *Genet. Med.* **2013**, *15*, 901–909. [CrossRef] [PubMed]

39. MacDonald, J.R.; Ziman, R.; Yuen, R.K.C.; Feuk, L.; Scherer, S.W. The database of genomic variants: A curated collection of structural variation in the human genome. *Nucleic Acids Res.* **2014**, *42*, D986–D992. [CrossRef] [PubMed]

40. Reiff, M.; Bernhardt, B.A.; Mulchandani, S.; Soucier, D.; Cornell, D.; Pyeritz, R.E.; Spinner, N.B. "What does it mean?": Uncertainties in understanding results of chromosomal microarray testing. *Genet. Med.* **2012**, *14*, 250–258. [CrossRef] [PubMed]

41. Reiff, M.; Giarelli, E.; Bernhardt, B.A.; Easley, E.; Spinner, N.B.; Sankar, P.L.; Mulchandani, S. Parents' perceptions of the usefulness of chromosomal microarray analysis for children with autism spectrum disorders. *J. Autism Dev. Disord.* **2015**, *45*, 3262–3275. [CrossRef] [PubMed]

42. Jez, S.; Martin, M.; South, S.; Vanzo, R.; Rothwell, E. Variants of unknown significance on chromosomal microarray analysis: Parental perspectives. *J. Community Genet.* **2015**, *6*, 343–349. [CrossRef] [PubMed]

43. Wilkins, E.J.; Archibald, A.D.; Sahhar, M.A.; White, S.M. "It wasn't a disaster or anything": Parents' experiences of their child's uncertain chromosomal microarray result. *Am. J. Med. Genet. A* **2016**, *170*, 2895–2904. [CrossRef] [PubMed]

44. Wenger, T.L.; Kao, C.; McDonald-McGinn, D.M.; Zackai, E.H.; Bailey, A.; Schultz, R.T.; Morrow, B.E.; Emanuel, B.S.; Hakonarson, H. The role of mGluR copy number variation in genetic and environmental forms of syndromic autism spectrum disorder. *Sci. Rep.* **2016**, *6*, 19372. [CrossRef] [PubMed]

45. Cox, D.M.; Butler, M.G. The 15q11.2 BP1–BP2 microdeletion syndrome: A review. *Int. J. Mol. Sci.* **2015**, *16*, 4068–4082. [CrossRef] [PubMed]

46. Pak, C.; Danko, T.; Zhang, Y.; Aoto, J.; Anderson, G.; Maxeiner, S.; Yi, F.; Wernig, M.; Südhof, T.C. Human neuropsychiatric disease modeling using conditional deletion reveals synaptic transmission defects caused by heterozygous mutations in NRXN1. *Cell Stem Cell* **2015**, *17*, 316–328. [CrossRef] [PubMed]

47. Prasad, A.; Merico, D.; Thiruvahindrapuram, B.; Wei, J.; Lionel, A.C.; Sato, D.; Rickaby, J.; Lu, C.; Szatmari, P.; Roberts, W.; et al. A discovery resource of rare copy number variations in individuals with autism spectrum disorder. *G3* **2012**, *2*, 1665–1685. [CrossRef] [PubMed]

International Journal of
Molecular Sciences

MDPI

Review

Delineating the Common Biological Pathways Perturbed by ASD's Genetic Etiology: Lessons from Network-Based Studies

Oded Oron and Evan Elliott *

Molecular and Behavioral Neurosciences Lab, Bar-Ilan University, Faculty of Medicine, 13215 Safed, Israel;
odedoron@gmail.com
* Correspondence: evan.elliott@biu.ac.il; Tel.: +972-72-264-4968

Academic Editor: Merlin G. Butler
Received: 28 February 2017; Accepted: 6 April 2017; Published: 14 April 2017

Abstract: In recent decades it has become clear that Autism Spectrum Disorder (ASD) possesses a diverse and heterogeneous genetic etiology. Aberrations in hundreds of genes have been associated with ASD so far, which include both rare and common variations. While one may expect that these genes converge on specific common molecular pathways, which drive the development of the core ASD characteristics, the task of elucidating these common molecular pathways has been proven to be challenging. Several studies have combined genetic analysis with bioinformatical techniques to uncover molecular mechanisms that are specifically targeted by autism-associated genetic aberrations. Recently, several analysis have suggested that particular signaling mechanisms, including the Wnt and Ca^{2+}/Calmodulin-signaling pathways are often targeted by autism-associated mutations. In this review, we discuss several studies that determine specific molecular pathways affected by autism-associated mutations, and then discuss more in-depth into the biological roles of a few of these pathways, and how they may be involved in the development of ASD. Considering that these pathways may be targeted by specific pharmacological intervention, they may prove to be important therapeutic targets for the treatment of ASD.

Keywords: ASD; autism; networks; genetics; Fragile-X Syndrome; Wnt; mTOR; Calmodulin; Calcium; NGF

1. The Genetic Basis of Autism

Autism Spectrum Disorder (ASD) is a developmental disorder characterized by persistent deficits in social communication, as well as restricted and repetitive patterns of behavior. It has been well characterized that genetic aberrations have a prominent role in the etiology of ASD [1]. Among the first studies to support a genetic etiology were twin studies published in the 1970s by Folstein and Rutter [2,3]. A recent meta-analysis by Rutter, encompassing a total of 6413 twins, showed that the heritability rate in families with an autistic proband was in the range of 64–91% [4]. Furthermore, it has been well-established that ASD overlaps at both the behavioral and genetic levels with other disorders such as social anxiety, Attention Deficit and Hyperactivity Disorder (ADHD), Intellectual Disability (ID), bipolar disorder and schizophrenia [5–8]. For example, Khazanda et al. discovered 23 genes, many of which are involved in circadian entrainment, that are associated with ASD, bipolar disorder and schizophrenia, therefore demonstrating shared genes and genetic pathways between various psychiatric disorders [7]. As such, common behavioral phenotypes of these disorders may have a root in shared genetic etiology.

A heterogeneous genetic etiology for ASD has been firmly established. Historically, some of the first genes that were successfully associated to autism were those responsible for syndromic forms of autism. Thus far, approximately 35 such syndromes have been documented [9]. A sub-group of these syndromic autisms are mendelian monogenic, and consistently develop due to rare mutations. For example, in the case of Fragile-X Syndrome (FXS), which develops due to aberrations in the Fragile-X Mental Retardation 1 (*FMR1*) gene, up to 30% of these male patients also develop ASD [10]. Cortical Dysplasia-focal Epilepsy (CDFE) syndrome is caused by mutations in Contactin Associated Protein 2 (*CNTNAP2*), where a subgroup of patients express autistic behavior [11]. Tuberous sclerosis is caused by mutations in either Tuberous Sclerosis 1/2 (*TSC1/TCS2*), where up to 61% of patients express autistic behaviors [12]. Rett syndrome, which includes autistic behavior, is caused by mutations in the Methyl CpG binding protein 2 (*MECP2*) [13]. Another group of syndromic autisms arise from Copy Number Variations (CNV) [14], which contain large duplication/deletion loci that usually encompass several genes, yet the causative gene is often not known. One such example is Phelan-McDermid syndrome where up to 9 Mb are deleted at Chr22q13 encompassing up to 130 genes [15]. The most notable gene in this CNV is *SHANK3/PROSAP2*, where mutations in this gene alone have been shown to be associated with ASD [16,17]. Additional CNVs include the 15q11-13 deletion or duplication syndromes [18], the chromosome 16p11.2 deletion or duplication syndromes [19,20] and the 2q23.1 microdeletion syndrome [7]. Most syndromic autisms, such as CDFE, TSC or the 2q23.1 microdeletion syndrome are considered extremely rare, however the highest discovery rate observed in FXS accounts for no more than 2% of autistic cases [9].

A wider search for ASD's genetic etiology started with linkage studies, which uncovered several candidate loci, including 2q, 3q 7q and 20p. However, it was primarily through Next Generation Sequencing (NGS) technologies that majority of genes were discovered, and by 2012, 10–20% of ASD cases had a known genetic link [9]. One approach for locating new genetic candidates involved in ASD development is to determine de-novo mutations in gene coding regions, usually found in sporadic familial cases. In 2011, O'Roak et al. conducted the first exome-wide sequencing study in ASD on 20 sporadic cases, revealing de-novo mutations in *FOXP1*, *GRIN2B*, *SCN1A* and *LAMC3* [21]. In 2012, in an effort to discover de-novo mutations, the Simons Simplex Collection conducted three large exome-sequencing trails which included approximately 750 families with affected and unaffected siblings. Many promising genes were identified, including *SCN2A*, *CHD8* and *NTNG1* [22–24]. By that time, it had become increasingly clear that genetic mutations involved in ASD do not fall into one particular biological category, but seem to be found in genes involved in several different biological systems. In recent years, whole genome sequencing studies are also beginning to appear, which aim to discover genetic aberrations in both coding and noncoding regions. One such study, by Yuen et al., found genetic aberrations in *STXBP1*, *UBE3A*, *KATNAL2*, *THRA*, *KCNQ4*, *MYH14*, *GJB6* and *COL11A1*, some of which have been previously associated with ASD and some with overlapping conditions such as hearing loss [25]. In addition, several Genome-Wide Association Studies (GWAS) emphasized the significance of common genetic variation to the inheritance of autistic traits [26].

To this day, over 800 genes have been identified and associated with ASD development [27]. Given this heterogeneous genetic reality, it has been hypothesized that the multitude of genetic aberrations are affecting specific molecular pathways that might be in common across the autistic spectrum, and are responsible for dysregulated neurodevelopment [28]. Discovering such common pathways has proven to be challenging. To address this difficulty, it is appropriate to perform network-based analysis of autism associated genes [29]. In the following chapters, we will discuss attempts at using network-based approaches to discover candidate signaling pathways that are effected by ASD-associated genetic aberrations, and how these signaling pathways may be involved in dysregulated neurodevelopment and the phenotypes of ASD.

2. Searching for Common Molecular Targets of Autism Mutations

2.1. Protein–Protein Network Analysis and Pathways Enrichment Tools

One approach to discovering molecular pathways that are commonly targeted by autism-associated mutations is to perform Protein-Protein Interaction (PPI) network analysis to identify groups of proteins within a given list of proteins that physically interact with each other. This analysis uses different databases that curate experimentally validated or predicted protein-protein interactions [29]. PPI also identifies hubs, highly interconnected proteins, which might prove to be central in the molecular pathway. However, this approach has some disadvantages and biases. First, tissue specific PPI databases are scarce, which are necessary to understand protein interactions in specific tissues, such as the brain. A very partial solution to this problem is to combine PPI databases with RNA expression from the specific tissue, therefore making sure that you only take into account proteins which are found in that specific tissue. Additional tools used to reveal biologically-relevant pathways within a gene list are gene ontology databases and signaling pathway enrichment tools. The Gene Ontology project initiated by the Gene Ontology Consortium (GOC) strives to provide a unifying vocabulary to genes and their biological roles [30]. In addition, there is a growing number of signaling pathway enrichment tools available for free or through commercial license [31]. There are several differences that should be considered when using these tools, such as method of data curation (manually and computerized), algorithms used to generate signaling pathway maps, types of pathway maps (e.g., disease specific, kinase signaling, hormone mediated) and breadth of knowledgebase. For example, in the next section, we mention studies which used KEGG and IPA pathway enrichment tools to generate pathways that might be affected in ASD. Due to the fact that IPA is a commercial tool, it is difficult to thoroughly compare it to other tools, and thus, reviews conducting such comparisons lack information about IPA. However, one important difference between KEGG and IPA is the size of their knowledgebase. KEGG mainly curates interaction and reactions of molecular pathways from the literature, while IPA also imports data from several external databases, which increases its knowledgebase substantially [32,33]. Therefore, it is necessary to integrate knowledge from several different pathway analysis tools in order to discover biological pathways that are commonly targeted by autism-associated mutations.

2.2. Autism-Associated Signaling Pathways Discovered Using Network Analysis and Pathway Enrichment Tools

As part of the effort conducted by the SSC in 2012, O'Roak et al. sequenced the exome of 209 families with sporadic cases of ASD, revealing 126 genes which had truncation or missense mutations, followed by PPI analysis, to determine if these genes form a biologically-relevant network [23]. They discovered one highly interconnected cluster of mutated genes, which included genes involved both in chromatin-binding and β-catenin function. An additional network analysis conducted with the IPA tool enriched 8 Wnt-signaling regulators, which are involved in the regulation of β-catenin, as we will explain later. Therefore, combination of exome sequencing with downstream network analysis revealed an autism-associated molecular pathway, the Wnt-β-catenin pathway. This was not the first study to suggest the involvement of Wnt-signaling in ASD [34,35], however it strongly demonstrated that network analysis may reveal pathways that are highly important for further research. In a 2014 follow up bioinformatics study which included the combined data for the O'Roak et al. study, and two other de-novo ASD mutation studies that were published in the same issue of *Nature*, a subset of these de-novo mutations were enriched for genes involved in chromatin-binding [36]. A separate study collected genome association data of rare CNVs from 6742 ASD patients, and constructed PPI networks based on the genes deleted or duplicated due to the CNVs. Calmodulin1, which is a major regulator of Calcium-signaling, was found as a central hub in the PPI network [37]. An additional network analysis of autism-associated genes has suggested a role for both Wnt and calcium pathways in ASD. Wen et al. performed KEGG pathway analysis on the SFARI annotated list of autism-associated genes. They found that these genes were particularly enriched for biological pathways including Calcium-signaling, Wnt-signaling, mTOR, and cellular adhesion molecules [38].

A different strategy to identify molecular pathways that are dysregulated by autism-associated mutations is to use publicly available data of gene expression patterns in the human brain to build gene expression networks and PPI networks that are found in specific brain regions, and then to probe if any of these networks are enriched for autism-associated genes, and are therefore likely to be perturbed by autism-related mutations. Lin et al. used such an approach to identify PPI networks in the brain that are effected by the autism-associated CNV at chromosome 16p11.2 [39]. Using published databases of gene expression and protein interaction data, the authors built protein interaction networks that appear in specific brain regions, and at specific developmental time points. The authors then discovered that four of these protein interaction networks would be disrupted by the CNV, particularly during mid-fetal stage and in early childhood. Gene ontology analysis of these protein networks showed enrichment for Wnt-signaling and NGF-signaling. A similar method was used to discover human brain protein networks dysregulated by rare or common variants associated with autism. Ben-David et al., built brain-region specific gene co-expression modules using WGCNA [40]. They found that a co-expression module that was highly effected by both rare and common autism-associated variants was enriched for genes involved in synaptic transmission and the Calmodulin-binding pathway. To summarize, these studies used different methodologies to uncover molecular pathways targeted by ASD mutations and implicated the Wnt-signaling and Ca^{2+}/Calmodulin-signaling [37–40]. Therefore, we will focus on defining these two specific pathways and how they may be related to the autism phenotype, followed by a short discussion of some of the other pathways identified.

3. Wnt-Signaling

3.1. Roles for Wnt-Signaling in Neurodevelopment and Adult Brain

Wnt-signaling pathways orchestrate cellular proliferation, polarity and differentiation; processes that are crucial for healthy tissue morphogenesis, especially in the embryonic stage [41]. Dysregulation of these pathways have been implicated in a wide variety of cancers, as well as in other pathologies, such as type II diabetes, osteoporosis and heart conditions [41,42]. In humans there are 19 Wnt glycoprotein ligands, which activate signal transduction pathways through binding to one of the 10 Frizzled receptors. As of today, there are two extensively researched Wnt pathways; a canonical pathway and a non-canonical pathway. In the canonical pathway the secreted Wnt glycoproteins bind to Frizzled receptors, as well as either the LRP6 or LRP5 co-receptors, to initiate a signaling cascade. The activated receptor recruits the scaffolding protein Dishevelled, which recruits and sequesters the Axin-GSK3β-APC destruction complex, thus preventing its role in phosphorylating β-catenin, leaving it stable for downstream activity. Stabilized β-catenin is transported to the nucleus where it fulfills its role as a co-transcription factor by binding to the TCF/LEF transcription factors, or it relocates to the cellular membrane where it maintains cell-cell adhesion complexes [41,42]. The two non-canonical Wnt pathways are independent of β-catenin, and include the Wnt/JNK pathway, which drives polarized cell movements such as neuronal crest migration, as well as the Wnt/Calcium pathway that has been shown to be involved in cardiac development [43].

Canonical Wnt-signaling has a pivotal role both in the developing and mature brain. Much evidence has indicated that during development the Wnt pathway regulates the balance between proliferation and differentiation of neuronal progenitor and precursor cells, as was recently reviewed by Noelanders and Vleminckx [44]. In transgenic mice with overexpression of stabilized β-catenin, neuronal precursor cells in the ventricular zone continue to divide beyond their natural timeframe, resulting in enlarged brains and deeper sulci folds that resemble higher mammals [45]. Concurrently, inhibition of β-catenin during embryonic development leads to premature differentiation of precursors to neurons [46]. These studies suggest that Wnt-signaling promotes proliferation of neuronal precursors and delays their differentiation. There is some evidence suggesting Wnt-signaling may inhibit proliferation in certain circumstances.

For example, as shown in developing zebrafish, induced activation of Wnt8 reduced the population of neuronal progenitor in the hypothalamus; reduction in neuronal progenitors was also induced by the canonical Wnt pathway activator, BIO [47]. These studies suggest that Wnt-signaling can regulate neuronal precursor proliferation in a manner that is either species-specific, or cell type-specific.

In addition to its roles in development, Wnt-signaling also affects Neuronal Stem Cell (NSC) proliferation and differentiation in the mature brain. One study of Wnt-signaling in NSCs used mice which express lacZ under the control of a TCF transcription response [48]. These mice received intraventricular infusions of the mitotic inhibitor Ara-C, which kills the dividing progenitor cells and decreases NSCs; followed with BrdU infusions, which allows the tracking of the regeneration of NSCs in the Subventricular Zone (SVZ). 30% of the BrdU positive cells displayed active Wnt-signaling (LacZ); and when Wnt-signaling was pharmacologically blocked, the number of neurospheres in the SVZ significantly decreased. Wnt-signaling has also been shown to induce NSC proliferation after injury [48]. Mice subjected to a stroke-inducing surgery (Pial Vessel Disruption, or PVD) displayed an increase in lacZ expressing neuronal stem cells of the SVZ after seven days. Other studies showed that Wnt-signaling is also involved in the differentiation of NSCs in the mature brain. For example, in adult NSCs of the SVZ, Wnt-signaling induces cell-cycle exit in the presence of Hipk1, an interactor of β-catenin. Furthermore, when overexpressing β-catenin and Hipk1, an increase of the cell-cycle inhibitor P16Ink4 was observed [49]. Therefore, Wnt has a role both in the proliferation and differentiation of NSCs cells in the adult brain.

Apart from its role in NSCs, Wnt-signaling has a positive developmental role in the maturation of dendrites and spines. The α-catenin/β-catenin/N-cadherin complex acts as a scaffold between intracellular actin and extracellular cadherin-dependent interactions [50,51]. This role is important for dendrite arborization in the developing brain, as demonstrated by the fact that overexpression of β-catenin, α-catenin and N-cadherin individually, and co-expression of β-catenin and N-cadherin in vitro, both induce increased dendritic branching [52]. In addition, β-catenin has also been implicated in spine pruning and maturation as part of the β-catenin/N-cadherin complex. By inducing a conditional knockout of β-catenin in the cerebral cortex and hippocampus of mice, spine density is increased and is comprised mostly from immature spines [53]. This suggests that pruning of immature spines may be dependent on β-catenin function.

In addition to the roles in neuronal differentiation and morphology, β-catenin has been suggested to have an additional role in neurotransmission through two distinct mechanisms. First, translocation of β-catenin to spines may increase the size of the post synaptic density (PSD) which leads to increased synaptic transmission [54]. This was shown by inducing a point mutation, which prevents the phosphorylation of Tyr654 in β-catenin, and results in its increased localization to spines. Considering the role of β-catenin in the α-catenin/β-catenin/N-cadherin complex, and the complex's importance in inducing cell-cell contacts [55], the authors deciphered if increased presence of β-catenin in spines might increase synaptic contacts, and affect other properties such as postsynaptic morphology and neurotransmission. Interestingly, they observed more intense and larger PSD-95 puncta and increased mEPSCs. Second, in the presynaptic neuron, β-catenin has been shown to have a role in recruiting vesicles to the presynaptic membrane: β-catenin has a PDZ domain used to recruit PDZ domain containing proteins such as Veli and Cadherin clusters [56]. Veli has been shown to create a tripartite complex with CASK and Mint1, which binds Munc18-1. This complex is essential for vesicle docking to the plasma membrane [57,58]. It has also been shown that regulation of synaptic vesicle release into the synaptic cleft is important for healthy development of synaptic plasticity in 4–8 day old rats [59]. In that regard, β-catenin is suggested to have an essential role. This has been shown by siRNA-mediated knockdown of synaptic β-catenin, which increased spillage of vesicle content into the synaptic cleft in nascent presynaptic terminals [60]. An additional study revealed that neuronal cultures treated with Wnt8A-conditioned media displayed increased PSD95 puncta, and that the co-receptor LRP6 selectively localizes to excitatory synapses where it takes part in regulating excitatory synaptic development [61]. These studies show that Wnt-signaling, and in particular β-catenin, is essential for normal brain development as well as adult brain performance.

3.2. How Wnt-Signaling May Be Involved in ASD

From the known roles of Wnt-signaling in neuronal differentiation, morphology, and neurotransmission, we can propose possible roles for dysregulated Wnt-signaling in the processes that are dysregulated in the brain of individuals diagnosed with ASD. Numerous studies have witnessed differences in cortical patterning or spine morphology in the brain of individuals with ASD. In some cases, it has been shown that cortical dendritic spine density is increased in the brain of individuals with autism [62]. This matches the phenotype of the β-catenin knockout mice, which displays increased spines [53]. However, it is not clear if the increase in spine density in the ASD brain is due to an increase in immature spines, as is in the case of the knockout mice. In addition, studies have shown dysregulation of cytoarchitecture in the cortex and in the architecture of the microcolumns in the cortex of individuals diagnosed with ASD [63,64]. A separate study, using the three dimensional imaging technique CLARITY, determined abnormal connections between axons in the brain of individuals with ASD [65]. Considering that dysregulation of Wnt-signaling affects development of normal cortical architecture, as well as neuronal morphology, it is possible that dysregulated Wnt-signaling is involved in these dysregulations in the ASD brain.

Experimentation in mouse models have also given more insight into the possible roles of Wnt in the development of ASD. Interestingly, a conditional knockdown of β-catenin in Parvalbumin (PV) neurons in mice induced increased repetitive behaviors and anxiety, decreased social interaction, and an increase in PV neuron density in the prefrontal cortex [66]. An additional Wnt-signaling component, Glycogen Synthase Kinase-3 (GSK3) was recently reviewed as a key driver of ASD development, as well as a therapeutic target for FXS [67,68]. GSK3 is hyperactive in several brain regions of the FXS mouse model [69] and impaired social preference and exaggerated anxiety during social interaction have been recorded in a GSK3 knockin mouse model [35]. So using this model, researchers inhibited GSK3 functioning in the hippocampus to explore how its hyperactivity contributes to FXS phenotypes. Both novel object detection and hippocampal learning, which are deficient in this mouse model, were rescued [70,71]. These studies suggest a direct link between Wnt-signaling and autism-like behavior in autism mouse models.

In addition, recent studies in mouse models have determined how autism-related mutations may induce autism-related behaviors through the dysregulation of Wnt-signaling. One such example is *CHD8*, a gene that was recently found to be strongly associated to ASD and encodes for a chromatin-binding protein. One of the first studies on CHD8 in the brain shows that it acts as a positive regulator of the Wnt pathway in the brain [72]. CHD8 binds to the promoter regions of Fzd1, Dvl3 and β-catenin. Knockdown of CHD8 during cortical development results in their downregulation, leading to the reduction of the TCF/LEF transcription factor family and ultimately defective brain development. By expressing a degradation-resistant β-catenin construct in CHD8-downregulated embryos, researchers rescued the aberrant dendritic arborization as well as increased spine density in the CHD8-downregulated mice. In addition, the study showed that inducing stable β-catenin restores the aberrant social and anxiogenic behaviors to the levels seen in control mice. The Shank scaffolding family is an additional group of high-risk ASD genes that have been suggested to interact with Wnt-signaling components. In a recent publication by Harris et al., a Shank-knockout drosophila model was used to study of the potential molecular pathways that the Shank family of genes regulate [73]. Surprisingly, the study revealed that the non-canonical Wnt-Frizzled Nuclear Import (FNI) pathway was affected. In this pathway, Wnt binding induces internalization of the Fz2 receptor, followed by proteolytic cleavage of the receptors' N-terminus to form Fz2-C, which translocates into the nucleus, where it regulates transcription necessary for synaptic development [74,75]. By knocking out Shank family genes, Harris et al. revealed reduction of the Fz2 internalization into the post-synaptic membrane, reduction of Fz2-C presence in the nucleus, and abnormal synaptic development. Therefore, there is accumulating evidence that autism-associated mutations affect abnormal cortical patterning, synaptic development, and autism-like behaviors through modulation of Wnt-signaling pathway.

4. Calcium and Calmodulin Signaling

4.1. Roles of Calcium-Signaling and the Calmodulin-Binding Pathway in the Brain

Shifts in Calcium (Ca^{2+}) concentrations have been shown to affect the function of several tissues and organs such as the heart, pancreas and components of the Central Nervous System (CNS). Therefore, it is of no surprise that disequilibrium in signaling leads to pathology [76]. In the brain, Ca^{2+} performs specific functions in the presynapse and postsynapse. In the presynapse, arrival of action potentials induce an increase of presynaptic Ca^{2+} levels through activation of the N and P/Q-voltage-gated channel. The influx of Ca^{2+} leads directly to the release of neurotransmitters [77]. However, the role of Ca^{2+} in postsynaptic signal transduction pathways has been found to be dysregulated in the autism genetic studies discussed previously.

Ca^{2+}-induced effects in postsynaptic neuronal function and excitability are mediated through its binding with the protein Calmodulin (CaM; Ca^{2+}-Modulated protein), which induces signaling to additional downstream CaM Kinases [77,78]. Glutamic acid induces Ca^{2+} influx at the postsynapse by binding to *N*-methyl-D-aspartate receptors (NMDARs) and metabotropic glutamate receptors such as mGlur1, which initiates the release of internal Ca^{2+} stores [77]. After binding Ca^{2+}, CaM may induce the activation of CaMKII or CaMKK2. CaMKII is a dodecamer holoenzyme which has 28 isoforms composed of four different subunits (CaMKIIα through δ), and resides mostly in spines and dendritic shafts of excitatory neurons [79–81]. The mandatory role of CaMKII in synaptic function has been well described [82]. Active CaM binds to CaMKII subunits and relieves them from autoinhibition, which leads to autophosphorylation and kinase activity [78]. CaMKIIα kinase activity is mandatory for long-term potentiation (LTP), the main electrophysiological determinant of experience-dependent synaptic strengthening, and is highly involved in behavioral processes, such as learning and memory. Deletion of *CAMKIIA*, or inhibiting the binding of CaM and Ca^{2+} to CaMKII, inhibits LTP induction [83]. CaMKIIβ has been shown to interact with cytoskeleton subunits such as F-actin, α-catinin and the PSD protein Densin-180 for the purpose of spine size regulation, as well as to affect long-term synaptic plasticity by binding and regulating receptors such as GluN2B and GluA1 [80,82]. Furthermore, accumulating evidence show that CaMKII phosphorylates the $GABA_A$ β, γ2 and α1 subunits, and regulates $GABA_A$ trafficking to the synaptic membrane [84]. Since GABA is the major inhibitory neurotransmitter, these findings highlight that CaMKII plays important roles in both excitatory and inhibitory neurotransmission.

CaM also activates CaMKK2, whose activity is crucial for spatial memory formation, and the downstream activation of additional kinases: CaMKI is a positive transducer of growth cone motility which is essential for neurite elongation and arborization [85]; phosphorylation of CaMKIV increases gene expression and protein synthesis, and is also necessary for contextual fear [86]. CaMKK2 also phosphorylates other kinases such as the AMP-activated Protein Kinase (AMPK) who's activity is essential for regulating the energy intake necessary for typical brain function [87]. Interestingly, *CAMKIG* and *CAMKIV* are among the genes which were enriched in the Calmodulin-binding pathway in the Ben-David et al. publication, which looked for common molecular pathways affected by rare and common variations in ASD [40]. Overall, this data provides compelling evidence of how Ca^{2+} signaling and the CaM pathway are involved in neurological functions by affecting a variety of synaptic characteristics, neurotransmission via excitatory and inhibitory receptor regulation and important biological functions such as LTP and LTM. Deficits in the CaM pathway and its branching cascades have the potential to be involved in many neuropsychiatric conditions due to their broad influence on many biological systems, and more specifically by the way it regulates neurotransmission and synaptic characteristics.

4.2. How Calcium-Signaling and the Calmodulin-Binding Pathway May Be Involved in ASD

Considering the central role of Ca^{2+} and CaM signaling in synaptic function and neuronal connectivity, it is reasonable to presume that dysregulation of this pathway could lead to autism-related symptoms. However, it is technically challenging to decipher if there are any dysregulation in synaptic functions such as LTP in humans diagnosed with ASD, while evidence for such dysregulation have been frequently observed in several ASD mouse models [88,89]. Therefore, our understanding of the possible role of Ca^{2+} signaling in autism is still at its infancy, compared to more established role of the Wnt pathway. Of great interest, one human study has used Transcranial Magnetic Stimulation (TMS) to study changes in long term potentiation-like synaptic plasticity in humans diagnosed with ASD [90]. In this study, the researchers performed TMS in cortical regions followed by motor-evoked potentials. Individuals with ASD did not show any changes in motor-evoked potentials after TMS, unlike neurotypical controls. This study suggests deficits in plasticity that resemble deficits in LTP. Additional human studies have verified similar deficits in neuronal network connectivity in ASD patients, as has been recently reviewed [91]. These studies have often found changes in electroencephalographic signals after different sensory stimuli in ASD patients. Overall, these studies suggest deficits in synaptic and network activity that may be related to Calcium-signaling.

While studies of Ca^{2+} signaling in the human brain remains challenging, recent studies have determined disturbances in Ca^{2+} signaling in cells derived from individuals with ASD. Agonist-evoked Ca^{2+} signaling has been shown to be dysfunctional in skin fibroblasts derived from individuals diagnosed with autism [92]. An elegant study was performed on induced Pluripotent Stem Cells (iPSC) derived from individuals diagnosed with Timothy syndrome [93], a syndromic autism where 80% of individuals are diagnosed with ASD. These iPSCs were differentiated into neurons in vitro and displayed dysregulated Ca^{2+} signaling and changes in activity-dependent gene transcription. While these studies suggest that Ca^{2+} signaling and CaM may be involved in the biology of ASD, technological improvements of Ca^{2+} imaging in the human brain, and more high-throughput studies in individuals diagnosed with ASD, are necessary to understand the role of Ca^{2+} in the specific behaviors and brain regions that are particularly relevant to ASD.

Animal models and in vitro studies have given some additional insights into how dysregulation of Ca^{2+} and Calmodulin-binding may be involved in abnormal neurodevelopment. For example, CaMKIV positively regulates the transcription of FMRP (Fragile-X Mental Retardation Protein), the causative gene of FXS [94]. A follow-up study found that a Single Nucleotide Polymorphism (SNP) in the gene CaMKIV (rs25925) is associated with higher risk for ASD development in a European cohort. This SNP appears to be located on a splicing factor binding site, and is predicted to alter the balance of CaMKIV isoforms [95]. In addition, CaMKIIα has been shown to regulate the activity of mGluR5, which is a potential target in FXS treatment [96]. It is still unclear how CaMKIIα perturb mGluR5 activity in FXS, however one proposed mechanism suggests that CaMKIIα is significantly elevated in the synapse of FXS mouse models, which might cause the hyperphosphorylation of the Homer 1 (H1) and 2 (H2) scaffolding proteins, resulting in their dissociation from mGluR5. This dissociation allows the short Homer 1 isoform, H1α, to bind mGluR5 and induce ligand-independent activity of the receptor [97]. Inhibition of mGluR5 in *FMR1* knockout mice improved learning and memory, which highlighted this therapy as a promising pharmaceutical treatment [98]. However, clinical trials that were designed to inhibit mGluR5 and its downstream pathways in FXS patients described only partial success, as extensively reviewed by Schaefer, Davenport and Erickson [99]. In one clinical study, the mGluR5 selective antagonist fenobam induced improvement in prepulse inhibition, but had no effects on the excessive impulsivity observed in FXS patients [100]. In another clinical study, treatment of a small cohort of FXS patients with Lithium—which mitigates signaling pathways activated via mGluR5 signaling—resulted in significant behavioral improvements such in hyperactivity and inappropriate speech, however induced only a tendency for improvement in irritability, lethargy and repetitive behaviors [101].

An additional syndromic autism linked with Ca^{2+} signaling dysregulation is Angelman Syndrome (AS), which is characterized by the deletion of the maternal allele of 15q13-11, including the gene *UBE3A*. In the *UBE3A* maternal allele null mouse model, disruptions in the autophosphorylation of CaMKIIα, which is essential for its kinase activity, is responsible for LTP deficits of the hippocampus [102]. CaMKIIα has also been found to interact directly with scaffolding proteins associated with ASD development. Specifically, using immunoprecipitation to pull-down CaMKII from different neuronal cell fractions of the forebrain, it was determined that CaMKII binds Shank3, Dlgap2 and Syngap1 in the synapse [103]. In summary, various studies have identified CaMKIV and CaMKIIα as the two main Ca^{2+} signaling enzymes that are most likely to be involved in ASD. On one hand the Ca^{2+} and Calmodulin-binding pathways may be interesting targets for therapeutic interventions. However, future studies in mouse models, and electrophysiology studies in humans, are necessary to understand how and why the Ca^{2+} signaling pathway is involved in ASD. Further studies are particularly needed to clarify the potential roles of Ca^{2+} signaling in social behaviors or repetitive behaviors, which form the core features of ASD.

5. Additional Signaling Pathways

Thus far, we have discussed the Wnt-signaling and Calmodulin-binding pathways as potential molecular mechanisms that are involved in ASD, due to the fact that these pathways have been found to be enriched for ASD-associated genes in multiple network-based analysis. In addition to these pathways, the mTOR-signaling and NGF-signaling pathways have also appeared in some of the discussed bioinformatic publications, although not quite as often as the Wnt and Calmodulin pathways. The mTOR pathway has already gained significant attention in connection to autism due to its significant involvement in syndromic autisms, as explained below. There is yet little known connection between NGF and ASD, however it is worth shortly considering the possible connections, considering its roles in brain development and function.

5.1. The PI3K/Mtor-Signaling Pathway

Mammalian Target of Rapamycin (mTOR) is a serine/threonine kinase, which is considered a central kinase in organism development. Therefore it is of no surprise that ablation of mTOR results in in-utero death a short time after the implantation stage of the embryo in the utcrus endometrium [104]. The mTOR pathway regulates brain development through two main cascades driven by two complexes: mTOR complex 1 (mTORC1) and mTOR complex 2 (mTORC2). In the developing brain each cascade seems to take part in unique developmental duties [105]. The mTORC2 cascade facilitates growth cone motility through its interaction with actin filaments. This role is essential for pathfinding dynamics of the neurite in the developing brain. Additionally, mTORC2 has an indirect influence on neuron size and morphology by regulating mTORC1 through RAC-alpha serine/threonine-protein kinase (AKT) activity. AKT activates mTORC1 which in turn phosphorylates p70 ribosomal protein S6 kinase (p70S6K) and eukaryotic Initiation Factor 4E (eIF4E)-binding protein (4EBP), which enhance a downstream cascade necessary for both protein synthesis and lipid synthesis. The products are used for plasma membrane expansion that is required for neurite elongation and arborization, and dendrite formation. This developmental mechanism driven by mTORC1 may also be directly activated by extracellular stimulation such as growth factors and neurotransmitters.

There is an abundance of publications linking mTOR-signaling to ASD, and it is considered to be one of the most promising converging signaling pathway candidates for ASD development. The involvement of mTOR-signaling in ASD and other neuropsychiatric conditions has been thoroughly reviewed previously [106]. However, unlike the previously described pathways in this review, and to the best of our knowledge, only one publication has described a link between the mTOR-signaling pathway and ASD using a network analysis-based approach [38]. Rather, the evidence for mTOR-signaling as a promising common ASD pathway candidate arise mostly from in vivo research into syndromic autisms such as Tuberous Sclerosis (TS), *PTEN*-related syndrome, Neurofibromatosis Type 1 (NF-1) and FXS [12,107–109]. In TS, the TSC1/2 complex has an inhibitory role on mTORC1, and knockdown of TSC2 results in over-activation of mTORC1 and mTORC2, leading to an increase in neuronal cell size of the fetal brain [110]. Interestingly, blocking mTOR activity in TSC2-haploinsufficient mice using rapamycin reversed social deficits, suggesting that mTOR-signaling has a role in the manifestation of overall autistic behaviors, and in TSC particularly [111]. PTEN is a phosphatase which also acts as a negative regulator of the mTOR-signaling pathway by reducing the activity of the PI3K/AKT pathway. As previously mentioned, AKT activates mTORC1 which is essential for protein synthesis of the developing neurite. Therefore, PTEN dysfunction might lead to increased protein synthesis and abnormal brain growth. Indeed, it has been shown that *PTEN* mutations in a subset of ASD cases are co-morbid with overgrowth and macrocephaly [112]. Mouse models with conditional knockout of *PTEN* in mature neurons of the cerebral cortex and hippocampus display axonal overgrowth, ectopic axonal projections, and abnormal synapses, as well as reduced social interaction, increased anxiety, and hyperexcitability [113]. In a recent publication by Cupolillo et al., conditional knockout of PTEN in cerebellar Purkinje cells led to a reduction in social interaction and repetitive behavior, paralleled with structural abnormalities in cerebellar axons and dendrites [114]. Neurofibromin (NF), a tumor suppressor through its GTPase-activating function, is an additional negative regulator of mTOR-signaling via TSC2 [115]. It is the key inducer of the familial cancer syndrome, NF-1, a pathology where autism-like social dysfunction has also been observed [108]. It is still unclear exactly how NF is specifically involved with the ASD-like phenotype observed in NF-1. However, considering that the disequilibrium of the negative regulation of mTOR-signaling both by TSC1/2 and PTEN has been shown to be impaired in syndromic autisms, it is possible to claim that NF has a role in ASD-like phonotype development through its regulation of mTOR. In the case of FXS, increased phosphorylation of mTOR and p70S6K was observed both in lymphocytes and brain samples of FXS patients, which suggest increased protein synthesis, which is the main avenue by which mTOR dysregulation is believe to induce ASD development [109].

In all of the examples given above, inability to downregulate mTOR-signaling is associated with the autism phenotype. In fact, studies determined that blocking mTOR through the use of Rapamycin improves social deficits in the TS and BTBR mouse models [111,116] which implies that Rapamycin might be a possible therapy for the social deficits in ASD.

5.2. Nerve Growth Factor

Nerve Growth Factor (NGF) is a primary neurotrophin for peripheral organ innervation and sensory neuron development. As the nervous system develops, target-organs secret NGF that is detected by elongating axons, and through receptor-mediated endocytosis it enhances survival, neurite outgrowth and synaptic plasticity and connectivity [117]. A role for NGF in the CNS was initially observed in the developing rat forebrain, as researchers injected exogenous NGF intraventricularly to neonatal rats, which resulted in an increase of Choline Acetyltransferase (ChAT) [118]. Further studies of NGF in the rat forebrain showed that NGF regulates cholinergic development and differentiation by binding Tropomyosin receptor kinase A (TrkA), which by a positive feedback loop, increases TrkA expression as well as ChAT in cholinergic neurons [119]. Since then, NGF, together with BDNF, have been shown to play an important role in orchestrating neuronal plasticity important for sociability in mice [117].

Given its important role in nervous system development, it is not surprising that dysregulation in NGF-signaling has been implicated in psychiatric disorders such as depression, schizophrenia and Alzheimer's disease [117]. Nevertheless, the number of publications linking NGF to ASD is scarce. In fact, some publications have shown that NGF levels in cerebral spinal fluid (CSF) and blood is typical in children with ASD [120,121]. However, more recent publications begin to present a different picture. A study analyzing Differential Alternative Splicing (DAS) in the blood mRNA of 2–4 year old boys diagnosed with ASD showed there was a significant difference in DAS for several genes involved in NGF-signaling, including the NGF receptors, Nerve Growth Factor Receptor (NGFR) and Neurotrophic Receptor Tyrosine Kinase 1 (NTRK1) [122]. A different study showed that SNPs in NTRK1 associated with ASD behavioral traits measured by the Empathy Quotient (EQ) and the Autism Spectrum Quotient (AQ) [123]. In an interesting study by Lu et al., researchers conducted a genome-wide Quantitative Trait Loci (QTL) study and found that several SNPs in NGF are significantly associated with deficits in non-verbal communication, which is an autistic trait [124]. The reasoning for the QTL approach was to focus on a specific trait and find the genetic loci associated to it and reduce some of the genetic heterogeneity that usually complicates ASD genetic research. Therefore, while there is scarce evidence for the role of NGF-signaling in ASD from animal studies, human genetic studies have actually found associations between the NGF pathway and behaviors that are dysregulated in ASD. Therefore, further research into the possible mechanistic roles for NGF in social behavior are needed [39].

6. Conclusions

ASD's elusive genetic etiology imposes a great challenge to our understanding of the disorder's pathology. The growing number of genes associated with ASD, both rare and common variants, and the fact that these genes are involved in a variety of biological processes, makes it difficult for the research community to find a specific target for therapy. Taking into account that there are currently no pharmacological agents that treat the core symptoms of ASD, there is great need to understand the biological pathways that are targeted by ASD-mutations, which can be novel pharmacological targets. Both Wnt and the Ca^{2+} signaling pathways that we discussed in-depth in this review are potential targets for pharmacological intervention. However, while these pathways have been well characterized, there is still a great need to understand exactly how dysregulation of these pathways are involved in the core characteristics of ASD, including social and repetitive behaviors. A more clear understanding of the specific roles of these pathways in specific brain region is also likely to shed some light into this issue.

Up until now, the only way to find common pathways affected by ASD-associated genes was by in-silico network analysis. However, with the creation of multiple mouse models based on these genetic aberrations, a current method can be to decipher common molecular dysregulations found in these multiple autism mouse models, and to correlate these dysregulations with the animal's behavioral and neurodevelopmental phenotypes. For example, Ellegood et al. revealed that autism mouse models can be clustered according to neuroanatomical differences such as changes in the volume of different brain regions [125]. In a recent review, Kim et al., compared publications on multiple mouse models, and searched for physiological dysregulations underlying the variety of repetitive behavior types observed in ASD mouse models. While Kim et al. deciphered that it is difficult to link repetitive behavior types to specific brain regions, they discovered a few common pathways that are often involved in ASD phenotypes [126]. One example is involvement of glutamatergic connections from the frontal cortex to the midbrain, involved in repetitive behaviors. Therefore, the spatial and temporal resolution of ASD-related molecular pathways cannot be determined by network analysis of genetic data, but rather through the analysis of in vivo models. Therefore, parallel investigation of molecular mechanisms dysregulated in multiple mouse models of autism genes is likely to reveal important mechanisms and novel therapeutic targets.

Acknowledgments: Our research is currently being supported by Israel Science Foundation grant 1047/12 and by a grant from Teva Pharmaceutical Industries.

Author Contributions: Both Oded Oron and Evan Elliott designed, wrote, and edited this review, according to own expertise. All authors have read and approved the final version of the manuscript.

Conflicts of Interest: The authors declare no conflict of interest.

Abbreviations

ADHD	Attention Deficit Hyperactivity Disorder
AKT	RAC-alpha serine/threonine-protein kinase
AMPK	AMP-activated Protein Kinase

AQ	Autism Spectrum Quotient
AS	Angelman Syndrome
ASD	Autism Spectrum Disorder
BDNF	Brain-derived Neurotrophic Factor
BIO	$(2'Z,3'E)$-6-Bromoindirubin-3'-oxime
CaM	Calmodulin
CAMK	Ca^{2+}/calmodulin-dependent protein kinase
CASK	Calcium/Calmodulin Dependent Serine Protein Kinase
CDFE	Cortical Dysplasia-focal Epilepsy
ChAT	Choline Acetyltransferase
CHD8	Chromodomain Helicase DNA Binding Protein 8
CNS	Central Nervous System
CNTNAP2	Contacting Associated Protein 2
CNV	Copy Number Variation
COL11A1	Collagen Type XI Alpha 1 Chain
DAS	Differential Alternative Splicing
Dlgap2	DLG Associated Protein 2
Dvl3	Dishevelled Segment Polarity Protein 3
eIF4E	Eukaryotic Translation Initiation Factor 4E
EQ	Empathy Quotient
FMR1	Fragile-X Mental Retardation 1
FNI	Frizzled Nuclear Import
FOXP1	Forkhead Box P1
FXS	Fragile-X Syndrome
Fzd1	Frizzled Class Receptor 1
GABA	Gamma-Aminobutyric
GJB6	Gap Junction Protein Beta 6
GOC	Gene Ontology Consortium
GRIN2B	Glutamate Ionotropic Receptor NMDA Type Subunit 2B
GSK3β	Glycogen Synthase Kinase 3 Beta
GWAS	Genome Wide Association Study
H1	Homer 1
H2	Homer 2
Hipk1	Homeodomain Interacting Protein Kinase 1
ID	Intellectual Deficiency
IPA	Ingenuity Pathway Analysis
iPSC	induced pluripotent stem cells
KATNAL2	Katanin Catalytic Subunit A1 Like 2
KCNQ4	Potassium Voltage-Gated Channel Subfamily Q Member 4
KEGG	Kyoto Encyclopedia of Genes and Genomes
LAMC3	Laminin Subunit Gamma 3
LEF	Lymphoid Enhancer Binding Factor
LRP5	LDL Receptor Related Protein 5
LRP6	LDL Receptor Related Protein 6
LTP	Long Term Potentiation
MECP2	Methyl CpG binding protein 2
mGluR5	Glutamate Metabotropic Receptor 5
mTOR	Mammalian Target of Rapamycin
mTORC1	Mammalian Target of Rapamycin complex 1
mTORC2	Mammalian Target of Rapamycin complex 2
MYH14	Myosin Heavy Chain 14
NF	Neurofibromin
NF-1	Neurofibromatosis Type 1

NGF	Nerve Growth Factor
NGFR	Nerve Growth Factor Receptor
NGS	Next Generations Sequencing
NMDAR	*N*-methyl-D-aspartate receptors
NSC	Neuronal Stem Cell
NTNG1	Netrin G1
NTRK1	Neurotrophic Receptor Tyrosine Kinase 1
p70S6K	p70 ribosomal protein S6 kinase
PI3K	Phosphatidylinositol-4,5-Bisphosphate 3-Kinase
PPI	Protein-Protein Interaction
PROSAP2	Proline Rich Synapse Associated Protein 2
PSD	Post Synaptic Density
PTEN	Phosphatase And Tensin Homolog
PV	Parvalbumin
PVD	Pial Vessel Disruption
QTL	Quantitative Trait Loci
SCN1A	Sodium Voltage-Gated Channel Alpha Subunit 1
SCN2A	Sodium Voltage-Gated Channel Alpha Subunit 2
SFARI	Simmons Foundation Autism Research Initiative
SHANK3	SH3 And Multiple Ankyrin Repeat Domains 3
SNP	Single Nucleotide Polymorphism
STXBP1	Syntaxin Binding Protein 1
SVZ	Subventricular Zone
Syngap1	Synaptic Ras GTPase Activating Protein 1
TCF	Transcription Factor
THRA	Thyroid Hormone Receptor, Alpha
TMS	Transcranial Magnetic Stimulation
TrkA	Tropomyosin receptor kinase A
TSC1	Tuberous Sclerosis 1
TSC2	Tuberous Sclerosis 2
UBE3A	Ubiquitin Protein Ligase E3A
WGCNA	Weighted Gene Co-expression Network Analysis
Wnt	Wingless-type

References

1. Grice, D.E.; Buxbaum, J.D. The Genetics of Autism Spectrum Disorders. *NeuroMol. Med.* **2006**, *8*, 451–460. [CrossRef]
2. Folstein, S.; Rutter, M. Infantile autism: A genetic study of 21 twin pairs. *J. Child Psychol. Psychiatry* **1977**, *18*, 297–321. [CrossRef] [PubMed]
3. Folstein, S.; Rutter, M. A Twin Study of Individuals with Infantile Autism. In *Autism: A Reappraisal of Concepts and Treatment*; Rutter, M., Schopler, E., Eds.; Springer: Boston, MA, USA, 1978; pp. 219–241.
4. Rutter, M. Heritability of autism spectrum disorders: A meta-analysis of twin studies. *J. Child Psychol. Psychiatry* **2016**, *57*, 585–595.
5. Taurines, R.; Schwenck, C.; Westerwald, E.; Sachse, M.; Siniatchkin, M.; Freitag, C. ADHD and autism: Differential diagnosis or overlapping traits? A selective review. *Atten. Deficit Hyperact. Disord.* **2012**, *4*, 115–139. [CrossRef] [PubMed]
6. Hollocks, M.J.; Howlin, P.; Papadopoulos, A.S.; Khondoker, M.; Simonoff, E. Differences in HPA-axis and heart rate responsiveness to psychosocial stress in children with autism spectrum disorders with and without co-morbid anxiety. *Psychoneuroendocrinology* **2014**, *46*, 32–45. [CrossRef] [PubMed]
7. Talkowski, M.E.; Mullegama, S.V.; Rosenfeld, J.A.; Van Bon, B.W.M.; Shen, Y.; Repnikova, E.A.; Gastier-Foster, J.; Thrush, D.L.; Kathiresan, S.; Ruderfer, D.M.; et al. Assessment of 2q23.1 microdeletion syndrome implicates MBD5 as a single causal locus of intellectual disability, epilepsy, and autism spectrum disorder. *Am. J. Hum. Genet.* **2011**, *89*, 551–563. [CrossRef] [PubMed]

8. Khanzada, N.; Butler, M.; Manzardo, A. GeneAnalytics Pathway Analysis and Genetic Overlap among Autism Spectrum Disorder, Bipolar Disorder and Schizophrenia. *Int. J. Mol. Sci.* **2017**, *18*, 527. [CrossRef] [PubMed]

9. Buxbaum, J.D.; Hof, P.R. *The Neuroscience of Autism Spectrum Disorders*; Elsevier Science: Amsterdam, The Netherlands, 2012.

10. Hagerman, R.; Hoem, G.; Hagerman, P. Fragile X and autism: Intertwined at the molecular level leading to targeted treatments. *Mol. Autism* **2010**, *1*, 12. [CrossRef] [PubMed]

11. Strauss, K.A.; Puffenberger, E.G.; Huentelman, M.J.; Gottlieb, S.; Dobrin, S.E.; Parod, J.M.; Stephan, D.A.; Morton, D.H. Recessive symptomatic focal epilepsy and mutant contactin-associated protein-like 2. *N. Engl. J. Med.* **2006**, *354*, 1370–1377. [CrossRef] [PubMed]

12. Vignoli, A.; La Briola, F.; Peron, A.; Turner, K.; Vannicola, C.; Saccani, M.; Magnaghi, E.; Scornavacca, G.F.; Canevini, M.P. Autism spectrum disorder in tuberous sclerosis complex: Searching for risk markers. *Orphanet J. Rare Dis.* **2015**, *10*, 154. [CrossRef] [PubMed]

13. Percy, A.K. Rett syndrome: Exploring the autism link. *Arch. Neurol.* **2011**, *68*, 985–989. [CrossRef] [PubMed]

14. Leppa, V.M.; Kravitz, S.N.; Martin, C.L.; Andrieux, J.; Le Caignec, C.; Martin-Coignard, D.; DyBuncio, C.; Sanders, S.J.; Lowe, J.K.; Cantor, R.M.; et al. Rare Inherited and De Novo CNVs Reveal Complex Contributions to ASD Risk in Multiplex Families. *Am. J. Hum. Genet.* **2016**, *99*, 540–554. [CrossRef] [PubMed]

15. Wilson, H.L. Molecular characterisation of the 22q13 deletion syndrome supports the role of haploinsufficiency of SHANK3/PROSAP2 in the major neurological symptoms. *J. Med. Genet.* **2003**, *40*, 575–584. [CrossRef] [PubMed]

16. Durand, C.M.; Betancur, C.; Boeckers, T.M.; Bockmann, J.; Chaste, P.; Fauchereau, F.; Nygren, G.; Rastam, M.; Gillberg, I.C.; Anckarsäter, H.; et al. Mutations in the gene encoding the synaptic scaffolding protein SHANK3 are associated with autism spectrum disorders. *Nat. Genet.* **2007**, *39*, 25–27. [CrossRef] [PubMed]

17. Wang, X.; Xu, Q.; Bey, A.L.; Lee, Y.; Jiang, Y.-H. Transcriptional and functional complexity of SHANK3 provides a molecular framework to understand the phenotypic heterogeneity of SHANK3 causing autism and SHANK3 mutant mice. *Mol. Autism* **2014**, *5*, 30. [CrossRef] [PubMed]

18. Ornoy, A.; Liza, W.F.; Ergaz, Z. Genetic syndromes, maternal diseases and antenatal factors associated with autism spectrum disorders (ASD). *Front. Neurosci.* **2016**, *10*, 1–21. [CrossRef] [PubMed]

19. De Anda, F.C.; Rosario, A.L.; Durak, O.; Tran, T.; Gräff, J.; Meletis, K.; Rei, D.; Soda, T.; Madabhushi, R.; Ginty, D.D.; et al. Autism spectrum disorder susceptibility gene TAOK2 affects basal dendrite formation in the neocortex. *Nat. Neurosci.* **2012**, *15*, 1022–1031. [CrossRef] [PubMed]

20. Golzio, C.; Willer, J.; Talkowski, M.E.; Oh, E.C.; Taniguchi, Y.; Jacquemont, S.; Reymond, A.; Sun, M.; Sawa, A.; Gusella, J.F.; et al. KCTD13 is a major driver of mirrored neuroanatomical phenotypes of the 16p11.2 copy number variant. *Nature* **2012**, *485*, 363–367. [CrossRef] [PubMed]

21. O'Roak, B.J.; Deriziotis, P.; Lee, C.; Vives, L.; Schwartz, J.J.; Girirajan, S.; Karakoc, E.; Mackenzie, A.P.; Ng, S.B.; Baker, C.; et al. Exome sequencing in spordic autism spectrum disorders identifies severe de novo mutations. *Nat. Genet.* **2011**, *43*, 585–589. [CrossRef] [PubMed]

22. Iossifov, I.; Ronemus, M.; Levy, D.; Wang, Z.; Hakker, I.; Rosenbaum, J.; Yamrom, B.; Lee, Y.H.; Narzisi, G.; Leotta, A.; et al. De Novo Gene Disruptions in Children on the Autistic Spectrum. *Neuron* **2012**, *74*, 285–299. [CrossRef] [PubMed]

23. O'Roak, B.J.; Vives, L.; Girirajan, S.; Karakoc, E.; Krumm, N.; Coe, B.P.; Levy, R.; Ko, A.; Lee, C.; Smith, J.D.; et al. Sporadic autism exomes reveal a highly interconnected protein network of de novo mutations. *Nature* **2012**, *485*, 246–250. [CrossRef] [PubMed]

24. Sanders, S.J.; Murtha, M.T.; Gupta, A.R.; Murdoch, J.D.; Raubeson, M.J.; Willsey, A.J.; Ercan-Sencicek, A.G.; DiLullo, N.M.; Parikshak, N.N.; Stein, J.L.; et al. De novo mutations revealed by whole-exome sequencing are strongly associated with autism. *Nature* **2012**, *485*, 237–241. [CrossRef] [PubMed]

25. Yuen, R.K.C.; Thiruvahindrapuram, B.; Merico, D.; Walker, S.; Tammimies, K.; Hoang, N.; Chrysler, C.; Nalpathamkalam, T.; Pellecchia, G.; Liu, Y.; et al. Whole-genome sequencing of quartet families with autism spectrum disorder. *Nat. Med.* **2015**, *21*, 185–191. [CrossRef] [PubMed]

26. Gaugler, T.; Klei, L.; Sanders, S.J.; Bodea, C.A.; Goldberg, A.P.; Lee, A.B.; Mahajan, M.; Manaa, D.; Pawitan, Y.; Reichert, J.; et al. Most genetic risk for autism resides with common variation. *Nat. Genet.* **2014**, *46*, 881–885. [CrossRef] [PubMed]

27. Basu, S.N.; Kollu, R.; Banerjee-Basu, S. AutDB: A gene reference resource for autism research. *Nucleic Acids Res.* **2009**, *37*, D832–D836. [CrossRef] [PubMed]

28. Geschwind, D.H. Autism: Many genes, common pathways? *Cell* **2008**, *135*, 391–395. [CrossRef] [PubMed]

29. Parikshak, N.N.; Gandal, M.J.; Geschwind, D.H. Systems biology and gene networks in neurodevelopmental and neurodegenerative disorders. *Nat. Rev. Genet.* **2015**, *16*, 441–458. [CrossRef] [PubMed]

30. The Gene Ontology Consortium; Ashburner, M.; Ball, C.A.; Blake, J.A.; Botstein, D.; Butler, H.; Michael Cherry, J.; Davis, A.P.; Dolinski, K.; Dwight, S.S.; Eppig, J.T.; et al. Gene Ontology: Tool for the unification of biology. *Nat. Genet.* **2000**, *25*, 25–29.

31. Chowdhury, S.; Sarkar, R.R. Comparison of human cell signaling pathway databases—Evolution, drawbacks and challenges. *Database* **2015**, *2015*. [CrossRef] [PubMed]

32. Ogata, H.; Goto, S.; Sato, K.; Fujibuchi, W.; Bono, H.; Kanehisa, M. KEGG: Kyoto Encyclopedia of Genes and Genomes. *Nucleic Acids Res.* **1999**, *27*, 29–34. [CrossRef] [PubMed]

33. Ingenuity Systems Ingenuity Pathway Analysis (IPA). Available online: https://www.qiagenbioinformatics.com/ (accessed on 13 April 2017).

34. Mines, M.A.; Yuskaitis, C.J.; King, M.K.; Beurel, E.; Jope, R.S. GSK3 Influences Social Preference and Anxiety-Related Behaviors during Social Interaction in a Mouse Model of Fragile X Syndrome and Autism. *PLoS ONE* **2010**, *5*, e9706. [CrossRef] [PubMed]

35. Okerlund, N.D.; Cheyette, B.N.R. Synaptic Wnt signaling—A contributor to major psychiatric disorders? *J. Neurodev. Disord.* **2011**, *3*, 162–174. [CrossRef] [PubMed]

36. Iossifov, I.; O'roak, B.J.; Sanders, S.J.; Ronemus, M.; Krumm, N.; Levy, D.; Stessman, H.A.; Witherspoon, K.; Vives, L.; Patterson, K.E.; et al. The contribution of de novo coding mutations to autism spectrum disorder. *November* **2014**, *13*, 216–221. [CrossRef] [PubMed]

37. Hadley, D.; Wu, Z.-L.; Kao, C.; Kini, A.; Mohamed-Hadley, A.; Thomas, K.; Vazquez, L.; Qiu, H.; Mentch, F.; Pellegrino, R.; et al. The impact of the metabotropic glutamate receptor and other gene family interaction networks on autism. *Nat. Commun.* **2014**, *5*. [CrossRef] [PubMed]

38. Wen, Y.; Alshikho, M.J.; Herbert, M.R. Pathway network analyses for autism reveal multisystem involvement, major overlaps with other diseases and convergence upon MAPK and Calcium signaling. *PLoS ONE* **2016**, *11*, e0153329. [CrossRef] [PubMed]

39. Lin, G.N.; Corominas, R.; Lemmens, I.; Yang, X.; Tavernier, J.; Hill, D.E.; Vidal, M.; Sebat, J.; Iakoucheva, L.M. Spatiotemporal 16p11.2 Protein Network Implicates Cortical Late Mid-Fetal Brain Development and KCTD13-Cul3-RhoA Pathway in Psychiatric Diseases. *Neuron* **2015**, *85*, 742–754. [CrossRef] [PubMed]

40. Ben-David, E.; Shifman, S. Networks of neuronal genes affected by common and rare variants in autism spectrum disorders. *PLoS Genet.* **2012**, *8*, e1002556. [CrossRef] [PubMed]

41. MacDonald, B.T.; Tamai, K.; He, X. Wnt/β-Catenin Signaling: Components, Mechanisms, and Diseases. *Dev. Cell* **2009**, *17*, 9–26. [CrossRef] [PubMed]

42. Polakis, P. Wnt signaling and cancer. *Genes Dev.* **2000**, *14*, 1837–1851. [CrossRef] [PubMed]

43. Rao, T.P.; Kühl, M. An updated overview on Wnt signaling pathways: A prelude for more. *Circ. Res.* **2010**, *106*, 1798–1806. [CrossRef] [PubMed]

44. Noelanders, R.; Vleminckx, K. How Wnt Signaling Builds the Brain. *Neuroscientist* **2016**. [CrossRef] [PubMed]

45. Chen, A.; Walsh, C. Regulation of Cerebral Cortical Size by Control of Cell Cycle Exit in Neural Precursors. *Science* **2002**, *297*, 365–369. [CrossRef] [PubMed]

46. Woodhead, G.J.; Mutch, C.A.; Olson, E.C.; Chenn, A. Cell-Autonomous beta-Catenin Signaling Regulates Cortical Precursor Proliferation. *J. Neurosci.* **2006**, *26*, 12620–12630. [CrossRef] [PubMed]

47. Duncan, R.N.; Xie, Y.; McPherson, A.D.; Taibi, A.V.; Bonkowsky, J.L.; Douglass, A.D.; Dorsky, R.I. Hypothalamic radial glia function as self-renewing neural progenitors in the absence of Wnt/β-catenin signaling. *Development* **2015**, *143*, 45–53. [CrossRef] [PubMed]

48. Piccin, D.; Morshead, C.M. Wnt signaling regulates symmetry of division of neural stem cells in the adult brain and in response to injury. *Stem Cells* **2011**, *29*, 528–538. [CrossRef] [PubMed]

49. Marinaro, C.; Pannese, M.; Weinandy, F.; Sessa, A.; Bergamaschi, A.; Taketo, M.M.; Broccoli, V.; Comi, G.; Götz, M.; Martino, G.; et al. Wnt signaling has opposing roles in the developing and the adult brain that are modulated by Hipk1. *Cereb. Cortex* **2012**, *22*, 2415–2427. [CrossRef] [PubMed]

50. Huber, O.; Krohn, M.; Kemler, R. A specific domain in alpha-catenin mediates binding to beta-catenin or plakoglobin. *J. Cell Sci.* **1997**, *110 Pt 1*, 1759–1765. [PubMed]

51. Huber, A.H.; Weis, W.I. The structure of the B-catenin/E-cadherin complex and the molecular basis of diverse ligand recognition by β-catenin. *Cell* **2001**, *105*, 391–402. [CrossRef]

52. Yu, X.; Malenka, R.C. Beta-catenin is critical for dendritic morphogenesis. *Nat. Neurosci.* **2003**, *6*, 1169–1177. [CrossRef] [PubMed]

53. Bian, W.-J.; Miao, W.-Y.; He, S.-J.; Qiu, Z.; Yu, X. Coordinated Spine Pruning and Maturation Mediated by Inter-Spine Competition for Cadherin/Catenin Complexes. *Cell* **2015**, *162*, 808–822. [CrossRef] [PubMed]

54. Murase, S.; Mosser, E.; Schuman, E.M. Depolarization drives β-catenin into neuronal spines promoting changes in synaptic structure and function. *Neuron* **2002**, *35*, 91–105. [CrossRef]

55. Adams, C.L.; Nelson, W.J.; Smith, S.J. Quantitative Analysis of Cadherin-Catenin-Actin Reorganization during Development of Cell-Cell Adhesion. *J. Cell Biol.* **1996**, *135*, 1899–1911. [CrossRef] [PubMed]

56. Bamji, S.X.; Shimazu, K.; Kimes, N.; Huelsken, J.; Birchmeier, W.; Lu, B.; Reichardt, L.F. Role of β-catenin in synaptic vesicle localization and presynaptic assembly. *Neuron* **2003**, *40*, 719–731. [CrossRef]

57. Butz, S.; Okamoto, M.; Südhof, T.C. A Tripartite Protein Complex with the Potential to Couple Synaptic Vesicle Exocytosis to Cell Adhesion in Brain. *Cell* **1998**, *94*, 773–782. [CrossRef]

58. Han, G.A.; Malintan, N.T.; Collins, B.M.; Meunier, F.A.; Sugita, S. Munc18-1 as a key regulator of neurosecretion. *J. Neurochem.* **2010**, *115*, 1–10. [CrossRef] [PubMed]

59. Bolshakov, V.Y.; Siegelbaum, S.A. Regulation of hippocampal transmitter release during development and long-term potentiation. *Science* **1995**, *269*, 1730–1734. [CrossRef] [PubMed]

60. Taylor, A.M.; Wu, J.; Tai, H.C.; Schuman, E.M. Axonal translation of beta-catenin regulates synaptic vesicle dynamics. *J. Neurosci.* **2013**, *33*, 5584–5589. [CrossRef] [PubMed]

61. Sharma, K.; Choi, S.Y.; Zhang, Y.; Nieland, T.J.F.; Long, S.; Li, M.; Huganir, R.L. High-throughput genetic screen for synaptogenic factors: Identification of LRP6 as critical for excitatory synapse development. *Cell Rep.* **2013**, *5*, 1330–1341. [CrossRef] [PubMed]

62. Hutsler, J.J.; Zhang, H. Increased dendritic spine densities on cortical projection neurons in autism spectrum disorders. *Brain Res.* **2010**, *1309*, 83–94. [CrossRef] [PubMed]

63. Stoner, R.; Chow, M.L.; Boyle, M.P.; Sunkin, S.M.; Mouton, P.R.; Roy, S.; Wynshaw-Boris, A.; Colamarino, S.A.; Lein, E.S.; Courchesne, E. Patches of disorganization in the neocortex of children with autism. *N. Engl. J. Med.* **2014**, *370*, 1209–1219. [CrossRef] [PubMed]

64. Casanova, M.F.; van Kooten, I.; Switala, A.E.; van Engeland, H.; Heinsen, H.; Steinbusch, H.W.M.; Hof, P.R.; Schmitz, C. Abnormalities of cortical minicolumnar organization in the prefrontal lobes of autistic patients. *Clin. Neurosci. Res.* **2006**, *6*, 127–133. [CrossRef]

65. Chung, K.; Wallace, J.; Kim, S.-Y.; Kalyanasundaram, S.; Andalman, A.S.; Davidson, T.J.; Mirzabekov, J.J.; Zalocusky, K.A.; Mattis, J.; Denisin, A.K.; et al. Structural and molecular interrogation of intact biological systems. *Nature* **2013**, *497*, 332–337. [CrossRef] [PubMed]

66. Dong, F.; Jiang, J.; McSweeney, C.; Zou, D.; Liu, L.; Mao, Y. Deletion of CTNNB1 in inhibitory circuitry contributes to autism-associated behavioral defects. *Hum. Mol. Genet.* **2016**, *25*, 2738–2751. [CrossRef] [PubMed]

67. Mines, M.A.; Jope, R.S. Glycogen synthase kinase-3: A promising therapeutic target for fragile X syndrome. *Front. Mol. Neurosci.* **2011**, *4*, 35. [CrossRef] [PubMed]

68. Caracci, M.O.; Avila, M.E.; de Ferrari, G.V. Synaptic Wnt/GSK3β-Signaling Hub in Autism. *Neural Plast.* **2016**, *2016*. [CrossRef] [PubMed]

69. Min, W.W.; Yuskaitis, C.J.; Yan, Q.; Sikorski, C.; Chen, S.; Jope, R.S.; Bauchwitz, R.P. Elevated glycogen synthase kinase-3 activity in Fragile X mice: Key metabolic regulator with evidence for treatment potential. *Neuropharmacology* **2009**, *56*, 463–472. [CrossRef] [PubMed]

70. Guo, W.; Murthy, A.C.; Zhang, L.; Johnson, E.B.; Schaller, E.G.; Allan, A.M.; Zhao, X. Inhibition of GSK3β improves hippocampusdependent learning and rescues neurogenesis in a mouse model of fragile X syndrome. *Hum. Mol. Genet.* **2012**, *21*, 681–691. [CrossRef] [PubMed]

71. Franklin, A.V.; King, M.K.; Palomo, V.; Martinez, A.; Mcmahon, L.L.; Jope, R.S. Glycogen synthase kinase-3 inhibitors reverse deficits in long- term potentiation and cognition in Fragile X mice. *Biol. Psychiatry* **2014**, *75*, 198–206. [CrossRef] [PubMed]

72. Durak, O.; Gao, F.; Kaeser-Woo, Y.; Rueda, R.; Martorell, A.; Nott, A.; Liu, C.; Watson, L.; Tsai, L.-H. Chd8 mediates cortical neurogenesis via transcriptional regulation of cell cycle and Wnt signaling. *Nat. Neurosci.* **2016**, *19*, 1477–1488. [CrossRef] [PubMed]

73. Harris, K.P.; Akbergenova, Y.; Cho, R.W.; Baas-Thomas, M.S.; Littleton, J.T. Shank Modulates Postsynaptic Wnt Signaling to Regulate Synaptic Development. *J. Neurosci.* **2016**, *36*, 5820–5832. [CrossRef] [PubMed]

74. Budnik, V.; Salinas, P.C. Wnt signaling during synaptic development and plasticity. *Curr. Opin. Neurobiol.* **2011**, *21*, 151–159. [CrossRef] [PubMed]

75. Mathew, D.; Ataman, B.; Chen, J.; Zhang, Y.; Cumberledge, S.; Budnik, V. Wingless Signaling at Synapses Is through Cleavage and Nuclear Import of Receptor DFrizzled2. *Science* **2005**, *310*, 1344–1347. [CrossRef] [PubMed]

76. Carafoli, E. Special issue: Calcium signaling and disease. *Biochem. Biophys. Res. Commun.* **2004**, *322*, 1097. [CrossRef]

77. Berridge, M.J.; Lipp, P.; Bootman, M.D. The versatility and universality of Calcium signalling. *Nat. Rev. Mol. Cell Biol.* **2000**, *1*, 11–21. [CrossRef] [PubMed]

78. Clapham, D.E. Calcium Signaling. *Cell* **2007**, *131*, 1047–1058. [CrossRef] [PubMed]

79. Bossuyt, J.; Bers, D.M. Visualizing CaMKII and CaM activity: A paradigm of compartmentalized signaling. *J. Mol. Med.* **2013**, *91*, 907–916. [CrossRef] [PubMed]

80. Okamoto, K.-I.; Narayanan, R.; Lee, S.H.; Murata, K.; Hayashi, Y. The role of CaMKII as an F-actin-bundling protein crucial for maintenance of dendritic spine structure. *Proc. Natl. Acad. Sci. USA* **2007**, *104*, 6418–6423. [CrossRef] [PubMed]

81. Jalan-Sakrikar, N.; Bartlett, R.K.; Baucum, A.J.; Colbran, R.J. Substrate-selective and calcium-independent activation of CaMKII by α-actinin. *J. Biol. Chem.* **2012**, *287*, 15275–15283. [CrossRef] [PubMed]

82. Hell, J.W. CaMKII: Claiming center stage in postsynaptic function and organization. *Neuron* **2014**, *81*, 249–265. [CrossRef] [PubMed]

83. Lisman, J.; Schulman, H.; Cline, H. The molecular basis of CaMKII function in synaptic and behavioural memory. *Nat. Rev. Neurosci.* **2002**, *3*, 175–190. [CrossRef] [PubMed]

84. Houston, C.M.; He, Q.; Smart, T.G. CaMKII phosphorylation of the GABA A receptor: Receptor subtype-and synapse-specific modulation. *J. Physiol.* **2009**, *58710*, 2115–2125. [CrossRef] [PubMed]

85. Wayman, G.A.; Kaech, S.; Grant, W.F.; Davare, M.; Impey, S.; Tokumitsu, H.; Nozaki, N.; Banker, G.; Soderling, T.R. Regulation of axonal extension and growth cone motility by calmodulin-dependent protein kinase I. *J. Neurosci.* **2004**, *24*, 3786–3794. [CrossRef] [PubMed]

86. Mizuno, K.; Ris, L.; Sánchez-Capelo, A.; Godaux, E.; Giese, K.P. Ca^{2+}/calmodulin kinase kinase alpha is dispensable for brain development but is required for distinct memories in male, though not in female, mice. *Mol. Cell. Biol.* **2006**, *26*, 9094–9104. [CrossRef] [PubMed]

87. Marcelo, K.L.; Means, A.R.; York, B. The Ca^{2+}/Calmodulin/CaMKK2 Axis: Nature's Metabolic CaMshaft. *Trends Endocrinol. Metab.* **2016**, *27*, 706–718. [CrossRef] [PubMed]

88. Yun, S.H.; Trommer, B.L. Fragile X mice: Reduced long-term potentiation and N-Methyl-D-Aspartate receptor-mediated neurotransmission in dentate gyrus. *J. Neurosci. Res.* **2011**, *89*, 176–182. [CrossRef] [PubMed]

89. Moretti, P.; Levenson, J.M.; Battaglia, F.; Atkinson, R.; Teague, R.; Antalffy, B.; Armstrong, D.; Arancio, O.; Sweatt, J.D.; Zoghbi, H.Y. Learning and memory and synaptic plasticity are impaired in a mouse model of Rett syndrome. *J. Neurosci.* **2006**, *26*, 319–327. [CrossRef] [PubMed]

90. Jung, N.H.; Janzarik, W.G.; Delvendahl, I.; Münchau, A.; Biscaldi, M.; Mainberger, F.; Bäumer, T.; Rauh, R.; Mall, V. Impaired induction of long-term potentiation-like plasticity in patients with high-functioning autism and Asperger syndrome. *Dev. Med. Child Neurol.* **2013**, *55*, 83–89. [CrossRef] [PubMed]

91. Modi, M.E.; Sahin, M. Translational use of event-related potentials to assess circuit integrity in ASD. *Nat. Rev. Neurol.* **2017**, *13*, 160–170. [CrossRef] [PubMed]

92. Schmunk, G.; Nguyen, R.L.; Ferguson, D.L.; Kumar, K.; Parker, I.; Gargus, J.J. High-throughput screen detects calcium signaling dysfunction in typical sporadic autism spectrum disorder. *Sci. Rep.* **2017**, *7*, 40740. [CrossRef] [PubMed]

93. Paşca, S.P.; Portmann, T.; Voineagu, I.; Yazawa, M.; Shcheglovitov, A.; Paşca, A.M.; Cord, B.; Palmer, T.D.; Chikahisa, S.; Nishino, S.; et al. Using iPSC-derived neurons to uncover cellular phenotypes associated with Timothy syndrome. *Nat. Med.* **2011**, *17*, 1657–1662. [CrossRef] [PubMed]

94. Wang, H.; Wu, L.-J.J.; Zhang, F.; Zhuo, M. Roles of calcium-stimulated adenylyl cyclase and calmodulin-dependent protein kinase {IV} in the regulation of {FMRP} by group I metabotropic glutamate receptors. *J. Neurosci.* **2008**, *28*, 4385–4397. [CrossRef] [PubMed]

95. Waltes, R.; Duketis, E.; Knapp, M.; Anney, R.J.L.; Huguet, G.; Schlitt, S.; Jarczok, T.A.; Sachse, M.; Kämpfer, L.M.; Kleinböck, T.; et al. Common variants in genes of the postsynaptic FMRP signalling pathway are risk factors for autism spectrum disorders. *Hum. Genet.* **2014**, *133*, 781–792. [CrossRef] [PubMed]

96. Pop, A.S.; Gomez-Mancilla, B.; Neri, G.; Willemsen, R.; Gasparini, F. Fragile X syndrome: A preclinical review on metabotropic glutamate receptor 5 (mGluR5) antagonists and drug development. *Psychopharmacology* **2014**, *231*, 1217–1226. [CrossRef] [PubMed]

97. Guo, W.; Ceolin, L.; Collins, K.A.; Perroy, J.; Huber Correspondence, K.M.; Huber, K.M. Elevated CaMKIIα and Hyperphosphorylation of Homer Mediate Circuit Dysfunction in a Fragile X Syndrome Mouse Model. *Cell Rep.* **2015**, *13*, 2297–2311. [CrossRef] [PubMed]

98. Michalon, A.; Bruns, A.; Risterucci, C.; Honer, M.; Ballard, T.M.; Ozmen, L.; Jaeschke, G.; Wettstein, J.G.; von Kienlin, M.; Künnecke, B. Chronic Metabotropic Glutamate Receptor 5 Inhibition Corrects Local Alterations of Brain Activity and Improves Cognitive Performance in Fragile X Mice. *Biol. Psychiatry* **2013**, *75*, 189–197. [CrossRef] [PubMed]

99. Schaefer, T.L.; Davenport, M.H.; Erickson, C.A. Emerging pharmacologic treatment options for fragile X syndrome. *Appl. Clin. Genet.* **2015**, *2015*, 75–93.

100. Berry-Kravis, E.; Hessl, D.; Coffey, S.; Hervey, C.; Schneider, A.; Yuhas, J.; Hutchison, J.; Snape, M.; Tranfaglia, M.; Nguyen, D.V.; et al. A pilot open label, single dose trial of fenobam in adults with fragile X syndrome. *J. Med. Genet.* **2009**, *46*, 266–271. [CrossRef] [PubMed]

101. Berry-Kravis, E.; Sumis, A.; Hervey, C.; Nelson, M.; Porges, S.W.; Weng, N.; Weiler, I.J.; Greenough, W.T. Open-label treatment trial of lithium to target the underlying defect in fragile X syndrome. *J. Dev. Behav. Pediatr.* **2008**, *29*, 293–302. [CrossRef] [PubMed]

102. Weeber, E.J.; Jiang, Y.-H.; Elgersma, Y.; Varga, A.W.; Carrasquillo, Y.; Brown, S.E.; Christian, J.M.; Mirnikjoo, B.; Silva, A.; Beaudet, A.L.; et al. Derangements of hippocampal calcium/calmodulin-dependent protein kinase II in a mouse model for Angelman mental retardation syndrome. *J. Neurosci.* **2003**, *23*, 2634–2644. [PubMed]

103. Baucum, A.J.; Shonesy, B.C.; Rose, K.L.; Colbran, R.J. Quantitative Proteomics Analysis of CaMKII Phosphorylation and the CaMKII Interactome in the Mouse Forebrain. *ACS Chem. Neurosci.* **2015**, *6*, 615–631. [CrossRef] [PubMed]

104. Murakami, M.; Ichisaka, T.; Maeda, M.; Oshiro, N.; Hara, K.; Edenhofer, F.; Kiyama, H.; Yonezawa, K.; Yamanaka, S. mTOR is essential for growth and proliferation in early mouse embryos and embryonic stem cells. *Mol. Cell. Biol.* **2004**, *24*, 6710–6718. [CrossRef] [PubMed]

105. Takei, N.; Nawa, H. mTOR signaling and its roles in normal and abnormal brain development. *Front. Mol. Neurosci.* **2014**, *7*, 28. [CrossRef] [PubMed]

106. Chen, J.A.; Peñagarikano, O.; Belgard, T.G.; Swarup, V.; Geschwind, D.H. The Emerging Picture of Autism Spectrum Disorder: Genetics and Pathology. *Annu. Rev. Pathol. Mech. Dis.* **2015**, *10*, 111–144. [CrossRef] [PubMed]

107. McBride, K.L.; Varga, E.A.; Pastore, M.T.; Prior, T.W.; Manickam, K.; Atkin, J.F.; Herman, G.E. Confirmation study of PTEN mutations among individuals with autism or developmental delays/mental retardation and macrocephaly. *Autism Res.* **2010**, *3*, 137–141. [CrossRef] [PubMed]

108. Plasschaert, E.; Descheemaeker, M.-J.; Van Eylen, L.; Noens, I.; Steyaert, J.; Legius, E. Prevalence of Autism Spectrum Disorder symptoms in children with neurofibromatosis type 1. *Am. J. Med. Genet. Part B* **2015**, *168B*, 72–80. [CrossRef] [PubMed]

109. Hoeffer, C.A.; Sanchez, E.; Hagerman, R.J.; Mu, Y.; Nguyen, D.V.; Wong, H.; Whelan, A.M.; Zukin, R.S.; Klann, E.; Tassone, F. Altered mTOR signaling and enhanced CYFIP2 expression levels in subjects with Fragile X syndrome. *Genes Brain Behav.* **2012**, *11*, 332–341. [CrossRef] [PubMed]

110. Tsai, V.; Parker, W.E.; Orlova, K.A.; Baybis, M.; Chi, A.W.S.; Berg, B.D.; Birnbaum, J.F.; Estevez, J.; Okochi, K.; Sarnat, H.B.; et al. Fetal Brain mTOR Signaling Activation in Tuberous Sclerosis Complex. *Cereb. Cortex* **2014**, *24*, 315–327. [CrossRef] [PubMed]

111. Sato, A.; Kasai, S.; Kobayashi, T.; Takamatsu, Y.; Hino, O.; Ikeda, K.; Mizuguchi, M. Rapamycin reverses impaired social interaction in mouse models of tuberous sclerosis complex. *Nat. Commun.* **2012**, *3*, 1292. [CrossRef] [PubMed]

112. Butler, M.G.; Dasouki, M.J.; Zhou, X.; Talebizadeh, Z.; Brown, M.; Takahashi, T.N.; Miles, J.H.; Wang, C.H.; Stratton, R.; Pilarski, R.; et al. Subset of individuals with autism spectrum disorders and extreme macrocephaly associated with germline PTEN tumour suppressor gene mutations. *J. Med. Genet.* **2005**, *42*, 318–321. [CrossRef] [PubMed]

113. Kwon, C.H.; Luikart, B.W.; Powell, C.M.; Zhou, J.; Matheny, S.A.; Zhang, W.; Li, Y.; Baker, S.J.; Parada, L.F. Pten Regulates Neuronal Arborization and Social Interaction in Mice. *Neuron* **2006**, *50*, 377–388. [CrossRef] [PubMed]

114. Cupolillo, D.; Hoxha, E.; Faralli, A.; De Luca, A.; Rossi, F.; Tempia, F.; Carulli, D. Autistic-Like Traits and Cerebellar Dysfunction in Purkinje Cell PTEN Knock-Out Mice. *Neuropsychopharmacology* **2015**, *41*, 1–27. [CrossRef] [PubMed]

115. Johannessen, C.M.; Reczek, E.E.; James, M.F.; Brems, H.; Legius, E.; Cichowski, K. The NF1 tumor suppressor critically regulates TSC2 and mTOR. *Proc. Natl. Acad. Sci. USA* **2005**, *102*, 8573–8578. [CrossRef] [PubMed]

116. Burket, J.A.; Benson, A.D.; Tang, A.H.; Deutsch, S.I. Rapamycin improves sociability in the BTBR T+itpr3tf/J mouse model of autism spectrum disorders. *Brain Res. Bull.* **2014**, *100*, 70–75. [CrossRef] [PubMed]

117. Berry, A.; Bindocci, E.; Alleva, E. NGF, Brain and Behavioral Plasticity. *Neural Plast.* **2012**, *2012*, 1–9. [CrossRef] [PubMed]

118. Gnahn, H.; Hefti, F.; Heumann, R.; Schwab, M.E.; Thoenen, H. NGF-Mediated increase of choline acetyltransferase (ChAT) in the neonatal rat forebrain: Evidence for a physiological role of NGF in the brain? *Dev. Brain Res.* **1983**, *9*, 45–52. [CrossRef]

119. Li, Y.; Holtzman, D.M.; Kromer, L.F.; Kaplan, D.R.; Chua-Couzens, J.; Clary, D.O.; Knüsel, B.; Mobley, W.C. Regulation of TrkA and ChAT expression in developing rat basal forebrain: Evidence that both exogenous and endogenous NGF regulate differentiation of cholinergic neurons. *J. Neurosci.* **1995**, *15*, 2888–2905. [PubMed]

120. Riikonen, R.; Vanhala, R. Levels of cerebrospinal fluid nerve-growth factor differ in infantile autism and Rett syndrome. *Dev. Med. Child Neurol.* **1999**, *41*, 148–152. [CrossRef] [PubMed]

121. Nelson, K.B.; Grether, J.K.; Croen, L.A.; Dambrosia, J.M.; Dickens, B.F.; Jelliffe, L.L.; Hansen, R.L.; Phillips, T.M. Neuropeptides and neurotrophins in neonatal blood of children with autism or mental retardation. *Ann. Neurol.* **2001**, *49*, 597–606. [CrossRef] [PubMed]

122. Stamova, B.S.; Tian, Y.; Nordahl, C.W.; Shen, M.D.; Rogers, S.; Amaral, D.G.; Sharp, F.R. Evidence for differential alternative splicing in blood of young boys with autism spectrum disorders. *Mol. Autism* **2013**, *4*, 30. [CrossRef] [PubMed]

123. Chakrabarti, B.; Dudbridge, F.; Kent, L.; Wheelwright, S.; Hill-Cawthorne, G.; Allison, C.; Banerjee-Basu, S.; Baron-Cohen, S. Genes related to sex steroids, neural growth, and social-emotional behavior are associated with autistic traits, empathy, and asperger syndrome. *Autism Res.* **2009**, *2*, 157–177. [CrossRef] [PubMed]

124. Lu, A.T.-H.; Yoon, J.; Geschwind, D.H.; Cantor, R.M. QTL replication and targeted association highlight the nerve growth factor gene for nonverbal communication deficits in autism spectrum disorders. *Mol. Psychiatry* **2013**, *18*, 226–235. [CrossRef] [PubMed]

125. Ellegood, J.; Anagnostou, E.; Babineau, B.A.; Crawley, J.N.; Lin, L.; Genestine, M.; DiCicco-Bloom, E.; Lai, J.K.Y.; Foster, J.A.; Peñagarikano, O.; et al. Clustering autism: Using neuroanatomical differences in 26 mouse models to gain insight into the heterogeneity. *Mol. Psychiatry* **2015**, *20*, 118–125. [CrossRef] [PubMed]

126. Kim, H.; Lim, C.-S.; Kaang, B.-K. Neuronal mechanisms and circuits underlying repetitive behaviors in mouse models of autism spectrum disorder. *Behav. Brain Funct.* **2016**, *12*, 3. [CrossRef] [PubMed]

International Journal of
Molecular Sciences

MDPI

Article

GeneAnalytics Pathway Analysis and Genetic Overlap among Autism Spectrum Disorder, Bipolar Disorder and Schizophrenia

Naveen S. Khanzada [1], Merlin G. Butler [1,2] and Ann M. Manzardo [1,*]

[1] Department of Psychiatry and Behavioral Sciences, University of Kansas Medical Center, Kansas City,
     KS 66160, USA; nkhanzada@kumc.edu (N.S.K.); mbutler4@kumc.edu (M.G.B.)
[2] Department of Pediatrics, University of Kansas Medical Center, Kansas City, KS 66160, USA
*   Correspondence: amanzardo@kumc.edu; Tel.: +1-913-588-6473

Academic Editor: William Chi-shing Cho
Received: 18 January 2017; Accepted: 23 February 2017; Published: 28 February 2017

Abstract: Bipolar disorder (BPD) and schizophrenia (SCH) show similar neuropsychiatric behavioral disturbances, including impaired social interaction and communication, seen in autism spectrum disorder (ASD) with multiple overlapping genetic and environmental influences implicated in risk and course of illness. GeneAnalytics software was used for pathway analysis and genetic profiling to characterize common susceptibility genes obtained from published lists for ASD (792 genes), BPD (290 genes) and SCH (560 genes). Rank scores were derived from the number and nature of overlapping genes, gene-disease association, tissue specificity and gene functions subdivided into categories (e.g., diseases, tissues or functional pathways). Twenty-three genes were common to all three disorders and mapped to nine biological Superpathways including Circadian entrainment (10 genes, score = 37.0), Amphetamine addiction (five genes, score = 24.2), and Sudden infant death syndrome (six genes, score = 24.1). Brain tissues included the medulla oblongata (11 genes, score = 2.1), thalamus (10 genes, score = 2.0) and hypothalamus (nine genes, score = 2.0) with six common genes (*BDNF, DRD2, CHRNA7, HTR2A, SLC6A3,* and *TPH2*). Overlapping genes impacted dopamine and serotonin homeostasis and signal transduction pathways, impacting mood, behavior and physical activity level. Converging effects on pathways governing circadian rhythms support a core etiological relationship between neuropsychiatric illnesses and sleep disruption with hypoxia and central brain stem dysfunction.

Keywords: mental illness; genetic profiling; GeneAnalytics molecular pathway analysis; circadian entrainment

---

1. Introduction

Severe neuropsychiatric disorders collectively present with similar behavioral, social, cognitive and perceptual disturbances including autism spectrum disorder (ASD). ASD includes classical autism, Asperger syndrome and pervasive developmental disorder with problems in social interaction and communication or repetitive behavior. Schizophrenia (SCH) presents with delusions, hallucinations, disorganized thinking and behavior with negative symptoms. Bipolar disorder (BPD) is considered a developmental disorder characterized by progressive cognitive impairment, residual symptoms, sleep disturbance, and emotional dysregulations with cycles of depression and mania. Schizophrenia and bipolar disorders share many common traits with ASD, including social and cognitive dysfunction and impaired ability to function, live and work independently [1]. Considerable overlap has been identified between the molecular mechanisms implicated in the etiology of schizophrenia, bipolar disorder, and autism, suggesting similar root causes [2]. Up to 30% of patients diagnosed with ASD during childhood will develop schizophrenia during adulthood [3]. Further, the presence of schizophrenia or bipolar disorder in first-degree relatives is a consistent and significant risk factor for ASD [4]. These three neuropsychiatric illnesses have complex inheritance patterns with >80% estimate for each disorder with multiple genetic and environmental factors influencing disease risk and course [2,4].

A recent large collaborative genetic study of families with schizophrenia and ASD showed significant overlap in candidate genes and susceptibility regions for both disorders using traditional karyotyping, genome-wide association studies (GWAS) and comparative genome hybridization (CGH) analyses by identifying chromosomal deletions and duplications in individuals with ASD [5–8]. ASD and SCH risk alleles appear to impact growth-signaling pathways with autism associated with loss of function in many genes [9–11], whereas schizophrenia tends to be associated with reduced function or activity of genes that up-regulate growth-related pathways [12–16].

Cytogenetic, linkage and association studies have identified common copy number variants (CNVs, deletions, duplications) between ASD and SCH which may produce dosage-dependent gain or loss of expression of genes contributing to phenotypic variation in presentation and course of illness. A large number of autism-specific CNVs have been found but with low recurrence (<1%) and they show a high level of genetic heterogeneity. For example, when the 15q11.2 BP1–BP2 region or 15q13.3 band contains a deletion or duplication in patients, then autism or a variety of neuropsychiatric traits including schizophrenia are identified [3,17–19]. Gene expression disturbances were found using postmortem cortical brain tissue from patients with autism, schizophrenia and bipolar disorder and they have shown a high correlation between the transcriptomes of ASD and schizophrenia, but not in BPD [5]. Hence, a large number of possible genetic and environmental factors do influence disease risk, expression and treatment.

Herein, we use the GeneAnalytics [20] program pathway analysis to further profile and characterize the underlying molecular architecture of clinically and etiologically relevant genes common to ASD [21], bipolar disorder [22] and schizophrenia [23] and associated diseases.

2. Results

The original gene lists reported in the literature included 792 genes for ASD [21], 290 genes for bipolar disorder [22] and 560 genes for schizophrenia [23], and of these, 23 genes were found in common in all three conditions (see Table 1). Functional analysis of the 23 genes identified from the submitted list of genes showed a high match for schizophrenia (17 genes, score = 15.1) with medium-match scores representing 25 other disorders including bipolar disorder (nine genes, score = 9.6) and autism spectrum disorder (10 genes, score = 9.1). Additional diseases identified were related to disorders of mental health including mood and personality disorders (see Table 2A). Tissues and cell types profiled for these 23 overlapping genes identified five common types of brain tissues which achieved high match scores (see Table 2B). These included the medulla oblongata (11 genes, score = 2.1), thalamus (10 genes, score = 2.0), hypothalamus (nine genes, score = 2.0), hippocampus (nine genes, score = 1.9) and cerebellum (eight genes, score = 1.9). Six (*BDNF, DRD2, CHRNA7, HTR2A, SLC6A3,* and *TPH2*) of the overlapping genes were matched to all five tissues types. Sixteen of the overlapping genes matched to a total of 36 Biological Superpathways with a total of nine Superpathways achieving significantly high match scores according to the GeneAnalytics pathway analysis and algorithm [20]. The Circadian entrainment pathway showed the highest match score involving 10 genes (score = 37.0), followed by Amphetamine addiction involving five genes (score = 24.3) and Sudden infant death syndrome (SIDS) susceptibility pathways (six genes, score = 24.1) (see Table 3).

Table 1. Twenty-three clinically relevant genes common to ASD, BPD and schizophrenia.

Gene Symbol	Gene Name	Chromosome Location
BDNF	Brain-derived neurotrophic factor	11p14.1
ANK3	Ankyrin 3	10q21.2
CACNA1C	Ca^2+ channel, voltage-dependent, L type, α1C subunit	12p13.33
CACNB2	Ca+ channel, voltage dependent, β2 subunit	10p12.33
CHRNA7	Cholinergic receptor, nicotinic, α7 (neuronal)	15q13.3
CNTNAP5	Contactin associated protein-like 5	2q14.3
CSMD1	CUB and sushi multiple domains 1	8p23.2
DISC1	Disruption in schizophrenia 1	1q42.2
DPP10	Dipeptidyl-peptidase 10 (non-functional)	2q14.1
DRD2	Dopamine receptor D2	11q23.2
FOXP2	Forkhead box P2	7q31.1
GSK3B	Glycogen synthase kinase 3β	3q13.33
HTR2A	5-Hydroxytryptamine (serotonin) receptor 2A, G-protein-coupled	13q14.2
MAOA	Monoamine oxidase A	Xp11.3
MTHFR	Methylenetetrahydrofolate reductase	1p36.22
NOS1AP	Nitric oxide synthase 1 (neuronal) adaptor protein	1q23.3
NRG1	Neuregulin 1	8p12
PDE4B	Phosphodiesterase 4B, CAMP-specific	1p31.3
SLC6A3	Solute carrier family 6 (neurotransmitter transporter, dopamine), member 3	5p15.33
SYN3	Synapsin III	22q12.3
TCF4	Transcription factor 4	18q21.1
TPH2	Tryptophan hydroxylase 2	12q21.1
ZNF804A	Zinc finger protein 804 A	2q32.1

Table 2. GeneAnalytics program mapping of diseases, tissues and cells that were significantly matched to 23 overlapping genes for autism spectrum disorder, bipolar disorder and schizophrenia.

A. Diseases	Genes Matched to Disease Type (Highmatch Score)	No. of Genes in Disease Type	Score
Schizophrenia	ANK3, BDNF, CACNA1C, CHRNA7, DISC1, NRG1, DRD2, GSK3B, HTR2A, MAOA, MTHFR, NOS1AP, PDE4B, SLC6A3, SYN3, TPH2, ZNF804A	249	15.1
	(Mediummatch scores > 6.0)		
Bipolar disorder	BDNF, DISC1, DRD2, GSK3B, HTR2A, MAOA, NRG1, SLC6A3, ZNF804A	39	9.6
Autism spectrum disorder	BDNF, CHRNA7, DISC1, DRD2, FOXP2, HTR2A, MAOA, MTHFR, SLC6A3, TPH2	103	9.1
Disease of mental health	BDNF, DISC1, DRD2, HTR2A, MAOA, NRG1, SLC6A3, ZNF804A	57	8.0
Attention deficit hyperactivity disorder	BDNF, CHRNA7, DRD2, HTR2A, MAOA, SLC6A3, TPH2, ZNF804A	63	7.8
Mood disorder	BDNF, CACNA1C, DISC1, DRD2, HTR2A, MAOA, TPH2	31	7.7
Psychotic disorder	BDNF, CHRNA7, DISC1, DRD2, HTR2A, NRG1, SLC6A3	37	7.5
Anxiety disorder	BDNF, DRD2, HTR2A, MAOA, SLC6A3, TPH2	21	7.1
Obsessive compulsive disorder	BDNF, DRD2, HTR2A, MAOA, SLC6A3, TPH2	29	6.8
Personality disorder	DRD2, HTR2A, MAOA, SLC6A3, TPH2	14	6.4
B. Tissues and Cells	**Genes Matched to Tissues and Cells**	**No. of Genes in Tissues And Cells**	**Score**
Medulla oblongata	BDNF, CHRNA7, DRD2, FOXP2, HTR2A, MAOA, NRG1, PDE4B, SLC6A3, TCF4, TPH2	2179	2.1
Thalamus	BDNF, CHRNA7, DRD2, HTR2A, MTHFR, PDE4B, SLC6A3, TCF4, TPH2	1736	2.0
Hypothalamus	BDNF, CHRNA7, DRD2, FOXP2, HTR2A, MTHFR, PDE4B, SLC6A3, TCF4, TPH2	1666	2.0
Hippocampus	ANK3, BDNF, CHRNA7, DISC1, DRD2, FOXP2, HTR2A, SLC6A3, TPH2	3335	1.9
Cerebellum	ANK3, BDNF, CHRNA7, DRD2, FOXP2, HTR2A, SLC6A3, TPH2	2609	1.9

Table 3. GeneAnalytics program mapping of superpathways with high match scores for 23 overlapping genes for autism spectrum disorder, bipolar disorder and schizophrenia.

Superpathways	Genes Matched to Superpathways	No. of Genes in Superpathways	Score
Circadian entrainment	*SLC6A3, GSK3B, HTR2A, MAOA, NOS1AP, PDE4B, TPH2, CACNA1C, CHRNA7, DRD2*	390	37.0
Amphetamine addiction	*SLC6A3, MAOA, BDNF, CACNA1C, DRD2*	87	24.3
SID susceptibility pathways	*HTR2A, MAOA, NOS1AP, TPH2, BDNF, CHRNA7*	185	24.1
Selective serotonin reuptake inhibitor pathways	*HTR2A, MAOA, TPH2*	29	17.5
Monoamine transport	*SLC6A3, MAOA, TPH2*	36	16.6
Transmission across chemical synapses	*SLC6A3, MAOA, SYN3, CACNB2, CHRNA7*	316	15.2
CREB pathways	*HTR2A, NRG1, BDNF, CACNA1C, CACNB2, CHRNA7*	562	14.9
Neurotransmitter clearance in the synaptic cleft	*SLC6A3, MAOA*	8	14.6
CAMP signaling pathways	*PDE4B, BDNF, CACNA1C, DRD2*	211	13.4

The remaining Superpathways identified did emphasize monoamine signaling and cellular re-uptake/transport. There was little intrinsic overlap between the 36 biological Superpathways with only one gene (*DRD2*) common to all nine Superpathways.

Examination of gene ontology pathways identified 17 genes that matched to a total of 32 GO-molecular functions but high match scores were found for only three molecular functions involving six genes: Serotonin binding with two genes (*HTR2, MAOA*) out of nine total pathway genes, score = 14.3; Dopamine binding with two genes (*SLC6A3, DRD2*) out of 10 total pathway genes, score = 14) and High Voltage-gated Calcium Channel Activity with two genes (*CACNA1C, CACNB2*) out of 10 total pathway genes, score = 14. The 23 overlapping genes mapped to a total of 55 GO-biological processes with high match scores identified for 16 biological processes involving 19 genes (see Table 4). These processes involved a variety of behavioral constructs including axon guidance, synaptic transmission, and particularly the activity of ion channels and dopamine homeostasis.

Table 4. GeneAnalytics program mapping of gene ontology (GO) biological processes with high match scores to 23 overlapping genes for autism spectrum disorder, bipolar disorder and schizophrenia.

GO-Biological Processes	Genes Matched to GO-Biological Processes	No. of Genes in GO-Biological Processes	Score
Startle response	*NRG1, CSMD1, DRD2*	20	19.1
Positive regulation of axon extension	*GSK3B, NRG1, DISC1*	30	17.4
Cellular calcium ion homeostasis	*HTR2A, CACNA1C, CHRNA7, DRD2*	107	17.2
Synaptic transmission	*SLC6A3, HTR2A, MAOA, CACNA1C, CACNB2, CHRNA7*	432	17.0
Dopamine catabolic process	*SLC6A3, MAOA*	5	16.0
Axon guidance	*GSK3B, NRG1, ANK3, BDNF, CACNA1C, CACNB2*	537	15.3
Synapse assembly	*NRG1, BDNF, DRD2*	52	15.0
Regulation of high voltage-gated calcium channel activity	*NOS1AP, PDE4B*	7	15.0
Regulation of potassium Ion transport	*ANK3, DRD2*	7	15.0
Response to hypoxia	*MTHFR, BDNF, CHRNA7, DRD2*	180	14.3
Negative regulation of synaptic transmission, glutamatergic	*HTR2A, DRD2*	9	14.3
Adenohypophysis development	*SLC6A3, DRD2*	9	14.3
Regulation of synaptic transmission, GABAergic	*SYN3, DRD2*	10	14.0
Behavioral response to ethanol	*CHRNA7, DRD2*	11	13.7
Regulation of dopamine secretion	*SLC6A3, CHRNA7*	11	13.7
Dopamine biosynthetic process	*HTR2A, DRD2*	12	13.4

A total of 106 phenotypes were mapped to the 23 overlapping genes with 36 phenotypes involving 18 genes having high match scores (see Table 5). The highest-matched phenotypes were behavioral despair (four genes, score = 24.8), hypoactivity (seven genes, score = 24.4), abnormal serotonin levels (four genes, score = 24.1) and abnormal response to novel objects (four genes, score = 22.2). Additionally, phenotypes impacted GABAergic neuron morphology, synaptic transmission and response to hypoxia and risk of death.

Table 5. GeneAnalytics profiling of high match score phenotypes to 23 overlapping genes for autism spectrum disorder, bipolar disorder and schizophrenia.

Phenotypes	Genes Matched to Phenotypes	No. of Genes	Score
Behavioral despair	GSK3B, CACNA1C, CSMD1, DISC1	28	24.8
Hypoactivity	MAOA, FOXP2, ANK3, BDNF, CACNA1C, CHRNA7, DRD2	314	24.4
Abnormal serotonin level	MAOA, FOXP2, TPH2, BDNF	32	24.1
Abnormal response to novel object	SLC6A3, FOXP2, TPH2, DISC1	44	22.2
Abnormal GABAergic neuron morphology	BDNF, CHRNA7, DRD2	11	21.7
Abnormal prepulse inhibition	NRG1, DISC1, DRD2	14	20.7
Abnormal social Investigation	MAOA, SYN3, CACNA1C, DISC1	14	20.7
Increase aggression towards males	MAOA, TPH2, BDNF	64	20.1
Small cerebellum	MTHFR, FOXP2, ANK3, DISC1	67	19.8
Decrease exploration in new environment	GSK3B, FOXP2, BDNF, CACNA1C	79	18.9
Abnormal CNS synaptic transmission	SLC6A3, BDNF, CACNA1C, DRD2	79	18.9
Premature death	SLC6A3, MTHFR, FOXP2, ANK3, TPH2, BDNF, CACNA1C, DRD2	830	18.6
Decrease startle reflex	FOXP2, CSMD1, DISC1, DRD2	84	18.6
Decrease anxiety-related response	HTR2A, CHRNA7, DISC1, DRD2	89	18.2
Increase dopamine level	SLC6A3, MAOA, FOXP2	31	17.2
Abnormal vocalization	FOXP2, CACNA1C, DRD2	37	16.5
Hyperactivity	SLC6A3, NRG1, BDNF, CSMD1, DISC1	272	16.3
Increased thigmotaxis	SLC6A3, CACNA1C, CSMD1	45	15.6
Abnormal serotonergic neuron morphology	TPH2, BDNF	6	15.4
Abnormal response to novel odor	SLC6A3, DRD2	6	15.4
Abnormal latent inhibition of conditioning	DISC1, DRD2	6	15.4
Impaired coordination	SLC6A3, FOXP2, BDNF, CACNA1C, DRD2	309	15.4
Abnormal learning/memory/conditioning	GSK3B, CACNA1C, DISC1	49	15.3
Limp posture	NRG1, DRD2	7	15.0
Decreased serotonin Level	TPH2, DISC1	7	15.0
Postnatal growth retardation	SLC6A3, MTHFR, FOXP2, TPH2, BDNF, DRD2	581	14.6
Abnormal pituitary gland physiology	SLC6A3, DRD2	8	14.6
Abnormal inhibitory postsynaptic currents	SLC6A3, SYN3, BDNF	60	14.4
Small nodose ganglion	NRG1, BDNF	9	14.3
Complete postnatal lethality	FOXP2, ANK3, TCF4, BDNF, CACNA1C	375	14.1
Small petrosal ganglion	NRG1, BDNF	10	14.0
Decreased somatotroph cell number	SLC6A3, DRD2	11	13.7
Abnormal grooming behavior	SLC6A3, MAOA	11	13.7
Abnormal excitatory postsynaptic currents	NRG1, SYN3, DRD2	72	13.6
Decreased left ventricle systolic pressure	MAOA, NRG1	12	13.4
Hunched posture	MAOA, BDNF, DRD2	78	13.3

3. Discussion

The molecular and genetic architecture of ASD, BPD and SCH with the identified 23 overlapping candidate susceptibility genes common to the three neuropsychiatric illnesses were analyzed to assess shared etiological factors and phenotypes to facilitate mechanistic understanding and potential development of new treatment approaches. Interestingly, the genetic architecture of the overlapping genes for the three disorders converged on brain structures (e.g., medulla oblongata, thalamus), neurotransmitter systems (e.g., dopamine, serotonin) and signal transduction pathways primarily involved in the regulation of circadian oscillations and sleep disturbances. GO-molecular processes were mapped to the neurotransmitter pathways (dopamine, andserotonin) and ion channels (high voltage–gated calcium channels) implicated in mood, addiction and psychotic disorders with regulatory function and expression in brain centers controlling Circadian entrainment.

The human thalamus, hypothalamus and hippocampus have been extensively targeted in relation to relaying and processing sensory and motor information, as well as regulating consciousness and sleep [24]. The central and peripheral circadian molecular clock is entrained through a complex and highly regulated molecular cascade in the suprachiasmatic nuclei (SCN) of the hypothalamus driven by the cyclical expression of *PER* and *CRY* [25–27]. The master lists of susceptibility genes for BPD and SCH contain *CLOCK*, an integral part of the Circadian entrainment pathway, and the identified ASD master gene list contains additional circadian-regulatory genes (e.g., *PER1*, *PER2*, *NPAS2*, *MTNR1A*, and *MTNR1B*). The master clock located in the suprachiasmatic nuclei of the hypothalamus synchronizes mainly by light signals, and releases glutamate and pituitary adenylate cyclase-activating polypeptide (PACAP) with the activation of signal transduction cascades, including nNOS activity, cAMP- and cGMP-dependent protein kinases. Additionally, multiple entrainment pathways converge to phosphorylate CREB and to activate *CLOCK* gene expression [25–27] (see Figure 1 for circadian pathways and related features).

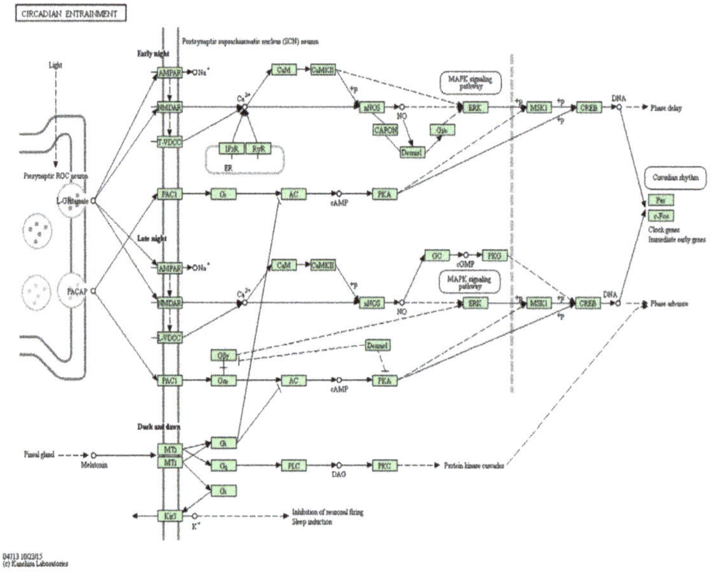

Figure 1. Circadian entrainment is an intrinsic, internal biological clock entrained by exogenous signals, such as endocrine and behavioral rhythms synchronized to environmental cues. The master clock located in the suprachiasmatic nuclei (SCN) of the hypothalamus synchronizes circadian oscillators in peripheral tissues. The main photic input to the suprachiasmatic nuclei comes from the retinal ganglion cells which use glutamate and PACAP, which leads to activation of AMPA and NMDA receptors. The release of glutamate and PACAP triggers the activation of signal transduction cascades including CamKII and nNOS activity, cAMP- and cGMP-dependent protein kinases and mitogen-activated protein kinase (MAPK). Also, melatonin affects non-photic entrainment by inhibiting light-induced phase shifts through inhibition of adenylate cyclase (AC). Additionally, multiple entrainment pathways converge to phosphorylate CREB and to activate *CLOCK* gene expression [28–31]. Solid and dotted lines indicate direct and indirect relationships, respectively.

Arousal and cortical responsiveness are modulated by a complex regulatory network including serotonergic feedback from the median raphe nuclei involved in the regulation of rapid eye movement (REM) sleep patterning [24]. Depletion of serotonin receptors and feedback are implicated in the pathology of Sudden infant death syndrome (SIDS), a secondary Superpathway identified in our study [32]. *BDNF*, a neurotrophic factor, and *CHRNA7* genes involved in the SIDS Superpathway were also implicated in the biological processes impacting the response to hypoxia and may reflect underlying vulnerability to neurological/physiological injury secondary to apnea, thereby influencing neurocognition and/or behavior. Mesencephalic dopaminergic neurons from the ventral tegmental area also project to the thalamic nuclei to directly modulate sleep induction and wakefulness through opposing effects of *DRD1* and *DRD2* receptors on adenylyl cyclase activity.

Neuronal activity and plasticity during brain development is sensitive to both neurotransmitter signaling of serotonin and dopamine and responsive to common genetic and environmental factors influencing brain maturation. Monoamine-sensitive periods also modulate select neurodevelopmental processes (e.g., neuron division, migration and dendritic connectivity) involved in the development of behavioral regulation and control as well as sleep patterns, which may reflect core etiological relationships linking neuropsychiatric illnesses to sleep disruption and hypoxia [28,33–38]. Additionally, the homeostatic value of sleep and circadian rhythmicity impact physiology and behavior in important ways, including the response to stress. Rhythmic circadian oscillations in glucocorticoid levels can modulate adult hippocampal growth and functioning through inhibitory effects of glucocorticoid hormones on neural stem cell and progenitor cell proliferation [39]. Thus, the disruption of circadian entrainment or loss of circadian regulation of glucocorticoid release could directly influence hippocampal neuroplasticity, learning and memory throughout life, which may play a role in psychopathology such as schizophrenia [40,41].

Our analyses are limited by the current status of research and availability of published literature reports on candidate genes as well as the reliability of the curated databases and integrated pathway analyses produced by the GeneAnalytics algorithms. Advances in genomic technology and bioinformatics will continue to identify and characterization new candidate genes, but not all identified genes will be equally important or certain to be causative. The relative contributions of any individual gene to the general disease prevalence must be assessed individually. Further, intrinsic bias in the curated literature may result from imbalances in the allocation of resources for study which may overemphasize some disease states or scientific disciplines over others (e.g., genetics of cancer over psychiatry). Nevertheless, the convergence of these model systems and overlaid genetic mechanisms provides relevant insight into the key macro systems involved in pathogenesis and the overlap of these three severe neuropsychiatric disorders.

4. Materials and Methods

We used recently published list of genes found to be clinically relevant and known to play a role in ASD [21], bipolar disorder [22] and schizophrenia [23] for molecular profiling and pathway analysis of genes common to all three neuropsychiatric conditions with similar features. GeneAnalytics (http://geneanalytics.genecards.org/ [20]) computer program and genomic databases are part of the GeneCards Suite developed by LifeMap Sciences (http://www.lifemapsc.com/products/genecards-suite-premium-tools/) and were used to map the resultant list of common genes to characterize molecular pathways, biological processes, molecular functions, phenotypes, tissues and cells, diseases and compounds affected by overlapping neuropsychiatric genes.

GeneAnalytics is powered by GeneCards, LifeMap Discovery, MalaCards and PathCards, which combine >100 archived data sources [20]. The databases contain gene lists for tissues and cells, diseases, phenotypes pathways and compounds curated from published literature reports to develop the best matched list of genes, scored and subdivided into their biological categories such as diseases or pathways. These applications are integrated with GeneCards human gene database, Malacards human disease database, PathCards, human biological pathways database, and LifeMap Discovery tissues and cells database in order to provide an extensive universe of data from human genes, proteins, cells, biological pathways, diseases and their relationships with integration valuable for research and discovery purposes.

Disease matching scores were derived based upon the number of overlapping genes found and the nature of the gene-disease associations. Tissues and cells were scored using a matching algorithm that weighs tissue specificity, abundance and function of the gene. Related pathways were then grouped into Superpathways to improve inferences and pathway enrichment, reduce redundancy and rank genes within a biological mechanism via the multiplicity of constituent pathways with the methodology and algorithm generated by the GeneAnalytics computer-based program. Superpathways were scored based upon transformation of the binomial p-value which was equivalent to a corrected p-value with significance defined at <0.0001.

5. Conclusions

Genetic overlap among autism spectrum disorder, bipolar disorder and schizophrenia was most strongly mapped to Superpathways and brain tissue types guiding Circadian entrainment. The results illustrate the converging effects of dopamine, serotonin and the signal transduction pathways involved in mood, behavior, cognition and impaired social functions; learning and memory are affected in these neuropsychiatric disorders which also guide brain development and sleep patterning disturbances. Thus, under-recognized sleep dysregulation as a common component of psychiatric illness appears to reflect the underlying molecular and genetic architecture of disease pathology in psychiatric illnesses. The convergence of pathways governing circadian rhythms supports the existence of a common core etiological relationship between neuropsychiatric illness and sleep disruption possibly related to central brain stem dysfunction impacting the presentation and underlying pathology and course of illness. This observation opens a new avenue to pursue for treatment modalities in order to change the clinical outcome and natural history of those affected with these relatively common mental health disorders in our current society.

Acknowledgments: We acknowledge support from the Consortium for Translational Research on Aggression and Drug Abuse and Dependence (ConTRADA) grant QB864900 from the University of Kansas and the National Institute of Child Health and Human Development (NICHD) grant HD02528. We thank Miwako Karikomi from Kanehisa Laboratories for granting the permission to publish the Kyoto Encyclopedia of Genes and Genomes (KEGG) pathways map image.

Author Contributions: Naveen Khanzada performed the literature review, compiled information for the GeneAnalytics analysis and primarily composed the manuscript. Merlin Butler contributed to the generation and interpretation of genetic data and the preparation of the manuscript. Ann Manzardo was responsible for the oversight of the study design, implementation, data analysis and interpretation, and manuscript preparation. All of the authors contributed to the final revision and accepted the manuscript.

Conflicts of Interest: The authors declare no conflict of interest.

Abbreviations

ASD	Autism spectrum disorder
BPD	Bipolar disorder
SCH	Schizophrenia
GWAS	Genome-wide association studies
CGH	Comparative genomic hybridization
CNV	Copy number variants
SIDS	Sudden infant death syndrome
SCN	Suprachiasmatic nuclei
PACAP	Pituitary adenylate cyclase-activating polypeptide
cAMP	Cyclic adenosine monophosphate
cGAMP	Cyclic guanosine monophosphate–adenosine monophosphate
nNOS	Nitric oxide synthase
CREB	cAMP response element-binding protein

References

1. American Psychiatric Association. *Diagnostic and Statistical Manual of Mental Disorders: DSMIV-TR*, 4th ed.; American Psychiatric Association: Washington, DC, USA, 2000.
2. Carroll, L.S.; Owen, M.J. Genetic overlap between autism, schizophrenia and bipolar disorder. *Genome Med.* **2009**, *1*, 102. [CrossRef] [PubMed]
3. Burbach, J.; Peter, H.; van der Zwaag, B. Contact in the genetics of autism and schizophrenia. *Trends Neurosci.* **2009**, *32*, 69–72. [CrossRef] [PubMed]
4. Sullivan, P.F.; Magnusson, C.; Reichenberg, A.; Boman, M.; Dalman, C.; Davidson, M.; Fruchter, E.; Hultman, C.M.; Lundberg, M.; Långström, N.; et al. Family history of schizophrenia and bipolar disorder as risk factors for autism. *Arch. Gen. Psychiatry* **2012**, *69*, 1099–1103. [CrossRef] [PubMed]
5. Sebat, J.; Lakshmi, B.; Malhotra, D.; Troge, J.; Lese-Martin, C.; Walsh, T.; Yamrom, B.; Yoon, S.; Krasnitz, A.; Kendall, J.; et al. Strong association of de novo copy number mutations with autism. *Science* **2007**, *316*, 445–449. [CrossRef] [PubMed]
6. Szatmari, P.; Paterson, A.D.; Zwaigenbaum, L.; Roberts, W.; Brian, J.; Liu, X.Q.; Vincent, J.B.; Skaug, J.L.; Thompson, A.P.; Senman, L.; et al. Autism Genome Project Consortium, Mapping autism risk loci using genetic linkage and chromosomal rearrangments. *Nat. Genet.* **2007**, *39*, 319–328. [CrossRef] [PubMed]
7. Marshall, C.R.; Noor, A.; Vincent, J.B.; Lionel, A.C.; Feuk, L.; Skaug, J.; Shago, M.; Moessner, R.; Pinto, D.; Ren, Y.; et al. Structural variation of chromosomes in autism spectrum disorder. *Am. J. Hum. Genet.* **2008**, *82*, 477–488. [CrossRef] [PubMed]
8. Kim, H.G.; Kishikawa, S.; Higgins, A.W.; Seong, I.S.; Donovan, D.J.; Shen, Y.; Lally, E.; Weiss, L.A.; Najm, J.; Kutsche, K.; et al. Disruption of neurexin 1 associated with autism spectrum disorder. *Am. J. Hum. Genet.* **2008**, *82*, 199–207. [CrossRef] [PubMed]
9. Belmonte, M.K.; Bourgeron, T. Fragile X syndrome and autism at the intersection of genetic and neural networks. *Nat. Neurosci.* **2006**, *9*, 1221–1225. [CrossRef] [PubMed]
10. Kwon, C.H.; Luikart, B.W.; Powell, C.M.; Zhou, J.; Matheny, S.A.; Zhang, W.; Li, Y.; Baker, S.J.; Parada, L.F. Pten regulates neuronal arborization and social interaction in mice. *Neuron* **2006**, *50*, 377–388. [CrossRef] [PubMed]
11. Hoeffer, C.A.; Tang, W.; Wong, H.; Santillan, A.; Patterson, R.J.; Martinez, L.A.; Tejada-Simon, M.V.; Paylor, R.; Hamilton, S.L.; Klann, E. Removal of FKBP12 enhances mTOR-Raptor interactions, LTP, memory, and perseverative/repetitive behavior. *Neuron* **2008**, *60*, 832–845. [CrossRef] [PubMed]
12. Cuscó, I.; Medrano, A.; Gener, B.; Vilardell, M.; Gallastegui, F.; Villa, O.; González, E.; Rodríguez-Santiago, B.; Vilella, E.; del Campo, M.; et al. Autism-specific copy number variants further implicate the phosphatidylinositol signaling pathway and the glutamatergic synapse in the etiology of the disorder. *Hum. Mol. Genet.* **2009**, *18*, 1795–1804. [CrossRef] [PubMed]
13. Emamian, E.S.; Hall, D.; Birnbaum, M.J.; Karayiorgou, M.; Gogos, J.A. Convergent evidence for impaired AKT1-GSK3β signaling in schizophrenia. *Nat. Genet.* **2004**, *36*, 131–137. [CrossRef] [PubMed]
14. Stopkova, P.; Saito, T.; Papolos, D.F.; Vevera, J.; Paclt, I.; Zukov, I.; Bersson, Y.B.; Margolis, B.A.; Strous, R.D.; Lachman, H.M. Identification of PIK3C3 promoter variant associated with bipolar disorder and schizophrenia. *Biol. Psychiatry* **2004**, *55*, 981–988. [CrossRef] [PubMed]
15. Kalkman, H.O. The role of the phosphatidylinositide 3-kinase-protein kinase B pathway in schizophrenia. *Pharmacol. Ther.* **2006**, *110*, 117–134. [CrossRef] [PubMed]
16. Krivosheya, D.; Tapia, L.; Levinson, J.N.; Huang, K.; Kang, Y.; Hines, R.; Ting, A.K.; Craig, A.M.; Mei, L.; Bamji, S.X.; et al. ErbB4-neuregulin signaling modulates synapse development and dendritic arborization through distinct mechanisms. *J. Biol. Chem.* **2008**, *283*, 32944–32956. [CrossRef] [PubMed]
17. Cox, D.M.; Butler, M.G. The 15q11.2 BP1–BP2 microdeletion syndrome: A review. *Int. J. Mol. Sci.* **2015**, *16*, 4068–4082. [CrossRef] [PubMed]
18. Burnside, R.D.; Pasion, R.; Mikhail, F.M.; Carroll, A.J.; Robin, N.H.; Youngs, E.L.; Gadi, I.K.; Keitges, E.; Jaswaney, V.L.; Papenhausen, P.R.; et al. Microdeletion/microduplication of proximal 15q11.2 between BP1 and BP2: A susceptibility region for neurological dysfunction including developmental and language delay. *Hum. Genet.* **2011**, *130*, 517–528. [CrossRef] [PubMed]

19. Ho, K.S.; Wassman, E.R.; Baxter, A.L.; Hensel, C.H.; Martin, M.M.; Prasad, A.; Twede, H.; Vanzo, R.J.; Butler, M.G. Chromosomal microarray analysis of consecutive individuals with autism spectrum disorders using an ultra-high resolution chromosomal microarray optimized for neurodevelopmental disorders. *Int. J. Mol. Sci.* **2016**, *17*, 2070. [CrossRef] [PubMed]

20. Ben-Ari, F.S.; Lieder, I.; Stelzer, G.; Mazor, Y.; Buzhor, E.; Kaplan, S.; Bogoch, Y.; Plaschkes, I.; Shitrit, A.; Rappaport, N.; et al. GeneAnalytics: An integrative gene set analysis tool for next generation sequencing, RNAseq and microarray data. *OMICS J. Integr. Biol.* **2016**, *20*, 139–151. [CrossRef] [PubMed]

21. Butler, M.G.; Rafi, S.K.; Manzardo, A.M. High-resolution chromosome ideogram representation of currently recognized genes for autism spectrum disorders. *Int. J. Mol. Sci.* **2015**, *16*, 6464–6495. [CrossRef] [PubMed]

22. Douglas, L.N.; McGuire, A.B.; Manzardo, A.M.; Butler, M.G. High-resolution chromosome ideogram representation of recognized genes for bipolar disorder. *Gene* **2016**, *586*, 136–147. [CrossRef] [PubMed]

23. Butler, M.G.; McGuire, A.B.; Masoud, H.; Manzardo, A.M. Currently recognized genes for schizophrenia: High-resolution chromosome ideogram representation. *Am. J. Med. Genet. B Neuropsychiatr. Genet. Part B* **2016**, *1B*, 181–202. [CrossRef] [PubMed]

24. Weber, F.; Chung, S.; Beier, K.T.; Xu, M.; Luo, L.; Dan, Y. Control of REM sleep by ventral medulla GABAergic neurons. *Nature* **2015**, *526*, 435–438. [CrossRef] [PubMed]

25. Reischl, S.; Kramer, A. Kinases and phosphatases in the mammalian circadian clock. *FEBS Lett.* **2011**, *585*, 1393–1399. [CrossRef] [PubMed]

26. Yin, L.; Wang, J.; Klein, P.S.; Lazar, M.A. Nuclear receptor Rev-Erbα is a critical lithium-sensitive component of the circadian clock. *Science* **2006**, *311*, 1002–1005. [CrossRef] [PubMed]

27. Zeidner, L.C.; Buescher, J.L.; Phiel, C.J. A novel interaction between Glycogen Synthase Kinase-3α (GSK-3α) and the scaffold protein receptor for activated C-Kinase 1 (RACK1) regulates the circadian clock. *Int. J. Biochem. Mol. Biol.* **2011**, *2*, 318–327. [PubMed]

28. Tarazi, F.I.; Tomasini, E.C.; Baldessarini, R.J. Postnatal development of dopamine D4-like receptors in rat forebrain regions: Comparison with D2-like receptors. *Brain Res. Dev. Brain Res.* **1998**, *110*, 227–233. [CrossRef]

29. Kanehisa, M.; Furumichi, M.; Tanabe, M.; Sato, Y.; Morishima, K. KEGG: New perspectives on genomes, pathways, diseases and drugs. *Nucleic Acids Res.* **2017**, *45*, D353–D361. [CrossRef] [PubMed]

30. Kanehisa, M.; Sato, Y.; Kawashima, M.; Furumichi, M.; Tanabe, M. KEGG as a reference resource for gene and protein annotation. *Nucleic Acids Res.* **2016**, *44*, D457–D462. [CrossRef] [PubMed]

31. Kanehisa, M.; Goto, S. KEGG: Kyoto Encyclopedia of Genes and Genomes. *Nucleic Acids Res.* **2000**, *28*, 27–30. [CrossRef] [PubMed]

32. Lavezzi, A.M.; Casale, V.; Oneda, R.; Weese-Mayer, D.E.; Matturri, L. Sudden infant death syndrome and sudden intrauterine unexplained death: Correlation between hypoplasia of raphé nuclei and serotonin transporter gene promoter polymorphism. *Pediatr. Res.* **2009**, *66*, 22–27. [CrossRef] [PubMed]

33. Severson, C.A.; Wang, W.; Pieribone, V.A.; Dohle, C.I.; Richerson, G.B. Midbrain serotoninergic neurons are central pH chemoreceptors. *Nat. Neurosci.* **2003**, *6*, 1139–1140. [CrossRef] [PubMed]

34. Gaspar, P.; Cases, O.; Maroteaux, L. The developmental role of serotonin: News from mouse molecular genetics. *Nat. Rev. Neurosci.* **2003**, *4*, 1002–1012. [CrossRef] [PubMed]

35. Haydon, P.G.; McCobb, D.P.; Kater, S.B. Serotonin selectively inhibits growth cone motility and synaptogenesis of specific identified neurons. *Science* **1984**, *226*, 561–564. [CrossRef] [PubMed]

36. Lauder, J.M. Ontogeny of the serotonergic system in the rat: Serotonin as a developmental signal. *Ann. N. Y. Acad. Sci.* **1990**, *600*, 297–313. [CrossRef] [PubMed]

37. McCarthy, D.; Lueras, P.; Bhide, P.G. Elevated dopamine levels during gestation produce region-specific decreases in neurogenesis and subtle deficits in neuronal numbers. *Brain Res.* **2007**, *1182*, 11–25. [CrossRef] [PubMed]

38. Popolo, M.; McCarthy, D.M.; Bhide, P. Influence of dopamine on precursor cell proliferation and differentiation in the embryonic mouse telencephalon. *Dev. Neurosci.* **2004**, *26*, 229–244. [CrossRef] [PubMed]

39. Fitzsimons, C.P.; Herbert, J.; Schouten, M.; Meijer, O.C.; Lucassen, P.J.; Lightman, S. Circadian and ultradian glucocorticoid rhythmicity: Implications for the effects of glucocorticoids on neural stem cells and adult hippocampal neurogenesis. *Front. Neuroendocrinol.* **2016**, *41*, 44–58. [CrossRef] [PubMed]

Int. J. Mol. Sci. **2017**, *18*, 527

40. Tam, S.K.; Pritchett, D.; Brown, L.A.; Foster, R.G.; Bannerman, D.M.; Peirson, S.N. Sleep and circadian rhythm disruption and recognition memory in schizophrenia. *Methods Enzymol.* **2015**, *552*, 325–349. [PubMed]

41. Iyer, R.; Wang, T.A.; Gillette, M.U. Circadian gating of neuronal functionality: A basis for iterative metaplasticity. *Front. Syst. Neurosci.* **2014**, *8*, 164. [CrossRef] [PubMed]

International Journal of
Molecular Sciences

MDPI

Article

Whole Exome Sequencing in Females with Autism Implicates Novel and Candidate Genes

Merlin G. Butler [1,*], Syed K. Rafi [1,†], Waheeda Hossain [1,†], Dietrich A. Stephan [2,†] and Ann M. Manzardo [1,†]

[1] Departments of Psychiatry & Behavioral Sciences, University of Kansas Medical Center, 3901 Rainbow Boulevard, MS 4015, Kansas City, KS 66160, USA; rafigene@yahoo.com (S.K.R.); whossain@kumc.edu (W.H.); amanzardo@kumc.edu (A.M.M.)

[2] Department of Human Genetics, University of Pittsburgh, Pittsburgh, PA 15260, USA; dstephan@pitt.edu

[*] Author to whom correspondence should be addressed; mbutler4@kumc.edu;
Tel.: +1-913-588-1873; Fax: +1-913-588-1305.

[†] These authors contributed equally to this work.

Academic Editor: William Chi-shing Cho

Received: 21 November 2014; Accepted: 31 December 2014; Published: 7 January 2015

Abstract: Classical autism or autistic disorder belongs to a group of genetically heterogeneous conditions known as Autism Spectrum Disorders (ASD). Heritability is estimated as high as 90% for ASD with a recently reported compilation of 629 clinically relevant candidate and known genes. We chose to undertake a descriptive next generation whole exome sequencing case study of 30 well-characterized Caucasian females with autism (average age, 7.7 ± 2.6 years; age range, 5 to 16 years) from multiplex families. Genomic DNA was used for whole exome sequencing via paired-end next generation sequencing approach and X chromosome inactivation status. The list of putative disease causing genes was developed from primary selection criteria using machine learning-derived classification score and other predictive parameters (GERP2, PolyPhen2, and SIFT). We narrowed the variant list to 10 to 20 genes and screened for biological significance including neural development, function and known neurological disorders. Seventy-eight genes identified met selection criteria ranging from 1 to 9 filtered variants per female. Five females presented with functional variants of X-linked genes (*IL1RAPL1*, *PIR*, *GABRQ*, *GPRASP2*, *SYTL4*) with cadherin, protocadherin and ankyrin repeat gene families most commonly altered (e.g., *CDH6*, *FAT2*, *PCDH8*, *CTNNA3*, *ANKRD11*). Other genes related to neurogenesis and neuronal migration (e.g., *SEMA3F*, *MIDN*), were also identified.

Keywords: whole exome sequencing; females; autism spectrum disorder; genetic variants; X chromosome inactivation

1. Introduction

Classical autism or autistic disorder belongs to a group of genetically heterogeneous conditions known as Autism Spectrum Disorders (ASD) characterized by three clinical features: (1) marked impairments in verbal and non-verbal communication; (2) impairments in social interactions; and (3) restricted repetitive behaviors and interests [1,2]. Additional findings include lack of eye contact, tactile defensiveness and sleep disturbances with diagnosis validation at a young age requiring test instruments such as Autism Diagnostic Observation Schedule (ADOS) and Autism Diagnostic Interview-Revised (ADI-R) [3–5]. Latest estimates reveal that ASD affects about 1 of every 68 children by 8 years of age with a predominant male to female (4:1) ratio [6]. This occurrence rate is higher than that of epilepsy or Down syndrome. Congenital anomalies, intellectual disability, growth retardation and/or seizures can occur in a subset of children presenting with features related to autism. Regression of skills is observed in about 30% of cases with ASD and about 60% show intellectual disabilities [7,8].

Small or large head size is seen in about 30% of affected children with ASD with extreme macrocephaly associated with mutation of the *PTEN* tumor suppressor gene [9]. Heritability studies to identify the contribution of genetic factors in autism have shown an estimate as high as 90%. Recognized single gene conditions such as fragile X syndrome, Rett syndrome or tuberous sclerosis account for less than 20% of all cases with ASD [10].

Standard routine chromosome studies in individuals with ASD have shown abnormalities of over one dozen chromosomes. Various cytogenetic findings are reported including deletions, duplications, translocations and inversions often involving the chromosome 15q11-q13 region or the 22q11.2 band [10,11]. More advanced chromosome microarray studies are more powerful in finding cytogenetic abnormalities than routine chromosome studies. High resolution microarrays have detected recurrent small submicroscopic deletions or duplications in individuals with ASD indicating the presence of hundreds of candidate and/or known ASD genes localized to each human chromosome. In addition, there are numerous submicroscopic copy number changes, more often of the deletion type, seen in greater than 20% of patients with ASD using microarray analysis [11]. Many of these chromosome abnormalities contain genes playing a causative role in ASD. In a recent review of genetic linkage data, candidate genes and genome-wide association studies along with further advances in genetic technology including high resolution DNA microarray and next generation sequencing have led to a compilation of 629 clinically relevant candidate and known genes for ASD [12].

Given the fact that females with ASD are historically understudied, we performed whole exome sequencing of well-characterized females with classical autism from multiplex families in the search for existing or potentially new candidate genes for autism. We utilized a cohort of affected females recruited by the Autism Genetic Research Exchange (AGRE), a gene bank housing data and biospecimens from over 2000 families (www.AGRE.autismspeaks.org). Most families had two or more affected children with autism. Identification of causative mutations (e.g., serotonin-related gene mutations and disturbed biology) could be important to guide selection of treatment options and medication use as well as to manage medical co-morbidities such as seizures, developmental regression (e.g., *MECP2* gene) or for cancer (e.g., *PTEN* gene). A cursory autism data base search revealed a large body of publications, particularly since 2008, linking autism to a wide range of genetic and environmental factors found only 3 (0.48%) clinically relevant ASD genes to be located on the Y chromosome while 68 (10.81%) clinically relevant ASD genes were recognized on the X-chromosome [12].

The preponderance of males with ASD may be attributable to the single X chromosome in males depriving the normal allelic pair of genes due to the XY sex chromosome constitution. Hence, sex chromosomes illustrate the most obvious genetic difference between men and women. All female mammals have two X chromosomes and achieve a balanced X chromosome gene expression with males by inactivating one of their X chromosomes, a process known as X chromosome inactivation (XCI) [13]. This process occurs randomly and very early in embryonic development. Once an X chromosome is "selected" for inactivation within a cell, then the same X chromosome remains inactivated in each subsequent daughter cell. Therefore, females have a mixture of cells with random expression of genes on a single X chromosome. Occasionally, XCI represents a nonrandom pattern or high skewness which is usually defined by at least 80% preferential inactivation of one of the two X chromosomes [14,15]. Skewed XCI appears to play a role in the increased incidence and presentation of diseases in females with known X-linked gene involvement such as Rett syndrome [16], X-linked intellectual disability [14], X-linked adrenoleukodystrophy [17] and possibly autism [15]. Skewed XCI may also reflect an early disruption in the developing embryo causing cell death followed by a small number of dividing cells repopulating the embryo [18–20]. XCI can be measured by using the androgen receptor (*AR*) gene located at chromosome Xq13 which contains a highly polymorphic CAG repeat region and normally inactivated on one of the X chromosomes in females. The *AR* gene is used most often to determine the XCI status [21,22]. Disorders with an imbalance in the number of males versus females affected as in autism become a likely condition to study XCI and its impact on X-linked genes in females. We suspect additional autosomal and/or sex chromosome based ASD gene(s) that either overwhelms the protective normal alleles of the second X chromosome among females, or preferentially silences them due to a plausible skewed X chromosome inactivation among neural cell lineages. Therefore, the whole exome sequence of females with ASD from multiplex families having more than one family member affected was analyzed in our study with next generation sequencing and bioinformatics with a novel machine-learning classification engine applied to identify the contributing ASD genes and the proportion of disease contributing alleles from the X chromosomes versus autosomes. The inactivation status of the two X chromosomes was determined to identify significant deviation from randomness, taking into consideration whether the X chromosome gene is known to escape inactivation, or known to act as a dominant gene.

2. Results

Whole exome sequencing from DNA of 30 females with autism spectrum disorders from AGRE identified between 100 and 300 genes showing genomic variants of novel or candidate genes for autism per subject with an initial classification score >0.5. To further narrow the list of putative disease causing genes, we increased the level of selection criteria by adding cutoff levels for other predictive parameters (GERP2, PolyPhen2 and SIFT) and specifically increased the classification score from >0.5 to a more stringent score of >0.7. For example, the initial list of genes and genomic variants for Subject HI2898 consisted of 245 genes when using a classification score >0.5 alone and then reduced to 22 genes when the cutoff levels of other predictive parameters (GERP2, PolyPhen2 and SIFT) were included at the chosen final selection criteria level. Finally, the classification score was raised to >0.7 to generate the master list of 14 genes. For the 30 females with autism, the number of genes identified based on the final selection criteria ranged between 8 to 23 genes with 10 females presenting with an X-linked gene in the list of selected genes meeting the final selection criteria. The putative candidate genes and genomic variants were then subjected to further screening for biological significance and relevance for neural development, function and for causation of known neurological disorders based on published medical literature. In so doing, we identified two potentially causative genes for autism for Subject HI2898. See Table 1 for the list of genes meeting or exceeding the established final selection criteria level and screening for biological significance. Identical genomic variants were recognized for several subjects even after increasing the final level of selection criteria required to reduce the number of candidate genes and considered artifactual findings. These genes with identical genomic variants included *KCNJ12*, *MLL3*, *OR9G1* and *PCMTD1* with different mutations reported in other disease states and appeared to reflect artifacts in the present analysis of autism.

We report genomic coordinates, previously recognized SNPs (and their ID numbers), reference base to alternate base sequences (e.g., G→A), variant call format (VCF) information, SNP effects (synonymous vs non- synonymous), SNP codons, and GERP2, PolyPhen2, SIFT, Blosum, Blosum62 and the classification scores for each genomic variant identified per affected female (Table 1). Our analysis identified a total of 79 different genes containing variants (both known and novel) consisting of 12 (15%) known candidate genes for ASD, 24 (30%) paralogues of known candidate ASD genes and 43 (54%) novel genes with functional roles in nerve cell growth, adhesion and neurodevelopment or function as putative candidate ASD genes (See Table 2). The identified candidate ASD genes and paralogues were distributed across 11 (37%) and 16 (53%) of the females with autism, respectively.

Four females (HI0555, HI0890, HI1402 and HI2126) possessed a single variant of significance in clinically relevant candidate or known ASD genes considered conclusive for causality. Seventy-seven percent of subjects (*N* = 23) possessed multiple variants with possible influences on causation of ASD. Two (9%) of the 23 females (7% of the total sample population) possessed more than five variants/mutations of known candidate ASD genes, paralogues or other functionally relevant genes with presumed influence over the development of ASD. Three (10%) of the total females (HI1143, HI1884 and HI2278) possessed a single gene variant with possible functional relevance toward the development of ASD but no known candidate ASD genes or paralogues identified to be considered conclusive.

Int. J. Mol. Sci. 2015, 16, 1312–1335

Table 1. Putative disease causing genes for autism identified using whole exome sequencing of females with autism spectrum disorder (ASD).

ID Number XCI	Gene Symbol (Category) *	Chromosome Position (Hg19)	Genomic Variant				VCF Allele Depth	Classification Score	GERP2	PolyPhen2 Score	SIFT Score	Blosum Score/Blosum62
			SNP Effect	Amino Acid	SNP Codon	SNP Function						
H10405 46%:54%	KCNC2 (P) [12]	12:75444895	Non-synon	p.F296C	tTt/tGt	Missense	162, 30	0.790	0	0.984	0	0.787/−2
	ASPM (F) [23]	1:197112823	Non-synon	p.R186G	Aga/Gga	Missense	69, 53	0.777	0.007	0.994	0	0.787/−2
	TAF3 (F) [24]	10:8006662	Non-synon	p.R396G	Cga/Gga	Missense	69, 60	0.740	0	0.956	0	0.787/−2
	FLRT2 (F) [25]	14:86089552	Non-synon	p.H564R	cAt/cGt	Missense	28, 29	0.736	0	0.958	0.02	0.787/0
	SYTL4 (F) [26]	X:99941091	Non-synon	p.H448D	Cat/Gat	Missense	36, 37	0.710	0.006	1	0.03	0.787/−1
H10555:58%:42%	SETD2 (A) [12,27,28]	3:47164293	Non-synon	p.K610O	aaG/aaT	Missense	23, 21	0.778	0	0.970	0	8.522/0
H10558 52%:48%	IFT122 (F) [29]	3:129238526	Non-synon	p.R986H	cGc/cAc	Missense	58, 61	0.795	0.008	0.989	0	0.074/0
	NUP98 (F) [30]	11:3756456	Non-synon	p.R519W	Cgg/Tgg	Missense	12, 9	0.793	0.006	0.999	0	24.449/−3
	ZKSCAN1 (F) [31]	7:99621487	Non-synon	p.R119C	Cgc/Tgc	Missense	8, 5	0.782	0.006	1	0	0.787/−3
	DNAH8 (F) [32]	6:38834433	Non-synon	p.M197V	Atg/Gtg	Missense	8, 14	0.783	0.009	0.902	0.01	4.182/1
	CENPJ (F) [33]	13:254487103	Non-synon	p.M20L	Atg/Ttg	Missense	45, 42	0.761	0.007	0.991	0	4.312/2
	TRIM5 (F) [34]	11:5878550	Non-synon	p.R127P	cGc/cCc	Missense	42, 24	0.742	0.0089	0.971	0	0.787/−2
	PCDH8 (A) [35]	13:53421358	Non-synon	p.A404V	gCg/gTg	Missense	13, 14	0.731	0	0.924	0	0.787/0
	CTNNA3 (A) [36]	10:68381521	Non-synon	p.M434V	Atg/Gtg	Missense	30, 17	0.723	0.009	0.942	0.01	4.183/1
	PPP1R18 (F) [37]	6:30652604	Non-synon	p.S397P	Tct/Cct	Missense	12, 6	0.703	0.005	0.958	0	0.787/−1
H10605 53%:47%	CCDC64 (A) [38]	12:120412012	Non-synon	p.A113T	Gcc/Acc	Missense	13, 24	0.794	0.006	0.96	0	4.774/−1
	DCHS1 (F) [39]	11:6662745	Non-synon Codon Change Plus Codon Insertion	p.L132LW	ctg/ctCTCg	None	11, 8	0.700	0	0.979	0.02	0/0
H10714 40%:60%	TRAK2 (F) [40]	2:202264189	Non-synon	p.R130G	Cga/Gga	Missense	31, 27	0.805	0.010	1	0	0.064/−2
	MPST (F) [41]	22:37425405	Utr_3 Prime	p.Null-1Null	Null	None	20, 21	0.731	0.008	1	0	0.787/−2
	FSTL3 (F) [42]	19:680331	Non-synon	p.R115H	cGc/cAc	Missense	7, 3	0.729	0.010	1	0.01	8.723/0
H10751 16%:84%	IST1 (F) [43,44]	16:71956522	Non-synon	p.P232R	cCc/cGc	Missense	63, 50	0.754	0.008	0.972	0.01	0.787/−2
	KIF27 (P) [43,44]	9:86474259	Non-synon	p.S921P	Tca/Cca	Missense	22, 25	0.730	0.009	0.970	0.01	0.787/−1
	PCDHGA1 (P) [45]	5:140870234	Non-synon	p.V475A	gTg/gCg	Missense	152, 141	0.713	0	0.995	0	0.787/0
H10765 77%:23%	SEMA3F (P) [46]	3:50225525	Non-synon	p.R679W	Cgg/Tgg	Missense	21, 7	0.793	0.006	1	0	0.787/−3
	SCN11A (F) [47]	3:38968409	Frame-shift	p.-166S5?	-/TCTT CACT	None	23, 17	0.748	0.008	0.982	0	0/0
	GPRASP2 (F) [48]	X:101972234	Non-synon	p.R812C	Cgt/Tgt	Missense	69, 51	0.706	0	0.989	0.01	0.787/−3
H10793 37%:64%	ELTD1 (F) [49]	1:79356850	Non-synon	p.C687R	Tgt/Cgt	Missense	50, 25	0.807	0.008	0.998	0	10.199/−3
	DENND3 (F) [50]	8:142161854	Non-synon	p.S252L	tCg/tTg	Missense	40, 20	0.806	0.006	0.992	0	0.787/−2
	MIDN (F) [51]	19:1250353	Non-synon	p.C19G	Tgc/Ggc	Missense	8, 3	0.793	0.008	0.998	0	17.850/−3
	RADIL (F) [52]	7:4917586	Non-synon	p.P61R	cCt/cGt	Missense	42, 27	0.774	0.004	1	0	14.302/−2
	EXOC7 (P) [53,54]	17:74097856	Non-synon	p.R71Q	cGg/cAg	Missense	36, 46	0.755	0.006	0.992	0.01	0.059/1
	PLCE1 (F) [55]	10:96064250	Non-synon	p.H1515Y	Cat/Tat	Missense	22, 14	0.737	0.008	0.999	0.02	7.078/2
	KDM5A (F) [56]	12:416817	Non-synon	p.A1244T	Gcc/Acc	Missense	42, 25	0.706	0	0.993	0	0.157/0
	ARAP2 (F) [57]	4:36115873	Non-synon	p.D1358N	Gat/Aat	Missense	70, 53	0.702	0.008	0.907	0.02	12.052/1

Table 1. *Cont.*

ID Number XCI	Gene Symbol (Category) *	Chromosome Position (Hg19)	Genomic Variant			SNP Function	VCF Allele Depth	Classification Score	GERP2	PolyPhen2 Score	SIFT Score	Blosum Score/Blosum62
			SNP Effect	Amino Acid	SNP Codon							
H10855 44%:56%	CHAC1 (F) [58]	15:41247844	Frame-shift	p.-177?	-/T	None	23, 19	0.941	0.009	0.997	0	0.787/0
	MIDN (F) [59]	19:1250353	Non-synon	p.C19G	Tgc/Ggc	Missense	3, 7	0.793	0.008	0.998	0	17.85/−3
	CDH6 (P) [60]	5:31323179	Non-synon	p.R712G	Aga/Gga	Missense	76, 81	0.749	0.003	0.988	0.01	0.787/−2
H10868 58%:42%	CHST3 (P) [61]	10:73767391	Non-synon	p.Y200C	tAc/tGc	Missense	43, 30	0.759	0	0.998	0.02	17.489/−2
	FBXO5 (P) [62]	6:153296090	Non-synon	p.H256R	cAt/cGt	Missense	103, 76	0.756	0.003	1	0	0.787/0
H10890/NI	BTAF1 (A) [63]	10:93719892	Non-synon	p.Y414F	tAt/tTt	Missense	29, 16	0.765	0.005	0.997	0	3/3
H11143/14%:86%	CLPTM1 (F) [64]	19:45476442	Non-synon	p.N94I	aAc/aTc	Missense	30, 27	0.744	0.007	1	0	0.787/−3
H11157 61%:39%	AMIGO1 (F) [65]	1:110050220	Non-synon	p.R438W	Cgg/Tgg	Missense	42, 28	0.806	0	0.994	0	0.787/−3
	PEX5 (P) [66]	12:7362354	Non-synon	p.R560C	Cgc/Tgc	Missense	71, 50	0.801	0.006	1	0	0.787/−3
	KIF23 (P) [67]	15:69718474	Non-synon	p.E266A	gAa/gCa	Missense	52, 48	0.775	0.007	0.998	0	0.787/−1
H11228 42%:58%	FOXA3 (P) [68]	19:46376266	Non-synon	p.G334R	Gga/Aga	Missense	30, 29	0.801	0.004	0.975	0	26.000/−2
	FRMD4A (F) [69]	10:13698719	Non-synon	p.S941L	tCg/tTg	Missense	6, 8	0.775	0	0.957	0	0.064/−2
H11305 73%:27%	PIR (F) [70]	X:15474053	Non-synon	p.E132G	gAa/gGa	Missense	32, 28	0.811	0.007	0.962	0	0.213/−2
	DNAH2 (F) [71]	17:7637815	Non-synon	p.I255T	aTa/aCa	Missense	25, 20	0.793	0.010	0.942	0	7.006/−1
H11375 17%:83%	LHX2 (F) [72]	9:126777468	Non-synon	p.A130P	Gct/Cct	Missense	50, 34	0.775	0.006	0.998	0	0.787/−1
	EPHA4 (P) [73]	2:222428829	Non-synon	p.A148T	Gct/Act	Missense	18, 25	0.744	0	0.977	0	0.787/0
H11402/85%:15%	IL1RAPL1 (A) [12,74]	X:29973282	Non-synon	p.P478Q	cCa/cAa	Missense	57, 42	0.761	0	0.999	0	7.684/−1
H11422 56%:44%	RTN4R (F) [75]	22:20230387	Non-synon	p.S89L	tCg/tTg	Missense	11, 18	0.778	0.004	0.998	0	0.064/−4
	DHX8 (F) [76]	17:41606990	Non-synon	p.R282H	cGc/cAc	Missense	4, 8	0.724	0.010	1	0	0.787/0
H11433 93%:7%	FAT2 (P) [77–79]	5:150947258	Non-synon	p.T411I	aCt/aTt	Missense	32, 29	0.757	0	0.992	0	6.276/−1
	FTM2 (F) [80,81]	20:42939631	Non-synon	p.R52H	cGc/cAc	Missense	6, 5	0.750	0.008	0.999	0.02	8.723/0
	RALGPS2 (P) [82,83]	1:179064186	Non-synon	p.V342M	Gtg/Atg	Missense	69, 73	0.734	0.010	0.970	0	0.787/1
H11739 31%:69%	ASB3 (F) [84]	2:53941656	Non-synon	p.S208N	aGc/aAc	Missense	53, 30	0.773	0.007	0.980	0	0.403/1
	AKAP6 (F) [85]	14:33291533	Non-synon	p.D1504P	gAt/gIt	Missense	43, 41	0.761	0	0.983	0.01	0.787/−3
	GRM4 (A) [86]	6:34100940	Non-synon	p.D111N	Gac/Aac	Missense	58, 51	0.726	0.005	0.993	0.02	0.787/1
H11884/48%:52%	TRAF3IP1 (F) [87]	2:239306221	Non-synon	p.M537I	aTg/aCg	Missense	13, 19	0.759	0.007	0.999	0.01	0.033/−1
H11954 23%:77%	DDX5 (P) [88]	17:62500102	Non-synon	p.S146C	tCt/tGt	Missense	88, 91	0.788	0.007	0.950	0	0.787/−1
	KIF5A (P) [89]	12:57974875	Non-synon	p.R891Q	cGg/cAg	Missense	25, 19	0.767	0.005	0.989	0	0.787/1
H12126/41%:59%	ANKRD11 (A) [90,91]	16:89351565	Non-synon	p.T461R	aCa/aGa	Missense	20, 11	0.767	0	0.961	0	0.787/−1
H12172 61%:39%	TNFRSF21 (F) [92]	6:47221105	Non-synon	p.W465R	Tgg/Agg	Missense	8, 16	0.805	0.004	1	0	10.610/−3
	LRSAM1 (F) [93,94]	9:130253524	Non-synon	p.L484M	Ctg/Atg	Missense	18, 23	0.796	0.010	0.935	0	0.787/−1
	CYFIP1 (A) [95]	15:22960872	Non-synon	p.D716N	Gac/Aac	Missense	28, 30	0.767	0.009	0.999	0	0.067/1

Table 1. *Cont.*

| ID Number XCI | Gene Symbol (Category) * | Chromosome Position (Hg19) | Genomic Variant | | | | VCF Allele Depth | Classification Score | GERP2 | PolyPhen2 Score | SIFT Score | Blosum Score/Blosum62 |
			SNP Effect	Amino Acid	SNP Codon	SNP Function						
HI2215 60%:40%	FPGT-TNNI3K (F) [96]	1:74836022	Non-synon	p.G673V	gGc/gTc	Missense	34, 53	0.802	0.009	0.998	0	0.787/−3
	MAP1A (P) [96]	15:43820377	Non-synon	p.P223S	Ccc/Tcc	Missense	65, 67	0.793	0	0.998	0	0.787/−1
	MAP1A (P) [96]	15:43819396	Non-synon	p.Y1908H	Tac/Cac	Missense	13, 21	0.782	0.004	0.994	0	0.787/2
HI2244 32%:68%	KDM5B (P) [97–99]	1:202724501	Non-synon	p.Y320K	aTa/aAa	Missense	52, 47	0.808	0.007	0.988	0	0.787/−4
	TRPM2 (P) [97–99]	21:45826549	Non-synon	p.I954F	Atc/Ttc	Missense	17, 16	0.786	0.008	0.997	0	12.615/−3
	RAI1 (A) [100]	17:17699284	Non-synon	p.P959A	Ccc/Gcc	Missense	37, 26	0.763	0	0.992	0	0.787/−1
HI2278 8%:92%	DCHS1 (F) [39]	11:6662745	Codon Change Plus Codon Insertion	p.L321LW	Ctg/ctCTCg	None	12, 11	0.700	0	0.979	0.02	0/0
HI2843 25%:75%	UBE2E2 (P) [101]	3:23631311	Non-synon	p.Y198H	Tac/Cac	Missense	64, 33	0.787	0	0.994	0	0.787/2
	PCDHB15 (P) [45]	5:140625618	Non-synon	p.R157W	Cgg/Tgg	Missense	33, 61	0.754	0.007	0.942	0	24.449/−3
	SCG2 (F) [102]	2:224463227	Non-synon	p.E257D	gaG/gaC	Missense	70, 64	0.750	0	0.993	0.01	0.787/2
HI2879 18%:82%	TRIM9 (F) [103]	14:51448554	Non-synon	p.R623Q	cGg/cAg	Missense	64, 69	0.774	0	1	0	0.787/1
	SOX7 (A) [104]	8:10583274	Non-synon	p.Y432H	Tac/Cac	Missense	25, 33	0.766	0	0.997	0	0.787/2
	ANKFN1 (P) [105]	17:54559849	Non-synon	p.G744R	Ggg/Agg	Missense	28, 16	0.715	0	0.995	0.04	26.000/−2
	DCHS1 (F) [39]	11:6662745	Codon Change Plus Codon Insertion	p.L321LW	ctg/ctCTCg	None	11, 12	0.700	0	0.979	0.02	0/0
HI2898 85%:15%	MPHOSPH8 (F) [106]	13:20242552	Non-synon	p.Y736C	tAc/tGc	Missense	39, 28	0.779	0.007	0.999	0	0.787/−2
	GABRQ (A) [107]	X:151819020	Non-synon	p.S292F	tCc/tTc	Missense	9, 16	0.749	0.008	0.963	0	0.787/−2

ID Number represents the AGRE identifier. For the 30 females with ASD in this table, the average age ± standard deviation was 7.7 ± 2.6 years and age range was 5–16 years. Final selection criteria: Classification Score > 0.7; GERP2 < 0.01; PolyPhen2 > 0.9; SIFT < 0.03. * (A) = Known clinically relevant gene for autism; (F) = Neurodevelopment functional gene for autism; (P) = Known paralogue of an autism gene; NI = not informative; XCI = X chromosome Inactivation (%); ? indicates an unspecified amino acid variation; [ ] represents literature citations for each gene description.

Table 2. Summary of Putative Disease Causing Gene Variants Identified by Exome Sequencing.

Gene Variant Category	Number of Subjects (%)	Number of Genes (%)	Range
Known ASD Genes	11 (37%)	12 (15%)	0–2
Paralogues of Known ASD Genes	16 (53%)	23 (30%)	0–2
Neurodevelopmental Function Genes	23 (77%)	43 (54%)	0–7

Six females (HI0605, HI2898, HI1739, HI2172, HI2244 and HI2879) possessed a single variant of a known candidate ASD gene meeting our final selection criteria for pathogenicity in addition to one or more paralogues of ASD genes or gene variants with functional influence over cell growth and neurodevelopment. Fourteen females (HI0714, HI0751, HI0765, HI0855, HI0868, HI1157, HI1228, HI1305, HI1375, HI1422, HI1433, HI1954, HI2215 and HI2843) possessed 2 to 4 variants of ASD gene paralogues or variants with functional roles of influence over ASD but without a conclusive or definable causal mutation. Variants of multiple putative candidate genes for ASD were identified for Subjects HI0405, HI0558 and HI0793.

Using XCI data, we found that eight of the 30 females had high skewness (XCI > 80%:20%) and an additional six females showed moderate skewness (XCI 65%:35% to 80%:20%). Two of the eight females with high skewness and two females with moderate skewness possessed an X-linked gene variant meeting the final selection criteria. When considering the 78 total genes identified meeting our final selection criteria, X-linked gene variants were disproportionally more likely to be found among the females with autism spectrum disorder who exhibited moderate to high level XCI skewness than females without XCI skewness ($OR = 6.4$, 95% CI: 0.68, 60.5, $p = 0.1$) but this relationship did not achieve statistical significance. Similarly, females who exhibited XCI skewness were more likely to possess an X-linked gene variant than females without XCI skewness, ($OR = 6.0$, 95% CI: 0.58, 61.8, $p = 0.132$).

3. Discussion

We chose females with ASD to study from the age of 5 to 16 years from multiplex families with more than one affected family member in order to increase the likelihood of identifying a single gene as a causative factor using a descriptive case study approach. The females were selected from the AGRE, a biorepository of specimens and clinical data from well-characterized families with children with ASD enrolled for genetic research studies. Females with ASD from multiplex families were also examined to determine whether an overabundance of X-linked genes could be identified contributing to ASD by using whole exome sequencing and XCI assays and if X-linked genes were more frequently disturbed in those females with XCI skewness.

3.1. Single Gene Variants

Our investigation identified four females (HI0555, HI0890, HI1402 and HI2126) with single gene variants of clinically relevant candidate or known genes for ASD with a high likelihood of causality. Subject HI0555 possessed a non-synonymous, missense mutation of the *SETD2* gene located on chromosome 3 with high likelihood to have a deleterious effect on gene expression and/or function. The *SETD2* encodes a Huntingtin-interacting protein B related to Huntington disease and known expression in the brain [27,28]. The *SETD2* gene is known as an autism susceptibility gene reported in autism [12] and was the highest rated and most likely candidate gene in this case. Subject HI0890 possessed a non-synonymous, missense mutation of the *BTAF1* gene located on chromosome 10 which is implicated as a helicase and sequence-specific binding transcription factor activity [63]. Subject HI2126 possessed a non-synonymous, missense mutation in a known candidate gene for ASD, *ANKRD11* which is located on chromosome 16 and likely to be causative in this case. The *ANKRD11* gene encodes an ankryin repeat domain-containing protein which inhibits ligand-dependent activation of transcription with mutations causing the KBG syndrome characterized by craniofacial features, short stature, skeletal anomalies, seizures and intellectual disability [90,91]. No physical examination or cognitive data were available from AGRE on this 12 year old female diagnosed with autism.

3.2. Involvement of Skewness of X Chromosome Inactivation and Putative Disease Causing Genes

High XCI skewness (>80%:20%) was observed for eight of our females with autism and two of these females possessed putative variants of X-linked genes (HI1402 and HI2898) which may contribute to the phenotype. Skewing of X-chromosome inactivation may increase expression of recessive disease causing variants on the X-chromosome. Subject HI1402 with an XCI status of 85%:15% indicating high skewness also possessed a non-synonymous, missense mutation of an X-linked gene (*IL1RAPL1*) which is a recognized candidate gene for ASD [12]. *IL1RAPL1* had the highest criteria selection score of the four genes meeting initial selection criteria and the most likely cause of ASD in this female. The *IL1RAPL1* gene encodes a protein with the highest expression in brain neurons and participates in the regulation of neurite outgrowth via interaction with neuronal calcium sensors thereby regulating synaptic formation and modulation of synaptic transmission [74].

Subject HI2898 with an XCI status of 85%:15% showing high skewness possessed a variant of the X-linked gene (*GABRQ*) found to be in the top two genes in the selection process. The *GABRQ* (gamma-aminobutyric acid A receptor theta) gene encodes a receptor protein for GABAergic neurotransmission in the mammalian central nervous system [107]. GABA is a major inhibitory neurotransmitter. *GABRQ* expression is distributed in the amygdala, hippocampus, anterior hypothalamus and cortex [107]. Subject HI2278 with an XCI status of 8%:92% possessed an X-linked gene (*PAK3*) in the top 25 genes but was excluded due to strict filtering requirements (*i.e.*, PolyPhen 2) and a correct classification score of >0.50. However, the *DCHS1* gene was identified which encodes a transmembrane cell adhesion molecule that belongs to the protocadherin superfamily and acts as a ligand for FAT4, another protocadherin protein. Both DCHS1 and FAT4 form an apically located adhesive complex in the developing brain [39].

The five remaining females (HI0751, HI1143, HI1375, HI1433 and HI2879) with high XCI skewness ranging from 93%:7% to 82%:18% showed no X-linked gene variants meeting the final selection criteria for pathogenicity. Other variations of possible clinical relevance in autosomal chromosomes were observed for these subjects. Subject HI1433 with an XCI status of 93%:7% indicating extreme skewness possessed a variant of the *FAT2* gene located on chromosome 5 which is a member of the cadherin-related FAT tumor suppressor homolog 2 (Drosophila) [77–79]. Subject HI2879 with an XCI status of 18%:82% showing high skewness possessed a variant of the *SOX7* (SRY-Box 7) gene which encodes a SOX protein, acting as a transcription factor regulating diverse developmental processes [104]. Based upon our investigation, XCI skewness in these females did not contribute to the presentation of autism.

Moderate skewness (65%:35% to 80%:20%) was observed for six females (HI0765, HI1305, HI1739, HI1954, HI2244 and HI2843). Two of these females possessed putative variants of X-linked genes that may have contributed to their phenotype. Subject HI1305 with an XCI status of 73%:27% possessed a variant for *PIR*, a highly conserved X-linked iron-binding nuclear protein gene which functions as a transcriptional co-regulator and contributes to regulation of cellular processes and may promote apoptosis when over expressed [70]. An associated disorder for this gene when disturbed is extratemporal epilepsy. Three genes were identified for Subject HI0765 with an XCI status of 77%:23% with moderate skewness including *GPRASP2* located on the X chromosome which encodes a protein that may regulate a variety of G-protein coupled receptors associated with autism spectrum disorders and schizophrenia [48], as an X-linked gene approaching high skewness becomes an important candidate for causation of ASD in this female. The *KCNC2* and *ASPM* genes were the highest rated and became likely candidates

3.3. Other Autosomal Putative Disease Causing Genes

The cadherin, protocadherin and ankyrin repeat gene families were the most commonly altered putative disease causing gene variants identified in our study. Subject HI0558 possessed a variant of *PCDH8*, a member of the cadherin superfamily of genes and functions in cell adhesion in a CNS-specific manner possibly playing a role in activity-induced synaptic reorganization underlying memory with down-regulation of dendritic spines, primarily in the hippocampal area [35]. In addition this affected female possessed a variant of the *CTNNA3* gene which is located on chromosome 10 and encodes an alpha-t-catenin which plays a role in functional cadherin-mediated cell adhesion. A possible association of this gene in Alzheimer disease has been proposed [36]. Subject HI0855 possessed a variant of *CDH6* located on chromosome 5 encoding a known cadherin playing a role in cell-cell adhesion and implicated in autism [60]. Subject HI0605 possessed a non-synonymous, missense mutation of the *CCDC64* (coiled-coil domain containing, 64), a recognized ASD gene on chromosome 12 [12,38]. CCDC64 is a component of the secretory vesicle machinery in developing neurons that acts as a regulator of neurite outgrowth in the early phase of neuronal differentiation which when disturbed is associated with neuronitis [38]. Subject HI1739 possessed a non-synonymous missense mutation of the *ASB3* gene located on chromosome 2 and related to the ankyrin repeat gene family known to play a role in brain development and function including autism [88]. *GRM4* (glutamate receptor, metabotropic, 4) is a known ASD gene involved with glutamate, a major excitatory neurotransmitter in the central nervous system [86]. Subject HI2172 possessed a non-synonymous, missense mutation of the *CYFIP1* (cytoplasmic FMRP-interacting protein 1) gene located on chromosome 15 which encodes a protein that interacts with the familial mental retardation protein (FMRP) that when disturbed, causes the fragile X syndrome by impacting on development and maintenance of neuronal structures [108,109]. Individuals with the 15q11.2 BP1-BP2 microdeletion or Burnside-Butler syndrome are known to have developmental and speech delay involving the *CYFIP1* gene with an increased rate of aberrant behavior and autism [95].

4. Experimental Section

4.1. Samples from Females with Autism

Thirty Caucasian females (average age, 7.7 ± 2.6 years; age range, 5 to 16 years) were selected with confirmed diagnosis of autism. We chose to undertake whole exome sequencing from well-characterized females with autism with a positive family history (e.g., affected brothers) classified as multiplex families with autism and having a high probability of causation due to gene disturbances. We selected affected females from the Autism Genetic Research Exchange (AGRE) repository (www.agre.autismspeaks.org) for a descriptive next generation whole exome sequencing case study. The family members with autism in the AGRE were recruited and enrolled after screening and diagnostic assessments were performed with autism-related testing instruments (e.g., ADOS). Medical examinations and neurology evaluations were undertaken to collect family, pregnancy and medical history data along with physical and anthropometric measures and neurological recordings. Blood was also collected at AGRE for lymphoblastoid cell line development, plasma storage and DNA isolation. Normal chromosome analysis (karyotype) and fragile X syndrome DNA testing were completed previously in all female subjects selected from AGRE. DNA from parents or other family members were not analyzed as a component of our study.

4.2. Whole Exome Sequencing

4.2.1. Exome Sequencing Methods

Genomic DNA (5 µg) samples were sent to the Silicon Valley Biosystems, Foster City, CA, USA for whole exome sequencing via paired-end next generation sequencing (NGS) approach using standard protocols with the Illumina HiSeq2000 platform (http://www.service.Illumina.com) and Agilent

SureSelect Human All Exon v4–51Mb (http://www.genomics.agilent.com). The primary sequence raw data produced by the sequencing phase of our study was aligned to reference, the variants called and functional significance of each variant determined and rank-ordered into a list of functional variants (in order of decreasing pathology) that were correlated with or causative for the autism phenotype. The raw data was in the form of *fastq* files, containing the sequence reads and quality scores of the NGS sequencing runs. The individual reads were initially aligned by mapping to the reference genome using Burrows-Wheeler aligner (BWA), a software program for mapping low-divergent genomic sequences of up to 100 base pairs against a large reference genome (http://www.bio-bwa.sourceforge.net). Variant calling procedures then identified genomic sequence variants from the NGS sequencing runs or the absence of variants in the sample of interest relative to a reference sequence. The platform used to produce variant calls was an amalgamation of tools and methods uniquely combined and specifically calibrated for enhanced performance incorporating *Genome Analysis Toolkit (GATK)* V2.39 (https://www.broadinstitute.org/gatk/). A number of intermediate steps were applied to improve the quality/accuracy of the alignment and variant calling results prior to the creation of the final list of variants. Sequence reads that were likely to represent duplicates were marked to be ignored by the downstream variant-calling tools. Additionally, realignment around known insertions and deletions was performed, and quality scores recalibrated based upon the number of reads. The list of variants was produced in the form of a .vcf file annotated to reflect the results of quality control filters. This established and calibrated alignment and variant calling platform performs well and can reliably predict calls for single insertion and deletion events (up to 50 bp with between 97% and 99.9% sensitivity and specificity). The average number of reads per generated fragment was 64 for each of the 30 females with autism.

4.2.2. Data Analysis

The resulting .vcf file contained annotated variants with scientific features, including sequence, gene-, variant-, and transcript-level annotations. Annotations include characteristics of the observed variant such as the genomic coordinates or the variant's genomic region (e.g., exonic, upstream *etc.*), predicted effects on the various transcripts (e.g., missense, frame-shift, *etc.*), observed frequencies of the variant in control populations and affected individuals, curated knowledge from both proprietary and public databases, and pathogenicity predictions (e.g., Polyphen and SIFT). The machine learning tool utilized both the 2012 April 1K Genome release reflecting the frequency at the position and variant levels and the European Exome Sequencing Project (ESP6500) (https://esp.gs.washington.edu/drupal/) at the variant level which reflects the frequency of variants as observed in the ESP6500 dataset of healthy individuals in generating the correct classification score. The regulatory significance of variants were derived from ENCODE data. A classification engine developed by Silicon Valley Biosystems was used to create a pathogenicity score based on a machine learning technique approach. The model is trained and validated using 150,000 gene polymorphisms and mutations from a large set of variants with known pathogenicity. The classification accuracy exceeds 99% and thus the score provides the probability that a particular variant has an effect on gene expression or protein structure. The classifier was trained using known disease-causing mutations and known benign polymorphisms, each of which was annotated with attributes such as location relative to splice junctions, type of non-synonymous amino acid substitution and allele frequency in the population. The classifier was validated on a similarly sized set of variants and was shown to perform with 93% sensitive and 97% specificity. Prioritization of putative functionally relevant variants based on the classification score was performed and these variants were additionally classified with other means of determining pathogenicity such as SIFT, PolyPhen2, Blosum and Blosum62. The variants that were the most highly predicted to be functionally significant were examined across the study cohort for clustering within and around genes hypothesized as relevant to the phenotype.

4.2.3. Gene Filtration/Selection Parameters

A high filter system was utilized whereby the indels and quality disqualified (QD) parameters were identified and removed based on the bioinformatics outlined in our study approach. The classification score (range 0 to 1) was generated as our primary delimiting factor in identifying novel or candidate genes for autism in the affected females. The classifier uses machine learning algorithms to predict a variant's functional impact on a gene with higher scores for genomic variants considered more pathogenic or disease-causing changes at the gene level. Thus, we applied an initial classification cutoff of >0.5 with final selection criteria >0.7. Genomic Evolutionary Rate Profiling (GERP2), an indicator of the prevalence of the genomic variants in the general population, was assigned a cutoff value of <0.01 and remained at <0.01 as final selection criteria. GERP2 identifies constrained elements in multiple alignments by quantifying substitution deficits. These deficits represented substitutions that would have occurred if the element was neutral DNA, but did not occur because the element was under functional constraint. The initial Polymorphism Phenotyping v2 (PolyPhen2) cutoff value was >0.85 with the final selection criteria >0.9 and the initial Sorts Intolerant From Tolerant (SIFT) cutoff valuewas <0.05 with a final selection criteria <0.03 were predictors of pathogenicity or disease causing changes. The Blosum (BLOcks SUbstitution Matrix) score reports the ratio between the frequency of an amino acid substitution in the Human Gene Mutation Database (HGMD) to the frequency of this amino acid substitution in known polymorphism data sets. Numbers greater than 1 indicate this variant was observed more in diseased individuals than in controls and less likely to be tolerated. We also used the Blosum62 matrix to score alignments between evolutionarily divergent protein sequences based on local alignments. Blosum62 is a log-odds score for each possible substitution of the 20 standard amino acids. Higher Blosum62 numbers represent more closely related species comparisons and provide a general indication of frequency of the substitution. Negative numbers indicate less frequent substitutions and a lower likelihood to be tolerated (e.g., negative values = unexpected or rare substitutions).

The list of disease causing genes were developed using the primary selection criteria (*i.e.*, classification score) and then adding cutoff levels for other predictive parameters (GERP2, PolyPhen2, and SIFT). The initial list of genes and genomic variants used the classification score >0.5 alone and then the cutoff levels of the other predictive parameters (GERP2, PolyPhen2 and SIFT) were included. The classification score was then raised to >0.7 to generate the final list of genes. These putative candidate genes and genomic variants were then subjected to further screening for biological significance for neural development, function and known neurological disorders using Online Mendelian Inheritance in Man (OMIM, http://www.ncbi.nlm.nih.gov/omim), GeneCards (http://www.genecards.org/), PubMed and other online websites and databases.

4.2.4. Clinical Relevance to ASD

Following the recommendations of the American College of Medical Genetics and Genomics (ACMGG) in reporting clinical exome and genome sequencing data, we analyzed only the primary findings, in depth, which is termed as the pathogenic alteration in a gene or genes relevant to the diagnostic indication for sequencing study, *i.e.*, causation of autism in females. Only clinically relevant candidate or known existing genes for autism were included for extended analysis. Incidental findings are those pathogenic or likely pathogenic alterations in genes that are disease associated or causing but not apparently relevant to a diagnostic indication for which sequencing was undertaken (*i.e.*, in our study on autism [12]) and were not further investigated in our study. We did not include mutations of recognized autosomal recessive genes causing human disorder (e.g., *GJB2* or *GJB6*, gene mutations causing hearing loss). It is estimated that about 1% of sequencing reports would include an incidental variant.

4.3. X Chromosome Inactivation in Females with Autism

Genomic DNA isolated from blood was used as a template for polymerase chain reaction (PCR) amplification, to identify the CAG polymorphic region of the *AR* gene. Prior to the PCR amplification, 200 ng of genomic DNA was digested with the methyl-sensitive restriction enzyme *Hpa*II [21]. Approximately 50 ng of digested or undigested genomic DNA was used as a template for PCR amplification to determine the peak height of the polymorphic PCR fragment of the *AR* gene using the following primers: forward 5' TCCAGAATCTGTTCCAGAGCGTGC 3' and reverse 5' GCTGTGAAGGTTGCTGTTCCTCAT 3' with the forward primer fluorescently labeled with 6-FAM. The lengths and peak heights of the resulting PCR fragments are determined with the use of capillary electrophoresis and fragment separation software with the ABI 3100 DNA sequencer (Applied Biosystems, Carlsbad, CA, USA) using established protocols [18,22].

The digestion process preferentially degrades activated (unmethylated) over inactivated (methylated) DNA. Undigested DNA is preferentially amplified and produces larger peak heights. The peak height values for digested DNA is normalized using peak height values for the undigested DNA for each subject. The percentage of XCI for each *AR* allele was then calculated using the following formula: $(d1/u1)/[(d1/u1) + (d2/u2)]$; $d1$ = peak height of digested DNA from the first allele and $u1$ = peak height of undigested DNA from the first allele; $d2$ = peak height of digested DNA from the second allele and $u2$ = peak height of undigested DNA from the second allele. Highly skewed XCI is defined as >80% calculated ratio for either one of the *AR* gene alleles in the digested DNA sample. To ensure the reproducibility of XCI results and equal amplification of both alleles, the digestion, PCR amplification, and genotyping were repeated up to three times in several samples.

XCI status for each subject was assigned to one of the three mutually exclusive categories: (i) randomly selected inactivation of either allele from each X chromosome (XCI = 50%:50% to 64%:36%); (ii) moderately skewed inactivation favoring 1 allele of one of the X chromosomes (XCI = 65%:35% to 80%:20%); and (iii) highly skewed inactivation of a single allele representing one X chromosome (XCI > 80%:20%). The relative frequency of random, moderate, and highly skewed XCI categories was determined for the females with autism. The binomial frequency distribution of X-linked gene variants (present or absent) among females with autism spectrum disorder exhibiting moderate and high levels of XCI skewness were approximated using Yates' chi-squire test and odds ratios were calculated with 95% confidence intervals.

The top genes chosen were based on the final selection criteria identified for each female and evaluated for biologic function and whether each gene was previously reported as a clinically relevant candidate or known gene for autism [12]. The X chromosome inactivation data and XCI status on each female were used to support whether an X-linked gene played a role in the causation of autism in the affected female if XCI skewness was found.

5. Conclusions

In summary, we shared our experience using a descriptive next generation whole exome sequencing case study approach examining 30 well-characterized Caucasian females with autism between 5 and 16 years of age recruited from multiplex families enrolled at the AGRE for research purposes. Interpretation of DNA findings are limited due to an inability to study other affected and non-affected family members to determine whether the gene variants were de-novo or inherited in origin for correlation with clinical phenotypes and lack of Sanger sequencing confirmation. Using strict selection criteria, four females (13%) were found to possess single gene variants of known candidate genes for ASD with a high likelihood of causality. We also identified multiple plausible candidate genes [some known (e.g., *CCDC64* in subject HI0605) and some novel (e.g., *CHAC1* in subject HI0855)] in 77% of the remaining females. In most females, we found more than one gene that could contribute to the causation of autism. This compares with a recent report of molecular findings among 2000 consecutive patients (primarily pediatric age) referred for clinical whole exome sequencing for evaluation of suspected genetic disorders at a large USA medical center in which 25% received a molecular diagnosis. About 60% of the diagnostic mutations were not previously reported [110]. By phenotypic category, 36% of those presenting with neurological involvement were found to have a molecular diagnosis while 20% of those from the non-neurological group had a diagnosis. We not only performed whole exome sequencing in our females with autism but also performed X chromosome inactivation (XCI) studies for evidence of X-linked genes in the females showing non-random XCI skewness. Moderate and high XCI skewness was seen in 14 of the 30 females and those with skewness were 4 times more likely to show a gene variant on the X chromosome.

Acknowledgments: The authors thank Carla Meister for manuscript preparation. Partial grant support was provided by the University of Kansas Medical Center and Center of Translational Science through the NIH National Center for Advancing Translational Sciences (Grant# UL1TR000001) and from NICHD (HD02528).

Author Contributions: Merlin G. Butler conceived the study, analyzed data and wrote the manuscript. Syed K. Rafi reviewed literature, assessed the function of gene variants and contributed to the content of the manuscript. Waheeda Hossain carried out DNA isolation and X-chromosome inactivation studies with data analysis. Dietrich A. Stephan carried out whole exome sequencing experiments, data analysis and interpretation and contributed to the content of the manuscript. Ann M. Manzardo carried out and summarized data analysis and interpretation and contributed to the text of the manuscript.

Conflicts of Interest: The authors declare no conflict of interest.

References

1. American Psychiatric Association. *Diagnostic and Statistical Manual of Mental Disorders*, 4th ed.; American Psychiatric Association Press: Washington, DC, USA, 2000.
2. Johnson, C.P.; Myers, S.M. The Council on Children With Disabilities. Identification and evaluation of children with autism spectrum disorders. *Pediatrics* **2007**, *120*, 1183–1215. [CrossRef] [PubMed]
3. Lord, C.; Risi, S.; Lambrecht, L.; Cook, E.H., Jr.; Leventhal, B.L.; DiLavore, P.C.; Pickles, A.; Rutter, M. The autism diagnostic observation schedule-generic: A standard measure of social and communication deficits associated with the spectrum of autism. *J. Autism Dev. Disord.* **2000**, *30*, 205–223. [CrossRef] [PubMed]
4. Le Couteur, A.; Lord, C.; Ruter, M. *Autism Diagnostic Interview-Reviewed (ADI-R)*; Western Psychological Services: Los Angeles, CA, USA, 2003.
5. Constantino, J.N.; Davis, S.A.; Todd, R.D.; Schindler, M.K.; Gross, M.M.; Brophy, S.L.; Metzger, L.M.; Shoushtari, C.S.; Splinter, R.; Reich, W. Validation of a brief quantitative measure of autistic traits: Comparison of the social responsiveness scale with the autism diagnostic interview-revised. *J. Autism Dev. Disord.* **2003**, *33*, 427–433. [CrossRef] [PubMed]
6. Developmental Disabilities Monitoring Network Surveillance Year 2010 Principal Investigators; Centers for Disease Control and Prevention (CDC). Prevalence of autism spectrum disorder among children aged 8 years—autism and developmental disabilities monitoring network, 11 sites, United States, 2010. *MMWR Surveill. Summ.* **2014**, *63*, 1–21.
7. Rapin, I.; Dunn, M. The neurology of autism: Many unanswered questions. *Eur. J. Neurol.* **1995**, *2*, 151–162. [CrossRef] [PubMed]
8. Geschwind, D.H.; Levitt, P. Autism spectrum disorders: Developmental disconnection syndromes. *Curr. Opin. Neurobio.* **2007**, *17*, 103–111. [CrossRef]
9. Butler, M.G.; Dasouki, M.J.; Zhou, X.P.; Talebizadeh, Z.; Brown, M.; Takahashi, T.N.; Miles, J.H.; Wang, C.H.; Stratton, R.; Pilarski, R.; et al. Subset of individuals with autism spectrum disorders and extreme macrocephaly associated with germline PTEN tumour suppressor gene mutations. *J. Med. Genet.* **2005**, *42*, 318–321. [CrossRef] [PubMed]
10. Herman, G.E.; Henninger, N.; Ratliff-Schaub, K.; Pastore, M.; Fitzgerald, S.; McBride, K.L. Genetic testing in autism: How much is enough? *Genet. Med.* **2007**, *9*, 268–274. [CrossRef] [PubMed]
11. Roberts, J.L.; Hovanes, K.; Dasouki, M.; Manzardo, A.M.; Butler, M.G. Chromosomal microarray analysis of consecutive individuals with autism spectrum disorders or learning disability presenting for genetic services. *Gene* **2014**, *535*, 70–78. [CrossRef] [PubMed]
12. Butler, MG; Rafi, S.K.; Manzardo, AM. Clinically relevant candidate and known genes for autism spectrum disorders (ASD) with representation on high resolution chromosome ideograms. *OA Autism* **2014**, *2*, 5–28.

13. Lyon, M.F. X-chromosome inactivation and human genetic disease. *Acta Paediatr.* **2002**, *91*, 107–112. [CrossRef]

14. Plenge, R.M.; Stevenson, R.A.; Lubs, H.A.; Schwartz, C.E.; Willard, H.F. Skewed X-chromosome inactivation is a common feature of X-linked mental retardation disorders. *Am. J. Hum. Genet.* **2002**, *71*, 168–173. [CrossRef] [PubMed]

15. Talebizadeh, Z.; Bittel, D.C.; Veatch, O.J.; Kibiryeva, N.; Butler, M.G. Brief report: Non-random X chromosome inactivation in females with autism. *J. Autism Dev. Disord.* **2005**, *35*, 675–681. [CrossRef] [PubMed]

16. Buyse, I.M.; Fang, P.; Hoon, K.T.; Amir, R.E.; Zoghbi, H.Y.; Roa, B.B. Diagnostic testing for Rett syndrome by DHPLC and direct sequencing analysis of the MECP2 gene: Identification of several novel mutations and polymorphisms. *Am. J. Hum. Genet.* **2000**, *67*, 1428–1436. [CrossRef] [PubMed]

17. Maier, E.M.; Kammerer, S.; Muntau, A.C.; Wichers, M.; Braun, A.; Roscher, A.A. Symptoms in carriers of adrenoleukodystrophy relate to skewed X inactivation. *Ann. Neurol.* **2002**, *52*, 683–688. [CrossRef] [PubMed]

18. Butler, M.G.; Theodoro, M.F.; Bittel, D.C.; Kuipers, P.J.; Driscoll, D.J.; Talebizadeh, Z. X-chromosome inactivation patterns in females with Prader-Willi syndrome. *Am. J. Med. Genet. A* **2007**, *143*, 469–475. [CrossRef]

19. Butler, M.G.; Sturich, J.; Myers, S.E.; Gold, J.A.; Kimonis, V.; Driscoll, D.J. Is gestation in Prader-Willi syndrome affected by the genetic subtype? *J. Assist. Reprod. Genet.* **2009**, *26*, 461–466. [CrossRef] [PubMed]

20. Cassidy, S.B.; Lai, L.W.; Erickson, R.P.; Magnuson, L.; Thomas, E.; Gendron, R.; Herrmann, J. Trisomy 15 with loss of the paternal 15 as a cause of Prader-Willi syndrome due to maternal disomy. *Am. J. Hum. Genet.* **1992**, *51*, 701–708. [PubMed]

21. Allen, R.C.; Zoghbi, H.Y.; Moseley, A.B.; Rosenblatt, H.M.; Belmont, J.W. Methylation of HpaII and HhaI sites near the polymorphic CAG repeat in the human androgen-receptor gene correlates with x chromosome inactivation. *Am. J. Hum. Genet.* **1992**, *51*, 1229–1239. [PubMed]

22. Bittel, D.C.; Theodoro, M.F.; Kibiryeva, N.; Fischer, W.; Talebizadeh, Z.; Butler, M.G. Comparison of X-chromosome inactivation patterns in multiple tissues from human females. *J. Med. Genet.* **2008**, *45*, 309–313. [CrossRef] [PubMed]

23. Bond, J.; Roberts, E.; Mochida, G.H.; Hampshire, D.J.; Scott, S.; Askham, J.M.; Springell, K.; Mahadevan, M.; Crow, Y.J.; Markham, A.F.; *et al.* ASPM is a major determinant of cerebral cortical size. *Nat. Genet.* **2002**, *32*, 316–320. [CrossRef]

24. Vermeulen, M.; Mulder, K.W.; Denissov, S.; Pim Pijnappel, W.W. M.; van Schaik, F.M.A.; Varier, R.A.; Baltissen, M.P.A.; Stunnenberg, H.G.; Mann, M.; Timmers, H.T.M. Selective anchoring of TFIID to nucleosomes by trimethylation of histone H3 lysine 4. *Cell* **2007**, *131*, 58–69. [CrossRef] [PubMed]

25. Lacy, S.E.; Bonnemann, C.G.; Buzney, E.A.; Kunkel, L.M. Identification of FLRT1, FLRT2, and FLRT3: A novel family of transmembrane leucine-rich repeat proteins. *Genomics* **1999**, *62*, 417–426. [CrossRef] [PubMed]

26. Wang, J.; Takeuchi, T.; Yokota, H.; Izumi, T. Novel rabphilin-3-like protein associates with insulin-containing granules in pancreatic beta cells. *J. Biol. Chem.* **1999**, *274*, 28542–28548. [CrossRef] [PubMed]

27. Faber, P.W.; Barnes, G.T.; Srinidhi, J.; Chen, J.; Gusella, J.F.; MacDonald, M.E. Huntingtin interacts with a family of WW domain proteins. *Hum. Mol. Genet.* **1998**, *7*, 1463–1474. [CrossRef] [PubMed]

28. Nagase, T.; Kikuno, R.; Hattori, A.; Kondo, Y.; Okumura, K.; Ohara, O. Prediction of the coding sequences of unidentified human genes. XIX. The complete sequences of 100 new cDNA clones from brain which code for large proteins *in vitro*. *DNA Res.* **2000**, *7*, 347–355. [CrossRef] [PubMed]

29. Claudio, J.O.; Liew, C.C.; Ma, J.; Heng, H.H.Q.; Stewart, A.K.; Hawley, R.G. Cloning and expression analysis of a novel WD repeat gene, WDR3, mapping to 1p12-p13. *Genomics* **1999**, *59*, 85–89. [CrossRef] [PubMed]

30. Iwamoto, M.; Asakawa, H.; Hiraoka, Y.; Haraguchi, T. Nucleoporin Nup98: A gatekeeper in the eukaryotic kingdoms. *Genes Cells* **2010**, *15*, 661–669. [CrossRef] [PubMed]

31. Tommerup, N.; Vissing, H. Isolation and fine mapping of 16 novel human zinc finger-encoding cDNAs identify putative candidate genes for developmental and malignant disorders. *Genomics* **1995**, *27*, 259–264. [CrossRef] [PubMed]

32. Chapelin, C.; Duriez, N.; Magnino, F.; Goossens, M.; Escudier, E.; Amselem, S. Isolation of several human axonemal dynein heavy chain genes: Genomic structure of the catalytic site, phylogenetic analysis and chromosomal assignment. *FEBS Lett.* **1997**, *412*, 325–330. [CrossRef] [PubMed]

33. Bond, J.; Roberts, E.; Springell, K.; Lizarraga, S.B.; Scott, S.; Higgins, J.; Hampshire, D.J.; Morrison, E.E.; Leal, G.F.; Silva, E.O.; *et al.* A centrosomal mechanism involving CDK5RAP2 and CENPJ controls brain size. *Nat. Genet.* **2005**, *37*, 353–355. [CrossRef]

34. Pertel, T.; Hausmann, S.; Morger, D.; Zuger, S.; Guerra, J.; Lascano, J.; Reinhard, C.; Santoni, F.A.; Uchil, P.D.; Chatel, L.; *et al.* TRIM5 is an innate immune sensor for the retrovirus capsid lattice. *Nature* **2011**, *472*, 361–365. [CrossRef]

35. Strehl, S.; Glatt, K.; Liu, Q.M.; Glatt, H.; Lalande, M. Characterization of two novel protocadherins (PCDH8 and PCDH9) localized on human chromosome 13 and mouse chromosome 14. *Genomics* **1998**, *53*, 81–89. [CrossRef] [PubMed]

36. Miyashita, A.; Arai, H.; Asada, T.; Imagawa, M.; Matsubara, E.; Shoji, M.; Higuchi, S.; Urakami, K.; Kakita, A.; Takahashi, H.; *et al.* Genetic association of CTNNA3 with late-onset Alzheimer's disease in females. *Hum. Mol. Genet.* **2007**, *16*, 2854–2869. [CrossRef]

37. Kao, S.C.; Chen, C.Y.; Wang, S.L.; Yang, J.J.; Hung, W.C.; Chen, Y.C.; Lai, N.S.; Liu, H.T.; Huang, H.; Chen, H.C.; *et al.* Identification of phostensin, a PP1 F-actin cytoskeleton targeting subunit. *Biochem. Biophys. Res. Commun.* **2007**, *356*, 594–598. [CrossRef] [PubMed]

38. Leidinger, P.; Backes, C.; Deutscher, S.; Schmitt, K.; Mueller, S.C.; Frese, K.; Haas, J.; Ruprecht, K.; Paul, F.; Stahler, C.; *et al.* A blood based 12-miRNA signature of Alzheimer disease patients. *Genome Biol.* **2013**, *14*. [CrossRef]

39. Cappello, S.; Gray, M.J.; Badouel, C.; Lange, S.; Einsiedler, M.; Srour, M.; Chitayat, D.; Hamdan, F.F.; Jenkins, Z.A.; Morgan, T.; *et al.* Mutations in genes encoding the cadherin receptor-ligand pair DCHS1 and FAT4 disrupt cerebral cortical development. *Nat. Genet.* **2013**, *45*, 1300–1308.

40. Grishin, A.; Li, H.; Levitan, E.S.; Zaks-Makhina, E. Identification of gamma-aminobutyric acid receptor-interacting factor 1 (TRAK2) as a trafficking factor for the K$^+$ channel Kir2.1. *J. Biol. Chem.* **2006**, *281*, 30104–30111. [CrossRef] [PubMed]

41. Billaut-Laden, I.; Rat, E.; Allorge, D.; Crunelle-Thibaut, A.; Cauffiez, C.; Chevalier, D.; Lo-Guidice, J.M.; Broly, F. Evidence for a functional genetic polymorphism of the human mercaptopyruvate sulfurtransferase (MPST), a cyanide detoxification enzyme. *Toxicol. Lett.* **2006**, *165*, 101–111. [CrossRef] [PubMed]

42. Hayette, S.; Gadoux, M.; Martel, S.; Bertrand, S.; Tigaud, I.; Magaud, J.P.; Rimokh, R. FLRG (follistatin-related gene), a new target of chromosomal rearrangement in malignant blood disorders. *Oncogene* **1998**, *16*, 2949–2954. [CrossRef] [PubMed]

43. Bajorek, M.; Morita, E.; Skalicky, J.J.; Morham, S.G.; Babst, M.; Sundquist, W.I. Biochemical analyses of human IST1 and its function in cytokinesis. *Mol. Biol. Cell* **2009**, *20*, 1360–1373. [CrossRef] [PubMed]

44. Klejnot, M.; Kozielski, F. Structural insights into human Kif7, a kinesin involved in hedgehog signalling. *Acta Crystallogr.* **2012**, *68*, 154–159.

45. Wu, Q.; Maniatis, T. A striking organization of a large family of human neural cadherin-like cell adhesion genes. *Cell* **1999**, *97*, 779–790. [CrossRef] [PubMed]

46. Tran, T.S.; Rubio, M.E.; Clem, R.L.; Johnson, D.; Case, L.; Tessier-Lavigne, M.; Huganir, R.L.; Ginty, D.D.; Kolodkin, A.L. Secreted semaphorins control spine distribution and morphogenesis in the postnatal CNS. *Nature* **2009**, *462*, 1065–1069. [CrossRef] [PubMed]

47. Zhang, X.Y.; Wen, J.; Yang, W.; Wang, C.; Gao, L.; Zheng, L.H.; Wang, T.; Ran, K.; Li, Y.; Li, X.; *et al.* Gain-of-function mutations in SCN11a cause familial episodic pain. *Am. J. Hum. Genet.* **2013**, *93*, 957–966. [CrossRef] [PubMed]

48. Piton, A.; Gauthier, J.; Hamdan, F.F.; Lafreniere, R.G.; Yang, Y.; Henrion, E.; Laurent, S.; Noreau, A.; Thibodeau, P.; Karemera, L.; *et al.* Systematic resequencing of X-chromosome synaptic genes in autism spectrum disorder and schizophrenia. *Mol. Psychiatry* **2011**, *16*, 867–880. [CrossRef] [PubMed]

49. Agrawal, A.; Pergadia, M.L.; Saccone, S.F.; Lynskey, M.T.; Wang, J.C.; Martin, N.G.; Statham, D.; Henders, A.; Campbell, M.; Garcia, R.; *et al.* An autosomal linkage scan for cannabis use disorders in the nicotine addiction genetics project. *Arch. Gen. Psychiatry* **2008**, *65*, 713–721. [CrossRef] [PubMed]

50. Yoshimura, S.; Gerondopoulos, A.; Linford, A.; Rigden, D.J.; Barr, F.A. Family-wide characterization of the DENN domain Rab GDP-GTP exchange factors. *J. Cell Biol.* **2010**, *191*, 367–381. [CrossRef] [PubMed]

51. Le Meur, N.; Martin, C.; Saugier-Veber, P.; Joly, G.; Lemoine, F.; Moirot, H.; Rossi, A.; Bachy, B.; Cabot, A.; Joly, P.; *et al.* Complete germline deletion of the *STK11* gene in a family with Peutz-Jeghers syndrome. *Eur. J. Hum. Genet.* **2004**, *12*, 415–418. [CrossRef] [PubMed]

52. Smolen, G.A.; Schott, B.J.; Stewart, R.A.; Diederichs, S.; Muir, B.; Provencher, H.L.; Look, A.T.; Sgroi, D.C.; Peterson, R.T.; Haber, D.A. A Rap GTPase interactor, RADIL, mediates migration of neural crest precursors. *Genes Dev.* **2007**, *21*, 2131–2136. [CrossRef] [PubMed]

53. Zuo, X.; Zhang, J.; Zhang, Y.; Hsu, S.C.; Zhou, D.; Guo, W. Exo70 interacts with the Arp2/3 complex and regulates cell migration. *Nat. Cell Biol.* **2006**, *8*, 1383–1388. [CrossRef] [PubMed]

54. Oeffner, F.; Moch, C.; Neundorf, A.; Hofmann, J.; Koch, M.; Grzeschik, K.H. Novel interaction partners of Bardet-Biedl syndrome proteins. *Cell Motil. Cytoskelet.* **2008**, *65*, 143–155. [CrossRef]

55. Lopez, I.; Mak, E.C.; Ding, J.; Hamm, H.E.; Lomasney, J.W. A novel bifunctional phospholipase C that is regulated by Galpha 12 and stimulates the Ras/mitogen-activated protein kinase pathway. *J. Biol. Chem.* **2001**, *276*, 2758–2765. [CrossRef] [PubMed]

56. Liefke, R.; Oswald, F.; Alvarado, C.; Ferres-Marco, D.; Mittler, G.; Rodriguez, P.; Dominguez, M.; Borggrefe, T. Histone demethylase KDM5A is an integral part of the core Notch-RBP-J repressor complex. *Genes Dev.* **2010**, *24*, 590–601. [CrossRef] [PubMed]

57. Chen, P.W.; Jian, X.; Yoon, H.Y.; Randazzo, P.A. ARAP2 signals through Arf6 and Rac1 to control focal adhesion morphology. *J. Biol. Chem.* **2013**, *288*, 5849–5860. [CrossRef] [PubMed]

58. Chi, Z.; Zhang, J.; Tokunaga, A.; Harraz, M.M.; Byrne, S.T.; Dolinko, A.; Xu, J.; Blackshaw, S.; Gaiano, N.; Dawson, T.M.; *et al.* Botch promotes neurogenesis by antagonizing Notch. *Dev. Cell* **2012**, *22*, 707–720. [CrossRef] [PubMed]

59. Tsukahara, M.; Suemori, H.; Noguchi, S.; Ji, Z.S.; Tsunoo, H. Novel nucleolar protein, midnolin, is expressed in the mesencephalon during mouse development. *Gene* **2000**, *254*, 45–55. [CrossRef] [PubMed]

60. Suzuki, S.; Sano, K.; Tanihara, H. Diversity of the cadherin family: Evidence for eight new cadherins in nervous tissue. *Cell Regul.* **1991**, *2*, 261–270. [PubMed]

61. Unger, S.; Lausch, E.; Rossi, A.; Megarbane, A.; Sillence, D.; Alcausin, M.; Aytes, A.; Mendoza-Londono, R.; Nampoothiri, S.; Afroze, B.; *et al.* Phenotypic features of carbohydrate sulfotransferase 3 (CHST3) deficiency in 24 patients: Congenital dislocations and vertebral changes as principal diagnostic features. *Am. J. Med. Genet. A* **2010**, *152A*, 2543–2549. [CrossRef]

62. Reimann, J.D.; Freed, E.; Hsu, J.Y.; Kramer, E.R.; Peters, J.M.; Jackson, P.K. Emi1 is a mitotic regulator that interacts with Cdc20 and inhibits the anaphase promoting complex. *Cell* **2001**, *105*, 645–655. [CrossRef] [PubMed]

63. Van Der Knaap, J.A.; van Den Boom, V.; Kuipers, J.; van Eijk, M.J.; van Der Vliet, P.C.; Timmers, H.T. The gene for human TATA-binding-protein-associated factor (TAFII) 170: Structure, promoter and chromosomal localization. *Biochem. J.* **2000**, *345*, 521–527. [CrossRef] [PubMed]

64. Yoshiura, K.; Machida, J.; Daack-Hirsch, S.; Patil, S.R.; Ashworth, L.K.; Hecht, J.T.; Murray, J.C. Characterization of a novel gene disrupted by a balanced chromosomal translocation t(2;19)(q11.2;q13.3) in a family with cleft lip and palate. *Genomics* **1998**, *54*, 231–240. [CrossRef] [PubMed]

65. Kuja-Panula, J.; Kiiltomaki, M.; Yamashiro, T.; Rouhiainen, A.; Rauvala, H. AMIGO, a transmembrane protein implicated in axon tract development, defines a novel protein family with leucine-rich repeats. *J. Cell Biol.* **2003**, *160*, 963–973. [CrossRef] [PubMed]

66. Dammai, V.; Subramani, S. The human peroxisomal targeting signal receptor, Pex5p, is translocated into the peroxisomal matrix and recycled to the cytosol. *Cell* **2001**, *105*, 187–196. [CrossRef] [PubMed]

67. Iwamori, T.; Iwamori, N.; Ma, L.; Edson, M.A.; Greenbaum, M.P.; Matzuk, M.M. TEX14 interacts with CEP55 to block cell abscission. *Mol. Cell. Biol.* **2010**, *30*, 2280–2292. [CrossRef] [PubMed]

68. Friedman, J.R.; Kaestner, K.H. The Foxa family of transcription factors in development and metabolism. *Cell Mol. Life Sci.* **2006**, *63*, 2317–2328. [CrossRef] [PubMed]

69. Yoon, D.; Kim, Y.J.; Cui, W.Y.; Van der Vaart, A.; Cho, Y.S.; Lee, J.Y.; Ma, J.Z.; Payne, T.J.; Li, M.D.; Park, T. Large-scale genome-wide association study of asian population reveals genetic factors in FRMD4A and other loci influencing smoking initiation and nicotine dependence. *Hum. Genet.* **2012**, *131*, 1009–1021. [CrossRef] [PubMed]

70. Licciulli, S.; Cambiaghi, V.; Scafetta, G.; Gruszka, A.M.; Alcalay, M. Pirin downregulation is a feature of AML and leads to impairment of terminal myeloid differentiation. *Leukemia* **2010**, *24*, 429–437. [CrossRef] [PubMed]

71. Pazour, G.J.; Agrin, N.; Walker, B.L.; Witman, G.B. Identification of predicted human outer dynein arm genes: Candidates for primary ciliary dyskinesia genes. *J. Med. Genet.* **2006**, *43*, 62–73. [CrossRef] [PubMed]

72. Chou, S.J.; O'Leary, D.D. Role for Lhx2 in corticogenesis through regulation of progenitor differentiation. *Mol. Cell. Neurosci.* **2013**, *56*, 1–9. [CrossRef] [PubMed]

73. Leighton, P.A.; Mitchell, K.J.; Goodrich, L.V.; Lu, X.; Pinson, K.; Scherz, P.; Skarnes, W.C.; Tessier-Lavigne, M. Defining brain wiring patterns and mechanisms through gene trapping in mice. *Nature* **2001**, *410*, 174–179. [CrossRef] [PubMed]

74. Piton, A.; Michaud, J.L.; Peng, H.; Aradhya, S.; Gauthier, J.; Mottron, L.; Champagne, N.; Lafreniere, R.G.; Hamdan, F.F.; team, S.D.; *et al.* Mutations in the calcium-related gene IL1RAPL1 are associated with autism. *Hum. Mol. Genet.* **2008**, *17*, 3965–3974. [CrossRef] [PubMed]

75. Sinibaldi, L.; De Luca, A.; Bellacchio, E.; Conti, E.; Pasini, A.; Paloscia, C.; Spalletta, G.; Caltagirone, C.; Pizzuti, A.; Dallapiccola, B. Mutations of the Nogo-66 receptor (RTN4R) gene in schizophrenia. *Hum. Mutat.* **2004**, *24*, 534–535. [CrossRef] [PubMed]

76. Ono, Y.; Ohno, M.; Shimura, Y. Identification of a putative RNA helicase (HRH1), a human homolog of yeast Prp22. *Mol. Cell. Biol.* **1994**, *14*, 7611–7620. [PubMed]

77. Katoh, Y.; Katoh, M. Comparative integromics on FAT1, FAT2, FAT3 and FAT4. *Int. J. Mol. Med.* **2006**, *18*, 523–528. [PubMed]

78. Neale, B.M.; Kou, Y.; Liu, L.; Ma'ayan, A.; Samocha, K.E.; Sabo, A.; Lin, C.F.; Stevens, C.; Wang, L.S.; Makarov, V.; *et al.* Patterns and rates of exonic de novo mutations in autism spectrum disorders. *Nature* **2012**, *485*, 242–245. [CrossRef] [PubMed]

79. Kenny, E.M.; Cormican, P.; Furlong, S.; Heron, E.; Kenny, G.; Fahey, C.; Kelleher, E.; Ennis, S.; Tropea, D.; Anney, R.; *et al.* Excess of rare novel loss-of-function variants in synaptic genes in schizophrenia and autism spectrum disorders. *Mol. Psychiatry* **2014**, *19*, 872–879. [CrossRef] [PubMed]

80. Kadereit, B.; Kumar, P.; Wang, W.J.; Miranda, D.; Snapp, E.L.; Severina, N.; Torregroza, I.; Evans, T.; Silver, D.L. Evolutionarily conserved gene family important for fat storage. *Proc. Natl. Acad. Sci. USA* **2008**, *105*, 94–99. [CrossRef] [PubMed]

81. Bai, S.W.; Herrera-Abreu, M.T.; Rohn, J.L.; Racine, V.; Tajadura, V.; Suryavanshi, N.; Bechtel, S.; Wiemann, S.; Baum, B.; Ridley, A.J. Identification and characterization of a set of conserved and new regulators of cytoskeletal organization, cell morphology and migration. *BMC Biol.* **2011**, *9*, 54. [CrossRef] [PubMed]

82. Liu, F.; Arias-Vasquez, A.; Sleegers, K.; Aulchenko, Y.S.; Kayser, M.; Sanchez-Juan, P.; Feng, B.J.; Bertoli-Avella, A.M.; van Swieten, J.; Axenovich, T.I.; *et al.* A genomewide screen for late-onset Alzheimer disease in a genetically isolated Dutch population. *Am. J. Hum. Genet.* **2007**, *81*, 17–31. [CrossRef] [PubMed]

83. Jin, J.; Smith, F.D.; Stark, C.; Wells, C.D.; Fawcett, J.P.; Kulkarni, S.; Metalnikov, P.; O'Donnell, P.; Taylor, P.; Taylor, L.; *et al.* Proteomic, functional, and domain-based analysis of *in vivo* 14-3-3 binding proteins involved in cytoskeletal regulation and cellular organization. *Curr. Biol.* **2004**, *14*, 1436–1450. [CrossRef] [PubMed]

84. Kile, B.T.; Viney, E.M.; Willson, T.A.; Brodnicki, T.C.; Cancilla, M.R.; Herlihy, A.S.; Croker, B.A.; Baca, M.; Nicola, N.A.; Hilton, D.J.; *et al.* Cloning and characterization of the genes encoding the ankyrin repeat and SOCS box-containing proteins asb-1, asb-2, asb-3 and asb-4. *Gene* **2000**, *258*, 31–41. [CrossRef] [PubMed]

85. Dodge-Kafka, K.L.; Soughayer, J.; Pare, G.C.; Michel, J.J.C.; Langeberg, L.K.; Kapiloff, M.S.; Scott, J.D. The protein kinase A anchoring protein mAKAP coordinates two integrated cAMP effector pathways. *Nature* **2005**, *437*, 574–578. [CrossRef] [PubMed]

86. Wu, S.; Wright, R.A.; Rockey, P.K.; Burgett, S.G.; Arnold, J.S.; Rosteck, P.R., Jr.; Johnson, B.G.; Schoepp, D.D.; Belagaje, R.M. Group III human metabotropic glutamate receptors 4, 7 and 8: Molecular cloning, functional expression, and comparison of pharmacological properties in RGT cells. *Brain Res. Mol. Brain Res.* **1998**, *53*, 88–97. [CrossRef] [PubMed]

87. Berbari, N.F.; Kin, N.W.; Sharma, N.; Michaud, E.J.; Kesterson, R.A.; Yoder, B.K. Mutations in Traf3ip1 reveal defects in ciliogenesis, embryonic development, and altered cell size regulation. *Dev. Biol.* **2011**, *360*, 66–76. [PubMed]

88. Davis, B.N.; Hilyard, A.C.; Lagna, G.; Hata, A. SMAD proteins control DROSHA-mediated microRNA maturation. *Nature* **2008**, *454*, 56–61. [CrossRef] [PubMed]

89. Goizet, C.; Boukhris, A.; Mundwiller, E.; Tallaksen, C.; Forlani, S.; Toutain, A.; Carriere, N.; Paquis, V.; Depienne, C.; Durr, A.; *et al.* Complicated forms of autosomal dominant hereditary spastic paraplegia are frequent in SPG10. *Hum. Mutat.* **2008**, *30*, E376–E385. [CrossRef]

90. Youngs, E.L.; Hellings, J.A.; Butler, M.G. ANKRD11 gene deletion in a 17-year-old male. *Clin. Dysmorphol.* **2011**, *20*, 170–171. [CrossRef] [PubMed]

91. Sirmaci, A.; Spiliopoulos, M.; Brancati, F.; Powell, E.; Duman, D.; Abrams, A.; Bademci, G.; Agolini, E.; Guo, S.; Konuk, B.; *et al.* Mutations in ANKRD11 cause KBG syndrome, characterized by intellectual disability, skeletal malformations, and macrodontia. *Am. J. Hum. Genet.* **2011**, *89*, 289–294. [CrossRef] [PubMed]

92. Mi, S.; Lee, X.; Hu, Y.; Ji, B.; Shao, Z.; Yang, W.; Huang, G.; Walus, L.; Rhodes, K.; Gong, B.J.; *et al.* Death receptor 6 negatively regulates oligodendrocyte survival, maturation and myelination. *Nat. Med.* **2011**, *17*, 816–821. [CrossRef] [PubMed]

93. Guernsey, D.L.; Jiang, H.; Bedard, K.; Evans, S.C.; Ferguson, M.; Matsuoka, M.; Macgillivray, C.; Nightingale, M.; Perry, S.; Rideout, A.L.; *et al.* Mutation in the gene encoding ubiquitin ligase LRSAM1 in patients with Charcot-Marie-Tooth disease. *PLoS Genet.* **2010**, *6*. [CrossRef]

94. Weterman, M.A.; Sorrentino, V.; Kasher, P.R.; Jakobs, M.E.; van Engelen, B.G.; Fluiter, K.; de Wissel, M.B.; Sizarov, A.; Nurnberg, G.; Nurnberg, P.; *et al.* A frameshift mutation in LRSAM1 is responsible for a dominant hereditary polyneuropathy. *Hum. Mol. Genet.* **2012**, *21*, 358–370. [CrossRef] [PubMed]

95. Burnside, R.D.; Pasion, R.; Mikhail, F.M.; Carroll, A.J.; Robin, N.H.; Youngs, E.L.; Gadi, I.K.; Keitges, E.; Jaswaney, V.L.; Papenhausen, P.R.; *et al.* Microdeletion/microduplication of proximal 15q11.2 between BP1 and BP2: A susceptibility region for neurological dysfunction including developmental and language delay. *Hum. Genet.* **2011**, *130*, 517–528. [CrossRef] [PubMed]

96. Pastuszak, I.; Ketchum, C.; Hermanson, G.; Sjoberg, E.J.; Drake, R.; Elbein, A.D. GDP-L-fucose pyrophosphorylase. Purification, cDNA cloning, and properties of the enzyme. *J. Biol. Chem.* **1998**, *273*, 30165–30174. [CrossRef] [PubMed]

97. Xu, C.; Li, P.P.; Cooke, R.G.; Parikh, S.V.; Wang, K.; Kennedy, J.L.; Warsh, J.J. TRPM2 variants and bipolar disorder risk: Confirmation in a family-based association study. *Bipolar Disord.* **2009**, *11*, 1–10. [CrossRef] [PubMed]

98. Bueno, M.T.; Richard, S. SUMOylation negatively modulates target gene occupancy of the KDM5B, a histone lysine demethylase. *Epigenetics* **2013**, *8*, 1162–1175. [CrossRef] [PubMed]

99. Catchpole, S.; Spencer-Dene, B.; Hall, D.; Santangelo, S.; Rosewell, I.; Guenatri, M.; Beatson, R.; Scibetta, A.G.; Burchell, J.M.; Taylor-Papadimitriou, J. PLU-1/JARID1B/KDM5B is required for embryonic survival and contributes to cell proliferation in the mammary gland and in ER+ breast cancer cells. *Int. J. Oncol.* **2011**, *38*, 1267–1277. [PubMed]

100. Carmona-Mora, P.; Walz, K. Retinoic acid induced 1, RAI1: A dosage sensitive gene related to neurobehavioral alterations including autistic behavior. *Curr. Genomics* **2010**, *11*, 607–617. [CrossRef] [PubMed]

101. Van Wijk, S.J.; Timmers, H.T. The family of ubiquitin-conjugating enzymes (E2s): Deciding between life and death of proteins. *FASEB J.* **2010**, *24*, 981–993. [CrossRef] [PubMed]

102. Kirchmair, R.; Hogue-Angeletti, R.; Gutierrez, J.; Fischer-Colbrie, R.; Winkler, H. Secretoneurin—A neuropeptide generated in brain, adrenal medulla and other endocrine tissues by proteolytic processing of secretogranin II (chromogranin C). *Neuroscience* **1993**, *53*, 359–365. [CrossRef] [PubMed]

103. Tanji, K.; Kamitani, T.; Mori, F.; Kakita, A.; Takahashi, H.; Wakabayashi, K. TRIM9, a novel brain-specific E3 ubiquitin ligase, is repressed in the brain of Parkinson's disease and dementia with Lewy bodies. *Neurobiol. Dis.* **2010**, *38*, 210–218. [CrossRef] [PubMed]

104. Takash, W.; Canizares, J.; Bonneaud, N.; Poulat, F.; Mattei, M.G.; Jay, P.; Berta, P. SOX7 transcription factor: Sequence, chromosomal localisation, expression, transactivation and interference with Wnt signalling. *Nucleic Acids Res.* **2001**, *29*, 4274–4283. [CrossRef] [PubMed]

105. Agrawal, A.; Lynskey, M.T.; Hinrichs, A.; Grucza, R.; Saccone, S.F.; Krueger, R.; Neuman, R.; Howells, W.; Fisher, S.; Fox, L.; *et al.* A genome-wide association study of DSM-IV cannabis dependence. *Addict. Biol.* **2011**, *16*, 514–518. [CrossRef] [PubMed]

106. Kokura, K.; Sun, L.; Bedford, M.T.; Fang, J. Methyl-H3K9-binding protein MPP8 mediates E-cadherin gene silencing and promotes tumour cell motility and invasion. *EMBO J.* **2010**, *29*, 3673–3687. [CrossRef] [PubMed]

107. Bonnert, T.P.; McKernan, R.M.; Farrar, S.; le Bourdelles, B.; Heavens, R.P.; Smith, D.W.; Hewson, L.; Rigby, M.R.; Sirinathsinghji, D.J.; Brown, N.; *et al.* Theta, a novel γ-aminobutyric acid type a receptor subunit. *Proc. Natl. Acad. Sci. USA* **1999**, *96*, 9891–9896. [CrossRef] [PubMed]

108. Kobayashi, K.; Kuroda, S.; Fukata, M.; Nakamura, T.; Nagase, T.; Nomura, N.; Matsuura, Y.; Yoshida-Kubomura, N.; Iwamatsu, A.; Kaibuchi, K. P140Sra-1 (specifically rac1-associated protein) is a novel specific target for rac1 small GTPase. *J. Biol. Chem.* **1998**, *273*, 291–295. [CrossRef] [PubMed]

109. Schenck, A.; Bardoni, B.; Moro, A.; Bagni, C.; Mandel, J.L. A highly conserved protein family interacting with the fragile X mental retardation protein (FMRP) and displaying selective interactions with fmrp-related proteins FXR1P and FXR2P. *Proc. Natl. Acad. Sci. USA* **2001**, *98*, 8844–8849. [CrossRef] [PubMed]

110. Yang, Y.; Muzny, D.M.; Xia, F.; Niu, Z.; Person, R.; Ding, Y.; Ward, P.; Braxton, A.; Wang, M.; Buhay, C.; *et al.* Molecular findings among patients referred for clinical whole-exome sequencing. *JAMA* **2014**, *312*, 1870–1879. [CrossRef] [PubMed]

International Journal of
Molecular Sciences

MDPI

Article

Whole Exome Sequencing for a Patient with Rubinstein-Taybi Syndrome Reveals *de Novo* Variants besides an Overt *CREBBP* Mutation

Hee Jeong Yoo [1,2,†], Kyung Kim [3,4,7,†], In Hyang Kim [1], Seong-Hwan Rho [5], Jong-Eun Park [1], Ki Young Lee [4,7,‡], Soon Ae Kim [6], Byung Yoon Choi [2,8] and Namshin Kim [3,*]

[1] Department of Psychiatry, Seoul National University Hospital, Seongnam, Gyeonggi 463-707, Korea; hjyoo@snu.ac.kr (H.J.Y.); iambabyvox@snu.ac.kr (I.H.K.); bulls18@snu.ac.kr (J.-E.P.)

[2] Department of Psychiatry, Seoul National University, College of Medicine, Seoul 110-744, Korea; twinwif2@snu.ac.kr

[3] Epigenomics Research Center, Genome Institute, Korea Research Institute of Bioscience and Biotechnology, Daejeon 305-806, Korea; kkyung412@gmail.com

[4] Department of Biomedical Informatics, Ajou University, School of Medicine, Suwon 443-749, Korea; kiylee@ajou.ac.kr

[5] Simulacre Modeling Group, Seoul 140-897, Korea; shrho@simulacre.re.kr

[6] Department of Pharmacology, Eulji University College of Medicine, Daejeon 301-746, Korea; sakim@eulji.ac.kr

[7] Department of Biomedical Science, Ajou University Graduate School of Medicine, Suwon 443-749, Korea

[8] Department of Otolaryngology, Seoul National University Hospital, Seongnam, Gyeonggi 463-707, Korea

[*] Author to whom correspondence should be addressed; n@rna.kr/deepreds@kribb.re.kr; Tel.: +82-42-879-8162; Fax: +82-42-879-8493.

[†] These authors contributed equally to this work.

[‡] Deceased.

Academic Editor: Merlin G. Butler

Received: 29 December 2014; Accepted: 28 February 2015; Published: 11 March 2015

Abstract: Rubinstein-Taybi syndrome (RSTS) is a rare condition with a prevalence of 1 in 125,000–720,000 births and characterized by clinical features that include facial, dental, and limb dysmorphology and growth retardation. Most cases of RSTS occur sporadically and are caused by *de novo* mutations. Cytogenetic or molecular abnormalities are detected in only 55% of RSTS cases. Previous genetic studies have yielded inconsistent results due to the variety of methods used for genetic analysis. The purpose of this study was to use whole exome sequencing (WES) to evaluate the genetic causes of RSTS in a young girl presenting with an Autism phenotype. We used the Autism diagnostic observation schedule (ADOS) and Autism diagnostic interview revised (ADI-R) to confirm her diagnosis of Autism. In addition, various questionnaires were used to evaluate other psychiatric features. We used WES to analyze the DNA sequences of the patient and her parents and to search for *de novo* variants. The patient showed all the typical features of Autism, WES revealed a *de novo* frameshift mutation in *CREBBP* and *de novo* sequence variants in *TNC* and *IGFALS* genes. Mutations in the *CREBBP* gene have been extensively reported in RSTS patients, while potential missense mutations in *TNC* and *IGFALS* genes have not previously been associated with RSTS. The *TNC* and *IGFALS* genes are involved in central nervous system development and growth. It is possible for patients with RSTS to have additional *de novo* variants that could account for previously unexplained phenotypes.

Keywords: Rubinstein-Taybi syndrome (RSTS); Autism spectrum disorder (ASD); *de novo* variants; Tenascin C (*TNC*) gene; insulin-like growth factor-binding protein; acid labile subunit (*IGFALS*) gene; CREB (cAMP response element binding protein) binding protein (*CREBBP*) gene

Int. J. Mol. Sci. **2015**, *16*, 5697–5713

1. Introduction

Rubinstein-Taybi syndrome (RSTS, OMIM 180849, 613684), also known as the broad thumb-hallux syndrome, is a rare condition with a prevalence of 1 in 125,000–720,000 births [1,2]. The syndrome is characterized by a group of well-defined clinical features including characteristic facial, dental, and limb dysmorphology, and growth retardation [3]. RSTS can also affect multiple internal organs including the heart, kidney, eyes, ears, and skin [4]. Patients are known to have an increased risk of developing non-cancerous and cancerous tumors, leukemia, and lymphoma [5]. Mental retardation and learning difficulties are other typical findings of RSTS. The average IQ of the patients is 35–50, although cognitive functioning outside this range has also been documented [6].

The diagnosis of RSTS is essentially a clinical diagnosis based on the features described above [7]. Most RSTS cases are sporadic and are caused by *de novo* mutations [8]. Cytogenetic or molecular abnormalities can be detected in only 55% of patients with RSTS [9]. Missense/non-sense mutations in genes encoding the CREB-binding protein (*CREBBP*) or the E1A-binding protein (*EP300*) are the most common, although micro-deletions in chromosome 16p13.3 and other rare mutations have also been reported as genetic causes of this condition [8,10–15]. Nevertheless, genetic analyses for identifying mutations in RSTS have often yielded inconsistent results, probably because of inadequacies in the methods used for the analyses, and many mutations remain unconfirmed. Previous studies examining genetic mutations in RSTS have mainly used fluorescence *in situ* hybridization (FISH) [16,17] or comparative genomic hybridization (CGH) array analyses [14]. Whole genome or exome sequencing (WES) methods have rarely been used. The purpose of this study was to use WES to evaluate the genetic causes of RSTS in a young girl presenting with RSTS and Autism spectrum disorder (ASD).

2. Results

2.1. Physical Characteristics

The typical dysmorphic features of RSTS were observed in the patient's face and mouth: Microcephaly (head circumference of 32 cm (approximately 3rd to 5th percentile) at birth and 47.8 cm (below the 3rd percentile) at her current age (6.75 years), highly arched eyebrows, long eyelashes, downward-slanting palpebral fissures, a broad nasal bridge, a beaked nose with the nasal septum extending well below the alae, a pouting lower lip, mild micrognathia, and dental crowding.

The patient had big toes that were short and broad, and mild clinodactyly of the fifth digits of both hands. The second, third, and fourth digits of the right hand had short distal phalanges, and the third and fourth digits of the left hand had broad distal phalanges. She also had a curly toe on the fourth digit of both feet and hallux valgus in the right foot. X-ray radiographs showed bilateral pes varus deformities and bilateral big-toe soft tissue thickening. There were abnormal accessory physes on the side of the distal phalanx, as well as other abnormalities in the physes of the big toes of both feet. Hand X-ray radiographs showed delayed carpal bone ossification in both hands and prominent thickening of the soft tissue of all fingers.

Marked growth retardation and poor weight gain were prominent in this patient from birth until her latest evaluation. Her height was 106.3 cm (below the 3rd percentile) and her weight was 17.2 kg (3rd percentile). A videofluoroscopic swallowing study revealed difficulty in swallowing. Furthermore, serum prealbumin and zinc levels were low. Other physical findings in the patient included anomalies of the eye (congenital cataract and nasolacrimal duct obstruction) and skin (pilomatrixoma on the left periauricular area, hypertrichosis on the forehead, anterior chest, and back, as well as nevus depigmentosus on the right calf). Gross hematuria had previously been found at five years of age, although a kidney ultrasound examination showed no abnormal findings. Results of cardiac evaluation were within normal limits. The characteristic physical features and growth curve of the patient are shown in Figure 1.

2.2. Neuropsychiatric and Behavioral Features

The patient showed pronounced developmental delay with no meaningful vocabulary. She also showed moderate mental retardation with a full-scale intelligence quotient (IQ) score of 37 on the Korean Wechsler Preschool and Primary Scale for Intelligence (K-WPPSI) and a score of 45 on the Leiter International Performance Scale. Her adaptive functioning level was equivalent to a child of approximately 3–5 years of age. She had shown the typical behavioral characteristics of Autism from early infancy, including abnormalities in communication and social interaction, along with repetitive behavior and restricted interests. Her scores surpassed the diagnostic thresholds on all categories of both the ADOS and ADI-R. She presented the distinctive mannerism of repetitively waving her hand whenever emotionally disturbed. Scores for other questionnaires also supported the diagnosis of Autism including a score of 15 on the social communication questionnaire (SCQ) and a score of 113 on the social component of the social responsiveness scale (SRS). Further behavioral characteristics included marked hyperactivity, short attention span, withdrawal, and poor motor coordination. Frequent snoring was reported, although the possibility of obstructive sleep apnea was deemed low. Brain magnetic resonance imaging (MRI) showed mild pachygyria. Electroencephalogram (EEG) showed intermittent diffuse high-amplitude semirhythmic 1–1.5-Hz delta activity and one episode of diffuse 1.5-Hz rhythmic, synchronous bifrontal δ/ζ activity lasting up to 20 s.

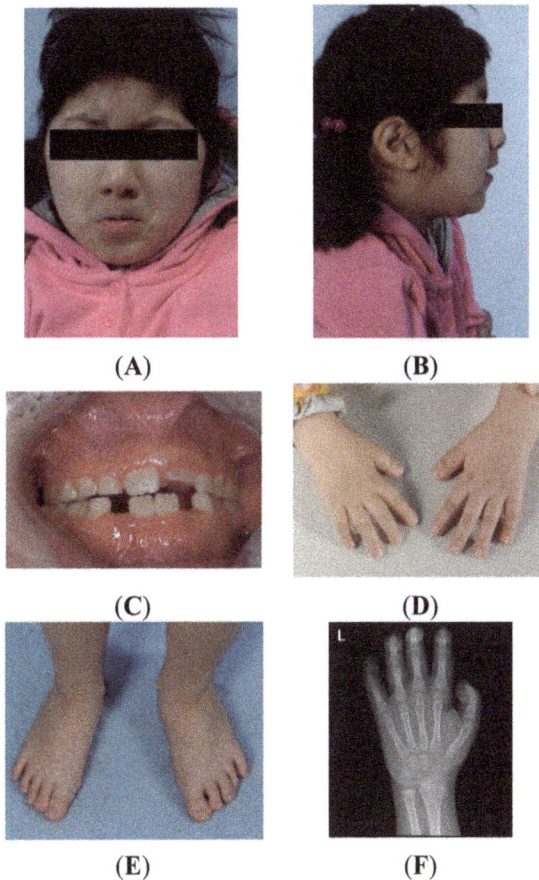

(A)   **(B)**

(C)   **(D)**

(E)   **(F)**

Figure 1. *Cont.*

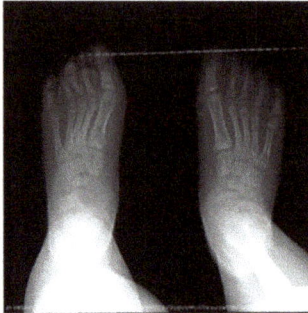

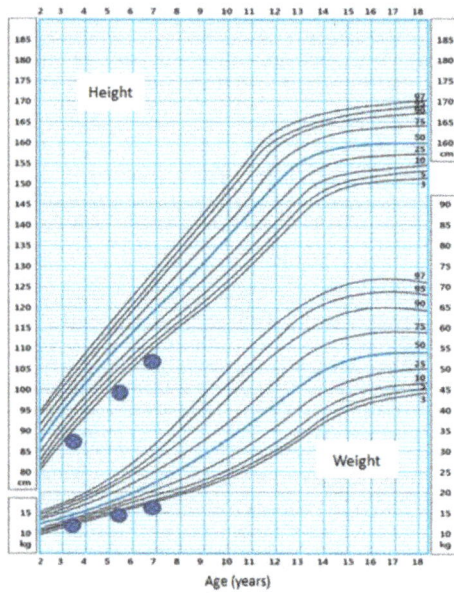

(G)   (H)

Figure 1. Physical characteristics of the proband. Patient's eyes are covered for reasons of confidentiality. Typical facial dysmorphic features can be observed in (**A,B**), including highly arched eyebrows, a broad nasal bridge, beaked nose with the nasal septum extending well below the alae, a pouting lower lip, and mild micrognathia. Dental crowding is shown in (**C**); The hand and feet abnormalities are shown in (**D,E**), showing short and broad big toes and thumbs, as well as mild clinodactyly on the 5th digit of both hands. The 2nd, 3rd, 4th digits of the right hand had short distal phalanges and the 3rd, 4th digits of the left hand had broad distal phalanges. The X-ray of the left hand shows delayed carpal bone ossification and prominent thickening of the soft tissue of all fingers (**F**); The X-ray of the feet shows bilateral pes varus deformities and abnormalities in the physes of big toe metatarsal bones (**G**); Growth curve (**H**) shows marked growth retardation in the proband (blue dot) superimposed on the normal growth curve of Korean girls of age 2 to 18 (black line). The growth curve shows the height and weight changes of the patient over time. These values never exceeded the third percentile since birth. Data for normal growth curve came from the Korea Center for Disease Control and Prevention.

2.3. Genetic Variants

Cytogenetic analyses showed normal karyotypes in the proband and both parents. Using in-depth statistics, we obtained a mean read-depth coverage of 120× over the targeted exome. Each exome contained more than 120,000 total variant calls. An average of 111 million individual paired-end reads of 100 bp were aligned to the human reference genome. More than 94% of targeted exon regions were covered with at least 10× depth, which is sufficient for variant discovery.

We found three *de novo* variants after filtering for all common variants in the dbSNP (except clinically associated variants) and an in-house control database comprising normal healthy individuals. We performed a visual inspection of read alignments using the Integrative Genomics Viewer (IGV) browser, and then removed "Benign" and "Likely Benign" variants after classification. A summary of the detected variants is described in Table S1.

Three *de novo* variants were validated by Sanger sequencing: a frameshift mutation in *CREBBP* (c.2199delG), and missense mutations in *TNC* (c.323G > A) and *IGFALS* (c.1415C > T). The results of family sequencing analyses are shown in Figure 2 and the properties of the *de novo* variants are summarized in Table 1 and Figure 2. The genetic variants of the proband and both parents are shown in Figure 3.

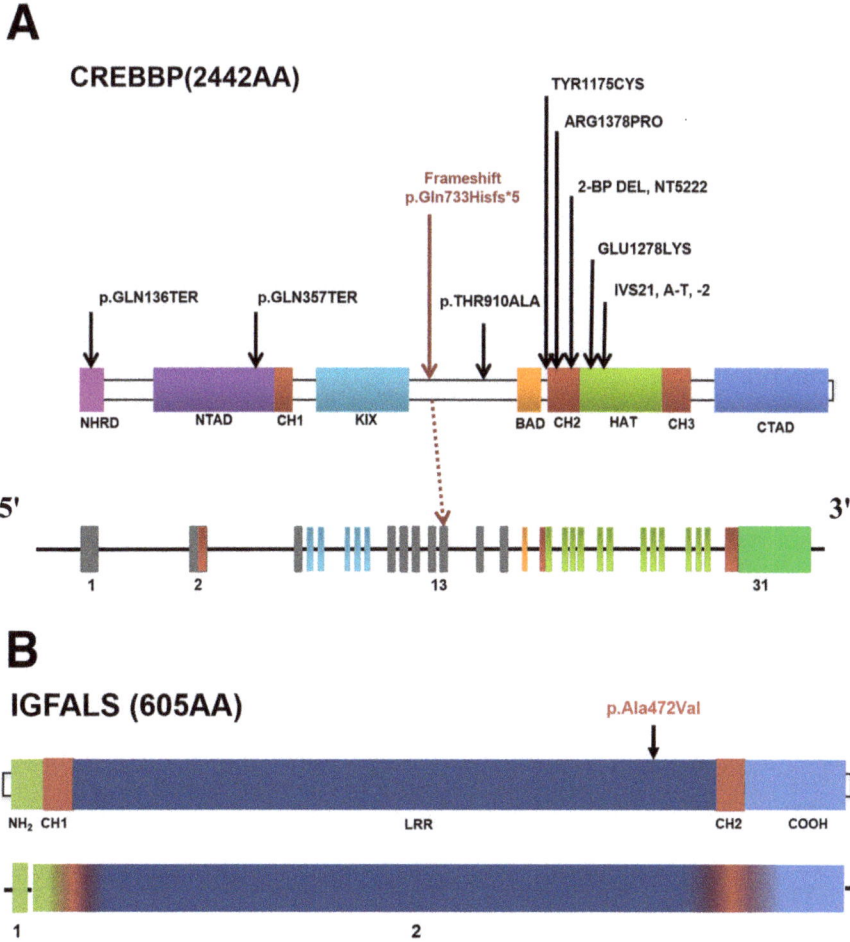

A

B

Figure 2. *Cont.*

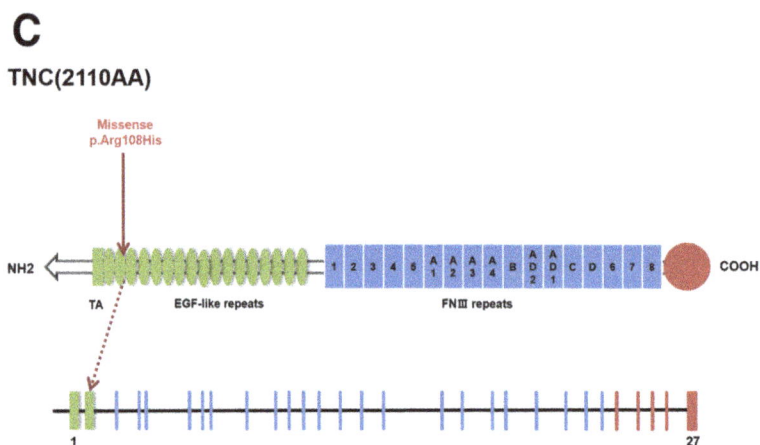

Figure 2. Schematic of *de novo* variants of *CREBBP*, *IGFALS*, and *TNC* in RSTS. (**A**) Genomic alterations and schematic drawing of proteins showing the predicted effects of selected missense mutations identified in *CREBBP*. The p.Thy1174Cys and p.Thr910Ala substitutions predicted by missense mutations do not affect known functional domains and are associated with a mild phenotype compared with the classic phenotype expected for mutations at other sites. The frameshift mutation p.Gln733Hisfs*5 is predicted to interrupt all of the following functional domains: BAD, CH2, HAT, CH3, and CTAD; (**B**) The *IGFALS* mutation site affects the LRR domain and (**C**) the *TNC* mutation one of the EGF-like repeats. The *de novo* variants are shown in red. CREBBP: CREB (cAMP response element-binding protein) binding protein; IGFALS: Insulin-like growth factor)-binding protein, acid labile subunit; TNC: Tenascin C; Protein domain names: (**A**) NHRD: Nuclear receptor-binding and receptor-interacting domain; NTAD: Amino-terminal transactivation domain; CH1: Cys/His-rich region; CREB/KIX: CREB-binding domain; BROMO/BRD: Bromo domain; HAT: Histone acetyltransferase domain; CH2: Cys/His-rich region 2; CH3: Cys/His-rich region 3; CTAD: C-terminal transactivation domain; (**B**) NH2: N-terminal; CH1: Cys-rich region 1; LRR: Leucine rich repeats; CH2: Cys-rich region 2; COOH: C-terminal; (**C**) TA: N-terminal tenascin assembly domain; EGF-like repeats: Epidermal growth factor-like repeats; FNIII repeats: Fibronectin type III-like repeats; FN Globe: C-terminal fibrinogen globe

Table 1. The three *de novo* variants validated by Sanger Sequencing.

Gene	*TNC*	*IGFALS*	*CREBBP*
Chromosome	chr9	chr16	chr16
Position	117852975	1841118	3823901
Reference Allele	C	G	C
Alternative Allele	T	A	–
Mutation Type	Missense	Missense	Frameshift
Sequence Variant	c.323G>A	c.1415C>T	c.2199delG
Protein Variant	p.Arg108His	p.Ala472Val	p.Gln733Hisfs*5
phyloP Score	2.369	0.598	1.492
GERP Score	1200.8	2536.4	476.2
SIFT Score	0	0.30	–
Mutation Assessor	1.87	0.995	–
Classification [1]	Likely pathogenic [2]	Variants of unknown significance	Pathogenic

[1] Variant Classification is based on the recommendations of Ambry Genetics, but pathogenicity of *de novo* variants in *TNC* and *IGFALS* are not certain; [2] *TNC* c.323G>A is reported in dbSNP as rs151119387, but MAF/MinorAlleleCount = 0.0002/1. Overall MAF in 1000 genomes, ESP6500, ExAc database is less than 0.0003. *TNC*: Tenascin C; *IGFALS*: Insulin-like growth factor-binding protein, acid labile subunit; *CREBBP*: CREB (cAMP response element-binding protein) binding protein; phyloP: Phylogenetic *p*-values; GERP: Genomic evolutionary rate profiling; SIFT: Sorting intolerant from tolerant.

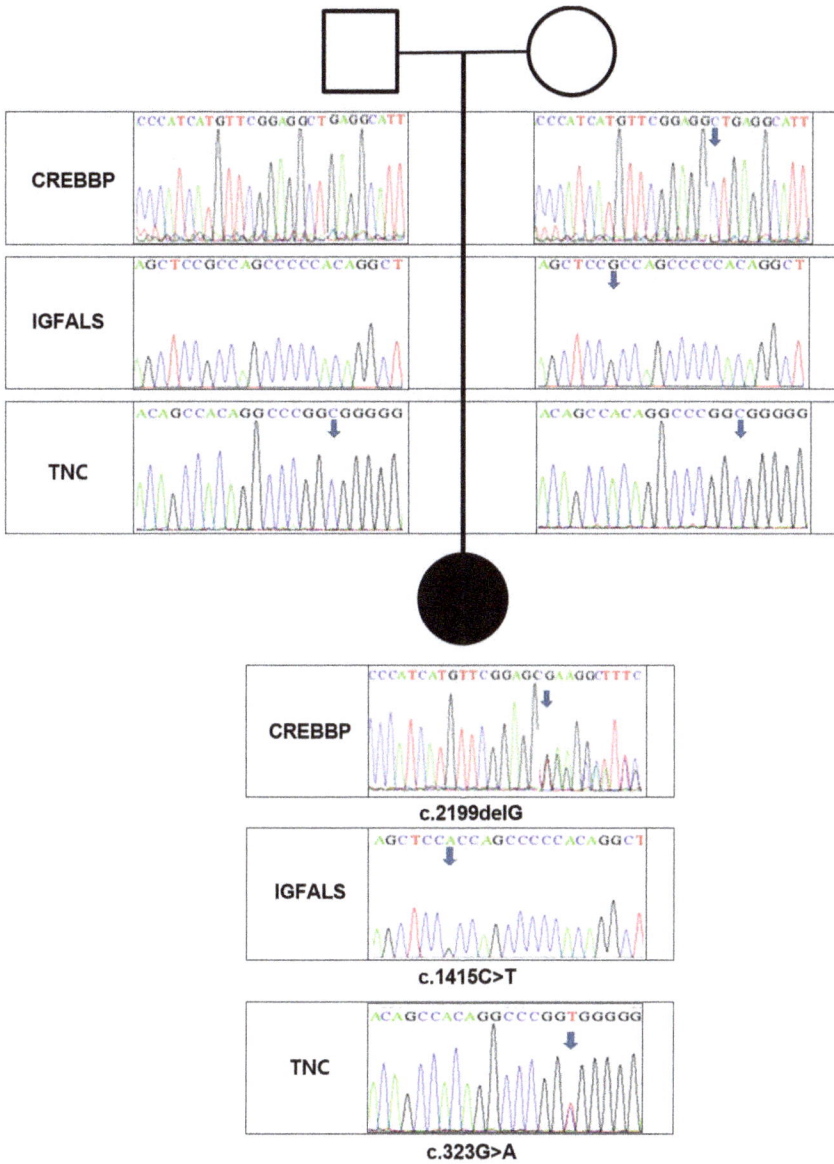

Figure 3. Sequences of *CREBBP, IGFALS,* and *TNC* for the family trio. A downward arrow indicates the mutated residue and nucleotides are shown in colored, single-letter codes. The black circle indicates the proband.

None of the observed sequence changes were observed in normal Korean subjects. The validation results from the normal population are shown in Figure 4.

CREBBP
chr16_3823901   GC/G

IGFALS
chr16_1841118   G/A

TNC
chr9_117852975   C/T

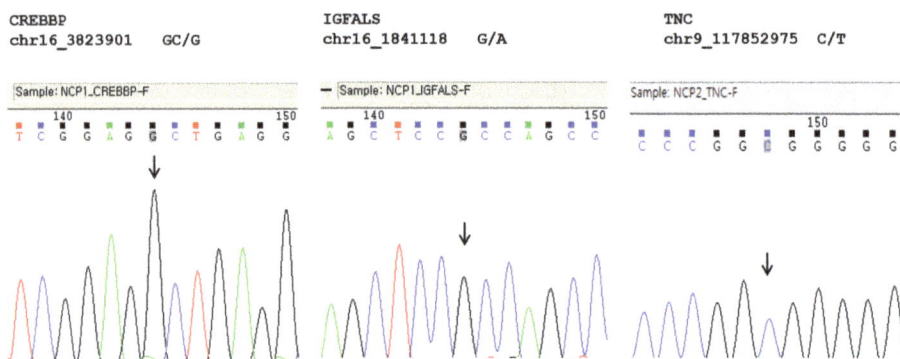

Figure 4. Validation results for *de novo* variants in *CREBBP*, *IGFALS*, and *TNC* in 80 normal Korean subjects. The downward arrow indicates the residue of *de novo* variants.

3. Discussion

We have shown that *de novo* variants in the *CREBBP*, *TNC*, and *IGFALS* genes are potential causes for RSTS by analyzing the whole genome of a female patient with RSTS. In WES analyses, we observed *de novo* variants in the *CREBBP*, *TNC*, and *IGFALS* genes. The girl showed typical morphological findings of RSTS, including abnormal features of the face, fingers, and toes, as well as growth retardation. Major characteristics of this patient were severe developmental/behavioral problems together with ASD with typical features that fulfilled the diagnostic criteria in multiple domains including communication, social interaction, specialized interests, and repetitive behavior.

Developmental/behavioral problems are common in patients with RSTS, with the most typical symptoms being moderate to severe mental retardation and language impairment [8,18]. There are no definitive reports regarding the specific frequency of psychiatric disorders in RSTS due to its low prevalence. In terms of autistic behavior, only a few reports have described this specific psychopathology associated with RSTS. In one relatively large-scale study exploring the socio-behavioral characteristics of RSTS, this syndrome was associated with behavioral problems such as a short attention span, motor stereotypies, and poor coordination [19]. These can also be part of the behavioral phenotype of ASD. The studies that have reported Autism-like phenotypes in patients with RSTS describe milder and atypical forms of ASD (*i.e.*, pervasive developmental disorder not otherwise specified; PDD NOS) in only a small proportion (8%) of the subjects [20–23].

Mutations in the *CREBBP* gene are one of the most commonly reported genetic etiologies of RSTS (Table S2). *CREBBP* mutations are found in 50%–60% of RSTS cases, whereas *EP300* (OMIM 602700) mutations are found in 5% of the cases [7,8]. RSTS is assumed to be a genetically heterogeneous disorder which is caused by mutations in the *CREBBP* gene as well as in other genes in up to 50% of cases [2]. The *CREBBP* mutations are heterogeneous, as 160 different mutations have been identified (Leiden Open Variation Database v.2.0 Build 35) [24]. The *CREBBP* gene is ubiquitously expressed and is involved in the transcriptional co-activation of many transcription factors [25]. *CREBBP* is also involved in multiple signaling pathways and in other cellular functions such as DNA repair, cell growth, differentiation, apoptosis, and tumor suppression [26]. The CREBBP protein plays an important role in regulating cell growth and division and is essential for normal fetal development. If one copy of *CREBBP* is deleted or mutated, cells make only half the normal amount of CREBBP protein. This reduction in the amount of this protein is thought to disrupt normal development before and after birth. The *CREBBP* mutation is known to be associated with more severe symptomatology than the *EP300* mutations [24]. Based on previous studies, the severe cognitive symptomatology of the subject of this study, including the ASD phenotype, could be attributed to the *CREBBP* mutation.

We can infer that the *de novo* sequence changes of the *TNC* and *IGFALS* genes, featuring as missense mutations observed in our study could potentially explain any additional phenotypes in patients with RSTS, but experimental validation is required to establish disease causation. Tenascin C (TNC), encoded by *TNC*, is a glycoprotein expressed in the extracellular matrix of various tissues during development, disease, or injury, as well as in restricted areas of the central nervous system involved in neurogenesis [27]. TNC expression changes during development until adulthood. In the developing central nervous system, it is involved in regulating the proliferation of both oligodendrocyte precursor cells and astrocytes [28]. TNC expression by radial glia precedes the onset of gliogenesis, and is thought to drive astrocyte differentiation. In the adult brain, TNC expression is downregulated in all areas except the hypothalamus and areas that maintain neurogenesis into adulthood [28]. Male-specific association signals have been found in the intronic region of the *TNC* gene [29]. A missense mutation c.5317G>A (p.V1773M) in exon 19 of the *TNC* gene was recently reported as a causative gene for nonsyndromic hearing loss in family-based exome sequencing and linkage analyses [30]. Based on the fact that pronounced-to-profound bilateral hearing loss or deafness is diagnosed more commonly in ASD than in the general population, the causality of deafness and ASD might have a partial overlap [31]. This study suggests that mutations in the *TNC* gene could contribute to sensory impairment in ASD, including hearing loss. The patient with RSTS in this study showed slight hearing problems (data now provided), but it is still not clear which specific aspect of the patient's phenotype can be attributed to the missense mutations in *TNC*.

To date, the presence of a genetic mutation in *CREBBP* has been linked with the behavioral phenotype in RSTS. For example, poor coordination is more prevalent in patients with RSTS and in subjects with genetic abnormalities in *CREBBP* than in those who have no abnormalities in this gene [19]. Severe cognitive impairments and autistic features have been reported in patients with large deletions in *CREBBP* [32]. This can be explained by the involvement of CREB-dependent gene transcription in long-term memory formation through signaling pathways that include multiple important kinases (protein kinase A, mitogen-activated protein kinase, and calcium/calmodulin-dependent kinase IV (CaMKIV)) in the nervous systems of many species [2]. The mental retardation and cognitive phenotype in patients with RSTS may be derived from impairments to the function of CREB-binding protein [2,13]. The role of the *CREBBP* mutation in long-term memory formation is supported by experiments showing that memory impairments in a $CREBBP^{+/-}$ mutant mouse model were ameliorated by phosphodiesterase 4 inhibitors [33]. Our observations suggest that *de novo* variants can have an additive effect and produce more severe and defined neurological/cognitive aberrations in subjects with RSTS.

The acid-labile subunit of insulin-like growth factor-binding protein (IGFALS) encoded by the *IGFALS* gene, is a serum protein that binds to IGFs. The IGFALS protein interacts with growth hormones to increase their half-life as well as their vascular localization [34]. *IGFALS*-dependent growth changes related to sex-specific body characteristics and bone size have been reported previously [35,36]. Defects in this gene cause acid-labile subunit deficiency which manifests as delayed and slow development during puberty [37]. The growth retardation in our patient, which has been present from birth until her current age, might be enhanced by the variants in the *IGFALS* gene.

One limitation of our study is that the analysis was restricted to only a single case. Validation of our results was also performed using relatively small samples from the general population. The precise function of the mutations and variants needs to be validated in future studies.

In conclusion, using WES, we observed *de novo* variants in the *CREBBP*, *TNC*, and *IGFALS* genes in a female patient with RSTS. To our knowledge, this is the first report suggesting that variants in the *TNC* and *IGFALS* genes can produce additional phenotypic effects. We assume that *de novo* variants in the *TNC* gene are related to the neurological phenotype and those in the *IGFALS* gene are related to growth retardation.

4. Methods

4.1. Ethics Statement

This study was approved by the Institutional Review Board (IRB) of the Seoul National University Bundang Hospital (Seongnam, Gyeonggi, Korea). Written informed consent was obtained from both parents of the patient.

4.2. Subject

The patient, a girl aged 6 years and 8 months, was examined in the child psychiatry clinic of the Seoul National University Bundang Hospital. She was thoroughly evaluated to assess problems related to delayed development and dysmorphology. A genetic assessment to diagnose RSTS, a possibility indicated during a previous diagnosis process, was also performed. The patient was the only child of a healthy non-consanguineous couple. Both biological parents were healthy with no clinically significant medical, developmental, or neuropsychological illnesses.

4.3. Clinical Evaluation

Thorough physical and laboratory evaluations were performed to confirm the diagnosis of RSTS and obtain detailed clinical and behavioral features. Physical, dental, ophthalmologic, and dermatologic examinations were performed. Chest and limb/digit X-ray radiography, urine and blood tests, a videofluoroscopic swallowing study, kidney ultrasonography, echocardiography, and electrocardiography were all performed as part of the laboratory analyses. Intelligence and adaptive functioning were assessed using the Korean version of the Wechsler preschool and primary scale of intelligence (K-WPPSI), the Leiter international performance scale [6,38], and the Vineland Adaptive Behavior scale (VABS) [39,40]. Two board-certified child psychiatrists evaluated the autistic symptoms. The Autism diagnostic observation schedule (ADOS), Module I [41,42], and the Autism diagnostic interview-revised (ADI-R) [43,44] were used, as well as the social communication questionnaire (SCQ) and social responsiveness scale (SRS) [45,46]. We measured the patients sleep quality using the Pittsburgh sleep quality index (PSQI) [47] and the "STOP" questionnaire (snoring, tiredness during daytime, observed apnea, high blood pressure) [48]. Brain magnetic resonance imaging (MRI) and electroencephalography (EEG) were performed to elucidate potential structural and functional abnormalities in the brain.

4.4. Genetic Analyses Using Bioinformatics Tools

Blood samples were drawn from the patient and both biological parents. Genomic DNA was extracted using the DNeasy Blood & Tissue kit (QIAGEN, Valencia, CA, USA). Raw reads in FASTQ format from exome sequencing were aligned to the hg19 reference genome from the UCSC genome browser using the Burrows–Wheeler Aligner (BWA) [49] with default parameters. Aligned reads were processed and polymerase chain reaction (PCR) duplicates were removed with SAMtools [50]. Single-nucleotide variants (SNVs) and insertions/deletions (indels) were identified using the Genome Analysis Toolkit (GATK) [51]. Regions near short indels were realigned using the IndelRealigner function in GATK according to default parameters. SNVs and indels affecting coding sequences or splicing sites were annotated by an in-house custom-made annotation system. All genomic changes were filtered against the Single Nucleotide Polymorphism Database (dbSNP; build 141) and the in-house control database comprising 54 Korean individuals.

We classified the *de novo* variants using the recommendations of Ambry Genetics (Scheme for AD and XD Mendelian disorders) into five categories: Benign, VLB (Variant, Likely benign), VUS (Variant, Unknown significance), VLP (Variant, Likely pathogenic), and Pathogenic mutation. Variants classified as VUS, VLP, and Pathogenic mutation were also confirmed using an IGV browser. To evaluate the mutations in the general population, we also applied Sanger sequencing for these mutations to 80 normal subjects belonging to the Korean ethnic group.

5. Conclusions

RSTS is a rare disorder characterized by facial, dental, and limb dysmorphology, growth retardation, and developmental disorders including intellectual disability and Autism spectrum disorder. Although many *de novo* mutations have been proposed as the genetic cause of RSTS, genetic analyses to identify mutations have often yielded inconsistent results due to inadequacies in the analysis methods, and many mutations are yet to be confirmed. The objective of this study was to investigate the genetic background of a young girl with RSTS presenting with an Autism phenotype, using whole exome sequencing (WES). We observed *de novo* variants in the *CREBBP*, *TNC*, and *IGFALS* genes. As far as the authors know, this is the first report to suggest that the *TNC* and *IGFALS* genes might contribute to clinical signs that are uncommon in RSTS patients.

Supplementary Materials: Supplementary materials can be found at http://www.mdpi.com/1422-0067/16/03/5697/s1.

Acknowledgments: A summary of this article was previously presented at the Annual Meeting of the American Society of Human Genetics, 2013, in Boston, MA, USA.
This work was supported by the Korea Healthcare Technology R&D project, Ministry of Health & Welfare, Republic of Korea [Grant Number A120029], Korean NRF grants (2011-0030049), and a grant from KRIBB Research Initiative Program.

Author Contributions: Hee Jeong Yoo and Seong-Hwan Rho recruited subjects. Hee Jeong Yoo and In Hyang Kim performed psychiatric and behavioral assessment of the subject. Namshin Kim, Hee Jeong Yoo and Ki Young Lee conceived and designed the experiments. Soon Ae Kim and Byung Yoon Choi performed the experiments. Kyung Kim and Namshin Kim analyzed the data. Kyung Kim, Hee Jeong Yoo, In Hyang Kim, Jong-Eun Park, and Namshin Kim wrote the paper.

Conflicts of Interest: The authors declare no conflict of interest.

References

1. Hennekam, R.C.; Stevens, C.A.; van de Kamp, J.J. Etiology and recurrence risk in Rubinstein-Taybi syndrome. *Am. J. Med. Genet. Suppl.* **1990**, *6*, 56–64. [PubMed]
2. Hallam, T.M.; Bourtchouladze, R. Rubinstein-Taybi syndrome: Molecular findings and therapeutic approaches to improve cognitive dysfunction. *Cell. Mol. Life Sci.* **2006**, *63*, 1725–1735. [CrossRef] [PubMed]
3. Rubinstein, J.H.; Taybi, H. Broad thumbs and toes and facial abnormalities. A possible mental retardation syndrome. *Am. J. Dis. Child.* **1963**, *105*, 588–608. [CrossRef] [PubMed]
4. Wiley, S.; Swayne, S.; Rubinstein, J.H.; Lanphear, N.E.; Stevens, C.A. Rubinstein-Taybi syndrome medical guidelines. *Am. J. Med. Genet. Part A* **2003**, *119A*, 101–110. [CrossRef]
5. Miller, R.W.; Rubinstein, J.H. Tumors in Rubinstein-Taybi syndrome. *Am. J. Med. Genet.* **1995**, *56*, 112–115. [CrossRef] [PubMed]
6. Hennekam, R.C.; Baselier, A.C.; Beyaert, E.; Bos, A.; Blok, J.B.; Jansma, H.B.; Thorbecke-Nilsen, V.V.; Veerman, H. Psychological and speech studies in Rubinstein-Taybi syndrome. *Am. J. Mental Retard.* **1992**, *96*, 645–660.
7. Hennekam, R.C. Rubinstein-Taybi syndrome. *Eur. J. Hum. Genet.* **2006**, *14*, 981–985. [CrossRef] [PubMed]
8. Roelfsema, J.H.; White, S.J.; Ariyurek, Y.; Bartholdi, D.; Niedrist, D.; Papadia, F.; Bacino, C.A.; den Dunnen, J.T.; van Ommen, G.J.; Breuning, M.H.; *et al.* Genetic heterogeneity in Rubinstein-Taybi syndrome: Mutations in both the *CBP* and *EP300* genes cause disease. *Am. J. Hum. Genet.* **2005**, *76*, 572–580. [CrossRef] [PubMed]
9. Bartsch, O.; Schmidt, S.; Richter, M.; Morlot, S.; Seemanova, E.; Wiebe, G.; Rasi, S. DNA sequencing of CREBBP demonstrates mutations in 56% of patients with Rubinstein-Taybi syndrome (RSTS) and in another patient with incomplete RSTS. *Hum. Genet.* **2005**, *117*, 485–493. [CrossRef] [PubMed]
10. Bentivegna, A.; Milani, D.; Gervasini, C.; Castronovo, P.; Mottadelli, F.; Manzini, S.; Colapietro, P.; Giordano, L.; Atzeri, F.; Divizia, M.T. Rubinstein-Taybi syndrome: Spectrum of *CREBBP* mutations in Italian patients. *BMC Med. Genet.* **2006**, *7*, 77. [CrossRef] [PubMed]
11. Negri, G.; Milani, D.; Colapietro, P.; Forzano, F.; Monica, M.D.; Rusconi, D.; Consonni, L.; Caffi, L.G.; Finelli, P.; Scarano, G.; *et al.* Clinical and molecular characterization of Rubinstein-Taybi syndrome patients carrying distinct novel mutations of the *EP300* gene. *Clin. Genet.* **2015**, *87*, 148–154. [CrossRef]

12. Spena, S.; Milani, D.; Rusconi, D.; Negri, G.; Colapietro, P.; Elcioglu, N.; Bedeschi, F.; Pilotta, A.; Spaccini, L.; Ficcadenti, A. Insights into genotype–phenotype correlations from *CREBBP* point mutation screening in a cohort of 46 Rubinstein-Taybi syndrome patients. *Clin. Genet.* **2014**. [CrossRef]

13. Petrij, F.; Giles, R.H.; Dauwerse, H.G.; Saris, J.J.; Hennekam, R.C.; Masuno, M.; Tommerup, N.; van Ommen, G.J.; Goodman, R.H.; Peters, D.J.; *et al.* Rubinstein-Taybi syndrome caused by mutations in the transcriptional co-activator CBP. *Nature* **1995**, *376*, 348–351. [CrossRef] [PubMed]

14. Tsai, A.C.-H.; Dossett, C.J.; Walton, C.S.; Cramer, A.E.; Eng, P.A.; Nowakowska, B.A.; Pursley, A.N.; Stankiewicz, P.; Wiszniewska, J.; Cheung, S.W. Exon deletions of the *EP300* and *CREBBP* genes in two children with Rubinstein-Taybi syndrome detected by aCGH. *Eur. J. Hum. Genet.* **2011**, *19*, 43–49. [CrossRef] [PubMed]

15. Kim, S.R.; Kim, H.J.; Kim, Y.J.; Kwon, J.Y.; Kim, J.W.; Kim, S.H. Cryptic microdeletion of the *CREBBP* gene from t(1;16) (p36.2;p13.3) as a novel genetic defect causing Rubinstein-Taybi syndrome. *Ann. Clin. Lab. Sci.* **2013**, *43*, 450–456. [PubMed]

16. Blough, R.I.; Petrij, F.; Dauwerse, J.G.; Milatovich-Cherry, A.; Weiss, L.; Saal, H.M.; Rubinstein, J.H. Variation in microdeletions of the cyclic AMP-responsive element-binding protein gene at chromosome band 16p13. 3 in the Rubinstein-Taybi syndrome. *Am. J. Med. Genet.* **2000**, *90*, 29–34. [CrossRef] [PubMed]

17. Gervasini, C.; Castronovo, P.; Bentivegna, A.; Mottadelli, F.; Faravelli, F.; Giovannucci-Uzielli, M.L.; Pessagno, A.; Lucci-Cordisco, E.; Pinto, A.M.; Salviati, L. High frequency of mosaic *CREBBP* deletions in Rubinstein–Taybi syndrome patients and mapping of somatic and germ-line breakpoints. *Genomics* **2007**, *90*, 567–573. [CrossRef] [PubMed]

18. Stevens, C.A.; Hennekam, R.C.; Blackburn, B.L. Growth in the Rubinstein-Taybi syndrome. *Am. J. Med. Genet. Suppl.* **1990**, *6*, 51–55. [PubMed]

19. Galera, C.; Taupiac, E.; Fraisse, S.; Naudion, S.; Toussaint, E.; Rooryck-Thambo, C.; Delrue, M.A.; Arveiler, B.; Lacombe, D.; Bouvard, M.P. Socio-behavioral characteristics of children with Rubinstein-Taybi syndrome. *J. Autism Dev. Disord.* **2009**, *39*, 1252–1260. [CrossRef] [PubMed]

20. Calì, F.; Failla, P.; Chiavetta, V.; Ragalmuto, A.; Ruggeri, G.; Schinocca, P.; Schepis, C.; Romano, V.; Romano, C. Multiplex ligation-dependent probe amplification detection of an unknown large deletion of the *CREB-binding protein* gene in a patient with Rubinstein-Taybi syndrome. *CEP* **2013**, *14025*, 220.

21. Hellings, J.A.; Hossain, S.; Martin, J.K.; Baratang, R.R. Psychopathology, GABA, and the Rubinstein-Taybi syndrome: A review and case study. *Am. J. Med. Genet.* **2002**, *114*, 190–195. [CrossRef] [PubMed]

22. Levitas, A.S.; Reid, C.S. Rubinstein-Taybi syndrome and psychiatric disorders. *J. Intellect. Disabil. Res.* **1998**, *42*, 284–292. [CrossRef] [PubMed]

23. Waite, J.; Moss, J.; Beck, S.R.; Richards, C.; Nelson, L.; Arron, K.; Burbidge, C.; Berg, K.; Oliver, C. Repetitive behavior in Rubinstein-Taybi syndrome: Parallels with Autism spectrum phenomenology. *J. Autism Dev. Disord.* **2014**. [CrossRef]

24. Lopez-Atalaya, J.P.; Valor, L.M.; Barco, A. Epigenetic factors in intellectual disability: The Rubinstein-Taybi syndrome as a paradigm of neurodevelopmental disorder with epigenetic origin. *Prog. Mol. Biol. Transl. Sci.* **2014**, *128*, 139. [PubMed]

25. Goodman, R.H.; Smolik, S. CBP/p300 in cell growth, transformation, and development. *Genes Dev.* **2000**, *14*, 1553–1577. [PubMed]

26. Marzuillo, P.; Grandone, A.; Coppola, R.; Cozzolino, D.; Festa, A.; Messa, F.; Luongo, C.; del Giudice, E.M.; Perrone, L. Novel cAMP binding protein-BP (*CREBBP*) mutation in a girl with Rubinstein-Taybi syndrome, GH deficiency, Arnold Chiari malformation and pituitary hypoplasia. *BMC Med. Genet.* **2013**, *14*, 28. [CrossRef] [PubMed]

27. Nies, D.E.; Hemesath, T.J.; Kim, J.-H.; Gulcher, J.R.; Stefansson, K. The complete cDNA sequence of human hexabrachion (Tenascin). A multidomain protein containing unique epidermal growth factor repeats. *J. Biol. Chem.* **1991**, *266*, 2818–2823. [PubMed]

28. Wiese, S.; Karus, M.; Faissner, A. Astrocytes as a source for extracellular matrix molecules and cytokines. *Front. Pharmacol.* **2012**, *3*, 120. [CrossRef] [PubMed]

29. Chang, S.C.; Pauls, D.L.; Lange, C.; Sasanfar, R.; Santangelo, S.L. Sex-specific association of a common variant of the *XG* gene with Autism spectrum disorders. *Am. J. Med. Genet. Part B* **2013**, *162*, 742–750. [CrossRef]

30. Zhao, Y.; Zhao, F.; Zong, L.; Zhang, P.; Guan, L.; Zhang, J.; Wang, D.; Wang, J.; Chai, W.; Lan, L.; *et al.* Exome sequencing and linkage analysis identified tenascin-C (*TNC*) as a novel causative gene in nonsyndromic hearing loss. *PLoS One* **2013**, *8*, e69549. [CrossRef] [PubMed]

31. Rosenhall, U.; Nordin, V.; Sandstrom, M.; Ahlsen, G.; Gillberg, C. Autism and hearing loss. *J. Autism Dev. Disord.* **1999**, *29*, 349–357. [CrossRef] [PubMed]

32. Schorry, E.; Keddache, M.; Lanphear, N.; Rubinstein, J.; Srodulski, S.; Fletcher, D.; Blough-Pfau, R.; Grabowski, G. Genotype–phenotype correlations in Rubinstein-Taybi syndrome. *Am. J. Med. Genet. Part A* **2008**, *146*, 2512–2519. [CrossRef]

33. Bourtchouladze, R.; Lidge, R.; Catapano, R.; Stanley, J.; Gossweiler, S.; Romashko, D.; Scott, R.; Tully, T. A mouse model of Rubinstein-Taybi syndrome: Defective long-term memory is ameliorated by inhibitors of phosphodiesterase 4. *Proc. Natl. Acad. Sci. USA* **2003**, *100*, 10518–10522. [CrossRef] [PubMed]

34. Leong, S.R.; Baxter, R.C.; Camerato, T.; Dai, J.; Wood, W.I. Structure and functional expression of the acid-labile subunit of the insulin-like growth factor-binding protein complex. *Mol. Endocrinol.* **1992**, *6*, 870–876. [PubMed]

35. Courtland, H.W.; DeMambro, V.; Maynard, J.; Sun, H.; Elis, S.; Rosen, C.; Yakar, S. Sex-specific regulation of body size and bone slenderness by the acid labile subunit. *J. Bone Miner. Res.* **2010**, *25*, 2059–2068. [CrossRef] [PubMed]

36. Domene, H.; Bengolea, S.; Jasper, H.; Boisclair, Y. Acid-labile subunit deficiency: Phenotypic similarities and differences between human and mouse. *J. Endocrinol. Investig.* **2004**, *28*, 43–46.

37. Domené, H.M.; Scaglia, P.A.; Lteif, A.; Mahmud, F.H.; Kirmani, S.; Frystyk, J.; Bedecarrás, P.; Gutiérrez, M.; Jasper, H.G. Phenotypic effects of null and haploinsufficiency of acid-labile subunit in a family with two novel *IGFALS* gene mutations. *J. Clin. Endocrinol. Metab.* **2007**, *92*, 4444–4450. [CrossRef] [PubMed]

38. Park, H.W.; Kwak, K.J.; Park, G.B. *Korean Wechsler Preschool and Primary Scale of Intelligence*; Special Education Publishing: Seoul, Korean, 2002.

39. Sparrow, S.; Balla, D.; Cichetti, D. *Vineland Adaptive Behavior Scales*; American Guidance Services: Circle Pines, MN, USA, 1984.

40. Aylward, G. *Practitioner's Guide to Developmental and Psychological Testing*; Springer: New York, NY, USA, 1994.

41. Lord, C.; Rutter, M.; DiLavore, P.D.; Risi, S. *Autism Diagnostic Observation Schedule*; Western Psychological Services: Los Angeles, CA, USA, 2001.

42. Lord, C.; Rutter, M.; Goode, S.; Heemsbergen, J.; Jordan, H.; Mawhood, L.; Schopler, E. Autism diagnostic observation schedule: A standardized observation of communicative and social behavior. *J. Autism Dev. Disord.* **1989**, *19*, 185–212. [CrossRef] [PubMed]

43. Le Couteur, A.; Lord, C.; Rutter, M. *The Autism Diagnostic Interview-Revised (ADI-R)*; Western Psychological Services: Los Angeles, CA, USA, 2003.

44. Lord, C.; Rutter, M.; Le Couteur, A. Autism diagnostic interview-revised: A revised version of a diagnostic interview for caregivers of individuals with possible pervasive developmental disorders. *J. Autism Dev. Disord.* **1994**, *24*, 659–685. [CrossRef] [PubMed]

45. Constantino, J.; Gruber, C.P. *Social Responsiveness Scale (SRS)*; Western Psychological Services: Los Angeles, CA, USA, 2005.

46. Rutter, M.; Bailey, A.; Lord, C. *Social Communication Questionnaire (SCQ)*; Western Psychological Services: Los Angeles, CA, USA, 2003.

47. Buysse, D.J.; Reynolds, C.F., III; Monk, T.H.; Berman, S.R.; Kupfer, D.J. The Pittsburgh sleep quality index: A new instrument for psychiatric practice and research. *Psychiatry Res.* **1989**, *28*, 193–213. [CrossRef] [PubMed]

48. Chung, F.; Yegneswaran, B.; Liao, P.; Chung, S.A.; Vairavanathan, S.; Islam, S.; Khajehdehi, A.; Shapiro, C.M. Stop questionnaire: A tool to screen patients for obstructive sleep apnea. *Anesthesiology* **2008**, *108*, 812–821. [CrossRef] [PubMed]

49. Li, H.; Durbin, R. Fast and accurate short read alignment with Burrows-Wheeler transform. *Bioinformatics* **2009**, *25*, 1754–1760. [CrossRef] [PubMed]

50. Li, H.; Handsaker, B.; Wysoker, A.; Fennell, T.; Ruan, J.; Homer, N.; Marth, G.; Abecasis, G.; Durbin, R.; Genome Project Data Processing Subgroup. The sequence alignment/map format and SAMtools. *Bioinformatics* **2009**, *25*, 2078–2079. [CrossRef] [PubMed]

51. McKenna, A.; Hanna, M.; Banks, E.; Sivachenko, A.; Cibulskis, K.; Kernytsky, A.; Garimella, K.; Altshuler, D.; Gabriel, S.; Daly, M.; *et al.* The genome analysis toolkit: A MapReduce framework for analyzing next-generation DNA sequencing data. *Genome Res.* **2010**, *20*, 1297–1303. [CrossRef] [PubMed]

52. Murata, T.; Kurokawa, R.; Krones, A.; Tatsumi, K.; Ishii, M.; Taki, T.; Masuno, M.; Ohashi, H.; Yanagisawa, M.; Rosenfeld, M,G.; *et al.* Defect of histone acetyltransferase activity of the nuclear transcriptional coactivator CBP in Rubinstein-Taybi syndrome. *Hum. Mol. Genet.* **2001**, *10*, 1071–1076. [CrossRef]

53. Bartsch, O.; Locher, K.; Meinecke, P.; Kress, W.; Seemanova, E.; Wagner, A.; Ostermann, K.; Rodel, G. Molecular studies in 10 cases of Rubinstein-Taybi syndrome, including a mild variant showing a missense mutation in codon 1175 of CREBBP. *J. Med. Genet.* **2002**, *39*, 496–501. [CrossRef] [PubMed]

54. Kalkhoven, E.; Roelfsema, J.H.; Teunissen, H.; den Boer, A.; Ariyurek, Y.; Zantema, A.; Breuning, M.H.; Hennekam, R.C.; Peters, D.J. Loss of CBP acetyltransferase activity by PHD finger mutations in Rubinstein-Taybi syndrome. *Hum. Mol. Genet.* **2003**, *12*, 441–450. [CrossRef] [PubMed]

55. Coupry, I.; Roudaut, C.; Stef, M.; Delrue, M.A.; Marche, M.; Burgelin, I.; Taine, L.; Cruaud, C.; Lacombe, D.; Arveiler, B. Molecular analysis of the *CBP* gene in 60 patients with Rubinstein-Taybi syndrome. *J. Med. Genet.* **2002**, *39*, 415–421. [CrossRef] [PubMed]

56. Kalkhoven, E. CBP and p300: HATs for different occasions. *Biochem. Pharmacol.* **2004**, *68*, 1145–1155. [CrossRef] [PubMed]

57. Udaka, T.; Samejima, H.; Kosaki, R.; Kurosawa, K.; Okamoto, N.; Mizuno, S.; Makita, Y.; Numabe, H.; Toral, J.F.; Takahashi, T.; *et al.* Comprehensive screening of CREB-binding protein gene mutations among patients with Rubinstein-Taybi syndrome using denaturing high-performance liquid chromatography. *Congenit. Anom. (Kyoto)* **2005**, *45*, 125–131. [CrossRef]

International Journal of
Molecular Sciences

MDPI

Article

Selection of Suitable Reference Genes for Analysis of Salivary Transcriptome in Non-Syndromic Autistic Male Children

Yasin Panahi [1], Fahimeh Salasar Moghaddam [2], Zahra Ghasemi [2], Mandana Hadi Jafari [1], Reza Shervin Badv [3,4], Mohamad Reza Eskandari [5,6] and Mehrdad Pedram [1,*]

[1]   Department of Genetics and Molecular Medicine, School of Medicine, Zanjan University of Medical Sciences (ZUMS), Zanjan 45139-56111, Iran; panahi.y@zums.ac.ir (Y.P.); mandana.hj@gmail.com (M.H.J.)
[2]   Department of Medical Biotechnology, School of Medicine, Zanjan University of Medical Sciences (ZUMS), Zanjan 45139-56111, Iran; f.moghaddam91@gmail.com (F.S.M.); zahra.ghasemi77@gmail.com (Z.G.)
[3]   Department of Pediatric Neurology, School of Medicine, Tehran University of Medical Sciences (TUMS), Tehran 14176-13151, Iran; badv@sina.tums.ac.ir
[4]   Children's Medical Center, Pediatric Center of Excellence, Tehran University of Medical Sciences (TUMS), Tehran 14176-13151, Iran
[5]   Metrowest CNS Research Center, Natick, MA 01760, USA; dr.eskandari@gmail.com
[6]   Department of Psychiatry, School of Medicine, Zanjan University of Medical Sciences (ZUMS), Zanjan 45139-56111, Iran
*   Correspondence: mpedram@zums.ac.ir; Tel./Fax: +98-24-3344-9553

Academic Editor: Stephen A. Bustin
Received: 9 August 2016; Accepted: 30 September 2016; Published: 12 October 2016

Abstract: Childhood autism is a severe form of complex genetically heterogeneous and behaviorally defined set of neurodevelopmental diseases, collectively termed as autism spectrum disorders (ASD). Reverse transcriptase quantitative real-time PCR (RT-qPCR) is a highly sensitive technique for transcriptome analysis, and it has been frequently used in ASD gene expression studies. However, normalization to stably expressed reference gene(s) is necessary to validate any alteration reported at the mRNA level for target genes. The main goal of the present study was to find the most stable reference genes in the salivary transcriptome for RT-qPCR analysis in non-syndromic male childhood autism. Saliva samples were obtained from nine drug naïve non-syndromic male children with autism and also sex-, age-, and location-matched healthy controls using the RNA-stabilizer kit from DNA Genotek. A systematic two-phased measurement of whole saliva mRNA levels for eight common housekeeping genes (HKGs) was carried out by RT-qPCR, and the stability of expression for each candidate gene was analyzed using two specialized algorithms, geNorm and NormFinder, in parallel. Our analysis shows that while the frequently used HKG *ACTB* is not a suitable reference gene, the combination of *GAPDH* and *YWHAZ* could be recommended for normalization of RT-qPCR analysis of salivary transcriptome in non-syndromic autistic male children.

Keywords: childhood autism; non-syndromic; transcriptome; saliva; reverse transcriptase quantitative real-time PCR (RT-qPCR); housekeeping genes (HKGs); reference gene; stability of expression; geNorm; NormFinder

1. Introduction

Autism spectrum disorders (ASD) are serious, lifelong pervasive neurodevelopmental disorders characterized by impairments in reciprocal social interaction, including verbal and non-verbal communication, and also repetitive stereotyped behavioral patterns [1]. Childhood autism, which is the most severe form of ASD with an early life onset, is typically diagnosed between the ages of 2 and

3 years [2] and a male to female ratio of about 4:1 [3]. There is no doubt that genetic components play a key role in the pathogenesis of ASD with a high heritability index. However, the genes that have been associated with ASD only relate to only a small portion of the cases [4].

Due to the wide range of genetic heterogeneity reported in ASD, it is quite possible that different causes and various underlying molecular mechanisms may lead to a common set of changes in the brain resulting in a similar behavioral profile [5,6]. Nonetheless, analysis of the transcriptome, as an intermediate phenotype, still could provide valuable insights for investigation of such complex cases. Careful comparison of the gene expression profiles in autistic patients with normal subjects could provide important clues for the development of autism and also the molecular mechanisms and crossroads involved. Reverse transcriptase quantitative real-time PCR (RT-qPCR) is a highly sensitive technique for investigation of gene expression profile, and it is also used for confirmation of microarray results [7,8]. However, accurate analysis of gene expression with RT-qPCR requires proper normalization to stably expressed internal control or reference gene(s) [9]. An important caveat to consider for selection of reference genes, which are typically selected from housekeeping genes (HKGs), is that expression of HKGs could vary significantly under different biological conditions including disease states, age, sex, drug treatment. Furthermore, the stability of expression for a particular HKG could also be dependent upon the tissue type/source used for RNA isolation and experimental design [10,11]. Therefore, selection of appropriate reference genes for a complex neurodevelopmental disorder such as autism requires very careful design and planning.

A review of the literature shows that no single gene or particular set of genes has been used consistently as reference in gene expression studies in ASD, using either the brain or blood as the main RNA source (Table 1) [12–34]. Surprisingly, as we will further outline in the discussion, almost none of these studies have embarked on a systematic validation of the reference gene(s) prior to the start of the experiment. In the present study, we: (1) propose using saliva as a readily available biofluid and an alternative noninvasive sampling source for transcriptome analysis in childhood autism; and (2) provide a systematic evaluation for the expression levels of eight commonly used HKGs in whole saliva samples in drug naïve non-syndromic autistic male children, as a highly important yet defined subset of ASD patients, and healthy sex-, age-, and location-matched controls.

Table 1. Internal control genes used in ASD gene expression studies.

Reference	RNA/Protein Source	Reference Gene(s)	Age (Year)	N	Sex
Purcell et al. (2001) [12]	Brain tissue Cerebellum	ACTIN [WB] GAPDH [RP]	5–54	10	M/F [1]
Fatemi et al. (2001) [13]	Brain tissue Cerebellum	ACTIN [WB]	23.5 ± 4.8	5	M
Araghi-Niknam et al. (2003) [14]	Brain tissue Cerebellum, Frontal Cortex	ACTIN [WB]	23.8 ± 4.9	5	M
Samaco et al. (2004) [15]	Brain tissue Frontal Cortex	*TFRC* and *ACTIN* [IF] *GAPDH, H1,* and *TFRC* [FISH]	2–32	13	M/F [8]
Hu et al. (2006) [16]	LCL	*18s rRNA* [qP]	6–16	10	M
Nishimura et al. (2007 [17])	LCL	*HPRT1* [qP]	AGRE	42	M
Garbett et al. (2008) [18]	Brain tissue Sup. temporal gyrus	*ACTB* [MA−qP]	4–30	6	M/F [2]
Gregg et al. (2008) [19]	Whole blood	*DDR1, RPL37A,* and *SMARC2* [MA−qP]	CHARGE	35	M/F [6]
Enstrom et al. (2009) [20]	PBL, Natural killer cells	*DDR1, RPL37A,* and *SMARC2* [MA−qP]	2.3–5.6	52	M
Hu et al. (2009a) [21]	LCL	*MDH1, ARF1, ACSL5* [MA−qP]	12.3 ± 3.7	116	M
Hu et al. (2009b) [22]	LCL	*18s rRNA* [MA−qP]	7.8 ± 3.4	20	NA
Sheikh et al. (2010) [23]	Brain tissue Frontal Cortex	ACTIN [WB]	8.9 ± 3.2	9	M/F [4]
Malik et al. (2011) [24]	PBL	ACTIN [WB, NB]	8.4 ± 0.3	6	NA
Kuwano et al. (2011) [25]	Whole blood	*GAPDH* and *HDAC1* [MA−qP]	26.7 ± 5.5	21	M/F [4]

Table 1. *Cont.*

Reference	RNA/Protein Source	Reference Gene(s)	Age (Year)	N	Sex
Ghahramani Seno et al. (2011) [26]	LCL	*TMEM32* [qP] *TUBB* [WB]	2–18	20	M/F [7]
Luo et al. (2012) [27]	LCL	*GAPDH* [MA–qP]	SSC	42	NA
Chow et al. (2012) [28]	Brain tissue Prefrontal Cortex	*ACTB, TBP*, and *RPL13A* [MA–qP]	2–56	20	M
Kong et al. (2012) [29]	Whole blood	*GAPDH* [MA–qP]	8.2 ± 3.0	170	M
Griesi-Oliveira et al. (2012) [30]	Dental pulp stem cells	*GAPDH, HPRT1, SDHA*, and *HMBS* [qP]	10	1	F
Anitha et al. (2012) [31]	Brain tissue Anterior cingulated gyrus, Motor Cortex, thalamus	*GAPDH* [WB] *B2M, HPRT1, RPL13A, GAPDH*, and *ACTB* [qP]	8–29	8	M/F [2]
Ginsberg et al. (2012) [32]	Brain tissue Occipital and cerebellar hemispheric cortices	*GAPDH* [qP]	2–60	9	M/F
Choi et al. (2014) [33]	Brain Tissue Cerebellum	*GAPDH* [qP]	4.5–82	29	M/F [8]
Nardone et al. (2014) [34]	Brain Tissue Anterior cingulated gyrus Prefrontal Cortex	*GAPDH, HPRT1, POLR2α*, and *SDHA* [qP]	18–51	13	M/F [2]

N, number of ASD samples; NA, not available; F [N], number of female ASD samples; Regular capitalized abbreviations under the Reference Gene(s) columns are indicative of proteins, whereas the italicized ones denote RNA or mRNA. Sample sources: AGRE, Autism Genetic Resource Exchange; CHARGE, Childhood Autism Risks from Genetics and Environment; SSC, Simon Simplex Collection. Cell lines: LCL, Lymphoblastoid Cell Line; PBL, Peripheral Blood Lymphocyte. Techniques: FISH, Fluorescence in situ Hybridization; IF, Immunofluorescence; MA-qP, Microarray-RT-qPCR; qP, Reverse transcriptase quantitative real-time PCR; NB, Northern Blot; RP, semi-quantitative RT-PCR; WB, Western Blot.

2. Results

2.1. Selection of Candidate Reference Genes and Initial Screening

The present study was prompted by an earlier experiment, during which saliva samples were obtained from five lab members and β-actin gene (*ACTB*) mRNA levels were measured for RNA quality analysis. It should be noted that *ACTB* has been used previously by other investigators for RNA quality and transcriptome analysis by RT-qPCR in saliva [35–38], and that it also has been used as a common internal control for gene expression studies in ASD (Table 1). Interestingly, while *ACTB* Cq values were stable in the salivary transcriptome derived from the lab members (i.e., adults), the Cq values varied when examined in five samples (one autistic and four healthy control subjects) taken from children (Table S1).

There were limited amounts of saliva samples taken from our clinical groups, collected within a two-year period, available for future studies. Thus, in order to systematically evaluate potential reference genes for normalization of gene expression levels in whole saliva collected from patients with non-syndromic childhood autism and the healthy controls, a two-phase analysis was performed. In the first phase, for initial screening, mRNAs levels of eight candidate HKGs (Table 2) were examined by RT-qPCR using equal amounts of RNA templates extracted from four saliva samples: two samples taken from autistic patients, and two from healthy controls. The RT-qPCR procedures were designed and performed in line with the Minimum Information for Publication of Quantitative Real-Time PCR Experiments (MIQE) guidelines [10]. The mean Cq values ± SD ($n = 2$–4) for each candidate gene are presented on Table 3 (see Table S2 for detailed individual average Cq values of replicate qPCR reactions ± SD).

Table 2. List of the housekeeping genes evaluated as candidate reference genes during the first phase.

Gene Name	Function	Gene Symbol	mRNA Accession No.	Amplicon [2] Length (bp)
18S ribosomal RNA	Ribosomal RNA Subunit	*18s rRNA*	M10098 [1]	99 *
β-actin	Cytoskeletal structural protein	*ACTB*	NM_001101	94
Glyceraldehyde-3-phosphate dehydrogenase	Glycolytic enzyme	*GAPDH*	NM_002046	142 *
Ribosomal protein L13a	Structural component of large subunit of Ribosome	*RPL13A*	NM_012423	223 *
Succinate dehydrogenase complex subunit A, flavoprotein	Electron transporter in the Krebs cycle	*SDHA*	NM_004168	154 *
Transferrin receptor	Cellular iron uptake	*TFRC*	NM_003234	134
Ubiquitin C	Protein degradation	*UBC*	NM_021009	192 *
Tyrosine 3 monooxygenase activation protein, zeta polypeptide	Signal transduction	*YWHAZ*	NM_003406	150 *

[1] Non-coding RNA; [2] The numbers with asterisks given for *18s rRNA, GAPDH, RPL13A, SDHA, UBC,* and *YWHAZ* do not reflect exact Amplicon sizes. They are indicative of the "context sequence length" [39] information provided by PrimerDesign Ltd. (Southampton, UK).

To analyze the stability of expression for the candidate HKGs, the Cq values of each data set were exported to a Microsoft Office Excel sheet and then imported into the geNorm (qbasePLUS software v2.2, Biogazelle, Ghent, Belgium) and also NormFinder (Molecular Diagnostic Labrotory, Department of Clinical Biochemistry Molecular Medicine, Aarhus University Hospital, Skejby, Aarhus, Denmark) software. GeNorm, which is a strong algorithm for small sample sizes and also the most widely used algorithm to determine the most stable reference gene, calculates an expression normalization factor for each sample set by geometric averaging of a number of user-defined candidate reference genes. For each candidate gene, the geNorm algorithm determines an internal control stability measure (M), which is defined as the average pairwise variation of a particular gene compared with all other test genes. The lowest M value indicates the most stable gene [40]. By contrast, NormFinder that is also a visual basic application for Microsoft Excel focuses on the expression variation of each gene with the least intra- and inter-group values in a model-based approach. In addition to giving a direct measure for the estimated expression variation, the NormFinder algorithm evaluates the systematic error introduced when using the HKG of interest as internal control [41].

Table 3. The mean Cq values ($\pm$SD, n = 2–4) of samples studied in phase I.

Gene Symbol	Control			Autism			Overall SD
	Mean	SD	N	Mean	SD	N	
18s rRNA	16.06	1.78	2	14.58	0.66	2	1.04
ACTB	25.87	0.38	2	29.18	1.77	2	2.34
GAPDH	26.27	0.29	2	27.01	0.44	2	0.52
RPL13A	31.09	0.89	2	32.72	NA	1	1.15
SDHA	33.73	NA	1	28.93	0.32	2	3.40
TFRC	31.14	1.73	2	33.22	1.58	2	1.47
UBC	30.22	0.62	2	32.23	0.59	2	1.42
YWHAZ	27.54	0.61	2	28.07	1.11	2	0.37

N, number of samples (from independent individuals) with valid signals; NA, not applicable; SD, standard deviation.

According to the geNorm algorithm, the M values below the cut-off value of 1.5 (the recommended threshold) are acceptable in the selection of internal controls for normalization in RT-qPCR [40]. As it can be seen from Figure 1A, the data output from geNorm for the first phase of evaluation indicated that glyceraldehyde-3-phosphate dehydrogenase gene (*GAPDH*) was the best HKG in our panel of candidate reference genes followed by succinate dehydrogenase complex subunit A, flavoprotein

(*SDHA*), tyrosine 3 monooxygenase activation protein, zeta polypeptide (*YWHAZ*), and ubiquitin C (*UBC*). *ACTB*, a traditionally common reference gene in normalization of RT-qPCR data in ASD studies (Table 1) and salivary transcriptome analysis [35,37,38], was ranked among the least stable genes. The number of reference genes needed for optimal normalization was determined by geNorm based on the average pairwise variation (Vn/n + 1) below the recommended cut-off point value of 0.15. At this phase, the geNorm output indicated that the combination of the best three reference genes would be sufficient for normalization of RT-qPCR data (Figure S1).

The ranking of the best candidate genes by NormFinder algorithm in phase I was somewhat similar to the geNorm output. *GAPDH*, *YWHAZ*, and *UBC* were ranked as the best candidates, respectively (Figure 1B). However, *SDHA* was demoted to the 6th rank, while *ACTB* was promoted two steps from low to the intermediate ranking. It should be noted that, in phase I, the clinical identifier value was not included in the NormFinder analysis due to the small number of samples, and thus analysis was performed as a single clinical group. In both algorithms, two of the candidate genes, encoding for ribosomal protein L13a (RPL13A) and transferrin receptor (TFRC), were ranked as the least stable HKGs based on their high variation of relative expression values.

2.2. Expression Profiling and Validation of Candidate Reference Genes

Three of the candidate HKGs that had the best scores in phase I based on both geNorm and NormFinder algorithms (*GAPDH*, *YWHAZ*, and *UBC*) were selected for the second phase of analysis. Despite its low-intermediate ranking in phase I, *ACTB* was also taken for further analysis because of its common usage as an internal control HKG in saliva [35,38,42] and ASD studies (Table 1). In this phase, the expression levels of the final four candidate HKGs were examined in saliva samples from a panel of nine autistic and healthy boys (average age: 47.6 ± 21.9 vs. 41.2 ± 26.7 months, respectively). The amplification efficiency for each gene and linear regression coefficient (R2) values were calculated by performing standard curves (Table 4). The scatter plot representations of the *Cq* values recorded for the final four candidate HKGs in the autistic and healthy clinical groups are shown in Figure 2 (Table S3 for detailed average *Cq* values for replicate qPCR reactions ±SD).

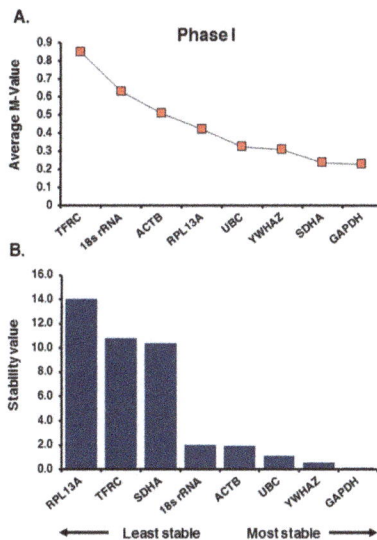

Figure 1. Expression stability analysis of the candidate reference genes in phase I. The stability of expression for 8 housekeeping genes tested in saliva is ranked (left to right) based on the least stable to the most stable gene by: (**A**) geNorm; and (**B**) NormFinder algorithms.

Table 4. Amplification efficiency data for candidate reference genes tested in phase II.

Gene Symbol	Slope	R^2	Efficiency (%)
ACTB	−3.400	0.981	96.8
GAPDH	−3.303	0.989	100.9
UBC	−3.747	0.974	84.8
YWHAZ	−3.398	0.984	97.2

Standard curves were used to calculate the reaction efficiency for each gene, using a five-point serial dilution made from pooled stock cDNAs. The amplification efficiency percentage for each gene was calculated using the equation $E = 10^{(-1/\text{Slope})} - 1$ multiplied by 100, where E stands for the efficiency and slope is the gradient of the best fit line. R^2: linear regression coefficient.

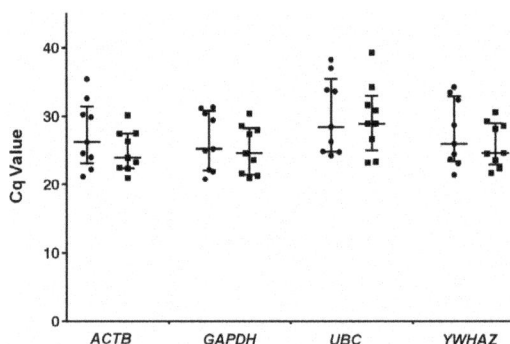

Figure 2. Scatter plot representations of the *Cq* values measured for the final four candidate genes. Filled squares and circles represent the control and autistic groups, respectively. The vertical bars indicate the median and interquartile range. The bottom and top of each bar represent the 25th and 75th percentiles, respectively, and the line bisecting each bar shows the median *Cq* value.

Noticeably, geNorm once again ranked *GAPDH* as the most stable HKG, and it ranked *ACTB* as the least stable gene (*M* values: 1.29 vs. 1.84, respectively). However, after the pairwise analysis, the combination of *GAPDH* and *YWHAZ* was selected as the best pair with an *M*-value of 1.02, followed by *UBC* and *ACTB* with *M*-values of 1.45 vs. 1.67, respectively (Figure 3A). An optimal number for reference genes was not determined by geNorm algorithm because the *V*-value was higher than recommended threshold (0.15). Similar to the geNorm analysis, NormFinder evaluated *GAPDH* as the most stable HKG, followed by *YWHAZ*, *UBC*, and *ACTB* (Stability values: 0.35, 0.60, 1.18, and 1.40, respectively). However, when the clinical subgroups were included in the analysis, the *GAPDH* and *YWHAZ* set was recommended as the best combination of two genes by NormFinder, followed by *UBC* (ranked third) and *ACTB*, ranked as the least stable gene once again (Figure 3B).

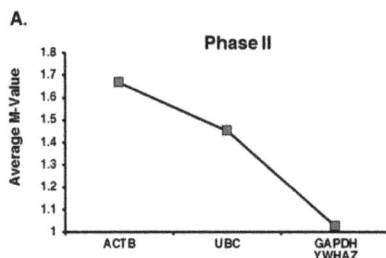

Figure 3. *Cont.*

B.

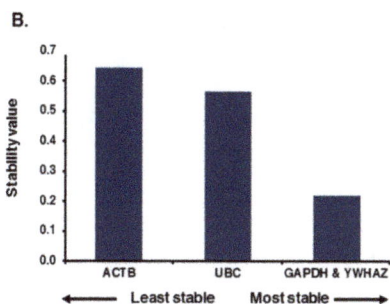

Figure 3. Expression stability analysis for the final four candidate reference genes in phase II. Expression of each candidate gene was examined in a total of 18 saliva samples (nine with childhood Autism and nine healthy controls) ranked compared with the rest by (**A**) geNorm and (**B**) NormFinder algorithms based on their stability of expression values; when the clinical group's identifier was included in the NormFinder analysis, *GAPDH* and *YWHAZ* were selected as the best genes combined, in agreement with the geNorm pair-wise ranking.

2.3. Relative Expression Levels of Candidate Reference Genes in Autistic vs. Healthy Subjects

Using *GAPDH*, the most stable HKG ranked by both geNorm and NormFinder algorithms, as reference, and employing an improved $2^{-\Delta Ct}$ method [43] (by taking into account the amplification efficiencies), the relative expression levels of other three genes were compared in the autistic children vs. the control group (Figure 4A). The results demonstrate that although *ACTB* shows a higher degree of variation compared with *UBC* and *YWHAZ* (*p*-values: 0.26, 0.45 and 0.82, respectively), as expected, there are no significant differences in the expression levels of any of these HKGs between the two clinical groups. By contrast, when *ACTB*, which was ranked as the least stable, was used as reference, the expression levels of *GAPDH*, *UBC*, and *YWHAZ* did not follow a normal distribution pattern, and in particular, the *UBC* expression levels and median values were significantly different (*p* = 0.03) between the autistic and control groups (Figure 4B).

The above observations are in support of the ranking validity of the candidate reference genes by geNorm and NormFinder algorithms, and they clearly demonstrate how selection of an inappropriate HKG as reference can bias interpretation of RT-qPCR data. Interestingly, examination of the *Cq* values obtained for the candidate HKGs using a third statistical algorithm applet, called BestKeeper [44], produced a similar ranking for the candidate reference genes, in particular for Phase II (Table S4). It should be noted that the BestKeeper algorithm works by calculating the SD for each gene of interest and Pearson's coefficient of correlation for each pair of the candidate genes. Thus, it would be best suited for our Phase II analysis, as BestKeeper cannot provide significant *p*-values for small sample sets (Table S4A,B). Nonetheless, based on the intra-variations for each candidate gene, the Phase I ranking by BestKeeper is not that far off from those produced by geNorm and NormFinder.

A.

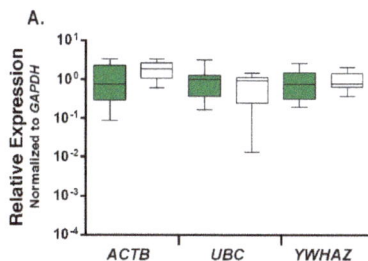

Figure 4. *Cont.*

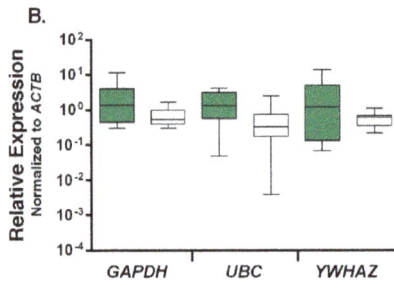

Figure 4. Comparison of the relative expression levels of candidate reference genes between the two groups. Colored and white boxes represent the autistic and control groups, respectively. The relative expression level for each gene was calculated by using either: (**A**) the *GAPDH* (glyceraldehyde-3-phosphate dehydrogenase); or (**B**) ACTB (β-actin) gene as the reference. Statistical analysis comparing the expression levels are done with Mann-Whitney *U* test (two-tailed), and the data are presented as the mean and interquartile range, with vertical bars representing the total range (**B**). *p*-values: (**A**) *ACTB* ($p = 0.26$), *UBC* ($p = 0.45$), *YWHAZ* ($p = 0.82$); and (**B**) *GAPDH* ($p = 0.26$), *UBC* ($p = 0.03$), *YWHAZ* ($p = 0.26$).

3. Discussion

To our knowledge, this is the first systematic evaluation of HKGs as reference for RT-qPCR analysis of salivary transcriptome in patients with autism reported so far. ASD encompass a heterogeneous set of complex and pervasive behaviorally-defined neurodevelopmental disorders, which brings up the possibility of different causes leading to similar brain deficits and behavioral patterns. In fact, there is a strong consensus on the involvement of various genetic and environmental components in the etiology of ASD [1,6]. Nonetheless, because of the high heritability reported in ASD, it should be quite plausible to search for a "common signature" at the level of gene expression profile. Thus, analysis of the transcriptome, as an intermediate phenotype for investigation, could provide insights into the molecular mechanism/s involved [5]. However, despite enormous number of research and concerted efforts at the international level during the past decade, as evident on Table 1, there is still an incoherent picture of a common gene expression profile signature/s for ASD, and the underlying molecular mechanisms are poorly understood.

A recent highly in depth meta-analysis of ASD gene expression data, taken from 12 independent studies in blood and brain tissues, by Ch'ng et al. conclude that constructing a somewhat consistent transcriptome signature might be possible in the studies using a small number of brain samples. By contrast, the outcomes of the studies using blood cells are highly heterogeneous and rather inconsistent [5]. As pointed out by the investigators, part of the inconsistency in these studies could be due to the variation in their inclusion and exclusion criteria, which include variations in the scope of the ASD patients studied, differences in gender and age groups, and differences in sampling and experimental procedures. A review of the information presented on Table 1 supports the notion of such variations in different ASD studies. Another important factor to note, however, is the inconsistency in the use of reference genes in different studies, even in cases where the type of tissue/cells used for transcriptome analysis were the same or very similar.

Selection of appropriate reference gene(s) plays an essential part in proper normalization of RT-qPCR assays and comparison of gene expression profiles between different samples and clinical groups. It is also essential that validity and stability of expression for the candidate or intended internal control HKG/s is experimentally examined for particular source/s of RNA, disease status, and experimental design prior to the start of each experiment [10,40,41]. Surprisingly, a careful examination of the ASD gene expression studies listed on Table 1, however, reveals that the HKGs used as internal controls were not systematically evaluated for their stability of expression for the particular tissue/cell type, and biological and experimental conditions. Perhaps, the only one exception on that list is the case report study by Griesi-Oliveira et al. using dental pulp stem cells for their main RNA source. Aside from the fact that the particular female patient in this study suffered from a craniofacial dysmorphism in addition to ASD phenotype, it should be noted that because of the very small and restricted sample size (only one patient and two controls from the same family), validation of the reference genes in this study has a very low power. It is also interesting to note that even with such a small and restricted sample size, when the investigators used the geNorm applet for selection of the reference genes, they ended up with a set of four HKGs in order to generate a normalization factor [30]. The study by Nardone et al., using postmortem brain tissues, used a set of four HKGs based on a geNorm evaluation of candidate reference genes by a separate group of investigators on suicide subjects [34,45]. Finally, the study by Garbett et al., using a small set (six; four male and two female subjects) of postmortem brain tissue samples, selected only one HKG, *ACTB*, for validation of their microarray results based on the previous works by the same lab [18] on small numbers of postmortem brain samples from schizophrenic and epileptic patients [46,47].

An ideal reference gene should maintain a stable mRNA level in all sample groups under different experimental conditions. It might not be possible to find a "universal" reference gene or set of genes in different tissues for various biological and experimental conditions. However, it is quite feasible to find the most appropriate HKGs for normalization of gene expression profiling if one limits the type and number of tissues tested under a well-defined experimental setting [40,41]. As mentioned several times already, ASD encompass a heterogeneous and complex set of neurodevelopmental disorders. In the present study, in an effort to limit the number of confounders and heterogeneity, we confined the scope of the ASD subjects to drug naïve non-syndromic autistic male children and age-, and sex-matched healthy children picked from the same neighborhoods as the patients (average ages: 47.6 ± 21.9 vs. 41.2 ± 26.7 months, respectively). Considering the severity of childhood autism and also the high ratio of male children affected, the autistic patients used in this study represent a highly important yet defined subset of ASD patients. We also used saliva as an alternative "noninvasive" sampling source for RNA instead of the blood. As the most available and portable biofluid, saliva contains a wealth of information regarding the physiological states of the body and is a good indicator of the plasma levels of hormones, drugs, immune system, and neurological conditions [48,49]. In recent years, saliva has been used a valid source of biomarkers and transcriptome analysis for a number of systemic human diseases including sleep disorder [38] and breast cancer [50].

Saliva has also been used as a DNA source for epigenetic analysis in bipolar disorder and schizophrenia [51], for genetic analysis and investigation of the impact of nutritional intake and environmental toxins in ASD [52]. Most recently, Frank Middleton and colleagues used the salivary microRNA profiles for identification of children with ASD [53]. Here, we present the first systematic evaluation of HKGs in the salivary transcriptome to be used for normalization of RT-qPCR data in ASD research. Based on our analysis of gene expression stability using geNorm and NormFinder algorithms in parallel, *GAPDH* was determined as the most stable HKG to be used as an internal control for salivary transcriptome analysis and/or validation of microarray results by RT-qPCR in male childhood autism. By contrast, *ACTB* was evaluated as unsuitable to be used as a reference gene. It should be noted, however, that *GAPDH* does not appear to be stable enough to be used as a single reference gene for normalization of RT-qPCR data in this setting. Both geNorm and Normfinder analyses recommend combination of *GAPDH* and *YWHAZ* as the best option. However, the V value given by the geNorm algorithm in our second phase of study is above the ideal 0.15. A simple interpretation of the results from geNorm's point of view would entails using all four candidate genes as internal controls. However, considering the high variation of *ACTB* expression, and significantly low levels of *UBC* expression between the autistic and control samples (Figure S1 and Figure 4), it would not be advisable to add either of them to the *GAPDH* and *YWHAZ* combination.

The poor evaluation of *ACTB* as a reference gene in saliva samples, in a sharp contrast to *GAPDH*, initially came to us as a surprise. However, a careful review of the literature revealed this to be a well-supported outcome by various lines of evidence. While *ACTB* was traditionally used as a reference gene in saliva [38], even pioneer labs specializing in saliva work have recently switched to using *GAPDH* instead [50,54]. In contrast to this notion, in a recent report by Pandit and colleagues on a high-yield RNA-extraction protocol for saliva, the investigators used *ACTB* in combination with the saliva-specific histatin 3 gene (*HTN3*) for validation of the quality and quantity of their isolated RNA samples. They also used *ACTB* alone as their reference HKG for analysis of 2 cell lines and a cohort of cancer patients and controls [35]. However, a review of the Cq values listed for *ACTB* in their Table 1, shows a very high variation in the Cq values (18 Cq differences for all samples, and 13 Cq differences within the clinical samples only) for *ACTB*, making it unsuitable to be used as a reference gene. Furthermore, it should be kept in mind that the saliva samples taken in our study were obtained from young children (average age for both groups: 44.4 ± 23.9 months) that are going through a highly significant growth period. A key developmental and cytoskeletal protein coding gene like *ACTB* may not be suited as a reference gene for this age period. In support of this line of reasoning, a recent investigation of HKGs in leukocyte subpopulations from children by Yu and colleagues found that *GAPDH* had the most stable expression in children (five samples: 1, 3, 6, 9, and 12 years old). By contrast, the ACTB protein levels were inconsistent. The investigators reported that *ACTB* mRNA levels were not significantly different across their samples and propose a difference at the level of post-translational regulation. However, there is no mention of the *ACTB* Cq values/variation provided in the article [55].

In contrast to our findings indicating *ACTB* as an unsuitable internal control for RT-qPCR analysis in childhood autism, the study by Garbett and colleagues gives *ACTB* a high praise as a reference gene for transcriptome analysis in postmortem brain tissue samples from ASD patients [18]. Aside from the fact that we have used a different source for RNA, a close inspection of the report by Garbett et al. reveals some interesting points. The authors provide several reasons for why they chose *ACTB* as the reference gene for RT-qPCR verification of their microarray results, including that it had been previously established as a stable HKG in the literature (referring to a study by Chen et al. on the primary human skin fibroblasts [56]); and in two previous studies on human postmortem brain tissues by the same lab, one in subjects with schizophrenia [46] and one in epilepsy [47]. Aside from the fact that we have used a different source of RNA in the present study compared with the studies noted above, the reasoning outlined by Garbett et al. for selection of their internal control HKG is not convincing. The fact that *ACTB* has been previously used as an endogenous reference gene in primary human skin fibroblasts (a different tissue type), and that it has been used in the brain tissue samples in schizophrenia and epilepsy (different disease types), are not valid justifications for using *ACTB* as a reference gene in brain tissue samples for ASD studies. Two recent systematic evaluations of HKGs, with both studies including *ACTB* and *GAPDH*, for RT-qPCR in postmortem human brain tissues came up with entirely different sets of reference genes. While Silberberg and colleagues found *TFRC* and *RPLP0* as the most stable reference genes in schizophrenia and bipolar disorders [57], Penna and colleagues found CYC1 and EIF4A2 as the best reference genes in Alzheimer's disease [58]. Finally, it is interesting to note that a review of the reference genes used in ASD studies shows a noticeable shift from *ACTB* to *GAPDH* during the past 15 years (Table 1).

4. Materials and Methods

4.1. Subjects and Sampling Procedure

Participants recruited to this study included nine drug naïve non-syndromic male children diagnosed with autism (47.6 ± 21.9 months) and nine age-, gender-, and location-matched typically developing healthy controls (41.2 ± 26.7 months). Children with autism were diagnosed according to the Diagnostic and Statistical Manual of Mental Disorders fourth edition, Text Revised (DSM-IV-TR) criteria by two qualified clinicians, a child neurologist and a child and adolescent psychiatrist. The diagnosis of autism was confirmed by Autism Diagnostic Interview-Revised (ADI-R), which is a standardized semi-structured diagnostic algorithm for autism based on the definitions set by DSM-IV and International Statistical Classification of Diseases and Related Health Problems 10th Revision (ICD-10) [59,60]. Exclusion criteria for patients included any other medical disorder and/or diagnosis including significant neurological problems such as depression, epileptic seizures, and genetic syndromes such as tuberous sclerosis, Angelman, and Fragile-X. Sex-matched healthy control children were picked from the same neighborhood locations as the autistic subjects. All healthy controls were screened by the Strengths and Difficulties Questionnaire (SDQ) [61]. The study was conducted in accordance with the Declaration of Helsinki, and it had the approval of the ZUMS Ethics Committee (ZUMS.REC.1392 97; 13 October 2013). All subjects had signed informed consent, provided by their parents, for inclusion before participating in the study.

Two mL of whole saliva was collected from each child who was able to provide sputum into the Oragene RNA collection kit RE-100 (DNA Genotek Inc., Ottawa, ON, Canada). Saliva sampling was supervised and collected by trained personnel according to the instructions provided in the collection kit. In the case of children unable to spit voluntarily, sampling was done by a 2-mL syringe from under the tongues and gingival crevices, mainly with the help of the participating parents and/or caregivers. Briefly, children were recommended to refrain from eating and drinking for 1 h prior to sampling and also wash their mouths with drinking water at least 15 min before providing a sample. Once collected, the tube containing the sample was covered by placing the cap securely and inverting the container vigorously by hand vortexing for approximately 10 s, in order to allow the saliva to mix well with the Oragene RNA solution. Samples were then stored at $-20\ ^\circ C$ and subsequently transferred to the central laboratory at the Zanjan University of Medical Science (ZUMS) for further processing.

4.2. Total RNA Preparation and cDNA Synthesis

Total RNA was extracted from whole saliva samples using a combination of two protocols: Oragene RNA collection kit RE-100 (DNA Genotek Inc., Ottawa, ON, Canada) and RNX-Plus solution (SinaClone BioScience, Tehran, Iran) with some modifications to the manufacturers' instructions. Briefly, the saliva collected in the Oragene RNA collection kit was incubated in a water bath at 50 $^\circ$C for 1 h. A 500-μL aliquot was then transferred into a 1.5-mL microcentrifuge tube and incubated in a water bath at 90 $^\circ$C for 15 min. Twenty μL of the Neutralizer solution was added to the sample and vortexed for a few seconds, incubated on ice for 10 min, and centrifuged at 15,000$\times$ *g* for 3 min. The clear supernatant was then transferred into a new microcentrifuge tube, 2 volumes of cold 95% ethanol was added to the sample and mixed vigorously by hand. The sample was incubated at $-20\ ^\circ$C for 30 min followed by centrifugation at 15,000$\times$ *g* for 3 min. The supernatant was discarded and the pellet was dissolved in 750 μL of RNX-Plus solution (SinaClone BioScience, Tehran, Iran) by vortexing, followed by incubation at room temperature for 5 min. A 200-μL volume of Chloroform was added to the solution and centrifuged for 15 min at 12,000$\times$ *g*. The upper phase was then transferred to a new microcentrifuge tube and an equal volume of isopropanol was added into the tube. The mixture was centrifuged for 15 min at 12,000$\times$ *g* and the resulting pellet was washed in 70% ethanol, air dried for 10 min, and dissolved in DEPC-treated ddH_2O. The quality and quantity of extracted RNA samples were analyzed by fiber optic spectrophotometry using a NanoDrop 2000c (Thermo Scientific, Waltham, MA, USA). While RNA quantity was measured using the A_{260} absorbance in DEPC-treated ddH_2O, the RNA purity was monitored by A_{260}/A_{280} absorbance ratios of sample dilutions in Tris pH 7.4. RNA samples with A_{260}/A_{280} ratios of 1.8–2.1 were used for cDNA in the study. The average concentration of the RNA samples used in the study was 220 $\pm$ 132 ng/μL. Potential residual genomic DNA was eliminated with RNase-free DNase I digestion (Thermo Scientific). Complementary DNA (cDNA) was synthesized from 500-ng total DNase I treated RNA per reaction using the PrimScriptTM Reagent Kit (Takara Inc., Shiga, Japan) following the manufacturer's instructions. Briefly, in a 10-μL reaction with 1$\times$ reaction buffer including MgCl$_2$ inside an RNase-free tube, the appropriate volume of RNA sample was first treated with 0.5 μL of DNase I (1 U/μL) by gently mixing the reaction and incubating it at 37 $^\circ$C for 30 min, followed by inactivation of DNase I with addition of 1 μL of 50 mM

EDTA and incubation at 65 °C for 10 min. This DNase I treated RNA sample (11 μL) was then used for cDNA synthesis in a 20-μL reaction containing 4 μL of 5× PrimeScript™ buffer, 1 μL of PrimeScript™ RT Enzyme mix I, 1 μL of 50 μM Oligo dT primer, 1 μL of 100 μM Random 6 mers, and 2 μL of RNase-free ddH$_2$O. The RT reaction was mixed well and incubated at 37 °C for 15 min, followed by inactivation of the RT enzyme at 85 °C for 5 s. The cDNA samples were kept at −20 °C for further use.

4.3. Candidate Reference Gene Selection

A total of 8 commonly used housekeeping genes with a wide range of biological functions were used in this study (see Table 2 for details). Six of the eight genes were picked from the internal control panel of genes offered by PrimerDesign Ltd. (Southampton, UK) with the following "anchor nucleotide" and "context sequence length" [39] information for the primer sets: *18s rRNA* (235, 99 bp), *GAPDH* (1087, 142 bp), *RPL13A* (727, 223 bp), *SDHA* (1032, 154 bp), *UBC* (452, 192 bp), and *YWHAZ* (2585, 150 bp). The remaining two genes included *ACTB*, and *TFRC* with the following forward and reverse primer sequences: ACTB_F, 5′-CGAGCACAGAGCCTCGCCTTTGCC-3′, ACTB_R, 5′-TGTCGACGACGAGCGCGGCGATAT-3′; and TFRC_F, 5′-ACCGGCACCATCAAGCT-3′, TFRC_R, 5′-TGATCACGCCAGACTTTGC-3′.

4.4. Quantitative Real-Time PCR

The mRNA levels of the potential reference genes were quantified using Micro Amp Optical 8-Cap Strips (Life Technology, Pittsburgh, PA, USA) with an ABI 7300 Real-Time PCR System (Applied Biosystems, Foster City, CA, USA). Each 20-μL reaction contained 10 μL of RealQ Plus 2× PCR Master Mix Green, High ROX TM (Ampliqon Inc., Odense, Denmark), 25 ng of cDNA, and 300 nM of forward and reverse primers. PCR amplifications were performed in triplicates with an initial enzyme activation step at 95 °C for 10 min, followed by 40 cycles of 95 °C for 15 s (template denaturation), and 60 °C for 1 min (annealing and data collection). No-template controls (NTC) were included in each run. The inter-run technical variation was calculated as SD for three HKGs Cq values for a fixed repeated cDNA sample dilutions: *ACTB*, ±0.13; *GAPDH*, ±0.15; *YWHAZ*, ±0.14.

4.5. Gene Expression Stability Analysis

Analysis of expression stability for the candidate reference genes was initially carried out by using two commonly used statistical algorithms, geNorm (qbasePLUS software v2.2, Biogazelle) [40] and NormFinder (an Excel Add-in, provided by Molecular Diagnostic Labrotory, Dept. of Molecular Medicine, Aarhus University Hospital, Skejby Sygehus, Denmark) [41], both of which are Visual Basic Applications for Microsoft Excel, in parallel. The raw Cq values in an Excel spread sheet were also fed into the BestKeeper applet (verison 1, http://www.gene-quantification.com/bestkeeper.html) [44] for additional analysis.

5. Conclusions

To our knowledge, the present study is the first systematic evaluation of candidate HKGs as internal controls for normalization of gene expression profiles in autistic children. Based on the analysis of eight commonly used HKGs by geNorm and NormFinder algorithms in parallel, a combination of the top two most stable candidate reference genes, *GAPDH* and *YWHAZ*, could be recommended for normalization of RT-qPCR analysis of salivary transcriptome in drug naïve non-syndromic autistic male children. We hope that the information presented here paves the way for: (1) systematic evaluation of reference genes in ASD studies under different experimental settings; and (2) using saliva as an alternative source of RNA, instead of the brain and blood, for transcriptome analysis in ASD research. As a readily-available and resourceful biofluid, saliva could open a noninvasive window to the complex molecular world of autism not only for diagnosis but also for surveillance of disease status and treatment.

Supplementary Materials: Supplementary materials can be found at www.mdpi.com/1422-0067/17/10/1711/s1.

Acknowledgments: This work was supported in part by Zanjan University of Medical Sciences (ZUMS) grant number A-12-534-1/-6-8. We would like to thank Khadijeh Babaei from the High Institute for Research and Education in Transfusion Medicine, Blood Transfusion Center, Tehran, for the preliminary set up of total RNA extractions from saliva. We are also thankful to Hamid Pezeshk (Department of Mathematics, Statistics, and Computer Science, University of Tehran) and Mohammad H. Rahbar (Department of Internal Medicine, UT Health Science Center at Houston) for their helpful comments and suggestions.

Author Contributions: Mehrdad Pedram and Yasin Panahi designed the study, and Mehrdad Pedram supervised the project. Yasin Panahi carried out the RT-qPCR experiments and did the geNorm and NormFinder analysis. Reza Shervin Badv and Mohammad Reza Eskandari were involved in the discussion and design of the scope of ASD patients for the study, and they also took care of the clinical examination of the patients, diagnosis, and clinical follow-ups. Fahimeh Salasar Moghaddam, Zahra Ghasemi, and Mandana Hadi Jafari did the saliva sample collection and processing. Yasin Panahi and Mehrdad Pedram analyzed the data. Yasin Panahi, Fahimeh Salasar Moghaddam, and Mandana Hadi Jafari wrote the initial draft of the manuscript, and Zahra Ghasemi provided the first draft of Table 1. Mehrdad Pedram wrote and revised the full manuscript for submission. All authors read and approved the final manuscript.

Conflicts of Interest: The authors declare no conflict of interest.

Abbreviations

ACTB, β-actin protein; *ACTB*, β-actin gene; ASD, Autism spectrum disorders; Cq, Quantification Cycle; *GAPDH*, Glyceraldehyde-3-phosphate dehydrogenase gene; HKG, House Keeping Gene; MIQE, Minimum Information for Publication of Quantitative Real-Time PCR Experiments; RT-qPCR, Reverse Transcriptase Quantitative real-time PCR; *UBC*, Ubiquitin C gene; *YWHAZ*, Tyrosine 3 monooxygenase activation, zeta polypeptide gene.

References

1. Lai, M.C.; Lombardo, M.V.; Baron-Cohen, S. Autism. *Lancet* **2014**, *383*, 896–910. [CrossRef]
2. Lord, C.; Risi, S.; DiLavore, P.S.; Shulman, C.; Thurm, A.; Pickles, A. Autism from 2 to 9 years of age. *Arch. Gen. Psychiatry* **2006**, *63*, 694–701. [CrossRef] [PubMed]
3. Werling, D.M.; Geschwind, D.H. Sex differences in autism spectrum disorders. *Curr. Opin. Neurol.* **2013**, *26*, 146–161. [CrossRef] [PubMed]
4. Betancur, C.; Coleman, M. Etiological Heterogeneity in Autism Spectrum Disorders: Role of Rare Variants. In *The Neuroscience of Autism Spectrum Disorders*; Joseph, D., Buxbaum, P.R.H., Eds.; Academic Press: Waltham, MA, USA, 2013; pp. 113–144.
5. Ch'ng, C.; Kwok, W.; Rogic, S.; Pavlidis, P. Meta-analysis of gene expression in autism spectrum disorder. *Autism Res.* **2015**, *8*, 593–608. [CrossRef] [PubMed]
6. Berg, J.M.; Geschwind, D.H. Autism genetics: Searching for specificity and convergence. *Genome Biol.* **2012**, *13*, 247–263. [CrossRef] [PubMed]
7. Morey, J.S.; Ryan, J.C.; van Dolah, F.M. Microarray validation: Factors influencing correlation between oligonucleotide microarrays and real-time PCR. *Biol. Proced. Online* **2006**, *8*, 175–193. [CrossRef] [PubMed]
8. Nolan, T.; Hands, R.E.; Bustin, S.A. Quantification of mrna using real-time RT-PCR. *Nat. Protoc.* **2006**, *1*, 1559–1582. [CrossRef] [PubMed]
9. Huggett, J.; Dheda, K.; Bustin, S.; Zumla, A. Real-time RT-PCR normalisation; strategies and considerations. *Genes Immun.* **2005**, *6*, 279–284. [CrossRef] [PubMed]
10. Bustin, S.A.; Benes, V.; Garson, J.A.; Hellemans, J.; Huggett, J.; Kubista, M.; Mueller, R.; Nolan, T.; Pfaffl, M.W.; Shipley, G.L.; et al. The MIQE guidelines: Minimum information for publication of quantitative real-time PCR experiments. *Clin. Chem.* **2009**, *55*, 611–622. [CrossRef] [PubMed]
11. Li, R.; Shen, Y. An old method facing a new challenge: Re-visiting housekeeping proteins as internal reference control for neuroscience research. *Life Sci.* **2013**, *92*, 747–751. [CrossRef] [PubMed]
12. Purcell, A.E.; Jeon, O.H.; Zimmerman, A.W.; Blue, M.E.; Pevsner, J. Postmortem brain abnormalities of the glutamate neurotransmitter system in autism. *Neurology* **2001**, *57*, 1618–1628. [CrossRef] [PubMed]
13. Fatemi, S.H.; Stary, J.M.; Halt, A.R.; Realmuto, G.R. Dysregulation of reelin and Bcl-2 proteins in autistic cerebellum. *J. Autism Dev. Disord.* **2001**, *31*, 529–535. [CrossRef] [PubMed]

14. Araghi-Niknam, M.; Fatemi, S.H. Levels of Bcl-2 and p53 are altered in superior frontal and cerebellar cortices of autistic subjects. *Cell. Mol. Neurobiol.* **2003**, *23*, 945–952. [CrossRef] [PubMed]
15. Samaco, R.C.; Nagarajan, R.P.; Braunschweig, D.; LaSalle, J.M. Multiple pathways regulate MeCP2 expression in normal brain development and exhibit defects in autism-spectrum disorders. *Hum. Mol. Genet.* **2004**, *13*, 629–639. [CrossRef] [PubMed]
16. Hu, V.W.; Frank, B.C.; Heine, S.; Lee, N.H.; Quackenbush, J. Gene expression profiling of lymphoblastoid cell lines from monozygotic twins discordant in severity of autism reveals differential regulation of neurologically relevant genes. *BMC Genom.* **2006**, *7*, 118–136. [CrossRef] [PubMed]
17. Nishimura, Y.; Martin, C.L.; Vazquez-Lopez, A.; Spence, S.J.; Alvarez-Retuerto, A.I.; Sigman, M.; Steindler, C.; Pellegrini, S.; Schanen, N.C.; Warren, S.T. Genome-wide expression profiling of lymphoblastoid cell lines distinguishes different forms of autism and reveals shared pathways. *Hum. Mol. Genet.* **2007**, *16*, 1682–1698. [CrossRef] [PubMed]
18. Garbett, K.; Ebert, P.J.; Mitchell, A.; Lintas, C.; Manzi, B.; Mirnics, K.; Persico, A.M. Immune transcriptome alterations in the temporal cortex of subjects with autism. *Neurobiol. Dis.* **2008**, *30*, 303–311. [CrossRef] [PubMed]
19. Gregg, J.P.; Lit, L.; Baron, C.A.; Hertz-Picciotto, I.; Walker, W.; Davis, R.A.; Croen, L.A.; Ozonoff, S.; Hansen, R.; Pessah, I.N. Gene expression changes in children with autism. *Genomics* **2008**, *91*, 22–29. [CrossRef] [PubMed]
20. Enstrom, A.M.; Lit, L.; Onore, C.E.; Gregg, J.P.; Hansen, R.L.; Pessah, I.N.; Hertz-Picciotto, I.; van de Water, J.A.; Sharp, F.R.; Ashwood, P. Altered gene expression and function of peripheral blood natural killer cells in children with autism. *Brain Behav. Immun.* **2009**, *23*, 124–133. [CrossRef] [PubMed]
21. Hu, V.W.; Sarachana, T.; Kim, K.S.; Nguyen, A.; Kulkarni, S.; Steinberg, M.E.; Luu, T.; Lai, Y.; Lee, N.H. Gene expression profiling differentiates autism case–controls and phenotypic variants of autism spectrum disorders: Evidence for circadian rhythm dysfunction in severe autism. *Autism Res.* **2009**, *2*, 78–97. [CrossRef] [PubMed]
22. Hu, V.W.; Nguyen, A.; Kim, K.S.; Steinberg, M.E.; Sarachana, T.; Scully, M.A.; Soldin, S.J.; Luu, T.; Lee, N.H. Gene expression profiling of lymphoblasts from autistic and nonaffected sib pairs: Altered pathways in neuronal development and steroid biosynthesis. *PLoS ONE* **2009**, *4*, e5775. [CrossRef] [PubMed]
23. Sheikh, A.; Li, X.; Wen, G.; Tauqeer, Z.; Brown, W.; Malik, M. Cathepsin D and apoptosis related proteins are elevated in the brain of autistic subjects. *Neuroscience* **2010**, *165*, 363–370. [CrossRef] [PubMed]
24. Malik, M.; Sheikh, A.M.; Wen, G.; Spivack, W.; Brown, W.T.; Li, X. Expression of inflammatory cytokines, Bcl-2 and cathepsin D are altered in lymphoblasts of autistic subjects. *Immunobiology* **2011**, *216*, 80–85. [CrossRef] [PubMed]
25. Kuwano, Y.; Kamio, Y.; Kawai, T.; Katsuura, S.; Inada, N.; Takaki, A.; Rokutan, K. Autism-associated gene expression in peripheral leucocytes commonly observed between subjects with autism and healthy women having autistic children. *PLoS ONE* **2011**, *6*, e24723. [CrossRef] [PubMed]
26. Ghahramani Seno, M.M.; Hu, P.; Gwadry, F.G.; Pinto, D.; Marshall, C.R.; Casallo, G.; Scherer, S.W. Gene and miRNA expression profiles in autism spectrum disorders. *Brain Res.* **2011**, *1380*, 85–97. [CrossRef] [PubMed]
27. Luo, R.; Sanders, S.J.; Tian, Y.; Voineagu, I.; Huang, N.; Chu, S.H.; Klei, L.; Cai, C.; Ou, J.; Lowe, J.K. Genome-wide transcriptome profiling reveals the functional impact of rare de novo and recurrent CNVs in autism spectrum disorders. *Am. J. Hum. Genet.* **2012**, *91*, 38–55. [CrossRef] [PubMed]
28. Chow, M.L.; Pramparo, T.; Winn, M.E.; Barnes, C.C.; Li, H.-R.; Weiss, L.; Fan, J.-B.; Murray, S.; April, C.; Belinson, H. Age-dependent brain gene expression and copy number anomalies in autism suggest distinct pathological processes at young versus mature ages. *PLoS Genet.* **2012**, *8*, e1002592. [CrossRef] [PubMed]
29. Kong, S.W.; Collins, C.D.; Shimizu-Motohashi, Y.; Holm, I.A.; Campbell, M.G.; Lee, I.-H.; Brewster, S.J.; Hanson, E.; Harris, H.K.; Lowe, K.R. Characteristics and predictive value of blood transcriptome signature in males with autism spectrum disorders. *PLoS ONE* **2012**, *7*, e49475. [CrossRef] [PubMed]
30. Griesi-Oliveira, K.; Moreira Dde, P.; Davis-Wright, N.; Sanders, S.; Mason, C.; Orabona, G.M.; Vadasz, E.; Bertola, D.R.; State, M.W.; Passos-Bueno, M.R. A complex chromosomal rearrangement involving chromosomes 2, 5, and X in autism spectrum disorder. *Am. J. Med. Genet. B Neuropsychiatr. Genet.* **2012**, *159*, 529–536. [CrossRef] [PubMed]
31. Anitha, A.; Nakamura, K.; Thanseem, I.; Yamada, K.; Iwayama, Y.; Toyota, T.; Matsuzaki, H.; Miyachi, T.; Yamada, S.; Tsujii, M.; et al. Brain region-specific altered expression and association of mitochondria-related genes in autism. *Mol. Autism* **2012**, *3*, 12–24. [CrossRef] [PubMed]

32. Ginsberg, M.R.; Rubin, R.A.; Falcone, T.; Ting, A.H.; Natowicz, M.R. Brain transcriptional and epigenetic associations with autism. *PLoS ONE* **2012**, *7*, e44736. [CrossRef] [PubMed]

33. Choi, J.; Ababon, M.R.; Soliman, M.; Lin, Y.; Brzustowicz, L.M.; Matteson, P.G.; Millonig, J.H. Autism associated gene, engrailed2, and flanking gene levels are altered in post-mortem cerebellum. *PLoS ONE* **2014**, *9*, e87208. [CrossRef] [PubMed]

34. Nardone, S.; Sams, D.S.; Reuveni, E.; Getselter, D.; Oron, O.; Karpuj, M.; Elliott, E. DNA methylation analysis of the autistic brain reveals multiple dysregulated biological pathways. *Transl. Psychiatry* **2014**, *4*, e433. [CrossRef] [PubMed]

35. Pandit, P.; Cooper-White, J.; Punyadeera, C. High-yield RNA-extraction method for saliva. *Clin. Chem.* **2013**, *59*, 1118–1122. [CrossRef] [PubMed]

36. Lee, Y.H.; Zhou, H.; Reiss, J.K.; Yan, X.; Zhang, L.; Chia, D.; Wong, D.T. Direct saliva transcriptome analysis. *Clin. Chem.* **2011**, *57*, 1295–1302. [CrossRef] [PubMed]

37. Park, N.J.; Li, Y.; Yu, T.; Brinkman, B.M.; Wong, D.T. Characterization of RNA in saliva. *Clin. Chem.* **2006**, *52*, 988–994. [CrossRef] [PubMed]

38. Seugnet, L.; Boero, J.; Gottschalk, L.; Duntley, S.P.; Shaw, P.J. Identification of a biomarker for sleep drive in flies and humans. *Proc. Natl. Acad. Sci. USA* **2006**, *103*, 19913–19918. [CrossRef] [PubMed]

39. Bustin, S.A.; Benes, V.; Garson, J.A.; Hellemans, J.; Huggett, J.; Kubista, M.; Mueller, R.; Nolan, T.; Pfaffl, M.W.; Shipley, G.L. Primer sequence disclosure: A clarification of the MIQE guidelines. *Clin. Chem.* **2011**, *57*, 919–921. [CrossRef] [PubMed]

40. Vandesompele, J.; de Preter, K.; Pattyn, F.; Poppe, B.; van Roy, N.; de Paepe, A.; Speleman, F. Accurate normalization of real-time quantitative RT-PCR data by geometric averaging of multiple internal control genes. *Genome Biol.* **2002**, *3*, 1–12. [CrossRef]

41. Andersen, C.L.; Jensen, J.L.; Orntoft, T.F. Normalization of real-time quantitative reverse transcription-PCR data: A model-based variance estimation approach to identify genes suited for normalization, applied to bladder and colon cancer data sets. *Cancer Res.* **2004**, *64*, 5245–5250. [CrossRef] [PubMed]

42. Lallemant, B.; Evrard, A.; Combescure, C.; Chapuis, H.; Chambon, G.; Raynal, C.; Reynaud, C.; Sabra, O.; Joubert, D.; Hollande, F.; et al. Reference gene selection for head and neck squamous cell carcinoma gene expression studies. *BMC Mol. Biol.* **2009**, *10*, 78–88. [CrossRef] [PubMed]

43. Livak, K.J.; Schmittgen, T.D. Analysis of relative gene expression data using real-time quantitative PCR and the $2^{-\Delta\Delta Ct}$ method. *Methods* **2001**, *25*, 402–408. [CrossRef] [PubMed]

44. Pfaffl, M.W.; Tichopad, A.; Prgomet, C.; Neuvians, T.P. Determination of stable housekeeping genes, differentially regulated target genes and sample integrity: Bestkeeper—Excel-based tool using pair-wise correlations. *Biotechnol. Lett.* **2004**, *26*, 509–515. [CrossRef] [PubMed]

45. Keller, S.; Sarchiapone, M.; Zarrilli, F.; Videtic, A.; Ferraro, A.; Carli, V.; Sacchetti, S.; Lembo, F.; Angiolillo, A.; Jovanovic, N.; et al. Increased BDNF promoter methylation in the Wernicke area of suicide subjects. *Arch. Gen. Psychiatry* **2010**, *67*, 258–267. [CrossRef] [PubMed]

46. Arion, D.; Sabatini, M.; Unger, T.; Pastor, J.; Alonso-Nanclares, L.; Ballesteros-Yanez, I.; Garcia Sola, R.; Munoz, A.; Mirnics, K.; DeFelipe, J. Correlation of transcriptome profile with electrical activity in temporal lobe epilepsy. *Neurobiol. Dis.* **2006**, *22*, 374–387. [CrossRef] [PubMed]

47. Arion, D.; Unger, T.; Lewis, D.A.; Levitt, P.; Mirnics, K. Molecular evidence for increased expression of genes related to immune and chaperone function in the prefrontal cortex in schizophrenia. *Biol. Psychiatry* **2007**, *62*, 711–721. [CrossRef] [PubMed]

48. Zhang, L.; Xiao, H.; Wong, D.T. Salivary biomarkers for clinical applications. *Mol. Diagn. Ther.* **2009**, *13*, 245–259. [CrossRef] [PubMed]

49. Saxena, V.; Yadev, N.; Juneja, V.; Singh, A.; Tiwari, U.; Santha, B. Saliva: A miraculous biofluid for early detection of disease. *J. Oral Health Community Dent.* **2013**, *7*, 64–68.

50. Zhang, L.; Xiao, H.; Karlan, S.; Zhou, H.; Gross, J.; Elashoff, D.; Akin, D.; Yan, X.; Chia, D.; Karlan, B.; et al. Discovery and preclinical validation of salivary transcriptomic and proteomic biomarkers for the non-invasive detection of breast cancer. *PLoS ONE* **2010**, *5*, e15573. [CrossRef] [PubMed]

51. Nohesara, S.; Ghadirivasfi, M.; Mostafavi, S.; Eskandari, M.R.; Ahmadkhaniha, H.; Thiagalingam, S.; Abdolmaleky, H.M. DNA hypomethylation of MB-COMT promoter in the DNA derived from saliva in schizophrenia and bipolar disorder. *J. Psychiatr. Res.* **2011**, *45*, 1432–1438. [CrossRef] [PubMed]

52. Rahbar, M.H.; Samms-Vaughan, M.; Ma, J.; Bressler, J.; Dickerson, A.S.; Hessabi, M.; Loveland, K.A.; Grove, M.L.; Shakespeare-Pellington, S.; Beecher, C.; et al. Synergic effect of GSTP1 and blood manganese concentrations in autism spectrum disorder. *Res. Autism Spectr. Disord.* **2015**, *18*, 73–82. [CrossRef] [PubMed]

53. Hicks, S.D.; Ignacio, C.; Gentile, K.; Middleton, F.A. Salivary mirna profiles identify children with autism spectrum disorder, correlate with adaptive behavior, and implicate asd candidate genes involved in neurodevelopment. *BMC Pediatr.* **2016**, *16*, 1–11. [CrossRef] [PubMed]

54. Zhang, L.; Farrell, J.J.; Zhou, H.; Elashoff, D.; Akin, D.; Park, N.H.; Chia, D.; Wong, D.T. Salivary transcriptomic biomarkers for detection of resectable pancreatic cancer. *Gastroenterology* **2010**, *138*, 941–947. [CrossRef] [PubMed]

55. Yu, H.R.; Kuo, H.C.; Huang, H.C.; Huang, L.T.; Tain, Y.L.; Chen, C.C.; Liang, C.D.; Sheen, J.M.; Lin, I.C.; Wu, C.C.; et al. Glyceraldehyde-3-phosphate dehydrogenase is a reliable internal control in western blot analysis of leukocyte subpopulations from children. *Anal. Biochem.* **2011**, *413*, 24–29. [CrossRef] [PubMed]

56. Chen, J.; Sochivko, D.; Beck, H.; Marechal, D.; Wiestler, O.D.; Becker, A.J. Activity-induced expression of common reference genes in individual cns neurons. *Lab. Investig.* **2001**, *81*, 913–916. [CrossRef] [PubMed]

57. Silberberg, G.; Baruch, K.; Navon, R. Detection of stable reference genes for real-time PCR analysis in schizophrenia and bipolar disorder. *Anal. Biochem.* **2009**, *391*, 91–97. [CrossRef] [PubMed]

58. Penna, I.; Vella, S.; Gigoni, A.; Russo, C.; Cancedda, R.; Pagano, A. Selection of candidate housekeeping genes for normalization in human postmortem brain samples. *Int. J. Mol. Sci.* **2011**, *12*, 5461–5470. [CrossRef] [PubMed]

59. Steinhausen, H.C.; Erdin, A. Abnormal psychosocial situations and ICD-10 diagnoses in children and adolescents attending a psychiatric service. *J. Child Psychol. Psychiatry* **1992**, *33*, 731–740. [CrossRef] [PubMed]

60. Lord, C.; Pickles, A.; McLennan, J.; Rutter, M.; Bregman, J.; Folstein, S.; Fombonne, E.; Leboyer, M.; Minshew, N. Diagnosing autism: Analyses of data from the autism diagnostic interview. *J. Autism Dev. Disord.* **1997**, *27*, 501–517. [CrossRef] [PubMed]

61. Goodman, R. The strengths and difficulties questionnaire: A research note. *J. Child. Psychol. Psychiatry* **1997**, *38*, 581–586. [CrossRef] [PubMed]

International Journal of
Molecular Sciences

MDPI

Article

Association Analysis of Noncoding Variants in Neuroligins 3 and 4X Genes with Autism Spectrum Disorder in an Italian Cohort

Martina Landini [1], Ivan Merelli [1], M. Elisabetta Raggi [2], Nadia Galluccio [1], Francesca Ciceri [2], Arianna Bonfanti [2], Serena Camposeo [3], Angelo Massagli [3], Laura Villa [2], Erika Salvi [4], Daniele Cusi [1,5], Massimo Molteni [2], Luciano Milanesi [1], Anna Marabotti [2,6] and Alessandra Mezzelani [1,*]

[1]   Institute of Biomedical Technologies, National Research Council, Via Fratelli Cervi 93, 20090 Segrate, Italy; landinimartina@gmail.com (M.L.); ivan.merelli@itb.cnr.it (I.M.); nadiagalluccio@gmail.com (N.G.); daniele.cusi@unimi.it (D.C.); luciano.milanesi@itb.cnr.it (L.M.)
[2]   Scientific Institute, IRCSS Eugenio Medea, 23842 Bosisio Parini, Italy; mariaelisabetta.raggi@bp.lnf.it (M.E.R.); francesca.ciceri@bp.lnf.it (F.C.); arianna.bonfanti@bp.lnf.it (A.B.); laura.villa@bp.lnf.it (L.V.); massimo.molteni@bp.lnf.it (M.M.); amarabotti@unisa.it (A.M.)
[3]   Scientific Institute, IRCSS Eugenio Medea, 72100 Brindisi, Italy; serena.camposeo@libero.it (S.C.); angelo.massagli@gmail.com (A.M.)
[4]   Department of Health Sciences, University of Milan, 20142 Milan, Italy; erika.salvi@unimi.it
[5]   Sanipedia srl, via Ariosto 21, 20091 Bresso, Italy
[6]   Department of Chemistry and Biology, University of Salerno, Via Giovanni Paolo II 132, 84084 Fisciano, Italy
*   Correspondence: alessandra.mezzelani@itb.cnr.it; Tel.: +39-02-2642-2606

Academic Editor: Merlin G. Butler
Received: 10 August 2016; Accepted: 12 October 2016; Published: 22 October 2016

Abstract: Since involved in synaptic transmission and located on X-chromosome, neuroligins 3 and 4X have been studied as good positional and functional candidate genes for autism spectrum disorder pathogenesis, although contradictory results have been reported. Here, we performed a case-control study to assess the association between noncoding genetic variants in *NLGN3* and *NLGN4X* genes and autism, in an Italian cohort of 202 autistic children analyzed by high-resolution melting. The results were first compared with data from 379 European healthy controls (1000 Genomes Project) and then with those from 1061 Italian controls genotyped by Illumina single nucleotide polymorphism (SNP) array 1M-duo. Statistical evaluations were performed using Plink v1.07, with the Omnibus multiple loci approach. According to both the European and the Italian control groups, a 6-marker haplotype on *NLGN4X* (rs6638575(G), rs3810688(T), rs3810687(G), rs3810686(C), rs5916269(G), rs1882260(T)) was associated with autism (odd ratio = 3.58, p-value = 2.58×10^{-6} for the European controls; odds ratio = 2.42, p-value = 6.33×10^{-3} for the Italian controls). Furthermore, several haplotype blocks at 5-, 4-, 3-, and 2-, including the first 5, 4, 3, and 2 SNPs, respectively, showed a similar association with autism. We provide evidence that noncoding polymorphisms on *NLGN4X* may be associated to autism, suggesting the key role of *NLGN4X* in autism pathophysiology and in its male prevalence.

Keywords: autism; genetics; neuroligins; SNPs; haplotype analysis; noncoding regions

1. Introduction

Autism spectrum disorder (ASD) is a complex neurodevelopmental disorder with an early onset, typically prior to age 3, characterized by impaired social interactions, absent or limited verbal communication, and stereotyped and restricted pattern of interests [1]. Considering the great variability of clinical presentations, the definition of ASD includes autism, Asperger's syndrome, and pervasive developmental disorder not otherwise specified (PDD-NOS) [2]. A combination of both genetic and environmental causes, such as pesticides, heavy metals, dysbiosis, and mycotoxins, has been suggested for ASD pathogenesis but, despite approximately 5%–15% of ASDs cases being due to identified genetic or chromosomal alterations, the ASD etiology is still largely unknown [3–8]. Considering the higher occurrence of ASD in males than in females, with a ratio of roughly 4:1, a role for the X-chromosome in the ASD etiology has been also suggested [7,9]. In this regard, cytogenetic abnormalities highlight the putative involvement of the loci from Xq12 to Xq21 and Xp22 in autism pathogenesis [10,11], as well as associations with structural variants that have been reported for some X-chromosomal genes, such as *NLGN4X* and *MECP2* genes [12–15]. Since located on the X-chromosome and involved in synaptic plasticity, both Neuroligin3 (*NLGN3*) and Neuroligin4X (*NLGN4X*), mapped at Xq13.1 and Xp22.31, respectively, were extensively studied as good positional and functional candidates for ASD predisposing. Neuroligins are postsynaptic cell-adhesion molecules that interact with the presynaptic cell-surface receptors neurexins. Neurexins contain a single transmembrane domain, and together with neuroligins they form Ca^{2+}-dependent neurexin/neuroligin complexes. These transsynaptic complexes play a crucial role in modulating neurotransmission and differentiation [16–18]. Several ASD genetic studies have identified mutations affecting neuroligin proteins and influencing synaptic function [19–21]. The involvement of neuroligins in ASD has been firstly confirmed in two Swedish families by a de novo missense mutation (R451C) in *NLGN3* and a frameshift mutation (1186insT) in *NLGN4X* causing premature protein termination (D396X), respectively associated with typical autism and Asperger's syndrome [12]. Both these mutations resulted in an intracellular retention of the mutant proteins, with a consistent loss of the synaptic activity, leading to neurodevelopmental defects and mental retardation [22,23]. In mouse models, autism-associated *NLGN3* mutations induced repetitive behaviors, and R451C mutant mice showed impaired social interactions, but enhanced spatial learning abilities [24–26]. A loss-of-function mechanism has also been suggested for the point mutation R87W in *NLGN4X*, found in two brothers with classical ASD [27]. Recently, missense variations in *NLGN3* (G426S) and *NLGN4X* (G84R, Q162K, and A283T) in Chinese ASD patients have been associated to ASD predisposing, by causing abnormal synaptic homeostasis [28]. Deletion of exons 4, 5, and 6 in *NLGN4X* have been also found in autistic children, suggesting that alternative splicing variants might lead to abnormal neuroligin function in ASD [29,30]. Moreover, several noncoding genetic variants have been specifically found in ASD patients [31–37]. All these variations often segregate into ASD families [12,13] and can also be associated with different cognitive phenotypes, such as intellectual and language disabilities [32,37], highlighting the role of neuroligins in the ASD pathogenesis. In this regard, a de novo base pair substitution (−335G>A) in the promoter region of *NLGN4X*, has been found in one autistic child with nonsyndromic mental retardation (NSMR) [32]. Since associated with an increased level of *NLGN4X* transcripts, this variation probably affects the binding sites of transcription factors in the mutated promoter sequence [32]. Four novel synonymous substitutions, specific to ASD, have been reported in the coding sequences of the *NLGN3* (p.K566K) and *NLGN4X* (p.G99G; p.I172I; p.F530F) [34]. Despite the uncertainty of the physiological and clinical relevance, they might affect the protein structures by altering splicing sites [34] or affecting gene regulation, as they are located in conserved regulatory regions, such as enhancer- and promoter-associated histone modification sites [33]. Moreover, a positive association with ASD has been identified in a homogeneous autistic Chinese cohort (made of patients with typical phenotype), for a common intronic variant in *NLGN3* (rs4844285(G)) [6]. A three-marker haplotype (rs11795613(A), rs4844285(G), rs4844286(T)), with a significant male bias, has been suggested, supporting the hypothesis that defects of the synapse might have a role in ASD pathogenesis [6]. In addition, three haplotype sets in *NLGN4X* (rs3810686(T), rs1882260(C)), (rs6638575(G), rs3810686(T), rs18882260(T)) and (rs6638575(A), rs3810686(C), rs18882260(C)), have been positively associated with nonspecific mental retardation (Intelligence Quotients IQs < 70) and social disability scores ≤ 8 [36] and genetic variants in *NLGN4X* also have a significant effect on male cognitive abilities, highlighting the role of neuroligins in psychiatric conditions [37].

However, contradictory results have been also provided in validating the relevance and frequency of neuroligin genetic variants in ASD [38–41]. Wermenter and coworkers failed to find an involvement of *NLGN3* and *NLGN4X* with ASD in a study group of 107 probands with autistic disorders at a high-functioning level [40], and similar results have been reported in the Quebec population, after screening for neuroligin mutations in 96 individuals diagnosed with ASD [41]. Lastly, no association was found between *NLGN3/NLGN4X* alterations and ASD in a Chinese cohort [38]. However, the genetic heterogeneity of ASD, as well as differences in the ethnicity background of ASD samples, make it difficult to elucidate the involvement of neuroligin genes in autism pathogenesis.

Considering all these evidences and starting from an Italian ASD sample set, we assessed the effects of three noncoding genetic variants in *NLGN3*, already reported as associated with ASD in a Chinese cohort [6], and six common single nucleotide polymorphisms (SNPs) in *NLGN4X*, only partially described in literature [36], with the final aim to validate the selected genetic variants as susceptibility loci for ASD.

2. Results

2.1. Single-Locus Analysis

Association analyses have been performed considering SNPs located in the intronic and 3' UTR regions of *NLGN3* and *NLGN4X*, considering both the European (EUR) population (1000 Genomes Project, Phase 2) [42] and the Italian population. Genotype counts of SNPs and results of the Hardy–Weinberg statistics, which do not present any significant disequilibrium, are shown in Tables S1 and S2. As shown in Table 1, a moderate statistical significance difference (odds ratio (OR) = 1.453, p = 0.0456) was identified in the single-locus analysis only for SNP rs3810688, with the minor allele T having frequency slightly lower in ASD cases in comparison to EUR controls (separate results for male and female analysis are provided in Tables S3 and S4). However, this association lost its significance after False Discovery Rate (FDR) correction for multiple testing (Table 1). As the number of samples was quite limited, Fisher's exact test has been chosen for comparing SNP allelic frequencies between cases and unrelated controls, always using the genetic data from the 1000 Genomes Project (Phase 2) as reference population [42]. We also verified that the allelic frequencies for all the SNPs analyzed do not statistically deviate in our control data set from those reported in the 1000 Genomes Project EUR population. The moderate statistical association of rs3810688(T) with autism disappeared when comparing ASD data with those from Italian controls.

Table 1. Single locus analysis for all the single nucleotide polymorphisms (SNPs) genotyped in *NLGN3* and *NLGN4X* (EUR control population).

Gene ID	SNP ID	Minor Allele	MAF Controls	MAF Case	*p*-Value	OR	95% CI	*p*-Corr
NLGN3	rs11795613	G	0.481	0.515	0.398	1.144	0.816–1.617	0.701
	rs4844285	A	0.4741	0.4895	0.701	1.064	0.755–1.496	0.701
	rs4844286	T	0.505	0.481	0.5396	0.908	0.646–1.280	0.701
NLGN4X	rs6638575	A	0.266	0.301	0.303	1.193	0.856–1.663	0.404
	rs3810688	T	0.2176	0.288	0.0456	1.453	1.018–2.074	0.152
	rs3810687	T	0.1086	0.1339	0.3369	1.269	0.805–2.000	0.404
	rs3810686	T	0.4466	0.4561	0.817	1.039	0.768–1.407	0.817
	rs5916269	A	0.1086	0.1339	0.3369	1.269	0.805–2.000	0.404
	rs1882260	C	0.2655	0.3347	0.0505	1.392	1.005–1.928	0.152

Statistical significance, for the single locus association analysis, was tested at $p < 0.05$, after Fisher's exact test. The *p*-values have been adjusted by false discovery rate (FDR) correction for multiple test analysis. MAF controls, frequency of the minor allele in controls (n = 379); MAF case, frequency of the minor allele in case (n = 202); OR, odds ratio; CI, confidence interval; *p*-corr, corrected *p*-value after FDR test.

2.2. Haplotype Analysis (EUR Control Population)

A 6-SNPs haplotype block in *NLGN4X* showed a strong significant association with ASD cases (odds ratio = 3.58, *p*-value = 2.58 × 10^{-6}), also after logistic regression and 100,000 permutation tests (empirical *p*-value < 1 × 10^{-5}, empirical *q*-value < 1 × 10^{-5}) (Table 2). This block (rs6638575(G), rs3810688(T), rs3810687(G), rs3810686(C), rs5916269(G), rs188260(T)) includes the SNP rs3810688(T), moderately associated with ASD in the single-locus analysis. Indeed, haplotypes on *NLGN4X* have been also analyzed as 5-, 4-, 3-, or 2-loci blocks, as reported in Table 2. A good statistical significance was obtained with blocks including the SNP rs3810688(T), especially together with the haplotype sets also including the SNPs rs3810687(G) and rs3810686(C)—both located in the 3′ UTR of *NLGN4X*—and with rs6638575(G), in the intron 5 of *NLGN4X* (Table 2). These variants are part of two 5-loci (rs6638575(G), rs3810688(T), rs3810687(G), rs381086(C), rs5916269(G)) and (rs3810688(T), rs3810687(G), rs381086(C), rs5916269(G), rs1882260(T)), two 4-loci (rs6638575(G), rs3810688(T), rs3810687(G), rs381086(C)) and (rs3810688(T), rs3810687(G), rs3810686(C), rs5916269(G)), two 3-loci (rs6638575(G), rs3810688(T), rs3810687(G)) and (rs3810688(T), rs3810687(G), rs3810686(C)) and two 2-loci (rs6638575(G), rs3810688(T)) and (rs3810688(T), rs3810687(G)) combinations, all with strong statistical association with ASD. Moreover, despite the single-locus analysis not showing any statistical difference for the SNPs rs1882260 and rs5916269, haplotype analysis showed that two 2-blocks (rs5916269(G), rs1882260(T)) and (rs3810686(C), rs5916269(G)), two 3-blocks (rs3810686(C), rs5916269(G), rs1882260(T)) and (rs3810687(G), rs3810686(C), rs5916269(G)) and one 4-block loci (rs3810687(G), rs3810686(C), rs5916269(G), rs1882260(T)), respectively, have also a very strong significant association with ASD (Table 2). Similar results have been obtained also comparing the ASD haplotype frequencies of *NLGN4X* with our unrelated controls, suggesting the reliability of the selected control population for the statistical analysis (EUR Phase 2, 1000 Genomes Project).

Table 2. Haplotype association analysis in *NLGN4X* (*EUR control population*).

N	SNPs in Haplotypes	Control Frequency	Case Frequency	Haplotype	*p*-Value	Emp. *p*-Value	Emp. *q*-Value
6	rs6638575-rs3810688-rs3810687-rs3810686-rs5916269-rs1882260	0.0335	0.1799	GTGCGT	2.58 × 10^{-6}	<1 × 10^{-5}	<1 × 10^{-5}
5	rs6638575-rs3810688-rs3810687-rs3810686-rs5916269	0.0805	0.1799	GTGCG	7.60 × 10^{-6}	<1 × 10^{-5}	<1 × 10^{-5}
5	rs3810688-rs3810687-rs3810686-rs5916269-rs1882260	0.0335	0.1799	TGCGT	2.58 × 10^{-6}	<1 × 10^{-5}	<1 × 10^{-5}
4	rs6638575-rs3810688-rs3810687-rs3810686	0.0805	0.1799	GTGC	7.6 × 10^{-6}	<1 × 10^{-5}	2 × 10^{-5}
4	rs3810688-rs3810687-rs3810686-rs5916269	0.0801	0.1793	TGCG	7.6 × 10^{-6}	<1 × 10^{-5}	2 × 10^{-5}
4	rs3810687-rs3810686-rs5916269-rs1882260	0.0943	0.1793	GCGT	1.3 × 10^{-5}	<1 × 10^{-5}	2 × 10^{-5}
3	rs6638575-rs3810688-rs3810687	0.3538	0.4463	GTG	8.16 × 10^{-6}	<1 × 10^{-5}	<1 × 10^{-5}
3	rs3810688-rs3810687-rs3810686	0.0701	0.1793	TGC	7.6 × 10^{-6}	<1 × 10^{-5}	<1 × 10^{-5}
3	rs3810687-rs3810686-rs5916269	0.4062	0.4445	GCG	8.16 × 10^{-4}	6.99 × 10^{-4}	3.2 × 10^{-3}
3	rs3810686-rs5916269-rs1882260	0.0943	0.1793	CGT	1.3 × 10^{-5}	<1 × 10^{-5}	3 × 10^{-4}
2	rs6638575-rs3810688	0.2109	0.2854	GT	1.08 × 10^{-4}	7 × 10^{-5}	8.3 × 10^{-4}
2	rs3810688-rs3810687	0.0943	0.1793	TG	1.3 × 10^{-5}	2 × 10^{-5}	1.5 × 10^{-4}
2	rs3810687-rs3810686	0.4062	0.4445	GC	8.16 × 10^{-4}	7.3 × 10^{-4}	5 × 10^{-3}
2	rs3810686-rs5916269	0.4062	0.4445	CG	8.16 × 10^{-4}	7.3 × 10^{-4}	5 × 10^{-3}
2	rs5916269-rs1882260	0.4314	0.6259	GT	2.4 × 10^{-8}	<1 × 10^{-5}	<1 × 10^{-5}

Haplotype combinations with significant *p*-values (*p* < 0.05), using the EUR control population, after permutation test, have been reported for all the SNPs located on *NLGN4X*. Empirical *p*-values and empirical *q*-values after 100,000 permutations are reported. Only haplotypes having case frequency >0.05 have been presented.

No statistically significant haplotypes have been found in the *NLGN3* after logistic regression and permutation analysis.

2.3. Haplotype Analysis (ITA control Population)

The analysis was then repeated comparing the frequencies of *NLGN4X* SNPs between ASD cases and 1061 healthy controls collected during the Hypergenes project [43]. The association with the 6-marker haplotype remained, although weaker than the EUR population (odds ratio = 2.42, *p*-value = 6.33×10^{-3}, empirical *p*-value = 6.79×10^{-3}, empirical *q*-value < 2.73×10^{-2}). A weaker association was also found for most of the 5-, 4-, 3- and 2-SNP haplotypes identified considering the EUR population. Data are reported in Table 3.

Table 3. Haplotype association analysis in *NLGN4X* (ITA control population).

N	SNPs in Haplotypes	Control Frequency	Case Frequency	Haplotype	*p*-Value	Empirical *p*-Value	Empirical *q*-Value
6	rs6638575-rs3810688-rs3810687-rs3810686-rs5916269-rs1882260	0.0678	0.1799	GTGCGT	6.33×10^{-3}	6.79×10^{-3}	2.73×10^{-3}
5	rs6638575-rs3810688-rs3810687-rs3810686-rs5916269	0.1349	0.1799	GTGCG	1.9×10^{-2}	1.87×10^{-2}	4.47×10^{-2}
5	rs3810688-rs3810687-rs3810686-rs5916269-rs1882260	0.0678	0.1799	TGCGT	6.33×10^{-3}	5.39×10^{-3}	2.78×10^{-2}
4	rs6638575-rs3810688-rs3810687-rs3810686	0.1189	0.1799	GTGC	1.66×10^{-2}	1.62×10^{-2}	4.41×10^{-2}
3	rs6638575-rs3810688-rs3810687	0.3782	0.4463	GTG	1.56×10^{-2}	1.29×10^{-2}	4.37×10^{-2}
3	rs3810688-rs3810687-rs3810686	0.0801	0.1793	TGC	1.17×10^{-2}	1.26×10^{-2}	4.78×10^{-2}
3	rs3810687-rs3810686-rs5916269	0.3360	0.4445	GCG	3.99×10^{-3}	4.10×10^{-3}	1.98×10^{-2}
2	rs3810687-rs3810686	0.3160	0.4445	GC	3.50×10^{-3}	3.20×10^{-3}	2.20×10^{-2}
2	rs3810686-rs5916269	0.3360	0.4445	CG	3.99×10^{-3}	3.70×10^{-3}	2.53×10^{-2}

Haplotype combinations with significant *p*-values (*p* < 0.05), using the ITA control population, after permutation test, have been reported for all the SNPs located on *NLGN4X*. Empirical *p*-values and empirical *q*-values after 100,000 permutations are reported. Only haplotypes having case frequency >0.05 have been presented.

2.4. Linkage Disequilibrium Analysis

The results for linkage disequilibrium (LD) analysis for the considered SNPs in *NLGN4X* are presented in Figure 1. Panel (a) represents the LD for SNPs in the EUR control populations; panel (b) shows the LD scores for the ITA control population patients; panel (c) reports the results achieved for ASD patients.

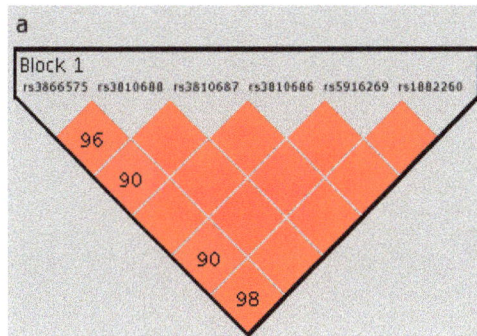

Figure 1. *Cont.*

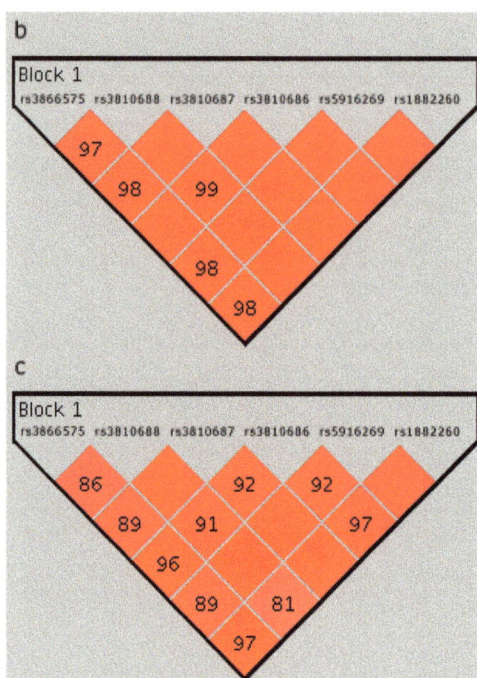

Figure 1. Linkage disequilibrium (LD) for the SNPs analyzed in *NLGN4X* calculated using Haploview: in panel (**a**) for the EUR population (1000 Genomes Project, Phase 2); in panel (**b**) for the Italian control population; and in panel (**c**) for the autistic patients enrolled in our study.

3. Discussion

According to our results, common noncoding variants in *NLGN4X* may play a role in ASD susceptibility. Starting from a relatively homogenous sample and a case-control study design, our results suggest that some SNP marker haplotype blocks, located in the 3′ UTR of *NLGN4X*, might be associated to ASD. After a first statistical analysis, referring to EUR 1000 Genomes Project population as control, the 3′ UTR of *NLGN3* did not show any association with ASD, nor at single SNP or at haplotype blocks. On the other hand, the 3′ UTR of *NLGN4X* displayed statistical association with the disease at single SNP rs3810688(T), which is the minor allele (Table 1), although this significance was lost after FDR correction. Indeed, although the frequency of this allele is higher in the ASD population in comparison to EUR and ITA controls, the allele is still present in 21% of EUR population, thus producing only a modest effect by itself.

To confirm the statistical association of the haplotype blocks with ASD, (Table 2), we repeated the same statistical analyses, referring to the ITA healthy population, and confirmed the association, although weaker, between some SNP marker haplotypes in the 3′ UTR of *NLGN4X* and ASD. Indeed, the haplotype analysis revealed a 6-SNPs haplotype block (rs6638575(G), −rs3810688(T), −rs3810687(G), −rs3810686(C), −rs5916269(G), −rs1882260(T)) with a significant association with ASD, even after correction by permutation both referring to EUR (Table 2) and ITA control cohort populations (Table 3). In particular, in the region spanning from SNPs rs6638575 to rs3810686, several multiloci haplotype blocks were found to be associated with ASD (Tables 2 and 3). Within the haplotype blocks containing rs3810687 and rs3810686, both SNPs always displayed the G allele. Interestingly, considering the LD analysis, we notice that the considered SNPs on *NLGN4X* are more in linkage than in control populations (Figure 1).

These results might imply that an approximately 4 kb region in the 3′ UTR of *NLGN4X*, spanning from SNPs rs66385775 to rs1882260, could be involved in autism susceptibility. This regulatory region in *NLGN4X* might have a role in modulating neuroligin biological function and, therefore, in causing or predisposing to neurological disorders. Since *NLGN4X* is involved in synaptic plasticity and X-linked, it might be a key gene in ASD pathogenesis and in its sex bias.

On the other hand, no associations were found for variants located on *NLGN3* at the single locus and haplotype analysis, despite Yu and coworkers (2011) reporting a positive association for the variants rs11795613(A), rs4844285(G), and rs4844286(T) with ASD [6]. Differences in sample ethnicity and size, as well as the ASD heterogeneity, might be taken into account for explaining such different results. Further studies, performed on a greater number of patients and with different genetic backgrounds, will be required to elucidate the role of *NLGN3* in autism susceptibility.

Nonetheless, our results are partially in agreement with Qi and collaborators (2009) that reported a positive association between SNPs in *NLGN4X* (rs6638575, rs3810686, rs1882260) and NSMR in a Chinese population [36]. In this regard, four putative haplotype combinations (rs3810686(T), rs1882260(T)), (rs3810686(C), −rs1882260(C)) and (rs6638575(G), rs3810686(T), rs18882260(T)) and (rs6638575(A), rs3810686(C), rs18882260(C)), in the sixth exon of *NLGN4X* have been reported as positively associated to NSMR in this Chinese cohort [36]. This genomic region, in fact, plays a crucial role for the biological function of *NLGN4X*, encoding for the transmembrane domain and postsynaptic density 95-disc large zone occludens-1 binding domain, both necessary for N-methyl aspartate receptors and downstream signal-transducing proteins. Furthermore, it has been reported that a target haplotype composed of the SNPs rs5916271(C) and rs6638575(A) had a significant effect in improving the general cognitive ability and especially verbal comprehension in male children, although further studies are required to highlight the role of these polymorphisms in ASD susceptibility [37].

As none of the analyzed SNPs on *NLGN4X* produce amino acid changes, two possible explanations should be taken into account for their association with ASD. First, considering the regulatory role of the 3′ UTR regions in modulating gene expression, the SNPs located in this region, correlated with ASD, might affect miRNA-binding sites, leading to impairments in *NLGN4X* expression. In this regard, the SNPs analyzed in this study might be located in or very close to putative miRNA regulatory regions, breaking or creating miRNA-binding sites with a possible effect on *NLGN4X* expression and thus on synaptic transmission. To this respect, it has been reported that a single nucleotide variant located in the 5′ UTR of *NLGN4X* is able to affect neuroligin expression [32], but further studies are required to validate if such a regulatory mechanism might also occur for the genetic variants in the 3′ UTR of neuroligin genes. Second, it might be also possible that other genetic variants, as putative predisposing loci or causative mutations, could exist in close linkage with the SNPs analyzed on *NLGN4X*, leading to differences in *NLGN4X* expression through alternative splicing or amino acid variations since each of these conditions can alter the biological function of *NLGN4X*.

Considering that males have higher predisposition in ASD etiology, an involvement of the X-chromosome in autism susceptibility has been suggested by genetic and cytogenetic studies [9–13] and several putative candidate genes, including neuroligins, have been largely investigated [7,8,23–37]. Nonetheless, taking advantages of the new sequencing technologies, further insights might be provided in discovering new X-linked genetic variants associated with ASD predisposition or its etiology.

Although the sample size of patients is quite limited for a genetic association study, statistical association between marker haplotype blocks located at 3′ UTR of *NLGN4X* and ASD was found referring both to EUR and ITA populations. Indeed, despite the stronger statistical significance found in reference to EUR controls, similar results were obtained in reference to ITA controls.

After all, considering the heterogeneity of ASD, in terms of symptoms, severity, and comorbidities, it sounds reasonable that several genes might be involved in ASD predisposition, as well as epigenetics or environmental factors. For this reason, a single genetic variant or haplotype, even if in a regulatory region or in a candidate gene, seems unlikely to be playing a major role in ASD predisposition. Thus, further functional and genetic studies coupled to these SNP profiles in a larger ASD cohort would be the next steps to elucidate the role of neuroligins and genetic variants in ASD etiology. Indeed, these SNPs could be the markers of other genetic variations or associated to specific phenotypes. Moreover, as gene–environment interaction has also been proposed for autism pathogenesis, further studies considering both the genetic and environmental risk factors should be designed.

As for clinical application, the relation between these genetic risk factors and symptoms or ASD characteristics could suggest molecular and epigenetic mechanisms triggering the disease and suggest possible preventive or therapeutic intervention.

In our study we found and discussed the association of several noncoding genetic variants in the *NLGN4X* gene with ASD in an Italian cohort. This implies that neuroligins and, therefore, impairments at the synaptic transmission, might play a role in ASD susceptibility and pathophysiology. To date, no other works have investigated the role of neuroligin noncoding variants in ASD susceptibility in an Italian autistic population, although Blasi and coworkers (2006) failed to find a significant association with ASD in neuroligin coding variants [39] when analyzing Italian ASD patients. Taking into account the contradictory results in validating the role of neuroligins in ASD predisposition as well as the complex etiology of this disease, our findings can be important for the characterization of this disease, although they need to be confirmed and replicated in different ethnic groups and to be associated with phenotypic and epigenetic data.

4. Materials and Methods

4.1. Subjects

The sample-sets consist of: 202 ASD patients (165 (81.7%) males; 37 (18.3%) females), recruited by IRCCS Eugenio Medea-La Nostra Famiglia in two different and far areas of Italy: Bosisio Parini (Lecco, Italy) in the North and Ostuni (BR, Italy) in the South. Patients had an age ranging between 2 and 12 years. Demographic characteristics are summarized in Table 4.

Table 4. Demographic data and diagnosis of autistic patients.

Patients: 202; Age: 2–12; Males: 165; Females: 37		
Patients from North Italy: n = 157	Males: 131, 83.44%	Young autism: 93 (71%) PDD-NOS: 34 (26%) Asperger's Syndrome: 4 (3%)
	Females: 26, 16.56%	Young autism: 18 (69.2%) PDD-NOS: 7 (27%) Asperger's Syndrome: 1 (3.8%)
Patients from South Italy: n = 45	Males: 34, 75.55%	Young autism: 24 (70.6%) PDD-NOS: 9 (26.5%) Asperger's Syndrome: 1 (2.9%)
	Females: 11, 24.45%	Young autism: 8 (72.7%) PDD-NOS: 3 (27.3%) Asperger's Syndrome: 0

The diagnosis for young autism (70.3%), Asperger's syndrome (3.5%), or PDD-NOS (26.2%) were performed according to DSM-IV TR (APA, 2000) and by the Autism Diagnostic Observation Scale (ADOS) and ADI-R. IQ, cognitive tests, and behavioral analyses were also done in all the patients, but 20, and are summarized in Table 5. Patients with genetic syndromes, epilepsy, and neuroradiological confirmed disorders were excluded.

One thousand sixty one healthy controls, collected during the Hypergenes Project [43] (European Network for Genetic-Epidemiological Studies; www.hypergenes.eu) were included in the study. These controls were from the same two areas of Italy from which patients were recruited and in similar proportion. The healthy subjects were genotyped at the University of Milan, using the Illumina

1M-duo array (San Diego, CA, USA) and imputed with Minimac software [44], a low-memory, fast, flexible, computationally efficient implementation of the Markov chain haplotyping method, using the haplotypes from the 1000 Genomes Project (March 2012) as reference [42].

Table 5. Cognitive and behavioral characteristics of autistic patients.

	IQ		Hyperactivity	Language and Communication
>70 *n* = 71 (35.1%)	Level 1: IQ > 101; *n* = 15 (7.4%) Level 2: 100 > IQ> 70; *n* = 56 (27.7%)		Level 0: *n* = 136; (67.3%)	Level 1: *n* = 7 (3.5%) Level 2: *n* = 48 (23.7%)
≤69 *n* = 131 (64.6%)	Level 3: 69 > IQ > 50; *n* = 79 (39.1%) Level 4: 49 > IQ > 35; *n* = 28 (13.9%)		Level 1: *n* = 53 (26.2%)	Level 3: *n* = 65 (32.2%) Level 4: *n* = 64 (31.7%)
	Level 5: 34 > IQ> 20; *n* = 17 (8.4%) Level 6: IQ < 19; *n* = 7 (3.5%)		Level 2: *n* = 13 (6.5%)	Level 5: *n* = 3 (1.5%) 15 patients not evaluated (7.4%)

The data relative to the SNPs considered in the statistical analyses were obtained directly from genotyping for rs5916269, rs3810686, rs3810687, and rs663857 and by imputation for rs1882260 and rs3810688.

The Ethical committees of IRCCS Eugenio Medea-La Nostra Famiglia and of Hypergenes project approved the study.

Moreover, genotype data from the 1000 Genomes Project (Phase 2) [42] were also used as control cases, considering the 379 individuals with EUR genotype.

4.2. Selection of SNPs

The NCBI Accession Numbers, used as reference sequences for *NLGN3* and *NLGN4X* were, respectively, NG_015874 and NG_008881. Relying on literature [6,36], a selection of SNPs located in the intronic or 3′ UTR regions of both *NLGN3* and *NLGN4X* were genotyped. Furthermore, three additional SNPs, located in the 3′ UTR of *NLGN4X*, have been included since they are located in a putative regulatory region and have a close map distance with the previously selected SNPs. In this regard, the inter-SNP distance, for the SNPs analyzed in *NLGN3* and *NLGN4X*, was less than 5 kb. All the studied SNPs have a minor allele frequency (MAF) >0.05—as reported by the 1000 Genomes Project (Phase 2) in the EUR population—and are listed in the Table 6. Linkage disequilibrium for SNPs considered in this study has been calculated using Haploview [45], and the most relevant LD scores for these data are displayed in Figure 1. Figure S1 represents the LD scores (derived from the 1000 Genomes Project EUR population) of all the SNPs located in *NLGN4X*, including those selected for our analyses and the related TAG SNPs.

Table 6. List of SNPs analyzed in *NLGN3* and *NLGN4X*.

Genes	SNPs	Alleles	MAF	Position in Gene	References
NLGN3	rs11795613 rs4844285 rs4844286	(A/G) (A/G) (T/G)	G: 0.49 A: 0.48 T: 0.49	Intron 1 Intron 2 Intron 2	Yu et al., *Behav. Brain Funct.* 2011, 7, 13 [6]
NLGN4X	rs6638575 rs3810686 rs1882260	(A/G) (T/C) (C/T)	A: 0.28 T: 0.44 C: 0.27	Intron 5 3′ UTR 3′ UTR	Qi et al., *Psychiatr. Genet.* 2009, 19, 1 [36]
	rs3810687 rs3810688 rs5916269	(T/G) (T/C) (A/G)	T: 0.11 T: 0.29 A: 0.11	3′ UTR 3′ UTR 3′ UTR	–

All the SNPs analyzed as well as their position along the *NLGN3* and *NLGN4X* genes and MAF, are listed. MAF is referred to that reported for the EUR population (1000 Genomes Project, Phase 2). According to literature, their association to autism or psychiatric conditions, has been also reported. SNP: Single Nucleotide Polymorphism; MAF: Minor Allele Frequency; UTR: Untranslated Region.

4.3. SNPs Screening and Genotyping

Genomic DNA was extracted from peripheral blood samples, using commercial kit (Macherey-Nagel GmbH & Co. KG, Düren, Germany). SNPs located on both *NLGN3* and *NLGN4X* were amplified and genotyped by high-resolution melting analysis (HRM) (Rotor-Gene® Q-Pure detection, Qiagen, Venlo, The Netherlands). For each sample a total of 20 ng of genomic DNA was amplified in a 20 µL reaction mixture Phusion Flash High Fidelity PCR Master Mix (Thermo Fisher Scientific, Waltham, MA, USA), Stati Uniti of primer mix (0.7 each) and 5% of the intercalating dye EvaGreen (Biotium Inc., Fremont, CA, USA), with 1.4, as final concentration. HRM analysis primers were designed by PRIMER3plus software (http://bioinfo.ut.ee/primer3-0.4.0/primer3/). All samples were analyzed in duplicates, and for each selected SNP three internal controls—resembling all three possible genotypes, previously characterized by Sanger-sequencing—have been included. The temperature raising for the HRM analysis was of 0.1 °C/s. Normalized melting curves and differential graphs have been compared for the genotype analysis. Primers and amplification details are shown in the Table S5.

4.4. Statistical Analysis

A first screening was performed comparing the data obtained from ASD patients with those from the European subset of the 1000 Genomes Project (EUR Phase 2 population). SNP association analyses were performed first using Plink v1.07 [46] to test for possible associations between all the SNPs identified in *NLGN3* and *NLGN4X* and ASD. The two-tailed Fisher's exact test was chosen to compare the polymorphisms' distributions (cases vs. controls) and testing their significance at $p < 0.05$, under the null hypothesis of no association between each SNP and ASD. The p-values have been adjusted by false discovery rate (FDR) correction for multiple test analysis.

Using the same control data, haplotype tests were also performed using Plink v1.07. Omnibus multiple loci analyses were carried out for all the SNPs included in this study regarding *NLGN3* and *NLGN4X*, using logistic regression under the null hypothesis of no association between haplotypes and ASD. Permutation tests (100,000 permutations) were used to correct the p-values. Haplotypes with p-values equal or below the α threshold of 0.05 were considered as statistically significant.

The same statistical analyses were performed on SNPs in *NLGN4X*, which revealed a significant correlation with ASD, considering two groups of Italian healthy controls recruited in the same two areas of Italy in which patients were recruited.

Supplementary Materials: Supplementary materials can be found at www.mdpi.com/1422-0067/17/10/1765/s1.

Acknowledgments: This work has been supported by the Italian Ministry of Health (Targeted Research Funding in Public Health—Young Researchers; grant no. GR-2009-1570296), by the Italian Ministry of Education and Research (MIUR) through the Flagship project InterOmics (PB05), by IRCCS "Eugenio Medea" through the "5XMILLE" funds 2008, by the FP7-Health-F4-2007-201550 "HYPERGENES". The Flagship project InterOmics (PB05) covers costs to publish in open access. We specially thank all the parents and the children involved in this study.

Author Contributions: Alessandra Mezzelani and Martina Landini conceived and designed the experiments; Martina Landini and Nadia Galluccio performed the experiments; Ivan Merelli, Erika Salvi, Martina Landini and Alessandra Mezzelani analyzed the data; M. Elisabetta Raggi, Francesca Ciceri, Arianna Bonfanti and Serena Camposeo recruited the patients; Daniele Cusi recruited the controls; Angelo Massagli, Laura Villa and Massimo Molteni made the diagnosis; Martina Landini wrote the paper; Luciano Milanesi is the responsible of the Italian Flagship-project InterOmics (PB05); Anna Marabotti was the Principal Investigator of the project funded by Italian Ministry of Health. All the authors critically revised the manuscript.

Conflicts of Interest: The authors declare no conflict of interest.

Abbreviations

ASD	Autism spectrum disorder
EUR	European
FDR	False discovery rate
HRM	High resolution melting

IQ	Intelligence Quotient
ITA	Italian
MAF	Minor allele frequency
NLGN	Neuroligin
NSMR	Non-syndromic mental retardation
PDD-NOS	Pervasive developmental disorder-not otherwise specified
SNP	Single nucleotide polymorphism

References

1. Abrahams, B.S.; Geschwind, D.H. Advances in autism genetics: On the threshold of a new neurobiology. *Nat. Rev.* **2008**, *9*, 341–355. [CrossRef] [PubMed]
2. Johnson, C.P. Myers SM: Identification and evaluation of children with autism spectrum disorder. *Pediatrics* **2007**, *120*, 188–197. [CrossRef] [PubMed]
3. London, E.; Etzel, R.A. The environment as an etiologic factor in autism: A new direction for research. *Environ. Health Perspect.* **2000**, *108*, 401–404. [CrossRef] [PubMed]
4. Tordjman, S.; Somogyi, E.; Coulon, N.; Kermarrec, S.; Cohen, D.; Bronsard, G.; Bonnot, O.; Weismann-Arcache, C.; Botbol, M.; Lauth, B.; et al. Gene × environment interactions in autism spectrum disorders: Role of epigenetic mechanisms. *Front. Psychiatr.* **2014**, *5*, 1–17. [CrossRef] [PubMed]
5. Mezzelani, A.; Landini, M.; Facchiano, F.; Raggi, M.E.; Villa, L.; Molteni, M.; de Santis, B.; Brera, C.; Caroli, A.M.; Milanesi, L.; et al. Environment, dysbiosis, immunity and sex-specific susceptibility: An evidence-based translational hypothesis for regressive autism pathogenesis. *Nutr. Neurosci.* **2014**, *1*, 1–17. [CrossRef] [PubMed]
6. Yu, J.; He, X.; Yao, D.; Li, Z.; Zhao, Z. A sex specific association of common variants of neuroligin genes (*NLGN3* and *NLGN4X*) with autism spectrum disorders in a Chinese Han cohort. *Behav. Brain Funct.* **2011**, *7*, 13–23. [CrossRef] [PubMed]
7. David, B.; Scherer, S.W. Genetic architecture in autism spectrum disorder. *Curr. Opin. Genet. Dev.* **2012**, *22*, 229–237.
8. Talkowsky, M.E.; Minikel, E.V.; Gusella, J.F. Autism spectrum disorder genetics: Diverse genes with diverse clinical outcomes. *Harv. Rev. Psychiatr.* **2014**, *22*, 65–74. [CrossRef] [PubMed]
9. Piton, A.; Gauthier, J.; Hamdam, F.F.; Lafreniere, R.G.; Yang, Y.; Henrion, E.; Laurent, S.; Noreau, A.; Thibodeau, P.; Karemera, L.; et al. Systematic resequencing of X-chromosome synaptic genes in autism spectrum disorder ad schizophrenia. *Mol. Psychiatry* **2011**, *16*, 867–880. [CrossRef] [PubMed]
10. Vorstman, J.A.; Staal, W.G.; van Daalen, E.; van Engeland, H.; Hochstenbach, P.F.; Franke, L. Identification of novel autism candidate regions through the analysis of reported cytogenetic abnormalities associated with autism. *Mol. Psychiatr.* **2006**, *11*, 18–28. [CrossRef] [PubMed]
11. Betancur, C. Etiological heterogeneity in autism spectrum disorders: More than 100 genetic and genomic disorders and still counting. *Brain Res.* **2011**, *1380*, 42–77. [CrossRef] [PubMed]
12. Jamain, S.; Quach, H.; Betancur, C.; Råstam, M.; Colineaux, C.; Gillberg, I.C.; Soderstrom, H.; Giros, B.; Leboyer, M.; Gillberg, C.; et al. Paris autism research international sibpair study: Mutations of the X-linked genes encoding neuroligins NLGN3 and NLGN4 are associated with autism. *Nat. Genet.* **2003**, *34*, 27–29. [CrossRef] [PubMed]
13. Laumonnier, F.; Bonnet-Brilhault, F.; Gomot, M.; Blanc, R.; David, A.; Moizard, M.P.; Raynaud, M.; Ronce, N.; Lemonnier, E.; Calvas, P.; et al. X-linked mental retardation and autism are associated with a mutation in the *NLGN4* gene, a member of the neuroligin family. *Am. J. Hum. Genet.* **2004**, *74*, 552–557. [CrossRef] [PubMed]
14. Shibayama, A.; Cook, E.H.; Feng, J.; Glanzmann, C.; Yan, J.; Craddock, N.; Jones, I.R.; Goldman, D.; Heston, L.L.; Sommer, S.S. MECP2 structural and 3' UTR variants in schizophrenia, autism and other psychiatric diseases; a possible association with autism. *Am. J. Med. Genet. B Neuropsychiatr. Genet.* **2004**, *128*, 50–53. [CrossRef] [PubMed]
15. Yan, J.; Oliveira, G.; Coutinho, A.; Yang, C.; Feng, J.; Katz, C.; Sram, J.; Bockholt, A.; Jones, I.R.; Craddock, N.; et al. Analysis of the neuroligin 3 and 4 genes in autism and other neuropsychiatric patients. *Mol. Psychiatr.* **2005**, *10*, 329–332. [CrossRef] [PubMed]

16. Chih, B.; Engelman, H.; Scheiffele, P. Control of excitatory and inhibitory synapse formation by neuroligins. *Science* **2005**, *307*, 1324–1328. [CrossRef] [PubMed]

17. Lisè, M.F.; El-Husseini, A. The neuroligin and neurexin families: From structure to function at the synapse. *Cell. Mol. Life Sci.* **2006**, *63*, 1833–1849. [CrossRef] [PubMed]

18. Varoqueaux, F.; Aramuni, G.; Rawson, R.L.; Mohrmann, R.; Missler, M.; Gottmann, K.; Zhang, W.; Südhof, T.C.; Brose, N. Neuroligins determine synapse maturation and function. *Neuron* **2006**, *51*, 741–754. [CrossRef] [PubMed]

19. Pettem, K.L.; Yokomaku, D.; Takahashi, H.; Ge, Y.; Craig, A.M. Interaction between autism-linked MDGAs and neuroligins suppresses inhibitory synapse development. *J. Cell Biol.* **2013**, *200*, 321–336. [CrossRef] [PubMed]

20. Südhof, T.C. Neuroligins and neurexins link synaptic function to cognitive disease. *Nature* **2008**, *455*, 903–911. [CrossRef] [PubMed]

21. Zoghbi, H.Y.; Bear, M.F. Synaptic dysfunction in neurodevelopmental disorders associated with autism and intellectual disabilities. *Cold Spring Harb. Perspect. Biol.* **2012**, *4*. [CrossRef] [PubMed]

22. Chih, B.; Afridi, S.K.; Clark, L.; Scheiffele, P. Disorder associated mutations lead to functional inactivation of neuroligins. *Hum. Mol. Genet.* **2004**, *13*, 1471–1477. [CrossRef] [PubMed]

23. Comoletti, D.; De Jaco, A.; Jennings, L.L.; Flynn, R.E.; Gaietta, G.; Tsigelny, I.; Ellisman, M.H.; Taylor, P. The Arg451Cys-neuroligin 3 mutation associated with autism reveals a defect in protein processing. *J. Neurosci.* **2004**, *24*, 4889–4893. [CrossRef] [PubMed]

24. Rothwell, P.E.; Fuccillo, M.V.; Maxeiner, S.; Hayton, S.J.; Gokce, O.; Lim, B.K.; Fowler, S.C.; Malenka, R.C.; Südhof, T.C. Autism-associated neuroligin 3 mutations commonly impair striatal circuits to boost repetitive behaviors. *Cell* **2014**, *158*, 198–212. [CrossRef] [PubMed]

25. Tabuchi, K.; Blundell, J.; Etherton, M.R.; Hammer, R.E.; Liu, X.; Powell, C.M.; Südhof, T.C. A neuroligin 3 mutation implicated in autism increases inhibitory synaptic transmission in mice. *Science* **2007**, *318*, 71–76. [CrossRef] [PubMed]

26. Jamain, S.K.; Hammerschmidt, R.K.; Granon, S.; Boretius, S.; Varoqueaux, F.; Ramanantsoa, N.; Gallego, J.; Ronnenberg, A.; Winter, D.; Frahm, J.; et al. Reduced social interaction and ultrasonic communication in a mouse model of monogenic heritable autism. *Proc. Natl. Acad. Sci. USA* **2008**, *105*, 1710–1715. [CrossRef] [PubMed]

27. Zhang, C.; Milunsky, J.M.; Newton, S.; Ko, J.; Zhao, G.; Maher, T.A.; Tager-Flusberg, H.; Bolliger, M.F.; Carter, A.S.; Boucard, A.A.; et al. A neuroligin 4 missense mutation associated with autism impairs neuroligin 4 folding and endoplasmic reticulum export. *J. Neurosci.* **2009**, *29*, 10843–10854. [CrossRef] [PubMed]

28. Hu, X.; Xiong, Z.; Zhang, L.; Liu, Y.; Lu, L.; Peng, Y.; Guo, H.; Zhao, J.; Xia, K.; Hu, Z. Variations analysis of *NLGN3* and *NLGN4X* gene in Chinese autism patients. *Mol. Biol. Rep.* **2014**, *41*, 4133–4140.

29. Talebizadeh, Z.; Lam, D.Y.; Theodoro, M.F.; Bittel, D.C.; Lushington, G.H.; Butler, M.G. Novel splice isoforms for NLGN3 and NLGN4 with possible implication in autism. *J. Med. Genet.* **2006**, *43*, e21. [CrossRef] [PubMed]

30. Lawson-Yuen, A.; Saldivar, J.S.; Sommer, S.; Picker, J. Familial deletion within *NLGN4* associated with autism and Tourette syndrome. *Eur. J. Hum. Genet.* **2008**, *16*, 614–618. [CrossRef] [PubMed]

31. Ylisaukko-oja, T.; Rehnström, K.; Auranen, M.; Vanhala, R.; Alen, R.; Kempas, E.; Ellonen, P.; Turunen, J.A.; Makkonen, I.; Riikonen, R.; et al. Analysis of four neuroligin genes as candidates for autism. *Eur. J. Hum. Genet.* **2005**, *13*, 1285–1292. [CrossRef] [PubMed]

32. Daoud, H.; Bonnet-Brilhault, F.; Vesdrine, S.; Demattei, M.V.; Vourc'h, P.; Bayou, N.; Andres, C.R.; Barthélémy, C.; Laumonnier, F.; Briault, S. Autism and nonsyndromic mental retardation associated with a de novo mutation in the *NLGN4X* gene promoter causing an increased expression level. *Soc. Biol. Psychiatry* **2009**, *66*, 906–910. [CrossRef] [PubMed]

33. Steinberg, K.M.; Ramachandran, D.; Patel, V.C.; Shetty, A.C.; Cutler, D.J.; Zwick, M.E. Identification of rare X-linked neuroligin variants by massively parallel sequencing in males with autism spectrum disorder. *Mol. Autism* **2012**, *3*, 8–20. [CrossRef] [PubMed]

34. Yanagi, K.; Kaname, T.; Wakui, K.; Hashimoto, O.; Fukushima, Y.; Naritomi, K. Identification of four novel synonymous substitutions in neuroligin 3 and neuroligin 4X in Japanase patients with autistic spectrum disorder. *Autism Res. Treat.* **2012**, *2012*, 724072–724077. [PubMed]

35. Volaki, K.; Pampanos, A.; Kitsiou-Tzeli, S.; Vrettou, C.; Oikonomakis, V.; Sofocleous, C.; Kanavakis, E. Mutation screening in the Greek population and evaluation of *NLGN3* and *NLGN4X* genes causal factors for autism. *Psychiatr. Genet.* **2013**, *23*, 198–203. [CrossRef] [PubMed]

36. Qi, H.; Xing, L.; Zhang, K.; Gao, X.; Zheng, Z.; Huang, S.; Guo, Y.; Zhang, F. Positive association of neuroigin-4 gene with non specific mental retardation in the Qinba mountains region of China. *Psychiatr. Genet.* **2009**, *19*, 1–5. [CrossRef] [PubMed]

37. Zhang, K.; Gao, X.; Qi, H.; Zheng, Z.; Zhang, F. Gender differencs in cognitive ability associated with genetic variants of *NLGN4*. *Neuropsychiatry* **2010**, *62*, 221–228.

38. Liu, Y.; Du, Y.; Liu, W.; Yang, C.; Liu, Y.; Wang, H.; Gong, X. Lack of association between *NLGN3*, *NLGN4*, *SHANK2* and *SHANK3* gene variants and autism spectrum disorder in a Chinese population. *PLoS ONE* **2013**, *8*, e56639. [CrossRef] [PubMed]

39. Blasi, F.; Bacchelli, E.; Pesaresi, G.; Carone, S.; Bailey, A.J.; Maestrini, E. International molecular genetic study of autism consortium (IMGAC): Absence of coding mutations in the X-linked genes neuroligin 3 and neuroligin 4 in individuals with autism from the IMGSAC collection. *Am. J. Med. Genet. B Neuropsychiatr. Genet.* **2006**, *141*, 220–221. [CrossRef] [PubMed]

40. Wermenter, A.K.; Kamp-Becker, I.; Strauch, K.; Schulte-Korne, G.; Remschmidt, H. No evidence for involvement of genetic variants in the X-linked neuroligin genes *NLGN3* and *NLGN4X* in probands with autism spectrum disorder on high functioning level. *Am. J. Med. Genet. B Neuropsychiatr. Genet.* **2008**, *147*, 535–537. [CrossRef] [PubMed]

41. Gauthier, J.; Bonnel, A.; St-Onge, J.; Karamera, L.; Laurent, S.; Mottron, L.; Fombonne, E.; Joober, R.; Rouleau, G.A. NLGN3/NLGN4 gene mutations are not responsible for autism in the Quebec population. *Am. J. Med. Genet. B Neuropsychiatr. Genet.* **2005**, *132*, 74–75. [CrossRef] [PubMed]

42. Altshuler, D.; Durbin, R.M.; Abecasis, G.R.; Bentley, D.R.; Chakravarti, A.; Clark, A.G.; Collins, F.S.; De La Vega, F.M.; Donnelly, P.; Egholm, M.; et al. A map of human genome variation from population-scale sequencing. *Nature* **2010**, *467*, 1061–1073.

43. Salvi, E.; Kutalik, Z.; Glorioso, N.; Benaglio, P.; Frau, F.; Kuznetsova, T.; Arima, H.; Hoggart, C.; Tichet, J.; Nikitin, Y.P.; et al. Genome wide association study using a high-density single nucleotide polymorphism array and case-control design identifies a novel essential hypertension susceptibility locus in the promoter region of endothelial NO synthase. *Hypertension* **2012**, *59*, 248–255. [CrossRef] [PubMed]

44. Howie, B.; Fuchsberger, C.; Stephens, M.; Marchini, J.; Abecasis, G.R. Fast and accurate genotype imputation in genome-wide association studies through pre-phasing. *Nat. Genet.* **2012**, *44*, 955–959. [CrossRef] [PubMed]

45. Barrett, J.C.; Fry, B.; Maller, J.; Daly, M.J. Haploview: Analysis and visualization of LD and haplotype maps. *Bioinformatics* **2005**, *21*, 263–265. [CrossRef] [PubMed]

46. Purcell, S.; Neale, B.; Todd-Brown, K.; Thomas, L.; Ferreira, M.A.R.; Bender, D.; Maller, J.; Sklar, P.; De Bakker, P.I.W.; Daly, M.J.; et al. PLINK: A toolset for whole-genome association and population-based linkage analysis. *Am. J. Hum. Genet.* **2007**, *81*, 559–575. [CrossRef] [PubMed]

International Journal of
Molecular Sciences

MDPI

Article

A Study of Single Nucleotide Polymorphisms of the *SLC19A1/RFC1* Gene in Subjects with Autism Spectrum Disorder

Naila Al Mahmuda [1,*], Shigeru Yokoyama [1], Jian-Jun Huang [1], Li Liu [1], Toshio Munesue [1], Hideo Nakatani [2], Kenshi Hayashi [3], Kunimasa Yagi [4], Masakazu Yamagishi [3] and Haruhiro Higashida [1,*]

[1] Research Center for Child Mental Development, Kanazawa University, Kanazawa 920-8640, Japan; shigeruy@med.kanazawa-u.ac.jp (S.Y.); jianjun453@163.com (J.-J.H.); liuli011258@sina.com (L.L.); munesue@med.kanazawa-u.ac.jp (T.M.)
[2] Division of Neuroscience, Kanazawa University Graduate School of Medical Science, Kanazawa 920-8640, Japan; nak@yd5.so-net.ne.jp
[3] Division of Cardiovascular Medicine, Kanazawa University Graduate School of Medical Science, Kanazawa 920-8640, Japan; kenshi@med.kanazawa-u.ac.jp (K.H.); myamagi@med.kanazawa-u.ac.jp (M.Y.)
[4] Medical Education Research Center, Kanazawa University Graduate School of Medical Science, Kanazawa 920-8640, Japan; diabe@med.kanazawa-u.ac.jp
* Correspondence: nlmahmuda@gmail.com (N.A.M.); haruhiro@med.kanazawa-u.ac.jp (H.H.); Tel.: +81-76-265-2457 (N.A.M. & H.H.); Fax: +81-76-234-4213 (N.A.M. & H.H.)

Academic Editor: Merlin G. Butler
Received: 19 February 2016; Accepted: 2 May 2016; Published: 19 May 2016

Abstract: Autism Spectrum Disorder (ASD) is a group of neurodevelopmental disorders with complex genetic etiology. Recent studies have indicated that children with ASD may have altered folate or methionine metabolism, suggesting that the folate–methionine cycle may play a key role in the etiology of ASD. *SLC19A1*, also referred to as reduced folate carrier 1 (*RFC1*), is a member of the solute carrier group of transporters and is one of the key enzymes in the folate metabolism pathway. Findings from multiple genomic screens suggest the presence of an autism susceptibility locus on chromosome 21q22.3, which includes *SLC19A1*. Therefore, we performed a case-control study in a Japanese population. In this study, DNA samples obtained from 147 ASD patients at the Kanazawa University Hospital in Japan and 150 unrelated healthy Japanese volunteers were examined by the sequence-specific primer-polymerase chain reaction method pooled with fluorescence correlation spectroscopy. $p < 0.05$ was considered to represent a statistically significant outcome. Of 13 single nucleotide polymorphisms (SNPs) examined, a significant p-value was obtained for AA genotype of one SNP (*rs1023159*, OR = 0.39, 95% CI = 0.16–0.91, $p = 0.0394$; Fisher's exact test). Despite some conflicting results, our findings supported a role for the polymorphism *rs1023159* of the *SLC19A1* gene, alone or in combination, as a risk factor for ASD. However, the findings were not consistent after multiple testing corrections. In conclusion, although our results supported a role of the *SLC19A1* gene in the etiology of ASD, it was not a significant risk factor for the ASD samples analyzed in this study.

Keywords: autism spectrum disorder; reduced folate carrier; single nucleotide polymorphism

1. Introduction

Autism spectrum disorder (ASD) is a devastating neurodevelopmental disorder with a complex biological basis and is thought to involve multiple and variable gene–environment interactions. ASD is characterized by social impairments, communication problems, and restricted repetitive behaviors [1].

Most candidate genes currently implicated in ASD are involved in neurodevelopmental pathways, social-emotional behavior, or sex or neuropeptide hormonal signaling [2].

The *SLC19A1* gene on human chromosome 21q22.3 [3] encodes one of the key enzymes in the folate metabolism pathway. *SLC19A1*, also referred to as reduced folate carrier 1 (RFC1), functions as a bidirectional anion exchanger, accepting folate cofactors and exporting various organic anions. *SLC19A1* has five exons that contain the total open reading frame (ORF) [4–6]. The ORF of human *SLC19A1* cDNA encodes a protein with 12 transmembrane domains and a single *N*-glycosylation site [3,7–9]. *SLC19A1* mRNA is detectable in all human tissues [10].

Recent studies indicated that children with ASD may have changed folate or methionine metabolism, suggesting that the folate–methionine cycle may play an important role in the etiology of ASD [11]. Many important genes, including *SLC19A1*, are involved in the folate metabolism pathway and their roles in human diseases, such as gastric and esophageal cancers, have been studied in depth [12,13]. A marginal association with ASD was identified for a 19-bp deletion in the dihydrofolate reductase (*DHFR*) gene (odds ratio (OR): 2.69; 95% CI: 1.00–7.28; $p < 0.05$), which is involved in folate metabolism [14]. Common variants of the decreased folate carrier (*RFC*) and methylene tetrahydrofolate reductase (*MTHFR*) genes conferred increased susceptibility to ASD, suggesting a potential etiological role of impaired folate-dependent one-carbon metabolism in susceptibility to ASD [15].

However, the findings for genes involved in folate transport have been inconsistent between reports. Although the largest study to date found an important association between the *SLC19A1* gene and ASD [15], a subsequent study failed to replicate this finding [16]. Other studies have not identified any mutations in genes included in folate transport in ASD populations [17–19].

Here, we hypothesized that genetic variants in *SLC19A1* may play a role in the pathways that are altered in ASD and can therefore be considered candidate genes for testing in ASD patients. We performed a case-control study of 13 genetic variations to assess the involvement of *SLC19A1* in ASD. The study was performed in a Japanese population, in which genetic variants of *CD38* and *BST-1/CD157* were reported to be associated with increased risk of ASD [20,21].

2. Results

Thirteen SNPs were analyzed in this study, five of which (*rs914232, rs3788205, rs1023159, rs944423,* and *rs9979087*) were located in the *SLC19A1* gene region; these were subjected to statistical analysis. The eight other SNPs (*rs1888533, rs11700708, rs12627639, rs2838965, rs6518253, rs9974061, rs9980967,* and *rs2838968*) were located in the adjacent region. Two SNPs (*rs9980967, rs9979087*) were excluded due to insufficient genotyping data. However, there were no significant associations between any of these SNPs and ASD, with the exception of *rs1023159*. As the results suggested a role ($p = 0.0394$; Table 1) of this polymorphism alone or in combination with others as a risk factor for ASD, this SNP was subjected to further analysis. No association was found after multiple testing corrections. Tests of Hardy–Weinberg equilibrium deviations were performed for each marker in two groups of case and control individuals, and polymorphisms showed evidence of deviation from Hardy–Weinberg equilibrium. The genotyping rate was above 95%. LD analysis of these SNPs identified three haplotype blocks, one of which (Block 1; Figure 1) consisted of two SNPs including one (*rs1023159*) with the lowest *p*-value ($p = 0.0394$; Table 1) among those analyzed.

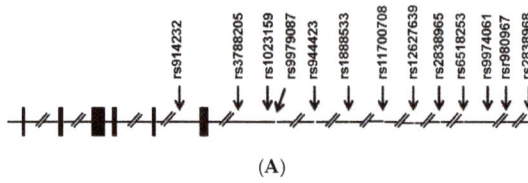

(A)

Figure 1. *Cont.*

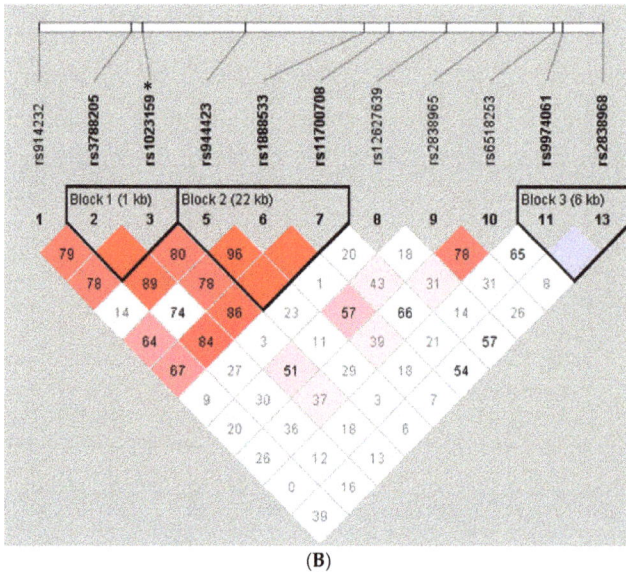

(B)

Figure 1. The genomic structure of *SLC19A1* (**A**). Bars, exons. Arrows, positions of single nucleotide polymorphisms (SNPs). Linkage disequilibrium plot of SNPs in the samples studied (**B**). Numbers in squares indicate D' values. Reference Number (rs) with asterisk indicates the SNP with $p < 0.05$. The blocks are defined following the four-gamete rule [22]. Explanation of color scheme: If $D' < 1$ and LOD (log of the likelihood odds ratio) <2, the cell color is white; if $D' = 1$ and LOD < 2, the cell color is blue; if $D' < 1$ and LOD $\geqslant 2$, the cell color is shades of pink/red; if $D' = 1$ and LOD $\geqslant 2$, the cell color is bright red.

Table 1. Genotype and allele frequencies of *rs1023159* at Kanazawa University Hospital for autism spectrum disorder (ASD).

rs1023159	Cases	Control	Odds Ratio (95% CI)	p
Genotype	(n = 144)	(n = 146)		
G/G	72 (50.0%)	62 (42.5%)	Reference	
A/G	63 (43.8%)	64 (43.8%)	0.85 (0.52–1.4)	0.5368
A/A	9 (6.3%)	20 (13.7%)	0.39 (**0.16–0.91**)	0.0394
Allele	(n = 288)	(n = 292)		
G	207 (71.9%)	188 (64.4%)	Reference	
A	81 (28.1%)	104 (35.6%)	0.71 (0.50–1.0)	0.0613

CI, confidence interval; *p*-values obtained by Fisher's exact test are given; $p < 0.05$ is indicated in bold.

3. Discussion

In this population-based case-control study, we investigated the relationship between polymorphisms in the *SLC19A1* gene and risk of ASD in a Japanese population. We identified no significant associations between SNPs of the *SLC19A1* gene and ASD, with the exception of one SNP, although the results eventually did not support a role of the SLC19A1 gene in the etiology of ASD in our sample.

We have also calculated the genotype and allele frequencies of *rs1023159* in Autism Genome Resources Exchange (AGRE) samples (Table 2). The frequency (15.5%) of the AA genotype in AGRE group of ASD cases, although with different ethnicity, was similar to the frequency (13.7%) observed in the group of Japanese controls.

Table 2. Genotype and allele frequencies of *rs1023159* in KU samples and AGRE samples for autism spectrum disorder (ASD).

rs1023159	KU	AGRE
Genotype	(*n* = 144)	(*n* = 191)
G/G	72 (50.0%)	63 (31.4%)
A/G	63 (43.8%)	104 (53.1%)
A/A	9 (6.3%)	30 (15.5%)
Allele	(*n* = 288)	(*n* = 394)
G	207 (71.9%)	230 (58.4%)
A	81 (28.1%)	164 (41.6%)

KU, Kanazawa University; AGRE, Autism Genome Resources Exchange.

Recent genetic studies recognized the contribution of the *SLC19A1* gene to neural tube defects (NTD) [23–28]. It was suggested that the maternal G allele may be a causative genetic risk factor for having a child with ASD independent of the child's genotype [29]. In case-control analysis, a significant increase in the *SLC19A1* G allele frequency was discovered among mothers of children with ASD, but not among affected children, and analysis of the *SLC19A1* A80G genotype within family trios discovered that the maternal G allele was allied with a significant increase in risk of ASD, whereas the inherited genotype of the children was not [29].

Evidence indicates that expression of *SLC19A1* in the intestine is subject to adaptive regulation in response to folate status [30]. Folic acid is the inactive, oxidized form of folate compounds that is important for many physiological systems of the body. Folate is the major one-carbon donor for *de novo* nucleotide synthesis for DNA replication and also for remethylation of homocysteine to methionine for essential methylation reactions [29]. The folate cycle interacts with the methionine cycle as well as the tetrahydrobiopterin construction and salvage pathways. Insufficiencies in folate can lead to anomalies in these pathways [31]. The methionine cycle is important for DNA methylation [15], a process that is important in regulating gene expression. Folate deficiency during various significant stages of fetal and infantile development upsets structural and functional alteration of the brain [32].

The reduced folate carrier is the principal mechanism by which folates and antifolates are delivered to mammalian cells and tissues [33]. As folate transport across cell membranes is mediated in part by the RFC, variants within this gene may affect the disease risk via an effect on folate and/or homocysteine levels [34]. Low levels of RFC could result in a number of pathophysiological states associated with folate deficiency, including cardiovascular disease, fetal anomalies, and neurological disorders [33]. Moretti *et al.* reported a 6-year-old girl with developmental delay, psychomotor regression, seizures, mental retardation, and autistic features associated with low cerebrospinal fluid (CSF) levels of 5-methyltetrahydrofolate, the biologically active form of folate in the CSF and blood [35]. Several studies reported considerably low CSF folate concentrations together with normal serum folate concentrations in children with autism [18,19,35–37].

SLC19A1 is situated on the CSF side of the choroid plexus, where it enables transport of concentrated folate into the CSF [11]. Taken together, variation in *SLC19A1* expression may involve both neuronal structures and metabolism in the Central Nervous System (CNS). Defective transport of folate into the CNS is related to cerebral folate deficiency (CFD), a neurological disorder that is important in diagnosis of children with unexplained neurodevelopmental symptoms, which suggests the possible involvement of the folate-methionine pathway in ASD [31]. Further, early-onset low-functioning autism with neurological deficits has been suggested as a characteristic of children with both autism and CFD [18,19,35,38].

The major limitation of this study was the small sample size, resulting in deviation from Hardy–Weinberg equilibrium and limited power (68%) to reliably detect the role of *SLC19A1* in ASD. We did not recognize any population stratification, admixture, and cryptic relation among the subjects in the present study, which may have contributed to the lack of association in this small sample. Another limitation was the lack of a replication cohort. Further studies with larger sample sizes and/or family-based association testing are needed to clarify the precise role of this gene in ASD. However, our findings were consistent with reports that *SLC19A1* may not contribute to genetic susceptibility to ASD in some populations.

4. Materials and Methods

4.1. Study Population

The study population consisted of 147 ASD subjects (113 males, 34 females; 15.6 ± 0.6 years old, mean ± s.e.m.) from the Outpatient Psychiatry Department of Kanazawa University Hospital, as described previously [20,39]. All subjects satisfied the Diagnostic and Statistical Mannual of Mental Disorders-IV (DSM-IV) criteria for pervasive developmental ailment and Childhood Autism Rating Scale. Two experienced child psychiatrists established the diagnosis of ASD in all patients based on semi-structured behavioral observations and conversations with the subjects and their parents. The interview structure and clinical records were described previously [20]. One of the following methods was used as an aid to evaluate the autism-specific behaviors and symptoms during interviews with parents: the Asperger Syndrome Diagnostic Interview [40], Autism Diagnostic Interview-Revised (ADI-R) [41], Pervasive Developmental Disorders Autism Society Japan Rating Scale [42], Diagnostic Interview for Social and Communication Disorders [43], or Tokyo Autistic Behavior Scale [44]. A total of 150 individuals were recruited as controls (115 males, 35 females; 23.8 ± 0.3 years old). All patients and controls were Japanese with no non-Japanese parents or grandparents. These controls were part of a stock used frequently for single nucleotide polymorphism (SNP) analysis of ion channels related to arrhythmia at Kanazawa University Heart Center. This study was approved by the ethics committee of Kanazawa University School of Medicine (July 2015), and all participants and/or their caregivers provided informed consent. The study protocol was performed in accordance with the Declaration of Helsinki.

4.2. Genotyping

Genomic DNA was extracted as described previously [39] from venous blood samples using a commercial kit (Wizard Genomic DNA Purification kit; Promega, Madison, WI, USA) or from nails using an ISOHAIR DNA extraction kit (Nippon Gene, Tokyo, Japan). The genomic DNA samples were subjected to whole-genome amplification, and SNPs were determined by the sequence-specific primer-polymerase chain reaction (SSP-PCR) method followed by fluorescence correlation spectroscopy as described by Bannai *et al.* [45]. We selected a set of tagging SNPs that capture common variations and linkage disequilibrium (LD) structures across the *SLC19A1* gene using the Tagger program incorporated with Haploview v4.2 software(Broad Institute of MIT and Harvard, Cambridge, MA, USA). The data source for tagging SNPs was the dbSNP database [46] and the HapMap genome browser, release 27 (operated by the National Institutes of Health (NIH), Bethesda, MD, USA) in the

JPT (Japanese individuals from Tokyo, Japan), CHB (Han Chinese individuals from Beijing, China), ASW (African ancestry in Southwest USA), and CEU (Utah residents of northern and western European ancestry) populations. Selection of tagging SNPs was based on pairwise tagging only and the minor allele frequency was ≥5% in any one of the different ethnicities.

4.3. Statistical Analysis

Genotype and allele frequencies were examined using a contingency table and Fisher's exact test (GraphPad Prism 6; GraphPad Software, San Diego, CA, USA), and $p < 0.05$ was taken to indicate statistical significance. We also used the method of Nyholt [47], which estimates an "effective number" of independent tests and then adjusts the smallest observed p-value using simulation based on this number of tests. In our samples, the estimated effective number for independent tests was 9 and the p-value was 0.005. The observed genotype frequency distributions were compared with those expected from the Hardy–Weinberg equilibrium and analyzed by the chi "χ" squared test.

Statistical power was calculated using the Genetic Power Calculator [48,49] assuming a population prevalence of 0.015 for ASD [50], and a D' value of 1 between the marker and disease with a false positive rate of 5%.

5. Conclusions

This study showed no evidence supporting a role of the *SLC19A1* gene in the etiology of ASD. The ethnic and cultural background may have influenced the results of our study. However, these findings warrant additional discussion and confirmation in subsequent studies. Further cellular and molecular studies are required to elucidate the precise role of this gene in ASD.

Acknowledgments: We acknowledge financial support from the Strategic Research Program for Brain Sciences from the Japan Agency for Medical Research and Development and Grants-in-Aid for Scientific Research (24590375 and 25461335) from the Japan Society for Promotion of Sciences. We thank AGRE for their cooperation.

Author Contributions: Haruhiro Higashida, Shigeru Yokoyama, Jian-Jun Huang and Li Liu conceived and designed the research. Shigeru Yokoyama and Naila Al Mahmuda performed experiments and analyzed data. Toshio Munesue, Kenshi Hayashi, Kunimasa Yagi, Masakazu Yamagishi and Hideo Nakatani contributed participant recruitment, clinical assessment and sample collections. Naila Al Mahmuda prepared the initial draft; and Shigeru Yokoyama, Naila Al Mahmuda, and Haruhiro Higashida revised the manuscript. All authors reviewed the final manuscript and approved its publication.

Conflicts of Interest: The authors declare no conflict of interest.

References

1. American Psychiatric Association. *Diagnostic and Statistical Mannual of Mental Disorders (DSM-V)*, 5th ed.; American Psychiatric Pub.: Washington, DC, USA, 2013.
2. Chakrabarti, B.; Dudbridge, F.; Kent, L.; Wheelwright, S.; Hill-Cawthorne, G.; Allison, C.; Banerjee-Basu, S.; Baron-Cohen, S. Genes related to sex steroids, neural growth, and social–emotional behavior are associated with autistic traits, empathy, and Asperger syndrome. *Autism Res.* **2009**, *2*, 157–177. [CrossRef] [PubMed]
3. Moscow, J.A.; Gong, M.; He, R.; Sgagias, M.K.; Dixon, K.H.; Anzick, S.L.; Meltzer, P.S.; Cowan, K.H. Isolation of a gene encoding a human reduced folate carrier (RFC1) and analysis of its expression in transport-deficient, methotrexate-resistant human breast cancer cells. *Cancer Res.* **1995**, *55*, 3790–3794. [PubMed]
4. Tolner, B.; Roy, K.; Sirotnak, F.M. Structural analysis of the human RFC-1 gene encoding a folate transporter reveals multiple promoters and alternatively spliced transcripts with 5′ end heterogeneity. *Gene* **1998**, *211*, 331–341. [CrossRef]
5. Williams, F.M.; Flintoff, W.F. Structural organization of the human reduced folate carrier gene: Evidence for 5′ heterogeneity in lymphoblast mRNA. *Somat. Cell Mol. Genet.* **1998**, *24*, 143–156. [CrossRef] [PubMed]
6. Zhang, L.; Wong, S.C.; Matherly, L.H. Structure and organization of the human reduced folate carrier gene1. *Biochim. Biophys. Acta* **1998**, *1442*, 389–393. [CrossRef]
7. Prasad, P.D.; Ramamoorthy, S.; Leibach, F.H.; Ganapathy, V. Molecular Cloning of the Human Placental Folate Transporter. *Biochem. Biophys. Res. Commun.* **1995**, *206*, 681–687. [CrossRef] [PubMed]

8.   Williams, F.M.R.; Flintoff, W.F. Isolation of a Human cDNA that Complements a Mutant Hamster Cell Defective in Methotrexate Uptake. *J. Biol. Chem.* **1995**, *270*, 2987–2992. [PubMed]

9.   Wong, S.C.; Proefke, S.A.; Bhushan, A.; Matherly, L.H. Isolation of human cDNAs that restore methotrexate sensitivity and reduced folate carrier activity in methotrexate transport-defective Chinese Hamster ovary cells. *J. Biol. Chem.* **1995**, *270*, 17468–17475. [CrossRef] [PubMed]

10. Ganapathy, V.; Smith, S.; Prasad, P. SLC19: The folate/thiamine transporter family. *Pflugers Arch.* **2004**, *447*, 641–646. [CrossRef] [PubMed]

11. Main, P.A.; Angley, M.T.; Thomas, P.; O'Doherty, C.E.; Fenech, M. Folate and methionine metabolism in autism: A systematic review. *Am. J. Clin. Nutr.* **2010**, *91*, 1598–1620. [CrossRef] [PubMed]

12. Wang, L.; Chen, W.; Wang, J.; Tan, Y.; Zhou, Y.; Ding, W.; Hua, Z.; Shen, J.; Xu, Y.; Shen, H. Reduced folate carrier gene G80A polymorphism is associated with an increased risk of gastroesophageal cancers in a Chinese population. *Eur. J. Cancer* **2006**, *42*, 3206–3211. [CrossRef] [PubMed]

13. Zhang, Z.; Xu, Y.; Zhou, J.; Wang, X.; Wang, L.; Hu, X.; Guo, J.; Wei, Q.; Shen, H. Polymorphisms of thymidylate synthase in the 5′- and 3′-untranslated regions associated with risk of gastric cancer in South China: A case-control analysis. *Carcinogenesis* **2005**, *26*, 1764–1769. [CrossRef] [PubMed]

14. Adams, M.; Lucock, M.; Stuart, J.; Fardell, S.; Baker, K.; Ng, X. Preliminary evidence for involvement of the folate gene polymorphism 19 bp deletion-DHFR in occurrence of autism. *Neurosci. Lett.* **2007**, *422*, 24–29. [CrossRef] [PubMed]

15. James, S.J.; Melnyk, S.; Jernigan, S.; Cleves, M.A.; Halsted, C.H.; Wong, D.H.; Cutler, P.; Bock, K.; Boris, M.; Bradstreet, J.J.; *et al.* Metabolic endophenotype and related genotypes are associated with oxidative stress in children with autism. *Am. J. Med. Genet. B Neuropsychiatr. Genet.* **2006**, *141B*, 947–956. [CrossRef] [PubMed]

16. Paşca, S.P.; Nemeş, B.; Vlase, L.; Gagyi, C.E.; Dronca, E.; Miu, A.C.; Dronca, M. High levels of homocysteine and low serum paraoxonase 1 arylesterase activity in children with autism. *Life Sci.* **2006**, *78*, 2244–2248. [CrossRef] [PubMed]

17. Gordon, N. Cerebral folate deficiency. *Dev. Med. Child Neurol.* **2009**, *51*, 180–182. [CrossRef] [PubMed]

18. Moretti, P.; Peters, S.U.; del Gaudio, D.; Sahoo, T.; Hyland, K.; Bottiglieri, T.; Hopkin, R.J.; Peach, E.; Min, S.H.; Goldman, D.; *et al.* Brief Report: Autistic Symptoms, Developmental Regression, Mental Retardation, Epilepsy, and Dyskinesias in CNS Folate Deficiency. *J. Autism Dev. Disord.* **2008**, *38*, 1170–1177. [CrossRef] [PubMed]

19. Ramaekers, V.T.; Blau, N.; Sequeira, J.M.; Nassogne, M.C.; Quadros, E.V. Folate receptor autoimmunity and cerebral folate deficiency in low-functioning autism with neurological deficits. *Neuropediatrics* **2007**, *38*, 276–281. [CrossRef] [PubMed]

20. Munesue, T.; Yokoyama, S.; Nakamura, K.; Anitha, A.; Yamada, K.; Hayashi, K.; Asaka, T.; Liu, H.-X.; Jin, D.; Koizumi, K.; *et al.* Two genetic variants of CD38 in subjects with autism spectrum disorder and controls. *Neurosci. Res.* **2010**, *67*, 181–191. [CrossRef] [PubMed]

21. Yokoyama, S.; Al Mahmuda, N.; Munesue, T.; Hayashi, K.; Yagi, K.; Yamagishi, M.; Higashida, H. Association Study between the CD157/BST1 Gene and Autism Spectrum Disorders in a Japanese Population. *Brain Sci.* **2015**, *5*, 188–200. [CrossRef] [PubMed]

22. Wang, N.; Akey, J.M.; Zhang, K.; Chakraborty, R.; Jin, L. Distribution of recombination crossovers and the origin of haplotype blocks: The interplay of population history, recombination, and mutation. *Am. J. Hum. Genet.* **2002**, *71*, 1227–1234. [CrossRef] [PubMed]

23. O'Leary, V.B.; Pangilinan, F.; Cox, C.; Parle-McDermott, A.; Conley, M.; Molloy, A.M.; Kirke, P.N.; Mills, J.L.; Brody, L.C.; Scott, J.M. Reduced folate carrier polymorphisms and neural tube defect risk. *Mol. Genet. Metab.* **2006**, *87*, 364–369. [CrossRef] [PubMed]

24. Pei, L.J.; Li, Z.W.; Zhang, W.; Ren, A.G.; Zhu, H.P.; Hao, L.; Zhu, J.H.; Li, Z. Epidemiological study on reduced folate carrier gene (RFC1 A80G) polymorphism and other risk factors of neural tube defects. *J. Peking Univ. Health Sci.* **2005**, *37*, 341–345.

25. Pei, L.J.; Zhu, H.P.; Li, Z.W.; Zhang, W.; Ren, A.G.; Zhu, J.H.; Li, Z. Interaction between maternal periconceptional supplementation of folic acid and reduced folate carrier gene polymorphism of neural tube defects. *Chin. J. Med. Genet.* **2005**, *22*, 284–287.

26. Zhang, T.; Lou, J.; Zhong, R.; Wu, J.; Zou, L.; Sun, Y.; Lu, X.; Liu, L.; Miao, X.; Xiong, G. Genetic variants in the folate pathway and the risk of neural tube defects: A meta-analysis of the published literature. *PLoS ONE* **2013**, *8*, e59570. [CrossRef] [PubMed]

27. Shang, Y.; Zhao, H.; Niu, B.; Li, W.I.; Zhou, R.; Zhang, T.; Xie, J. Correlation of polymorphism of *MTHFRs* and *RFC-1* genes with neural tube defects in China. *Birth Defects Res. A Clin. Mol. Teratol.* **2008**, *82*, 3–7. [CrossRef] [PubMed]

28. De Marco, P.; Calevo, M.G.; Moroni, A.; Merello, E.; Raso, A.; Finnell, R.H.; Zhu, H.; Andreussi, L.; Cama, A.; Capra, V. Reduced folate carrier polymorphism (80A→G) and neural tube defects. *Eur. J. Hum. Genet.* **2003**, *11*, 245–252. [CrossRef] [PubMed]

29. James, S.J.; Melnyk, S.; Jernigan, S.; Lehman, S.; Seidel, L.; Gaylor, D.W.; Cleves, M.A. A functional polymorphism in the reduced folate carrier gene and DNA hypomethylation in mothers of children with autism. *Am. J. Med. Genet. B Neuropsychiatr. Genet.* **2010**, *153B*, 1209–1220. [CrossRef] [PubMed]

30. Said, H.M.; Chatterjee, N.; Haq, R.U.; Subramanian, V.S.; Ortiz, A.; Matherly, L.H.; Sirotnak, F.M.; Halsted, C.; Rubin, S.A. Adaptive regulation of intestinal folate uptake: Effect of dietary folate deficiency. *Am. J. Physiol. Cell Physiol.* **2000**, *279*, C1889–C1895. [PubMed]

31. Frye, R.E.; Rossignol, D.A. Cerebral Folate Deficiency in Autism Spectrum Disorders. *Autism Sci. Dig. J. Autsmone* **2011**. Available online: http://www.autismone.org/content/cerebral-folate-deficiency-autism-spectrum-disorders-richard-frye-md-phd-and-daniel-rossigno (accessed on 19 May 2016).

32. Ramaekers, V.; Sequeira, J.M.; Quadros, E.V. Clinical recognition and aspects of the cerebral folate deficiency syndromes. *Clin. Chem. Lab. Med.* **2013**, *51*, 497–511. [CrossRef] [PubMed]

33. Matherly, L.; Hou, Z.; Deng, Y. Human reduced folate carrier: Translation of basic biology to cancer etiology and therapy. *Cancer Metastasis Rev.* **2007**, *26*, 111–128. [CrossRef] [PubMed]

34. Stanislawska-Sachadyn, A.; Mitchell, L.E.; Woodside, J.V.; Buckley, P.T.; Kealey, C.; Young, I.S.; Scott, J.M.; Murray, L.; Boreham, C.A.; McNulty, H.; *et al.* The reduced folate carrier (*SLC19A1*) c.80G>A polymorphism is associated with red cell folate concentrations among women. *Ann. Hum. Genet.* **2009**, *73 Pt 5*, 484–491.[CrossRef] [PubMed]

35. Moretti, P.; Sahoo, T.; Hyland, K.; Bottiglieri, T.; Peters, S.; del Gaudio, D.; Roa, B.; Curry, S.; Zhu, H.; Finnell, R.H.; *et al.* Cerebral folate deficiency with developmental delay, autism, and response to folinic acid. *Neurology* **2005**, *64*, 1088–1090. [CrossRef] [PubMed]

36. Lowe, T.L.; Cohen, D.J.; Miller, S.; Young, J.G. Folic acid and B12 in autism and neuropsychiatric disturbances of childhood. *J. Am. Acad. Child Psychiatry* **1981**, *20*, 104–111. [CrossRef]

37. Ramaekers, V.T.; Rothenberg, S.P.; Sequeira, J.M.; Opladen, T.; Blau, N.; Quadros, E.V.; Selhub, J. Autoantibodies to folate receptors in the cerebral folate deficiency syndrome. *N. Eng. J. Med.* **2005**, *352*, 1985–1991. [CrossRef] [PubMed]

38. Ramaekers, V.T.; Sequeira, J.M.; Blau, N.; Quadros, E.V. A milk-free diet downregulates folate receptor autoimmunity in cerebral folate deficiency syndrome. *Dev. Med. Child Neurol.* **2008**, *50*, 346–352. [CrossRef] [PubMed]

39. Ma, W.-J.; Hashii, M.; Munesue, T.; Hayashi, K.; Yagi, K.; Yamagishi, M.; Higashida, H.; Yokoyama, S. Non-synonymous single-nucleotide variations of the human oxytocin receptor gene and autism spectrum disorders: A case-control study in a Japanese population and functional analysis. *Mol. Autism* **2013**, *4*, 22. [CrossRef] [PubMed]

40. Gillberg, C.; Gillberg, C.; Råstam, M.; Wentz, E. The Asperger Syndrome (and High-Functioning Autism) Diagnostic Interview (ASDI): A Preliminary Study of a New Structured Clinical Interview. *Autism* **2001**, *5*, 57–66. [CrossRef] [PubMed]

41. Lord, C.; Rutter, M.; le Couteur, A. Autism Diagnostic Interview-Revised: A revised version of a diagnostic interview for caregivers of individuals with possible pervasive developmental disorders. *J. Autism Dev. Disord.* **1994**, *24*, 659–685. [CrossRef] [PubMed]

42. Autism Society Japan. *Pervasive Developmental Disorders Autism Society Japan Rating Scale (PARS)*; Spectrum Publishing Company: Tokyo, Japan, 2006.

43. Wing, L.; Leekam, S.R.; Libby, S.J.; Gould, J.; Larcombe, M. The diagnostic interview for social and communication disorders: Background, inter-rater reliability and clinical use. *J. Child Psychol. Psychiatry* **2002**, *43*, 307–325. [CrossRef] [PubMed]

44. Kurita, H.; Miyake, Y. The Reliability and Validity of the Tokyo Autistic Behaviour Scale. *Psychiatry Clin. Neurosci.* **1990**, *44*, 25–32. [CrossRef]

45. Nishida, N.; Tanabe, T.; Takasu, M.; Suyama, A.; Tokunaga, K. Further development of multiplex single nucleotide polymorphism typing method, the DigiTag2 assay. *Anal. Biochem.* **2007**, *364*, 78–85. [CrossRef] [PubMed]

46. dbSNP: Short Genetic Variations. Available online: http://www.ncbi.nlm.nih.gov/SNP/ (accessed on 24 June 2015).

47. Nyholt, D.R. A simple correction for multiple testing for SNPs in linkage disequilibrium with each other. *Am. J. Hum. Genet.* **2004**, *74*, 765–769. [CrossRef] [PubMed]

48. Purcell, S.; Cherny, S.S.; Sham, P.C. Genetic power calculator: Design of linkage and association genetic mapping studies of complex traits. *Bioinformatics (Oxf. Engl.)* **2003**, *19*, 149–150. [CrossRef]

49. Genetic Power Calculator. Available online: http://pngu.mgh.harvard.edu/purcell/gpc/cc2.html (accessed on 27 June 2015).

50. Developmental Disabilities Monitoring Network Surveillance Year 2010 Principal Investigators; Centers for Disease Control and Prevention (CDC). Prevalence of autism spectrum disorder among children aged 8 years—Autism and developmental disabilities monitoring network, 11 sites, United States, 2010. *MMWR Surveill. Summ.* **2014**, *63*, 1–21.

International Journal of
Molecular Sciences

MDPI

Communication

Association between *IRS1* Gene Polymorphism and Autism Spectrum Disorder: A Pilot Case-Control Study in Korean Males

Hae Jeong Park [1], Su Kang Kim [1], Won Sub Kang [2], Jin Kyung Park [2], Young Jong Kim [2], Min Nam [3], Jong Woo Kim [2] and Joo-Ho Chung [1,*]

[1] Kohwang Medical Research Institute, School of Medicine, Kyung Hee University, Seoul 02447, Korea; hjpark17@gmail.com (H.J.P.); skkim7@khu.ac.kr (S.K.K.)
[2] Department of Neuropsychiatry, School of Medicine, Kyung Hee University, Seoul 02447, Korea; menuhin@hanmail.net (W.S.K.); parkdawit@naver.com (J.K.P.); jimmypage@nate.com (Y.J.K.); psyjong@gmail.com (J.W.K.)
[3] Seoul Metropolitan Eunpyeong Hospital, Seoul 06801, Korea; passion17@hanmail.net
* Correspondence: jhchung@khu.ac.kr; Tel.: +82-2-961-0281

Academic Editors: Merlin G. Butler and Katalin Prokai-Tatrai
Received: 11 May 2016; Accepted: 25 July 2016; Published: 29 July 2016

Abstract: The insulin-like growth factor (IGF) pathway is thought to play an important role in brain development. Altered levels of IGFs and their signaling regulators have been shown in autism spectrum disorder (ASD) patients. In this study, we investigated whether coding region single-nucleotide polymorphisms (cSNPs) of the insulin receptor substrates (*IRS1* and *IRS2*), key mediators of the IGF pathway, were associated with ASD in Korean males. Two cSNPs (rs1801123 of *IRS1*, and rs4773092 of *IRS2*) were genotyped using direct sequencing in 180 male ASD patients and 147 male control subjects. A significant association between rs1801123 of *IRS1* and ASD was shown in additive ($p = 0.022$, odds ratio (OR) = 0.66, 95% confidence interval (CI) = 0.46–0.95) and dominant models ($p = 0.013$, OR = 0.57, 95% CI = 0.37–0.89). Allele frequency analysis also showed an association between rs1801123 and ASD ($p = 0.022$, OR = 0.66, 95% CI = 0.46–0.94). These results suggest that *IRS1* may contribute to the susceptibility of ASD in Korean males.

Keywords: autism spectrum disorder; insulin receptor substrate; single nucleotide polymorphism; insulin-like growth factor

---

1. Introduction

Autism spectrum disorder (ASD) is a neurodevelopmental disorder characterized by deficits in social communication, impaired reciprocal social interaction, and repetitive patterns of behaviors or interests [1]. ASD has been reported to occur in approximately six out of 1000 births, affecting males and females in a ratio of 4:1 [2]. ASD has been found throughout the world across all racial, ethnic and social backgrounds. Although the cause of ASD remains elusive, accumulating evidence suggests that genetic factors play a prominent role in this disease. Heritability is estimated to be above 90% [3]. Inherited copy number variations (CNVs) and chromosomal abnormalities have shown to contribute to genetic vulnerability to ASD [4,5]. In addition, recent multiple genome-wide association studies suggest several common single-nucleotide polymorphisms (SNPs) as markers of ASD [6–8].

Previous studies have revealed the importance of the insulin-like growth factor (IGF) pathway in the development and maintenance of the central nervous system (CNS). During both development and adulthood, increased IGF1 and IGF2 levels were associated with increased neuronal complexity and impaired learning, suggested that IGFs support the activity-dependent neuronal plasticity underlying cognitive processes [9,10]. Moreover, recent studies have highlighted a significant role for IGF1 in complex social interactions [11,12]. It has been reported that levels of IGF1, IGF2, and IGF binding protein 3 (IGFBP3) were significantly increased in patients with ASD [13]. Furthermore, both cross-sectional and longitudinal studies have reported that a subset of ASD patients show age-dependent brain overgrowth [14,15]. Brain overgrowth at an early age could be caused by conditional overexpression of IGF1 in the brain, which is responsible for significant increases in brain volume during the embryonic and early postnatal period [16]. Indeed, the significant correlation between head circumference and IGF1 levels was also shown in ASD patients, but not in the controls [13]. On the other hand, other studies reported lower IGF1 levels in the cerebrospinal fluid (CSF) of autistic patients [17,18]. In addition, it was proposed that a reduced peripartum level of IGF1 due to genetic, epigenetic, or environmental factors may be a sentinel biomarker of increased probability of the later development of autism [19].

Although little is known about the biological function of the IGF pathway molecules in ASD, given the previous reports, we speculated that IGFs and their signaling regulator genes might be candidate genes involved in ASD. However, to our knowledge, there have not been any studies on the possible genetic association of IGFs or IGF signaling regulator genes with ASD. In this study, we focused on key molecules of the IGF pathway, insulin receptor substrate 1 (*IRS1*) and *IRS2*, which are the major cytosolic substrates of the IGF receptors and mediators for the downstream pathway processes [20,21].

Insulin has been regarded as primarily a metabolic signal, while IGFs has been implicated as an important mitogen and cell differentiation factor [22,23]. The IRS family contains several members (IRS1-6), of which IRS1 and IRS2 have been most widely studied. IRS1 and IRS2 regulate body weight control and glucose homeostasis [24]. They could also control body growth and peripheral insulin action. Thus, they have been suggested as markers of an active IGF pathway within tumors [25,26], although they are involved in insulin signaling. Indeed, polymorphisms of *IRS1* and/or *IRS2* have shown the significant associations with diabetes, glucose levels [27], and obesity [28], as well as with cancers, along with IGF signaling regulator genes [29–31]. Herein, we investigated the association of the coding region single-nucleotide polymorphisms (cSNP) of *IRS1* and *IRS2*, active markers of the IGF pathway, with ASD in Korean males.

2. Results

Two cSNPs of *IRS1* and *IRS2* were polymorphic, and the genotype distributions of the SNPs were in Hardy-Weinberg equilibrium (HWE) ($p > 0.05$; data not shown). We calculated the power of the sample size to verify our data using a genetic power calculator [32]. Considering a two-fold genotype relative risk, the sample powers of the SNPs were 0.900 (rs1801123, number of effective samples for 80% power = 148) and 0.967 (rs4773092, $n = 110$), respectively ($\alpha = 0.05$). In addition, the sample powers of rs1801123 were 0.841 ($n = 175$) and 0.761 ($n = 213$) for a 1.9- and 1.8-fold relative risk, respectively. The sample powers of rs4773092 were 0.938 ($n = 128$) for a 1.9-fold relative risk, 0.889 ($n = 154$) for a 1.8-fold relative risk, and 0.889 ($n = 189$) for a 1.7-fold relative risk. Therefore, the results of our study had a significant power and sample size to detect the genotype relative risks up to 1.9-fold on rs1801123 and 1.8-fold on rs4773092.

As shown in Table 1, rs1801123 of *IRS1* was associated with ASD in additive (AG vs. GG vs. AA, $p = 0.022$, odds ratio (OR) = 0.66, 95% confidence interval (CI) = 0.46–0.95) and dominant models ($p = 0.013$, OR = 0.57, 95% CI = 0.37–0.89). The frequency of the genotypes containing the G allele (AG/GG, 36.7%) was decreased in the ASD patients compared to the control subjects (50.3%). In allele frequency analysis, we also found that rs1801123 was associated with ASD ($p = 0.022$, OR = 0.66, 95% CI = 0.46–0.94). The frequency of the G allele was lower in ASD patients (21.1%) than in control subjects (28.9%). This significance remained after the Bonferroni correction.

Interestingly, when we analyzed the differences between patients with autistic disorder and healthy individuals, rs1801123 of IRS1 showed a statistically more significant association (Table 2).

The association was revealed in the additive (p = 0.0037, OR = 0.56, 95% CI = 0.37–0.83) and dominant models (p = 0.0041, OR = 0.50, 95% CI = 0.31–0.81). Allele frequency analysis also revealed a stronger association between rs1801123 and autistic disorder (p = 0.004, OR = 0.56, 95% CI = 0.38–0.84). The frequencies of the AG/GG genotypes (33.6% and 50.3% in patients with autistic disorder and control subjects, respectively) and the G allele (18.6% and 28.9%) were more remarkably decreased in patients with autistic disorder compared to control subjects.

Table 1. Multiple logistic regression analysis of *IRS1* and *IRS2* polymorphisms in autism spectrum disorder (ASD) patients and control subjects.

SNP	Model/Allele	Genotype	Control	ASD	OR (95% CI)	p
			n (%)	*n* (%)		
rs1801123	Additive	AA	73 (49.7)	114 (63.3)	1	
Ala804Ala		AG	63 (42.9)	56 (31.1)		
IRS1		GG	11 (7.5)	10 (5.6)	0.66 (0.46–0.95)	**0.022**
	Dominant	AA	73 (49.7)	114 (63.3)	1	
		AG/GG	74 (50.3)	66 (36.7)	0.57 (0.37–0.89)	**0.013**
	Recessive	AA/AG	136 (92.5)	170 (94.4)	1	
		GG	11 (7.5)	10 (5.6)	0.73 (0.30–1.76)	0.48
	Allele	A	209 (71.1)	284 (78.9)	1	
		G	85 (28.9)	76 (21.1)	0.66 (0.46–0.94)	**0.022**
rs4773092	Additive	AA	41 (27.9)	51 (28.3)	1	
Cys816Cys		AG	76 (51.7)	95 (52.8)		
IRS2		GG	30 (20.4)	34 (18.9)	0.96 (0.70–1.32)	0.8
	Dominant	AA	41 (27.9)	51 (28.3)	1	
		AG/GG	106 (72.1)	129 (71.7)	0.98 (0.60–1.59)	0.93
	Recessive	AA/AG	117 (79.6)	146 (81.1)	1	
		GG	30 (20.4)	34 (18.9)	0.91 (0.53–1.57)	0.73
	Allele	A	158 (53.7)	197 (54.7)	1	
		G	136 (46.3)	163 (45.3)	0.96 (0.70–1.31)	0.8

Bold characters represent statistically significant values (p < 0.025). ASD, autism spectrum disorder. 1—It is a statistical reference in our genetic analysis.

Table 2. Multiple logistic regression analysis of *IRS1* and *IRS2* polymorphisms in patients with autistic disorder and control subjects.

SNP	Model/allele	Genotype	Control	Autistic Disorder	OR (95% CI)	p
			n (%)	*n* (%)		
rs1801123	Additive	AA	73 (49.7)	91 (66.4)	1	
Ala804Ala		AG	63 (42.9)	41 (29.9)		
IRS1		GG	11 (7.5)	5 (3.6)	0.56 (0.37–0.83)	**0.0037**
	Dominant	AA	73 (49.7)	91 (66.4)	1	
		AG/GG	74 (50.3)	46 (33.6)	0.50 (0.31–0.81)	**0.0041**
	Recessive	AA/AG	136 (92.5)	132 (96.3)	1	
		GG	11 (7.5)	5 (3.6)	0.47 (0.16–1.38)	0.16
	Allele	A	209 (71.1)	223 (81.4)	1	
		G	85 (28.9)	51 (18.6)	0.56 (0.38–0.84)	**0.004**
rs4773092	Additive	AA	41 (27.9)	36 (26.3)	1	
Cys816Cys		AG	76 (51.7)	70 (51.1)		
IRS2		GG	30 (20.4)	31 (22.6)	1.08 (0.77–1.51)	0.64
	Dominant	AA	41 (27.9)	36 (26.3)	1	0.76
		AG/GG	106 (72.1)	101 (73.7)	1.09 (0.64–1.83)	
	Recessive	AA/AG	117 (79.6)	106 (77.4)	1	0.65
		GG	30 (20.4)	31 (22.6)	1.14 (0.65–2.01)	
	Allele	A	158 (53.7)	142 (51.8)	1	
		G	136 (46.3)	132 (48.2)	0.96 (0.70–1.31)	0.8

Bold characters represent statistically significant values (p < 0.025). 1—It is a statistical reference in our genetic analysis.

3. Discussion

In our study, we found that rs1801123 of *IRS1* was significantly associated with ASD in Korean males. The G allele of rs1801123 contributed to a decreased risk of ASD and, particularly, the contribution was potently shown in patients with autistic disorder.

The IGF pathway plays an important role in regulating cell proliferation, differentiation and apoptosis, and, thus, IGFs and their signaling regulators have been studied in growth-, weight gain-, and obesity-related diseases [20,28,33–35]. IRS1 and IRS2 are key mediators of the IGF pathway [20,21]. Binding of IGFs to IGF receptors phosphorylates IRSs and triggers downstream cascades such as MAPK and PI3K/AKT signaling, which finally leads to cell proliferation and differentiation [20,21]. Thus, IRS1 and IRS2 together with IGFs and IGFRs have been involved in obesity, birth weight, diabetes mellitus, insulin sensitivity and cancer, showing the genetic associations of their polymorphisms [28–31,33,34].

In ASD patients, increased head growth, and particularly brain overgrowth in early life, has been reported with or without higher weights and body mass indexes (BMIs) Thus, several studies have suggested the involvement of growth-related hormones such as IGFs and their regulators, which lead to increased head growth and higher weights and BMIs, in the pathophysiology of autistic disorder/ASD [13,17,18]. Indeed, Mills et al. [13] reported increased levels of IGF1, IGF2, IGFBP3 and GHBP in the plasma of autistic disorder/ASD patients, and also showed a positive correlation between IGF1 level and head circumference in autistic disorder/ASD patients.

On the other hands, IGFs and their regulators have been also reported to play a role in growth retardation. Indeed, transgenic mice lacking *IRS1* showed prenatal and postnatal growth retardation [36, 37]. In mice lacking *IRS2*, growth retardation was also observed, although it was minimal compared to mice lacking *IRS1* [36]. Moreover, in the brain, IGFs and their regulators are essential factors for normal brain growth and development, as well as synaptogenesis and myelination [16,19,38]. They directly affect the rate that oligodendrocytes promote myelination, and thus factors which relatively reduce the production or availability of IGFs could retard normal nerve programming [19]. Indeed, in early laboratory embryos, the addition of IGF-receptor inhibitors blocked the normal formation of midbrain neurons [39]. *IGF1* knockout mice had defective neurologic development [40]. Moreover, other studies on the relationship between IGF1 level and autistic patients reported lower IGF1 levels in the cerebrospinal fluid (CSF) of autistic patients, although the IGF1 levels of patients were compared to abnormal controls instead of normal controls due to ethical reasons that do not allow researchers to obtain CSF from normal control subjects [17,18]. Thus, relatively low activities of IGF pathway molecules such as IGF1, IRS1 and IRS2 may play a role as risk factors in the pathophysiology of autistic disorder/ASD, leading to the growth and development retardation. Hence, it is controversial which one among excessive activations and reduced availabilities of IGF-related factors is involved in the pathology of autistic disorder/ASD.

In the present study, we found that rs1801123 of *IRS1* was associated with ASD, and the association was shown more strongly in patients with autistic disorder. In particular, our results showed that the frequency of the minor G allele of rs1801123 was decreased in patients with autistic disorder/ASD; thus, the G allele of rs1801123 may contribute to a decreased risk of autistic disorder/ASD as a protective factor. In a previous study, carriers of the minor allele of rs1801123 (TG/GG) were reported to be associated with higher fasting plasma glucose and insulin levels [27]. Furthermore, the G allele of rs1801123 was associated with an increased risk of breast cancer in women carrying the *BRCA1* mutation [30], although in a recent study of the same group using a large set of *BRCA1* and *BRCA2* mutation carriers, its lack of association was revealed [41]. Moreover, the G allele of rs1801123 was significantly associated with lymph node involvement in estrogen-receptor-positive primary invasive breast cancer patients, who were treated with surgery and tamoxifen [29]. These reports indicated that the minor allele of rs1801123 may be involved in the increased product and ability of the IGFs, along with the activation of the insulin-related signaling pathway. Therefore, the decreased frequency of the minor allele of rs1801123 in autistic disorder/ASD patients in our study may contribute to relatively low activation of the IGF pathway molecules. Taken together, we postulated that the activation of IGF pathway molecules may be reduced in autistic disorder/ASD patients, although it is controversial as mentioned above. Also, the G allele of rs1801123 may play a role as a protective factor against the decreased activity of the IGF pathway in autistic disorder/ASD. Further studies are needed to determine how rs1801123 and the IGF pathway affect the pathophysiology of autistic disorder/ASD.

Our study is the first pilot to report an association of the *IRS1* with autistic disorder/ASD. The limitation of our study is that only one SNP of each *IRS1* and *IRS2* was selected and analyzed. Replication studies are needed to determine the association between the *IRS1* and autistic disorder/ASD, as well as the lack of association of *IRS2*, analyzing more polymorphisms in addition to rs1801123 and rs4773092. Moreover, as shown in our sample power analysis, our results have statistical confidence, only assuming a genotype relative risk up to 1.9-fold on rs1801123. Thus, the relatively small sample size limits the generalizability of the findings from the present study. Our findings are preliminary and need to be validated in further studies with larger sample sizes. Our work provides evidence that the *IRS1* gene may play a role in the pathophysiology of ASD.

4. Experimental Section

4.1. Subjects

One hundred eighty male ASD patients (mean age $\pm$ standard deviation (SD), 15.5 $\pm$ 4.8 years) and 147 healthy male individuals (39.9 $\pm$ 5.8 years) were enrolled in this study. ASD patients were diagnosed with ASDs by well-trained psychiatrists, child and adolescent specialists according to Diagnostic and Statistical Manual of Mental Disorders, 4th ed (DSM-IV) criteria [42], using available historical information from interviews and clinical records. Each ASD patient was also evaluated using the Childhood Autism Rating Scale (CARS) [43], one of the most widely used instrument to evaluate the developmental degree of autism, applying cut-off score of 30. The average of CARS score was 38.6 $\pm$ 5.8 (mean $\pm$ SD). The ASD group consisted of 137 patients with autistic disorder, 11 patients with Asperger's disorder, and 32 patients with Pervasive Developmental Disorder-Not Otherwise Specified (PDD-NOS). A summary of clinical characteristics of ASD patients is provided in Table 3. The healthy adult controls were recruited from subjects who visited the hospital for routine health checkups. Controls were investigated to determine whether they or their first-degree relatives had psychiatric disturbances or previous psychiatric treatment through personal interviews. Only unaffected subjects with no psychiatric disorder or family history were included in this study.

Table 3. Clinical characteristics of ASD patients and control subjects.

Characteristics	ASDs	Control
Total no. of subject	180	147
Age (mean $\pm$ SD, years)	15.5 $\pm$ 4.8	39.9 $\pm$ 5.8
CARS score	38.6 $\pm$ 5.8	
Autistic disorder (n = 137)	41.1 $\pm$ 4.3	
Asperger's disorder (n = 11)	30.7 $\pm$ 0.4	
PDD-NOS (n = 32)	30.9 $\pm$ 1.0	

ASD, autism spectrum disorder; CARS, Childhood Autism Rating Scale; PDD-NOS, Pervasive Developmental Disorder-Not Otherwise Specified.

All the ASD patients and control subjects were of Korean background. The present study was conducted in accordance with the guidelines of the Helsinki Declaration and was approved by the Ethics Review Committee of Medical Research Institute, Kyung Hee University Medical Center on 15 September 2004 (2004-09-15). Written informed consents were obtained from the parents or guardians of ASD patients and control subjects.

4.2. Single-Nucleotide Polymorphism (SNP) Selection and Genotyping

Of SNPs in the *IRS1* and *IRS2*, cSNPs were targeted and selected from the National Center for Biotechnology Information SNP database [44]. Some cSNPs alter a functionally important amino acid residue, and these are of interest for their potential links with phenotype. Other cSNPs may prove useful for their potential links to functional cSNPs via linkage disequilibrium mapping [45,46]. We selected common SNPs with a minor allele frequency of >0.1 in Chinese and Japanese populations, excluding SNPs without data on genotype frequency. Finally, we selected two cSNP (rs1801123 (Ala804Ala) of *IRS1*, and rs4773092 (Cys816Cys) of *IRS2*).

Genomic DNA was extracted from the whole blood of each subject using the High Pure PCR Template Preparation kit (Roche, Mannheim, Germany) following the manufacturer's protocol. SNP genotyping was conducted with direct sequencing using the following primers for each SNP: rs1801123 in *IRS1* (sense, 5′-TCCTACTACTCATTGCCAAGATC-3′; antisense, 5′-CTATTGGTCTGA GCAGCTGTGT-3′), and rs4773092 in *IRS2* (sense, 5′-ATGTGGTGCGGTTCCAAGCTGT-3′; antisense, 5′-GCCAAAGTCGATGTTGATGTACT-3′). The PCR products were sequenced using the ABI PRISM 3730XL analyzer (PE Applied Biosystems, Foster City, CA, USA), and sequence data were then analyzed using SeqManII software (DNASTAR Inc., Madison, WI, USA).

4.3. Statistical Analysis

SNPStats [47] and SPSS 18.0 software (SPSS Inc., Chicago, IL, USA) were used to analyze the genetic data and the HWE. The association between SNP genotypes and ASD were estimated by computing the ORs and their 95% CIs with logistic regression analyses. In the logistic regression analysis for each SNP, the following models were used: codominant inheritance (that is, where the relative hazard differed between subjects with one minor allele and those with two minor alleles), dominant inheritance (subjects with one or two minor alleles had the same relative hazard for the disease), or recessive inheritance (subjects with two minor alleles were at increased risk of the disease). The chi-square test was used to compare allele frequencies between groups. To avoid chance findings due to multiple testing, a Bonferroni correction was applied by lowering the significance levels to $p = 0.025$ ($p = 0.05/2$) for two SNPs.

Acknowledgments: This work was supported by a grant from Kyung Hee University (Seoul, Korea) in 2012 (KHU-20121738).

Author Contributions: Joo-Ho Chung designed and directed the whole project. Won Sub Kang, Jin Kyung Park, Young Jong Kim, Min Nam and Jong Woo Kim collected the blood samples from ASD patients and control subjects. Hae Jeong Park and Su Kang Kim performed the experiments, collected the results, and analyzed the data. Joo-Ho Chung and Hae Jeong Park discussed and interpreted the data and results. Hae Jeong Park wrote the first draft of the manuscript. All authors contributed to and have approved the final manuscript.

Conflicts of Interest: The authors declare no conflict of interest.

References

1. Tanguay, P.E. Pervasive developmental disorders: A 10-year review. *J. Am. Acad. Child Adolesc. Psychiatry* **2000**, *39*, 1079–1095.
2. Nicholas, J.S.; Charles, J.M.; Carpenter, L.A.; King, L.B.; Jenner, W.; Spratt, E.G. Prevalence and characteristics of children with autism-spectrum disorders. *Ann. Epidemiol.* **2008**, *18*, 130–136.
3. Burmeister, M.; McInnis, M.G.; Zollner, S. Psychiatric genetics: Progress amid controversy. *Nat. Rev. Genet.* **2008**, *9*, 527–540.
4. Nowakowska, B.A.; de Leeuw, N.; Ruivenkamp, C.A.; Sikkema-Raddatz, B.; Crolla, J.A.; Thoelen, R.; Koopmans, M.; den Hollander, N.; van Haeringen, A.; van der Kevie-Kersemaekers, A.M.; et al. Parental insertional balanced translocations are an important cause of apparently de novo CNVs in patients with developmental anomalies. *Eur. J. Hum. Genet.* **2012**, *20*, 166–170.
5. Carter, M.T.; Nikkel, S.M.; Fernandez, B.A.; Marshall, C.R.; Noor, A.; Lionel, A.C.; Prasad, A.; Pinto, D.; Joseph-George, A.M.; Noakes, C.; et al. Hemizygous deletions on chromosome 1p21.3 involving the DPYD gene in individuals with autism spectrum disorder. *Clin. Genet.* **2011**, *80*, 435–443.
6. Anney, R.; Klei, L.; Pinto, D.; Regan, R.; Conroy, J.; Magalhaes, T.R.; Correia, C.; Abrahams, B.S.; Sykes, N.; Pagnamenta, A.T.; et al. A genome-wide scan for common alleles affecting risk for autism. *Hum. Mol. Genet.* **2010**, *19*, 4072–4082.

7. Weiss, L.A.; Arking, D.E. Gene Discovery Project of Johns Hopkins and the Autism Consortium. A genome-wide linkage and association scan reveals novel loci for autism. *Nature* **2009**, *461*, 802–808. [PubMed]
8. Wang, K.; Zhang, H.; Ma, D.; Bucan, M.; Glessner, J.T.; Abrahams, B.S.; Salyakina, D.; Imielinski, M.; Bradfield, J.P.; Sleiman, P.M.; et al. Common genetic variants on 5p14.1 associate with autism spectrum disorders. *Nature* **2009**, *459*, 528–533.
9. Chen, D.Y.; Stern, S.A.; Garcia-Osta, A.; Saunier-Rebori, B.; Pollonini, G.; Bambah-Mukku, D.; Blitzer, R.D.; Alberini, C.M. A critical role for IGF-II in memory consolidation and enhancement. *Nature* **2011**, *469*, 491–497.
10. O'Kusky, J.R.; Ye, P.; D'Ercole, A.J. Insulin-like growth factor-I promotes neurogenesis and synaptogenesis in the hippocampal dentate gyrus during postnatal development. *J. Neurosci.* **2000**, *20*, 8435–8442.
11. Schaevitz, L.R.; Moriuchi, J.M.; Nag, N.; Mellot, T.J.; Berger-Sweeney, J. Cognitive and social functions and growth factors in a mouse model of Rett syndrome. *Physiol. Behav.* **2010**, *100*, 255–263.
12. Moy, S.S.; Nadler, J.J.; Young, N.B.; Nonneman, R.J.; Grossman, A.W.; Murphy, D.L.; D'Ercole, A.J.; Crawley, J.N.; Magnuson, T.R.; Lauder, J.M. Social approach in genetically engineered mouse lines relevant to autism. *Genes Brain Behav.* **2009**, *8*, 129–142.
13. Mills, J.L.; Hediger, M.L.; Molloy, C.A.; Chrousos, G.P.; Manning-Courtney, P.; Yu, K.F.; Brasington, M.; England, L.J. Elevated levels of growth-related hormones in autism and autism spectrum disorder. *Clin. Endocrinol.* **2007**, *67*, 230–237.
14. Courchesne, E.; Carper, R.; Akshoomoff, N. Evidence of brain overgrowth in the first year of life in autism. *JAMA* **2003**, *290*, 337–344.
15. Courchesne, E.; Karns, C.M.; Davis, H.R.; Ziccardi, R.; Carper, R.A.; Tigue, Z.D.; Chisum, H.J.; Moses, P.; Pierce, K.; Lord, C.; et al. Unusual brain growth patterns in early life in patients with autistic disorder: An MRI study. *Neurology* **2001**, *57*, 245–254.
16. Popken, G.J.; Hodge, R.D.; Ye, P.; Zhang, J.; Ng, W.; O'Kusky, J.R.; D'Ercole, A.J. In vivo effects of insulin-like growth factor-I (IGF-I) on prenatal and early postnatal development of the central nervous system. *Eur. J. Neurosci.* **2004**, *19*, 2056–2068.
17. Riikonen, R. Neurotrophic factors in the pathogenesis of Rett syndrome. *J. Child Neurol.* **2003**, *18*, 693–697.
18. Vanhala, R.; Turpeinen, U.; Riikonen, R. Low levels of insulin-like growth factor-I in cerebrospinal fluid in children with autism. *Dev. Med. Child Neurol.* **2001**, *43*, 614–616.
19. Steinman, G.; Mankuta, D. Insulin-like growth factor and the etiology of autism. *Med. Hypotheses* **2013**, *80*, 475–480.
20. Laban, C.; Bustin, S.A.; Jenkins, P.J. The GH-IGF-I axis and breast cancer. *Trends Endocrinol. Metab.* **2003**, *14*, 28–34.
21. Jones, J.I.; Clemmons, D.R. Insulin-like growth factors and their binding proteins: Biological actions. *Endocr. Rev.* **1995**, *16*, 3–34.
22. White, M.F. Insulin signaling in health and disease. *Science* **2003**, *302*, 1710–1711.
23. Burks, D.J.; White, M.F. IRS proteins and β-cell function. *Diabetes* **2001**, *50*, S140–S145.
24. Withers, D.J.; Burks, D.J.; Towery, H.H.; Altamuro, S.L.; Flint, C.L.; White, M.F. IRS-2 coordinates IGF-1 receptor-mediated beta-cell development and peripheral insulin signalling. *Nat. Genet.* **1999**, *23*, 32–40.
25. Werner, H.; Le Roith, D. The insulin-like growth factor-I receptor signaling pathways are important for tumorigenesis and inhibition of apoptosis. *Crit. Rev. Oncog.* **1997**, *8*, 71–92.
26. Van Obberghen, E.; Baron, V.; Scimeca, J.C.; Kaliman, P. Insulin receptor: Receptor activation and signal transduction. *Adv. Second Messenger Phosphoprot. Res.* **1993**, *28*, 195–201.
27. Feng, X.; Tucker, K.L.; Parnell, L.D.; Shen, J.; Lee, Y.C.; Ordovas, J.M.; Ling, W.H.; Lai, C.Q. Insulin receptor substrate 1 (IRS1) variants confer risk of diabetes in the Boston Puerto Rican Health Study. *Asia Pac. J. Clin. Nutr.* **2013**, *22*, 150–159. [PubMed]
28. Lautier, C.; El Mkadem, S.A.; Renard, E.; Brun, J.F.; Gris, J.C.; Bringer, J.; Grigorescu, F. Complex haplotypes of *IRS2* gene are associated with severe obesity and reveal heterogeneity in the effect of Gly1057Asp mutation. *Hum. Genet.* **2003**, *113*, 34–43.
29. Winder, T.; Giamas, G.; Wilson, P.M.; Zhang, W.; Yang, D.; Bohanes, P.; Ning, Y.; Gerger, A.; Stebbing, J.; Lenz, H.J. Insulin-like growth factor receptor polymorphism defines clinical outcome in estrogen receptor-positive breast cancer patients treated with tamoxifen. *Pharmacogen. J.* **2014**, *14*, 28–34.

30. Neuhausen, S.L.; Brummel, S.; Ding, Y.C.; Singer, C.F.; Pfeiler, G.; Lynch, H.T.; Nathanson, K.L.; Rebbeck, T.R.; Garber, J.E.; Couch, F.; et al. Genetic variation in insulin-like growth factor signaling genes and breast cancer risk among BRCA1 and BRCA2 carriers. *Breast Cancer Res.* **2009**, *11*, R76.

31. Slattery, M.L.; Samowitz, W.; Curtin, K.; Ma, K.N.; Hoffman, M.; Caan, B.; Neuhausen, S. Associations among IRS1, IRS2, IGF1, and IGFBP3 genetic polymorphisms and colorectal cancer. *Cancer Epidemiol. Biomark. Prev.* **2004**, *13*, 1206–1214.

32. Genetic Power Calculator. Available online: http://pngu.mgh.harvard.edu/~purcell/gpc/cc2.html (accessed on 28 July 2016).

33. Rasmussen, S.K.; Urhammer, S.A.; Hansen, T.; Almind, K.; Moller, A.M.; Borch-Johnsen, K.; Pedersen, O. Variability of the insulin receptor substrate-1, hepatocyte nuclear factor-1α (*HNF-1α*), *HNF-4α*, and *HNF-6* genes and size at birth in a population-based sample of young Danish subjects. *J. Clin. Endocrinol. Metab.* **2000**, *85*, 2951–2953.

34. Almind, K.; Frederiksen, S.K.; Bernal, D.; Hansen, T.; Ambye, L.; Urhammer, S.; Ekstrom, C.T.; Berglund, L.; Reneland, R.; Lithell, H.; et al. Search for variants of the gene-promoter and the potential phosphotyrosine encoding sequence of the insulin receptor substrate-2 gene: Evaluation of their relation with alterations in insulin secretion and insulin sensitivity. *Diabetologia* **1999**, *42*, 1244–1249.

35. Almind, K.; Bjorbaek, C.; Vestergaard, H.; Hansen, T.; Echwald, S.; Pedersen, O. Aminoacid polymorphisms of insulin receptor substrate-1 in non-insulin-dependent diabetes mellitus. *Lancet* **1993**, *342*, 828–832.

36. Withers, D.J.; Gutierrez, J.S.; Towery, H.; Burks, D.J.; Ren, J.M.; Previs, S.; Zhang, Y.; Bernal, D.; Pons, S.; Shulman, G.I.; et al. Disruption of IRS-2 causes type 2 diabetes in mice. *Nature* **1998**, *391*, 900–904.

37. Tamemoto, H.; Kadowaki, T.; Tobe, K.; Yagi, T.; Sakura, H.; Hayakawa, T.; Terauchi, Y.; Ueki, K.; Kaburagi, Y.; Satoh, S.; et al. Insulin resistance and growth retardation in mice lacking insulin receptor substrate-1. *Nature* **1994**, *372*, 182–186.

38. Anlar, B.; Sullivan, K.A.; Feldman, E.L. Insulin-like growth factor-I and central nervous system development. *Horm. Metab. Res.* **1999**, *31*, 120–125.

39. Onuma, T.A.; Ding, Y.; Abraham, E.; Zohar, Y.; Ando, H.; Duan, C. Regulation of temporal and spatial organization of newborn GnRH neurons by IGF signaling in zebrafish. *J. Neurosci.* **2011**, *31*, 11814–11824.

40. Beck, K.D.; Powell-Braxton, L.; Widmer, H.R.; Valverde, J.; Hefti, F. Igf1 gene disruption results in reduced brain size, CNS hypomyelination, and loss of hippocampal granule and striatal parvalbumin-containing neurons. *Neuron* **1995**, *14*, 717–730.

41. Ding, Y.C.; McGuffog, L.; Healey, S.; Friedman, E.; Laitman, Y.; Paluch-Shimon, S.; Kaufman, B.; Swe, B.; Liljegren, A.; Lindblom, A.; et al. A nonsynonymous polymorphism in IRS1 modifies risk of developing breast and ovarian cancers in BRCA1 and ovarian cancer in BRCA2 mutation carriers. *Cancer Epidemiol. Biomark. Prev.* **2012**, *21*, 1362–1370.

42. Association Psychiatric Association. *Diagnostic and Statistical Manual of Mental Disorders*, 4th ed.; American Psychiatric Press: Washington, DC, USA, 1994.

43. Schopler, E.; Reichier, R.J.; Renner, B.R. *Childhood Autism Rating Scale*; Western Psychological Services: Los Angeles, CA, USA, 1988.

44. NCBI databases. Available online: http://www.ncbi.nlm.nih.gov/SNP (accessed on 28 July 2016).

45. Collins, F.S.; Guyer, M.S.; Chakravarti, A. Variations on a theme: Cataloging human DNA sequence variations. *Science* **1997**, *278*, 1580–1581.

46. Garg, K.; Green, P.; Nickerson, D.A. Identification of candidate coding region single nucleotide polymorphisms in 165 human genes using assembled expressed sequence tags. *Genome Res.* **1999**, *9*, 1087–1092.

47. ICO. Available online: http://bioinfo.iconcologia.net/index.php (accessed on 28 July 2016).

Chapter III:
Other

International Journal of
Molecular Sciences

MDPI

Review

Modulation of the Genome and Epigenome of Individuals Susceptible to Autism by Environmental Risk Factors

Costas Koufaris and Carolina Sismani *

Department of Cytogenetics and Genomics, the Cyprus Institute of Neurology and Genetics, P.O. Box 3462, Nicosia 1683, Cyprus; costask@cing.ac.cy

* Author to whom correspondence should be addressed; csismani@cing.ac.cy; Tel.: +357-22-392-696; Fax: +357-22-392-793.

Academic Editor: Merlin G. Butler

Received: 27 November 2014; Accepted: 8 April 2015; Published: 20 April 2015

Abstract: Diverse environmental factors have been implicated with the development of autism spectrum disorders (ASD). Genetic factors also underlie the differential vulnerability to environmental risk factors of susceptible individuals. Currently the way in which environmental risk factors interact with genetic factors to increase the incidence of ASD is not well understood. A greater understanding of the metabolic, cellular, and biochemical events involved in gene x environment interactions in ASD would have important implications for the prevention and possible treatment of the disorder. In this review we discuss various established and more alternative processes through which environmental factors implicated in ASD can modulate the genome and epigenome of genetically-susceptible individuals.

Keywords: xenobiotic; immune; endocrine; epigenome; gut microbes; oxidative stress; transposable elements; endocrine disruptors; transgenerational; gene environment interactions; seizures; genotype

1. Introduction

Autism spectrum disorders (ASD) are a collection of neurodevelopmental disorders typically diagnosed in the first three years of life, characterized by deficits in verbal and nonverbal communication, repetitive behaviors, and impairments in social interactions [1]. ASD is caused by the genetic or epigenetic disruption of genes that are essential for normal neurodevelopment, with hundreds of candidate susceptibility genes identified so far. Genetic variations associated with ASD include large chromosomal abnormalities, copy number variants (CNV), indels/deletions, and single nucleotide variants (SNV). The genetic variants implicated in ASD are either inherited from parents to affected individuals or form *de novo* in the patients. Inherited genetic factors (both common and rare variants) have been estimated to explain ~40% of ASD risk [2], while *de novo* mutations are thought to contribute to 15%–20% of cases [3]. Recurrent mutations found in ASD patients implicate disruption of certain biological pathways as particularly important in ASD, such as post-synaptic density [4], phosphoinositide 3-kinase (PI3K)-mammalian target of rapamycin (mTOR) [5], and ubiquitin processing [6].

Strong evidence now also implicates epigenetic dysregulation with ASD development. For example, the epigenetic regulator methyl CpG binding protein 2 (*MECP2*) is well established as being implicated in ASD [7], while altered DNA and histone epigenetic markers have been identified in brain tissue and immune-cell derived cells from autistic patients [8–11]. The biological significance of the observed epigenetic alterations in ASD is supported by the observation that they affect the expression of genes implicated in ASD such as *MECP2*, retinoic acid related orphan receptor A (*RORA*), and engrailed-2 (*EN-2*) [8–11].

Of particular interest to both researchers and the general public is the substantial evidence linking exposure to environmental agents with the incidence of ASD. Epidemiological studies have associated diverse prenatal and perinatal environmental exposures to an increased risk of ASD [12–14], while discordance between monozygotic twins in the incidence of the disorder also supports the importance of environmental influences [15]. Additionally, the largest genetic study reported so far also indicated a large influence of environmental factors on the development of ASD [2]. A sensible hypothesis, based on the importance of both genetic and environmental factors on ASD, is that the interactions between these two components influence the incidence of the disorder. Indeed, a number of recent reviews have discussed the evidence from genetic, epidemiological, and animal studies that support the influence of gene x environment interactions on ASD [16–19]. However, our understanding of the biological mechanisms underlying genotype-environment interactions in ASD is currently inadequate.

2. Disruption of the Genome and Epigenome of Genetically Susceptible Individuals by ASD Environmental Risk Factors

Environmental ASD risk factors are those non-genetic factors that can influence the development of the disorder in genetically-susceptible individuals. These have been suggested to include dietary factors, maternal diabetes, pre-and-perinatal stress, parental age, medications, zinc deficiencies, supplements, pesticides, and infections. Ultimately the environmental risk factors act by disrupting the genome/epigenome of developing neurons. This review will not aim to provide an overview of the numerous proposed environmental ASD risk factors. Rather here we will focus on how environmental factors can modulate the genome and epigenome of individuals susceptible to develop ASD. An increased understanding of this gene-environment interaction would have great implications for the prevention and treatment of the disorder. For example, identifying genetically vulnerable individuals would allow targeted strategies for reducing their exposure to ASD risk factors. Taking into account genotype-environment interactions could also improve the consistency and interpretation of association and epidemiological studies, which often fail to replicate. Finally, elucidating the mechanisms of genotype-environment interactions could also inform the identification of novel therapeutic targets in ASD patients.

2.1. Defective Xenobiotic Metabolism of ASD Environmental Risk Factors

The initial interaction between environmental toxins and biological organisms involves xenobiotic metabolic enzymes, specialized systems responsible for detoxifying and removing harmful chemicals from the body. Although the liver is considered the primary organ where xenobiotic metabolism takes place, detoxification in the brain is also known to occur [20]. Xenobiotic metabolic enzymes are also involved with the biochemical processing of neurotoxicants and environmental ASD risk factors. For example, 2,2',4,4'-tetrabromodiphenyl ether (BDE-47), a main component of the flame retardant polybrominated diphenyl ethers (PBDE) linked to ASD [21], is metabolised to its more harmful metabolites by action of CYP2B6 [22]. It is also the case that several environmental risk factors associated with ASD risk are established mutagens, possibly acting by increasing the incidence of mutations in key neurodevelopmental genes [23]. The majority of mutagenic compounds have to be metabolically converted from non-mutagenic to mutagenic forms by the action of cytochrome p450 (CYP450) enzymes, thus are also strongly affected by the functioning of xenobiotic metabolic activity.

Xenobiotic enzymes display extensive variation in copy number between individuals and ethnic groups. Additionally, SNVs are also known to affect the activity of xenobiotic enzymes. Polymorphisms in xenobiotic enzymes are an important source of individual variability in the response to environmental agents. Indeed, such a mode of action is widely accepted to influence individual susceptibility to environmental carcinogens [24]. A reasonable hypothesis then is that genetic variability that affects the metabolism of environmental ASD risk factors can render subsets of the population particularly vulnerable to environmental chemicals that modulate the genome or epigenome. However, so far few studies have examined the association between polymorphisms affecting xenobiotics metabolic enzymes and ASD incidence. One interesting case involves paraoxonase (*PON1*), a gene involved in the removal from the body of organophosphate pesticides. D'Amelio *et al.* [25] reported that variants of the *PON1* gene were associated with increased incidence of ASD in USA populations, but not in Italian populations. The authors interpreted these observations as relating to the higher levels of organophosphate to which the USA population is exposed. A second study reported an association between *PON1* polymorphisms and neurodevelopment in children exposed in utero to organophosphate pesticides [26]. Five studies have examined the association between *PON1* polymorphisms and ASD. These studies have had mixed results, with only three out of the five studies finding an association between the genetic variants and the disorder [27].

A study by Serajee *et al.* [28] examined the association between polymorphisms affecting six xenobiotic metabolism genes and ASD in 196 families and reported a significant association for a polymorphism affecting the metal-regulatory transcription factor 1 (*MTF1*). Buyske *et al.* [29] examined the association between deletions of the glutathione *S*-transferase M1 (*GSTM1*) allele and autism. This enzyme is involved in the detoxification of electrophilic compounds. In this study a significant association was found for homozygous deletion of *GSTM1* and ASD risk. In conclusion, genetic variability affecting the functioning of xenobiotic enzymes can render individuals more sensitive to environmental ASD risk factors. Taking into account exposures to environmental factors and variants affecting xenobiotic enzymes might potentially improve the design and interpretation of epidemiological studies, as demonstrated by D'Amelio *et al.* [25].

2.2. Increased Sensitivity to Endocrine-Disrupting Chemicals

Increasing evidence indicates that defects in the physiological sex hormone-dependent processes can influence ASD development [30,31]. Diagnosis of ASD is four times more frequent in males compared to females, which might be partly attributed to hormonal involvement in the development of the disease. Fetal testosterone affects brain organization [32] and investigation of amniotic fluid found elevated levels of sex steroid hormones in boys diagnosed with autism [33]. Importantly for our purposes, diverse environmental agents (including plasticizers, flame retardants, drugs, and pesticides) are acknowledged to be able to block or mimic normal hormone signaling by affecting the synthesis of hormones or by interacting with their receptors. Such environmental agents, termed endocrine-disruptor chemicals (EDC), have been implicated with the development of diverse human diseases, including ASD [34]. In support of the influence of EDC on ASD, epidemiological studies have linked gestational exposure to EDC with behavioral problems relating to ASD development [35,36].

Vulnerability to EDC exposures is known to be highly dependent on individual genotype. For example, disruption of male reproductive development in juvenile mice following EDC exposure varied more than 16-fold between mouse strains [37]. Individual variability in sensitivity to EDC can arise by various means, for example different basal levels of sex hormones or binding affinity of hormone receptors. Genetic variants affecting the sex hormones signaling axis in the developing brain could render individuals more susceptible to low EDC doses that induce abnormal brain development and ASD. One study has reported an association between polymorphisms of the androgen receptor and ASD [38]. A second recent study examined the association between 29 single nucleotide polymorphisms (SNP) in genes relating to sex steroids and ASD-like traits belonging to a subset (*n* = 1771) from The Child and Adolescent Twin Study in Sweden (CATSS) [39]. The authors reported two SNPs to be associated with autism-like traits, one located in the 3'UTR of the estrogen receptor and the other a non-synonymous variant affecting the 3-oxo-5-α-steroid 4-dehydrogenase 2 (SRD5A2) enzyme that is involved in testosterone metabolism. Another interesting observation is the reduced expression of RORA, a gene involved in the transcriptional regulation of genes that convert male to female hormones, in the brain and lymphocytes of ASD patients [9]. Nevertheless there is inadequate understanding at present of how abnormal sex hormone signaling can disrupt normal neurodevelopment and neurobiology in order to contribute to ASD.

2.3. Increased Susceptibility to Agents that Induce Oxidative Stress

Oxidative stress refers to an imbalance between the generation and removal of reactive oxygen species (ROS) in biological systems. Accumulating evidence supports that oxidative stress is also implicated with the development of ASD [40–42]. ROS can directly attack and damage DNA, thus increasing the incidence of mutations that can potentially contribute to ASD. A second possible consequence of oxidative stress is destabilization of the epigenome. In conditions of oxidative stress the enhanced requirement for the antioxidant glutathione can impair synthesis of *S*-adenosyl methionine (SAM), the major methyl donor used for DNA methylation, thus diminishing the available pool of methyl donors used for methylating DNA. The subsequent loss of normal methylation patterns results in the loss of normal regulation and aberrant gene expression. In addition, DNA hypomethylation can also lead to genomic instability [43]. A third detrimental consequence of ROS is the induction of mitochondrial dysfunction, which can initiate vicious cycles of increased ROS generation from the damaged mitochondria. Mitochondrial dysfunction can affect biological functions of the organelle that are important in normal neuronal functioning, such as regulation of neurotransmission [44]. The brain is considered to be particularly vulnerable to ROS due to the fact that it contains non-replicating cells that can become permanently dysfunctional or undergo necrotic or apoptotic cell death.

Increased oxidative stress has been reported in autistic patients [45] and a more oxidized microenvironment is associated with a more favourable development [40]. Additionally, mitochondrial disease and markers of mitochondrial dysfunction are much higher in ASD population compared to control populations [41]. A recent study has also found decreased tryptophan metabolism in lymphoblastoid cells from ASD patients [42]. Aberrant tryptophan metabolism is important because this amino acid is the precursor of molecules that are important for mitochondrial energy generation and anti-oxidant defenses [46]. Tryptophan metabolism is involved in the generation of NAD$^+$ involved in mitochondrial energy generation through oxidative phosphorylation. Tryptophan is also ultimately involved in the generation of melatonin, which acts as a scavenger of free radicals [46].

Importantly, diverse types of environmental factors can act to either augment or protect against oxidative stress. Subsets of individuals with inherently higher rates of ROS formation or reduced anti-oxidative capabilities would therefore be more vulnerable to environmental conditions that favor a pro-oxidant microenvironment that can contribute to ASD development. Recently published studies have reported that oxidative stress induces more prominent mitochondrial dysfunction in lymphoblastoid cells derived from subsets of ASD patients [47,48]. Damage in the mitochondria can also lead to even greater oxidative stress as ROS are released from the damaged mitochondria. In an interesting recent study it has been reported that exposure of lymphoblasts to bisphenol A results in greater levels of oxidative stress and mitochondrial dysfunction in ASD individuals compared to unaffected siblings [49].

At present the greatest effort to identify the underlying genetic basis for an increased vulnerability of ASD patients to environmental agents has been carried out in relation to ROS and mitochondrial dysfunction [40–42,42,45,45,47,47–49]. Work in animal models also supports that genes implicated with ASD affect ROS and mitochondria. Examples are conditional phosphatase and tensin homolog (*Pten*) haplo-insufficient mice which display aberrant social behavior coupled with mitochondrial dysfunction [50] and neuroligin knockout *Caenorhabditis elegans* that show both abnormal behavior and increased sensitivity to oxidative stress [51]. However, since these two proteins have multiple biological functions it is not possible to exclude that loss of these genes contributes to ASD development through alternative mechanisms. Association studies have revealed an increased frequency of genetic variants affecting Methylenetetrahydrofolate reductase (*MTHFR*) [52] and superoxide dismutase 1 (*SOD1*) [53], enzymes that affect cellular anti-oxidative machinery, in ASD patients. MTHFR is a crucial enzyme of the folate pathway and is involved in the generation of the anti-oxidant glutathione and in generating methyl units that are used for repairing genetic and epigenetic damage caused by ROS. Intriguingly, a SNP that results in reduced activity of the MTHFR was associated with greater incidence of autism, but only in countries with low levels of folate fortification during pregnancy [51]. SOD1 is a member of the superoxide dismutase family that are important participants in antioxidant defense mechanisms. Researchers have recently reported the presence of rare variants within the non-coding potentially regulatory regions of the *SOD1* gene. Although not verified experimentally, the authors consider a plausible hypothesis to be that these variants will affect the regulation of this gene and subsequently the ability of individuals to respond to oxidative stress [53].

A number of recent studies have also identified increased frequencies of genetic variants that affect mitochondrial functions in ASD patients. Smith *et al.* [54] found an increased presence of CNV affecting genes involved in mitochondrial oxidative phosphorylation. Nava *et al.* [55] identified mutations affecting trimethyllysine dioxygenase (*TMLHE*), an enzyme catalyzing the first step of carnitine biosynthesis, in ASD patients. Carnitine is a molecule that is essential for mitochondrial fatty acid metabolism, as well as having anti-oxidant functions. Dysfunction in the mitochondrial aspartate/glutamate carrier (*AGC1*) involved in calcium homeostasis has also been reported in autism [56]. Another alternative mechanism that could render individuals more sensitive to environmental inducers of oxidative stress is the presence of persistent inflammation and immune dysregulation. In these individuals the immune system would result in the continuous generation of excess ROS that will act to drain the available anti-oxidative defenses of cells, rendering them more vulnerable to environmental agents [57].

2.4. Susceptibility to Epigenomic Dysregulation

Epigenetic markers such as DNA methylation and histone modifications regulate gene expression in eukaryotic cells. The proper epigenetic marking of DNA is crucial for the normal development of human tissues and organs. Studies of ASD patients have found aberrant patterns of epigenetic regulation, both in the brain and in lymphoblastoid cells [8–11,58]. Aberrant epigenetic mechanisms can contribute to ASD development by causing abnormal gene expression patterns or due to increased genomic damage occurring subsequent to DNA hypomethylation. By their very nature epigenetic mechanisms are at the interface between the genome and environmental factors. A variety of environmental factors can affect the epigenome, such as exposure to heavy metals and levels of dietary folate. Genetic factors that influence the interaction between environmental factors and the epigenome could therefore be important in determining the risk of ASD developing in exposed individuals. In one interesting study it was shown that mice with a truncated version of *Mecp2*, a regulator of the epigenome in neuronal cells, demonstrated social behavioral defects when exposed prenatally to the organic pollutant PBDE. Decreased sociability was associated with reduced global DNA methylation in the female but not in the male mice [21]. As mentioned previously, SNPs that reduce activity of MTHFR have been associated with increased risk of ASD in countries with low fortification [52]. One of the effects of reduced MTHFR activity is lower generation of methyl groups by the folate cycle that can be used for methylating DNA. An increased propensity towards a dysregulated epigenome could also arise due to an inherently higher baseline level of ROS, which use one-carbon units for the generation of glutathione, in ASD patients (discussed in Section 2.3).

An alternative mechanism that would render individuals more susceptible to environmental modulators of the epigenome has been proposed by La Salle [16]. According to this hypothesis large-scale genomic differences between individuals in the size of their repetitive regions can affect their vulnerability to environmental factors that influence the epigenome. It is being increasingly recognized that large-scale genomic differences exist between individuals. Polymorphic regions in the human region include duplications, deletions, inversions, microsatellite repeats, ribosomal DNA repeats, and interstitial telomeric repeats. Since the repetitive regions of the genome act as major drains of methyl donors, it is suggested that individuals with greater genomic size of repetitive elements will be more vulnerable to environmental factors that interact with the DNA methylome [16].

2.5. Hyperactive Transposable Elements

Transposable elements are DNA sequences that can mobilize themselves to new regions of the genome, thus acting as mutagens. Long interspersed element-1 (L1) is the most abundant autonomous transposable element in the human genome. Importantly, the L1 retrotransposition is active in neural progenitor cells [59]. Additionally, L1 retrotransposition is enhanced by a large variety of chemical treatments [60]. Intriguingly, reduced activity of Mecp2, a gene implicated in Rett syndrome and ASD, is associated with a higher occurrence of L1 retrotransposition [61]. Therefore, a propensity to a hyperactive mobility of L1 elements in neurons in response to environmental exposures is a plausible mechanism that can drive the incidence of the pathogenic mutations that in turn drive the development of the disease. The precise mechanisms by which environmental agents activate L1 retrotransposition are diverse and probably differ between environmental agents. For example, heterocyclic amines activate L1 mobility through activation of ligand-bound transcription factors [62] and alcohol affecting methylation [63]. As of yet, clear experimental evidence for an increased susceptibility to L1 retrotransposition hyperactivity caused by environmental agents in ASD remains to be reported.

2.6. Increased Genomic Instability

Genomic rearrangements lead to the formation of CNV through deletions, duplications, inversions, and complex rearrangements. Importantly, genes involved in neurodevelopmental disorders including ASD are often flanked by segmental duplications or repetitive elements which drive genomic rearrangements and disruption of gene function [64]. An interesting observation is that ASD patients possess significantly elevated CNV load, even when removing rare and pathogenic events [65]. This finding suggests that beyond the sampling bias this population is also characterized by an increased rate of genomic rearrangements [65]. Environmental agents such as ionizing radiation and genotoxic chemicals are known to be able to initiate genomic rearrangements which subsequently lead to CNV formation. An increased genomic instability in individuals following exposure to such environmental agents could contribute to ASD by enhancing the rate of genomic rearrangements that involve genes implicated in the disorder. Increased genomic instability could be the result of defects in the machinery involved in DNA repair and recombination [65]. So far one study has suggested a link between defective DNA repair and ASD development. Fanconi-associated nuclease 1 (*FAN1*) is a relatively recently identified nuclease involved in the repair of DNA inter-strand cross-links. In a recent study rare variants affecting the FAN1 have been suggested to be candidate drivers of ASD [66]. A decreased ability to repair DNA damage caused by environmental agents could also arise due to the presence in the genome of susceptible individuals of regions with significant sequence homology that flank genes important for normal neurodevelopment. Such genomic regions can act as substrates to allow improper chromosomal recombination (non-allelic homologous recombination or microhomology-mediated break-induced replication) to occur, resulting in loss or gain of the intervening genes.

2.7. Abnormal Immune Activation

Several studies have reported abnormalities in the peripheral immune system of ASD individuals and a number of genes that are frequently mutated in ASD relate to immune function [67]. Additionally recent epidemiological reports link infections in early pregnancy to an increased incidence of autism [12]. Increased neuronal inflammation is also observed in the central nervous system of ASD patients [67]. Consequently, it is firmly established that the immune system is dysregulated in autism and it has been suggested that this could be a contributing factor to the development of the disorder [67,68]. Cytokines receptors are expressed in the brain and cytokines released by immune cells are able to affect both the development and function of the neuronal system. Imbalanced cytokine release can therefore disrupt physiological neurodevelopment [69]. Prenatal maternal inflammation can also disrupt global and gene-specific methylation patterns in the brain [70]. Chronic inflammation can also be a source of ROS contributing to oxidative stress [40,41] which has been implicated in ASD (see Section 2.3).

Environmental exposures that trigger immune responses include infections, toxins that impact the immune system, and allergens. An increased propensity to immune dysregulation predisposing to ASD could be due to inappropriate activation of immune responses, prolonged and persistent immune responses, and autoimmunity [67,68]. Abnormal immune activation can also occur due to defective clearance of environmental toxicants by xenobiotic metabolic enzymes. In one interesting study Ashwood *et al.* [71] showed that peripheral blood mononuclear cells from control and autistic individuals resulted in qualitatively different profiles of secreted cytokines following bacterial lipopolysaccharide (LPS) exposure when pre-treated with environmental toxins PBDEs. This experiment suggested an inherently different immune response of ASD individuals to PBDE exposure compared to controls. Evidence in support of genotype-environment interactions for the immune system is also available from animal studies. Mouse research has shown a link between maternal immune activation and ASD behavior [67]. In a recent interesting study comparing two mice strains, C57BL/6J and Black and Tan BRachyury *T* + *tf*/J (BTBR), it was shown that the immune responses and their propensity for the development of ASD-like phenotypes in the offspring differed following exposure to viral mimics, with the BTBR strain displaying a greater susceptibility [72]. Consequently, this study supports the importance of gene x environment interactions in the development of immune system dysregulation and ASD. The genetic factors associated with the greater susceptibility of the BTBR strain were not clarified in this study, although Disrupted in Schizophrenia 1 (*Disc1*), a gene that this strain lacks and is involved in immunological responses, has been suggested as a potential candidate. In another recent study researchers showed that haploinsufficiency of tuberous sclerosis complex (*TSC*), a gene for which mutations are associated with a greatly increased risk of ASD development, interacted with infection to disrupt social behavior in adult mice [73]. In this case specific gene x environment interactions and their effects on normal behavior, at least in adult mice, were demonstrated.

2.8. Gut Microbiota

Gut microbiota are the collection of thousands of microbial species that are found within the intestine. The gut microbiota affects multiple biological systems such as gut permeability, immune system, and metabolism. Gut microbiota can also affect neuronal development, functioning, and signaling by interacting with the immune system, neural, and hormonal pathways [74]. As discussed in Section 2.7 abnormal immune system activation has been associated with ASD development. The gut microbiota also induce the generation of ROS in epithelial cells which can contribute to immune dysfunction and to oxidative stress [75]. Studies have reported differences in the gut microbiota between ASD patients and control cohorts [76,77]. In an intriguing recent study it was reported that in a mouse model of ASD altering the gut microbiota ameliorates behavioral abnormalities [78]. Importantly the gut microbiome is highly flexible and is affected by a person's age, geographic region, environment, health, and genotype [79]. Studies in mice support that the genotype of animals is a strong determinant of the type of microbial species present in the animals guts [80,81]. Differences in the composition of the gut microbiota in humans could therefore potentially render them more susceptible to ASD environmental factors through distinct mechanisms e.g., increased sensitivity to environmental agents that disrupt the immune system. Currently we are not aware of any published studies that have investigated whether an individual's gut microbiota can render them more or less susceptible to environmental risk factors for autism.

2.9. Transgenerational Environmental Effects

The transmission of the effects of environmental exposures across generations has been suggested to be involved in various human diseases, including autism and other behavioral disorders [82,83]. There is considerable debate about the mechanisms by which transgenerational information can be transmitted, although miRNA, DNA methylation, and histone modification are considered as primary candidates. Although highly speculative at present, it is possible that defects in epigenetic machinery could render individuals more susceptible to transgenerational epigenetic defects caused by environmental exposures. In one study gestational exposure to bisphenol A was linked to transgenerational behavioral abnormalities in mice [82]. A second study found that exposure to valproic acid *in utero* was associated with abnormal gut microbiome across multiple generation, as well as with immune and neurological abnormalities [83,84]. By causing transgenerational epigenetic effects, environmental factors could interact with genetic factors to promote ASD in individuals that were not exposed to the environmental agents directly.

2.10. Exacerbation of Environmental Effects by Seizures

Seizures are the result of abnormal electrical activity in the brain and can result in loss of consciousness and involuntary muscle contractions. Importantly, there is evidence for a link between seizures and ASD. There is a much higher prevalence of epilepsy in ASD children compared to controls [85]. Seizures are also associated with the exacerbation of ASD clinical features [86,87]. The interaction between seizures and autism could be important for two different reasons. First, seizures can modulate the incidence and clinical features of ASD associated with a given genotype. Second, seizures could render individuals more susceptible to environmental agents that disrupt normal neurodevelopment to promote ASD development. One proposed mechanism by which seizures can affect ASD is the downregulation of the Fragile X mental retardation protein (*FMRP*). It has long been known that Fragile X syndrome (FXS) is associated with increased incidence of ASD [88]. In recent years post-mortem examination of brain tissue from idiopathic ASD patients has also found deregulated FMRP [89,90]. Interesting research has found that seizures disrupt the normal functioning of the FMRP. In a rat model of seizures it has been found that this results in abnormal FMRP phosphorylation, localization, and function of the protein [91].

3. Conclusions

So far substantially more research effort has been invested into cataloging the presence of genetic and epigenetic aberrations in ASD individuals compared to how and why these emerge in the first place. We have presented here a number of processes through which environmental ASD risk factors can modulate the genome and epigenome of genetically-susceptible individuals (Table 1). Some of these mechanisms have currently more supportive evidence (for example oxidative stress and immune dysregulation) while others are more speculative (for example transgenerational inheritance). Crucially, these different mechanisms do not act in isolation but can interact to enhance their detrimental effects. In Figure 1 we present examples of how different genetic and environmental factors and biological processes can interact to produce the genetic and epigenetic aberrations that promote ASD development. For example we have previously discussed the potential contribution of oxidative stress to ASD (Section 2.3). Defective one-carbon metabolism or dysfunctional mitochondria can increase the susceptibility of individuals to pro-oxidant diet or toxin exposures. Aberrant immune responses can also contribute to oxidative stress through the generation of ROS as a consequence of inflammation. Altered gut microbiota which disrupt the immune system could also possibly contribute to the generation of a state of oxidative stress. An integrative analysis of the multiple mechanisms of gene x environment interactions will be required to understand the development of ASD.

The susceptibility of individuals to environmental ASD risk factors is restricted to the early stages of life, especially during embryonic and fetal life, when the developing brain is uniquely sensitive to environmental agents [92,93]. The biological mechanisms described here should therefore only be relevant to ASD if they affect normal neurodevelopment *in utero* or possibly during very early post-natal stages. This observation raises the possibility that the genetically-susceptible individuals on which environmental risk factors act may not be the ASD patients themselves but their parents. For example an increased mutation rate as a consequence of environmental exposures (due to an inherently greater genomic instability or oxidative stress) could act to increase the presence of pathogenic mutations in the gametes and offspring. In this case mutations occurring in somatic tissues post-natally will not be relevant to ASD development. For many of the proposed mechanisms it is not clear at present whether environmental factors act on the ASD patients themselves or on their parents, a question of potential clinical significance.

It is important to remember when interpreting studies examining the association between exposure to environmental agents and ASD development that correlation does not necessarily imply causation. This is especially the case when the mechanisms through which environmental factors increase ASD risk are of a more speculative nature. Moreover, it is possible that environmental agents associated with ASD act to increase the incidence of the disorder through different mechanisms than the ones considered to be the primary candidates. Finally, in cases where environmental agents may act through multiple distinct mechanisms it can be difficult to determine which are the most relevant. For example, in conditions of low dietary folate MTHFR variants that affect the activity of the enzyme could increase the incidence of ASD by affecting oxidative stress, the epigenome or mutation rates (or all three). Uncovering the implicated mechanism could be of great therapeutic importance since epigenetic aberrations, unlike genetic aberrations, are in principle reversible.

Table 1. Mechanisms modulating the genome and epigenome of susceptible individuals.

Mechanisms	Effects on Genome and Epigenome	Environmental Factors	Genetic Factors
Oxidative stress	DNA damage; Mitochondria dysfunction; Disruption of the epigenome; Deregulated gene expression	Pro-oxidant toxicants; Diet low in one-carbon donors; Immunogens	Defective one-carbon metabolism; Defective anti-oxidant defenses; Abnormal immune activation; mitochondrial dysfunction
Abnormal immune activation	Autoantibody generation; CNS inflammation; Cytokine secretion; Gut permeability	Infections; Allergens; Immunogens	Defective xenobiotic metabolism; Autoimmune diseases; Hypersensitivity to immunogens
Genomic instability	Increased incidence of mutations predisposing to autism	DNA damaging agents; Diet low in one-carbon donors	Defective DNA repair; Defective xenobiotic metabolism; Genomic architecture
Epigenome dysregulation	Loss of normal gene regulation; Genomic instability	Heavy metals and toxins; CNS inflammation; Diet low in one-carbon donors	Defective one-carbon metabolism; Genomic architecture
Altered gut microbiome	Gut permeability; Neurotransmitter release; Immune dysfunction	Infections; Diet composition	Interaction of immune system with microbiota
Hyperactive transposable elements	Mutations affecting genes implicated in ASD	Chemicals activating retrotransposition	Reduced Mecp2 activity
Transgenerational inheritance	Abnormal gene expression patterns during neurodevelopment in unexposed generations	Perinatal stress; Endocrine disruptors; Effects on gut microbiome	Mechanisms are currently controversial

Elucidating the relevance of the candidate mechanisms by which environmental agents affect the genome and epigenome will require the usage of diverse research strategies. Family and population association studies that specifically examine genotype-environment interactions can provide strong evidence for the importance of such mechanisms in the disorder and identify genetic factors that increase individual susceptibility. We have previously discussed how variants affecting *MTHFR* have been associated with ASD risk according to folate fortification [52]. Another recent study reported that variants in the *MET* gene interact with air pollution exposure to increase ASD risk [94]. However, association studies examining genotype-environment interactions have low statistical power, necessitating very large study populations. Consequently, such studies are difficult to set up. A second limitation is that even when statistically significant gene x environmental factor interactions are detected by association studies, it can still be difficult to pinpoint the biological mechanisms by which these occur.

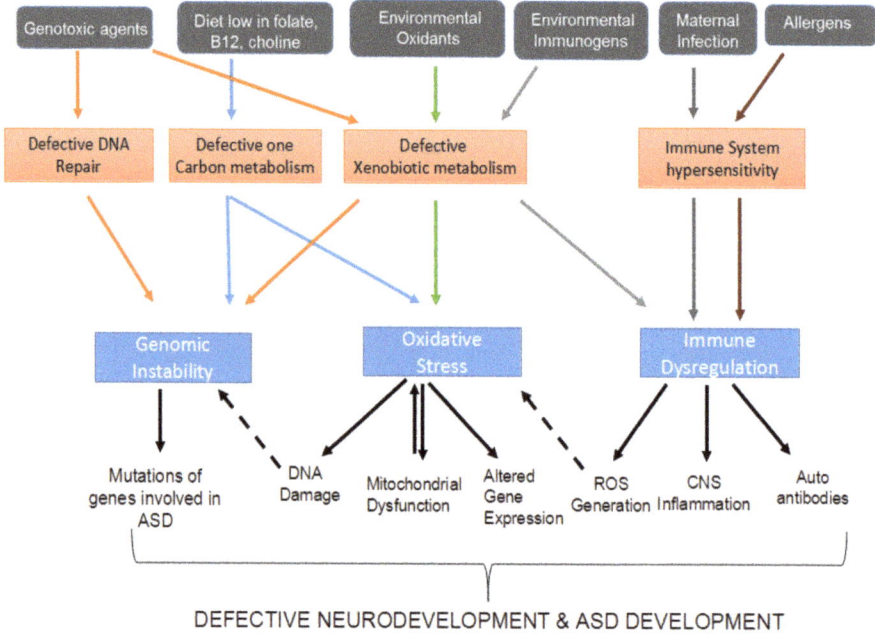

Figure 1. The figure illustrates the complex interactions between environmental and genetic factors that modulate the genome and epigenome of susceptible individuals. Environmental factors are shown in grey boxes, genetic factors in orange boxes, and the biological mechanism in blue boxes. The biological mechanisms shown in this figure are abnormal immune activation, oxidative stress, and genomic instability. The solid, colored arrows indicate interactions between environmental and genetic factors that eventually contribute to genomic instability, oxidative stress or immune activation. Orange colored lines indicate the effects of genotoxic agents, blue lines the effects of diets that are low in sources of one-carbon units, green lines indicate environmental agents, light gray lines indicate environmental immunogens, dark gray maternal infection, and red lines indicate allergens. Solid black lines link genomic instability, oxidative stress, and immune dysregulation to processes which contribute to ASD development. Dashed lines indicate interactions between biological processes involved in ASD e.g., DNA damage caused by oxidative stress also contributes to genomic instability and ROS generated by the immune system contribute to oxidative stress. As can be seen not only are multiple genotype-environmental interactions implicated in modulating the genome and epigenome of susceptible individuals, but interactions also occur between the different biological mechanisms.

Highly useful for studies investigating the mechanisms of genotype-environment interactions in ASD are the repositories of biomaterials and phenotypic and genotypic data for ASD patients and unaffected relatives, such as the AGRE and SFARI depositories. Patient-derived lymphoblastoid cells can be highly informative as gene expression profiles of immune cells reflect to a surprising extent those of human tissues, including the brain [95]. Additionally, the utilization of induced pluripotent stem cells (iPSC) is becoming an increasingly more powerful methodology for simulating neurodevelopmental disorders in patients. The lymphoblastoid and iPSC from ASD patients and controls can be used to investigate differences in the vulnerabilities of ASD patients to environmental factors. Cell line based investigations can be performed much cheaper and quicker than association studies that require the recruitment of large numbers of patients. An additional advantage is that cell lines can be easily genetically manipulated, something that can be used for examining the biological mechanisms by which environmental agents interact with genetic factors. An example of a recent study using this approach to investigate a potential inherent susceptibility of autistic individuals was conducted by Main *et al.*, 2013 [96]. In this study the researchers compared the effects of treatment with hydrogen peroxide and *S*-nitroprusside on lymphoblastoid cell lines from six ASD-normal sibling pairs. They reported finding that the autistic individuals were more sensitive to necrotic death but not to DNA damage induced by this treatments. Although innovative this study was limited in its use of a very small sample size and in not examining the epigenetic susceptibilities of the cell lines derived from ASD individuals (for example by determining disruption of global and gene specific patterns of DNA and histone methylation).

A big limitation of *in vitro* methods is that they are not applicable to the study of neurodevelopmental or behavioural processes. Consequently a variety of animal models have been developed for use in ASD research, with different advantages and disadvantages. Lower order species such as *C. elegans* and zebrafish can be genetically manipulated relatively quickly and at a lower cost, while using mammalian models is trickier and more expensive, but can better replicate human neurobiology and complex social behavior. Animal models also provide powerful tools for investigating genotype X environment interactions in ASD. For example, studies have demonstrated that neuroligin-deficient *C. elegans* are more vulnerable to oxidative stress [51] and hippocampal-slices from Mecp2 mice are more sensitive to hypoxia [97]. A large number of animal models for ASD have already been developed, which can be used in the future for studies investigating gene X environment interactions.

An increased understanding of genotype-environment interactions in ASD is important for identifying key genetic variants and environmental risk factors, reducing the exposure of vulnerable individuals to environmental risk factors, increasing the power and efficacy of epidemiological and clinical traits, and could facilitate the development of personalized treatments. A priority of autism research in the coming years should therefore be to increase the investment of time and resources into elucidating the nature, extent, and importance of genotype-environment interactions in autism. This will require the systematic application of research approaches that will facilitate progress in the field of genotype-environment interactions in ASD.

Acknowledgments: The authors of this manuscript have been supported by the Cyprus Institute of Neurology and Genetics.

Author Contributions: Costas Koufaris drafted the manuscript and Carolina Sismani revised the manuscript.

Conflicts of Interest: The authors declare no conflict of interest.

References

1. Geschwind, D.H. Advances in autism. *Annu. Rev. Med.* **2009**, *60*, 367–380. [CrossRef] [PubMed]
2. Hallmayer, J.; Cleveland, S.; Torres, A.; Phillips, J.; Cohen, B.; Torigoe, T.; Miller, J.; Fedele, A.; Collins, J.; Smith, K.; *et al.* Genetic heritability and shared environmental factors among twin pairs with autism. *Arch. Gen. Psychiatry* **2011**, *68*, 1095–1102. [CrossRef] [PubMed]
3. Devlin, B.; Scherer, S.W. Genetic architecture in autism spectrum disorder. *Curr. Opin. Genet. Dev.* **2012**, *22*, 229–237. [CrossRef] [PubMed]
4. Leblond, C.S.; Nava, C.; Polge, A.; Gauthier, J.; Huguet, G.; Lumbroso, S.; Giuliano, F.; Stordeur, C.; Depienne, C.; Mouzat, K.; *et al.* Meta-analysis of SHANK mutations in autism spectrum disorders: A gradient of severity in cognitive impairments. *PLoS Genet.* **2014**, *10*, e1004580. [CrossRef] [PubMed]
5. Clipperton-Allen, A.E.; Page, D.T. Pten haploinsufficient mice show broad brain overgrowth but selective impairments in autism-relevant behavioral tests. *Hum. Mol. Genet.* **2014**, *23*, 3490–3505. [CrossRef] [PubMed]
6. Tsai, N.P.; Wilkerson, J.R.; Guo, W.; Maksimova, M.A.; DeMartino, G.N.; Cowan, C.W.; Huber, K.M. Multiple autism-linked genes mediate synapse elimination via proteasomal degradation of a synaptic scaffold PSD-95. *Cell* **2012**, *151*, 1581–1594. [CrossRef] [PubMed]

7. Shepherd, G.M.; Katz, D.M. Synaptic microcircuit dysfunction in genetic models of neurodevelopmental disorders: Focus on *Mecp2* and *Met. Curr. Opin. Neurobiol.* **2011**, *21*, 827–833. [CrossRef] [PubMed]

8. James, S.J.; Shpyleva, S.; Melnyk, S.; Pavliv, O.; Pogribny, I.P. Elevated 5-hydroxymethylcytosine in the Engrailed-2 (EN-2) promoter is associated with increased gene expression and decreased MeCP2 binding in autism cerebellum. *Transl. Psychiatry* **2014**, *4*, e460. [CrossRef] [PubMed]

9. Nguyen, A.; Rauch, T.A.; Pfeifer, G.P.; Hu, V.W. Global methylation profiling of lymphoblastoid cell lines reveals epigenetic contributions to autism spectrum disorders and a novel autism candidate gene, *RORA*, whose protein product is reduced in autistic brain. *FASEB J.* **2010**, *24*, 3036–3051. [CrossRef] [PubMed]

10. Ladd-Acosta, C.; Hansen, K.D.; Briem, E.; Fallin, M.D.; Kaufmann, W.E.; Feinberg, A.P. Common DNA methylation alterations in multiple brain regions in autism. *Mol. Psychiatry* **2014**, *19*, 862–871. [CrossRef] [PubMed]

11. James, S.J.; Shpyleva, S.; Melnyk, S.; Pavliv, O.; Pogribny, I.P. Complex epigenetic regulation of the Engrailed-2 (EN-2) homeobox gene in the autism cerebellum. *Transl. Psychiatry* **2013**, *3*, e232. [CrossRef] [PubMed]

12. Atladóttir, H.O.; Thorsen, P.; Østergaard, L.; Schendel, D.E.; Lemcke, S.; Abdallah, M.; Parner, E.T. Maternal infection requiring hospitalization during pregnancy and autism spectrum disorders. *J. Autism Dev. Disord.* **2010**, *40*, 1423–1430. [CrossRef] [PubMed]

13. Roberts, A.L.; Lyall, K.; Hart, J.E.; Laden, F.; Just, A.C.; Bobb, J.F.; Koenen, K.C.; Ascherio, A.; Weisskopf, M.G. Perinatal air pollutant exposures and autism spectrum disorder in the children of Nurses' Health Study II participants. *Environ. Health Perspect.* **2013**, *121*, 978–984. [PubMed]

14. Surén, P.; Roth, C.; Bresnahan, M.; Haugen, M.; Hornig, M.; Hirtz, D.; Lie, K.K.; Lipkin, W.I.; Magnus, P.; Reichborn-Kjennerud, T.; *et al.* Association between maternal use of folic acid supplements and autism risk in children. *JAMA* **2013**, *309*, 570–577. [CrossRef] [PubMed]

15. Herbert, M.R. Contributions of the environment and environmentally vulnerable physiology to autism spectrum disorders. *Curr. Opin. Neurol.* **2010**, *23*, 103–110. [CrossRef] [PubMed]

16. LaSalle, J.M. A genomic-point-of-view on environmental factors influencing the human brain methylome. *Epigenetics* **2011**, *6*, 862–869. [CrossRef] [PubMed]

17. Chaste, P.; Leboyer, M. Autism risk factors: Genes, environment, and gene-environment interactions. *Dialogues Clin. Neurosci.* **2012**, *14*, 281–292. [PubMed]

18. LaSalle, J.M. Epigenomic strategies at the interface of genetic and environmental risk factors for autism. *J. Hum. Genet.* **2013**, *58*, 396–401. [CrossRef] [PubMed]

19. Stamou, M.; Streifel, K.M.; Goines, P.E.; Lein, P.J. Neuronal connectivity as a convergent target of gene × environment interactions that confer risk for Autism Spectrum Disorders. *Neurotoxicol. Teratol.* **2013**, *36*, 3–16. [CrossRef] [PubMed]

20. Miksys, S.; Tyndale, R.F. Cytochrome P450-mediated drug metabolism in the brain. *J. Psychiatry Neurosci.* **2013**, *38*, 152–163. [CrossRef] [PubMed]

21. Woods, R.; Vallero, R.O.; Golub, M.S.; Suarez, J.K.; Ta, T.A.; Yasui, D.H.; Chi, L.-H.; Kostyniak, P.J.; Pessah, I.N.; Berman, R.F.; *et al.* Long lived epigenetic interactions between perinatal PBDE exposure and Mecp22308 mutation. *Hum. Mol. Genet.* **2012**, *21*, 2399–2411. [CrossRef] [PubMed]

22. Feo, M.L.; Gross, M.S.; McGarrigle, B.P.; Eljarrat, E.; Barceló, D.; Aga, D.S.; Olson, J.R. Biotransformation of BDE-47 to potentially toxic metabolites is predominantly mediated by human CYP2B6. *Environ. Health Perspect.* **2013**, *121*, 440–446. [CrossRef] [PubMed]

23. Kinney, D.K.; Barch, D.H.; Chayka, B.; Napoleon, S.; Munir, K.M. Environmental risk factors for autism: Do they help cause de novo genetic mutations? *Med. Hypotheses* **2010**, *74*, 102–106. [CrossRef] [PubMed]

24. Turesky, R.J. The role of genetic polymorphisms in the metabolism of carcinogenic heterocyclic amines. *Curr. Drug Metab.* **2004**, *5*, 169–180. [CrossRef] [PubMed]

25. D'Amelio, M.; Ricci, I.; Sacco, R.; Liu, X.; D'Agruma, L.; Muscarella, L.A.; Guarnieri, V.; Militerni, R.; Bravaccio, C.; Elia, M.; *et al.* Paraoxonase gene variants are associated with autism in North America, but not in Italy: Possible regional specificity in gene-environment interactions. *Mol. Psychiatry* **2005**, *10*, 1006–1016. [CrossRef] [PubMed]

26. Eskenazi, B.; Huen, K.; Marks, A.; Harley, K.G.; Bradman, A.; Barr, D.B.; Holland, N. PON1 and neurodevelopment in children from the CHAMACOS study exposed to organophosphate pesticides in utero. *Environ. Health Perspect.* **2010**, *118*, 1775–1781. [CrossRef] [PubMed]

27. Rossignol, D.A.; Genuis, S.J.; Frye, R.E. Environmental toxicants and autism spectrum disorders: A systematic review. *Transl. Psychiatry* **2014**, *4*, e360. [CrossRef] [PubMed]
28. Serajee, F.J.; Nabi, R.; Zhong, H.; Huq, M. Polymorphisms in xenobiotic metabolism genes and autism. *J. Child Neurol.* **2004**, *19*, 413–417. [PubMed]
29. Buyske, S.; Williams, T.A.; Mars, A.E.; Stenroos, E.S.; Ming, S.X.; Wang, R.; Sreenath, M.; Factura, M.F.; Reddy, C.; Lambert, G.H.; *et al.* Analysis of case-parent trios at a locus with a deletion allele: Association of GSTM1 with autism. *BMC Genet.* **2006**, *7*, 8. [CrossRef] [PubMed]
30. Baron-Cohen, S.; Knickmeyer, R.C.; Belmonte, M.K. Sex differences in the brain: Implications for explaining autism. *Science* **2005**, *310*, 819–823. [CrossRef] [PubMed]
31. Werling, D.M.; Geschwind, D.H. Sex differences in autism spectrum disorders. *Curr. Opin. Neurol.* **2013**, *26*, 146–153. [CrossRef] [PubMed]
32. Lombardo, M.V.; Ashwin, E.; Auyeung, B.; Chakrabarti, B.; Taylor, K.; Hackett, G.; Bullmore, E.T.; Baron-Cohen, S. Fetal testosterone influences sexually dimorphic grey matter in the brain. *J. Neurosci.* **2012**, *32*, 674–680. [CrossRef] [PubMed]
33. Baron-Cohen, S.; Auyeung, B.; Nørgaard-Pedersen, B.; Hougaard, D.M.; Abdallah, M.W.; Melgaard, L.; Cohen, A.S.; Chakrabarti, B.; Ruta, L.; Lombardo, M.V. Elevated fetal steroidogenic activity in autism. *Mol. Psychiatry* **2015**, *20*, 369–376. [CrossRef] [PubMed]
34. Kajta, M.; Wójtowicz, A.K. Impact of endocrine-disrupting chemicals on neural development and the onset of neurological disorders. *Pharmacol. Rep.* **2013**, *65*, 1632–1639. [CrossRef] [PubMed]
35. Miodovnik, A.; Engel, S.M.; Zhu, C.; Ye, X.; Soorya, L.V.; Silva, M.J.; Calafat, A.M.; Wolff, M.S. Endocrine disruptors and childhood social impairments. *Neurotoxicology* **2011**, *32*, 261–267. [CrossRef] [PubMed]
36. Braun, J.M.; Kalkbrenner, A.E.; Just, A.C.; Yolton, K.; Calafat, A.M.; Sjödin, A.; Hauser, R.; Webster, G.M.; Chen, A.; Lanphear, B.P. Gestational exposure to endocrine-disrupting chemicals and reciprocal social, repetitive, and stereotypic behaviors in 4- and 5-year-old children: The HOME study. *Environ. Health Perspect.* **2014**, *122*, 513–520. [PubMed]
37. Spearow, J.L.; Doemeny, P.; Sera, R.; Leffler, R.; Barkley, M. Genetic variability in susceptibility to endocrine disruptors in mice. *Science* **1999**, *285*, 1259–1261. [CrossRef] [PubMed]
38. Henningsson, S.; Jonsson, L.; Ljunggren, E.; Westberg, L.; Gillberg, C.; Råstam, M.; Anckarsäter, H.; Nygren, G.; Landén, M.; Thuresson, K.; *et al.* Possible association between the androgen receptor gene and autism spectrum disorder. *Psychoneuroendocrinology* **2009**, *34*, 752–761. [CrossRef] [PubMed]
39. Zettergren, A.; Jonsson, L.; Johansson, D.; Melke, J.; Lundström, S.; Anckarsäter, H.; Lichtenstein, P.; Westberg, L. Associations between polymorphisms in sex steroid related genes and autistic-like traits. *Psychoneuroendocrinology* **2013**, *38*, 2575–2584. [CrossRef] [PubMed]
40. Frye, R.E.; Delatorre, R.; Taylor, H.; Slattery, J.; Melnyk, S.; Chowdhury, N.; James, S.J. Redox metabolic abnormalities in autistic children associated with mitochondrial disease. *Transl. Psychiatry* **2013**, *3*, e273. [CrossRef] [PubMed]
41. Rossignol, D.A.; Frye, R.E. Mitochondrial dysfunction in autism spectrum disorders: A systematic review and meta-analysis. *Mol. Psychiatry* **2012**, *17*, 290–314. [CrossRef] [PubMed]
42. Boccuto, L.; Chen, C.-F.; Pittman, A.R.; Skinner, C.D.; McCartney, H.J.; Jones, K.; Bochner, B.R.; Stevenson, R.E.; Schwartz, C.E. Decreased tryptophan metabolism in patients with autism spectrum disorders. *Mol. Autism* **2013**, *4*, 16. [CrossRef] [PubMed]
43. Li, J.; Harris, R.A.; Cheung, S.W.; Coarfa, C.; Jeong, M.; Goodell, M.A.; White, L.D.; Patel, A.; Kang, S.-H.; Shaw, C.; *et al.* Genomic hypomethylation in the human germline assosiates with selective structural mutability in the human genome. *PLoS Genet.* **2012**, *8*, e1002692. [CrossRef] [PubMed]
44. Li, Z.; Okamoto, K.-I.; Hayashi, Y.; Sheng, M. The importance of dendritic mitochondria in the morphogenesis and plasticity of spines and synapses. *Cell* **2004**, *119*, 873–887. [CrossRef] [PubMed]
45. James, S.J.; Melnyk, S.; Jernigan, S.; Cleves, M.A.; Halsted, C.H.; Wong, D.H.; Cutler, P.; Bock, K.; Boris, M.; Bradstreet, J.J.; *et al.* Metabolic endophenotype and related genotypes are associated with oxidative stress in children with autism. *Am. J. Med. Genet. Part B* **2006**, *141B*, 947–956. [CrossRef]
46. Schwartz, C.E. Aberrant tryptophan metabolism: The unifying biochemical basis for autism spectrum disorders? *Biomark. Med.* **2014**, *8*, 313–315. [CrossRef] [PubMed]

47. Rose, S.; Frye, R.E.; Slattery, J.; Wynne, R.; Tippett, M.; Pavliv, O.; Melnyk, S.; James, S.J. Oxidative stress induces mitochondrial dysfunction in a subset of autism lymphoblastoid cell lines in a well-matched case control cohort. *PLoS ONE* **2014**, *9*, e85436. [CrossRef] [PubMed]

48. Rose, S.; Frye, R.E.; Slattery, J.; Wynne, R.; Tippett, M.; Melnyk, S.; James, S.J. Oxidative stress induces mitochondrial dysfunction in a subset of autistic lymphoblastoid cell lines. *Transl. Psychiatry* **2014**, *4*, e377. [CrossRef] [PubMed]

49. Kaur, K.; Chauhan, V.; Gu, F.; Chauhan, A. Bisphenol A induces oxidative stress and mitochondrial dysfunction in lymphoblasts from children with autism and unaffected siblings. *Free Radic. Biol. Med.* **2014**, *76C*, 25–33. [CrossRef] [PubMed]

50. Napoli, E.; Ross-Inta, C.; Wong, S.; Hung, C.; Fujisawa, Y.; Sakaguchi, D.; Angelastro, J.; Omanska-Klusek, A.; Schoenfeld, R.; Giulivi, C. Mitochondrial dysfunction in Pten haplo-insufficient mice with social deficits and repetitive behavior: Interplay between Pten and p53. *PLoS ONE* **2012**, *7*, e42504. [CrossRef] [PubMed]

51. Hunter, J.W.; Mullen, G.P.; McManus, J.R.; Heatherly, J.M.; Duke, A.; Rand, J.B. Neuroligin-deficient mutants of C. elegans have sensory processing deficits and are hypersensitive to oxidative stress and mercury toxicity. *Dis. Model. Mech.* **2010**, *3*, 366–376. [CrossRef] [PubMed]

52. Pu, D.; Shen, Y.; Wu, J. Association between *MTHFR* gene polymorphisms and the risk of autism spectrum disorders: A meta-analysis. *Autism Res.* **2013**, *6*, 384–392. [CrossRef] [PubMed]

53. Kovač, J.; Macedoni Lukšič, M.; Trebušak Podkrajšek, K.; Klančar, G.; Battelino, T. Rare single nucleotide polymorphisms in the regulatory regions of the superoxide dismutase genes in autism spectrum disorder. *Autism Res.* **2014**, *7*, 138–144. [CrossRef] [PubMed]

54. Smith, M.; Flodman, P.L.; Gargus, J.J.; Simon, M.T.; Verrell, K.; Haas, R.; Reiner, G.E.; Naviaux, R.; Osann, K.; Spence, M.A.; *et al.* Mitochondrial and Ion Channel gene alterations in autism. *Biochim. Biophys. Acta* **2012**, *1817*, 1796–1802. [CrossRef] [PubMed]

55. Nava, C.; Lamari, F.; Héron, D.; Mignot, C.; Rastetter, A.; Keren, B.; Cohen, D.; Faudet, A.; Bouteiller, D.; Gilleron, M.; *et al.* Analysis of the chromosome X exome in patients with autism spectrum disorders identified novel candidate genes, including TMLHE. *Transl. Psychiatry* **2012**, *2*, e179. [CrossRef] [PubMed]

56. Napolioni, V.; Persico, A.M.; Porcelli, V.; Palmieri, L. The mitochondrial aspartate/glutamate carrier AGC1 and calcium homeostasis: Physiological links and abnormalities in autism. *Mol. Neurobiol.* **2011**, *44*, 83–92. [CrossRef] [PubMed]

57. Rossignol, D.A.; Frye, R.E. A review of research trends in physiological abnormalities in autism spectrum disorders: Immune dysregulation, inflammation, oxidative stress, mitochondrial dysfunction and environmental toxicant exposures. *Mol. Psychiatry* **2012**, *17*, 389–401. [CrossRef] [PubMed]

58. Zhu, L.; Wang, X.; Li, X.-L.; Towers, A.; Cao, X.; Wang, P.; Bowman, R.; Yang, H.; Goldstein, J.; Li, Y.-J.; *et al.* Epigenetic dysregulation of SHANK3 in brain tissues from individuals with autism spectrum disorders. *Hum. Mol. Genet.* **2014**, *23*, 1563–1578. [CrossRef] [PubMed]

59. Muotri, A.R.; Chu, V.T.; Marchetto, M.C.N.; Deng, W.; Moran, J.V.; Gage, F.H. Somatic mosaicism in neuronal precursor cells mediated by L1 retrotransposition. *Nature* **2005**, *435*, 903–910. [CrossRef] [PubMed]

60. Terasaki, N.; Goodier, J.L.; Cheung, L.E.; Wang, Y.J.; Kajikawa, M.; Kazazian, H.H.; Okada, N. *In vitro* screening for compounds that enhance human L1 mobilization. *PLoS ONE* **2013**, *8*, e74629. [CrossRef] [PubMed]

61. Muotri, A.R.; Marchetto, M.C.N.; Coufal, N.G.; Oefner, R.; Yeo, G.; Nakashima, K.; Gage, F.H. L1 retrotransposition in neurons is modulated by MeCP2. *Nature* **2010**, *468*, 443–446. [CrossRef] [PubMed]

62. Okudaira, N.; Okamura, T.; Tamura, M.; Iijma, K.; Goto, M.; Matsunaga, A.; Ochiai, M.; Nakagama, H.; Kano, S.; Fujii-Kuriyama, Y.; *et al.* Long interspersed element-1 is differentially regulated by food-borne carcinogens via the aryl hydrocarbon receptor. *Oncogene* **2013**, *32*, 4903–4912. [CrossRef] [PubMed]

63. Wilhelm-Benartzi, C.S.; Houseman, E.A.; Maccani, M.A.; Poage, G.M.; Koestler, D.C.; Langevin, S.M.; Gagne, L.A.; Banister, C.E.; Padbury, J.F.; Marsit, C.J. In utero exposures, infant growth, and DNA methylation of repetitive elements and developmentally related genes in human placenta. *Environ. Health Perspect.* **2012**, *120*, 296–302. [CrossRef] [PubMed]

64. Mefford, H.C.; Eichler, E.E. Duplication hotspots, rare genomic disorders, and common disease. *Curr. Opin. Genet. Dev.* **2009**, *19*, 196–204. [CrossRef] [PubMed]

65. Girirajan, S.; Johnson, R.L.; Tassone, F.; Balciuniene, J.; Katiyar, N.; Fox, K.; Baker, C.; Srikanth, A.; Yeoh, K.H.; Khoo, S.J.; *et al.* Global increases in both common and rare copy number load associated with autism. *Hum. Mol. Genet.* **2013**, *22*, 2870–2880. [CrossRef] [PubMed]

66. Ionita-Laza, I.; Xu, B.; Makarov, V.; Buxbaum, J.D.; Roos, J.L.; Gogos, J.A.; Karayiorgou, M. Scan statistic-based analysis of exome sequencing data identifies FAN1 at 15q13.3 as a susceptibility gene for schizophrenia and autism. *Proc. Natl. Acad. Sci. USA* **2014**, *111*, 343–348. [CrossRef] [PubMed]

67. Goines, P.; van de Water, J. The immune system's role in the biology of autism. *Curr. Opin. Neurol.* **2010**, *23*, 111–117. [CrossRef] [PubMed]

68. Onore, C.; Careaga, M.; Ashwood, P. The role of immune dysfunction in the pathophysiology of autism. *Brain. Behav. Immun.* **2012**, *26*, 383–392. [CrossRef] [PubMed]

69. Deverman, B.E.; Patterson, P.H. Cytokines and CNS development. *Neuron* **2009**, *64*, 61–78. [CrossRef] [PubMed]

70. Basil, P.; Li, Q.; Dempster, E.L.; Mill, J.; Sham, P.C.; Wong, C.C.; McAlonan, G.M. Prenatal maternal immune activation causes epigenetic differences in adolescent mouse brain. *Transl. Psychiatry* **2014**, *2*, e434. [CrossRef]

71. Ashwood, P.; Schauer, J.; Pessah, I.N.; van de Water, J. Preliminary evidence of the *in vitro* effects of BDE-47 on innate immune responses in children with autism spectrum disorders. *J. Neuroimmunol.* **2009**, *208*, 130–135. [CrossRef] [PubMed]

72. Schwartzer, J.J.; Careaga, M.; Onore, C.E.; Rushakoff, J.A.; Berman, R.F.; Ashwood, P. Maternal immune activation and strain specific interactions in the development of autism-like behaviors in mice. *Transl. Psychiatry* **2013**, *3*, e240. [CrossRef] [PubMed]

73. Ehninger, D.; Sano, Y.; de Vries, P.J.; Dies, K.; Franz, D.; Geschwind, D.H.; Kaur, M.; Lee, Y.-S.; Li, W.; Lowe, J.K.; *et al.* Gestational immune activation and Tsc2 haploinsufficiency cooperate to disrupt fetal survival and may perturb social behavior in adult mice. *Mol. Psychiatry* **2012**, *17*, 62–70. [CrossRef] [PubMed]

74. Cryan, J.F.; Dinan, T.G. Mind-altering microorganisms: The impact of the gut microbiota on brain and behaviour. *Nat. Rev. Neurosci.* **2012**, *13*, 701–712. [CrossRef] [PubMed]

75. Jones, R.M.; Mercante, J.W.; Neish, A.S. Reactive oxygen production induced by the gut microbiota: Pharmacotherapeutic implications. *Curr. Med. Chem.* **2012**, *19*, 1519–1529. [CrossRef] [PubMed]

76. Finegold, S.M.; Molitoris, D.; Song, Y.; Liu, C.; Vaisanen, M.-L.; Bolte, E.; McTeague, M.; Sandler, R.; Wexler, H.; Marlowe, E.M.; *et al.* Gastrointestinal microflora studies in late-onset autism. *Clin. Infect. Dis.* **2002**, *35*, S6–S16. [CrossRef] [PubMed]

77. Parracho, H.M.R.T.; Bingham, M.O.; Gibson, G.R.; McCartney, A.L. Differences between the gut microflora of children with autism spectrum disorders and that of healthy children. *J. Med. Microbiol.* **2005**, *54*, 987–991. [CrossRef] [PubMed]

78. Hsiao, E.Y.; McBride, S.W.; Hsien, S.; Sharon, G.; Hyde, E.R.; McCue, T.; Codelli, J.A.; Chow, J.; Reisman, S.E.; Petrosino, J.F.; *et al.* Microbiota modulate behavioral and physiological abnormalities associated with neurodevelopmental disorders. *Cell* **2013**, *155*, 1451–1463. [CrossRef] [PubMed]

79. Holmes, E.; Li, J.V.; Marchesi, J.R.; Nicholson, J.K. Gut microbiota composition and activity in relation to host metabolic phenotype and disease risk. *Cell Metab.* **2012**, *16*, 559–564. [CrossRef] [PubMed]

80. McKnite, A.M.; Perez-Munoz, M.E.; Lu, L.; Williams, E.G.; Brewer, S.; Andreux, P.A.; Bastiaansen, J.W.M.; Wang, X.; Kachman, S.D.; Auwerx, J.; *et al.* Murine gut microbiota is defined by host genetics and modulates variation of metabolic traits. *PLoS ONE* **2012**, *7*, e39191. [CrossRef] [PubMed]

81. Parks, B.W.; Nam, E.; Org, E.; Kostem, E.; Norheim, F.; Hui, S.T.; Pan, C.; Civelek, M.; Rau, C.D.; Bennett, B.J.; *et al.* Genetic control of obesity and gut microbiota composition in response to high-fat, high-sucrose diet in mice. *Cell Metab.* **2013**, *17*, 141–152. [CrossRef] [PubMed]

82. Wolstenholme, J.T.; Edwards, M.; Shetty, S.R.J.; Gatewood, J.D.; Taylor, J.A.; Rissman, E.F.; Connelly, J.J. Gestational exposure to bisphenol a produces transgenerational changes in behaviors and gene expression. *Endocrinology* **2012**, *153*, 3828–3838. [CrossRef] [PubMed]

83. Lim, J.P.; Brunet, A. Bridging the transgenerational gap with epigenetic memory. *Trends Genet.* **2013**, *29*, 176–186. [CrossRef] [PubMed]

84. De Theije, C.G.M.; Wopereis, H.; Ramadan, M.; van Eijndthoven, T.; Lambert, J.; Knol, J.; Garssen, J.; Kraneveld, A.D.; Oozeer, R. Altered gut microbiota and activity in a murine model of autism spectrum disorders. *Brain. Behav. Immun.* **2014**, *37*, 197–206. [CrossRef] [PubMed]

85. Viscidi, E.W.; Triche, E.W.; Pescosolido, M.F.; McLean, R.L.; Joseph, R.M.; Spence, S.J.; Morrow, E.M. Clinical characteristics of children with epilepsy and co-occuring autism. *PLoS ONE* **2013**, *8*, e67797. [CrossRef] [PubMed]

86. Van Eeghen, A.M.; Pulsifer, M.B.; Merker, V.L.; Neumeyer, A.M.; van Eeghen, E.E.; Thibert, R.L.; Cole, A.J.; Leigh, F.A.; Plotkin, S.R.; Thiele, E.A. Understanding relationships between autism, intelligence, and epilepsy: A cross-disorder approach. *Dev. Med. Child Neurol.* **2013**, *55*, 146–153. [CrossRef] [PubMed]

87. Hara, H. Autism and epilepsy: A retrospective follow-up study. *Brain Dev.* **2007**, *29*, 486–490. [CrossRef] [PubMed]

88. Hagerman, R.; Hoem, G.; Hagerman, P. Fragile X and autism: Intertwined at the molecular level leading to targeted treatments. *Mol. Autism* **2010**, *1*, 12. [CrossRef] [PubMed]

89. Fatemi, S.H.; Folsom, T.D. Dysregulation of fragile $\times$ mental retardation protein and metabotropic glutamate receptor 5 in superior frontal cortex of individuals with autism: A postmortem brain study. *Mol Autism.* **2011**, *2*, 1–11. [CrossRef] [PubMed]

90. Rustan, O.G.; Folsom, T.D.; Yousefi, M.K.; Fatemi, S.H. Phosphorylated fragile X mental retardation protein at serine 499, is reduced in cerebellar vermis and superior frontal cortex of subjects with autism: Implications for fragile X mental retardation protein-metabotropic glutamate receptor 5 signaling. *Mol. Autism.* **2013**, *4*, 41. [CrossRef] [PubMed]

91. Bernard, P.B.; Castano, A.M.; O'Leary, H.; Simpson, K.; Browning, M.D.; Benke, T.A. Phosphorylation of FMRP and alterations of FMRP complex underlie enhanced mLTD in adult rats triggered by early life seizures. *Neurobiol. Dis.* **2013**, *59*, 1–17. [CrossRef] [PubMed]

92. Grandjean, P.; Landrigan, P.J. Developmental neurotoxicity of industrial chemicals. *Lancet* **2006**, *368*, 2167–2178. [CrossRef] [PubMed]

93. Miodovnik, A. Environmental neurotoxicants and the developing brain. *Mt. Sinai J. Med.* **2011**, *78*, 58–77. [CrossRef] [PubMed]

94. Volk, H.E.; Kerin, T.; Lurmann, F.; Hertz-Picciotto, I.; McConnell, R.; Campbell, D.B. Autism spectrum disorder: Interaction of air pollution with the MET receptor tyrosine kinase gene. *Epidemiology* **2014**, *25*, 44–47. [CrossRef] [PubMed]

95. Kohane, I.S.; Valtchinov, V.I. Quantifying the white blood cell transcriptome as an accessible window to the multiorgan transcriptome. *Bioinform. Oxf. Engl.* **2012**, *28*, 538–545. [CrossRef]

96. Main, P.A.E.; Thomas, P.; Esterman, A.; Fenech, M.F. Necrosis is increased in lymphoblastoid cell lines from children with autism compared with their non-autistic siblings under conditions of oxidative and nitrosative stress. *Mutagenesis* **2013**, *28*, 475–484. [CrossRef] [PubMed]

97. Fischer, M.; Reuter, J.; Gerich, F.J.; Hildebrandt, B.; Hägele, S.; Katschinski, D.; Müller, M. Enhanced hypoxia susceptibility in hippocampal slices from a mouse model of rett syndrome. *J. Neurophysiol.* **2009**, *101*, 1016–1032. [CrossRef] [PubMed]

International Journal of
Molecular Sciences

MDPI

Article

Novel Systems Modeling Methodology in Comparative Microbial Metabolomics: Identifying Key Enzymes and Metabolites Implicated in Autism Spectrum Disorders

Colin Heberling [1,2,]* and Prasad Dhurjati [2]

[1] Department of Biotechnology, Johns Hopkins University, Rockville, MD 20850, USA

[2] Department of Chemical and Biomolecular Engineering, University of Delaware, Newark, DE 19716, USA; dhurjati@udel.edu

* Author to whom correspondence should be addressed; cheb@udel.edu; Tel.: +1-302-690-1003.

Academic Editor: Merlin G. Butler

Received: 14 January 2015; Accepted: 24 March 2015; Published: 22 April 2015

Abstract: Autism spectrum disorders are a group of mental illnesses highly correlated with gastrointestinal dysfunction. Recent studies have shown that there may be one or more microbial "fingerprints" in terms of the composition characterizing individuals with autism, which could be used for diagnostic purposes. This paper proposes a computational approach whereby metagenomes characteristic of "healthy" and autistic individuals are artificially constructed via genomic information, analyzed for the enzymes coded within, and then these enzymes are compared in detail. This is a text mining application. A custom-designed online application was built and used for the comparative metabolomics study and made publically available. Several of the enzyme-catalyzing reactions involved with the amino acid glutamate were curiously missing from the "autism" microbiome and were coded within almost every organism included in the "control" microbiome. Interestingly, there exists a leading hypothesis regarding autism and glutamate involving a neurological excitation/inhibition imbalance; but the association with this study is unclear. The results included data on the transsulfuration and transmethylation pathways, involved with oxidative stress, also of importance to autism. The results from this study are in alignment with leading hypotheses in the field, which is impressive, considering the purely *in silico* nature of this study. The present study provides new insight into the complex metabolic interactions underlying autism, and this novel methodology has potential to be useful for developing new hypotheses. However, limitations include sparse genome data availability and conflicting literature experimental data. We believe our software tool and methodology has potential for having great utility as data become more available, comprehensive and reliable.

Keywords: autism; metabolomics; comparative; microbial; computational; bioinformatics; biomarkers; gut; gastrointestinal; microbiome

1. Introduction

Autism spectrum disorders are a category of mental illnesses characterized by social cognitive impairments and stereotyped behaviors [1]. Autistic individuals often have trouble fitting into society and put a financial burden on their families, lifelong. Autism diagnosis has been steadily on the rise in the past decade or so, affecting one in every 88 children according to a source in 2012 [1] and one in every 68 children from current data (2015) from the Centers for Disease Control and Protection [2]. There are most likely many adults suffering from autism that have never been diagnosed. This is because there is as of yet no clear method to diagnose autism; diagnosis is purely based on making qualitative observations of an individual. This is why there is a pressing need to find reliable methods for autism diagnosis and for treatment, as well.

Recent research suggests that individuals with autism often suffer from gastrointestinal dysfunction, as well [3,4]. Rather than focusing on the illness by way of human genetics, many scientists are now exploring the impact of microbial genetics. There exists a full isolated ecosystem of microbial lifeforms that inhabit the human gastrointestinal tract. These microbes have a profound impact on the health and disease of the human host and in general have a symbiotic co-existence with the host [5]. Microbial species or strain composition is believed to largely contribute to homeostasis or abnormality in humans. While each person's microbial "fingerprint" is unique, there are specific patterns seen in those that are healthy and those that have specific illnesses [6,7]. Remarkably, autism has been correlated with the overgrowth of certain types of bacteria, such as certain species of *Clostridia* [8,9] and *Desulfovibrio* [9,10], and there is some preliminary evidence that *Sutterella* may be implicated, as well [11]. Data from the pyrosequencing study by Finegold *et al.* [9] suggested that several other organisms may be implicated in autism, as well, but these results were less significant than those implicating *Clostridia* and *Desulfovibrio*.

Nonetheless, one organism, *Akkermansia*, was chosen from among these organisms to be included in the study, to see how much of an impact these less significant organisms might have. *Akkermansia* is part of the *Verrucomicrobia* phylum. According to a study by Williams *et al.* [12] where live intestinal biopsies were taken to measure bacterial composition, *Verrucomicrobia* made up approximately 1% of the total bacterial composition in individuals with autism and only 0.5% in healthy age-matched controls. This was another leading factor for the decision to include *Akkermansia* in the present study.

As one might expect, as bacterial composition has an effect on health and disease, the molecular metabolites that these microbes produce are the "tools" that carry out these biological changes. For example, individuals with autism have been cited as having sulfur metabolic deficiencies, to suffer from elevated oxidative stress and to have trouble detoxifying xenobiotic compounds and heavy metals [13–15]. It is therefore important to analyze the metabolome of these individuals and that of healthy individuals (for control) in order to identify useful biomarkers and perhaps even gain a greater understanding of the disorder. Knowledge may be gained about other related illnesses, as well, such as others related to gastrointestinal dysfunction.

Metagenomics is the study of the genetic content contained in whole microbial ecosystems isolated from natural environments [16,17]. In contrast, the more traditional genomics is the study of only DNA from a single, isolated species or strain of organism. Metagenomics therefore brings with it new challenges in bioinformatics. This is because before regular genomics can be applied to each organism, the key organisms need to be identified by aligning sample DNA reads to reference sequences contained in curated databases [18]. Of course, the environment that this study focuses on is the human gut microbiome. This study does not use the conventional methods of metagenomics, but rather simplifies the problem by constructing artificial metagenomes based on genomes of organisms already known to be contained in the target environment. There are numerous pieces of literature already describing the composition of microbes contained in autistic individuals and "healthy" individuals. The present study constructs a metagenomics "model", rather than conducting a pure metagenomics study.

2. Methods

The literature suggests that there are distinct microbiomes characteristic of a "healthy" individual and that of an autistic individual [4]. Therefore, the first step was to choose the microorganisms to represent these microbiomes. It was the intended goal to be as comprehensive as possible with this step, so that all possible enzymes and metabolites that are possible to exist in urinary or stool samples are covered. Choosing the correct microorganisms can be tricky, as sometimes there are conflicting data or interpretations of data in different literature [19]. Another quandary is whether to include the more common organisms in both microbiomes studied or to only include the "problem" organisms; that is, the microorganisms thought to cause the main differences between the two microbiomes. In the latter case, the autistic microbiome could be represented with just a few species of microbe. The end goal is to find the key enzymes and metabolites expressed that are entirely different in the autistic individual. Of course, autism is a complex disorder with multiple different pathways for pathogenesis, so the autistic microbiome chosen will try to represent all cases of autism arising from gut microbiological origin.

Only those microorganisms thought to be represented in significantly higher amounts (proportionally) in the guts of autistic individuals were chosen for the autistic microbiome, with the core microorganisms representative of a healthy human gut for the control microbiome. However, there is still no consensus on what defines a typical "healthy" gut microbiome, so the "control" microbiome was chosen based on a few key steps. First, those microbes that have been reported to be in decreased quantities in autism were included in the "control" microbiome [9,12,20]. Next, those reported to have no change in composition were included. Finally, other organisms that are normally found in the human gut, but that had very little discussion in related autism literature were included in the "control" microbiome. To facilitate this final step, we have unpublished work where the MEtaGenome Analyzer (MEGAN software) [21,22] was used to analyze three sets of metagenomic data from the Human Microbiome Project's collection of human stool samples [6,23]. These design criteria may seem confusing at first if we take into consideration that those microorganisms that are decreased in autism are nonetheless still implicated in autism, just like those organisms that are found in increased amounts. However, our design considerations are predicated on the assumption that those organisms that are found in decreased numbers in autism are in fact usually beneficial for the host in otherwise "healthy" microbiomes. There is at least some evidence for two such cases: *Lactobacillus* species and *Bifidobacterium* species [24]. We can therefore justify stratifying gut microbes into two groupings based on a simple binary distinction: those microbes that are found in increased numbers in autism and those microbes that are not found in increased numbers in autism. Comparing these two groups should reveal key differences in expressed enzymes. Thus, finding enzymes that are expressed in the "autism" group and not in the "control" group may identify metabolites indicative only of our "problem" organisms and thus may serve as potential biomarkers for autism. On the other hand, finding enzymes that are expressed by the "control" group and not the "autism" group may also identify key biomarkers, except in this case, clinicians may be able to look for decreased quantities of such molecules instead.

Enzyme Commission (EC) numbers were used for the comparative analysis. EC numbers are a standardized way to identify enzymes, much like CAS numbers (Chemical Abstracts Service) for common chemicals. GenBank files for each microbe were downloaded from the National Center for Biotechnology Information (NCBI) [25]. The limitations of this study occurred at this stage. Not all microbes known have a completely sequenced genome uploaded to NCBI, and even less still have their protein products annotated with EC numbers. This greatly limited which organisms could be used in the study. The following list organizes microorganisms into three categories: those included in the first software query, those additional organisms that were able to be included in Query 2 after *de novo* EC number annotations were made (more detail later) and those organisms that were desired to be included, but in the end were not.

The usual approach for metagenomics presents significant bottlenecks for analysis and data storage. This is why the idea came about that there may be another approach that is more efficient in very specific circumstances. Instead of using metagenomic data, this study attempts to construct an artificial metagenome by carefully choosing the microorganisms to include in the model and using the curated database of GenBank files to analyze the full spectra of enzymes coded in the genomic DNA of these organisms. This study is differentiated from a proteomics study, because we are only interested in proteins that have been assigned a standardized Enzyme Commission (EC) number, so that there is no ambiguity between gene products. Two enzymes from two different organisms may have the same EC number, but slightly different amino acid sequences or slightly different protein names in the GenBank annotation files. Using EC numbers will allow automated programs to match identical enzymes correctly. The enzymes are correlated with their associated chemical reactions and the implicated metabolites. This method of intentionally leaving out some information and looking at the bigger picture on more of a systems level could be classified as systems biology [26].

A personal computer was used for all programming and analysis. A simple, personally-designed online application was used for all analysis. This software can be found at the web link [27]. Full instructions on how to use the software can be found in a link to the readme file on the main program interface linked above. The software utilizes a relational database system using MySQL to store and query information from the GenBank annotation files from NCBI and information from an EC number database [28]. The EC number data file was parsed for inserting data into a personally-designed database table. The following example (Figure 1) shows the type of information contained in the data file that was parsed for EC numbers.

```
ID    1.8.1.19
DE    Sulfide dehydrogenase.
CA    Hydrogen sulfide + (sulfide)(n) + NADP(+) = (sulfide)(n+1) + NADPH.
CF    Flavoprotein; Iron-sulfur.
CC    -!- In the archaeon Pyrococcus furiosus the enzyme is involved in the
CC        oxidation of NADPH which is produced in peptide degradation.
CC    -!- The enzyme also catalyzes the reduction of sulfur with lower
CC        activity.
DR    Q8U195, SUDHA_PYRFU;  Q8U194, SUDHB_PYRFU;
//
```

Figure 1. Example data extracted from The ENZYME Database in 2000, Bairoch [28].

The web application allows its user to choose two groups of microbiomes for metabolomics comparison. The program was designed with the intent to be used for any microbial metabolomics study between two microbiomes, two small alterations of the same microbiome or even simply between two single organisms. The application allows the user to upload a new genome file in GenBank format. Initially, the application could only accept GenBank files with EC number annotations present, but now, there does not need to be any EC number annotations in the original file. However, inserting EC number annotations with automation has certain limitations. Database queries fall into three categories for each comparison: no matches, one unique match or more than one match. In the event that there is more than one match, manual curation will be necessary. In the meantime, the program instead prints the first three matches to the GenBank file and makes a designation within the file and within the associated database that says that the information may be less reliable than unique matches or the data that was already annotated in the original file.

The application returns output directly to the screen within the web browser. The output consists of a data table, where each row contains columns showing the GI accession number (from NCBI) of the organism that the data come from, the organism's common name, an EC number, the predominant name associated with that EC number, the associated biochemistry, a "reliability factor" and a keyword describing which microbiome the data are associated with (default "Microbiome 1" and "Microbiome 2", or custom names supplied by the user on the form page). The table is broken up into three major sections: those enzymes that are identical between the two microbiomes and those enzymes that are unique for each microbiome. For this study (and likely others that might benefit from this software package), the sub-tables showing the unique results are most useful. An example partial result (Figure 2) might be as follows.

In Figure 2, GI accession is the unique identifier for that GenBank record within NCBI, and the reliability column differentiates between new annotations with multiple EC number matches (reliability = 0) and new annotations with unique matches and old annotations (reliability = 1).

Organisms in Microbiome1:

187426706 Akkermansia muciniphila ATCC BAA-835

Organisms in Microbiome2:

149935097 Bacteroides vulgatus ATCC 8482
291526581 Eubacterium rectale DSM 17629

Received search term query ".

Similar coded enzymes: 619 match(es).

GI Accession	Organism Name	EC_number	Enzyme Name	Biochemistry	Reliability	Microbiome Name
187426706	Akkermansia muciniphila ATCC BAA-835	1.3.1.12	Prephenate dehydrogenase.	Prephenate + NAD(+) = 4-hydroxyphenylpyruvate + CO(2) + NADH.	0	Microbiome1
291526581	Eubacterium rectale DSM 17629	1.3.1.12	Prephenate dehydrogenase.	Prephenate + NAD(+) = 4-hydroxyphenylpyruvate + CO(2) + NADH.	1	Microbiome2
187426706	Akkermansia muciniphila ATCC BAA-835	6.5.1.2	DNA ligase (NAD(+)).	NAD(+) + (deoxyribonucleotide)(n) + (deoxyribonucleotide)(m) = AMP + beta-nicotinamide D-ribonucleotide + (deoxyribonucleotide)(n+m).	1	Microbiome1
291526581	Eubacterium rectale DSM 17629	6.5.1.2	DNA ligase (NAD(+)).	NAD(+) + (deoxyribonucleotide)(n) + (deoxyribonucleotide)(m) = AMP + beta-nicotinamide D-ribonucleotide + (deoxyribonucleotide)(n+m).	1	Microbiome2

Figure 2. Example extract from the original software results page.

Three queries total were made with the web application. The first involved the organisms from the top section of Table 1. The second added the five organisms from the middle section of Table 1. In the third query, we tried to filter our results to just the organisms that are thought to be most important in each microbiome. These microbiomes consisted of those in Table 2. Table 1 is also available in the Supplemental Material as Table S1.

Table 1. Microorganisms chosen for the comparative metabolomics study. Organisms included in Query 2 had *de novo* Enzyme Commission (EC) number annotations.

Query	"Healthy" Microbiome	"Autism" Microbiome
Included in Query 1	Bacteroides fragilis Bifidobacterium longum Ruminococcus bromii Roseburia intestinalis Faecalibacterium prausnitzii Eubacterium rectale Alistipes putredinis Alistipes shahii Weissella koreensis Prevotella ruminicola Odoribacter splanchnicus Methanobrevibacter smithii Clostridium leptum Lactobacillus acidophilus	Desulfovibrio desulfuricans Clostridium perfringens Clostridium difficile Akkermansia muciniphila

Table 1. *Cont.*

Query	"Healthy" Microbiome	"Autism" Microbiome
Additional organisms includedin Query 2	Bacteroides thetaiotaomicron Bacteroides caccae	Sutterella wadsworthensis Clostridium bolteae Bacteroides vulgatus
Desired, but not included	Bacteroides finegoldii Bacteroides ovatus Bacteroides uniformis Bifidobacterium adolescentis Bifidobacterium pseudolongum Eubacterium siraeum Dorea Veillonella Turicibacter Barnesiella intestinihominis Odoribacter laneus Dialister invisus	Desulfovibrio piger Desulfovibrio intestinalis Clostridium butyricum Clostridium paraputrificum Clostridium subterminale Clostridium tertium Clostridium bifermentans Clostridium glycolicum Bacteroides stercoris Parabacteroides distasonis Parabacteroides merdae Paraprevotella xylaniphila Eubacterium eligens Prevotella oulorum

Table 2. Selected organisms for Query 3 of the Metabolomics Software.

	Organisms in Control:
392623967	Bacteroides caccae CL03T12C61
60495220	Bacteroides fragilis ATCC 25285 = NCTC 93
29342100	Bacteroides thetaiotaomicron VPI-5482
666001751	Bifidobacterium longum BXY01
488447870	Lactobacillus acidophilus La-14
148552872	Methanobrevibacter smithii ATCC 35061; PS; DSMZ
	Organisms in Autism:
149935097	Bacteroides vulgatus ATCC 8482
480704622	Clostridium bolteae 90A9
110676061	Clostridium perfringens ATCC 13124
219869941	Desulfovibrio desulfuricans ATCC 27774
115252745	Peptoclostridium difficile 630
512689910	Sutterella wadsworthensis HGA0223

The organisms chosen for each query were input into the program and the results copied and pasted into an Excel file (available as the Supplemental Material, Tables S2–S7: each query is separated into two worksheets, one for complete raw data, and one for filtered data of interest). The "unique" results were looked through manually and considered on a case by case basis, with literature data in mind. Special attention was paid to those enzymes and metabolites found in the methionine and cysteine metabolic pathways and those involved with oxidative stress [13] and amino acid metabolism [29–31].

3. Results

The full results will not be reproduced here, but can easily be reproduced using the online software and the methods detailed above. The full results for each query are available as the Supplemental Material, Tables S2, S4, and S6. Additionally, the full results have been filtered for the most meaningful results and pasted within the Supplemental Material, Tables S3, S5, and S7. The first query resulted in the most meaningful results. It had 19,959 identical enzyme matches between the two microbiomes tested, 834 unique enzymes for the "control" microbiome and 161 unique enzymes for the "autism" microbiome. Gene copies are included in this count, so these numbers are a bit inflated. The identical enzymes are meaningless for this study, so we direct our attention to the tables of "unique" enzymes. Any meaningful results must be found manually. Table 3 below shows the differences in match statistics between each query. The majority of the results found with Query 1 were also found in Queries 2 and 3, with possibly a few minor differences. Judging by the quantity of the results, these minor differences were assumed to be near negligible. Therefore, when comparing Queries 1 and 2, we can come up with an approximate difference in the number of unique results by subtracting the statistics. Therefore, for instance, Query 2 only had 19 new results unique to the "control" microbiome and 34 new results unique to the "autism" microbiome, compared to the results of Query 1. Thus, one can see that Queries 2 and 3 would have less novel results overall than Query 1, given that Query 1 was conducted first.

Table 3. Comparing match statistics between web application queries.

Query	Similar Enzymes	Control Unique	Autism Unique
1	19,959	834	161
2	41,035	853	195
3	14,451	356	387

From an experimental study on metabolic biomarkers in autistic individuals [29], we know that many amino acids are found in significantly lower abundance *vs.* controls, including glycine, serine, threonine, alanine, histidine, glutamine, glutamate and the organic acid, taurine. Antioxidants (especially glutathione) are also in much lower abundance. Another study [30] cites reduced glutathione as being in much lower abundance and another type of antioxidant, thioredoxins, as being in much higher abundance. James *et al.* [13] claims that there is a metabolic bottleneck in many autistic individuals, where the conversion of methionine to cysteine is inhibited, leading to diminished glutathione formation. This information was kept in mind when analyzing the data. The relevance of these findings will be discussed in more detail in the Discussion Section.

In Tables 4–11 we present the most meaningful results obtained from the data mining. The majority of the results come from the first web application query. First, for the enzymes unique to the "control" microbiome, we present enzymes indicative of glutamate metabolism. Some of the entries below contain notation, such as "×3" after an organism name. This denotes that that organism has that number of gene copies coding for that particular enzyme (the first table with this notation is Table 6 below).

Table 4. Acetylornithine transaminase expression ("control" unique).

2.6.1.11	Acetylornithine Transaminase	N(2)-Acetyl-L-ornithine + 2-Oxoglutarate = N-Acetyl-L-glutamate5-semialdehyde + L-Glutamate
Expressed by the following organisms:		
291516108	Alistipes shahii WAL 8301	
291516108	Alistipes shahii WAL 8301	
60495220	Bacteroides fragilis ATCC 25285 = NCTC 93	
291526581	Eubacterium rectale DSM 17629	
291526581	Eubacterium rectale DSM 17629	
295102938	Faecalibacterium prausnitzii L2/6	
148552872	Methanobrevibacter smithii ATCC 35061; PS; DSMZ	
324314063	Odoribacter splanchnicus DSM 220712	
294473972	Prevotella ruminicola Bryant 23	
291541371	Roseburia intestinalis XB6B4	

Table 5. Phosphoserine transaminase expression ("control" unique).

2.6.1.52	Phosphoserine Transaminase	(1) *O*-Phospho-L-serine + 2-Oxoglutarate = 3-Phosphonooxypyruvate + L-Glutamate (2) 4-Phosphonooxy-L-threonine + 2-Oxoglutarate = (3*R*)-3-Hydroxy-2-oxo-4-phosphonooxybutanoate + L-Glutamate
Expressed by the following organisms:		
167660682	*Alistipes putredinis* DSM 17216	
291516108	*Alistipes shahii* WAL 8301	
60495220	*Bacteroides fragilis* ATCC 25285 = NCTC 93	
295102938	*Faecalibacterium prausnitzii* L2/6	
324314063	*Odoribacter splanchnicus* DSM 220712	
294473972	*Prevotella ruminicola* Bryant 23	
291541371	*Roseburia intestinalis* XB6B4	
291543183	*Ruminococcus bromii* L2-63	

Table 6. 2-oxoglutarate synthase expression ("control" unique).

1.2.7.3	2-Oxoglutarate Synthase	2-Oxoglutarate + CoA + 2 Oxidized Ferredoxin = Succinyl-CoA + CO(2) +2 Reduced Ferredoxin + 2 H(+)
Expressed by the following organisms (multiple gene copies present):		
291516108	*Alistipes shahii* WAL 8301 ($\times$5)	
148552872	*Methanobrevibacter smithii* ATCC 35061; PS; DSMZ ($\times$5)	
324314063	*Odoribacter splanchnicus* DSM 220712 ($\times$3)	

Table 7. Glutamate synthase expression ("control" unique).

1.4.1.14 Glutamate Synthase (NADH)	2 L-Glutamate + NAD(+) = L-Glutamine + 2-Oxoglutarate + NADH
Expressed by the following organisms:	
291526581	*Eubacterium rectale* DSM 17629 ($\times$3)
295102938	*Faecalibacterium prausnitzii* L2/6 ($\times$4)
291541371	*Roseburia intestinalis* XB6B4 ($\times$3)
291543183	*Ruminococcus bromii* L2-63 ($\times$3)

There were several enzymes of interest coded by *Lactobacillus acidophilus* alone, many involving methionine and cysteine metabolism. Arginase produces ornithine from arginine, a precursor to glutamate. Mercury (II) reductase is included, because sulfur metabolism is involved with detoxifying toxic heavy metals, such as mercury. NADH peroxidase is an antioxidant, and the enzymes associated with methionine metabolism are ultimately associated with glutathione formation, another antioxidant. See Table 8 for these results.

Table 8. Enzymes of interest coded by Lactobacillus acidophilus: Query 1.

2.1.1.10	Homocysteine *S*-methyltransferase.	*S*-methyl-L-methionine + L-homocysteine = 2 L-methionine.
2.1.1.14	5-methyltetrahydropteroyltriglutamate homocysteine *S*-methyltransferase.	5-methyltetrahydropteroyltri-L-glutamate + L-homocysteine = tetrahydropteroyltri-L-glutamate + L-methionine.
3.5.3.1	Arginase.	L-arginine + H_2O = L-ornithine + urea.
4.2.1.22	Cystathionine beta-synthase.	L-serine + L-homocysteine = L-cystathionine + H_2O.
4.4.1.1	Cystathionine gamma-lyase.	L-cystathionine + H_2O = L-cysteine + NH_3 + 2-oxobutanoate.
1.16.1.1	Mercury(II) reductase.	$Hg + NADP^+ + H^+ = Hg^{2+} + NADPH$.
2.1.1.176	16S rRNA (cytosine(967)-C(5))-methyltransferase.	*S*-adenosyl-L-methionine + cytosine(967) in 16S rRNA = *S*-adenosyl-L-homocysteine + 5-methylcytosine(967) in 16S rRNA.
1.11.1.1	NADH peroxidase.	$NADH + H_2O_2 = NAD^+ + 2 H_2O$.

Other enzymes of interest were antioxidants superoxide reductase (1.15.1.2), coded by *Faecalibacterium prausnitzii*, and glutathione peroxidase (1.11.1.9), coded by *Prevotella ruminicola*. Several others exhibited metabolic pathways associated with glutamate, cysteine and methionine and a few for some amino acids of less interest, such as serine or histidine.

The following (Table 9) are some enzymes of interest from the "autism" microbiome. Interestingly, *Clostridium perfringens* is the only organism studied (among both microbiomes) that codes for glutamate: cysteine ligase and glutathione synthase, the two enzymes needed for glutathione formation.

Table 9. Enzymes of interest coded by the "autism" microbiome; Query 1.

110676061	Clostridium perfringens ATCC 13124	4.1.1.50	Adenosylmethionine decarboxylase.	S-adenosyl-L-methionine = S-adenosyl 3-(methylthio)propylamine + CO_2.
115252745	Peptoclostridium difficile 630	4.1.1.50	Adenosylmethionine decarboxylase.	S-adenosyl-L-methionine = S-adenosyl 3-(methylthio)propylamine + CO_2.
110676061	Clostridium perfringens ATCC 13124	4.1.1.22	Histidine decarboxylase.	L-histidine = histamine + CO_2.
110676061	Clostridium perfringens ATCC 13124	6.3.2.2	Glutamate–cysteine ligase.	ATP + L-glutamate + L-cysteine = ADP + phosphate + gamma-L-glutamyl-L-cysteine.
110676061	Clostridium perfringens ATCC 13124	6.3.2.3	Glutathione synthase.	ATP + gamma-L-glutamyl-L-cysteine + glycine = ADP + phosphate +glutathione.
219869941	Desulfovibrio desulfuricans ATCC 27774	2.6.1.44	Alanine–glyoxylate transaminase.	L-alanine + glyoxylate = pyruvate + glycine.
219869941	Desulfovibrio desulfuricans ATCC 27774	1.8.99.3	Hydrogen sulfite reductase.	$(O_3S.S.SO_3)^{2-}$ + acceptor + 2 H_2O + OH^- = 3 HSO_3^- + reduced acceptor.
115252745	Peptoclostridium difficile 630	4.4.1.11	Methionine gamma-lyase.	L-methionine + H_2O = methanethiol + NH_3 + 2-oxobutanoate.
115252745	Peptoclostridium difficile 630	1.8.1.2	Sulfite reductase (NADPH).	H_2S + 3 $NADP^+$ + 3 H_2O = sulfite + 3 NADPH.
115252745	Peptoclostridium difficile 630	5.4.3.5	D-ornithine 4,5-aminomutase.	D-ornithine = (2R,4S)-2,4-diaminopentanoate.
115252745	Peptoclostridium difficile 630	5.1.1.12	Ornithine racemase.	L-ornithine = D-ornithine.

It was found that *Desulfovibrio desulfuricans* and *Clostridium difficile* uniquely coded for enzymes that involved the use of thioredoxins, including sarcosine reductase, betaine reductase and glycine reductase.

Queries 2 and 3 had much less novel results in comparison. We will look at these results first, then take a look at why adding new EC number annotations to the files did not change much in terms of the overall results.

Query 2 only resulted in one new meaningful result (Table 10): the conversion of S-adenosyl-methionine to S-adenosyl-homocysteine, part of the transsulfuration pathways.

Table 10. SAM (S-adenosyl-methionine) conversion to SAH (S-adenosyl-homocysteine).

219869941	Desulfovibrio desulfuricans ATCC 27774	2.1.1.77	Protein-L-isoaspartate (D-aspartate) O-methyltransferase	S-adenosyl-L-methionine + protein L-isoaspartate = S-adenosyl-L-homocysteine + protein L-isoaspartate α-methyl ester.

Query 3 also resulted in only one major meaningful result, as well as two more that are at least noteworthy, involved with cyanide metabolism (Table 11). Of course, cyanide is highly toxic to humans, so the fact that *Clostridium difficile* is the only organism studied that can metabolize it is interesting, but its relevance to the present study is inconclusive. Also of note was that *Bifidobacterium* and *Lactobacillus* of the control microbiome uniquely coded for 2-haloacid dehalogenase, and *Lactobacillus* uniquely coded for haloalkane dehalogenase. Again, upon searching the literature in regards to these results, the interpretation is inconclusive. The relevance to the present study cannot be determined.

Table 11. Enzymes of interest coded by the "autism" microbiome in Query 3.

115252745	Peptoclostridium difficile 630	2.8.1.2	3-mercaptopyruvate sulfurtransferase.	3-mercaptopyruvate + cyanide = pyruvate + thiocyanate.
115252745	Peptoclostridium difficile 630	2.8.1.1	Thiosulfate sulfurtransferase.	Thiosulfate + cyanide = sulfite + thiocyanate.
115252745	Peptoclostridium difficile 630	4.4.1.8	Cystathionine beta-lyase.	L-cystathionine + H_2O = L-homocysteine + NH_3 + pyruvate.
666001751	Bifidobacterium longum BXY01	3.8.1.9	(R)-2-haloacid dehalogenase.	(R)-2-haloacid + H_2O = (S)-2-hydroxyacid + halide.
666001751	Bifidobacterium longum BXY01	3.8.1.2	(S)-2-haloacid dehalogenase.	(S)-2-haloacid + H_2O = (R)-2-hydroxyacid + halide.
666001751	Bifidobacterium longum BXY01	3.8.1.10	2-haloacid dehalogenase (configuration-inverting).	(1) (S)-2-haloacid + H_2O = (R)-2-hydroxyacid + halide. (2) (R)-2-haloacid + H_2O = (S)-2-hydroxyacid + halide.
488447870	Lactobacillus acidophilus La-14	3.8.1.2	(S)-2-haloacid dehalogenase.	(S)-2-haloacid + H_2O = (R)-2-hydroxyacid + halide.
488447870	Lactobacillus acidophilus La-14	3.8.1.5	Haloalkane dehalogenase.	1-haloalkane + H_2O = a primary alcohol + halide.

In order to analyze why some organisms provided more information than others, we compared the number of EC number annotations within each GenBank file to the total CDSs (coding sequences) within each file. For the new organisms added for Queries 2 and 3, we can see that very few annotations were actually added compared to the total CDSs. "Good" coverage seems to lie between 12% and 25% when taking the ratio of EC number annotations to total CDSs, and all five of the newly annotated files fall well short of this mark. However, Bifidobacterium had greatly improved coverage after updating EC number annotations, rising from <4% to >25%. Despite this, very few new results were obtained. Table 12 shows these statistics. This table is included in the Supplemental Material as Table S8 as well. It may seem strange at first that there are so few EC number annotations compared to the number of annotated CDSs, but keep in mind that enzymes are only a fraction of the organisms' proteome (e.g., non-catalytic proteins, such as inter-membrane proteins), and some of the organisms' genes code for non-translated RNA transcripts, such as tRNAs. There also seems to be a dearth in annotation of EC numbers in general.

Table 12. GenBank file statistic comparisons between the number of EC number annotations and coding sequences (CDSs).

Organism Name	Original File			Updated EC Annotations				% Increase
	Total CDSs	Total EC Numbers	%	Total EC Numbers	Total "Multi-EC" Designations	Actual EC Total	%	
Akkermansia muciniphila	2138	286	13.38	401	34	367	17.17	28.32
Alistipes putredinis	659	92	13.96	110	6	104	15.78	13.04
Alistipes shahii	2563	548	21.38	603	17	586	22.86	6.93
Bacillus subtilis	4140	912	22.03	941	8	933	22.54	2.30
Bacteroides fragilis	4406	366	8.31	377	5	372	8.443	1.64
Bifidobacterium longum	1903	68	3.57	672	189	483	25.38	610.29
Clostridium difficile	3902	1015	26.01	1,035	6	1029	26.37	1.38
Clostridium leptum	602	100	16.61	114	4	110	18.27	10.00
Clostridium perfringens	2878	504	17.51	539	10	529	18.38	4.96
Desulfovibrio desulfuricans	2356	292	12.39	392	32	360	15.28	23.29
Escherichia coli	4967	612	12.32	1542	285	1257	25.31	105.39
Eubacterium rectale	2898	636	21.95	696	16	680	23.46	6.92
Faecalibacterium prausnitzii	2756	586	21.26	641	17	624	22.64	6.48
Lactobacillus acidophilus	1876	486	25.91	585	31	554	29.53	13.99

Table 12. *Cont.*

Organism Name	Original File			Updated EC Annotations				% Increase
	Total CDSs	Total EC Numbers	%	Total EC Numbers	Total "Multi-EC" Designations	Actual EC Total	%	
Methanobrevibacter smithii	1795	414	23.06	454	12	442	24.62	6.76
Odoribacter splanchnicus	3498	515	14.72	603	27	576	16.47	11.84
Prevotella ruminicola	2791	480	17.20	519	12	507	18.17	5.62
Roseburia intestinalis	3630	709	19.53	793	17	776	21.38	9.45
Ruminococcus bromii	1811	467	25.79	496	9	487	26.89	4.28
Weissella koreensis	1335	93	6.97	288	61	227	17	144.09
Bacteroides caccae	3441	0	0.00	76	19	57	1.656	
Bacteroides thetaiotaomicron	4787	0	0.00	413	125	288	6.016	
Bacteroides vulgatus	4065	0	0.00	267	80	187	4.6	
clostridium bolteae	5830	0	0.00	737	227	510	8.748	
Sutterella wadsworthensis	2433	0	0.00	136	39	97	3.987	

4. Discussion

The present study successfully identified several key enzymes associated with autism spectrum disorders using a bioinformatics data mining approach, by comparing the metabolomes of two distinct microbiomes. We must compare the results to published experimental data in order to evaluate the impact of this study, such as those found in [29–33].

We expected to find key biomarkers unique to *Desulfovibrio* and *Clostridia* species indicative of autism, but instead, the key was based on identifying enzymes that were missing from these organisms; hence, a decreased abundance of such enzymes could be used as diagnostic biomarkers for autism. Several different amino acids have previously been reported to be in lower abundance in autistic individuals, but in this study, the amino acid that stood out the most was glutamate. Several organisms in the "control" microbiome coded for enzymes associated with glutamate metabolism, but were curiously missing from the "autism" microbiome. Glutamate is an integral part of glutathione, a tripeptide made of glutamate, cysteine and glycine. It was known before that glutathione was down-regulated in autism, but researchers focused their attention on an inhibition of cysteine metabolism to be the culprit.

Upon closer inspection of glutamate's association with autism, we find that there is a hypothesis that does not involve oxidative stress at all. Glutamate is the human body's major excitatory neurotransmitter, which works in opposition to gamma-aminobutyric acid (GABA) [34,35]. Contrary to the Ming *et al.* study [29], these studies, as well as Adams *et al.* [31] reported higher quantities of glutamate *vs.* neurotypical controls. Tevarst van Elst *et al.* [34] actually reviewed two different hypotheses: that a hypoglutamatergic condition is related to autism and also that a hyperglutamatergic condition is related to autism. Regardless of which one it is, the authors theorize that an imbalance in the neurological excitation/inhibition chemical signaling in the central nervous system is thought to be associated with autism. Furthermore, glutamate decarboxylase is the enzyme responsible for converting glutamate into GABA. Loss of the gene coding for GABA in host neurons has been shown to lead to symptoms characteristic of autism [36]. Going back to the results of our software (*i.e.*, Query 2), we found that glutamate decarboxylase is expressed by organisms from both the control and autism groups. Therefore, what then is the association of our research to the hypotheses on glutamate and autism? That remains to be determined, but we must not ignore the fact that we might not have heard of this field of research had not our software pointed in that direction. Even with limited data availability, it seems that our software can be useful in developing new hypotheses and revealing literature that was previously unknown to us.

After completing our analysis, we came across another study where *Akkermansia muciniphila* was found in decreased amounts in individuals with autism [37]. According to the authors, *Akkermansia* is integral to the host's gut health; the gut mucus layer is reduced as *Akkermansia* composition within the gut is depleted. This contradicts the Finegold *et al.* pyrosequencing study. In light of these two conflicting studies, it seems that we may have been correct in assuming that the involvement of *Akkermansia* in autism is questionable at best. We still included it in the study, because theoretically, it should serve as a basis for comparison. If we think that certain *Clostridia* species and sulfate reducers are indicative of autism, then we may be biased towards results that agree with that assumption. If we include a questionable case, such as *Akkermansia,*in the "autism" group analysis, then we may better be able to rule out false positives. Regarding this inclusion, no significant results were found in Queries 1 or 2, leading to the exclusion of *Akkermansia* from Query 3. Query 3 had quite similar results as Queries 1 and 2, which lends credence to the idea that those organisms still included in the "autism" group for Query 3 are more likely to be involved in autism pathogenesis.

Hydrogen sulfide has been cited as a toxic metabolite produced by *Desulfovibrio* species [10], but data on this metabolite were inconclusive. It was found that *Desulfovibrio* is not the only organism (in either microbiome) that can metabolize hydrogen sulfide. What we did notice is that *Clostridium difficile* codes for enzymes that metabolize alternative pathways for ornithine and methionine, precursors to glutamate and cysteine, respectively, with the enzymes ornithine racemase, D-ornithine 4,5-aminomutase, adenosylmethionine decarboxylase and methionine gamma-lyase. Nearly every organism studied codes for enzymes using ferredoxins and thioredoxins, alternative antioxidants to glutathione, which also use a sulfhydryl electron acceptor for reducing power. It is possible that thioredoxins were found in much higher abundance in autistic individuals [28], because thioredoxins had to take over from glutathione as the body's dominant antioxidant. However, if glutathione is usually dominant, then there must be something not as chemically favorable associated with ferredoxins and thioredoxins or perhaps they are just not sufficient to fill in the antioxidative gap that glutathione usually fills. That *Clostridium difficile* can uniquely use ornithine and methionine for other purposes other than glutathione formation provides evidence for this hypothesis. Interestingly, the final two catalytic steps in glutathione formation are uniquely coded by *Clostridium perfringens*, part of the "autism" microbiome. However, many of the precursor steps are uniquely coded by *Lactobacillus acidophilus*, part of the "control" microbiome. Glutathione is still existent in autistic individuals, so perhaps *Clostridium perfringens* allows for some extra utility in glutathione formation; or maybe this is just coincidence.

Quite remarkable was the fact that nearly every reaction of the transsulfuration and transmethylation metabolic pathways was coded by the organisms studied, but that each microbiome ("control" and "autism") could not express the pathways in their entirety alone. This suggests some sort of interdependence between the two microbiomes on completing these pathways. The control microbiome supplies glutamate metabolism, and cysteine and glutathione formation is split between the two groups of organisms. Based on this information, it seems that organisms, such as abnormal *Clostridia* species and *Desulfovibrio*, may be products of the sulfur metabolic deficiency found in so many autistic individuals and, therefore, may be essential in filling a unique metabolic niche. However, because of the known toxic byproducts of these organisms, they may still be causing gastrointestinal inflammation and subsequent neuro-inflammation associated with "leaky gut syndrome", leading to regressive autism [15,31]. Therefore, even though these organisms may be causing some major ill effects, the matter becomes complicated by the necessity of their presence in the host. The sulfur metabolic deficiencies and lack of proper anti-oxidation therefore seem to lie at the heart of the problems that lead to regressive autism. If treatment is possible, it would therefore be more likely to entail dietary supplements that restore the metabolic deficiencies to normal conditions [13], rather than eradicating the proposed offending organisms from the host's gut. James *et al.* [13] reported some improvements in symptoms with specific dietary supplements, and a study by Adams *et al.* [38] provides some more evidence of this. The Adams *et al.* study included adult subjects with autism, as well.

It is important to note that there were some crucial limitations to this study. Besides the obvious, that this study is purely computational, the dependence on having annotation data for EC numbers greatly limits the scope of the study. When the first query to the web application was completed and analyzed, it was thought that including more organisms of interest (with newly annotated EC numbers) would greatly strengthen the study. This, in fact, hardly had any impact at all on the initial study. These files had so few annotations, that they did not give good representations of the metabolome of these organisms. Thus, we can only surmise that there may be other more meaningful results possible with better data. It is possible that this study could be improved by creating new gene annotations altogether by using the publically available BLAST tool for sequence alignment of microbial genomic DNA and then annotating these genes with translated gene products and EC numbers. Also of note is that the two *Clostridium* species studied most deeply in this context (*C. perfringens* and *C. difficile*) have not, to our knowledge, been specified as part of the "autism" microbiome, but they happened to be two of the most widely available and comprehensive genomes of *Clostridia* that could appear in the human gut. Therefore, they were chosen as "representatives" of the *Clostridium* genus, and thus, associated results should be interpreted with scrutiny. It appears that this type of study has not been attempted previously, yielding mixed results. This provides substantial evidence that the present approach is novel and may be a step in a new, potentially useful direction, but its novelty severely limits its impact, because of a lack of resources. It is surmised that as data on microbial genomics and data on gut bacteria associated with autism patients improves, so too will the utility of our novel software and our comparative modeling approach. Future work could include full *de novo* genomic sequencing of gut microbes and validation through next-generation sequencing devices, followed by comprehensive annotation with accurate EC numbers. On NCBI's server, there is a noted difference in genomic comprehensiveness between microorganisms, where key gut microbes are covered definitively less than model organisms, such as *Escherichia coli* or *Bacillus subtilis*. We need to be able to have more studies on gut microbes in their natural habitat and their interactions with host cells in order to have more experimental data to fine-tune our metabolomics model. It also remains to be determined what would be the best way of evaluating data significance from this type of study.

Nevertheless, we still believe that substantial results came about from this study. We have not seen this type of comparative analysis conducted with autism in mind before, and we are proud to say that our computational approach was successful, given some of its parallels to previous experimental studies. Some experimental studies were not able to be validated with this approach, which is to be expected. As stated earlier, this software was designed with the intent to be used for many other applications, besides autism, hence making the software publically available online. Many other human illnesses and conditions are thought to be associated with the human gut microbiome [5], and we believe that these applications might benefit from our approach, as well. Microbiomes in other environments may also benefit; much potential could be lost if the software's use were limited to only the human gut.

Appendix Supplemental Materials

Supplementary materials can be found at http://www.mdpi.com/1422-0067/16/04/8949/s1.

Acknowledgments: Colin Heberling would like to acknowledge Thomas Koval and the Johns Hopkins Advanced Academic Programs (AAP) graduate program for guidance and for the opportunity to work on this project, as well as Joshua Orvis for instruction in web development as applied to bioinformatics and for his guidance on an early prototype of the online software.

Author Contributions: Prasad Dhurjati provided a mentoring and supervisory role throughout this project and contributed to its conceptual design. Colin Heberling had a large part in the project's conceptual design and contributed fully to its execution and analysis.

Conflicts of Interest: The authors declare no conflict of interest.

References

1. Baio, J. Prevalence of Autism Spectrum Disorders. In *Autism and Developmental Disabilities Monitoring Network, 14 Sites, United States, 2008*; Centers for Disease Control and Prevention, Surveillance Summaries: Atlanta, GA, USA, 2012; Volume 61, SS03, pp. 1–19.
2. CDC Features—Ten Things to Know About New Autism Data. Available online: http://www.cdc.gov/Features/dsAutismData/ (accessed on 25 February 2015).

3. Parracho, H.M.R.T.; Bingham, M.O.; Gibson, G.R.; McCartney, A.L. Differences between the gut microflora of children with autistic spectrum disorders and that of healthy children. *J. Med. Microbiol.* **2005**, *54*, 987–991. [CrossRef] [PubMed]

4. Adams, J.B.; Johansen, L.J.; Powell, L.D.; Quig, D.; Rubin, R.A. Gastrointestinal flora and gastrointestinal status in children with autism—Comparisons to typical children and correlation with autism severity. *BMC Gastroenterol.* **2011**, *11*, 22. [CrossRef] [PubMed]

5. Kranich, J.; Maslowski, K.M.; Mackay, C.R. Commensal flora and the regulation of inflammatory and autoimmune responses. *Semin. Immunol.* **2011**, *23*, 139–145. [CrossRef] [PubMed]

6. Human Microbiome Project Consortium. Structure, function and diversity of the healthy human microbiome. *Nature* **2012**, *486*, 207–214.

7. Turnbaugh, P.J.; Hamady, M.; Yatsunenko, T.; Cantarel, B.L.; Duncan, A.; Ley, R.E.; Sogin, M.L.; Jones, W.J.; Roe, B.A.; Affourtit, J.P.; *et al.* A core gut microbiome in obese and lean twins. *Nature* **2009**, *457*, 480–485. [CrossRef]

8. Finegold, S.M.; Downes, J.; Summanen, P.H. Microbiology of regressive autism. *Anaerobe* **2012**, *18*, 260–262. [CrossRef] [PubMed]

9. Finegold, S.M.; Dowd, S.E.; Contcharova, V.; Liu, C.; Henley, K.E.; Wolcott, R.D.; Youn, E.; Summanen, P.H.; Granpeesheh, D.; Dixon, D.; *et al.* Pyrosequencing study of fecal microflora of autistic and control children. *Anaerobe* **2010**, *16*, 444–453. [CrossRef]

10. Finegold, S. Desulfovibrio species are potentially important in regressive autism. *Med. Hypotheses* **2011**, *77*, 270–274. [CrossRef] [PubMed]

11. Wang, L.; Christophersen, C.T.; Sorich, M.J.; Gerber, J.P.; Angley, M.T.; Conlon, M.A. Increased abundance of *Sutterlla* spp. and *Ruminococcus torques* in feces of children with autism spectrum disorder. *Mol. Autism* **2013**, *4*, 42. [CrossRef] [PubMed]

12. Williams, B.L.; Hornig, M.; Buie, T.; Bauman, M.L.; Paik, M.C.; Wick, I.; Bennett, A.; Jabado, O.; Hirschberg, D.L.; Lipkin, W.I. Impaired carbohydrate digestion and transport and mucosal dysbiosis in the intestines of children with autism and gastrointestinal disturbances. *PLoS ONE* **2011**, *6*. [CrossRef] [PubMed]

13. James, S.J.; Cutler, P.; Melnyk, S.; Jernigan, S.; Janak, L.; Gaylor, D.W.; Neubrander, J.A. Metabolic biomarkers of increased oxidative stress and impaired methylation capacity in children with autism. *Am. J. Clin. Nutr.* **2004**, *80*, 1611–1617. [PubMed]

14. Alberti, A.; Pirrone, P.; Elia, M.; Waring, R.H.; Romano, C. Sulphation deficit in "Low-Functioning" autistic children: A pilot study. *Biol. Psychiatry* **1999**, *46*, 420–424. [CrossRef] [PubMed]

15. Heberling, C.A.; Dhurjati, P.S.; Sasser, M. Hypothesis for a systems connectivity model of autism spectrum disorder pathogenesis: Links to gut bacteria, oxidative stress, and intestinal permeability. *Med. Hypotheses* **2013**, *80*, 264–270. [CrossRef] [PubMed]

16. Nealson, K.; Venter, J.C. Metagenomics and the global ocean survey: What's in it for us, and why should we care? *ISME J.* **2007**, *1*, 185–190. [CrossRef] [PubMed]

17. Rusch, D.B.; Halpern, A.L.; Sutton, G.; Heidelberg, K.B.; Williamson, S.; Yooseph, S.; Wu, D.; Eisen, J.A.; Hoffman, J.M.; Remington, K.; *et al.* The Sorcerer II Global Ocean sampling expedition: Northwest Atlantic through Eastern Tropical Pacific. *PLoS Biol.* **2007**, *5*. [CrossRef]

18. The Human Microbiome Jumpstart Reference Strains Consortium. A catalog of reference genomes from the human microbiome. *Science* **2010**, *328*. [CrossRef]

19. Fernell, E.; Fagerburg, U.L.; Hellström, P.M. No evidence for a clear link between active intestinal inflammation and autism based on analyses faecal calprotectin and rectal nitric oxide. *Acta Paediatr.* **2007**, *96*, 1076–1079. [CrossRef] [PubMed]

20. Finegold, S.M.; Molitoris, D.; Song, Y.; Liu, C.; Vaisanen, M.L.; Bolte, E.; McTeague, M.; Sandler, R.; Wexler, H.; Marlowe, E.M.; *et al.* Gastrointestinal microflora studies in late-onset autism. *Clin. Infect. Dis.* **2002**, *35* (Suppl. 1), S6–S16. [CrossRef]

21. Heberling, C. Human gut microbiome functional analysis—Exploring the capabilities of the MEGAN Software. April 2013; unpublished.

22. Huson, D.H.; Auch, A.F.; Qi, J.; Schuster, S.C. MEGAN analysis of metagenomic data. *Genome Res.* **2007**, *17*, 377–386. [CrossRef] [PubMed]

23. Human Microbiome Project Consortium. A framework for human microbiome research. *Nature* **2012**, *486*, 215–221.

24. Lin, M.Y.; Chang, F.J. Antioxidative effect of intestinal bacteria *Bifidobacterium longum* ATCC 15708 and *Lactobacillus acidophilus* ATCC 4356. *Dig. Dis. Sci.* **2000**, *45*, 1617–1622. [CrossRef] [PubMed]

25. *The National Center for Biotechnology Information*; U.S. National Library of Medicine: 8600 Rockville Pike, Bethesda, MD, USA, 2014. Available online: http://www.ncbi.nlm.nih.gov/ (accessed on 25 September 2014).

26. Dhurjati, P.; Mayadevan, R. Systems biology: The synergistic interplay between biology and mathematics. *Can. J. Chem. Eng.* **2008**, *86*, 127–141. [CrossRef]

27. Heberling, C. Comparative Microbial Metabolomics. Available online: http://cheberling-homepc.com/comparative_microbial_metabolomics/gene_search_form.cgi (accessed on 12 October 2014).

28. Bairoch, A. The ENZYME database in 2000. *Nucleic Acids Res.* **2000**, *28*, 304–305. [CrossRef] [PubMed]

29. Ming, X.; Stein, T.P.; Barnes, V.; Rhodes, N.; Guo, L. Metabolic perturbance in autism spectrum disorders: A metabolomics study. *J. Proteome Res.* **2012**, *11*, 5856–5862. [PubMed]

30. Al-Yafee, Y.A.; Al-Ayadhi, L.Y.; Haq, S.H.; El-Ansary, A.E. Novel metabolic biomarkers related to sulfurdependent detoxification pathways in autistic patients of Saudi Arabia. *BMC Neurol.* **2011**, *11*, 139. [CrossRef] [PubMed]

31. Adams, J.B.; Audhya, T.; McDonough-Means, S.; Rubin, R.A.; Quig, D.; Geis, E.; Gehn, E.; Loresto, M.; Mitchell, J.; Atwood, S.; *et al.* Nutritional and metabolic status of children with autism *vs.* neurotypical children, and the association with autism severity. *Nutr. Metab.* **2011**, *8*, 34. [CrossRef]

32. Yap, I.K.S.; Angley, M.; Veselkov, K.A.; Holmes, E.; Lindon, J.C.; Nicholson, J.K. Urinary metabolic phenotyping differentiates children with autism from their unaffected siblings and age-matched controls. *J. Proteome Res.* **2010**, *9*, 2996–3004. [CrossRef] [PubMed]

33. Wang, L.; Angley, M.T.; Gerber, J.P.; Sorich, M.J. A review of candidate urinary biomarkers for autism spectrum disorder. *Biomarkers* **2011**, *16*, 537–552. [CrossRef] [PubMed]

34. Van Tebartz, E.L.; Maier, S.; Fangmeier, T.; Endres, D.; Mueller, G.T.; Nickel, K. Disturbed cingulate glutamate metabolism in adults with high-functioning autism spectrum disorder: Evidence in support of the excitatory/inhibitory imbalance hypothesis. *Mol. Psychiatry* **2014**, *19*, 1314–1325. [CrossRef] [PubMed]

35. El-Ansary, A.; Al-Ayadhi, L. GABAergic/glutamatergic imbalance relative to excessive neuroinflammation in autism spectrum disorders. *J. Neuroinflamm.* **2014**, *11*. [CrossRef]

36. Zhang, K.; Hill, K.; Labak, S.; Blatt, G.J.; Soghomonian, J.-J. Loss of glutamic acid decarboxylase (Gad67) in Gpr88-expressing neurons induces learning and social behavior deficits in mice. *Neuroscience* **2014**, *275*, 238–247. [CrossRef] [PubMed]

37. Wang, L.; Christophersen, C.T.; Sorich, M.J.; Gerber, J.P.; Angley, M.T.; Conlon, M.A. Low relative abundances of the mucolytic bacterium *Akkermansia muciniphila* and *Bifidobacterium* spp. in feces of children with autism. *Appl. Environ. Microbiol.* **2011**, *77*, 6718–6721. [CrossRef] [PubMed]

38. Adams, J.B.; Audhya, T.; McDonough-Means, S.; Rubin, R.A.; Quig, D.; Geis, E.; Gehn, E.; Loresto, M.; Mitchell, J.; Atwood, S.; *et al.* Effect of a vitamin/mineral supplement on children and adults with autism. *BMC Pediatr.* **2011**, *11*, 111. [CrossRef]

International Journal of
Molecular Sciences

MDPI

Article

Variability of Creatine Metabolism Genes in Children with Autism Spectrum Disorder

Jessie M. Cameron [1], Valeriy Levandovskiy [1], Wendy Roberts [2,3], Evdokia Anagnostou [2,3], Stephen Scherer [1,4,5], Alvin Loh [2,6] and Andreas Schulze [1,2,7,*]

[1] Genetics and Genome Biology, Peter Gilgan Center for Research and Learning, Toronto, ON M5G 0A4, Canada; jessie.cameron@sickkids.ca (J.M.C.); Valeriy.Levandovskiy@sickkids.ca (V.L.); Stephen.Scherer@sickkids.ca (S.S.)
[2] Department of Paediatrics, University of Toronto, Toronto, ON M5S 1A1, Canada; wendy.roberts@sickkids.ca (W.R.); Evdokia.Anagnostou@sickkids.ca (E.A.); Alvin.Loh@SurreyPlace.on.ca (A.L.)
[3] Holland Bloorview Kids Rehabilitation Hospital, 150 Kigour Rd, Toronto, ON M4G 1R8, Canada
[4] The Centre for Applied Genomics and Genetics and Genome Biology, the Hospital for Sick Children, Toronto, ON M5G 1X8, Canada
[5] McLaughlin Centre and Department of Molecular Genetics, 686 Bay Street, 13th Floor, Peter Gilgan Center for Research and Learning, Toronto, ON M5G 0A4, Canada
[6] Surrey Place Center, 2 Surrey Place, Toronto, ON M5S 2C2, Canada
[7] Department of Biochemistry, University of Toronto, Toronto, ON M5S 1A8, Canada
* Correspondence: andreas.schulze@sickkids.ca; Tel.: +1-(416)-813-7654 (ext. 301480)

Received: 5 May 2017; Accepted: 25 July 2017; Published: 31 July 2017

Abstract: Creatine deficiency syndrome (CDS) comprises three separate enzyme deficiencies with overlapping clinical presentations: arginine:glycine amidinotransferase (*GATM* gene, glycine amidinotransferase), guanidinoacetate methyltransferase (*GAMT* gene), and creatine transporter deficiency (*SLC6A8* gene, solute carrier family 6 member 8). CDS presents with developmental delays/regression, intellectual disability, speech and language impairment, autistic behaviour, epileptic seizures, treatment-refractory epilepsy, and extrapyramidal movement disorders; symptoms that are also evident in children with autism. The objective of the study was to test the hypothesis that genetic variability in creatine metabolism genes is associated with autism. We sequenced *GATM*, *GAMT* and *SLC6A8* genes in 166 patients with autism (coding sequence, introns and adjacent untranslated regions). A total of 29, 16 and 25 variants were identified in each gene, respectively. Four variants were novel in *GATM*, and 5 in *SLC6A8* (not present in the 1000 Genomes, Exome Sequencing Project (ESP) or Exome Aggregation Consortium (ExAC) databases). A single variant in each gene was identified as non-synonymous, and computationally predicted to be potentially damaging. Nine variants in *GATM* were shown to have a lower minor allele frequency (MAF) in the autism population than in the 1000 Genomes database, specifically in the East Asian population (Fisher's exact test). Two variants also had lower MAFs in the European population. In summary, there were no apparent associations of variants in *GAMT* and *SLC6A8* genes with autism. The data implying there could be a lower association of some specific *GATM* gene variants with autism is an observation that would need to be corroborated in a larger group of autism patients, and with sub-populations of Asian ethnicities. Overall, our findings suggest that the genetic variability of creatine synthesis/transport is unlikely to play a part in the pathogenesis of autism spectrum disorder (ASD) in children.

Keywords: autism spectrum disorder; creatine deficiency syndrome; *glycine amidinotransferase*; *guanidinoacetate methyltransferase*; *solute carrier family 6 member 8*; genetic variability

1. Introduction

Autism spectrum disorder (ASD) can be present in children with inborn errors of metabolism, the latter having a prevalence of 1 in every 800 live births [1]. Less than 5% of ASD cases can be attributed to routinely-screened-for inborn errors of metabolism [2,3]. Creatine deficiency syndrome (CDS) is an inherited metabolic disorder that is not included in routine newborn screening panels or in standard ASD diagnostic work up. The presence of autistic symptoms in patients with CDS suggests that genes in the creatine metabolic pathway may play a part in the neurobiology of ASD and may represent a treatable cause of ASD [4]. We recently ascertained the prevalence of CDS in children with non-syndromic autism in a large, prospective, multicentre study by screening for creatine metabolites in urine and sequencing *GAMT*, *GATM* and *SLC6A8* genes (glycine amidinotransferase, guanidinoacetate methyltransferase and solute carrier family 6 member 8) for pathogenic mutations (443 children with ASD). The estimated prevalence of CDS was less than 7 in 1000 and there was no obvious correlation between pathogenic CDS gene mutations and ASD, Therefore, the chances of a child with ASD having a CDS were very low [5].

In the study presented here, we investigate the association of autism with non-CDS disease causing variants in the three creatine metabolism genes. Whereas non-synonymous nucleotide changes resulting in amino acid changes are likely to result in pathogenic mutations, synonymous changes that do not alter the amino acid are often attributable to polymorphisms. It has been shown that single nucleotide polymorphisms (SNPs) can explain up to 40% of the variance in liability to ASD [6,7]; thus common genetic polymorphisms can be a major risk factor for autism.

The clinical presentation of CDS is characterized by developmental delays and cognitive decline, intellectual disability, impaired speech and language, autistic behaviour, epileptic seizure activity, treatment-refractory epilepsy, and extrapyramidal movement disorders [8–11]. Significant behavioural, social and communication challenges seen in ASD can be caused by neurodevelopmental disabilities of complex aetiology. The symptoms in patients with CDS and those with ASD can overlap.

In patients with CDS, the impaired synthesis or transport of creatine leads to depletion of creatine/phosphocreatine in the brain. The creatine–phosphocreatine system plays an essential role in cellular energy homeostasis in the central nervous system. This system acts as a temporal and spatial energy buffer, energy transducer and general regulator of cellular energetics [12]. Creatine has been shown to act as a neuroprotective agent [13]; it protects rat hippocampal neurons against glutamate and amyloid beta peptide toxicity [14]; it seems to function as a potent natural survival and neuroprotective factor for γ-aminobutyric acid (GABA)-ergic neurons in a model for Huntington disease [15] and of dopaminergic neurons in a model for Parkinson's disease [16]. In two studies, creatine supplementation was convincingly shown to improve concentration [17], as well as short-term memory and learning [18] in healthy human subjects. Due to these protective functions of creatine the deficiency of creatine could play a part in the neurobiology of ASD.

CDS includes three separate deficiencies: arginine:glycine amidinotransferase (AGAT), guanidinoacetate methyltransferase (GAMT), and creatine transporter (CrT) [4,8,19]. AGAT and GAMT deficiencies are autosomal recessive inherited biosynthesis defects of creatine. The AGAT gene (NCBI Gene ID 2628, official nomenclature: *GATM*) located on chromosome 15q15.3 is 16.8 Kb in size and contains nine exons, which encode 424 amino acids. The *GAMT* gene (Gene ID 2593) located on chromosome 19p13.3 is 4.46 Kb in size and has six exons encoding 237 amino acids. Dysfunction of the CrT leads to impaired intra-cellular creatine uptake and has X-linked inheritance. The CrT gene (Gene ID 6535, *SLC6A8*) located on Xq28 spans 8.4 Kb and consists of 13 exons encoding 635 amino acids [4]. To date, there are 6, 29, and 88 pathogenic mutations described throughout the *GATM*, *GAMT*, and *SLC6A8* genes, respectively [20]. The common denominator of all these diseases is the virtually complete absence of creatine and phosphocreatine in the brain that can be assessed with in vivo magnetic resonance spectroscopy (MRS). The diagnosis can be made in spot urine determining the concentration of creatine and guanidinoacetate normalized for creatinine, in 24-h urine measuring the excretion of creatine, guanidinoacetate, and creatinine, and by measuring the concentration of these metabolites in blood [4,21,22]. Among CDS, GAMT deficiency, while presenting as a spectrum from mild to severe, has the most severe phenotype that includes severe intellectual deficits. Treatment-refractory epilepsy and extrapyramidal movement disorders are exclusively seen in GAMT patients [4,8,10,23,24]. The few AGAT-deficient patients described so far seem to have a less severe presentation with developmental delays and impaired communication and social contact. Mild seizures have been reported in only some of the cases [9,25,26]. Males with X-linked CrT deficiency present with impairments of speech and language and may have a seizure disorder. They usually show behavioural abnormalities and develop intellectual disability. While most of the female carriers have a normal appearance, some may have mild learning difficulties. Treatment with supplementation of creatine can fully (AGAT) or partly (GAMT) replenish brain creatine and attenuate the clinical course in AGAT and GAMT-deficient patients. In contrast, creatine supplementation has no effect on brain creatine and the clinical course in male CrT patients [4]. There is an even gender distribution in GAMT [10] and likely in AGAT deficiencies, whereas predominantly males are affected with the CrT defect. The latter is an X-linked condition and female carriers are mostly asymptomatic [8].

Because of the overlap of symptoms between CDS and ASD and because of the potential role of the creatine-phosphocreatine homeostasis in autism, we hypothesize that comparing the ASD population to the general population may reveal an altered frequency of genetic variants in the three CDS genes (*GATM*, *GAMT* and *SLC6A8*) that are associated with autism.

2. Results

A total of 166 subjects enrolled in the study, with 71 from the prospective group enrolled in Toronto and 95 from the Genome Canada study sample. Of these, 32 were females, and 134 were males. Ethnicities were described as European "EUR", Asian "ASN", African "AFR" or Admixed/Unknown, "MIX/UNK" (Supplementary Table S1).

2.1. Novel/Rare Variants Observed in CDS Genes in ASD Patients

Sequence data was compared with NCBI reference genes, and genetic variations were compared to three variation databases: the 1000 Genomes phase 3 dataset, the National Heart, Lung and Blood Institute Grand Opportunity Exome Sequencing Project (NHLBI GO ESP) and the Exome Aggregation Consortium (ExAC). A number of variants were identified, some of which are novel (not in the variant databases) or rare (present at allele frequencies less than 1% or 0.01) (Tables 1–3, Figures 1–3).

2.1.1. GATM Gene Variants

A total of 29 variants were identified, of which four were present in coding regions (Figure 1, Table 1). Of these, two were rare variants: one was synonymous and one was non-synonymous. Synonymous variants are presumed to not have any potential pathogenicity, unless they are likely to affect splicing. The non-synonymous variant (c.700G > C, p.Asp234His) was not present in 1000 Genomes or ESP, but was present in the ExAC database at a minor allele frequency of 0.00001651 (2/121104 chromosomes, mixed ethnic population) (Table 1). The variant was identified as heterozygous in one individual in the ASD population, resulting in a minor allele frequency (MAF) of 0.003 (1/332 chromosomes). Computational software (SIFT [27], PolyPhen-2 [28] and MutationTaster [29]) all predict this variant to be damaging, and therefore pathogenic if homozygous (Table 1). Further functional testing would be needed to confirm the true pathogenicity of this variant.

Table 1. Summary of *GATM* (glycine amidinotransferase) gene sequencing results. Variants identified in the autisim spectrum disorder (ASD) population are noted, and minor allele frequency (MAF) calculated. MAF of variants present in 1000 Genomes, Exome Sequencing Project (ESP) and Exome Aggregation Consortium (ExAC) databases are shown for comparison. [1] Total number of samples = 166 (332 alleles); [2] MAF = minor allele frequency; the number of alleles in which variant was found /total number of alleles; [3] MAC = minor allele count; the number of times the minor allele was observed in the sample population of chromosomes; [4] Polymorphic variant: (>0.01 MAF in one database); [5] Rare variant: (>0.01 MAF in at least one published database); [6] Novel variant: Not present in any published database; coding sequence (exons) are in bold. SNP: single nucleotide polymorphism; UTR: untranslated region; and SIFT: 'sorting tolerant from intolerant' algorithm.

			Autism Population (n = 166, Alleles = 332 [1])			Databases			
GATM Exon/Intron	DNA Change/Protein Change [1]	SNP ID	Homozygous/ Heterozygous Change [1]	Number of Observed Alleles	MAF [2]	1000 Genomes (Phase 3) MAF, MAC [3]	ESP Report (July 2013) MAF, MAC	ExAC (January 2015) MAF, MAC	Comments
5UTR	c. – 200C > T	rs7164139	Homo: 20 Hetero: 78	118	0.355	0.444 2223/5008	x	0.5533 83/150	polymorphic variant [4]
5UTR	c. – 140C > T	rs533626184	Hetero: 1	1	0.003	0.0022 11/5008	x	x	rare variant
5UTR	c. – 30T > G	rs8024550	Hetero: 5	5	0.015	0.144 720/5008	0.073 651/8926	0.07742 155/2002	polymorphic variant
intron 1	c.70 – 77C > T	rs12437887	Homo: 20 Hetero: 76	116	0.349	0.443 2221/5008	x	x	polymorphic variant
intron 1	c.70 – 38G > T	rs12437840	Homo: 23 Hetero: 73	119	0.358	0.443 2219/5008	0.270 3509/12992	0.3827 24153/63112	polymorphic variant
exon 2	**c.282G > A p.=**	**rs141223762**	**Hetero: 1**	**1**	**0.003**	**NA**	**0.000154 2/12992**	**0.0001153 14/121390**	**rare variant [5]**
intron 2	c.289 – 76T > C	rs540536879	Hetero: 3	3	0.009	0.000998 5/5008	x	x	rare variant
intron 2	c.289 – 34T > C	rs74009633	Hetero: 1	1	0.003	0.0096 48/5008	0.005 65/12982	0.001947 193/99144	rare variant
intron 2	c.289 – 24G > A	rs145644806	Hetero: 1	1	0.003	0.0002 1/5008	0.000154 2/12988	x	rare variant
exon 3	**c.330A > T p.Q110H**	**rs1288775**	**Homo: 29 Hetero: 74**	**132**	**0.398**	**0.619 3098/5008**	**0.435 5651/12992**	**0.581 70416/121104**	**polymorphic variant**
intron 3	c.485-11_dupT		Homo: 1	2	0.006	x	x	x	novel variant [6]
exon 5	**c.700G > C**		**Hetero: 1**	**1**	**0.003**	**NA**	**x**	**0.00001651**	**rare variant**

Table 1. *Cont.*

GATM Exon/Intron	DNA Change/Protein Change	SNP ID	Autism Population (n = 166, Alleles = 332[1])			Databases			Comments
			Homozygous/ Heterozygous Change[1]	Number of Observed Alleles	MAF[2]	1000 Genomes (Phase 3) MAF, MAC[3]	ESP Report (July 2013) MAF, MAC	ExAC (January 2015) MAF, MAC	SIFT: deleterious (score: 0); PolyPhen-2: probably damaging (score: 0.986); MutationTaster: disease-causing (p-value: 1.0)
	p.D234H							2/121104	
intron 5	c.813 + 46C > G	rs150282769	Hetero: 1	1	0.003	0.0002 1/5008	0.000847 11/12992	0.0004 46/118404	rare variant
intron 6	c.978 + 43T > C	rs9972405	Hetero: 1	1	0.003	×	×	×	novel variant
intron 6	c.979 − 316A > G		Homo: 1 Hetero: 38	40	0.120	0.0998 500/5008	×	×	polymorphic variant
intron 6	c.979-49_979-51delTAA	rs200176845	Homo: 1	2	0.006	0.0100 50/5008	0.0137 171/12480	0.00228 248/108652	rare variant
intron 6	c.979 − 39T > C	rs57369693	Hetero: 1	1	0.003	×	×	×	novel variant
intron 7	c.1043 − 254A > G		Homo: 1 Hetero: 38	40	0.120	0.0998 500/5008	×	×	polymorphic variant
intron 8	c.1159 + 39A > G	rs113129788	Hetero: 1	1	0.003	0.0002 1/5008	×	0.000019 2/105236	rare variant
intron 8	c.1160 − 46C > T	rs201589362	Hetero: 2	2	0.006	0.0002 1/5008	×	0.00006987 7/100180	rare variant
exon 9	c.1252T > C p.=	rs1145086	Homo: 46 Hetero: 80	172	0.518	0.2823 1414/5008	0.534 6937/12992	0.5329 64562/121146	polymorphic variant
3UTR	c.*27C > G	rs200143728	Hetero: 1	1	0.003	NA	×	0.0004716 56/118754	rare variant
3UTR	c.*125G > A	rs143689218	Hetero: 2	2	0.006	0.0086 43/5008	×	×	rare variant
3UTR	c.*411G > A	rs1049503	Hetero: 1	1	0.003	×	×	×	novel variant
3UTR	c.*600A > G		Homo: 23 Hetero: 74	120	0.361	0.4507 2257/5008	×	×	polymorphic variant
3UTR	c.*715T > C	rs1049508	Homo: 27 Hetero: 69	123	0.370	0.618 3094/5008	×	×	polymorphic variant
3UTR	c.*734_*735insCA	rs35410548	Homo: 46 Hetero: 79	171	0.515	0.718 3594/5008	×	×	polymorphic variant
3UTR	c.*913G > A	rs17618637	Hetero: 10	10	0.030	0.0553 277/5008	×	×	polymorphic variant
3UTR	c.*940C > T	rs1049518	Homo: 54	108	0.325	0.718 3594/5008	×	×	polymorphic variant

Table 2. Summary of guanidinoacetate methyltransferase (GAMT) gene sequencing results. Variants identified in the ASD population are noted, and minor allele frequency (MAF) calculated. MAF of variants present in 1000 Genomes, ESP and Exome Aggregation Consortium (ExAC) databases are shown for comparison. [1] Total number of samples = 166 (332 alleles); [2] MAF = minor allele frequency; the number of alleles in which variant was found/total number of alleles; [3] MAC = minor allele count; the number of times the minor allele was observed in the sample population of chromosomes; [4] Polymorphic variant: (>0.01 MAF in one database); [5] Rare variant: ≤0.01 MAF in at least one published database; coding sequence (exons) are in bold; and dbSNP: Single Nucleotide Polymorphism database.

| GAMT Exon/Intron | DNA Change/Protein Change [1] | SNP ID | Autism Population (n = 166, Alleles = 332 [1]) | | | 1000 Genomes (Phase 3) MAF, MAC [3] | Databases | | Comments |
			Homozygous/ Heterozygous Change [1]	Number of Observed Alleles	MAF [2]		ESP Report (July 2013) MAF, MAC	ExAC (January 2015) MAF, MAC	
intron 1	c.182 − 173G > A	rs112975707	Hetero: 6	6	0.018	0.0491 246/5008	x	x	polymorphic variant [4]
exon 2	**c.227C > T** p.S76L	rs150338273	Hetero: 2	2	**0.006**	NA	0.000616 8/12984	0.000411 29/70554	rare variant [5] SIFT: deleterious (score: 0.03); PolyPhen-2: possibly damaging (score: 0.66); Mutation Taster: disease-causing (p-value: 0.986)
intron 2	c.327 + 69T > G	rs266808	Hetero: 4	4	0.012	0.0451 226/5008	x	x	polymorphic variant
exon 3	**c.348G > A** **p.=**	rs117884619	**Hetero: 1**	1	**0.003**	**0.0104 52/5008**	x	0.003969 431/108600	rare variant, benign allele in dbSNP
intron 3	c.391 + 47A > G	rs73515058	Hetero: 12	12	0.036	0.0865 433/5008	0.0852 1106/12988	0.06853 5944/86734	polymorphic variant
intron 4	c.460 − 31G > A	rs55776826	Homo: 2 Hetero: 41	45	0.136	0.1132 567/5008	0.153 1989/13006	0.1294 15585/120422	polymorphic variant
intron 5	c.571 − 60C > T	rs266809	Homo: 1 Hetero: 6	8	0.024	0.0731 366/5008	x	x	polymorphic variant
intron 5	c.571 − 6G > A	rs2074899	Hetero: 22	22	0.066	0.1088 545/5008	0.02 265/13000	0.06984 8254/118186	polymorphic variant
exon 6	**c.626C > T** **p.T209M**	rs17851582	**Hetero: 23**	23	**0.069**	**0.0365 183/5008**	0.071 926/13004	0.07554 8958/118584	**polymorphic variant**
3UTR	c.*146A > G	rs659455	Homo: 1 Hetero: 5	7	0.021	0.0765 383/5008	x	x	polymorphic variant
3UTR	c.*151T > C	rs659460	Homo: 1 Hetero: 5	7	0.021	0.0761 381/5008	x	x	polymorphic variant
3UTR	c.*276C > T	rs266810	Homo: 1 Hetero: 5	7	0.021	0.0761 381/5008	x	x	polymorphic variant
3UTR	c.*311C > G	rs266811	Homo: 1 Hetero: 5	7	0.021	0.0761 381/5008	x	x	polymorphic variant
3UTR	c.*369C > A	rs75762821	Hetero: 3	3	0.009	0.006 28/5008	x	x	rare variant
3UTR	c.*388C > T	rs266812	Homo: 1 Hetero: 5	7	0.021	0.0757 379/5008	x	x	polymorphic variant
3UTR	c.*406A > G	rs266813	Homo: 2 Hetero: 15	19	0.057	0.274 1372/5008	x	x	polymorphic variant

Table 3. Summary of solute carrier family 6 member 8 (SLC6A8) gene sequencing results. Variants identified in the ASD population are noted, and minor allele frequency (MAF) calculated. MAF of variants present in 1000 Genomes, ESP and Exome Aggregation Consortium ExAC databases are shown for comparison. [1] Total number of samples = 166 (134 males, 32 females, 198 alleles); [2] MAF = minor allele frequency; the number of alleles in which variant was found/total number of alleles; [3] MAC = minor allele count; the number of times the minor allele was observed in the sample population of chromosomes; [4] Polymorphic variant: (>0.01 MAF in one database); [5] Rare variant: ≤0.01 MAF in at least one published database; and [6] Novel variant: Not present in any published database; coding sequence (exons) are in bold.

SLC6A8 Exon/Intron	DNA Change/Protein Change	SNP ID	Autism Pop (n = 166, Alleles = 198 [1])			Databases			Comments
			Homozygous/ Heterozygous Change [1]	No. Observed Alleles	MAF [2]	1000 Genomes (Phase 3) MAF, MAC [3]	ESP Report (July 2013) MAF, MAC	ExAC (January 2015) MAF, MAC	
5UTR	c. − 5 A > G	rs384573	Homo: 166	196	1.000	NA	x	1 6172/6172	polymorphic variant [4]
intron 1	c.262 + 26T > C	rs192387453	Homo: 25M Hetero: 6F	31	0.158	0.151 570/3775	0.121 1249/10352	0.1084 5029/46406	polymorphic variant
intron 1	c.263 − 95G > A	rs6643763	Homo: 1M	1	0.005	x	x	x	novel variant [6]
intron 2	c.394 + 88G > A		Homo: 21M Hetero: 7F	28	0.143	0.102 385/3775	x	x	polymorphic variant
intron 2	c.394 + 108G > A		Homo: 1M	1	0.005	x	x	x	novel variant [5]
exon 5	**c.813C > T p.=**	**rs138064933**	**Homo: 1M**	**1**	**0.005**	**0.001 2/3775**	**0.0027 29/10561**	**0.003346 280/83683**	**rare variant [5]**
intron 5	c.913 − 40T > C	rs187505163	Homo: 1M	1	0.005	0.009 34/3775	0.014 148/10563	0.003969 339/85417	rare variant
intron 6	c.1016 + 41dupTGCCC	rs371888321	Homo: 1M	1	0.005	x	x	x	novel variant
intron 6	c.1016 + 56del TGCCC		Homo: 1M	1	0.005	x	x	x	novel variant
intron 7	c.1141 + 18G > A	rs187400676	Hetero: 1F	1	0.005	0.006 22/3775	0.000284 3/10563	0.004194 365/87034	rare variant
intron 7	c.1141 + 37G > A	rs2071028	Homo: 20M Hetero: 4F	24	0.122	0.153 576/3775	0.127 1341/10563	0.1014 8811/86854	polymorphic variant
intron 7	c.1141 + 87A > G	rs41302172	Homo: 25 (21M + 4F)	32	0.163	0.101 383/3775	x	x	polymorphic variant
intron 7	c.1142 − 130C > T	rs1411015652	Hetero: 3F Homo: 3M	3	0.015	0.021 81/3775	x	x	polymorphic variant
intron 7	c.1142 − 69G > A		Homo: 1M	1	0.005	x	x	x	novel variant
intron 7	c.1142 − 35G > A	rs201555047	Hetero: 1F	2	0.010	0.006 21/3775	0.00396 39/9845	0.008994 117/13009	rare variant
exon 8	**c.1162G > A**	**rs374163604**	**Homo: 1M**	**1**	**0.005**	**0.0003**	**0.0002886**	**0.00007051**	**rare variant**

Table 3. *Cont.*

SLC6A8 Exon/Intron	DNA Change/Protein Change	SNP ID	Autism Pop (n = 166, Alleles = 198 [1])			Databases			Comments
			Homozygous/Heterozygous Change [1]	No. Observed Alleles	MAF [2]	1000 Genomes (Phase 3) MAF, MAC [3]	ESP Report (July 2013) MAF, MAC	ExAC (January 2015) MAF, MAC	
	p.A388T					1/3775	3/10394	1/14182	SIFT: deleterious (score: 0.02); PolyPhen-2: probably damaging (score: 0.969); MutationTaster: disease-causing (p-value: 1.0)
intron 8	c.1255 − 44delG	rs34035058	Homo: 1M	1	0.005	NA	0.000688 7/10180	0.0002874 18/62628	rare variant
intron 8	c.1255 − 31C > T	rs193175235	Homo: 2M	2	0.010	0.011 42/3775	0.00559 59/10560	0.004952 378/76334	rare variant
intron 9	c.1392 + 31T > C	rs183780161	Homo: 1M	1	0.005	0.003 10/3775	0.00578 61/10556	0.001411 123/87147	rare variant
intron 10	c.1495 + 38C > T	rs200729826	Homo: 2M	2	0.010	0.021 78/3775	0.0000947 1/10562	0.00997 858/86058	rare variant
intron 10	c.1496 − 17G > A	rs375265267	Homo: 1M	1	0.005	NA	0.000189 2/10563	0.0000347 3/86591	rare variant
intron 10	c.1496 − 8C > T	rs376038235	Hetero: 1F	1	0.005	0.002 9/3775	x	0.001025 89/86861	rare variant
intron 11	c.1596 + 21G > A	rs73633747	Homo: 1M	1	0.005	0.019 70/3775	0.0186 194/10422	0.005616 445/79237	rare variant
intron 12	c.1768 − 3C > T	rs150207268	Hetero: 1F	1	0.005	0.002 9/3775	0.00398 42/10554	0.0006573 254/38642	rare variant
3UTR	c.*207G > C	rs6571290	Homo: 21M Hetero: 7F	28	0.143	0.194 731/3775	x	x	polymorphic variant

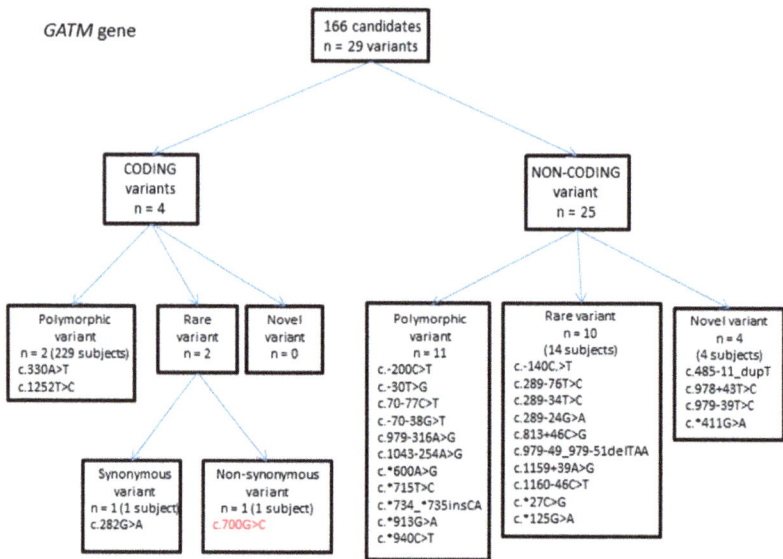

Figure 1. Flowchart summarizing variants identified in the glycine amidinotransferase (*GATM*) gene.

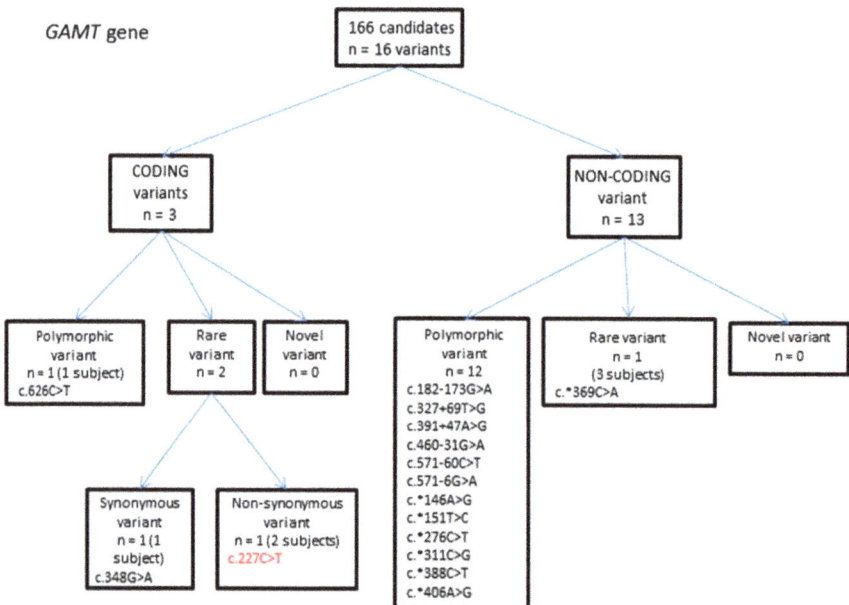

Figure 2. Flowchart summarizing variants identified in the guanidinoacetate methyltransferase (*GAMT*) gene.

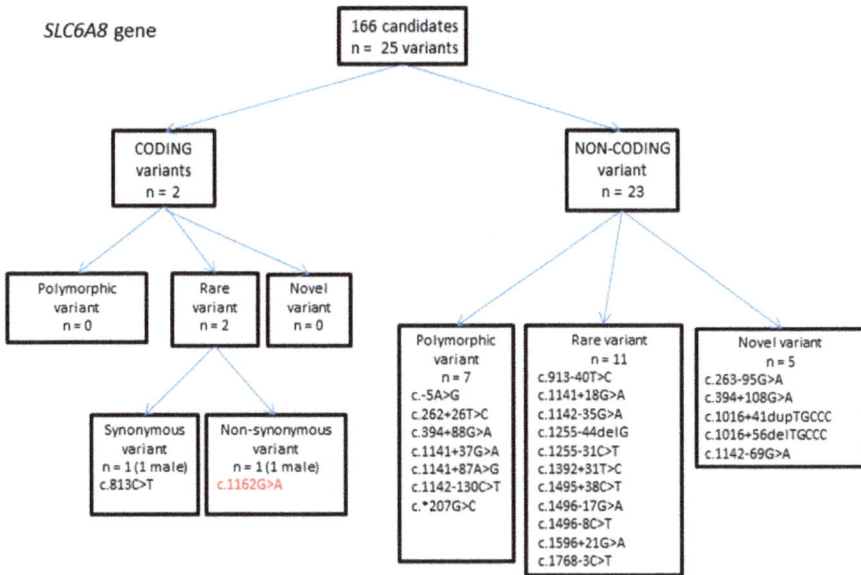

Figure 3. Flowchart summarizing variants identified in the solute carrier family 6 member 8 (*SLC6A8*) gene.

There were several differences seen in the MAFs calculated for *GATM* variants identified in the ASD patient group compared to those from ethnic populations in 1000 Genomes database (European, East Asian, African and Mixed/Unknown) (Supplementary Table S2). Figure 4 shows the *p*-values from the Fisher's exact test (plotted as -log (p)) and the variants identified. Fifteen of these differences were initially found to be significant (Fisher's exact test, $p = 0.01$), and after running the Benjamini–Hochberg procedure, 11 of these values remained above the threshold of significance. Nine of these values represented a significant difference between the allele frequencies seen in our autism population and the East Asian population of the 1000 Genomes database: SNPs rs7164139 (c. $-$ 200C > T, $p = 0.000005$), rs12437887 (c.70 $-$ 77C > T, $p = 0.00002$), rs12437840 (c.70 $-$ 38G > T, $p = 0.00002$), rs1288775 (c.330A > T, $p = 0.000037$), rs1145086(c.1252T > C, $p = 0.000114$), rs1049503 (c.*600A > G, $p = 0.0003$), rs1049508 (c.*715T > C, $p = 0.000037$), rs35410548 (c.*734_*735insCA, $p = 0.000499$) and rs1049518 (c.*940C > T, $p = 0$), Figure 4). All nine variants had a significantly lower MAF than the 1000 Genomes East Asian population database (Supplementary Table S2). It is possible that some of these SNPs could be inherited as a single haplotype (adjacent SNPs that are inherited together). To further investigate these nine variants, we compared the data from our Asian population with the 1000 Genomes South Asian population (Table 4). This data suggested that none of the variants were significantly different.

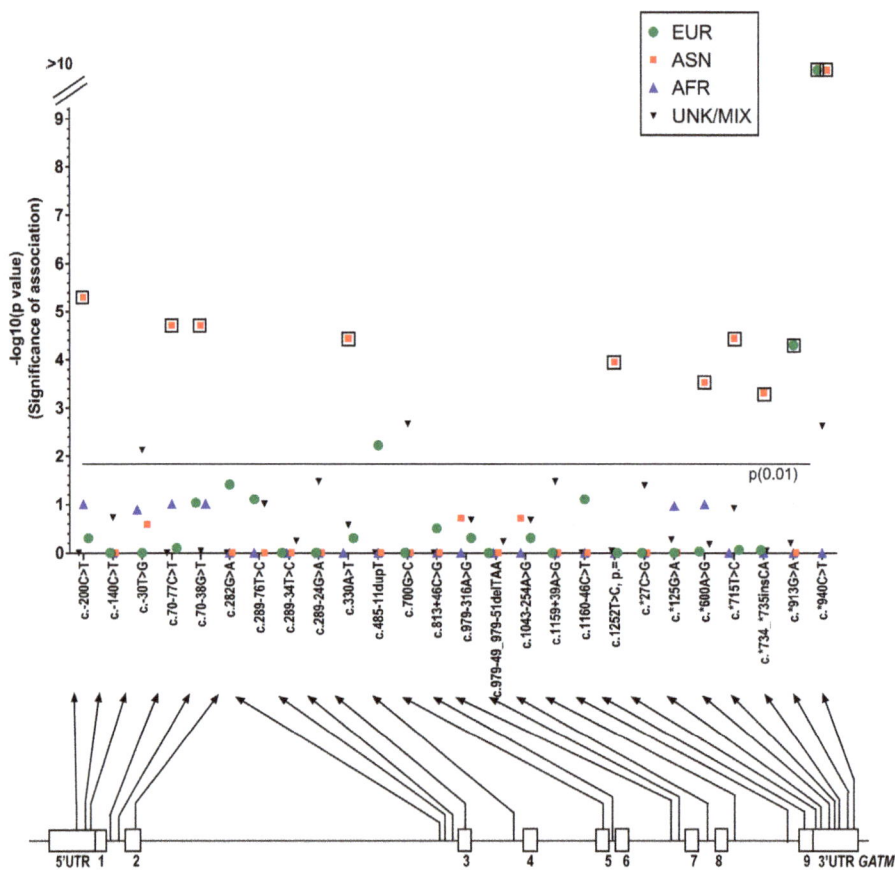

Figure 4. Manhattan plot showing the *p*-values from the Fisher's exact test (significance of association, plotted as -log (*p*)) for each SNP sequenced on the glycine amidinotransferase (*GATM*) gene. *p*-values are shown for European, East Asian, African, and Unknown/Admixed populations. The nominal statistical threshold for *p* = 0.01 is shown, and *p*-values that are still significant after running the Benjamin–Hochburg procedure are indicated with boxes. The location of each SNP is shown on the *GATM* gene below the x-axis. EUR: European; ASN: East Asian; AFR: African; and UNK/MIX: unknown/admixed populations.

Two variants in the European cohort show significant variation from the 1000 Genomes European database population: rs17618637 (c.*913G > A, *p* = 0.00005) and rs1049518 (c.*940C > T, *p* = 0) both in the 3′UTR (untranslated region) of the *GATM* gene. Both had lower MAFs than in the database (Supplementary Table S2).

Table 4. Summary of glycine amidinotransferase (GATM) gene variants that showed significance when compared to East Asian population, with additional data from 1000 Genomes South Asian population. [1] Total number of autism cohort Asian samples = 17 (n = 34 alleles); [2] Number of minor alleles seen in population; [3] Number of major alleles seen in the population; [4] Minor allele frequency; the number of alleles in which variant was found/total number of alleles.

GATM (AGAT) DNA Change/Protein Change	SNP ID	ASN n = 34 [1]			East ASN (1000 Genomes Phase 3) n = 1008				South ASN (1000 Genomes Phase 3) n = 978			
		Minor [2]	Major [3]	MAF [4]	Minor	Major	MAF	Fisher's Exact Test ASN	Minor	Major	MAF	Fisher's Exact Test ASN
c.−200C > T	rs7164139	15	19	0.441	811	197	0.805	0.000005	351	627	0.359	0.365138
c.70−77C > T	rs12437887	16	18	0.471	813	195	0.807	0.00002	351	627	0.359	0.205191
c.70−38G > T	rs12437840	16	18	0.471	813	195	0.807	0.00002	351	627	0.359	0.205191
c.330A > T, p.Q110H	rs1288775	17	17	0.500	825	183	0.818	0.000037	451	527	0.461	0.727522
c.1252T > C, p.=	rs1145086	22	12	0.647	906	102	0.899	0.000114	621	357	0.635	1
c.*600A > G	rs1049503	18	16	0.529	813	195	0.807	0.0003	384	594	0.393	0.112623
c.*715T > C	rs1049508	17	17	0.500	825	183	0.818	0.000037	451	527	0.461	0.727522
c.*734_*735insCA	rs35410548	23	11	0.676	906	102	0.899	0.000499	621	357	0.635	0.718426
c.*940C > T	rs1049518	18	16	0.529	906	102	0.899	0	621	357	0.635	0.21126

2.1.2. *GAMT* Gene Variants

A total of 16 variants were identified, of which three were in coding regions (Figure 2, Table 2). Two of these were rare variants, and one of these was non-synonymous. This variant (c.227C > T), observed in a heterozygous state in two individuals, had an MAF of 0.000616 in the ESP database (8/12984 chromosomes) and 0.000411 in the ExAC database (29/70554 chromosomes) (Table 2). In our ASD population, the variant was observed in 2/332 chromosomes, resulting in an MAF of 0.006. SIFT, PolyPhen-2 and MutationTaster all predicted the variant to be damaging (Table 2).

There were some differences seen in the MAFs calculated for *GAMT* variants identified in the ASD patient group compared to those from ethnic populations in the 1000 Genomes database (Supplementary Table S3). Figure 5 shows the *p*-values from Fisher's exact test (plotted as −log (*p*)) and the variants identified. Only three of these differences were significant (Fisher's exact test, *p* = 0.01), and after running the Benjamini–Hochberg procedure only one of these remained significant (r266813).

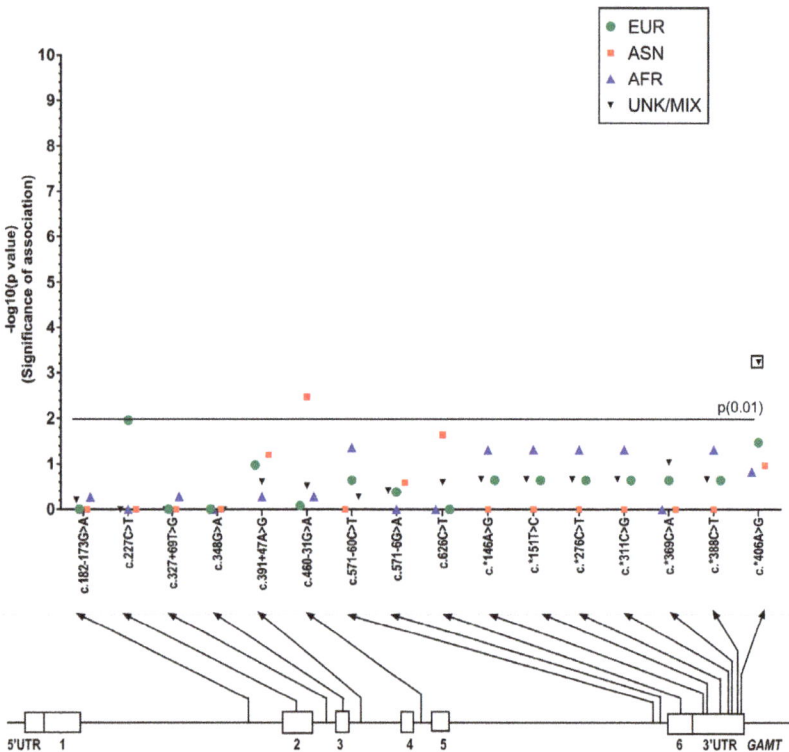

Figure 5. Manhattan plot showing the *p*-values from Fisher's exact test (significance of association, plotted as -log (*p*)) for each SNP sequenced on the guanidinoacetate methyltransferase (*GAMT*) gene. *p*-values are shown for European, East Asian, African, and Unknown/Admixed populations. The nominal statistical threshold for *p* = 0.01 is shown, and *p*-values that are still significant after running the Benjamin–Hochburg procedure are indicated with boxes. The location of each SNP is shown on the *GAMT* gene below the x-axis. EUR: European; ASN: East Asian; AFR: African; UNK/MIX: Unknown/Admixed populations.

2.1.3. *SLC6A8* Gene Variants

In total 25 variants were identified, of which two were in coding regions (Figure 3, Table 3). Of these, one resulted in a rare, non-synonymous coding change in exon 8: c.1162G > A. This was identified as hemizygous in a male individual, which would imply the variant would cause disease if found to be pathogenic, since the *SLC6A8* gene is on the X chromosome. The MAF was 0.005 (1/198 alleles), and allele frequencies for the 1000 Genomes, ESP and ExAC databases were 0.0003 (1/3775 alleles), 0.0002886 (3/10394 alleles) and 0.00007051 (1/14182 alleles) respectively. SIFT, PolyPhen-2 and MutationTaster all suggest the variant is damaging, and if so, the variant could potentially be disease causing. However, the variant is listed in the Leiden Open Variation Database v2 [30] and is shown by one study to be benign. The variant was identified in a patient with mental retardation, but fibroblasts from the patient were confirmed to have normal creatine uptake. The variant was seen in 2/1900 patients with mental retardation [31]. Two more rare variants were also identified as non-pathogenic by other studies: c.813C > T, a synonymous variant in exon 5 [32], and c.1142 − 35G > A in intron 7, as well as several of the non-coding polymorphic variants [33].

There were some differences seen in the MAFs calculated for *SLC6A8* variants identified in the ASD patient group compared to those from ethnic populations in 1000 Genomes database (Supplementary Table S4). Figure 6 shows the *p*-values from Fisher's exact test (plotted as -log (*p*)) and the variants identified. Only three of these differences were significant (Fisher's exact test, *p* = 0.01), but after running the Benjamini–Hochberg procedure, these values fell below the threshold of significance.

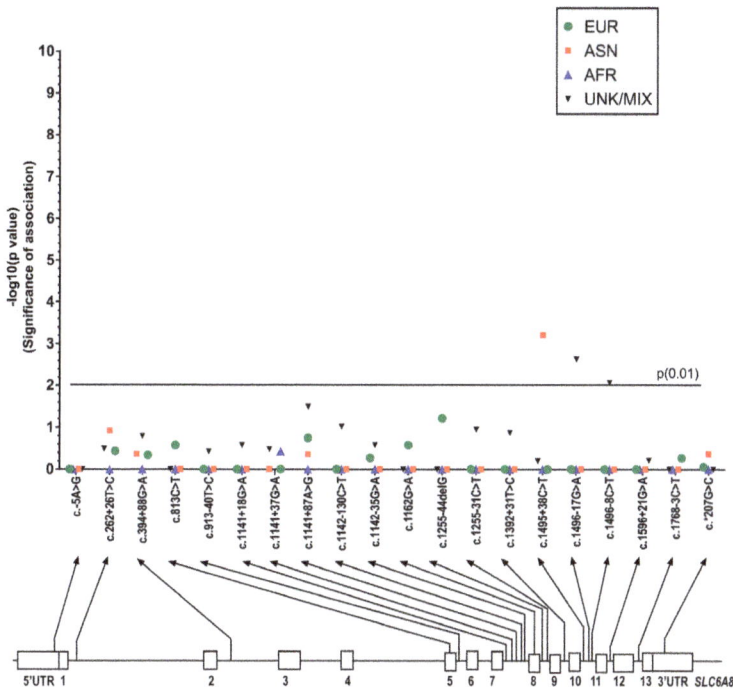

Figure 6. Manhattan plot showing the *p*-values from Fisher's exact test (significance of association, plotted as -log (p)) for each SNP sequenced on the solute carrier family 6 member 8 (*SLC6A8*) gene. *p*-values are shown for European, East Asian, African, and Unknown/Admixed populations. The nominal statistical threshold for *p* = 0.01 is shown. The location of each SNP is shown on the *SLC6A8* gene below the *x*-axis. EUR: European; ASN: East Asian; AFR: African; UNK/MIX: Unknown/Admixed populations.

3. Discussion

Three rare, non-synonymous genetic variants were identified in coding regions in *GATM*, *GAMT* and *SLC6A8* genes in four individuals with ASD. The variants were heterozygous in *GATM* and *GAMT*, implying possible carrier status for a pathogenic mutation; and hemizygous for *SLC6A8* in a single male, suggesting creatine transporter deficiency if the variant is proven pathogenic. All three variants were predicted as being damaging using SIFT, PolyPhen-2 and MutationTaster. No creatine transporter defect had been demonstrated in males with the hemizygous *SLC6A8* c.1162G > A variant. This variant was subsequently proven to be benign, based on its presence in two patients with mental retardation and no creatine transporter defect [31]. The variants in *GATM* and *GAMT* can only be proven as truly pathogenic if functional studies are performed. No homozygous or compound-heterozygous variants classified as "damaging" were identified in the ASD population, suggesting the prevalence of CDS in autism patients in our sample is zero. We have recently published another study confirming the prevalence to be zero, as no cases of CDS were identified in 443 children with ASD (screening of patients was done using both metabolic and molecular methods) [5]. Schiff et al. [34] screened urine from 203 children with non-syndromic autism and found none to be affected with CDS. The only other genetic screen was carried out by Newmeyer et al [35]. They screened 100 males with ASD for mutations in the *SLC6A8* gene, and reported one affected child. However, this male patient had the c.1162G > A variant. He was later further investigated and had a normal urine creatine/creatinine ratio, suggesting the variant was benign.

Statistical Differences between MAFs in CDS Genes in ASD and the General Population

Fisher's exact test was used to determine the statistical significance of any differences identified in the allele frequencies of the autism cohort compared to the 1000 Genomes database (divided into four population groups).

There were some differences seen in the MAFs calculated for *GATM*, *GAMT* and *SLC6A8* variants identified in the ASD patient group, but the only statistically significant results were seen in the *GATM* gene. These included variants in non-coding regions (3' and 5'UTRs, introns) as well as changes in exons. The Asian autism cohort was compared to the East Asian data in 1000 Genomes initially. Nine variants were identified at a statistically lower frequency in our Asian autism cohort compared to the 1000 Genomes database. The data for these nine variants was then compared to the South Asian data in 1000 Genomes, which further suggested that the MAFs were not significantly different. We do not know the diversity of ethnicities within our Asian cohort and so cannot determine if our population is skewed towards one or another subpopulation. This potentially interesting finding would need further investigation with Asian subpopulations. Two variants in the European autism cohort also presented at a statistically lower allele frequency than in the 1000 Genomes database.

4. Materials and Methods

A prospective group consisting of 71 subjects was enrolled in Toronto, and a retrospective sample of 95 subjects was collected as part of the Canadian Autism Genome Project (families with only one autistic child). The former was also part of the prevalence study of CDS in ASD [5]. Supplementary Table S1 shows the cohort divided by ethnicity and sex.

Participants received a diagnosis of ASD from clinician experts based on diagnostic and statistical manual of mental disorders, 4th edition (DSM-IV) criteria for ASD (autism, Asperger's, or pervasive developmental disorder—not otherwise specified (PDD-NOS)) with the diagnosis confirmed by the autism diagnostic observation schedule (ADOS-G) [36]. An additional group of children with moderate to severe ASD eligible for a publicly-funded Intensive Behavioural Intervention Therapy program for children with ASD was recruited. Children included from this subgroup also underwent assessment by clinician experts, however, rather than the ADOS, their assessment included observation with the Childhood Autism Rating Scale (CARS) [37] and the DSM-IV checklist. Individuals were excluded if they had Rett syndrome, Childhood disintegrative disorder, severe bilateral visual impairment, or severe bilateral hearing impairment.

4.1. Study Design and Measurement

To investigate the genetic variability in genes involved in creatine metabolism in children with ASD, Sanger-based DNA sequencing chemistry was used (ABI-3730) to sequence the three genes, *GATM* (AGAT), *GAMT*, and *SLC6A8* (CrT). The entire gene was sequenced including 3' and 5'UTRs, coding regions and flanking intronic segments. Examination of raw sequence data was completed manually for missense, nonsense, or small insertion/deletion events, by aligning sequences with NCBI reference genes. Genetic variations were compared to sequence data in Alamut (Alamut Visual version 2.6, Interactive Biosoftware, Rouen, France) and three variation databases: the 1000 Genomes phase 3 dataset [38], Exome Sequencing Project [39] and the Exome Aggregation Consortium [40]. Fisher's exact test was used to calculate "goodness-of-fit" between the allele frequencies in published databases (primarily the 1000 Genomes phase 3 dataset and the autistic patient population, based on ethnicity. If ethnicity was not known, or admixed, then allele frequencies were compared to the combined allele frequency for all ethnicities. Significant variation between allele frequencies for the two populations was noted when $p < 0.01$.

To reduce the false discovery rate, and help reduce type 1 errors (false positives) we applied the Benjamini–Hochberg procedure. The variants for each ethnic population for each gene (excluding the novel variants) were ranked in ascending p-values. Each individual p-value's Benjamini–Hochberg critical value was calculated using the formula $(i/m)Q$ (with i = the individual p-value's rank; m = total number of tests; and Q = the false discovery rate (0.01)). The highest p-value that was also smaller than the critical value was noted, and that value and all values with lower p-values were considered significant. These variants are identified on the Manhattan plots.

4.2. Molecular Genetics

DNA was prepared from blood or lymphoblasts; PCR products were amplified using the primers listed in Supplementary Table S5, and sequenced by TCAG (Toronto Center of Applied Genomics, Toronto, ON, Canada). *GATM* gDNA was amplified using Herculase II Fusion Taq polymerase (Agilent Tech., Mississauga, ON, Canada). *GAMT* gDNA was amplified using Hotstar Taq polymerase (Qiagen, Toronto, ON, Canada). In some cases, a second nested amplification was carried out on the first PCR product; some of these primers had a GC clamp at their 5' end. *SLC6A8* was amplified using Herculase II Fusion Taq polymerase, Hotstar Taq polymerase or TaKaRa La Taq polymerase (Takara Bio Inc., Shiga, Japan). Some of these primers had an M13 clamp at their 5' end (G. Salomons, personal communication).

4.3. Ethics, Consent and Permissions

The study was approved by the Research Ethics Boards of The Hospital of Sick Children, Surrey Place Center, and Holland Bloorview Kids Rehabilitation. Subjects enrolled in the study after parents completed written informed consent.

5. Conclusions

We hypothesized that genetic variability (polymorphisms) in the three genes associated with CDS could impact the health of children and result in an autistic phenotype. This hypothesis was suggested by the fact that rare variants have been shown to have a cumulative effect, contributing to ASD risk (narrow sense heritability); and that CSD and ASD have overlapping phenotypes suggesting some similar disease mechanisms.

Our findings suggest there could be a lower association of some specific *GATM* gene variants in Asians and Europeans with autism compared to the 1000 Genomes database (East Asian population and all Europeans), observations that would need to be corroborated in a larger group of autism patients in which ethnic sub-populations are known. Variants in *GATM* have not been associated with autism previously, nor has the chromosomal region of 15q21.1. There are no genes overlapping *GATM* that have been associated with autism.

By sequencing the three CSD genes, the findings demonstrate that genetic variability in genes of biosynthesis (AGAT and GAMT) and transport of creatine (CrT) is unlikely to play a part in the pathogenesis of ASD in children. Through this work, we have corroborated results determined in our prior prevalence study in which none of the 443 children with ASD was found to be affected with CDS [5]. Both studies taken together suggest that, despite the possibility that there is a shared feature that might cause the separate but similar ASD and CDS autistic symptoms, we should look elsewhere for causation of ASD.

Supplementary Materials: Supplementary materials can be found at www.mdpi.com/1422-0067/18/8/1665/s1.

Acknowledgments: This project was supported by the Health Resources and Services Administration (HRSA) of the U.S. Department of Health and Human Services (HHS) under cooperative agreement UA3 MC11054—Autism Intervention Research Network on Physical Health. This information or content and conclusions are those of the author and should not be construed as the official position or policy of, nor should any endorsements be inferred from the HRSA, HHS or the U.S. Government. This work was conducted through the Autism Speaks Autism Treatment Network serving as the Autism Intervention Research Network on Physical Health.

Author Contributions: Andreas Schulze designed the study, carried out data collection, data analysis, drafted and revised the manuscript. Andreas Schulze is the principal investigator in this project. Jessie M. Cameron wrote the manuscript and carried out data analysis. Valeriy Levandovskiy carried out laboratory analyses, participated in data analysis and revised the manuscript. Wendy Roberts, Evdokia Anagnostou, Stephen Scherer, and Alvin Loh contributed to study conceptualization, data collection, data analysis and revised the manuscript. All authors reviewed the article and approved its publication.

Conflicts of Interest: Scherer holds patents for autism biomarkers; the other authors have indicated they have no potential conflicts of interest to disclose.

Abbreviations

ADOS-G	Autism Diagnostic Observation Schedule
AGAT	Arginine:glycine Amidinotransferase
ASD	Autism Spectrum Disorder
bp	Base Pair
CDS	Creatine Deficiency Syndrome
CARS	Childhood Autism Rating Scale
CrT	Creatine Transporter
ExAC	Exome Aggregation Consortium
GAMT	Guanidinoacetate Methyltransferase
MAC	Minor allele Count
MAF	Minor allele Frequency
NHLBI GO ESP	National Heart, Lung and Blood Institute Grand Opportunity Exome Sequencing Project
PDD-NOS	Pervasive Developmental Disorder–Not Otherwise Specified
SNP	Single Nucleotide Polymorphism
UTR	Untranslated Region

References

1. Pampols, T. Inherited metabolic rare disease. *Adv. Exp. Med. Biol.* **2010**, *686*, 397–431. [PubMed]
2. Ververi, A.; Vargiami, E.; Papadopoulou, V.; Tryfonas, D.; Zafeiriou, D.I. Clinical and laboratory data in a sample of Greek children with autism spectrum disorders. *J. Autism Dev. Disord.* **2012**, *42*, 1470–1476. [CrossRef] [PubMed]
3. Manzi, B.; Loizzo, A.L.; Giana, G.; Curatolo, P. Autism and metabolic diseases. *J. Child Neurol.* **2008**, *23*, 307–314. [CrossRef] [PubMed]
4. Schulze, A. Creatine deficiency syndromes. *Mol. Cell. Biochem.* **2003**, *244*, 143–150. [CrossRef] [PubMed]

5. Schulze, A.; Bauman, M.; Tsai, A.C.; Reynolds, A.; Roberts, W.; Anagnostou, E.; Cameron, J.; Nozzolillo, A.A.; Chen, S.; Kyriakopoulou, L.; et al. Prevalence of Creatine Deficiency Syndromes in Children with Nonsyndromic Autism. *Pediatrics* **2016**, *137*, 1–9. [CrossRef] [PubMed]

6. Gaugler, T.; Klei, L.; Sanders, S.J.; Bodea, C.A.; Goldberg, A.P.; Lee, A.B.; Mahajan, M.; Manaa, D.; Pawitan, Y.; Reichert, J.; et al. Most genetic risk for autism resides with common variation. *Nat. Genet.* **2014**, *46*, 881–885. [CrossRef] [PubMed]

7. Klei, L.; Sanders, S.J.; Murtha, M.T.; Hus, V.; Lowe, J.K.; Willsey, A.J.; Moreno-De-Luca, D.; Yu, T.W.; Fombonne, E.; Geschwind, D.; et al. Common genetic variants, acting additively, are a major source of risk for autism. *Mol. Autism* **2012**, *3*. [CrossRef] [PubMed]

8. Schulze, A. Creatine deficiency syndromes. *Handb. Clin. Neurol.* **2013**, *113*, 1837–1843. [PubMed]

9. Leuzzi, V.; Mastrangelo, M.; Battini, R.; Cioni, G. Inborn errors of creatine metabolism and epilepsy. *Epilepsia* **2013**, *54*, 217–227. [CrossRef] [PubMed]

10. Stockler-Ipsiroglu, S.; van Karnebeek, C.; Longo, N.; Korenke, G.C.; Mercimek-Mahmutoglu, S.; Marquart, I.; Barshop, B.; Grolik, C.; Schlune, A.; Angle, B.; et al. Guanidinoacetate methyltransferase (*GAMT*) deficiency: Outcomes in 48 individuals and recommendations for diagnosis, treatment and monitoring. *Mol. Genet. Metab.* **2014**, *111*, 16–25. [CrossRef] [PubMed]

11. Van de Kamp, J.M.; Betsalel, O.T.; Mercimek-Mahmutoglu, S.; Abulhoul, L.; Grunewald, S.; Anselm, I.; Azzouz, H.; Bratkovic, D.; de Brouwer, A.; Hamel, B.; et al. Phenotype and genotype in 101 males with X-linked creatine transporter deficiency. *J. Med. Genet.* **2013**, *50*, 463–472. [CrossRef] [PubMed]

12. Wallimann, T.; Wyss, M.; Brdiczka, D.; Nicolay, K.; Eppenberger, H.M. Intracellular compartmentation, structure and function of creatine kinase isoenzymes in tissues with high and fluctuating energy demands: The 'phosphocreatine circuit' for cellular energy homeostasis. *Biochem. J.* **1992**, *281*, 21–40. [CrossRef] [PubMed]

13. Almeida, L.S.; Salomons, G.S.; Hogenboom, F.; Jakobs, C.; Schoffelmeer, A.N. Exocytotic release of creatine in rat brain. *Synapse* **2006**, *60*, 118–123. [CrossRef] [PubMed]

14. Brewer, G.J.; Wallimann, T.W. Protective effect of the energy precursor creatine against toxicity of glutamate and beta-amyloid in rat hippocampal neurons. *J. Neurochem.* **2000**, *74*, 1968–1978. [CrossRef] [PubMed]

15. Andres, R.H.; Ducray, A.D.; Huber, A.W.; Perez-Bouza, A.; Krebs, S.H.; Schlattner, U.; Seiler, R.W.; Wallimann, T.; Widmer, H.R. Effects of creatine treatment on survival and differentiation of GABA-ergic neurons in cultured striatal tissue. *J. Neurochem.* **2005**, *95*, 33–45. [CrossRef] [PubMed]

16. Andres, R.H.; Huber, A.W.; Schlattner, U.; Perez-Bouza, A.; Krebs, S.H.; Seiler, R.W.; Wallimann, T.; Widmer, H.R. Effects of creatine treatment on the survival of dopaminergic neurons in cultured fetal ventral mesencephalic tissue. *Neuroscience* **2005**, *133*, 701–713. [CrossRef] [PubMed]

17. Watanabe, A.; Kato, N.; Kato, T. Effects of creatine on mental fatigue and cerebral hemoglobin oxygenation. *Neurosci. Res.* **2002**, *42*, 279–285. [CrossRef]

18. Rae, C.; Digney, A.L.; McEwan, S.R.; Bates, T.C. Oral creatine monohydrate supplementation improves brain performance: A double-blind, placebo-controlled, cross-over trial. *Proc. Biol. Sci.* **2003**, *270*, 2147–2150. [CrossRef] [PubMed]

19. Stromberger, C.; Bodamer, O.A.; Stockler-Ipsiroglu, S. Clinical characteristics and diagnostic clues in inborn errors of creatine metabolism. *J. Inherit. Metab. Dis.* **2003**, *26*, 299–308. [CrossRef] [PubMed]

20. Comeaux, M.S.; Wang, J.; Wang, G.; Kleppe, S.; Zhang, V.W.; Schmitt, E.S.; Craigen, W.J.; Renaud, D.; Sun, Q.; Wong, L.J. Biochemical, molecular, and clinical diagnoses of patients with cerebral creatine deficiency syndromes. *Mol. Genet. Metab.* **2013**, *109*, 260–268. [CrossRef] [PubMed]

21. Stöckler, S.; Holzbach, U.; Hanefeld, F.; Marquardt, I.; Helms, G.; Requart, M.; Hanicke, W.; Frahm, J. Creatine deficiency in the brain: A new, treatable inborn error of metabolism. *Pediatr. Res.* **1994**, *36*, 409–413. [CrossRef] [PubMed]

22. Schulze, A.; Hess, T.; Wevers, R.; Mayatepek, E.; Bachert, P.; Marescau, B.; Knopp, M.V.; De Deyn, P.P.; Bremer, H.J.; Rating, D. Creatine deficiency syndrome caused by guanidinoacetate methyltransferase deficiency: Diagnostic tools for a new inborn error of metabolism. *J. Pediatr.* **1997**, *131*, 626–631. [CrossRef]

23. Mercimek-Mahmutoglu, S.; Stoeckler-Ipsiroglu, S.; Adami, A.; Appleton, R.; Araujo, H.C.; Duran, M.; Ensenauer, R.; Fernandez-Alvarez, E.; Garcia, P.; Grolik, C.; et al. GAMT deficiency: Features, treatment, and outcome in an inborn error of creatine synthesis. *Neurology* **2006**, *67*, 480–484. [CrossRef] [PubMed]

24. Gordon, N. Guanidinoacetate methyltransferase deficiency (GAMT). *Brain Dev.* **2009**, *32*, 79–81. [CrossRef] [PubMed]

25. Battini, R.; Leuzzi, V.; Carducci, C.; Tosetti, M.; Bianchi, M.C.; Item, C.B.; Stöckler-Ipsiroglu, S.; Cioni, G. Creatine depletion in a new case with AGAT deficiency: Clinical and genetic study in a large pedigree. *Mol. Genet. Metab.* **2002**, *77*, 326–331. [CrossRef]

26. Nouioua, S.; Cheillan, D.; Zaouidi, S.; Salomons, G.S.; Amedjout, N.; Kessaci, F.; Boulahdour, N.; Hamadouche, T.; Tazir, M. Creatine deficiency syndrome. A treatable myopathy due to arginine-glycine amidinotransferase (AGAT) deficiency. *Neuromuscul. Disord.* **2013**, *23*, 670–674. [CrossRef] [PubMed]

27. Ng, P.C.; Henikoff, S. Predicting deleterious amino acid substitutions. *Genome Res.* **2001**, *11*, 863–874. [CrossRef] [PubMed]

28. Adzhubei, I.A.; Schmidt, S.; Peshkin, L.; Ramensky, V.E.; Gerasimova, A.; Bork, P.; Kondrashov, A.S.; Sunyaev, S.R. A method and server for predicting damaging missense mutations. *Nat. Methods* **2010**, *7*, 248–249. [CrossRef] [PubMed]

29. Schwarz, J.M.; Cooper, D.N.; Schuelke, M.; Seelow, D. MutationTaster2: Mutation prediction for the deep-sequencing age. *Nat. Methods* **2014**, *11*, 361–362. [CrossRef] [PubMed]

30. Fokkema, I.F.; Taschner, P.E.; Schaafsma, G.C.; Celli, J.; Laros, J.F.; den Dunnen, J.T. LOVD v.2.0: The next generation in gene variant databases. *Hum. Mutat.* **2011**, *32*, 557–563. [CrossRef] [PubMed]

31. Betsalel, O.T.; Rosenberg, E.H.; Almeida, L.S.; Kleefstra, T.; Schwartz, C.E.; Valayannopoulos, V.; Abdul-Rahman, O.; Poplawski, N.; Vilarinho, L.; Wolf, P.; et al. Characterization of novel SLC6A8 variants with the use of splice-site analysis tools and implementation of a newly developed LOVD database. *Eur. J. Hum. Genet.* **2011**, *19*, 56–63. [CrossRef] [PubMed]

32. Clark, A.J.; Rosenberg, E.H.; Almeida, L.S.; Wood, T.C.; Jakobs, C.; Stevenson, R.E.; Schwartz, C.E.; Salomons, G.S. X-linked creatine transporter (SLC6A8) mutations in about 1% of males with mental retardation of unknown etiology. *Hum. Genet.* **2006**, *119*, 604–610. [CrossRef] [PubMed]

33. Rosenberg, E.H.; Almeida, L.S.; Kleefstra, T.; de Grauw, R.S.; Yntema, H.G.; Bahi, N.; Moraine, C.; Ropers, H.H.; Fryns, J.P.; Degrauw, T.J.; et al. High prevalence of SLC6A8 deficiency in X-linked mental retardation. *Am. J. Hum. Genet.* **2004**, *75*, 97–105. [CrossRef] [PubMed]

34. Schiff, M.; Benoist, J.F.; Aissaoui, S.; Boepsflug-Tanguy, O.; Mouren, M.C.; de Baulny, H.O.; Delorme, R. Should metabolic diseases be systematically screened in nonsyndromic autism spectrum disorders? *PLoS ONE* **2011**, *6*, e21932. [CrossRef]

35. Newmeyer, A.; deGrauw, T.; Clark, J.; Chuck, G.; Salomons, G. Screening of male patients with autism spectrum disorder for creatine transporter deficiency. *Neuropediatrics* **2007**, *38*, 310–312. [CrossRef] [PubMed]

36. Lord, C.; Risi, S.; Lambrecht, L.; Cook, E.H., Jr.; Leventhal, B.L.; DiLavore, P.C.; Pickles, A.; Rutter, M. The autism diagnostic observation schedule-generic: A standard measure of social and communication deficits associated with the spectrum of autism. *J. Autism Dev. Disord.* **2000**, *30*, 205–223. [CrossRef] [PubMed]

37. Schopler, E.; Reichler, R.J.; DeVellis, R.F.; Daly, K. Toward objective classification of childhood autism: Childhood Autism Rating Scale (CARS). *J. Autism Dev. Disord.* **1980**, *10*, 91–103. [CrossRef] [PubMed]

38. The International Genome Sample Resource (IGSR). 1000 Genomes Phase 3 Dataset. Available online: http://www.1000genomes.org (accessed on 20 May 2015).

39. National Heart Lung and Blood Institute (NHLBI). Exome Sequencing Project (ESP). Available online: http://evs.gs.washington.edu/EVS/ (accessed on 20 May 2015).

40. Exome Aggregation Consortium (ExAC). ExAC Browser (Beta). Available online: http://exac.broadinstitute.org/ (accessed on 20 May 2015).

International Journal of
Molecular Sciences

MDPI

Article

Increased Force Variability Is Associated with Altered Modulation of the Motorneuron Pool Activity in Autism Spectrum Disorder (ASD)

Zheng Wang [1,2,3], MinHyuk Kwon [1,2,3], Suman Mohanty [4], Lauren M. Schmitt [1,2,3], Stormi P. White [4], Evangelos A. Christou [5] and Matthew W. Mosconi [1,2,3,*]

[1]   Schiefelbusch Institute for Life Span Studies, University of Kansas, 1000 Sunnyside Ave., Lawrence, KS 66045, USA; zhengwang@ku.edu (Z.W.); minhyuk.kwon@marquette.edu (M.K.); lmschmitt@ku.edu (L.M.S.)
[2]   Clinical Child Psychology Program, University of Kansas, 1000 Sunnyside Ave., Lawrence, KS 66045, USA
[3]   Kansas Center for Autism Research and Training (K-CART), University of Kansas Medical School, Overland Park, KS 66213, USA
[4]   Center for Autism and Developmental Disabilities, University of Texas Southwestern Medical Center, Dallas, TX 75390, USA; smohanty3@humana.com (S.M.); Stormi.White@UTsouthwestern.edu (S.P.W.)
[5]   Department of Applied Physiology and Kinesiology, University of Florida, Gainesville, FL 32611, USA; eachristou@hhp.ufl.edu
*   Correspondence: mosconi@ku.edu; Tel.: +1-785-864-3350

Academic Editor: Merlin G. Butler
Received: 21 February 2017; Accepted: 22 March 2017; Published: 25 March 2017

Abstract: Force control deficits have been repeatedly documented in autism spectrum disorder (ASD). They are associated with worse social and daily living skill impairments in patients suggesting that developing a more mechanistic understanding of the central and peripheral processes that cause them may help guide the development of treatments that improve multiple outcomes in ASD. The neuromuscular mechanisms underlying force control deficits are not yet understood. Seventeen individuals with ASD and 14 matched healthy controls completed an isometric index finger abduction test at 60% of their maximum voluntary contraction (MVC) during recording of the first dorsal interosseous (FDI) muscle to determine the neuromuscular processes associated with sustained force variability. Central modulation of the motorneuron pool activation of the FDI muscle was evaluated at delta (0–4 Hz), alpha (4–10 Hz), beta (10–35 Hz) and gamma (35–60 Hz) frequency bands. ASD patients showed greater force variability than controls when attempting to maintain a constant force. Relative to controls, patients also showed increased central modulation of the motorneuron pool at beta and gamma bands. For controls, reduced force variability was associated with reduced delta frequency modulation of the motorneuron pool activity of the FDI muscle and increased modulation at beta and gamma bands. In contrast, delta, beta, and gamma frequency oscillations were not associated with force variability in ASD. These findings suggest that alterations of central mechanisms that control motorneuron pool firing may underlie the common and often impairing symptoms of ASD.

Keywords: autism spectrum disorder (ASD); index finger abduction; force variability; motorneuron pool; first dorsal interosseus (FDI) muscle; decomposition-based electromyography (dEMG)

---

1. Introduction

Sensorimotor impairments are common in autism spectrum disorder (ASD) [1,2]. Disrupted sensorimotor developments may be among the earliest emerging symptoms of ASD, and they are associated with increased severity of social-communication, cognitive, and daily living impairments [3–6]. Sensorimotor deficits also represent an important target for determining physiological processes disrupted in ASD. Specifically, sensorimotor processes are supported by central and peripheral nervous system mechanisms that are relatively well-understood. By determining patterns of sensorimotor deficits in ASD, and clarifying their underlying physiology, studies of sensorimotor control in patients may provide new insights into neurobiological processes associated with the disorder.

Multiple types of sensorimotor abnormalities have been identified in individuals with ASD, including reduced eye movement accuracy [7,8], postural instability [9,10], increased gait variability [11,12], and atypical handwriting [13]. Reduced ability to control force output also has been repeatedly documented in studies of ASD. Specifically, studies of grip force control in ASD have suggested reduced strength [6,14,15], increased sustained force variability, and reduced force accuracy [16,17]. Force control is essential for everyday tasks requiring manual dexterity (e.g., writing, feeding, and buttoning clothes), and thus determining the physiological processes associated with these deficits may identify new targets for treatments aimed at increasing daily living skills and functional independence.

Force control involves excitatory cortical commands relayed from primary motor cortex (M1) to spinal motor neurons which innervate skeletal muscle fibers [18]. Motor neurons and muscle fibers collectively form individual motor units which work together with other motor units within a motorneuron pool to generate force and maintain target force production. Force production during voluntary contractions involves increasing the number of motor units recruited and their rate of discharge action potentials [19–21]. During increases in force production, slow-twitch motor units are recruited early and followed by a gradual recruitment of fast-twitch motor units (the "size principle") [22]. During sustained force production, recruited motor units show an increase in their discharge rate over time to support a constant level of force output [19,23].

Motorneuron pool firing is coordinated by descending central commands generated at multiple frequencies to dynamically ensure precise force production [18]. Simultaneous recordings of brain and muscle during slow isometric force production have identified a low frequency 0–4 Hz delta oscillation "common drive" generated neocortically that modulates motorneuron pool activation [20,24]. The central origin of this drive is not yet determined, though delta oscillations are seen in premotor and supplementary motor areas during force production [25]. Alpha rhythms from 4 to 10 Hz also are observed during healthy individuals' slow phasic arm movements and isometric force production [26–28].

Beta band (10–35 Hz) frequency modulation of the motorneuron pool coincides with a default mode of cortical innervation from contralateral M1 during tasks involving low to moderate force contractions [29–33]. Increased beta modulation of the motorneuron pool is associated with greater force accuracy in healthy individuals suggesting that strengthened M1 cortical communication to skeletal muscles supports motor precision [34,35]. Gamma frequency (35–60 Hz) modulation of the motorneuron pool activity has been documented during individuals' maximal voluntary force production with its origins likely located in basal ganglia and frontal cortex [30]. Consistent with this hypothesis, studies have reported reduced gamma modulation in untreated individuals with Parkinson's disease along with a gradual recovery of gamma synchronization after patients received dopaminergic treatment [30,31,36]. Increased gamma power also is associated with reduced variability of sustained force production [31,37,38].

Defining alterations in the central modulation of the motorneuron pool activity associated with the increased force variability previously documented in ASD [16,17,39–41] may offer insights into both the musculophysiological processes that underpin motor deficits, and key central and peripheral processes that are disrupted. To address this critical issue, we applied Delsys decomposition-based quantitative electromyography (dEMG; Delsys, Inc., Boston, MA, USA) recording during participants' isometric index finger abduction test. The Delsys system collects four-channel surface EMG (sEMG) signal from participants' hand, from which specific skeletal muscle generates force. sEMG time series were then decomposed offline into distinct motor unit action potential trains using Delsys' decomposition algorithms (v42) [42,43] to allow us evaluating motorneuron pool activity of the muscle, thus further reveal mechanisms of the central modulation of skeletal muscles at the periphery.

Three study aims were pursued in the present study. First, whereas deficits in controlling grip force may reflect impairments in coordinating force production across multiple effectors, we tested whether individuals with ASD show deficits controlling sustained force produced by one finger in isolation. Second, we examined motorneuron pool firing properties and central modulation of the motorneuron pool at different frequency bands (delta: 0–4 Hz, alpha: 4–10 Hz, beta: 10–35 Hz, and gamma: 35–60 Hz) during isometric index finger abduction in order to characterize neuromuscular properties underlying increased force variability in ASD (Figure 1). Third, we compared the relationships between force production and motorneuron pool activity in both individuals with ASD and healthy controls to determine whether force output is controlled by separate neuromuscular processes in patients. We hypothesized that patients would show increased force variability when attempting to maintain a constant level of force by abducting their index finger. We also predicted that increased force variability in ASD would be associated with increased motorneuron pool discharge rate, increased motorneuron pool discharge rate variability, and atypical central modulation of the motorneuron pool at multiple frequency bands, especially delta, beta and gamma as these frequencies are highly associated with force production and variability in healthy individuals. Consistent with previous findings, we also expected that healthy controls would show a positive association between force variability and 0–4 Hz modulation of the motorneuron pool, and inverse relationships between force variability and modulation of the motorneuron pool at 10–35 and 35–60 Hz. For individuals with ASD, we predicted that these associations would be attenuated suggesting reduced central organization of motorneuron pool activation. Based on prior studies showing that force control impairments may be associated with core symptoms of ASD [15–17,39], we also hypothesized impaired force production and modulation of the motorneuron pool activity in individuals with ASD would be related to the severity of their ASD symptoms, including clinically rated social-communication abnormalities [4,5].

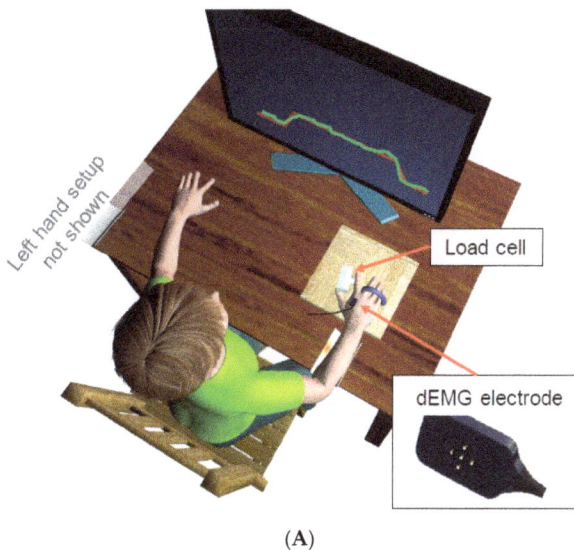

(A)

Figure 1. *Cont.*

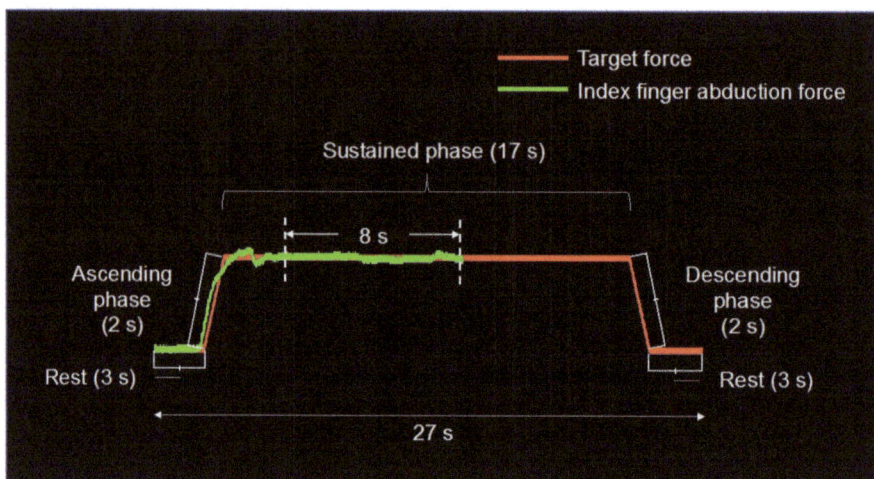

Figure 1. A participant pressed against a load cell with an abductive movement of his right index finger while viewing and tracing the red trapezoidal target template displayed on the monitor (**A**); the participant's middle, ring and little fingers were isolated and restricted to move on the hand plate. Delsys surface EMG (sEMG) electrode was attached to the back of the participant's hand in alignment with his first dorsal interosseus (FDI) muscle fibers to record muscle activity during each trial. The left index finger was used for left-handed individuals during the test (setup not shown). The duration of each trial was 27 s (**B**). The red trapezoidal target template includes a 2-s ascending phase, a 17-s sustained phase, and a 2-s descending phase of force production as well as two 3-s rest periods before and after each trial. Participant's index finger abduction force was displayed on the monitor as a light green line proceeding with time from left to right with its upward displacement representing force increase and downward movement representing force reduction. For each trial, time series of an 8-s force and its corresponding sEMG time series within the sustained phase of force production were randomly selected for evaluation of participants' behavioral performance (i.e., mean force and standard deviation of force) and motorneuron pool activity of the FDI muscle (i.e., mean discharge rate, discharge rate variability and normalized power of the motorneuron pool activity).

2. Results

2.1. Isometric Index Finger Abduction Force and Variability

Healthy controls and individuals with ASD showed similar levels of maximum voluntary contraction (MVC) ($F_{1,30} = 0.007$, $p = 0.935$; CNT = 22.500 N, SE = 2.198 N; ASD = 22.747 N, SE = 2.055 N) and mean sustained force ($F_{1,30} = 0.009$, $p = 0.924$; Figure 2A) during the test of isometric index finger abduction. Compared to healthy controls, individuals with ASD showed increased sustained force variability ($F_{1,30} = 6.641$, $p = 0.015$; Figure 2B).

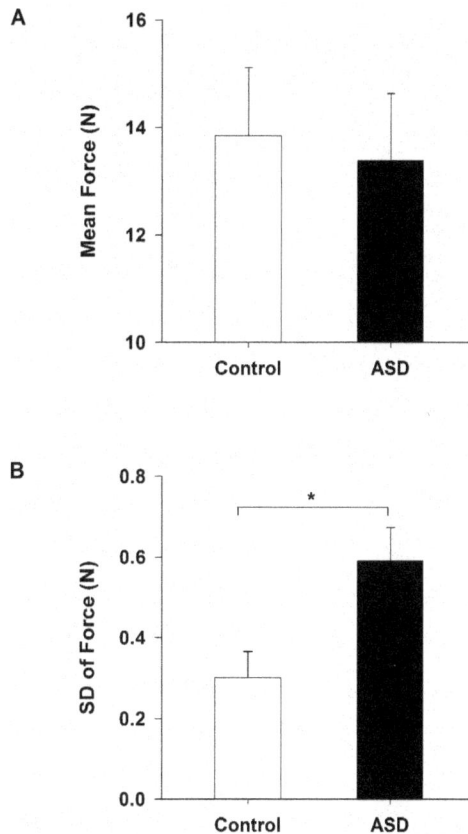

Figure 2. Isometric index finger abduction mean force (**A**); and standard deviation of force (**B**) in healthy controls and individuals with autism spectrum disorder (ASD). Both groups showed similar levels of mean force, although individuals with ASD showed a greater level of sustained force variability than healthy controls. * represents between group significance at the alpha level of 0.05. Error bars represent standard error.

2.2. decomposition-Based Electromyography (dEMG) Assessments of Motorneuron pool Activituy of the First Dorsal Interosseous (FDI) Muscle

Individuals with ASD and healthy controls showed similar FDI muscle motorneuron pool mean discharge rate ($F_{1,30} = 3.316$, $p = 0.079$; Figure 3A) and discharge rate variability ($F_{1,30} = 1.023$, $p = 0.320$; Figure 3B). For normalized power of the motorneuron pool, the main effect of frequency band ($F_{1.557,45.157} = 59.379$, $p = 0.000$) and the interactive effect of frequency band and group ($F_{1.557,45.157} = 5.173$, $p = 0.015$) were significant (Figure 3C). Normalized beta power (10–35 Hz) was significantly greater than power at other frequencies, whereas gamma power (35–60 Hz) was significantly lower than other frequency bands (delta (0–4 Hz): 26.976% ± 1.945%; alpha (4–10 Hz): 21.807% ± 1.091%; beta (10–35 Hz): 39.467% ± 1.097%; gamma (35–60 Hz): 11.750% ± 0.737%). Normalized power was significantly lower in individuals with ASD compared to controls at delta band (ASD-Control = −9.787% ± 3.889% , $p = 0.018$), while patients showed greater normalized power than controls at beta (10–35 Hz: ASD-Control = 5.546% ± 2.194%, $p = 0.017$) and gamma (35–60 Hz: ASD-Control = 3.445 ± 1.474%, $p = 0.027$) frequencies. No differences between groups were found for alpha band power.

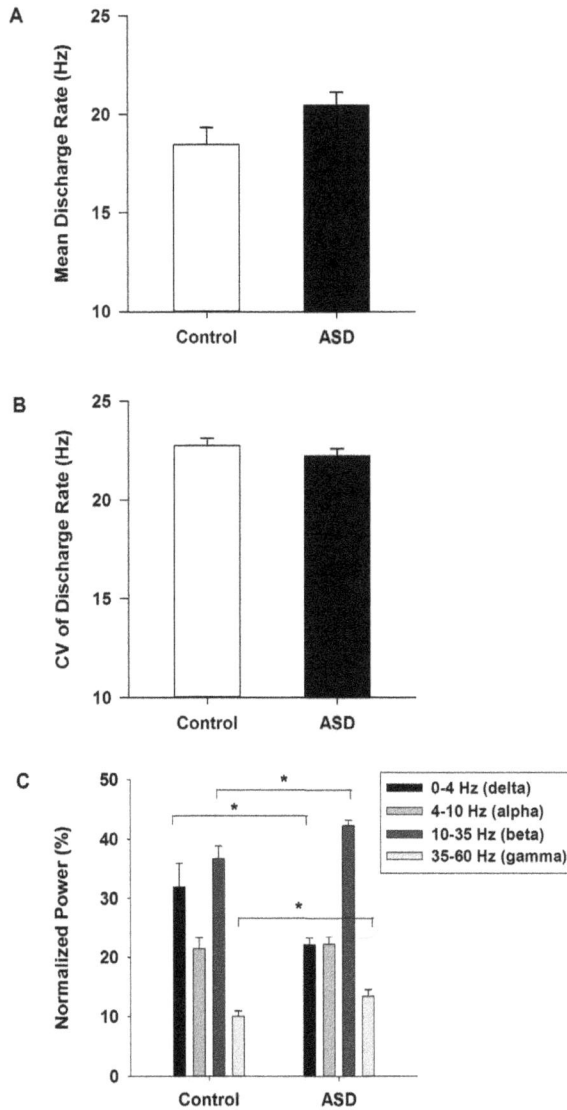

Figure 3. Mean discharge rate (**A**); coefficient of variation of mean discharge rate (**B**); and normalized power (**C**) of the first dorsal interosseous (FDI) muscle motorneuron pool for healthy controls and individuals with ASD. No between group differences were observed for mean discharge rate or discharge rate variability. Individuals with ASD showed greater normalized power at discharge rate of 10–35 (beta) and 35–60 (gamma) Hz, while they also showed lower normalized power at 0–4 (delta) Hz compared to healthy controls. * represents between group significance at the level of 0.05. Error bars represent standard error.

2.3. Relationship between Force Performance and Modulation of the FDI Muscle Motorneuron Pool Activity

For control participants, increased sustained force variability was associated with increased modulation of the motorneuron pool at delta band (0–4 Hz; Figure 4A) and reduced modulation at frequencies of beta (10–35 Hz; Figure 4C) and gamma (35–60 Hz; Figure 4D) bands. For individuals with ASD, the relationships between force variability and modulation of the motorneuron pool at different frequency bands were not significant. The relationships between sustained force variability and modulation of the motorneuron pool at alpha band (4–10 Hz) were not significant for healthy controls or individuals with ASD (Figure 4B).

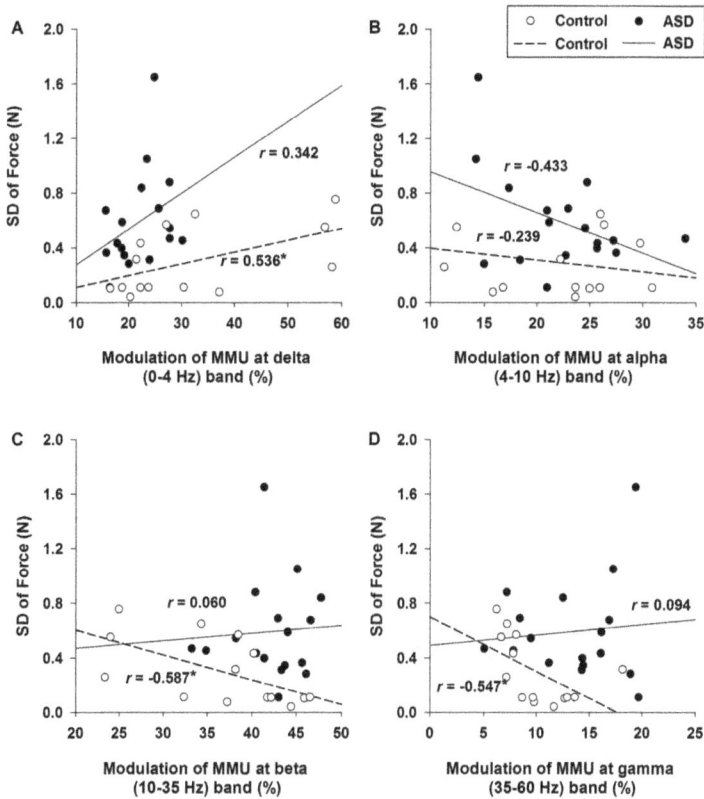

Figure 4. Relationships between the standard deviation of force and motorneuron pool activity of the FDI muscle at frequency bands of: delta, 0–4 Hz (**A**); alpha, 4–10 Hz (**B**); beta, 10–35 Hz (**C**); and gamma, 35–60 Hz (**D**). Behavioral–neuromuscular correlations were observed not significant for individuals with ASD at any frequency bands. Increased force variability was significantly associated with increased modulation of the motorneuron pool firing at 0–4 Hz in healthy controls. Force variability reduction was also significantly associated with increased modulation of motor units at beta (10–35 Hz) and gamma (35–60 Hz) frequency bands in healthy participants. * represents *p* value less than 0.05 and the absolute value of the correlation coefficient (|*r*|) greater than 0.5.

2.4. Demographic and Clinical Correlations

Neither force variability nor neuromuscular measurements were associated with intelligence quotient (IQ) scores for healthy controls or individuals with ASD (Table 1). For healthy controls, greater MVC was associated with higher full-scale and performance IQs (Figure 5A). Increased age was associated with reduced force variability for healthy controls (Figure 5B). For individuals with ASD, greater MVCs were associated with increased age but not IQ scores. Higher clinically rated social-communication deficits were associated with lower MVCs (Figure 5C) and greater motorneuron pool mean discharge rate for individuals with ASD (Figure 5D). No correlations were observed between clinically rated restricted, repetitive behaviors ratings and either force or neuromuscular measurements in individuals with ASD.

Table 1. Relationships between force and decomposition-based electromyography (dEMG) measurements with demographic, cognitive and clinical dimensions.

Control (n = 14)	Age	FSIQ	PIQ	VIQ		
MVC	0.399	0.590 *	0.640 *	0.477		
SD force	−0.576 *	0.172	0.033	0.270		
Mean discharge rate	0.462	−0.175	−0.067	−0.235		
ASD (n = 17)	Age	FSIQ	PIQ	VIQ	ADOS.soc.com	ADOS.rrb
MVC	0.787 **	0.296	0.310	0.230	−0.709 **	−0.058
SD force	0.143	−0.300	−0.108	−0.453	0.026	0.329
Mean discharge rate	−0.470	−0.388	−0.332	−0.397	0.674 **	0.1474

MVC: maximum voluntary contraction; SD Force: standard deviation of force; FSIQ: full scale IQ; PIQ: performance IQ; VIQ: verbal IQ; ADOS.sco.com: ADOS-2 social-communication algorithm total; ADOS.rrb: ADOS-2 restricted and repetitive behavior algorithm total; Statistical significance at * α = 0.05; ** α = 0.01 and r > |0.5|.

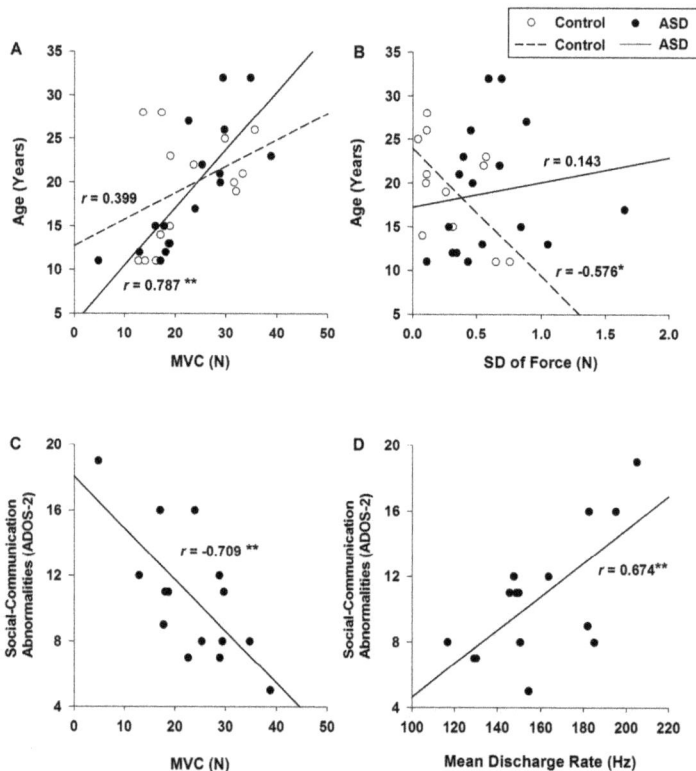

Figure 5. Relationships between age and isometric index finger abduction maximum voluntary contraction (MVC) (**A**) and standard deviation of force (**B**) for both groups. Relationships between clinical ratings of social-communication abnormalities on Autism diagnostic observation schedule-2 (ADOS-)2 and measures of MVC (**C**) and mean discharge rate of motorneuron pool activity (**D**) in individuals with ASD. * represents p value less than 0.05, ** represents p-value less than 0.01 and the absolute value of the correlation coefficient (|r|) greater than 0.5.

3. Discussion

The present study adds to the growing literature documenting deficits of voluntary hand movements and force production in ASD. Our findings also provide new evidence that motorneuron pool activity power during isometric index finger force production is abnormal in ASD at delta (0–4 Hz), beta (10–35 Hz), and gamma (35–60 Hz) frequency bands. Results further indicate that neuromuscular oscillations at these frequency bands are tightly linked to force control in healthy individuals, but that the central-peripheral communication processes that support the attenuation of force output variability in ASD are distinct and less organized. Taken together, alterations of central modulation of the motorneuron pool activity of the FDI muscle during constant force production may represent a key neurophysiological deficit related to both motor impairment and ASD symptoms.

3.1. Altered Force Production in Autism Spectrum Disorder (ASD)

Our findings were consistent with prior studies showing increased variability of manual motor output in ASD during precision gripping [16,17,39], writing [13], object lifting [40,41] and the use of simple finger gestures [44]. As the task of isometric index finger abduction only involves the FDI muscle, our study suggests that failure to precisely adjust force control and motor output in response to visual feedback is evidenced even when actions are restricted to a single muscle group as opposed to requiring coordination across different effectors or muscles. Given that manual motor deficits appear to be associated with increased severity of social-communication symptoms and daily living skills in ASD [3–6], our findings indicate that the compromised ability of patients to adjust motor output online in response to visual feedback may serve as a component of multiple key clinical issues and functional outcomes.

3.2. Altered Motorneuron Pool Activation during Force Control in ASD

Relative to controls, individuals with ASD showed similar levels of FDI muscle motorneuron pool discharge rate (Figure 3A) and discharge rate variability (Figure 3B). While these findings suggest intact central-to-peripheral modulation of skeletal muscles, it remains possible that impairments exist during recruitment of fast-twitch motor units and when increasing the firing rate of low-twitch motor units as has been seen in other developmental disorders [45]. Our findings of increased beta (10–35 Hz) and gamma (35–60 Hz) modulation of the motorneuron pool in individuals with ASD (Figure 3C) indicate atypical central modulation of the FDI muscle during isometric index finger abduction. Increased beta and gamma power may be attributable to increased central noise and/or compensatory modulation processes used to achieve specific motor goals. Increased inherent noise of central oscillators likely would involve elevation of signal power across different frequency bands. As our results showed a reduction of 0–4 Hz delta power in ASD and similar power in ASD and controls at higher frequencies from 4 to 10 Hz, it is likely that increased modulation of the motorneuron pool in ASD reflects compensatory modulation processes used to achieve specific force production.

Motorneuron pool oscillations within the delta band (0–4 Hz) represent a default mode of neural processing with the reductions seen as individuals engage in skilled actions, and associated with greater sustained motor precision [20,24]. At relatively higher force levels, such as the 60% MVC target studied here, output variability increases as delta power is increased due to greater muscle sensitivity to low frequency modulations [21,38,46,47]. In contrast, power of beta (10–35 Hz) and gamma (35–60 Hz) frequency oscillations are increased after visuomotor skill learning and during tasks involving greater attentional and cognitive demands [34,48,49]. Beta oscillations (10–35 Hz) represent co-activation of a large scale central network involving primary sensorimotor and inferior posterior parietal cortex, particularly during visuomotor tests requiring low and medium levels of force production [50,51]. Gamma (35–60 Hz) frequency oscillations, on the other hand, channel primary motor cortex as well as basal ganglia innervation during high level force production and slow phasic movements [26–28,30,31,36]. The relationship between reduced force variability and greater modulation of the motorneuron pool at beta and gamma bands thus reflects direct cortical communication to skeletal muscles of the hand that facilitates more precise motor output.

The nature of these compensatory processes remains unclear. Consistent with prior studies, we found that increased modulation of the motorneuron pool at beta (10–35 Hz) and gamma (35–60 Hz) as well as decreased modulation at delta (0–4 Hz) were highly associated with force variability reduction in healthy individuals [21,38,46,47]. These associations were not evident for individuals with ASD, suggesting the neurophysiological processes involved in central modulation of the motorneuron pool activity and control of force output are distinct from healthy individuals. These processes appear sufficient to allow patients to produce a similar level of MVC and mean force as healthy individuals, though they are not sufficient to stabilize motor output during continuous activity. Findings that individuals with ASD utilize unique neurophysiological processes during basic sensorimotor tasks are consistent with prior studies showing that prefrontal-striatal brain systems are more involved in basic movements in ASD than for controls, whereas cerebellar-cortical brain systems typically dedicated to controlling simple visuomotor actions are less involved in movement control in ASD [52,53]. Direct measurements of how modulation of the motorneuron pool varies across different force levels in individuals with ASD may be informative for determining central-to-peripheral mechanisms of force control and hand dexterity deficits in patients.

Unlike with controls, we did not find any relationship in ASD between modulation of motorneuron pool at beta and gamma bands and sustained force variability suggesting that excessive central oscillations were needed to support patients maintaining the target force level, but that these central processes had no effect on the precision of motor output. These findings suggest that central modulations may not be organized in ASD in the same manner as they are in health. For example, it is possible that, for individuals with ASD, fast-twitch and fatigable motor units are not recruited [45] or are recruited earlier than slow-twitch, fatigue resistant motor units. Alternatively, changes in the precise relation between firing rate and the mechanical twitch properties of motor units may also impair force production. In particular, when motor unit firing rates drop to the point where partial fusion of muscle twitches is reduced, muscle contractions become less efficient and more effort must be expended to achieve a force goal. Such increased effort may allow a target force to be reached, but it also likely leads to an increase in the variability of the force output [54].

3.3. Neuromotor Deficits, Demographic Characteristics and Clinical Symptoms in ASD

MVC production was more strongly associated with age in ASD compared to controls suggesting that increases in strength during development likely are delayed in ASD. This finding may help explain inconsistencies of prior studies showing both reductions in manual strength [6,14,15] and relatively intact manual strength [16,39]. It also is possible that we did not find MVC differences whereas grip strength has been shown to be impaired in ASD because gripping involves co-contraction of agonist and antagonist hand muscles, central modulation of which might be disrupted in individuals with ASD. It has been documented that motor units from different motorneuron pools respond to central modulation in a synergetic manner during co-contraction due to the fact that these motor units activate as a group according to the task goal and the advantage of this synergetic modulation is to reduce the computational load of central processing [25,55,56]. As our study showed atypical modulation of the motorneuron pool of a single muscle, it is possible that each individual's hand muscles are modulated in atypical and non-synergetic ways with the resultant effects being augmented during contraction of multiple muscles or co-contractions involving both agonist and antagonist muscles.

Increased age was associated with greater force variability reduction (Figure 5B) in healthy controls, but not in individuals with ASD (Figure 5A), suggesting persistent deficits in controlling force variability in patients. We also found that MVC reductions (Figure 5C) and greater mean discharge rates (Figure 5D) are associated with more severe clinically-rated social-communication abnormalities in ASD. Previous studies also have shown that social-communication symptom severity is related to different aspects of force control deficits in ASD, including reduced sustained force accuracy [39], lower complexity of force outputs [16], and greater target force overshooting during the ascending phase of force production [17]. Together, these findings suggest that social-communication and motor deficits in ASD may reflect common neurodevelopmental mechanisms [4,5], including central processes involved in modulating sensory-motor output. For example, increased motor variability in ASD has been shown to emerge early in development [44], and be linked to failures to understand the movements of others [57] and develop age-appropriate social and cognitive abilities [4–6]. Clarifying the timing and course of atypical modulation of motorneuron pool activity in ASD will be important for developing a more mechanistic understanding of motor, social-communication, and cognitive disturbances in patients and their dysmaturation.

3.4. Study Limitation

While the present study documents several novel findings useful for understanding neuromuscular processes underpinning force control deficits in ASD, our results should be considered in the context of multiple limitations. First, our sample spans a relatively broad age range. The small sample size may have contributed to insufficient power for characterizing the developmental trajectory of hand MVC and force variability increases in ASD. Further, the high functioning individuals may show less force variability as opposed to patients with lower IQ scores. Second, some of the participants with ASD in our sample may have shown comorbid conditions that are common in this disorder (e.g., Attention deficit hyperactivity disorder (ADHD) and depression). Systematic studies of contributions of these comorbid conditions to force variability increase as well as atypical central modulation of the motorneuron pool at different frequency bands in ASD are needed. Third, interpretations of our current findings, particularly the central origins of delta, alpha, beta and gamma frequency bands will need to be tested by integrating simultaneous measures of central oscillations using electroencephalogram (EEG) or magnetoencephalography (MEG). Such studies, in combination with EMG recording, will allow us to better understand altered central-to-peripheral mechanisms related to both motor and the defining symptoms of ASD. Lastly, antihypertensive and antidepressant medications have unclear effects on psychomotor functioning, with studies documenting both improvement and decline [58–61]. However, it is unlikely performance was impacted in participants taking either of these medications given that these medications appear to have minimal effect on basic motor functioning and peripheral processes [62].

4. Materials and Methods

4.1. Participants

Seventeen individuals with ASD and 14 healthy controls matched on age, gender, IQ and handedness (Table 2) performed an isometric index finger abduction test at 60% of their MVC. Tests of 20% and 40% MVC also were administered, though off-line observation of the surface-based EMG signals showed insufficient signal-to-noise ratio; thus, data analyses were conducted only for the 60% MVC condition. IQ was assessed using the Wechsler Abbreviated Scales of Intelligence [63], and handedness was determined using the Edinburgh questionnaire [64]. Individuals with ASD were recruited through community advertisements and local clinical programs. Diagnoses of ASD were confirmed using the ADOS-2 [65] and based on expert clinical opinion using DSM-5 criteria [66]. When possible, the Autism Diagnostic Inventory-Revised (ADI-R) [67] also was used to establish an ASD diagnosis. As parents of several adults with ASD in our study were not available, the ADI-R could only be conducted on 6/17 patients.

Table 2. Demographic characteristics (mean ± SD) of healthy controls and participants with autism spectrum disorder (ASD).

Demographic Characteristics	Control (*n* = 14)	ASD (*n* = 17)	*t*	*p*
Age (yr)	19.57 ± 6.24	18.95 ± 7.14	0.067	0.798
Range	11–28	11–32		
% Male *	85.7 (12/14)	94.14 (16/17)	0.576	0.425
% Right-handed *	92.9 (13/14)	88.23 (15/17)	0.653	0.422
Verbal IQ	112.62 ± 17.74	107.63 ± 17.14	0.589	0.449
Range	82–140	71–126		
Performance IQ	112.69 ± 13.68	106.81 ± 17.68	0.965	0.335
Range	85–133	79–129		
Full-scale IQ	114.77 ± 16.41	108.31 ± 18.34	0.975	0.449
Range	82–138	78–131		

* Chi-square (χ^2) statistics.

Participants with ASD were excluded if they had a known genetic or metabolic disorder associated with ASD (e.g., Fragile-X syndrome, Rett syndrome, and Tuberous sclerosis) or history of non-febrile seizures. Healthy controls were recruited from the community and were required to have a score of 8 or lower on the Social Communication Questionnaire (SCQ) [68]. Control participants were excluded for current or past history of psychiatric or neurological disorders, family history of ASD in first-, second- or third-degree relatives, or a history in first-degree relatives of a developmental or learning disorder, psychosis, or obsessive compulsive disorder.

No participants were taking medications known to affect sensorimotor control at the time of testing, including antipsychotics, stimulants, or anticonvulsants [62]. Seven individuals with ASD were taking antidepressant medication and two were taking antihypertensive medication at the time of testing. No participant had a history of head injury, birth injury, or seizure disorder. After a complete description of the study, written informed consent was obtained from adult participants, and informed parental consent and written assent were obtained for individuals aged less than 18 years. Study procedures were approved by the Institutional Review Board at Children's Medical Center Dallas (IRB # 062011-010) on 29 April 2012. Participants who are 18 years of age or older provided written consent and minors provided assent in addition to written consent from their legal guardian.

4.2. Apparatus and Procedures

Participants were seated in a darkened room facing a 27-inch Dell (Dell Inc., Dallas, TX, USA) LCD monitor (resolution: 1920 × 1080; refresh rate: 120 Hz) located 60 cm in front of them. They sat with their shoulder abducted at 45°, elbow flexed and forearm resting on a customized arm brace (Figure 1A). The arm brace was clamped to a table to keep participants' arm position stable throughout the test. Participants' dominant hand (i.e., the left hand was used for left handed individuals) was pronated and laid flat with digits comfortably extended on a hand plate with their middle, ring and little fingers isolated and restricted from moving. This setup only allows isometric index finger abduction at the metacarpophalangeal joint in the horizontal plane, which is a movement exclusively involving contractions of the FDI muscle [19,36]. Participants used the index finger of their dominant hand to press against a precision load cell (capacity: 100 lbf (≈445 N), Miniature Beam Load Cell, Interface Inc., Berwyn, PA, USA) that was securely attached to the hand plate and connected to a Bagnoli-16 surface EMG (sEMG) System (Delsys, Inc., Boston, MA, USA). Participants' index finger abduction force recorded from the load cell was sampled at a rate of 20 KHz using the Bagnoli-16 sEMG System (Delsys, Inc., Boston, MA, USA).

Prior to testing, each individual's index finger MVC was measured for their dominant hand. Participants completed three separate 5-s trials in which they were instructed to press against the load cell with as much force as possible by abducting their index finger. The amount of force they generated was displayed on the monitor as a red bar moving upwards with increased force. Participants rested for 1-min between consecutive trials to minimize the effect of muscle fatigue. The average maximum force across trials was calculated as the estimate of each participant's index finger MVC [16,17,39].

Prior to sEMG sensor attachment, individuals' skin over the FDI muscle was shaved and cleansed with rubbing alcohol to remove oil, debris and dead skin cells. A reference electrode was taped over the lateral epicondyle of participants' humerus. Subsequently, a specialized sEMG sensor (dEMG^{TM}, Delsys Inc.) was placed on the back of participants' hand in alignment with their FDI muscle. The sEMG electrode consists of five non-invasive probes (0.5 mm diameter of each) with four arranged in a square and the fifth probe located in the center of the square at a fixed distance of 3.6 mm from each of the surrounding four probes. Pairwise signal subtraction of these probes results in four differential sEMG channels from the FDI muscle. The sEMG time series was then amplified at 1 K, sampled at 20 KHz, and band-pass filtered at 20–450 Hz. Four-channel sEMG time series were decomposed offline into distinct motor unit action potential trains using Delsys decomposition algorithms (v42) [42,43] to evaluate the motorneuron pool activity of the FDI muscle.

During the test, participants viewed a red trapezoidal template displayed on the monitor with a target plateau set at 60% of their own MVC. Participants were instructed to accurately trace the template by adjusting the amount of force generated by abducting their index finger (Figure 1B). Participants' finger force was displayed as a light green line on the monitor that moved from left to right over time, upward with increased force, and downward with decreased force. Participants adjusted their finger force to keep the light green line as close as possible to the red trapezoidal template throughout the trial. Each trial was 27 s in duration. The red trapezoidal template consisted of three distinct phases, including: (1) a 2-s ascending phase during which participants gradually increased their force; (2) a 17-s sustained phase during which participants attempted to maintain a constant level of force; and (3) a 2-s descending phase during which participants slowly decreased their force. Two 3-s rest phases in addition to a 1 min break were administered before and after each trial to quantify baseline noise of the sEMG signal. During these rest phases, participants kept their index finger away from the load cell. Participants completed a block of three trials for each target force level, as well as practice trials at 30% MVC in order to confirm that they understood task instructions. The test consisted of 3 trials (×3 force levels) alternated with 1-min rest blocks. Total testing time including the electrode attachment, MVC testing, practice trials and task trials lasted 15–20 min.

4.3. Data Processing and Analyses

The trial on which the force trace best followed the trapezoidal template was selected for data analyses [38,69]. For each selected trial, the initial and final 5-s of force data was removed from analyses in order to limit variable effects related to initiating sustained force production and fatigue at the end of trials (Figure 1B). Thus, the middle 8-s of abduction force and its corresponding sEMG time series were analyzed off-line using custom-written Matlab programs and Delsys decomposition algorithms (v42) [42,43], respectively.

4.3.1. Force Data

Each raw force trace was digitally filtered using a 4th order low-pass Butterworth filter at a cutoff frequency of 20 Hz and detrended afterwards. Participants' behavioral performance was quantified using the mean and standard deviation of their 8-s index finger abduction force at 60% MVC.

4.3.2. sEMG Data and Decomposition Procedures for Motor Units' Activities

The sEMG time series of each trial was decomposed into distinct motor units. As shown in Figure 6, the action potentials of identified motor units were displayed in order from the smallest to the largest waveforms. To determine the accuracy of the decomposition procedure, a Decompose-Synthesize-Decompose-Compare (DSDC) test was conducted on the action potential train of each motor unit to reduce the incidence of false identification, which further increased the accuracy of validated wave forms [43,70]. Motor units were retained only when their firing was less than 10% of the false identification rate [38,70]. Among those retained motor units, we identified a range of 11–36 and 8–38 motor units for healthy participants and individuals with ASD, respectively.

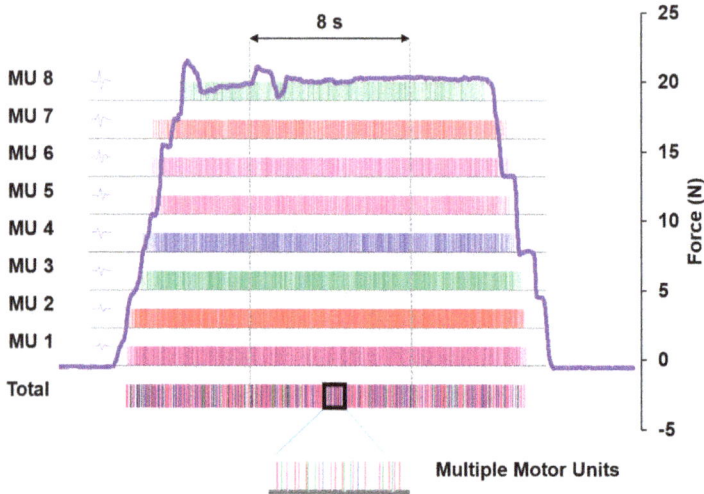

Figure 6. A representative index finger abduction force trace overlaid on eight identified and validated motor units' action potential firing trains of the FDI muscle. Eight motor units' action potential patterns were orderly displaced on the left from the smallest waveform located at the bottom to the largest at the top. The total action potential firing train representing motorneuron pool activituy of the FDI muscle was derived by summarizing action potential firing trains of all eight selected motor units. The summarized action potential train thus represents the motorneuron pool activity of the FDI muscle during a trial. The double-sided arrow shows an 8-s period during which force and eight motor unit action potential firing trains were selected for behavioral and FDI muscle activity assessments.

In order to compare the same number of motor units across groups, the smallest number of validated motor units across participants ($n = 8$) was selected from each trial to represent motorneuron pool activation of the FDI muscle and included in analyses of the mean discharge rate of the motorneuron pool. For those trials with a total number of retained motor units greater than eight, we selected eight motor units based on the following procedures: (1) the smallest and largest motor units were identified as the first (MU1) and the last (MU8) recruited motor units; (2) assigning motor units in the middle of the scale as the fourth (MU4) and fifth (MU5) recruited motor units; (3) assigning motor units in the middle of MU1 and MU4 as the second (MU2) and third (MU3) recruited motor units; and (4) assigning motor units in the middle of MU5 and MU8 as the sixth (MU6) and seventh (MU7) recruited motor units (for detailed procedures, see Appendix A). We then retained eight selected motor units for each trial with the wave forms of their action potentials evenly distributed across the motorneuron pool to ensure an unbiased comparison between individuals with ASD and healthy controls [38]. Motorneuron pool activity of the FDI muscle was derived by summarizing action potential trains of these eight selected motor units.

For each trial, mean discharge rate of the motorneuron pool was examined as the average of the inter-spike intervals. The discharge rate variability of the motorneuron pool was quantified as the coefficient of variation (CV) of the inter-spike intervals calculated using the formula below:

$$\text{CV of Discharge Rate} = \frac{\text{Standar deviation of interspike unterval}}{\text{Mean discharge rate}} \times 100\% \tag{1}$$

Modulation of the motorneuron pool activity was quantified using power spectrum analyses in the frequency domain. Inter-spike intervals of the motorneuron pool were initially transformed into a continuous time series by interpolating each trial [38]. Then, finite Fourier transformation was applied to quantify the power spectrum of the motorneuron pool [38,49,71] with frequencies separated into four bands: delta (0–4 Hz), alpha (4–10 Hz), beta (10–35 Hz) and gamma (35–60 Hz). For each frequency band, the normalized power spectrum of the motorneuron pool was calculated using the formula below:

$$\text{Normalized Power (\%)} = \frac{\left(\sum \text{power(specific frequency band)}/\text{Hz}\right)}{\sum \text{power}(0-60\ \text{Hz})} \times 100\% \tag{2}$$

Each specific frequency bin refers to each of the four frequency bands; thus, normalized power was derived for delta (0–4 Hz), alpha (4–10 Hz), beta (10–35 Hz) and gamma (35–60 Hz) frequency bands.

4.4. Clinical Measures

The ADOS-2 was used to confirm participants' diagnosis and rate ASD symptoms based on the observation of each participant's behavior [65]. The ADOS-2 is a semi-structured assessment of play, social abilities, communication skills, and imaginative use of materials performed with each individual with ASD by an examiner trained to research reliability. For the ADOS-2, higher scores reflect more severe abnormality in a given domain.

4.5. Statistical Analyses

A student *t*-test was used to compare index finger MVCs between individuals with ASD and healthy controls. A series of one-way ANOVAs were conducted to examine between-group (ASD vs. Control) differences on mean sustained force, standard deviation of force, mean discharge rate and discharge rate variability of the FDI muscle motorneuron pool. A 2 (group) × 4 (frequency band: delta (0–4 Hz), alpha (4–10 Hz), beta (10–35 Hz) and gamma (35–60 Hz)) fixed effect repeated measure ANOVA was applied to identify between-group differences in motorneuron pool activations of the FDI muscle across the power spectrum. For all analyses, results were interpreted as significant if $p < 0.05$. Where Mauchly's test indicated a violation of sphericity, the Greenhouse–Geisser estimate was used to provide a conservative test of ANOVA main and interaction effects.

To determine the relationships between FDI muscle motorneuron pool activity at each frequency band and force variability, Pearson correlations were conducted separately for individuals with ASD and healthy controls. Pearson correlation coefficients also were used to examine the relationships between individuals' muscle strength (i.e., index finger abduction MVC), force and dEMG measurements found to be different between groups and age, IQ and clinical ratings of ASD severity from the ADOS-2. Results were interpreted as significant if alpha was less than 0.05 and the absolute value of the correlation coefficient ($|r|$) was greater than 0.5.

5. Conclusions

Our results demonstrate that individuals with ASD show atypical central modulation of the motorneuron pool of a single hand muscle during isometric index finger abduction. Increased central modulations at beta (10–35 Hz) and gamma (35–60 Hz) frequency bands as well as delta (0–4 Hz) band reduction were all associated with lower force variability in healthy participants, although these relationships were all attenuated in patients. These findings suggest a lack of central communication to skeletal muscles in ASD as well as less organized motor unit recruitment at the periphery. An emerging literature has indicated that early motor developmental abnormalities are among the earliest signs of ASD [3,4,6], combined with our findings that neuromuscular impairments are associated with age and clinically rated social-communication abnormalities in ASD, these results suggest that studies of the development of force control and underlying neuromuscular properties in ASD may provide important insights into neurodevelopmental mechanisms that cause ASD and the emergence of sensorimotor and other core symptoms during childhood.

Acknowledgments: This study was supported by NIMH K23 Research Career Development Award (MH092696), the Kansas Center for Autism Research and Training (KCART) Research Investment Council Strategic Initiative Grant to Mosconi, and the NICHD U54 Kansas Intellectual and Developmental Disabilities Research Center Award (HD090216-01) to John Colombo.

Author Contributions: Matthew W. Mosconi and Suman Mohanty are responsible for the conception and design of the study; Lauren M. Schmitt and Stormi P. White performed ADI-R, ADOS-2 and IQ diagnostic tests for individuals with ASD and healthy controls and assisted in the interpretation of clinical data; Lauren M. Schmitt administered the testing protocol to participants; MinHyuk Kwon prepared the Matlab scoring program and scored the raw data; MinHyuk Kwon and Zheng Wang performed statistical analyses; Evangelos A. Christou, MinHyuk Kwon, Matthew W. Mosconi and Zheng Wang interpreted the experimental results; MinHyuk Kwon and Zheng Wang prepared figures and drafted the manuscript; each author edited the manuscript; and Evangelos A. Christou, Matthew W. Mosconi and Zheng Wang revised the manuscript. All authors have approved the final version of the manuscript.

Conflicts of Interest: The authors declare no conflict of interest.

Abbreviations

ASD	Autism spectrum disorder
ADOS-2	Autism diagnostic observation schedule-2
ADI-R	Autism diagnostic inventory-Revised
sEMG	Surface electromyography
dEMG	Decomposition-based quantitative electromyography
MVC	Maximum voluntary contraction
FDI muscle	First dorsal interosseus muscle
Motor unit	A single motorneuron and the muscle fiber that it innervates
Motorneuron pool	A motorneuron pool consists of all individual motorneurons that innervate a single muscle
Motor unit recruitment or Modulation of motor units	The CNS is responsible for the orderly recruitment of motorneurons through two distinct ways: spatial and temporal recruitment. Spatial recruitment activates more motor units to produce greater force. Temporal recruitment, or rate coding, deals with the frequency or activation rate of motor units firing
Size principle	Henneman's size principle [22] explains spatial recruitment of motor units, in which motor units are recruited from smallest to largest based on the amount of force production. For smaller force, slow twitch, low-force, fatigue-resistant muscle fibers are activated prior to the recruitment of the fast twitch, high-force, less fatigue-resistant muscle fibers
Motor unit discharge rate	Motor unit discharge rate describes temporal recruitment of motor units represented by spike firing frequency or rate of action potentials

Appendix A. Motor Unit Selection

MU1 and MU8 always represent the smallest and largest motor units in the motorneuron pool of FDI muscle, respectively. If the total number of recruited MUs is greater than 11, we used formulas below to identify MU2 through MU7 [38]:

$$MU2 = round(\frac{1 \times (a-1)}{7})$$

$$MU3 = round(\frac{2 \times (a-1)}{7})$$

$$MU4 = round(\frac{3 \times (a-1)}{7})$$

$$MU5 = round(\frac{4 \times (a-1)}{7})$$

$$MU6 = round(\frac{5 \times (a-1)}{7})$$

$$MU7 = round(\frac{6 \times (a-1)}{7})$$

where *a* stands for the total number of MUs identified using Delsys decomposition algorithms (v42) [67,68]; and *round* stands for rounding up the element within parentheses to its nearest integer. If the total number of identified MUs is 9, 10 and 11, the above listed formulas cannot be used due to overlapping assignments of specific MUs. For example, it is possible that the smallest MU will be assigned as both MU1 and MU2. If this happens, the following manual selection chart will apply:

9	MU8		10	MU8		11	MU8
8	MU7		9	MU7		10	MU7
7	MU6		8	MU6		9	MU6
6	MU5		7			8	MU5
5	MU4		6	MU5		7	
4			5	MU4		6	MU4
3	MU3		4	MU3		5	
2	MU2		3	MU2		4	MU3
1	MU1		2			3	MU2
			1	MU1		2	
						1	MU1

Light blue boxes represent MU1 and MU8, which are always the first and the last MU in the motorneuron pool, respectively; and dark blue boxes represent manually selected MUs with their order assigned on their right.

References

1. Fournier, K.A.; Hass, C.J.; Naik, S.K.; Lodha, N.; Cauraugh, J.H. Motor Coordination in Autism Spectrum Disorders: A Synthesis and Meta-Analysis. *J. Autism Dev. Disord.* **2010**, *40*, 1227–1240. [CrossRef] [PubMed]

2. Ming, X.; Brimacombe, M.; Wagner, G.C. Prevalence of motor impairment in autism spectrum disorders. *Brain Dev.* **2007**, *29*, 565–570. [CrossRef] [PubMed]

3. Loh, A.; Soman, T.; Brian, J.; Bryson, S.E.; Roberts, W.; Szatmari, P.; Smith, I.M.; Zwaigenbaum, L. Stereotyped motor behaviors associated with autism in high-risk infants: A pilot videotape analysis of a sibling sample. *J. Autism Dev. Disord.* **2007**, *37*, 25–36. [CrossRef] [PubMed]

4. Landa, R.J. Diagnosis of autism spectrum disorders in the first 3 years of life. *Nat. Clin. Pract. Neurol.* **2008**, *4*, 138–147. [CrossRef] [PubMed]

5. Hannant, P.; Tavassoli, T.; Cassidy, S. The Role of Sensorimotor Difficulties in Autism Spectrum Conditions. *Front. Neurol.* **2016**, *7*, 124. [CrossRef] [PubMed]

6. Travers, B.G.; Bigler, E.D.; Duffield, T.C.; Prigge, M.D.; Froehlich, A.L.; Lange, N.; Alexander, A.L.; Lainhart, J.E. Longitudinal development of manual motor ability in autism spectrum disorder from childhood to mid-adulthood relates to adaptive daily living skills. *Dev. Sci.* **2016**. [CrossRef] [PubMed]

7. Takarae, Y. Oculomotor abnormalities parallel cerebellar histopathology in autism. *J. Neurol. Neurosurg. Psychiatry* **2004**, *75*, 1359–1361. [CrossRef] [PubMed]

8. Schmitt, L.M.; Cook, E.H.; Sweeney, J.A.; Mosconi, M.W. Saccadic eye movement abnormalities in autism spectrum disorder indicate dysfunctions in both cerebellum and brainstem. *Mol. Autism* **2014**, *5*, 47. [CrossRef] [PubMed]

9. Minshew, N.J.; Sung, K.; Jones, B.L.; Furman, J.M. Underdevelopment of the postural control system in autism. *Neurology* **2007**, *63*, 2056–2061. [CrossRef]

10. Wang, Z.; Hallac, R.R.; Conroy, K.C.; White, S.P.; Kane, A.A.; Collinsworth, A.L.; Sweeney, J.A.; Mosconi, M.W. Postural orientation and equilibrium processes associated with increased postural sway in autism spectrum disorder (ASD). *J. Neurodev. Disord.* **2016**, *8*, 43. [CrossRef] [PubMed]

11. Hallett, M.; Lebiedowska, M.K.; Thomas, S.L.; Stanhope, S.J.; Benckla, M.B.; Rumsey, J. Locomotion of autistic adults. *Arch. Neurol.* **1993**, *50*, 1304–1308. [CrossRef] [PubMed]

12. Kindregan, D.; Gallagher, L.; Gormley, J. Gait deviations in children with autism spectrum disorders: A review. *Autism Res. Treat.* **2015**, *2015*, 741480. [CrossRef] [PubMed]

13. Fuentes, C.T.; Mostofsky, S.H.; Bastian, A.J. Children with autism show specific handwriting impairments. *Neurology* **2009**, *73*, 1532–1537. [CrossRef] [PubMed]

14. Hardan, A.Y.; Kilpatrick, M.; Keshavan, M.S.; Minshew, N.J. Motor Performance and Anatomic Magnetic Resonance Imaging (MRI) of the Basal Ganglia in Autism. *J. Child Neurol.* **2003**, *18*, 317–324. [CrossRef] [PubMed]

15. Kern, J.K.; Geier, D.A.; Adams, J.B.; Troutman, M.R.; Davis, G.; King, P.G.; Young, J.L.; Geier, M.R. Autism severity and muscle strength: A correlation analysis. *Res. Autism Spectr. Disord.* **2011**, *5*, 1011–1015. [CrossRef]

16. Mosconi, M.W.; Mohanty, S.; Greene, R.K.; Cook, E.H.; Vaillancourt, D.E.; Sweeney, J.A. Feedforward and feedback motor control abnormalities implicate cerebellar dysfunctions in autism spectrum disorder. *J. Neurosci.* **2015**, *35*, 2015–2025. [CrossRef] [PubMed]

17. Wang, Z.; Magnon, G.C.; White, S.P.; Greene, R.K.; Vaillancourt, D.E.; Mosconi, M.W. Individuals with autism spectrum disorder show abnormalities during initial and subsequent phases of precision gripping. *J. Neurophysiol.* **2015**, *113*, 1989–2001. [CrossRef] [PubMed]

18. Grosse, P.; Cassidy, M.J.; Brown, P. EEG-EMG, MEG-EMG and EMG-EMG frequency analysis: Physiological principles and clinical applications. *Clin. Neurophysiol.* **2002**, *113*, 1523–1531. [CrossRef]

19. De Luca, C.J.; LeFever, R.S.; McCue, M.P.; Xenakis, A.P. Behaviour of human motor units in different muscles during linearly varying contractions. *J. Physiol.* **1982**, *329*, 113–128. [CrossRef] [PubMed]

20. De Luca, C.J.; LeFever, R.S.; McCue, M.P.; Xenakis, A.P. Control scheme governing concurrently active human motor units during voluntary contractions. *J. Physiol.* **1982**, *329*, 129–142. [CrossRef] [PubMed]

21. Moritz, C.T.; Barry, B.K.; Pascoe, M.A.; Enoka, R.M. Discharge rate variability influences the variation in force fluctuations across the working range of a hand muscle. *J. Neurophysiol.* **2005**, *93*, 2449–2459. [CrossRef] [PubMed]

22. Henneman, E. Organization of the motorneuron pool: The size principle. In *Medical Physiology*, 14th ed.; Mountcastle, V.B., Ed.; Mosby: St. Louis, MO, USA, 1980; pp. 718–741.

23. Seki, K.; Marusawa, M. Firing rate modulation of human motor units in different muscles during isometric contraction with various forces. *Brain Res.* **1996**, *719*, 1–7. [CrossRef]

24. De Luca, C.J.; Erim, Z. Common drive of motor units in regulation of muscle force. *Trends Neurosci.* **1994**, *17*, 299–305. [CrossRef]

25. De Luca, C.J.; Erim, Z. Common drive in motor units of a synergistic muscle pair. *J. Neurophysiol.* **2002**, *87*, 2200–2204. [CrossRef] [PubMed]

26. Vallbo, A.B.; Wessberg, J. Organisation of motor output in slow finger movements in man. *J. Physiol.* **1993**, *469*, 673–691. [CrossRef] [PubMed]

27. Wessberg, J.; Vallbo, A.B. Coding of pulsatile motor output by human muscle afferents during slow finger movements. *J. Physiol.* **1995**, *485*, 271–282. [CrossRef] [PubMed]

28. Marsden, C.D. Origins of normal and pathological tremor. In *Movement Disorders: Tremor*; Findley, L.J., Capideo, R., Eds.; MacMillan Press: London, UK, 1984; pp. 37–84.

29. Conway, B.A.; Halliday, D.M.; Farmer, S.F.; Shahani, U.; Maas, P.; Weir, A.I.; Rosenberg, J.R. Synchronization between motor cortex and spinal motorneuronal pool during the performance of a maintained motor task in man. *J. Physiol.* **1995**, *489*, 917–924. [CrossRef] [PubMed]

30. Salenius, S.; Portin, K.; Kajola, M.; Salmelin, R.; Hari, R. Cortical control of human motorneuron firing during isometric contraction. *J. Neurophysiol.* **1997**, *77*, 3401–3405. [PubMed]

31. Chakarov, V.; Naranjo, J.R.; Schulte-Monting, J.; Omlor, W.; Huethe, F.; Kristeva, R. β-Range EEG-EMG Coherence With Isometric Compensation for Increasing Modulated Low-Level Forces. *J. Neurophysiol.* **2009**, *102*, 1115–1120. [CrossRef] [PubMed]

32. Brown, P. Cortical drives to human muscle: The Piper and related rhythms *Prog. Neurobiol.* **2000**, *60*, 97–108. [CrossRef]

33. Engel, A.K.; Fries, P. Beta-band oscillations-signalling the status quo? *Curr. Opin. Neurobiol.* **2010**, *20*, 156–165. [CrossRef] [PubMed]

34. Witte, M.; Patino, L.; Andrykiewicz, A.; Hepp-Reymond, M.C.; Kristeva, R. Modulation of human corticomuscular β-range coherence with low-level static forces. *Eur. J. Neurosci.* **2007**, *26*, 3564–3570. [CrossRef] [PubMed]

35. Kristeva, R.; Patino, L.; Omlor, W. β-Range cortical motor spectral power and corticomuscular coherence as a mechanism for effective corticospinal interaction during steady-state motor output. *Neuroimage* **2007**, *36*, 785–792. [CrossRef] [PubMed]

36. Brown, P. Muscle sounds in Parkinson's disease. *Lancet* **1997**, *349*, 533–535. [CrossRef]
37. Tracy, B.L.; Maluf, K.S.; Stephenson, J.L.; Hunter, S.K.; Enoka, R.M. Variability of motor unit discharge and force fluctuations across a range of muscle forces in older adults. *Muscle Nerve* **2005**, *32*, 533–540. [CrossRef] [PubMed]
38. Park, S.H.; Kwon, M.; Solis, D.; Lodha, N.; Christou, E.A. Motor control differs for increasing and releasing force. *J. Neurophysiol.* **2016**, *115*, 2924–2930. [CrossRef] [PubMed]
39. Neely, K.A.; Mohanty, S.; Schmitt, L.M.; Wang, Z.; Sweeney, J.A.; Mosconi, M.W. Motor Memory Deficits Contribute to Motor Impairments in Autism Spectrum Disorder. *J. Autism Dev. Disord.* **2016**. [CrossRef] [PubMed]
40. David, F.J.; Baranek, G.T.; Giuliani, C.A.; Mercer, V.S.; Poe, M.D.; Thorpe, D.E. A Pilot Study: Coordination of Precision Grip in Children and Adolescents with High Functioning Autism. *Pediatr. Phys. Ther.* **2009**, *21*, 205–211. [CrossRef] [PubMed]
41. David, F.J.; Baranek, G.T.; Wiesen, C.; Miao, A.F.; Thorpe, D.E. Coordination of precision grip in 2–6 years-old children with autism spectrum disorders compared to children developing typically and children with developmental disabilities. *Front. Integr. Neurosci.* **2012**, *6*, 122. [PubMed]
42. De Luca, C.J.; Adam, A.; Wotiz, R.; Gilmore, L.D.; Nawab, S.H. Decomposition of surface EMG signals. *J. Neurophysiol.* **2006**, *96*, 1646–1657. [CrossRef] [PubMed]
43. Nawab, S.H.; Chang, S.S.; De Luca, C.J. High-yield decomposition of surface EMG signals. *Clin. Neurophysiol.* **2010**, *121*, 1602–1615. [CrossRef] [PubMed]
44. Anzulewicz, A.; Sobota, K.; Delafield-Butt, J.T. Toward the Autism Motor Signature: Gesture patterns during smart tablet gameplay identify children with autism. *Sci. Rep.* **2016**, *6*, 31107. [CrossRef] [PubMed]
45. Rose, J.; McGill, K.C. Neuromuscular activation and motor-unit firing characteristics in cerebral palsy. *Dev. Med. Child Neurol.* **2005**, *47*, 329–336. [CrossRef] [PubMed]
46. Yoshitake, Y.; Shinohara, M. Low-frequency component of rectified EMG is temporally correlated with force and instantaneous rate of force fluctuations during steady contractions. *Muscle Nerve* **2013**, *47*, 577–584. [CrossRef] [PubMed]
47. Moon, H.; Kim, C.; Kwon, M.; Chen, Y.T.; Onushko, T.; Lodha, N.; Christou, E.A. Force control is related to low-frequency oscillations in force and surface EMG. *PLoS ONE* **2014**, *9*, e109202. [CrossRef] [PubMed]
48. Perez, M.A.; Lundbye-Jensen, J.; Nielsen, J.B. Changes in corticospinal drive to spinal motorneurones following visuo-motor skill learning in humans. *J. Physiol.* **2006**, *573*, 843–855. [CrossRef] [PubMed]
49. Onushko, T.; Baweja, H.S.; Christou, E.A. Practice improves motor control in older adults by increasing the motor unit modulation from 13 to 30 Hz. *J. Neurophysiol.* **2013**, *110*, 2393–2401. [CrossRef] [PubMed]
50. Classen, J.; Gerloff, C.; Honda, M.; Hallett, M. Integrative visuomotor behavior is associated with interregionally coherent oscillations in the human brain. *J. Neurophysiol.* **1998**, *79*, 1567–1573. [PubMed]
51. Brovelli, A.; Ding, M.; Ledberg, A.; Chen, Y.T.; Nakamura, R.; Bressler, S.L. Beta oscillations in a large-scale sensorimotor cortical network: Directional influences revealed by Granger causality. *Proc. Natl. Acad. Sci. USA* **2004**, *101*, 9849–9854. [CrossRef] [PubMed]
52. Takarae, Y.; Minshew, N.J.; Luna, B.; Sweeney, J.A. Atypical involvement of frontostriatal systems during sensorimotor control in autism. *Psychiatry Res.* **2007**, *156*, 117–127. [CrossRef] [PubMed]
53. Allen, G.; Muller, R.A.; Courchesne, E. Cerebellar function in autism: Functional magnetic resonance image activation during a simple motor task. *Biol. Psychiatry* **2004**, *56*, 269–278. [CrossRef] [PubMed]
54. Hu, X.; Suresh, A.K.; Li, X.; Rymer, W.Z.; Suresh, N.L. Impaired motor unit control in paretic muscle post stroke assessed using surface electromyography: A preliminary report. *Conf. Proc. IEEE Eng. Med. Biol. Soc.* **2012**, *2012*, 4116–4119. [PubMed]
55. De Luca, C.J.; Mambrito, B. Voluntary control of motor units in human antagonist muscles: Coactivation and reciprocal activation. *J. Neurophysiol.* **1987**, *58*, 525–542. [PubMed]
56. Negro, F.; Holobar, A.; Farina, D. Fluctuations in isometric muscle force can be described by one linear projection of low-frequency components of motor unit discharge rates. *J. Physiol.* **2009**, *587*, 5925–5938. [CrossRef] [PubMed]
57. Cook, J.L.; Blakemore, S.J.; Press, C. Atypical basic movement kinematics in autism spectrum conditions. *Brain* **2013**, *136*, 2816–2824. [CrossRef] [PubMed]
58. Kalra, L.; Swift, C.G.; Jackson, H.D. Psychomotor performance and antihypertensive treatment. *Br. J. Clin. Pharmacol.* **1994**, *37*, 165–172. [CrossRef] [PubMed]

59. Arora, E.; Khajuria, V.; Tandon, V.R.; Sharma, A.; Choudhary, N. Comparative evaluation of aliskiren, ramipril, and losartan on psychomotor performance in healthy volunteers: A preliminary report. *Perspect. Clin. Res.* **2014**, *5*, 190–194. [PubMed]

60. Dumont, G.J.; de Visser, S.J.; Cohen, A.F.; van Gerven, J.M.; Biomarker Working Group of the German Association for Applied Human, P. Biomarkers for the effects of selective serotonin reuptake inhibitors (SSRIs) in healthy subjects. *Br. J. Clin. Pharmacol.* **2005**, *59*, 495–510. [CrossRef] [PubMed]

61. Schrijvers, D.; Maas, Y.J.; Pier, M.P.; Madani, Y.; Hulstijn, W.; Sabbe, B.G. Psychomotor changes in major depressive disorder during sertraline treatment. *Neuropsychobiology* **2009**, *59*, 34–42. [CrossRef] [PubMed]

62. Reilly, J.L.; Lencer, R.; Bishop, J.R.; Keedy, S.; Sweeney, J.A. Pharmacological treatment effects on eye movement control. *Brain Cogn.* **2008**, *68*, 415–435. [CrossRef] [PubMed]

63. Wechsler, D. *Wechsler Abbreviated Scale of Intelligence*, 1st ed.; Psychological Corporation: New York, NY, USA, 1999.

64. Oldfield, R.C. The assessment and analysis of handedness: The Edinburgh inventory. *Neuropsychologia* **1971**, *9*, 97–113. [CrossRef]

65. Lord, C.; Rutter, M.; DiLavore, P.C.; Risi, S.; Gotham, K.; Bishop, S. *Autism Disgnostic Observation Schedule*, 2nd ed.; Western Psychological Services: Torrance, CA, USA, 2012.

66. Association, A.P. *Diagnostic and Statistical Manual of Mental Disorders*, 5th ed.; American Psychiatric Association: Washington, DC, USA, 2013.

67. Lord, C.; Rutter, M.; Le Couteur, A. Autism Diagnostic Interveiw-Revised: A revised version of a diagnostic interview for caregivers of individuals with possible pervasive developmental disorders. *J. Autism Dev. Disord.* **1994**, *24*, 659–685. [CrossRef] [PubMed]

68. Rutter, M.; Bailey, A.; Lord, C. *The Social Communication Questionnaire: Manual*, 1st ed.; Western Psychological Services: Torrance, CA, USA, 2003.

69. Taylor, A.M.; Christou, E.A.; Enoka, R.M. Multiple features of motor unit activity influence force fluctuations during isometric contractions. *J. Neurophysiol.* **2003**, *90*, 1350–1361. [CrossRef] [PubMed]

70. Kline, J.C.; De Luca, C.J. Error reduction in EMG signal decomposition. *J. Neurophysiol.* **2014**, *112*, 2718–2728. [CrossRef] [PubMed]

71. Mottram, C.J.; Christou, E.A.; Meyer, F.G.; Enoka, R.M. Frequency modulation of motor unit discharge has task-dependent effects on fluctuations in motor output. *J. Neurophysiol.* **2005**, *94*, 2878–2887. [CrossRef] [PubMed]

International Journal of
Molecular Sciences

MDPI

Article

Pharmacogenetics Informed Decision Making in Adolescent Psychiatric Treatment: A Clinical Case Report

Teri Smith *, Susan Sharp, Ann M. Manzardo and Merlin G. Butler

Department of Psychiatry and Behavioral Sciences, University of Kansas Medical Center, 3901 Rainbow Boulevard, Kansas City, Kansas 66160, USA; ssharp@kumc.edu (S.S.); amanzardo@kumc.edu (A.M.); mbutler4@kumc.edu (M.G.B.)

* Author to whom correspondence should be addressed; tsmith2@kumc.edu;
  Tel.: +1-913-588-6487, Fax: +1-913-588-6414.

Academic Editor: Kenji Hashimoto

Received: 19 December 2014; Accepted: 12 February 2015; Published: 20 February 2015

Abstract: Advances made in genetic testing and tools applied to pharmacogenetics are increasingly being used to inform clinicians in fields such as oncology, hematology, diabetes (endocrinology), cardiology and expanding into psychiatry by examining the influences of genetics on drug efficacy and metabolism. We present a clinical case example of an adolescent male with anxiety, attention deficit hyperactivity disorder (ADHD) and autism spectrum disorder who did not tolerate numerous medications and dosages over several years in attempts to manage his symptoms. Pharmacogenetics testing was performed and DNA results on this individual elucidated the potential pitfalls in medication use because of specific pharmacodynamic and pharmacokinetic differences specifically involving polymorphisms of genes in the cytochrome p450 enzyme system. Future studies and reports are needed to further illustrate and determine the type of individualized medicine approach required to treat individuals based on their specific gene patterns. Growing evidence supports this biological approach for standard of care in psychiatry.

Keywords: pharmacogenetics; cytochrome p450 enzymes; psychotropic medications; psychiatry; behavior; autism; clinical case

1. Introduction

Pharmacogenetics is the field of study that examines the influence of genetics on drug efficacy and tolerability often based on the cytochrome p450 enzymes involved in drug metabolism encoded by a large family of protein-coding genes [1]. Cytochrome p450 (CYP450) enzymes are present in most bodily tissues primarily positioned within the inner mitochondrial membrane or endoplasmic reticulum of cells. They are perhaps best known for their function in the metabolism of potentially toxic compounds including metabolic byproducts (e.g., bilirubin) and drugs but play an important role in the biosynthesis and metabolism of lipids, steroids, including hormones, and select vitamins [1]. PharmacoGENETICS can be discriminated from the related field of "PharmacoGENOMICS" which considers the broader influences of inheritance on gene products, expression and function outside the limited scope of the cytochrome p450 enzyme system, primarily found in the liver. Natural variation in sensitivity and action of cytochrome p450 enzymes contributes to variations in drug response and side effect profiles of direct relevance to medical management of drug therapy. For example, about one in every 15 individuals show an exaggerated response to standard doses of beta blockers, a class of medications commonly prescribed for hypertension and metabolized by the CYP450 enzyme system and when disturbed leads to side effects [2,3].

CYP450 enzymes are encoded by genes with the wild-type allele occurring in most individuals; however, an extensive or normal metabolizer receives two copies of the wild-type allele. The presence of other allele variants usually indicates reduced or no CYP450 enzyme activity. Individuals with two copies of a variant allele of one of the genes encoding a CYP450 enzyme are considered poor metabolizers while individuals with one wild-type allele and one variant allele have significantly reduced enzymatic activity. Individuals who inherit multiple copies of the wild-type allele are generally extensive metabolizers or ultrametabolizers and degrade drugs quickly [2,3].

Despite the relatively large number and broad function of the cytochrome p450 enzyme superfamily with over 50 members, 90 percent of all drugs are metabolized by just six different enzymes (CYP1A2, CYP3A5, CYP2C19, CYP2D6, CYP3A4 and CYP3A5) which significantly limit the number of genetic targets needed for screening and enhancing clinical utility [2]. Clinical testing platforms tailored for medical specialties are now available and typically provide interpretive services for treatment and dosage recommendations based upon the genetic testing results. Interpretive insight is particularly important for clinicians and hospitals that may be unfamiliar or apprehensive about the application of novel technologies in diagnosis and treatment in changing medical practice.

Pharmacogenetics have been used successfully to optimize treatment in cardiology, diabetes, and oncology [4,5], and has been particularly helpful for psychiatric medications which account for 20% of the 121 pharmacogenetic markers currently recognized by the US Food and Drug Administration [6]. Pharmacogenetics testing typically targets single nucleotide polymorphisms (SNPs) of the top six CYP450 enzyme genes which are now utilized in many areas of medicine including psychiatry for drug selection and adjustment to improve efficacy and reduce medication adverse events [7]. Potential financial and personal costs of adverse drug events, including deaths, and effectiveness in treatment and reducing psychiatric costs may be helped by providing better informed decision making when prescribing psychotropic drugs as outlined in a recent report by Durham [8]. The American Medical Association has made similar suggestions regarding how pharmacogenetics can reduce healthcare costs by decreasing the number of adverse drug reactions and number of medications used by patients to yield more effective therapies [9]. The following detailed clinical case history and report is presented as an example of how pharmacogenetics and the practice of psychiatry can interface now and in the future. In this case, pharmacogenetics testing provided more comprehensive and informed psychiatric care decision-making and evaluation as well as therapeutic outcomes beneficial for the patient and immediate family.

2. Results and Discussion

2.1. Clinical Case Report

Our 12-year-old Caucasian male was first evaluated by his current psychiatrist in April 2010 after years of seeking care from other psychiatrists and psychotherapists during his childhood due to disruptive behaviors. His history included psychiatric hospitalization for 2 days while on fluoxetine at 10 years of age and occupational therapy for sensory issues at 12 years of age. He had received a variety of treatments from psychotherapists. His past diagnoses included Attention Deficit Hyperactivity Disorder (ADHD), combined type, since 7 years of age, Anxiety Disorder-Not Otherwise Specified (NOS), learning problems, Obsessive Compulsive Disorder, chronic bed-wetting at a younger age, Pervasive Developmental Disorder-NOS (Autism Spectrum Disorder), Sensory Integration Disorder and drug induced mania. He reported no drug allergies, food sensitivities or intolerances.

Many different psychotropic medications had been prescribed, many of which were not tolerated or helpful. At the time of his initial evaluation, he was taking oxcarbazepine (total 375 mg), lamotrigine (175 mg), fluvoxamine (75 mg) and risperidone (0.5 mg) which were of limited to no benefit in controlling his behavior. He had experienced irritability, tantrums, impulsivity, distractibility, fidgetiness and obsessive thoughts without rituals, rigidity to change, sensory sensitivities and oppositional behaviors for much of his life. He slept well and had a normal diet. Over the next two years (2010–2012), he was prescribed atomoxetine (18 mg) and stimulants (dexamfetamine (30–60 mg), dextroamphetamine (15 mg). Risperidone was replaced with aripiprazole (5–10 mg). In February 2011 his laboratory findings showed a normal hematogram, a normal comprehensive metabolic panel, normal thyroid studies and normal fasting lipid levels. A fluoxetine (5–15 mg) trial showed minimal benefit, and he did not tolerate doses higher than 15 mg. Citalopram 10 mg was added in April 2012.

He was evaluated at 14 years of age by a clinical geneticist on our team (MGB). His height was 159 cm (15%), weight was 53.8 kg (50%) and head circumference was 53.1 cm (5%). He was non-dysmorphic and no single gene disorder was identified. The family history was positive for obsessive compulsive disorder, anxiety, irritability and anger control problems, ADHD, learning problems, heart problems, bipolar disorder, stuttering and drug and alcohol addiction. A chromosomal microarray test was normal without recognized deletions or duplications in the genome. Additional laboratory studies showed a normal comprehensive metabolic panel, hematogram, cortisol (morning), C-reactive protein, insulin (fasting), somatomedin C, but his total testosterone levels were low (80 ng/dL with reference range 270–1070 ng/dL). An X-ray bone age study showed delayed bone development. A referral was then made to an endocrinologist who prescribed testosterone cypionate injections (200 mg/mL). He was also taking over the counter (OTC) supplements, fish oil and lactobacillus probiotics at this time along with OTC flaxseed oil and ω-3 fatty acids. Veema liquid vitamins were added in January 2012. All OTC supplements were given as suggested by label. There was no perceived benefit from aripiprazole (2.5 mg) and it was discontinued. He then experienced trouble sleeping with irritability and reportedly visualized shadows. Aripiprazole (2.5 mg) was again prescribed and citalopram was increased to 15 mg. Psychotherapy was suggested, but he became more hyperactive, irritable, and angry. Aripiprazole was decreased to 2 mg while citalopram was decreased to 10 mg and testosterone injections continued. In April 2013, aripiprazole was decreased to 1 mg and hypnotherapy with relaxation techniques were offered but with only small positive effects. Aripiprazole was then discontinued. He began to take OTC amino acid supplements, salmon oil, melatonin for sleep and Empower Plus Vitamin with mineral supplements as suggested by label. He was taking 10 mg of citalopram prescribed by his psychiatrist but began to experience a crawling sensation of his skin. A low dose of diphenhydramine was suggested. His mother was advised to discuss with the pharmacist the use of OTC supplements and their possible side effects.

By June 2013, he was having frequent panic attacks and tactile sensory sensitivities which often triggered explosive episodes. His citalopram was reduced to 7.5 mg and buspirone was considered as a possible next step for pharmacotherapy. By August 2013, he was off all medications other than Empower Plus Vitamins and experienced increased symptoms of inattention and anxiety about school. He was sleeping better, but became stressed prior to attending school and wearing street clothes. He saw a new psychotherapist in August 2013 at 15 years of age for constructive methods to deal with his anxiety and sensory issues. He was a high school sophomore who did well academically (A's and B's). He participated in cross country and swimming but experienced anger dysregulation at home. Wearing street clothes bothered him extremely causing mood irritability by the time he arrived home from school.

He reported taking the following prescription medications in the past: Fluvoxamine (25–100mg), divalproex (125 mg), clonidine (0.5 mg), guanfacine (1 mg), sertraline (25 mg), aripiprazole (2–10 mg), oxcarbazepine (375 mg), risperidone (0.5 mg), lamotrigine (175 mg), lisdexamfetamine (30–60 mg), dextroamphetamine (15 mg), citalopram (10 mg), fluoxetine (5–15 mg), atomoxetine (18 mg), quetiapine (50 mg), imipramine (25 mg), testosterone, and alprazolam (0.5 mg). He felt hopelessness at times and did not understand why this was happening to him. He was polite and had a supportive family. He was socially awkward and friendships were limited. He had problems transitioning from one activity to another. At this point, his mother was giving him OTC supplements including amino acids, salmon oil, Empower vitamins, inositol when anxious, vitamin D3, choline, and probiotics as suggested by label.

At 15 years of age, he continued to have frequent "melt-downs" in the morning and could not tolerate tactile sensory stimulation. He had anxiety attacks during which he screamed and protested loudly. He became shaky, hot, and sweaty. He did not eat well or sleep normally. On September 30, 2013 escitalopram (5 mg) was prescribed and he immediately recorded less anxiety and his sensory issues improved. He requested an increase in dosage of escitalopram immediately because of the positive effect. He then began to use coping skills more efficiently such as mindfulness, listening to music, and deep breathing as well as behavioral strategies for coping with his symptoms. At about this time, he contracted a sinus infection and was using Zyrtec and escitalopram. His anger problems became much worse and his sensory problems increased. His therapist consulted with an occupational therapist in October 2013 and he began to use exercise and joint compressions along with other behavioral strategies to decrease sensory sensitivities. The patient's parents observed when the patient recently used OTC Nyquil Cold Medicine which contains acetaminophen, dextromethorphan and doxylamine succinate that it not only helped his cold symptoms, but his anxiety, sensory sensitivities, and other related behavioral problems lessened. When he discontinued this OTC medication, his anxiety, panic, and nail picking behaviors increased. Pharmacogenetics testing was then suggested due to his multiple episodes of behavioral problems that were not controlled over time with the use of several different medications and dosages.

Over time, escitalopram continued to help his anxiety, but sensory issues were still present. Escitalopram was increased to 10 mg on October 28, 2013 and behavioral improvement was notable with many days without melt-downs. Over the following few months he was dealing better with wearing specific clothing (e.g., long pants) and with transitions. He was better motivated to improve his grades from Bs to As and his overall school performance. He began to wear dress clothes (usually he would wear only soft, loose clothing). He wanted to socialize more by making new friends and to improve his social skills. ADHD symptoms lessened while on the increased level of escitalopram. He felt he could now take charge of his own self-soothing but could think through issues *versus* over-reacting to trivial frustrations. His social life improved and escitalopram was increased to 20 mg in June 2014. The family saw a correlation between inadequate eating and sleeping and the patient's over-reactions. They devised a plan to remind him to eat regularly and to encourage him to obtain sufficient sleep. His sensory issues improved and he was much kinder to others. He now had a job and was able to tolerate wearing rough-textured fabric pants and a T-shirt. He excelled at work and was given a promotion.

2.2. Pharmacogenetics

The DNA-based pharmacogenetics Genecept assay testing (Genomind, Chalfont, PA, USA) examines polymorphisms from 10 separate genes with three genes encoding cytochrome p450 enzymes related to medication metabolism (*CYP2D6, CYP2C19, CYP3A4*) and 7 additional genes consisting of neurotransmitter receptors (*5HT2C, DRD2*) and transporters (*SLC6A4*), enzymes (*COMT, MTHFR*) and ion channel function (*CACNA1C, ANK3*) involved with pharmacodynamics of drug activity and interaction. The three liver cytochrome p450 enzymes selected are major metabolizers of psychiatric medications and their gene variants are determined to have clinically relevant impacts on drug interaction and metabolism in the clinical setting. The assay contains a C/C gene variation of a serotonin receptor [5-hyroxytryptamine receptor 2C (5HT2C)] that has been associated with increased weight gain with atypical antipsychotic therapy [10–14]. The *SLC6A4* gene codes for a presynaptic serotonin transporter protein (SERT) responsible for serotonin reuptake and targeted by most selective serotonin reuptake inhibitors (SSRIs). The *SLC6A4* gene product can produce a long (L) and short (S) length variant with different clinical significance. Possession of two S variants is associated with a poor or slow response to SSRIs or with adverse events [10,15].

The DRD2 receptor is a target of most neuroleptics which act to block signaling of the neurotransmitter dopamine. The *DRD2* variant selected (-141C Ins/Del) is a variation in the promoter region of the gene that reduces *DRD2* gene expression and responsiveness along with potential adverse events when using atypical antipsychotic medications. The *COMT* gene codes for catechol-*O*-methyl-transferase which is an enzyme responsible for the majority of dopamine metabolism. Dopamine is critical for memory, judgment, attention, and other executive functions and strongly linked to multiple neuropsychiatric disorders [16]. A valine (Val) to methionine (Met) amino acid substitution is produced by a polymorphism at codon 158 due to a nucleotide G to A transition which results in approximately 40% reduction in COMT enzymatic activity [17]. The Val/Val substitution leads to elevated enzyme activity causing increased dopamine degradation (producing low dopamine) while the Met/Met substitution produces a 3 fold reduction in enzyme activity and reduced dopamine metabolism relative to Val/Val [18]. The Val/Val substitution is associated with a hypodopaminergic state and lower executive function and implicated in susceptibility to schizophrenia, panic disorder and anorexia nervosa.

A methylenetetrahydrofolate reductase (*MTHFR*) C/T gene variation has been shown to slow the conversion of folate or folic acid to methylfolate, a precursor to serotonin, norepinephrine, and dopamine synthesis. This variant impacts monoamine and catecholamine production [19–26] and associated with depression. L-methylfolate has shown efficacy as an adjunctive therapy in individuals with Selective Serotonin Reuptake Inhibitor resistant major depression [21–25]. The MTHFR enzyme metabolizes homocysteine and if not broken down effectively, can build up in the blood stream and lead to health concerns related to cardiovascular disease [26].

Molecular transport and regulation of intracellular calcium levels are important in neurological development and function with pathology linked to numerous neuropsychiatric disorders including depression, schizophrenia and bipolar disorder [27]. A variant (G to A) of the α-1C subunit of the L-type voltage gated calcium channel gene (*CACNA1C*) influences the threshold for activation and duration of channel opening leading to excessive calcium influx into the cell and neuronal hypersensitivity to activating stimuli [27]. The variation predicts poor clinical response to current pharmacotherapy. Another ion channel function related gene is Ankyrin-G (ANK3) which encodes a protein located at the nodes of Ranvier and neurons responsible for the generation of action potentials. It is important for the function and maintenance of voltage dependent sodium channels. Modest evidence links a common T to G transition with schizophrenia in this specific ankyrin gene family member [27].

2.3. DNA-based Pharmacogenetic Results

The Genecept assay results identified in our clinical case involved three known significant variations in the 10 genes tested which impact on several pharmacologic substrates (see Table 1). Our clinical case was found to have a *5HT2C* C/C gene variation of a serotonin receptor which is associated with satiety signaling in the hypothalamus and hence, serotonin has a potent satiety signal function and thus 5HT2C antagonism can lead to increased food intake [10–14]. Although the weight of our clinical case was within normal limits, this finding suggested that caution be used when prescribing atypical antipsychotics such as risperidone. Our clinical case also showed a *MTHFR* C/T gene variation that suggested reduced enzymatic activity associated with a reduced conversion of folic acid to methylfolate. As methylfolate is a precursor to serotonin, norepinephrine, and dopamine, this gene variant would indicate a possible reduced production of these peptides [19–26].

Variants of *MTHFR* have been related to increased risk for depression and L-methylfolate has shown efficacy as an adjunctive therapy. It was recommended that our clinical case should take folic acid supplements or L-methylfolate to help in the conversion of homocysteine and health concerns related to cardiovascular disease [26]. Interestingly, there was a maternal family history of heart disease that may be associated with this gene variation and homocysteine levels.

Table 1. Pharmacologic substrates, inhibitors and inducers of cytochrome P450 (CYP2D6) of relevant psychotropic drugs.

Substrate	Inhibitors	Inducers
Acetaminophen	Amiodarone	Dexamethasone
Amphetamine-Dextroamphetamine	Bupropion	Rifampin
Aripiprazole	Celecoxib	
Atomoxetine	Chlorpheniramine	
Clonidine	Chlorpromazine	
Codeine *	**Citalopram**	
Dextromethorphan	Clozapine	
Duloxetine	Cocaine	
Fluoxetine	Desipramine	
Fluvoxamine	**Diphenhydramine**	
Haloperidol	Duloxetine	
Iloperidone	**Fluoxetine**	
Methadone	Halofantrine	
Methamphetamine	Haloperidol	
Mirtazapine	Hydroxychloroquine	
Nefazodone	Imatinib	
Olanzapine	**Imipramine**	
Paroxetine	Levomepromazine	
Phenothiazines	Methadone	
Risperidone *	Metoclopramide	
Sertraline	Mibefradil	
Tricyclic antidepressants (TCAs)	Moclobemide	
Tramadol	Nelfinavir	
Venlafaxine *	Norfluoxetine	
Vortioxetine	Paroxetine	
	Perphenazine	
	Quinidine	
	Ranitidine	
	Ritonavir	
	Sertraline	
	Terbinafine	
	Thioridazine	
	Tranylcypromine	

Phenothiazines include: Chlorpromazine, fluphenazine, perphenazine, promethazine, thioridazine; tricyclic antidepressants include: Amitriptyline, amoxapine, clomipramine, desipramine, doxepin, imipramine, nortriptyline, protriptyline, trimipramine; * Metabolized to active compound; ω-3 fatty acids can inhibit CYP2D6 activity at high doses [28]; Compounds prescribed for our clinical case are indicated in **bold**. Table revised from literature supplied by Genomind, LLC (www.genomind.com).

Additionally, our clinical case had *CYP2D6*4/*5* gene allele variation that indicated significant reduction in enzyme activity. The *4 variation represents a G to A transition at the first nucleotide of exon 4 of one allele while the *5 variation represents a deletion of the second allele [29]. This is likely to put the patient at risk for significantly reduced hepatic degradation of targeted drugs and higher plasma levels of drugs that are typically processed by this enzyme thereby increasing the risk for drug interactions and reduced effectiveness of medications such as risperidone [13,30–53]. Caution should be used when prescribing medications that require this enzyme for metabolic break down. It would be important to avoid prescribing any inhibitors of CYP2D6, as well, which includes other medications that may lower further the enzymatic activity. On the other hand, inducers of CYP2D6 would increase the metabolic activity of CYP2D6. Table 1 lists psychotropic drugs known to be processed by CYP2D6, as well as inhibitors and inducers of this enzyme activity [54,55].

Many drugs prescribed for our patient (aripiprazole, dextroamphetamine, fluoxetine, fluvoxamine) were dependent on normal CYP2D6 enzyme activity and metabolism for degradation while other drugs such as risperidone require conversion to a therapeutic agent using this enzyme. Disruption of CYP2D6 function may partially or completely explain problems experienced by our clinical case when using these drugs. Some of the medications used were also inhibitors of CYP2D6 (e.g., citalopram, sertraline) which are expected to further suppress the reduced activity (see Table 2). Also, dextomethoraphan (DM) is an ingredient found in Nyquil Cold Medicine and used as a cough suppressant. It is a non-psychotropic medication substrate of CYP2D6 and excreted by the kidneys. It has a half-life of 2 to 4 h (for those with extensive (normal) metabolism) but 24 h for those individuals categorized as poor metabolizers (as found in our clinical case). Directions for use of Nyquil Cold Medication are not to exceed 4 doses in 24 h and with this recommended dosage, the amount of DM would be expected to be increased in the blood stream in those with poor metabolism. DM acts as an NMDA receptor antagonist at high doses and produces dissociative states similar to what is seen by other dissociative anesthetics such as ketamine and phencyclidine. When exceeding label-specified maximum dosages, dextromethorphan can thus act as a dissociative hallucinogen including visual field disturbances, distorted bodily perception and excitement.

Table 2. Pharmacogenetic test results with drug interactions from our clinical case.

Pharmacogenetic Target	Variant Functional Impact	Compounds Prescribed
CYP2D6	Poor cytochrome p450 metabolism	Acetaminophen, Aripiprazole, Atomoxetine, Citalopram, Dextroamphetamine, Dextromethorphan, Fluoxetine, Fluvoxamine, Risperidone *, Sertraline
5HT2C	Reduced affinity for serotonin	Fluoxetine, Fluvoamine, Sertraline
MTHFR	Reduced activity (low monoamine and catecholamine production)	Methyl/folate-related agents (vitamins)

* Metabolized to active compound.

As a slow or poor metabolizer, our clinical case would likely experience an increased prolonged sedative effect which was noted by his parents during the time he was using Nyquil Cold Medication [56,57]. Of interest, escitalopram is not dependent on CYP2D6 for metabolism but is dependent on both CYP3A4 and CYP2C19 [53], different p450 enzymes which were found to be normal by gene polymorphism testing in our clinical case. Ultimately, escitalopram, which was prescribed, was found to be the most effective medication to date for treating his behavioral problems and supported by his pharmacogenetic testing results.

3. Experimental Section

Saliva was collected from our patient in June 2014 when he was 16 years of age. DNA was isolated from saliva samples and sent for pharmacogenetics Genecept DNA-based assay commercially available from Genomind (Chalfont, PA, USA). The Genecept assay examines polymorphisms of selected genes for CYP450 hepatic enzymes which metabolize drugs as well as other genes related to neurotransmitters, their function, receptors and enzymes implicated in psychiatric disorders, and responsiveness to psychiatric medications. The list includes ten separate genes encoding cytochrome p450 enzymes with three related to the metabolism of pharmaceutical agents commonly used in psychiatry (*CYP2D6, CYP2C19, CYP3A4*) and seven additional genes encoding neurotransmitter receptors (*5HT2C, DRD2*), transporters (*SLC6A4*), enzymes (*COMT, MTHFR*) and ion channel function (*CACNA1C, ANK3*) involved in pharmacodynamic aspects of psychiatric medications.

4. Conclusions

It is clear in this clinical case that certain recommendations for care could be made due to pharmacogenetic findings. An approach would be to select a different class or drug that has a similar function but metabolized by a different CYP450 enzyme. The dosage or frequency of drug of administration of the specific medication could be adjusted to account for the metabolic disturbance (poor or ultrametabolizer). One should also consider other prescribed drugs or over the counter that may induce or inhibit the specific CYP450 enzyme (Table 1 shows a list of medications/agents that either

Int. J. Mol. Sci. **2015**, *16*, 4416–4428

induce or inhibit CYP2D6 activity). For example, eicosapentaenoic acid (EPA) is an ω-3 fatty acid and acts as an inhibitor of CYP2D6 [28] at higher doses as are antihistamines such as diphenhydramine [58]. Dietary considerations, such as the addition of folic acid or methyl group donor compound can decrease homocysteine blood levels and thus lower the negative impact from the *MTHFR* gene variant seen in our clinical case.

In summary, the history of our clinical case illustrates how pharmacogenetics and psychiatry can potentially interface to provide more informed decision making regarding use of psychotropic medications. Pharmacogenetics testing is new to the field of psychiatry and in similar situations may appear to be more humane and likely save years of personal distress associated with trial and error drug prescription without taking into consideration an individual's specific drug metabolism pattern. It would also likely be more cost effective to prescribe psychotropic medications with the more informed decision making provided by this type of genetic testing. The cost of testing (approximately $400) is low when compared to the financial costs of many other treatments utilized by this clinical case over several years. The current report involving only one clinical case is limited in generalizability and more clinical reports are needed to help guide treatment for providers. However, if similar experiences and clinical case studies are found then pharmacogenetic testing should become standard of care in mental health treatment and decision making when using psychotropic medications. Sharing these experiences would be encouraged by the authors.

Acknowledgments: We thank the family of the patient whose case history is presented in this paper. We acknowledge the support from the NICHD (HD02528) grant. The authors thank Carla Meister for manuscript preparation.

Author Contributions: Teri Smith and Merlin G. Butler conceived this study; Teri Smith, Susan Sharp, Ann Manzardo and Merlin G. Butler analyzed data and wrote the manuscript; Susan Sharp, Teri Smith and Merlin G. Butler evaluated the clinical information; and Teri Smith, Susan Sharp, Ann Manzardo and Merlin G. Butler reviewed the literature and contributed to the content of the manuscript.

Conflicts of Interest: The authors declare no conflict of interest.

References

1. Lee, J.W.; Aminkeng, F.; Bhavsar, A.P.; Shaw, K.; Carleton, B.C.; Hayden, M.R.; Ross, C.J. The emerging era of pharmacogenomics: Current successes, future potential, and challenges. *Clin. Genet.* **2014**, *86*, 21–28. [CrossRef] [PubMed]

2. Lynch, T.; Price, A. The effect of cytochrome p450 metabolism on drug response, interactions, and adverse effects. *Am Fam Physician* **2007**, *76*, 391–396. [PubMed]

3. Yang, X.; Zhang, B.; Molony, C.; Chudin, E.; Hao, K.; Zhu, J.; Gaedigk, A.; Suver, C.; Zhong, H.; Leeder, J.S.; *et al.* Systematic genetic and genomic analysis of cytochrome p450 enzyme activities in human liver. *Genome Res.* **2010**, *20*, 1020–1036. [CrossRef] [PubMed]

4. Weng, L.; Zhang, L.; Peng, Y.; Huang, R.S. Pharmacogenetics and pharmacogenomics: A bridge to individualized cancer therapy. *Pharmacogenomics* **2013**, *14*, 315–324. [CrossRef] [PubMed]

5. Toomula, N.; Hima Bindu, K.; Sathish Kumar, D.; Kumar, A. Pharmacogenomics—Personalized treatment of cancer, diabetes and cardiovascular diseases. *J. Pharmacogenomics Pharmacoproteomics* **2011**, *2*, 107.

6. Hamilton, S.P. The promise of psychiatric pharmacogenomics. *Biol. Psychiatry* **2015**, *77*, 29–35. [CrossRef] [PubMed]

7. Zandi, P.P.; Judy, J.T. The promise and reality of pharmacogenetics in psychiatry. *Psychiatr. Clin. N. Am.* **2010**, *33*, 181–224. [CrossRef]

8. Durham, D. Utilizing Pharmacogenetics in psychiarty: The time has come. *Mol. Diagn. Ther.* **2014**, *18*, 117–119. [CrossRef] [PubMed]

9. American Medical Association website resource. Available online: http://www.ama-assn.org/ama/pub/physician-resources/medical-science/genetics-molecular-medicine/current-topics/pharmacogenomics.page (accessed on 19 December 2014).

10. Reynolds, G.P.; Zhang, Z.J.; Zhang, X.B. Association of antipsychitic drug-induced weight gain with a 5-HT2C receptor gene polymorphism. *Lancet* **2002**, *359*, 2086–2087. [CrossRef] [PubMed]

11. Mulder, H.; Franke, B.; van der-Beek vander, A.A.; Arends, J.; Wilmink, F.W.; Scheffer, H.; Egberts, A.C. The association between HTR2C gene polymorphisms and the metabolic syndrome in patients with schizophrenia. *J. Clin. Psychopharmacol.* **2007**, *27*, 338–343. [CrossRef] [PubMed]

12. Reynolds, G.P. Pharmacogenetic aspects of antipsychotic drug-induced weight gain—A critical review. *Clin. Psychopharmacol. Neurosci.* **2012**, *10*, 71–77. [CrossRef] [PubMed]

13. Altar, C.A.; Hornberger, J.; Shewade, A.; Cruz, V.; Garrison, J.; Mrazek, D. Clinical validity of cytochrome P450 metabolism and serotonin gene variants in psychiatric pharmacotherapy. *Int. Rev. Psychiatry* **2013**, *25*, 509–533. [CrossRef] [PubMed]

14. Reynolds, G.P.; Hill, M.J.; Kirk, S.L. The 5HT2C receptor and antipsychotic induced weight grain–mechanisms and genetics. *J. Psychopharmacol.* **2006**, *20*, 15–18. [CrossRef] [PubMed]

15. Serretti, A.; Kato, M.; de Ronchi, D.; Kinoshita, T. Meta-analysis of serotonin transporter gene promoter polymorphism (5-HTTLPR) association with selective serotonin reuptake inhibitor efficacy in depressed patients. *Mol. Psychiatry* **2007**, *12*, 247–257. [PubMed]

16. Cohen, B.M.; Carlezon, W.A., Jr. Can't get enough of that dopamine. *Am. J. Psychiatry* **2007**, *164*, 543–546. [CrossRef] [PubMed]

17. Sim, S.C.; Kacevska, M.; Ingelman-Sundberg, M. Pharmacogenomics of drug-metabolizing enzymes: A recent update on clinical implications and endogenous effects. *Pharmacogenomics J.* **2013**, *13*, 1–11. [CrossRef] [PubMed]

18. Lachman, H.M.; Papolos, D.F.; Saito, T.; Yu, Y.M.; Szumlanski, C.L.; Weinshilboum, R.M. Human catechol-*O*-methyltransferase pharmacogenetics: Description of a functional polymorphism and its potential application to neuropsychiatric disorders. *Pharmacogenetics* **1996**, *6*, 243–250. [CrossRef] [PubMed]

19. Heisler, L.K; Zhou, L.; Bajwa, P.; Hsu, J.; Tecott, L.H. Serotonin 5HT2C receptors regulate anxiety-like behavior. *Genes Brain Behav.* **2007**, *6*, 491–496. [CrossRef] [PubMed]

20. Alex, K.D.; Yavanian, G.J.; McFarlane, H.G.; Pluto, C.P.; Pehak, E.A. Modulation of dopamine release by striatal 5-HT2C receptors. *Synapse* **2007**, *55*, 242–251. [CrossRef]

21. Stahl, S.M. L-methylfolate: A vitamin for your monoamines. *J. Clin. Psychiatry* **2008**, *69*, 1352–1353. [CrossRef] [PubMed]

22. Gilbody, S.; Lewis, S.; Lightfoot, T. Methylenetetrahydrofolate reductase (MTHFR) genetic polymorphisms and psychiatric disorders: A huge review. *Am. J. Epidemiol.* **2007**, *165*, 1–13. [CrossRef] [PubMed]

23. Wu, Y.L.; Ding, X.X.; Sun, Y.H.; Chen, J.; Zhao, X.; Jiang, Y.H.; Lv, X.L.; Wu, Z.Q. Association between MTHFR C677T polymorphism and depression: An updated meta-analysis of 26 studies. *Prog. Neuropsychopharmacol. Biol. Psychiatry* **2013**, *46*, 78–85. [CrossRef] [PubMed]

24. Ginsberg, L.D.; Oubre, A.Y.; Daoud, Y.A. L-methylfolate plus SSRI or SNRI from treatment initiation compared to SSRI or SNRI monotherapy in a major depressive episode. *Innov. Clin. Neurosci.* **2011**, *8*, 19–28. [PubMed]

25. Papkostas, G.I.; Shelton, R.C.; Zajecka, J.M.; Etemad, B.; Rickels, K.; Clain, A.; Baer, L.; Dalton, E.D.; Sacco, G.R.; Shoenfeld, D.; *et al.* L-methylfolate as adjunctive therapy for SSRI-resistant major depression: Results of two randomized, double-blind, parallel-sequential trials. *Am. J. Psychiatry* **2012**, *169*, 1267–1274. [CrossRef] [PubMed]

26. Frosst, P.; Blom, H.J.; Milos, R.; Goyette, P.; Sheppard, C.A.; Matthews, R.G.; Boers, G.J.; den Heijer, M.; Kluijtmans, L.A.; van den Heuvel, L.P.; *et al.* A candidate genetic risk factor for vascular disease: A common mutation in methylenetetrahydrofolate reductase. *Nat. Genet.* **1995**, *10*, 111–113. [CrossRef] [PubMed]

27. Yoshimizu, T.; Pan, J.Q.; Mungenast, A.E.; Madison, J.M.; Su, S.; Ketterman, J.; Ongur, D.; McPhie, D.; Cohen, B.; Perlis, R.; *et al.* Functional implications of a psychiatric risk variant within CACNA1C in induced human neurons. *Mol. Psychiatry.* **2014**. [CrossRef]

28. Yao, H.T.; Chang, Y.W.; Lan, S.J.; Chen, C.T.; Hsu, J.T.; Yeh, T.K. The inhibitory effect of polyunsaturated fatty acids on human CYP enzymes. *Life Sci.* **2006**, *79*, 2432–2440. [CrossRef] [PubMed]

29. Gough, A.C.; Miles, J.S.; Spurr, N.K.; Moss, J.E.; Gaedigk, A.; Eichelbaum, M.; Wolf, C.R. Identification of the primary gene defect at the cytochrome p450 CYP2D locus. *Nature* **1990**, *347*, 773–776. [CrossRef] [PubMed]

30. Nichols, A.I.; Focht, K.; Jiang, Q.; Preskorn, S.H.; Kane, C.P. Pharmacokinetics of venlafaxine extended release 75 mg and desvenlafaxine 50 mg in healthy CYP2D6 extensive and poor metabolizers: A randomized, open-label, two-period, parallel-group, crossover study. *Clin. Drug Investig.* **2011**, *48*, 155–167. [CrossRef]

31. Zhou, S.F. Polymorphism of human cytochrome P450 2D6 and its clinical significance: Part I. *Clin. Pharmacokinet.* **2009**, *48*, 689–723. [CrossRef] [PubMed]
32. Grasmader, K.; Verwohlt, P.L.; Rietschel, M.; Dragicevic, A.; Muller, M.; Hiemke, C.; Freymann, N.; Zobel, A.; Maier, W.; Rao, M.L. Impact of polymorphism of human cytochrome-450 isoenzymes 2C9, 2C19, and 2D6 on plasma concentrations and clinical effects of antidepressants in a naturalistic clinical setting. *Eur. J. Clin. Pharmacol.* **2004**, *60*, 329–336. [PubMed]
33. Zhou, S.F. Polymorphism of human cytochrome P450 2D6 and its clinical significance: Part II. *Clin. Pharmacokinet.* **2009**, *48*, 761–804. [CrossRef] [PubMed]
34. Samer, C.F.; Lorenzini, K.I.; Rollason, V.; Daali, Y.; Desmeules, J.A. Applications of CYP450 testing in the clinical setting. *Mol. Diagn. Ther.* **2013**, *17*, 165–184. [CrossRef] [PubMed]
35. Haertter, S. Recent examples on the clinical relevance of the CYP2D6 polymorphism and endogenous functionality of CYP2D6. *Drug Metab. Drug Interact.* **2013**, *28*, 209–216. [CrossRef]
36. De Leon, J.; Susce, M.T.; Pan, R.M.; Fairchild, M.; Koch, W.H.; Wedlund, P.J. The CYP2D6 poor metabolizer phenotype may be associated with risperidone adverse drug reactions and discontinuation. *J. Clin. Psychiatry* **2005**, *66*, 15–27. [CrossRef] [PubMed]
37. Wantanabe, J.; Suzuki, Y.; Fukui, N.; Sugai, T.; Ono, S.; Inoue, Y.; Someya, T. Dose-dependent effect of the CYP2D6 genotype on the steady-state fluvoxamine concentration. *Ther. Drug Monit.* **2008**, *30*, 705–708. [CrossRef] [PubMed]
38. Whyte, E.M.; Romkes, M.; Mulsant, B.H.; Kirshne, M.A.; Begley, A.E.; Reynolds, C.F., III; Pollock, B.G. CYP2D6 genotype and venlafaxine-XR concentrations in depressed elderly. *Int. J. Geriatr. Psychiatry* **2006**, *21*, 542–549. [CrossRef] [PubMed]
39. Lobello, K.W.; Preskorn, S.H.; Guico-Pabia, C.J.; Jiang, Q.; Paul, J.; Nichols, A.L.; Patroneva, A.; Ninan, P.T. Cytochrome P450 2D6 predicts antidepressant efficacy of venlafaxine: A secondary analysis of 4 studies in major depressive disorder. *J. Clin. Psychiatry* **2010**, *71*, 1482–1487. [CrossRef] [PubMed]
40. Van der Weide, J.; van Baalen-Benedek, E.H.; Kootstra-Ros, J.E. Mataloic ratios of psychotropics as indications of cytochrome P450 2D6/2C19 genotype. *Ther. Drug Monit.* **2005**, *27*, 478–483. [CrossRef] [PubMed]
41. Sauer, J.M.; Ring, B.J.; Witcher, J.W. Clinical pharmacokinetics of atomoxetine. *Clin. Pharmacokinet.* **2005**, *44*, 571–590. [CrossRef] [PubMed]
42. Wu, A.H. Drug metabolizing enzyme activities *versus* genetic variances for drug of clinical pharmacogenomics relevance. *Clin. Proteomics* **2011**, *8*, 12. [CrossRef] [PubMed]
43. Desmarais, J.R.; Looper, K.J. Interactions between tamoxifen and antidepressants via cytochrome P450 2D6. *J. Clin. Psychiatry* **2009**, *70*, 1688–1697. [CrossRef] [PubMed]
44. Anzenbacher, P.; Anzenbacherova, E. Cytochromes P450 and metabolism of xenobiotics. *Cell Mol. Life Sci.* **2001**, *58*, 737–747. [CrossRef] [PubMed]
45. Mannheimer, B.; von Bahr, C.; Pettersson, H.; Eliasson, E. Impact of multiple inhibitors or substrates of cytochrome P450 on plasma risperidone levels in patients on polypharmacy. *Ther. Drug Monit.* **2008**, *30*, 565–569. [CrossRef] [PubMed]
46. Jeppesen, U.; Gram, L.F.; Vistisen, K.; Loft, S.; Poulsen, H.E.; Brosen, K. Dose-dependent inhibition of CYP1A2, CYP2C19 and CYP2D6 by citalopram, fluoxetine, fluvoxamine and paroxetine. *Eur. J. Clin. Pharmacol.* **1996**, *51*, 73–78. [CrossRef] [PubMed]
47. DeLeon, J.; Armstrong, S.C.; Cozza, K.L. The dosing of atypical antipsychotics. *Psychosomatics* **2005**, *46*, 262–273.
48. DeLeon, J. Psychopharmacology: Atypical antipsychotic dosing: The effect of co-medication with anticonvulsants. *Psychiatr. Serv.* **2004**, *55*, 125–128. [CrossRef] [PubMed]
49. Sistonen, J.; Sajantila, A.; Lao, O.; Corander, J.; Barbujani, G.; Fuselli, S. CYP2D6 worldwide genetic variation shows high frequency of altered activity variants and no continental structure. *Pharmacogenet. Genomics* **2007**, *17*, 93–101. [PubMed]
50. Gaedigk, A.; Gotschall, R.R.; Forbes, N.S.; Simon, S.D.; Kearns, G.L.; Leeder, J.S. Optimization of cytochrome P4502D6 (CYP2D6) phenotype assignment using a genotyping algorithm based on allele frequency data. *Pharmacogenetics* **1999**, *9*, 669–682. [CrossRef] [PubMed]
51. Swen, J.J.; Nijenhuis, M.; de Boer, A.; Grandia, L.; Maitland-van der Zee, A.H.; Mulder, H.; Rongen, G.A.; van Schaik, R.H.; Schalekamp, T.; Touw, D.J.; *et al.* Pharmacogenetics: From bench to byte—An update of guidelines. *Clin. Pharmacol. Ther.* **2011**, *89*, 662–673. [CrossRef] [PubMed]

52. Pratt, V.M.; Zehnbauer, B.; Wilson, J.A.; Baak, R.; Babic, N.; Bettinotti, M.; Buller, A.; Butz, K.; Campbell, M.; Civalier, C.; *et al.* Characterization of 107 genomic DNA reference materials for CYP2D6, CYP2C19, CYP2C9, VKORC1, and UGT1A1: A GeT-RM and Association for Molecular Pathology collaborative project. *J. Mol. Diagn.* **2010**, *12*, 835–846. [CrossRef] [PubMed]

53. Huezo-Diaz, P.L.; Perroud, N.; Spencer, E.P.; Smith, R.; Sim, S.; Virding, S.; Uher, R.; Gunasinghe, C.; Gray, J.; Campbell, D.; Hauser, J.; *et al.* CYP2C19 genotype predicts steady state escitalopram concentration in GENDEP. *J. Psychopharmacol.* **2012**, *26*, 398–407. [CrossRef] [PubMed]

54. Genomind Literature Review, Version 3.0. Available online: http://www.geneceptassay.com/Content/LitReview/GNOMD_Lit_Review_LATEST.pdf (accessed 9 December 2014).

55. Genomind Assay Report (Sample). Available online: https://www.genomind.com/wp-content/uploads/2014/09/Sample-10-Gene-Report-Jul-2014-Lit-Sum-V-3.01.pdf (accessed 10 December 2014).

56. Zawertailo, L.A.; Kaplan, H.L.; Busto, U.E.; Tyndale, R.F.; Sellers, E.M. Psychotropic effects of dextromethorphan are altered by the CYP2D6 polymorphism: A pilot study. *J. Clin. Psychopharmacol.* **1998**, *18*, 332–337. [CrossRef] [PubMed]

57. Manap, R.A.; Wright, C.E.; Gregory, A.; Rostami-Hodjegan, A.; Meller, S.T.; Kelm, G.R.; Lennard, M.S.; Tucker, G.T.; Morice, A.H. The antitussive effect of dextromethorphan in relation to CYP2D6 activity. *Br. J. Clin. Pharmacol.* **1999**, *48*, 382–387. [CrossRef] [PubMed]

58. Hamelin, B.A.; Bouayad, A.; Methot, J.; Jobin, J.; Desgagnes, P.; Poirier, P.; Allaire, J.; Dumesnil, J.; Turgeon, J. Significant interaction between the nonprescription antihistamine diphenhydramine and the CYP2D6 substrate metoprolol in healthy men with high or low CYP2D6 activity. *Clin. Pharmaco. Ther.* **2000**, *67*, 466–477. [CrossRef]

International Journal of
Molecular Sciences

MDPI

Article

Parents' Attitudes toward Clinical Genetic Testing for Autism Spectrum Disorder—Data from a Norwegian Sample

Jarle Johannessen [1,5,*,†], Terje Nærland [1,4,†], Sigrun Hope [1,6], Tonje Torske [10], Anne Lise Høyland [7,8], Jana Strohmaier [9], Arvid Heiberg [3], Marcella Rietschel [9], Srdjan Djurovic [1,3,11] and Ole A. Andreassen [1,2]

[1] NORMENT, KG Jebsen Centre for Psychosis Research, University of Oslo, Oslo 0424, Norway; ternae@ous-hf.no (T.N.); sighop@ous-hf.no (S.H.); srdjan.djurovic@medisin.uio.no (S.D.); ole.andreassen@medisin.uio.no (O.A.A.)
[2] Division of Mental Health and Addiction, Oslo University Hospital, Oslo 0315, Norway
[3] Department of Medical Genetics, Oslo University Hospital, Oslo 0424, Norway; arvhei@ous-hf.no
[4] NevSom, Department of Rare Disorders and Disabilities, Oslo University Hospital, Oslo 0424, Norway
[5] Autism Society Norway, Oslo 0609, Norway
[6] Department of Neurohabilitation, Oslo University Hospital, Oslo 0424, Norway
[7] Regional Centre for Child and Youth Mental Health and Child Welfare, Department of Mental Health, Faculty of Medicine and Health Sciences, Norwegian University of Science and Technology, Trondheim 7491, Norway; anne.lise.hoyland@ntnu.no
[8] Department of Pediatrics, St. Olavs Hospital, Trondheim University Hospital, Trondheim 7006, Norway
[9] Department of Genetic Epidemiology in Psychiatry, Central Institute of Mental Health, Medical Faculty Mannheim, University of Heidelberg, Mannheim 68159, Germany; Jana.Strohmaier@zi-mannheim.de (J.S.); Marcella.Rietschel@zi-mannheim.de (M.R.)
[10] Division of Mental Health and Addiction, Vestre Viken Hospital Trust, Drammen 3004, Norway; tonje.torske@vestreviken.no
[11] Department of Clinical Science, University of Bergen, Bergen 5021, Norway
[*] Correspondence: jarle@autismeforeningen.no; Tel.: +47-23054573
[†] The authors contributed equally to this work.

Academic Editor: Merlin G. Butler
Received: 27 February 2017; Accepted: 13 May 2017; Published: 18 May 2017

Abstract: Clinical genetic testing (CGT) of children with autism spectrum disorder (ASD) may have positive and negative effects. Knowledge about parents' attitudes is needed to ensure good involvement of caregivers, which is crucial for accurate diagnosis and effective clinical management. This study aimed to assess parents' attitudes toward CGT for ASD. Parent members of the Norwegian Autism Society were given a previously untested questionnaire and 1455 answered. Linear regression analyses were conducted to evaluate contribution of parent and child characteristics to attitude statements. Provided it could contribute to a casual explanation of their child's ASD, 76% would undergo CGT. If it would improve the possibilities for early interventions, 74% were positive to CGT. Between 49–67% agreed that CGT could have a negative impact on health insurance, increase their concern for the child's future and cause family conflicts. Parents against CGT (9%) were less optimistic regarding positive effects, but not more concerned with negative impacts. The severity of the children's ASD diagnosis had a weak positive association with parent's positive attitudes to CGT (p-values range from <0.001 to 0.975). Parents prefer that CGT is offered to those having a child with ASD (65%), when the child's development deviates from normal (48%), or before pregnancy (36%). A majority of the parents of children with ASD are positive to CGT due to possibilities for an etiological explanation.

Int. J. Mol. Sci. **2017**, *18*, 1078

Keywords: ASD; autism; Asperger syndrome; clinical genetic testing; parents; attitudes; ethics; genetic counselling

1. Introduction

Autism spectrum disorder (ASD) is a set of heterogeneous conditions, characterized by early-onset difficulties in social communication and unusually restricted, repetitive behavior and interests [1]. The worldwide prevalence is about 1%, with a high global disease burden [2]. In Norway, the cumulative incidence for 11-year-olds has recently been estimated to be 0.8% [3].

ASD is among the most heritable neurodevelopmental conditions [4]. The recurrence risk for a sibling of an affected child is about 10–20%, which increases to 30–50% if the child has more than one sibling with ASD [5]. The etiology is still largely unknown, although it is clear that environmental factors are involved in addition to genetic susceptibility [1]. The etiology of ASD is very heterogeneous [6], and the search for genes that can explain the pathology is ongoing.

Recent advances in genetic technology have increased the diagnostic yield to 30–40% in clinical testing [7], and genetic testing is increasingly being used in the clinics, although not as a diagnostic tool [8]. In Norway, clinical genetic testing (CGT) is performed upon request by specialists in medical genetics, neurology or pediatrics and requires good phenotypic description. In cases of developmental delay and intellectual disability, as well as at least some features of ASD, chromosome 16p11.2 deletion syndrome is specifically tested for. Other copy number variants (CNVs) of diagnostic yield are emerging. A recent study on the use of ultra-high resolution chromosomal microarray analysis (CMA) found 15q11.2 BP1–BP2 microdeletion as the most common cytogenetic finding in those with ASD presenting for genetic services [9].

Meaningful interventions exist for ASD, but approved medications that target the core symptoms are absent. Finding effective treatments is regarded as the most important goal of ASD research [10], and the identification of a genetic etiology is a first step in the development of an individualized medical approach for these patients [5].

Towards the achievements of a new era of precision medicine [11], essential non-technological issues must be clarified. These issues include respecting the patients' autonomy and privacy and their need for protection from discriminations [12]. Good involvement of caregivers is needed to improve the accuracy of diagnosis, and to enable effective clinical management, tailored education [13,14], and genetic counseling, which involves complicated issues [15–17] that require a high degree of ethical reflection [15] informed by up-to-date knowledge about technological advancements [18,19] as well as of the views of those involved.

According to a recent report, professionals believe that genetic testing could improve the possibility for early intervention, enable the prevention of specific comorbid diseases, or relieve guilt [20]. Clarifying the role of specific genetic factors may contribute to parents' understanding of why their child has ASD [5]. Clinical relevance of genetic factors may include impact on the diagnostic process [15,21,22], on pharmacological and behavioral treatment [23], on adequate planning of actions [15] and on preparing parents for a life with an ASD child [22].

However, knowledge of specific genetic factors may also be potentially harmful, depending on the characteristics and roles of the genes involved, the family situation and clinical relevance of the genetic factors. A test showing that a child inherited a disease-causing gene may induce feelings of guilt and worry while a de novo mutation may relieve the parents from feelings of responsibility [21,24]. Genetic testing may also cause family conflicts [25]. Feelings of guilt and family conflicts are common in monogenic disorders [26,27]. The likelihood of recurrence of ASD in the family involved, i.e., the risk of getting another child with ASD [28–30], is also relevant for whether knowledge of specific genetic factors may be beneficial or harmful.

Current knowledge of parents´ attitudes toward CGT for ASD is limited. One study [31] found that more than 90% of the parents agreed that genetic testing is useful in health care, and that 86% of the parents are interested in finding out if genetic factors are causal in their child's ASD. Another study [32] shows that 80% of parents were interested in a genetic risk assessment test for ASD in an undiagnosed younger sibling. Both studies found that parents held a belief that the genetic test result could allow for closer monitoring and earlier access to evaluation and intervention. However, these studies included a small number of parents. Early identification is important for access to specialized evidence-based interventions [33], and prolonged diagnostic process is associated with stress in parents [34]. Parental stress or psychological well-being is not significantly different in parents of children across the ASD sub-diagnoses of infantile autism and Asperger syndrome [35]. However, we previously found very small significant differences in attitudes toward genetic research between parents of children with autism and parents of children with Asperger syndrome [36].

Perceived recurrence risk is found to influence reproductive decisions [37]. A survey of parents of children with fragile X syndrome (FXS) found that 59% indicated that the diagnosis affected their decision to have another child greatly while the rest was less affected. Those who indicated they were unaffected had already decided not to have more children [38,39]. A review found that concerns about increased risk of having a child with a given genetic condition influences decisions to undergo prenatal testing [40]. To the extent that these findings are applicable to ASD, they suggest that the reproductive decisions of parents of ASD individuals may be influenced by their perceived risk [39].

Genetic test results are also believed to reinforce stigmatization and discrimination, both by society in general and by insurance companies [25]. The stigma attached to mental illness is claimed to always lead to stress through discrimination in academic settings, in work, and in public, although often unintended [41]. Advocates for ASD fear the possible outcomes of genetic prenatal testing or treatment and argue that aggressively seeking interventions devalues autistic traits and tendencies [42]. Adults with ASD fear that people with ASD traits eventually will be eliminated through prenatal testing and selective abortion [42]. Despite genetic non-discrimination acts in most western countries [21,25], people with a disease mutation may be offered less favorable health insurance [25,43]. In some countries, it is explicitly allowed to use genetic risk information when the insurance payment exceeds a certain level [44,45].

To the best of our knowledge, most findings with respect to attitudes in the field are generated from small samples or samples that may be skewed due to selection bias or self-reference. Thus, there is a need for larger empirical investigations on parents' attitudes and outlooks of CGT in ASD. The objective of the current study was to describe and explore parents' attitudes to CGT for ASD. Our first hypothesis was that (1) parents would be positive toward CGT for ASD, due to learning more about their child's genetic risk and etiology of confirmed ASD. We also hypothesized that (2) small differences in attitudes to CGT exist between parents of children with autism and parents of children with Asperger syndrome. Health or life insurance discrimination is a potential negative effect of CGT for ASD risk. Thus, the third hypothesis was that (3) discrimination from insurance companies would be a reason to be opposed to CGT. Early diagnosis is important for early access to specialized evidence-based interventions and our fourth hypothesis was that (4) parents of children with ASD would prefer CGT for ASD risk to be offered as early as possible, i.e., during pregnancy. As a corollary, we also hypothesized that (5) parents of children with ASD would prefer CGT for ASD risk to be offered to all those interested, i.e., those doubting they are able to care for a child with ASD.

2. Results

2.1. Participants: Sociodemographic

As seen in Table 1, mean age of the parents was 46.7 years, and the mean age of their children with ASD was 16.5 years. Of the children with ASD, 81% were male and 19% were female, 49% had

Asperger syndrome and 51% had infantile autism or atypical autism (infantile autism 38%, atypical autism 13%).

Table 1. Demographics: Responding parents and their children with ASD.

Responding Parents n = 1444 [a]	Mean Age, (SD), Age Range, [count], %
Age: mean (SD) {range}	46.7 (8.6) {22–87}
Children with ASD of the responding parents: [count]	[1432] [b]
Age of the children with ASD: mean (SD) {range}	16.5 (7.6) {3–58}
Male gender of the children with ASD [c]	80.5%
ASD sub-diagnoses of the children with ASD of the responding parents [d]	
Asperger syndrome	49.3%
Infantile autism	37.6%
Other ASD	13.1%

Missing information: [a] n = 11; [b] n = 23; [c] n = 29; [d] n = 39; ASD: Autism spectrum disorder, SD: Standard deviation.

2.2. Possible Positive Effects of Clinical Genetic Testing

Proportions of participants agreeing to statements about expected positive and negative effects of CGT for ASD are presented in Table 2. Regression model summaries are listed in Table S1a,b.

Table 2. Parents' opinions on possible effects of CGT for ASD (n = 1455).

Positive Effects	Agree
Causal Explanation [a] I would only take a genetic test if it could help explain the cause of ASD in me or in my child.	76.4%
Intervention Planning [b] Early diagnosis based on a genetic test will improve the possibilities of planning for good interventions and facilitations.	74.8%
Treatment Relevance [c] I would only take a genetic test if it has a direct relevance for treatment or intervention.	53.1%
Recurrence Prevention [d] I would take a genetic test if it gave me the possibility to prevent having more children with ASD.	40.0%
Family Planning [e] Genetic testing is important for family planning.	24.1%
Negative Effects	**Agree**
Insurance Discrimination [f] A genetic test showing ASD risk will cause discrimination from insurance companies.	66.8%
Parental Concern [g] Early diagnosis based on a genetic test will increase the parents' concern for the future health and development of the child.	56.1%
Family Conflicts [h] Genetic testing may cause family conflicts.	48.9%

Missing information: [a] n = 33; [b] n = 74; [c] n = 42; [d] n = 36; [e] n = 38; [f] n = 68; [g] n = 71; [h] n = 76; CGT: Clinical genetic testing; ASD: Autism spectrum disorder.

Three quarters of the participants would only do CGT if it could contribute to explaining the etiology of ASD in their child (76%) and if the CGT could improve the possibilities of planning for good interventions and treatments (75%). Half of the respondents would do CGT only if it has a direct relevance for treatment or intervention (53%). A total of 40% would do CGT if it enabled them to prevent having further children with ASD, and 24% agreed that CGT is important for family planning.

Some parent and child characteristics had a small but significant influence on opinions regarding positive effects of clinical genetic testing. Parents of children with autism differed from parents of children with Asperger syndrome on how CGT is important for etiological explanation, for recurrence prevention and for family planning. Parents of children with autism parents agreed significantly more to the statements than parents of children with Asperger syndrome in all three models ($p < 0.001$). In a family planning model, age of parent contributed significantly ($p = 0.007$). Agreement to the family planning statement increases by age of the parent. The variances explained in the significant models are, however, small (p-values range from <0.001 to 0.009; R^2 range from 0.018 to 0.081). Regression model summaries are presented in Table S1a.

2.3. Possible Negative Effects of Clinical Genetic Testing

Two thirds (67%) of the parents agreed that a genetic test result showing risk for ASD will cause insurance company discrimination, and over half (56%) had increased concern for the child's future as a consequence of CGT. Half (49%) of the parents agreed that genetic testing may cause family conflicts. We found no significant association between age, gender or ASD sub-diagnosis and opinions of insurance discrimination, parental concern or family conflicts (p-values range from 0.055 to 0.404) (Table S1b).

2.4. Management of Clinical Genetic Testing: Who and When?

The proportions of participants' opinions concerning who should be offered CGT for ASD and when to offer CGT are listed in Table 3. Regression model summaries are listed in Table S2a,b.

Table 3. Parents' opinions on management of CGT ($n = 1330$).

Who Should Be Offered Testing	Agree
Parents of children with ASD Those already having a child with ASD	64.9%
Doubting able to care Those doubting ability to care for a child with ASD	15.6%
Worried about the fetus Those anxious/worried that there is something wrong with the fetus	14.7%
No pregnant women	11.3%
All pregnant women	9.5%
When Should Testing Be Offered	**Agree**
Development deviates When a child shows behavioral or developmental difficulties	47.7%
Before pregnancy	35.6%
During pregnancy	8.6%
Immediately after birth	8.1%

CGT: Clinical genetic testing; ASD: Autism spectrum disorder.

On opinions about who should be offered CGT for ASD risk, 65% of the respondents agreed to those who already have an ASD child, 16% agreed to those doubting they are able to care for a child with ASD and 15% agreed to those worried about the fetus. A total of 11% of the respondents agreed that CGT should be offered to no pregnant women, and 10% agreed to all pregnant women.

Regression analyses of the parent and child characteristics (see Table S2a) show small but significant contributions from ASD sub-diagnoses in the parents of children with an ASD model and in those worried about the fetus model. Parents of children with autism agree more than parents of children with Asperger syndrome that parents of children with ASD and those worried about the fetus should be offered CGT for ASD risk ($p < 0.001$).

On items about time of testing, 48% agreed to do CGT when development deviates, 36% before pregnancy, 9% during pregnancy and 8% immediately after birth. Regression analyses of the parent and child characteristics (see Table S2b) show small but significant contributions from ASD sub-diagnosis in the before pregnancy model ($p < 0.001$) and the during pregnancy model ($p = 0.015$). Parents of children with autism parents agreed more than parents of children with Asperger syndrome in both models. The variances explained in the significant models are, however, small ($p = 0.002$; $R^2 = 0.017$ and 0.018).

2.5. Participants Opposing Clinical Genetic Testing

One option on the question "When is the best time to offer genetic testing for ASD risk?" was "I am against genetic testing for ASD", to which 9% of the participants agreed. Of those opposing CGT, 31% agreed that they would do testing only if it could help explain the cause of ASD, 42% if it has directly treatment-relevant consequences, and 35% if that early diagnosis based on CGT will improve the possibilities of planning for good interventions and facilitations (Table 4).

Parents opposing CGT differ significantly from parents not explicitly opposing all statements (p-values range from <0.001 to <0.01). We found the largest differences in scores on the statements of positive effects and the smallest differences on negative effects. See Table 4 for details.

Table 4. Parents in favor versus parents opposed CGT.

Possible Positive Effects	Agree N	In Favor of CGT	Opposed CGT	MD	Sig.
Causal Explanation I would take a genetic test only if it could help explain the cause of ASD in me or in my child.	1422	80.7%	30.9%	1.9	**
Recurrence Prevention I would take a genetic test if it gave me the possibility to prevent getting more children with ASD.	1419	43.3%	4.9%	1.8	**
Intervention Planning Early diagnosis based on a genetic test will improve the possibilities of planning for good interventions and facilitations.	1381	78.6%	35.0%	1.4	**
Family Planning Genetic testing is important for family planning.	1417	26.3%	0.8%	1.4	**
Treatment Relevance I would take a genetic test only if it has directly treatment-relevant consequences for me or my child.	1413	54.1%	41.8%	0.7	**
Possible negative effects					
Family Conflicts Genetic testing may cause family conflicts.	1379	47.3%	66.7%	-0.9	**
Parental Concern Early diagnosis based on a genetic test will increase the parents' concern for the future health and development of the child.	1384	54.9%	68.9%	-0.7	**
Insurance Discrimination A genetic test showing ASD risk will cause discrimination from insurance companies.	1387	66.0%	74.4%	-0.4	*

CGT: Clinical genetic testing; ASD: Autism spectrum disorder; MD: Difference in means; Sig.: Significance probability; ** $p < 0.001$, * $p < 0.01$.

3. Discussion

The main result of this study is that parents of ASD individuals have a clearly positive view on CGT for ASD. Etiological explanation is the most frequent effect of CGT that our respondents agreed to in our survey. This seems to confirm our hypothesis #1. The second positive effect was improvement in the possibilities of planning for good interventions and facilitations. These results are in line with findings from a study of professionals who also believe that genetic testing implies benefits for the affected families, e.g., a better possibility for early intervention, prevention of specific comorbid diseases, or relief of guilt [20]. The current findings seem to provide mixed support for routine CGT for ASD, and could help in designing such programs.

Nine percent of our respondents seem to be clearly against CGT. However, this seems mostly due to doubts about the positive effects and less due to fears of negative effects. The large agreement to positive effects of CGT among the responders (Table 4) may indicate that they will become in favor of CGT when clinical utility and treatment implications improve. Taken together, these findings seem to indicate rather robust positive attitudes toward CGT for ASD among parents of children with ASD. Furthermore, the findings also suggest that there are very small differences between parents of children with autism and parents of children with Asperger syndrome, supporting our hypothesis #2.

Skepticism against beneficial effects of CGT rather than fear of negative consequences seems to be a general finding of the current analysis. The parents' attitudes could be interpreted as being cautious and focused on preventing negative consequences of introducing technological innovations into clinical practice prematurely, e.g., by giving inaccurate etiological diagnoses [46]. Their opinions seems to be in line with the recent guidelines of EuroGentest and the European Society of Human Genetics (ESHG) for diagnostic next-generation sequencing, which states that insufficiently validated tests present a threat to patients, and their use in a clinical diagnostic setting is unacceptable [47].

Discrimination from insurance companies is the negative effect of CGT most frequently stated in our sample. The difference between those in favor and those against CGT is small (see Table 4), supporting our hypothesis #3. Fear of discrimination from insurance companies may come as a surprise in the Norwegian context of our study. In Norway, genetic testing is covered by the compulsory state health insurance, and discrimination on the basis of genetic constitution is prohibited by law [48], as it is in most other western countries [21,25]. The UN Universal Declaration of the Human Genome and Human Rights also prohibit all forms of discrimination based on genetic characteristics [49]. However, the use of voluntary private health insurance (VPHI) has recently increased also among Norwegians, but VPHI typically do not cover treatments for psychiatric conditions [50]. Knowledge of risk is integral to the concept of insurance premiums [44] and some European countries explicitly allow the use of genetic risk information when the insurance payment exceeds a certain level, which usually applies for life insurance, illness/health insurance or insurance to protect income in the case of being disabled and unable to work [44,45]. Having a child with ASD is associated with long term sick leave, not being in the labor force and low income [51], and it seems prudent to be aware of issues of possible discriminations from health and life insurance companies both in relation to parents and to people with ASD.

While only 8% agree that it is best to offer CGT during pregnancy (i.e., prenatal diagnosis), half (48%) of the parents in our sample prefer CGT to be offered when a child shows behavioral or developmental challenges and more than a third (36%) before pregnancy. Hypothesis #4, saying that parents would prefer CGT as early as possible, i.e., during pregnancy, is thus refuted. In the event of a prenatal test for ASD, this result may indicate that the Norwegian practice of direct access to prenatal testing for Trisomies (21, 18, 13) for women 38 years or older, or in later pregnancies when one has a child with a severe disability, may be less relevant for parents of individuals with ASD. This may have consequences for genetic counselling practices of parents of children with ASD. Effective counselling may depend on professional exposure to individuals with ASD. In a Down syndrome (DS) context, it is claimed that what genetic counselors believe to be the most salient information to discuss with parents differs based on whether they practice in the prenatal or postnatal setting [52], which may be a point to consider in ASD as well.

We found that more than half (65%) of the parents of children with ASD prefer CGT to be offered to those already having a child with ASD. This contradicts our hypothesis #5 that parents would prefer all those interested, i.e., those doubting that they are able to care for a child with ASD, to be offered CGT. Fifteen percent of the respondents support our hypothesis #5, and 36% prefer CGT before pregnancy. Although not specified, testing before pregnancy is plausibly interpreted as carrier testing in parents, which is not likely in a complex disorder such as ASD. This may reflect a lack of knowledge in parents, but may as well be due to the questionnaire that asked to imagine the existence of a test saying "something about the risk of ASD". However, parents' recurrence risk perceptions have shown to be inaccurately high and to influence reproductive decisions [39], and our finding may be interpreted in line with this.

We found that 40% in our sample would do a CGT if it gave them the possibility to prevent having more children with ASD. This may be interpreted as our sample preferring CGT when a child shows behavioral or developmental challenges with the purpose of augmenting development, or for preventive purposes for those parents that already have children with ASD. In comparison, a study of parents of children with fragile X (FX) found that 59% agreed that the FX-test influenced their decision to have another child [38]. It is surprising that only 24% agree that CGT is important for family planning (Table 2) when 40% would do a CGT if it enabled them to prevent having further children with ASD (Table 2). According to the World Health Organization (WHO), "family planning allows people to attain their desired number of children and determine the spacing of pregnancies. It is achieved through use of contraceptive methods and the treatment of infertility" [53]. This is also the official standard use of the term in Norway. However, the term may have been interpreted as very different from preventive measures, perhaps because this was addressed explicitly in another questionnaire statement. Family planning may have been interpreted as a question of organizing the family life and about interventions in the parents' home as almost every child and teenager with ASD in Norway lives with their parents. Regarding genetic counselling and reproductive issues, this may warrant a focus on experienced and knowledgeable parents and collaboration between them and professionals, in addition to the current predominance of a "teaching model" that may not be optimal in supporting patients to make decisions relevant to their health care [54]. These results also seem to underscore the need for recurrence risk counselling.

Ethical concerns regarding genetic testing in a reproductive setting may vary with cultural, religious and societal value bases both in parents and professionals [55]. A collaborative climate that respects the autonomy of individuals and the values of the communities and groups to which they belong seems necessary [56]. It is also important to cover these ethical issues further than the scope of the present study, which may be limited by empirical foundations for such inquiries. Relevant issues for ethical inquiries include how society values people with ASD and their parents. Advocates for ASD fear the possible outcomes of prenatal genetic testing or treatment. They argue that aggressively seeking interventions devalues ASD traits and tendencies [42], and adults with ASD fear that people with these traits will eventually be eliminated through prenatal genetic testing and selective abortion [42].

Three quarters of our sample agreed that CGT is relevant for intervention planning. This seems in line with results from a study of the impact of CMA on clinical management [57]. Treatment plans targeting associated medical conditions and planned additional assessments were directly impacted for more than 40% of individuals with neurodevelopmental disorders, including ASD [58,59]. Furthermore, the current parents' views seem in line with recent findings that genetic testing may shorten the 'diagnostic odyssey' and provide a causal explanation, indicate a recurrence risk and enable access to appropriate early interventions and tailoring the care of affected individuals [8,32,59,60]. Interestingly, a recent study summarizing the results of over four years of real-world clinical ultra-high resolution CMA testing optimized for neurodevelopmental disorders found direct correlation between a higher rate of detected abnormalities and age in the ASD cohort, which suggests that earlier use of CMA and perhaps other genetic testing methods may be important for early intervention [9].

Overall, characteristics of the parents or their children seem to have only small influences on the informants' attitudes toward CGT of ASD. This underscores the general positive attitude toward the use of molecular genetic technology in this group of patients. Some differences were, however, present in parents of children with ASD. Whether the child had autism or Asperger syndrome influences the attitudes. Parents of children with autism are more optimistic about the clinical opportunities of genetic testing for ASD than parents of children with Asperger syndrome by rating the positive effects higher and the negative effects lower. The difference between Asperger syndrome and autism may be regarded as difference in disease severity. In contrast, no association between various FX severities and opinions about screening in caregivers have been found [61].

The stigma attached to mental illness is claimed to always lead to stress through discrimination in academic settings, in work, and in public, although often unintended [41]. For instance, in a future case of neonatal genetic screening for ASD, a child with a susceptibility genotype whose behavioral presentation is considered "odd" may be stigmatized as "autistic" despite not meeting the diagnostic criteria. Conversely, a child with a susceptibility genotype may be excused for "inappropriate" behavior [21]. In either case, less fortunate interventions or treatments may be initiated. Parents are also in danger of being blamed for their children's condition when the condition is inherited. It is important to avoid terminology that may be misused and cause unjust stigmatizing and blaming, as happened with Kanner's early characterization of parents [42]. If the ethical implications of genetic identities become uncertain as research gains more knowledge about how genes and environments interrelate [42], it is clear that empirical studies of attitudes, hopes and fears of all stakeholders, at pace with the technological advancements, are necessary.

Limitations

There is limited knowledge about the attitudes toward CGT among parents of children with ASD. Although the current study provides important new information about the attitudes toward CGT for ASD and collected information from a large numbers of parents, it has limitations. One limitation is a lack of comprehensive demographic information of the participants, and little information about disease severity or genetic origin of the ASD in their children. Our findings are also limited to opinions about the specific items of positive and negative effects stated in our questionnaire. The lack of explicit questions about attitudes toward prenatal diagnosis and the related issues of stigma and offence of such practice is also a limitation. The low number of parents that think CGT should be offered during pregnancy may indicate that this is an issue that should have been addressed in more detail.

Another potential limitation is that the previously untested questionnaire was inspired from a questionnaire about malignant melanoma, which is a disease quite different from ASD. Malignant melanoma is a deadly skin cancer disease while ASD is a disability. There are important differences between what is perceived as disease and what is perceived as disability. Disabilities, and particularly those involving mental capacities, raise existential issues of identity in a sense that issues of diseases do not. To determine identities through genetic testing is more controversial than to determine diseases. This difference may have made the questions generated from malignant melanoma studies less suitable for capturing attitudes toward CGT for ASD. A further limitation is that the questionnaire did not specify whether the questions concerned testing parents or testing the child, where it could mean either, which makes the accuracy of some of our findings uncertain.

It is also possible that our sample is biased by being recruited from an interest organization. Our large sample may not be representative of attitudes of parents of children with ASD in general. Despite it being the only ASD organization in Norway, with a high number of members, a significant proportion of parents of children with ASD are not members of the interest organization.

4. Materials and Methods

4.1. Participants

In collaboration with the Norwegian patient organization for ASD "Autism Society Norway", a questionnaire was sent by email or post to all of the parent members of the organization. The organization has no formal requirement for membership of parents, but 3539 members were registered as parents to one or more children with ASD. In order to reach as many as possible, two approaches were used. Members registered with email addresses were contacted by email, which contained a link to a web-based Response Form Questionnaire from the University of Oslo. Members without email addresses received paper mail with the questionnaire and a prepaid return envelope. Email reminders were sent to those not responding, while no paper mail reminders were sent. Answers were anonymous. A total of 3539 invitations were sent, $n = 1990$ by email and $n = 1549$ by paper letters. A total of $n = 1455$ responded (41%), $n = 917$ (46%) through email, and $n = 538$ (35%) by paper mail. The survey was closed after two reminders. No financial incentive or other rewards were provided.

4.2. Questionnaire

The questionnaire investigated attitudes toward clinical use of a potential genetic test for ASD. It was emphasized that, currently, a predictive genetic test for ASD is not available. The participants were asked to imagine the existence of a clinical test that could say "something" about the risk of ASD.

Some of the questions concerning attitudes were based on a survey of attitudes toward genetic research in families with a risk of malignant melanoma [62], others were composed and tested by our multi-professional team and selected representatives of the target group. One of the questions was also in line with a qualitative study of awareness and attitudes among parents of children with ASD [25]. The questionnaire as a whole has not been previously tested and the overall psychometric properties are unknown. Although difficult to establish, the validity seems high at face value. Our explorations of how responder characteristics influence the ratings provide some indications of the validity of the questionnaire. The answers were given on a 5-point Likert scale with alternatives "strongly disagree", "disagree", "neither disagree or agree", "agree" and "strongly agree", in addition to "Don´t know" or "Have no meaning" options [63].

The questions concerning attitudes were part of a questionnaire with 45 items in four sections. It took 10–15 min to complete the questionnaire. The first section of the questionnaire asked about age and gender of the parents. The second section asked about the parents' firstborn child with ASD diagnosis, the child´s ASD diagnosis (infantile autism or Asperger syndrome), the child´s age when first diagnosed with ASD, and the child´s comorbid somatic and psychiatric diagnoses. It was not asked how many children with ASD the participants have. International Statistical Classification of Diseases and Related Health Problems (ICD) criteria are used in Norway. The third section of the questionnaire asked about experiences with the specialist health services. The fourth section asked about attitudes toward genetic research and CGT. The questions about attitudes toward genetic research have been previously reported [36].

With a total of $n = 1455$ members responding, we estimate the response rate to be approximately 50%. Because the information to be collected was anonymized, the Regional Committee for Medical and Health Research Ethics (REC) concluded that the study did not require REC approval.

4.3. Data Handling and Analyses

SPSS version 23 (IBM, Armonk, NY, USA) was used for statistical analyses. The paper responses were scanned and read using ABBYY FlexiCapture 10 (ABBYY Europe, Munich, Germany) and subsequently imported into a SPSS file, while the online responses were imported directly.

First, descriptive statistics were performed. Percentages of agreement and disagreement to the attitude statements were obtained by dichotomizing the Likert-scale variables in the following way: Scores 1 ("Totally disagree") and 2 ("Somewhat disagree") were combined to a "disagree"—score, scores 4 ("Somewhat agree") and 5 ("Totally agree") were combined to an "agree"—score. Score 3 ("neither/nor") was merged with the "have no meaning"—and the "don't know"—scores in order to form an "NA"—score labelling the "no distinct negative or positive attitude available".

Forced entry linear regression analyses were conducted both on the original 5-point attitude variables and on the binary attitude variables [64] concerning management of CGT to analyze the effect of respondent age and gender, child age, child gender, and child ASD diagnose on the attitude items. Due to multiple comparisons, we used Bonferroni corrected significance thresholds, setting the significance limits to $p = 0.01$ per test (0.05/5 tests). Cases with missing values were excluded listwise in the regression analyses and deleted on an analysis-by-analysis basis in the independent sample *t*-tests.

5. Conclusions

Parents of children with ASD seem to be clearly positive toward CGT. Only one of ten is opposed to CGT and the reason is mainly skepticism about the possible benefits, not because they fear possible harm. This suggests that parents of children with ASD seem well informed, have a practical attitude to CGT, and have similar attitudes as health care personnel and ASD experts.

Few parents seem interested in prenatal diagnosis and family planning. They seem rather to prefer CGT when a child shows behavioral or developmental challenges with the purpose of augmenting development, or for preventive purposes for those that already have children with ASD. Further research is necessary in order to draw direct clinical implications of these results. However, the current results may form the basis for such testing procedures in the future when they will be more informative. It is important to avoid initiating premature CGT procedures.

Supplementary Materials: Supplementary materials can be found at www.mdpi.com/1422-0067/18/5/1078/s1.

Acknowledgments: We are thankful to the participants in the study and the Autism Society Norway. We are also thankful to Berge Solberg for valuable comments and suggestions. The project was supported by the Research Council of Norway (Grant Nos. 213694 and 223273) and the KG Jebsen Stiftelsen, and has received funding from the European Community's Seventh Framework Programme (FP7/2007–2013) under Grant No. 602450. This paper reflects only the author's views and the European Union is not liable for any use that may be made of the information contained therein. This study is part of the BUPgen Study group and the research network NeuroDevelop.

Author Contributions: Terje Nærland and Ole A. Andreassen designed the study, Jarle Johannessen and Terje Nærland collected and analysed data and wrote the first draft together with Ole A. Andreassen, Sigrun Hope, Marcella Rietschel, Tonje Torske, Anne Lise Høyland, Jana Strohmaier, Arvid Heiberg and Srdjan Djurovic participated in interpreting data. All authors contributed to the final version of the paper.

Conflicts of Interest: The authors declare no conflict of interest.

Abbreviations

ASD	Autism spectrum disorder
CGT	Clinical genetic testing
CMA	Chromosomal microarray
CNV	Copy number variant
DS	Down syndrome
ESHG	European Society of Human Genetics
FX	Fragile X
FXS	Fragile X syndrome
ICD	International Statistical Classification of Diseases and Related Health Problems
REC	Regional Committee for Medical and Health Research Ethics
VPHI	Voluntary private health insurance
WHO	World Health Organization

References

1. Lai, M.C.; Lombardo, M.V.; Baron-Cohen, S. Autism. *Lancet* **2014**, *383*, 896–910. [CrossRef]
2. Baxter, A.J.; Brugha, T.S.; Erskine, H.E.; Scheurer, R.W.; Vos, T.; Scott, J.G. The epidemiology and global burden of autism spectrum disorders. *Psychol. Med.* **2015**, *45*, 601–613. [CrossRef] [PubMed]
3. Surén, P.; Bakken, I.J.; Aase, H.; Chin, R.; Gunnes, N.; Lie, K.K.; Magnus, P.; Reichborn-Kjennerud, T.; Schjølberg, S.; Øyen, A.-S.; et al. Autism spectrum disorder, ADHD, epilepsy, and cerebral palsy in norwegian children. *Pediatrics* **2012**, *130*, e152–e158. [CrossRef] [PubMed]
4. Xu, L.; Carpenter-Aeby, T.; Aeby, V.G.; Lu, W. An integrated model of emotions, attitudes, and intentions associated with undergoing autism genetic testing. *Psychol. Cogn. Sci. Open J.* **2015**, *1*, 29–38. [CrossRef]

5. Schaefer, G. Clinical genetic aspects of ASD spectrum disorders. *Int. J. Mol. Sci.* **2016**, *17*, 180. [CrossRef] [PubMed]

6. Yuen, R.K.C.; Thiruvahindrapuram, B.; Merico, D.; Walker, S.; Tammimies, K.; Hoang, N.; Chrysler, C.; Nalpathamkalam, T.; Pellecchia, G.; Liu, Y.; et al. Whole-genome sequencing of quartet families with autism spectrum disorder. *Nat. Med.* **2015**, *21*, 185–191. [CrossRef] [PubMed]

7. Schaefer, G.B.; Mendelsohn, N.J. Clinical genetics evaluation in identifying the etiology of autism spectrum disorders: 2013 guideline revisions. *Genet. Med.* **2013**, *15*, 399–407. [CrossRef] [PubMed]

8. Heil, K.M.; Schaaf, C.P. The genetics of autism spectrum disorders—A guide for clinicians. *Curr. Psychiatry Rep.* **2013**, *15*, 334–342. [CrossRef] [PubMed]

9. Ho, K.; Wassman, E.; Baxter, A.; Hensel, C.; Martin, M.; Prasad, A.; Twede, H.; Vanzo, R.; Butler, M. Chromosomal microarray analysis of consecutive individuals with autism spectrum disorders using an ultra-high resolution chromosomal microarray optimized for neurodevelopmental disorders. *Int. J. Mol. Sci.* **2016**, *17*, 2070. [CrossRef] [PubMed]

10. Fischbach, R.L.; Harris, M.J.; Ballan, M.S.; Fischbach, G.D.; Link, B.G.; Charman, T.; Leake, A.R.; Sivberg, B. Is there concordance in attitudes and beliefs between parents and scientists about autism spectrum disorder? *Autism* **2015**, *6*, 207–214. [CrossRef] [PubMed]

11. Collins, F.S.; Varmus, H. A new initiative on precision medicine. *N. Engl. J. Med.* **2015**, *372*, 793–795. [CrossRef] [PubMed]

12. Julia, S.; Soulier, A.; Leonard, S.; Sanlaville, D.; Vigouroux, A.; Keren, B.; Heron, D.; Till, M.; Chassaing, N.; Bouneau, L. Ethical issues raised by the clinical implementation of new diagnostic tools for genetic diseases in children: Array comparative genomic hybridization (ACGH) as a case study. *J. Clin. Med. Res.* **2015**, *6*, 2.

13. Polyak, A.; Girirajan, S. A need for precision medicine to enable tailored special education. *Am. J. Med. Genet.* **2016**, *171*, 300. [CrossRef] [PubMed]

14. Polyak, A.; Kubina, R.M.; Girirajan, S. Comorbidity of intellectual disability confounds ascertainment of autism: Implications for genetic diagnosis. *Am. J. Med. Genet.* **2015**, *168*, 600–608. [CrossRef] [PubMed]

15. Hens, K.; Peeters, H.; Dierickx, K. The ethics of complexity. Genetics and autism, a literature review. *Am. J. Med. Genet.* **2016**, *171B*, 305–316. [CrossRef] [PubMed]

16. Burke, W.; Laberge, A.M.; Press, N. Debating clinical utility. *Public Health Genom.* **2010**, *13*, 215–223. [CrossRef] [PubMed]

17. Freitag, C.M. The genetics of autistic disorders and its clinical relevance: A review of the literature. *Mol. Psychiatry* **2007**, *12*, 2–22. [CrossRef] [PubMed]

18. McMahon, W.M.; Baty, B.J.; Botkin, J. Genetic counseling and ethical issues for autism. *Am. J. Med. Genet.* **2006**, *142C*, 52–57. [CrossRef] [PubMed]

19. Finucane, B.; Myers, S.M. Genetic counseling for autism spectrum disorder in an evolving theoretical landscape. *Curr. Genet. Med. Rep.* **2016**, *4*, 147–153. [CrossRef] [PubMed]

20. Hens, K.; Peeters, H.; Dierickx, K. Genetic testing and counseling in the case of an autism diagnosis: A caregivers perspective. *Eur. J. Hum. Genet.* **2016**, *59*, 452–458. [CrossRef] [PubMed]

21. Marchant, G.E.; Robert, J.S. Genetic testing for autism predisposition: Ethical, legal and social challenges. *Houst J. Health Law Policy* **2009**, *9*, 203–235.

22. Walsh, P.; Elsabbagh, M.; Bolton, P.; Singh, I. In search of biomarkers for autism: Scientific, social and ethical challenges. *Nat. Rev. Neurosc.* **2011**, *12*, 603–612. [CrossRef] [PubMed]

23. Lintas, C.; Persico, A.M. Autistic phenotypes and genetic testing: State-of-the-art for the clinical geneticist. *J. Med. Genet.* **2009**, *46*, 1–8. [CrossRef] [PubMed]

24. Scherer, S.W.; Dawson, G. Risk factors for autism: Translating genomic discoveries into diagnostics. *Hum. Genet.* **2011**, *130*, 123–148. [CrossRef] [PubMed]

25. Chen, L.S.; Xu, L.; Huang, T.Y.; Dhar, S.U. Autism genetic testing: A qualitative study of awareness, attitudes, and experiences among parents of children with autism spectrum disorders. *Genet. Med.* **2013**, *15*, 274–281. [CrossRef] [PubMed]

26. Rew, L.; Kaur, M.; McMillan, A.; Mackert, M.; Bonevac, D. Systematic review of psychosocial benefits and harms of genetic testing. *Issues Ment. Health Nurs.* **2010**, *31*, 631–645. [CrossRef] [PubMed]

27. Rew, L.; Mackert, M.; Bonevac, D. A systematic review of literature about the genetic testing of adolescents. *J. Spec. Pediatr. Nurs.* **2009**, *14*, 284–294. [CrossRef] [PubMed]

28. Gronborg, T.K.; Hansen, S.N.; Nielsen, S.V.; Skythhe, A.; Parner, E.T. Stoppage in autism spectrum disorders. *J. Autism. Dev. Disord.* **2015**, *45*, 3509–3519. [CrossRef] [PubMed]
29. Lauritsen, M.B.; Pedersen, C.B.; Mortensen, P.B. Effects of familial risk factors and place of birth on the risk of autism: A nationwide register-based study. *J. Child Psychol. Psychiatry* **2005**, *46*, 963–971. [CrossRef] [PubMed]
30. Ozonoff, S.; Young, G.S.; Carter, A.; Messinger, D.; Yirmiya, N.; Zwaigenbaum, L.; Bryson, S.; Carver, L.J.; Constantino, J.N.; Hutman, T.; et al. Recurrence risk for autism spectrum disorders: A baby siblings research consortium study. *Pediatrics* **2011**, *128*, e488–e495. [CrossRef] [PubMed]
31. Cuccaro, M.L.; Czape, K.; Alessandri, M.; Lee, J.; Deppen, A.R.; Bendik, E.; Dueker, N.; Nations, L.; Pericak-Vance, M.; Hahn, S. Genetic testing and corresponding services among individuals with autism spectrum disorder (ASD). *Am. J. Med. Genet.* **2014**, *164A*, 2592–2600. [CrossRef] [PubMed]
32. Narcisa, V.; Discenza, M.; Vaccari, E.; Rosen-Sheidley, B.; Hardan, A.Y.; Couchon, E. Parental interest in a genetic risk assessment test for autism spectrum disorders. *Clin. Pediatr.* **2013**, *52*, 139–146. [CrossRef] [PubMed]
33. Zwaigenbaum, L.; Bauman, M.L.; Stone, W.L.; Yirmiya, N.; Estes, A.; Hansen, R.L.; McPartland, J.C.; Natowicz, M.R.; Choueiri, R.; Fein, D.; et al. Early identification of autism spectrum disorder: Recommendations for practice and research. *Pediatrics* **2015**, *136*, S10–S40. [CrossRef] [PubMed]
34. Golfenshtein, N.; Srulovici, E.; Medoff-Cooper, B. Investigating parenting stress across pediatric health conditions: A systematic review. *Compr. Child. Adolesc. Nurs.* **2016**, *39*, 41–79. [CrossRef] [PubMed]
35. Lai, W.W.; Goh, T.J.; Oei, T.P.; Sung, M. Coping and well-being in parents of children with autism spectrum disorders (ASD). *J. Autism. Dev. Disord.* **2015**, *45*, 2582–2593. [CrossRef] [PubMed]
36. Johannessen, J.; Nærland, T.; Bloss, C.; Rietschel, M.; Strohmaier, J.; Gjevik, E.; Heiberg, A.; Djurovic, S.; Andreassen, O.A. Parents' attitudes toward genetic research in autism spectrum disorder. *Psychiatr. Genet.* **2016**, *26*, 74–80. [CrossRef] [PubMed]
37. Jones, M.B.; Szatmari, P. Stoppage rules and genetic studies of autism. *J. Autism. Dev. Disord.* **1988**, *18*, 31–40. [CrossRef] [PubMed]
38. Bailey, D.B.; Skinner, D.; Sparkman, K.L. Discovering fragile X syndrome: Family experiences and perceptions. *Pediatrics* **2003**, *111*, 407–416. [CrossRef] [PubMed]
39. Selkirk, C.G.; McCarthy Veach, P.; Lian, F.; Schimmenti, L.; Leroy, B.S. Parents' perceptions of autism spectrum disorder etiology and recurrence risk and effects of their perceptions on family planning: Recommendations for genetic counselors. *J. Genet. Couns.* **2009**, *18*, 507–519. [CrossRef] [PubMed]
40. Sivell, S.; Elwyn, G.; Gaff, C.L.; Clarke, A.J.; Iredale, R.; Shaw, C.; Dundon, J.; Thornton, H.; Edwards, A. How risk is perceived, constructed and interpreted by clients in clinical genetics, and the effects on decision making: Systematic review. *J. Genet. Couns.* **2008**, *17*, 30–63. [CrossRef] [PubMed]
41. Adler, B.A.; Minshawi, N.F.; Erickson, C.A. Evolution of autism: From Kanner to the DSM-V. In *Handbook of Early Intervention for Autism Spectrum Disorders: Research, Policy, and Practice*; Tarbox, J., Dixon, D.R., Sturmey, P., Matson, J.L., Eds.; Springer: New York, NY, USA, 2014; pp. 3–19.
42. Silverman, C. *Understanding Autism: Parents, Doctors, and the History of a Disorder*; Princeton University Press: Princeton, NJ, USA, 2012.
43. Campbell, E.; Ross, L.F. Parental attitudes and beliefs regarding the genetic testing of children. *Community Genet.* **2005**, *8*, 94–102. [CrossRef] [PubMed]
44. Soini, S. Genetic testing legislation in Western Europe—A fluctuating regulatory target. *J. Community Genet.* **2012**, *3*, 143–153. [CrossRef] [PubMed]
45. Kvale, H.; Gudmundsdóttir, L.H.; Stoll, J.; Faber, B.A. *Legislation on Biotechnology in the Nordic Countries—An Overview 2016*; NordForsk: Oslo, Norway, 2016.
46. Ashley, E.A. Towards precision medicine. *Nat. Rev. Genet.* **2016**, *17*, 507–522. [CrossRef] [PubMed]
47. Matthijs, G.; Souche, E.; Alders, M.; Corveleyn, A.; Eck, S.; Feenstra, I.; Race, V.; Sistermans, E.; Sturm, M.; Weiss, M.; et al. Guidelines for diagnostic next-generation sequencing. *Eur. J. Hum. Genet.* **2016**, *24*, 2–5. [CrossRef] [PubMed]
48. Bioteknologiloven. Lov om Humanmedisinsk Bruk av Bioteknologi m.m. (Bioteknologiloven). [Act of 5 December 2003 No. 100 Relating to the Application of Biotechnology in Human Medicine, etc]. Available online: http://app.uio.no/ub/ujur/oversatte-lover/data/lov-20031205-100-eng.pdf (accessed on 16 April 2017).

49. UN General Assembly. Universal declaration on the human genome and human rights. In *Resolution 53/152 of 9 December 1998*; United Nations Human Rights Office of the High Commissioner: Geneva, Switzerland, 1998.

50. Alexandersen, N.; Anell, A.; Kaarboe, O.; Lehto, J.S.; Tynkkynen, L.-K.; Vrangbaek, K. The development of voluntary private health insurance in the nordic countries. *Nordic J. Health Econ.* **2016**, *4*, 68–83. [CrossRef]

51. McEvilly, M.; Wicks, S.; Dalman, C. Sick leave and work participation among parents of children with autism spectrum disorder in the stockholm youth cohort: A register linkage study in Stockholm, Sweden. *J. Autism. Dev. Disord.* **2015**, *45*, 2157–2167. [CrossRef] [PubMed]

52. Acharya, K. Prenatal testing for intellectual disability: Misperceptions and reality with lessons from down syndrome. *Dev. Disabil. Res. Rev.* **2011**, *17*, 27–31. [CrossRef] [PubMed]

53. World Health Organization. Family Planning/Contraception. Available online: http://who.int/mediacentre/factsheets/fs351/en/ (accessed on 2 February 2017).

54. Skirton, H.; Cordier, C.; Ingvoldstad, C.; Taris, N.; Benjamin, C. The role of the genetic counsellor: A systematic review of research evidence. *Eur. J. Hum. Genet.* **2015**, *23*, 452–458. [CrossRef] [PubMed]

55. MacKenzie, M.J.; Baumeister, R.F. Meaning in life: Nature, needs, and myths. In *Meaning in Positive and Existential Psychology*; Batthyany, A., Russo-Netzer, P., Eds.; Springer: New York, NY, USA, 2014; pp. 25–37.

56. Blasimme, A.; Vayena, E. Becoming partners, retaining autonomy: Ethical considerations on the development of precision medicine. *BMC Med. Ethics* **2016**, *17*, 67–75. [CrossRef] [PubMed]

57. Tammimies, K.; Marshall, C.R.; Walker, S.; Kaur, G.; Thiruvahindrapuram, B.; Lionel, A.C.; Yuen, R.K.C.; Uddin, M.; Roberts, W.; Weksberg, R.; et al. Molecular diagnostic yield of chromosomal microarray analysis and whole-exome sequencing in children with autism spectrum disorder. *JAMA* **2016**, *314*, 895–903. [CrossRef] [PubMed]

58. Henderson, L.B.; Applegate, C.D.; Wohler, E.; Sheridan, M.B.; Hoover-Fong, J.; Batista, D.A.S. The impact of chromosomal microarray on clinical management: A retrospective analysis. *Genet. Med.* **2014**, *16*, 1–8. [CrossRef] [PubMed]

59. Tammimies, K.; Falck-Ytter, T.; Bölte, S. Quo vadis clinical genomics of ASD? *Autism* **2016**, *20*, 259–261. [CrossRef] [PubMed]

60. Reiff, M.; Giarelli, E.; Bernhardt, B.A.; Easley, E.; Spinner, N.B.; Sankar, P.L.; Mulchandani, S. Parents' perceptions of the usefulness of chromosomal microarray analysis for children with autism spectrum disorders. *J. Autism. Dev. Disord.* **2015**, *45*, 3262–3275. [CrossRef] [PubMed]

61. Bailey, D.B.; Bishop, E.; Raspa, M.; Skinner, D. Caregiver opinions about fragile X population screening. *Genet. Med.* **2012**, *14*, 115–121. [CrossRef] [PubMed]

62. Branstrom, R.; Kasparian, N.A.; Affleck, P.; Tibben, A.; Chang, Y.-m.M.; Azizi, E.; Baron-Epel, O.; Bergman, W.; Chan, M.; Davies, J.; et al. Perceptions of genetic research and testing among members of families with an increased risk of malignant melanoma. *Eur. J. Cancer* **2012**, *48*, 3052–3062. [CrossRef] [PubMed]

63. Bishop, P.A.; Herron, R.L. Use and misuse of the likert item responses and other ordinal measures. *Int. J. Exerc. Sci.* **2015**, *8*, 297–302. [PubMed]

64. Hellevik, O. Linear versus logistic regression when the dependent variable is a dichotomy. *Qual. Quant.* **2009**, *43*, 59–74. [CrossRef]

MDPI AG

St. Alban-Anlage 66

4052 Basel, Switzerland

Tel. +41 61 683 77 34

Fax +41 61 302 89 18

http://www.mdpi.com

IJMS Editorial Office

E-mail: ijms@mdpi.com

http://www.mdpi.com/journal/ijms